21ˢᵗ Century Astronomy's new Animations help students master core concepts.

Developed specifically for the Second Edition of *21st Century Astronomy*, these lessons animate 18 core concepts central to astronomy and use interactivity to test students' mastery.

Animations illustrate specific sections of the text

A.	The Celestial Sphere and the Ecliptic	Text Section 2.2
B.	The Earth Spins and Revolves	Text Sections 2.2 and 2.3
C.	The Moon's Orbit: Eclipses and Phases	Text Section 2.4 and 2.5
D.	Kepler's Laws	Text Section 3.2
E.	Newton's Laws and Universal Gravitation	Text Section 3.5
F.	Light as a Wave, Light as a Photon	Text Section 4.2, 4.4
G.	Doppler Effect	Text Section 4.4
H.	Atomic Energy Levels and the Bohr Model	Text Section 4.4
I.	Atomic Energy Levels and Light Emission and Absorption	Text Section 4.4
J.	Geometric Optics and Lenses	Text Section 5.1
K.	Solar System Formation	Text Section 6.1–6.4
L.	Processes That Shape the Planets	Text Section 7.2, 7.3, 7.5, 7.7
M.	Tides and the Moon	Text Section 10.2, 10.3
N.	Solar Spectrum	Text Section 13.3
O.	The H-R Diagram	Text Section 13.5
P.	The Solar Core	Text Section 14.2
Q.	Star Formation	Text Sections 15.3, 15.4
R.	Hubble's Law	Text Sections 20.2, 20.3, 20.4

There are a variety of ways to access the Animations

 Available free from the StudySpace student website at **www.wwnorton.com/astro21**

 Integrated into SmartWork homework assignments

Or ready for classroom presentation as a part of the PowerPoint resources on the Norton Media Library CD-ROM

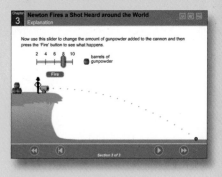

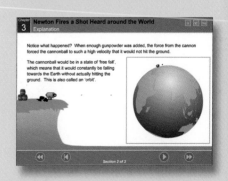

SECOND EDITION

21ST CENTURY

ASTRONOMY

THE SOLAR SYSTEM

SECOND EDITION

THE SOLAR SYSTEM

21ST CENTURY
ASTRONOMY

Jeff Hester
ARIZONA STATE UNIVERSITY

David Burstein
ARIZONA STATE UNIVERSITY

George Blumenthal
UNIVERSITY OF CALIFORNIA—
SANTA CRUZ

Ronald Greeley
ARIZONA STATE UNIVERSITY

Bradford Smith
UNIVERSITY OF HAWAII—MANOA

Howard G. Voss
ARIZONA STATE UNIVERSITY

 W. W. NORTON & COMPANY • NEW YORK • LONDON

W. W. Norton & Company has been independent since its founding in 1923, when William Warder Norton and Mary D. Herter Norton first published lectures delivered at the People's Institute, the adult education division of New York City's Cooper Union. The Nortons soon expanded their program beyond the Institute, publishing books by celebrated academics from America and abroad. By mid-century, the two major pillars of Norton's publishing program—trade books and college texts—were firmly established. In the 1950s, the Norton family transferred control of the company to its employees, and today—with a staff of 400 and a comparable number of trade, college, and professional titles published each year—W. W. Norton & Company stands as the largest and oldest publishing house owned wholly by its employees.

Printed in the United States of America

Second Edition

Editor: Leo Wiegman
Associate Editor: Sarah England
Editorial Assistant: Lisa Rand
Managing Editor—College: Marian Johnson
Senior Project Editor: Kim Yi
Copy Editor: Meg McDonald
Developmental Editor: Brian Loehr
Science Media Editor: April Lange
Supplements Editor: Karen Misler
Director of Manufacturing—College: Roy Tedoff
Art Director: Rubina Yeh
Photo Researcher: Kelly Mitchell
Illustrations: J/B Woolsey Associates/Penumbra
Composition and layout: Brad Walrod/High Text Graphics, Inc.
Manufacturing: VonHoffmann Company

ISBN-13: 978-0-393-93009-2 (pbk.)
ISBN-10: 0-393-93009-2 (pbk.)

The Library of Congress has cataloged the one-volume edition as follows:
21st century astronomy / Jeff Hester . . . [et al.].—2nd ed.
    p. cm.
  Includes index.
  ISBN-13: 978-0-393-92443-5 (pbk.)
  ISBN-10: 0-393-92443-2 (pbk.)
  1. Astronomy. I. Hester, John Jeffrey. II. Title: Twenty-first century astronomy.
  QB45.2.A14 2006
  520—dc22   2006047240

W. W. Norton & Company, Inc., 500 Fifth Avenue, New York, N.Y. 10110
www.wwnorton.com
W. W. Norton & Company Ltd., Castle House, 75/76 Wells Street, London W1T 3QT

1 2 3 4 5 6 7 8 9 0

Jeff Hester dedicates this book to Vicki, to whom he is still married, even after "The Book"; to his children—may they learn about rushing in where the wise fear to tread; to his colleagues and graduate students, who too often were left to pick up the slack as this project became all consuming; and finally to the many students, past and future, who make it all worthwhile.

David Burstein wishes to thank Gail, Jon, and Liz for their love and help.

George Blumenthal gratefully thanks his wife, Kelly Weisberg, and his children, Aaron and Sarah Blumenthal, for their support during this project.

Ronald Greeley dedicates this book to his wife, Cindy.

Bradford Smith dedicates this Second Edition to his patient and understanding wife, Diane.

Howard G. Voss dedicates this book to his wife, Helen Ann, who cheerfully sacrificed much so that he could attend to his teaching and other professional work. Helen Ann passed away at the time of the completion of this edition. She will be greatly missed by Howard, their family, and her many friends. She lived a life of selfless service even despite being ill most of that life.

Brief Contents

CHAPTERS 13 AND 15–20 ARE NOT INCLUDED IN THIS ALTERNATE EDITION.

EPILOGUE We Are Stardust in Human Form

Contents

CHAPTER 4   Light

PART II The Solar System

6.4   The Inner Disk Is Hot, but the Outer Disk Is Cold   170
Rock, Metal, and Ice   172

Foundations 6.1   **Thinking about Energy in Different Ways   173**

Solid Planets Gather Atmospheres   174

6.5   A Tale of Eight Planets   174

6.6   There Is Nothing Special about Our Solar System   177
The Search for Extrasolar Planets   177
Planetary Systems Seem to Be Commonplace   178

Connections 6.1   **Origins—Choosing the Right Kind of Planet   180**

Summary   181
Seeing the Forest through the Trees   181
Key Terms   182
Student Questions   182

CHAPTER 7   The Terrestrial Planets and Earth's Moon

7.1   How Are Planets the Same, and How Are They Different?   185

7.2   Four Main Processes Shape Our Planet   186

7.3   Impacts Help Shape the Evolution of the Planets   188

Connections 7.1   **Origins—Where Have All the Dinosaurs Gone?   190**

Calibrating a Cosmic Clock   192

7.4   The Interiors of the Terrestrial Planets Tell Their Own Tale   193
We Can Probe the Interior of Earth in Many Ways   193

Foundations 7.1   **Determining the Ages of Rocks   194**

Building a Model of Earth   195
The Moon Was Born from Earth   197

Foundations 7.2   **Pressure and Weight   197**

The Evolution of Planetary Interiors Depends on Heating and
        Cooling   198
Most Planets Generate Their Own Magnetic Fields   199

7.5   Tectonism—How Planetary Surfaces Evolve   201

Excursions 7.1   **Paleomagnetism: A Ticker-Tape Record of Plate Tectonics   202**

Plate Tectonics Is Driven by Convection   202

Connections 7.2   **Convection: From the Stovetop to Interiors of Stars   204**

Tectonism on Other Planets Is Different from on Earth   205

7.6   Igneous Activity: A Sign of a Geologically Active Planet   208
Terrestrial Volcanism Is Related to Tectonism   208
Volcanism Also Occurs Elsewhere in the Solar System   209

**7.7   Gradation: Wearing Down the High Spots and Filling in the
        Low   212**

Summary   216
Seeing the Forest through the Trees   216
Key Terms   217
Student Questions   218

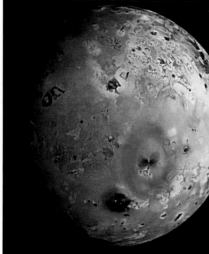

CHAPTER 10   Gravity Is More Than Kepler's Laws

CHAPTER 11   Planetary Moons and Rings,
and Dwarf Planets

PART III Stars and Stellar Evolution

CHAPTER 15   Star Formation and the Interstellar Medium

CHAPTER 16   Stars in the Slow Lane

Preface

What purpose does an introductory course in astronomy serve at today's college and university campuses? Some intrepid students would boldly answer this question, "To satisfy my lifelong curiosity about astronomy, physics, and planetary science!" Thank goodness such students exist. Their intent faces and interested questions get us through those occasional class periods when we think to ourselves, "Boy, am I glad I didn't have to *sit through* the lecture I just gave!"

We know most students take "Stars for Poets" to satisfy a science requirement. When faced with the choices at hand—biology (too squishy), physics (turn the catalog page quickly), chemistry (too smelly), geology (what's geology?)—many students opt for astronomy as the least of the available evils. Looking overhead at night, they see the stars and think, "Big Dipper, Little Dipper, North Star...sure, I can do stars!" It is of course ironic that astronomy turns out to be a mixture of physics, chemistry, and geology, with a bit of math and biology thrown in for good measure.

Students who study introductory astronomy typically come from diverse backgrounds; often their introductory astronomy course is the only formal exposure to science they will have in their lives! When viewed in this way, teaching introductory astronomy seems like a big responsibility, and so it is. As instructors, everything we do should turn on the answer to basic questions such as these: What should students carry away from this course that will still be with them 20 years from now? How should this course change the students who take it? And how can we, as scientists and educators, facilitate those changes within our students?

A thoughtful response to these questions does *not* lead to a course that is primarily about memorization of facts. At best such facts will enter students' short-term memory just long enough to be regurgitated on an exam. At worst students might come to think that science really is nothing more than memorizing information.

Facts are essential to reasoning and understanding in any course. Yet many students arrive at our doors with the idea that *learning* and *memorization* are synonymous. They confuse *knowing* with the ability to call an object by name or recite a few relevant pieces of information about it, as if stars once gathered into constellations and named were of no further interest. If an introductory science course does nothing else, it should encourage students to think of their brains as muscles that grow strong only with exercise and training. An introduction to science should be an introduction to what it means to think deeply about the universe, finding patterns and relationships within the world that go beyond the specifics of a particular object or setting, and applying those patterns and relationships broadly. An introductory science course should first and foremost be about understanding the world through the eyes of a scientist.

One of the authors of this text, Howard Voss, enjoys telling of a time when he was teaching an introductory physics course for nonscientists. One afternoon after a class discussing Newton's laws, a student came up to him in tears. Howard braced for the worst, but instead of a complaint about the difficulty of the course, he was met by a triumphant "I get it!" The student's tears were tears of joy. She had come to see the *sense* of Newton's laws. Not only did she know what they say, but she had come to understand something of what they mean and of how they can be used to make sense of the world.

In that moment of insight into Newton's laws, Howard's student caught a glimpse of what knowledge really is—of what it means to understand something rather than just know about it—and her discovery of that world of understanding was an emotional experience. This is the essence of an introductory science course.

An introductory astronomy course should also show students something of the epistemology of science: Knowledge is arrived at as we constantly challenge and test our ideas. The universe is not capricious; it behaves according to consistent, explicable rules. All scientific knowledge is provisional. How can it be that in science we never really prove things to be true, but instead only fail to show that they are false? Because that is how we give what is *really* true precedence over what we would *like* to be true. It is precisely because science acknowledges lack of certainty that the pinnacle of human knowledge is a well-tested, well-corroborated scientific theory. Here is the key to the difference between science and the myriad pseudoscientific pretenders that would borrow the credibility of science by adopting its trappings but forgo its intellectual rigor. We will have succeeded if a student comes to recognize the difference between science and pseudoscience, and if 20 years from now a former student still chuckles inwardly at the naiveté of the phrase "It's only a theory."

There is a strong practical side to what we teach about the process of science. Thinking about the world as a scientist means learning to sort the information from the opinion or happenstance. Physical scientists tend not to buy lottery tickets, call psychic help lines, or consume the many other forms of modern-day snake oil. Our students are served well if we pass on a healthy dose of our practical skepticism. A democratic society, which in the long run can fare no better than its citizens can think, is served also.

Yet the real reason it matters that students come to see the world through the eyes of a scientist is the sheer beauty of the view. The majesty of the universe in which we live is staggering. Simply staring at a deep image of distant galaxies or the glow of a newly formed star or a picture of Jupiter's swirling Great Red Spot can be a mind-boggling experience. Add a dose of quantum mechanics or the idea of the tortured fabric of spacetime around a black hole, and we are into territory that would awe Aristotle. Science—this marvelous expression of human curiosity and passion and reason—has shattered the mental cage in which we have lived since our ancestors first looked to the sky and wondered about what they saw there. It is extraordinary to face these vistas surrounded not by speculation but by the meticulously constructed edifice of science. Finally, to live during that moment in history when we first recognized our own origins in the Big Bang that occurred some 14 billion years ago—*that* is something to write home about!

Science is a vital, exciting, ongoing expression of our humanity. While an introductory astronomy book should teach students about what science is, showing them what science has revealed and giving them powerful tools for thinking about the world, it should also share with them something of the passion scientists feel for the endeavor to which we have devoted our lives.

The book that meets these goals would have to be very different from most textbooks. Rather than presenting a blocked-out set of facts and descriptions, that book would need to tell a story. On a small scale, it would need to tell the story of specific ideas. Understanding does not come from a recitation of facts. Understanding comes from thinking carefully and critically about things and how they work. Helping a student understand a concept as a scientist means guiding that student through the concept, making heavy use of examples and analogies, and tying the concept back to everyday phenomena and experiences to which the stu-

dent can relate. That is what the text we envisioned would have to do.

So rather than the expository style of traditional textbooks, this book tries for the explanatory style of a good teacher. Rather than stating that "the altitude of the north celestial pole above the horizon equals the observer's latitude" and leaving it at that, we talk about what the sky looks like over the course of the day as viewed from the North Pole and why. Then we invite the student to follow along as we head south, seeing how things change along the way.

On a larger scale, *21st Century Astronomy* tells a web of different stories. Some of these stories have to do with what science is and how it is done. Why did Newton choose the form that he did for his universal law of gravitation? What are the fundamental differences between Kepler's empirical "laws" and Newton's theoretical derivation of the same relationships? And if Einstein was "right," why wasn't Newton "wrong"? Other stories explore the human side of science, such as how science has led us to think beyond the box that evolution built for us. Representing science fairly and honestly even means telling those stories that might make some students uncomfortable, like the story of why our understanding of the evolving universe and our place in it is science, while "creation science" is not.

We tell the story of motion, the story of light, the story of matter and energy, the story of Earth and our planetary system, the story of the Sun and stars, the story of life, and the story of the universe as narratives, with one idea flowing into the next and each idea connected to the whole. That is the way humans learn best. Knowledge and understanding are interwoven, with each idea and insight given meaning by how it fits into the whole. None of the stories in *21st Century Astronomy* exists in isolation. The stories are threads in a grand intellectual tapestry, woven from the recurrent themes of the physics of matter, energy, radiation, and motion.

Our writing style for *21st Century Astronomy* fits this vision of science. The writing you will find here is far less formal than that found in most textbooks. We allow ourselves to be conversational, to pursue the occasional digression, to be irreverent or provocative or idiosyncratic when it suits our taste, and to allow the sense of passion and wonder that we feel for the subject to show through from time to time. Our hope is to catch the feel of a pleasant, stimulating conversation with a student during office hours.

This brings us to the first of many tensions that were with us throughout the writing and editing of *21st Century Astronomy*—the question of who is sitting across the desk, listening to what we have to say. There is a very real pull in the market in the direction of simplification. Books filled cover to cover with the latest images from NASA missions and simple, factual descriptions look pretty and are doubtless much easier to teach out of than more conceptually challenging treatments of the material. The problem is that such texts are so accessible that what they contain sometimes seems hardly worth accessing at all. For the most part, students pass through such courses untouched, the course forgotten the moment they walk out the door.

Plenty of books can be adapted to a course aimed at the lowest common denominator. That is not the slice of the market we chose to target. Instead we hope that *21st Century Astronomy* will appeal to instructors who, like us, have found that a serious nonscience student in today's colleges and universities is capable of thinking more deeply about more conceptually challenging material than is often assumed. Frankly, with the story that *21st Century Astronomy* has to tell—a story that deals head-on with some of the most fundamental, fascinating, and far-reaching questions humans have ever asked themselves—few students can resist being drawn in if approached in the right way.

Another significant tension in the book is between a traditional organization of topics that may be more comfortable for instructors and an organization that more accurately represents the way that scientists think and work today. True to our purpose, we chose an approach that reflects the state of our science. For example, we organized the Solar System section around a theme of comparative planetology rather than the more traditional "this is Tuesday, so it must be Mars" approach. We pulled the discussion of tidal interactions, orbital resonances, chaos, and similar phenomena into a separate chapter (Chapter 10) titled "Gravity Is More Than Kepler's Laws." We posited the existence of a rotating, collapsing interstellar cloud and a rotating protoplanetary disk and discussed the formation of the Solar System in Chapter 6 *before* discussing the planets themselves. This allowed us to look at the planets from the outset within the context of how they formed. Enough examples—you get the idea.

Even so, there is flexibility in the text. For example, we feel the flow of material through the early chapters discussing physical principles works well. We have tried to include enough discussion of the human and social aspects of science to break up the treatment of conceptually difficult material. On the other hand, some instructors may like to show their students a bit more astronomy and planetary science before hitting the physics. Such instructors might want to start with Chapter 6 on the formation of the Solar System and then pull in material about orbits and radiation as it is needed in what follows. Another approach is to start with stars in Chapter 13, pull in radiation and orbits as needed to understand the properties of stars, then insert the Solar System as an extended excursion into the topic of the formation of low-mass stars. While we wrote the book to tell a complete, well-integrated story, we tried to allow for different paths tailored to the needs and tastes of individual instructors.

Astronomy is one of the most rapidly advancing fields in modern science, and as such presents a continuing challenge for textbook authors to keep up with the most recent discoveries, interpretations, and definitions. For example,

in August 2006, shortly before this Second Edition was ready to go to press, the International Astronomical Union (IAU) announced new definitions for a planet. A "classical" planet is now defined as a celestial body that (a) is in orbit around the Sun, (b) has sufficient mass for its self-gravity to overcome rigid body forces so that it assumes a hydrostatic equilibrium (nearly round) shape, and (c) has cleared the neighborhood around its orbit. Definition (c) removes Pluto from the list of classical planets, a distinction it held since its discovery in 1930. Pluto is now a member of a new class of planets called "dwarf planets." A dwarf planet has the general characteristics of a classical planet, except that it has not cleared smaller bodies from the neighboring regions around its orbit. Ceres, formerly called an asteroid, and the recently discovered distant body Eris are also dwarf planets. The IAU defines all other objects orbiting the Sun collectively as "small solar system bodies." This includes most asteroids, comets and most objects orbiting beyond Neptune. These new criteria represent fundamental and historic changes in how we define the major bodies in our Solar System.

We should point out, however, that the decision by the IAU to demote Pluto created a firestorm of controversy. Many professional astronomers and planetary scientists expressed strong disagreement with the action taken by the IAU, and cries of disappointment were heard from the lay public, including disillusioned school children. Even so, it was our decision to accept the ruling by the IAU and the majority astronomers and thereby bring this Second Edition of *21st Century Astronomy* completely up to date. Through the heroic efforts of our editors and production team we have been able to do this.

In closing we should say a few words about who we are. Among the authors of this text you will find a group of accomplished and well-known scientists who have been at the forefront of many of the most exciting and significant events in astronomy and planetary science in the second half of the 20th century. The authors of this text include members of scientific teams that built three of the cameras that have flown on the Hubble Space Telescope, team members and leaders of many of the major planetary missions including the *Voyager* and *Galileo* missions to the giant planets and the *Mars Rover* mission, and a former president of the American Association of Physics Teachers. We have been involved in significant fundamental research on topics ranging from planetary geology, to the origin and evolution of the Solar System, to the formation and evolution of stars, to the structure and dynamics of the interstellar medium, to the nature of galaxies, and to the origin of structure in the universe. There are remarkably few fields discussed in this book to which one or more of the authors have not contributed in some important way over the years. For us, 21st century astronomy is not a textbook. Rather it is the life we live. We hope you find value in our attempt to share with you what *we* see when we look up at the sky at night.

Acknowledgments

Production of a book like this is a far larger enterprise than any of us imagined when we jumped on board. We would like to thank the teachers who reviewed portions of the First Edition manuscript along the way: Jeff Adams, William Andersen, Barbara Anthony-Twarog, Allen Armstrong, Keith Ashman, Edward Baron, David Baum, Dwight Beery, William Bittle, John Blake, Anita Corn, John Cummins, Kathy Eastwood, Terry Ellis, Randy Emmons, Thomas English, Eric Feigelson, Simonetta Frittelli, Martin Gaskell, Billy Graves, Kim Griest, Erick Guerra, Javier Hasbrun, Paul Heckert, Roger Hewins, Paul Hinds, Eric Hintz, David Hufnagel, Andrew Ingersoll, Adam Johnston, Khondkar Karim, Frank Kowalski, Claud Lacy, John Laird, Kenneth LaSota, Irene Little-Marenin, Bruce Margon, Bradley Matson, George McGill, Kenneth McLaughlin, Karen Meech, Zdzislaw Musielak, Robert O'Connell, Aileen O'Donoghue, Charles Peterson, Cynthia Peterson, Randy Phelps, Carlton Pryor, George Rosensteel, Anthony Russo, Carl Rutledge, Stephen Schneider, William Schoenfeld, Michael Sitko, Tim Slater, Larry Smith, Larry Sromovsky, Thomas Statler, Christine Staver, Curtis Struck, Donald Terndrup, Suzanne Willis, Louis Winkler, and George Wolf.

We also thank the following colleagues who served as Second Edition reviewers: Scott Atkins, University of South Dakota; Peter A. Becker, George Mason University; Timothy C. Beers, Michigan State University; Juan Cabanela, Minnesota State University–Moorhead; Judith Cohen, California Institute of Technology; Paul Hintzen, California State University–Long Beach; Paul Hodge, University of Washington; William A. Hollerman, University of Louisiana at Lafayette; Adam Johnston, Weber State University; Laura Kay, Barnard College; Bill Keel, University of Alabama; Leslie Looney, University of Illinois at Urbana–Champaign; Norm Markworth, Stephen F. Austin State University; Scott Miller, Pennsylvania State University; Ata Sarajedini, University of Florida; Paul P. Sipiera, William Rainey Harper College; Donald Terndrup, Ohio State University; Richard Williamon, Emory University.

There are many at W. W. Norton & Company without whom this project would not have come together. The authors would like to thank our editors, Stephen Mosberg, who first looked across the table at us and said, "We will find a way"; John Byram, who stepped into the First Edition midstream and held it together through difficult times; and Leo Wiegman, who spearheaded the Second Edition. Roby Harrington gave us enough rope to hang ourselves, then helped us out of trouble when we tried to do just that. Among those in the trenches were Brian Loehr, Developmental Editor; Lisa Rand, Editorial Assistant; Sarah England, Associate Editor; Kim Yi, Project Editor; Meg McDonald, Copy Editor; April Lange, Science Media Editor; Karen Misler, Supplements Editor; Sally Whiting, Emedia Assistant; Neil Hoos, Manager of Photo Permissions; Kelly Mitchell, Photo Researcher; Marian Johnson, Managing Editor—College Books; Roy Tedoff, Director of Manufacturing—College; Rubina Yeh, Art Director and designer of this book; Brad Walrod, Layout Artist; Science Technologies for the Animations; Sapling Systems for SmartWork; and our ancillary authors, Scott Miller, Ann Schmiedekamp, Don Terndrup, David Wood. Finally, we would like to thank John Woolsey and Kelly Paralis of J/B Woolsey Associates and Penumbra, who became true collaborators on the project rather than simply artists drawing to spec.

Jeff Hester
David Burstein
George Blumenthal
Ronald Greeley
Bradford Smith
Howard G. Voss

The Features of *21st Century Astronomy*

The overall themes and approach of the First Edition remain, emphasizing foundational physical science concepts and stressing scientific literacy for nonmajors by showing the process of scientific discovery. The following features support students as they start chapters, while they read, and as they finish and prepare for the next topic.

Key Concepts boxes at the beginning of each chapter call attention to the main issues and topics that will be discussed in the chapter.

All truths are easy to understand once they are discovered. The point is to discover them.

GALILEO GALILEI (1564–1642)

Robotic rovers *Spirit* and *Opportunity* have roamed the surface of Mars.

CHAPTER 5

The Tools of the Astronomer

5.1 The Optical Telescope— An Extension of Our Eyes

Do you remember the first time you saw the Moon up close? Perhaps your family or a neighbor had a backyard telescope like the one shown in **Figure 5.1**. Or the Moon might have been the featured attraction during a visit to your local planetarium. With that first view came recognition of what the Moon really is—a nearly planetary world covered with craters and vast lava-flooded basins. You might also have been treated to a breathtaking look at the Orion Nebula,

FIGURE 5.1 Amateur astronomers with their telescope. There are several hundred thousand amateur astronomers in the United States alone.

KEY CONCEPTS

In the previous chapter we learned how our understanding of the physical and chemical properties of distant planets, stars, and galaxies comes to us in the form of electromagnetic radiation. But this information must first be collected and processed before it can be analyzed and converted to useful knowledge. Here we will learn about the tools astronomers use to capture and scrutinize that information. We will find that:

* Telescopes of various types collect radiation over the entire range of the electromagnetic spectrum—from gamma rays to radio signals.
* Optical telescopes come in two basic types, refractors and reflectors, but all of the larger astronomical telescopes are reflectors.
* Telescope resolution increases with aperture; image size is proportional to a telescope's focal length.
* Earth's atmosphere distorts telescopic images and prevents large parts of the electromagnetic spectrum from reaching the ground. Telescopes in orbit overcome this problem.
* The most effective way to study the planets and moons of our Solar System is to go there.

133

Key Idea statements—one-sentence summaries, appear throughout each chapter to draw students' attention to fundamental concepts as they read. They are also a useful scanning tool for review.

98 **Chapter 4** Light

1 A passing ripple causes a cork to bob up and down.

Electric field Charged oscillations particles

Electromagnetic wave

2 The alternating electric field of a passing electromagnetic wave causes a charged particle to oscillate.

FIGURE 4.7 (a) When waves moving across the surface of water reach a cork, they cause the cork to bob up and down. (b) Similarly, a passing electromagnetic wave causes an electric charge to wiggle in response to the wave.

spectrum of the light. On the long-wavelength (and therefore low-frequency) end of the visible spectrum is red light. A **micrometer**, or **micron**, is the unit often used for measuring the wavelength of visible light. A micrometer is a millionth (1/10[6]) of a meter. The abbreviation used for a micron

The spectrum of visible light is seen as the colors of the rainbow.

is μm (where μ is the Greek letter "mu"). Another commonly used unit is the nanometer, abbreviated nm. A nanometer is one billionth (1/10[9]) of a meter. The wavelengths of the light we perceive as red fall between about 600 and 700 nm.

[1] Astronomers conventionally use nanometers (nm) when referring to wavelengths at visible and shorter wavelengths, micrometers or "microns" (μm) in the infrared, and centimeters (cm) and meters (m) in the microwave and radio regions of the electromagnetic spectrum.

FIGURE 4.8 The visible part of the electromagnetic spectrum is laid out in all its glory in the colors of this rainbow.

That frequency corresponds to 580 *trillion* wave crests passing by each second!

When we say "visible light," what we mean is "the light that the light-sensitive cells in our eyes respond to." But this is not the whole range of possible wavelengths for elec-

Visible light is only one small segment of the electromagnetic spectrum.

tromagnetic radiation. Radiation can have wavelengths that are much shorter or much longer than our eyes can perceive. The whole range of different wavelengths of light is collectively referred to as the **electromagnetic spectrum**.

Follow along in **Figure 4.9** as we take a tour of the electromagnetic spectrum, beginning with visible light and working our way to shorter and longer wavelengths. We start at the blue end of the visible spectrum. Beyond this short-wavelength, high-frequency side of the visible spectrum, there is light that is "bluer than blue" or actually "more violet than violet." This light, with wavelengths between 40 and 350 nm, is called **ultraviolet (UV) radiation**. You can remember what ultraviolet light is just by looking at the name. The prefix *ultra-* means "extreme," so ultraviolet light is light that is more "extremely" violet than violet. It is important to remember that ultraviolet light is fundamentally no different from visible light, any more than high C on a piano is fundamentally different from middle C.

As we go to shorter wavelengths of light (and so to higher frequencies), we pass through the ultraviolet part of the spectrum. At a wavelength shorter than 40 nm, or

4.2 Our Picture of Light Evolved with Time 99

4×10^{-9} m, we stop calling radiation ultraviolet light and instead start calling it X-rays. This distinction comes for historical reasons. When X-rays were discovered in the last part of the 19th century, they were given the name "X" by their discoverer, **Wilhelm Conrad Roentgen** (1845–1923), to indicate they were "a new kind of ray." As we continue to even shorter wavelengths, we come to another somewhat arbitrary break. Electromagnetic radiation with the very shortest wavelengths (less than about 10^{-11} m) is referred to as **gamma rays**. Again the reasons are historical. Gamma rays (or γ-rays) were first discovered as a type of radiation given off by radioactive material. It was only later that their true nature became known.

So far we have been considering ever shorter wavelengths and higher frequencies. In principle, there is no limit to this process. We can conceive of gamma rays of arbitrarily short wavelengths and arbitrarily high frequencies (even though practical considerations eventually come into play). We can also go in the other direction. Just as there is light that is more violet than violet, there is also light that is "redder than red." Such light, covering wavelengths longer than about 700 nm and shorter than 500 μm (5×10^{-4} m), is referred to as **infrared (IR) radiation**. Again the key to remembering what infrared light is comes from looking at the word itself. *Infra-* is a prefix that means "below." *Infrared* light is light that has a frequency that is lower than (below) that of red light. When the wavelength of light gets longer than this, we start calling it **microwave radiation**. The longest-wavelength (and therefore lowest-frequency) electromagnetic

At the other end of the visible spectrum is violet light, which is the bluest of blue light. The shortest-wavelength violet light that our eyes can see has a wavelength of around 350 nm. Stretched out between the two, literally in a rainbow, is the rest of the visible spectrum. The colors in the visible spectrum in order of decreasing wavelength can be remembered as a name: "Roy G. Biv," which stands for

Red Orange Yellow Green Blue Indigo Violet.

The human eye is most sensitive to light in the green to yellow part of the spectrum. This light has a wavelength of around 500 to 550 nm. Green light with a wavelength of 520 nm has a frequency of

$$f = \frac{c}{\lambda} = \frac{3.00 \times 10^8 \frac{m}{s}}{520 \times 10^{-9} m} = \frac{5.8 \times 10^{14}}{s}$$

or 5.8×10^{14} Hz

FIGURE 4.9 By convention the electromagnetic spectrum is broken into loosely defined regions ranging from gamma rays to radio waves.

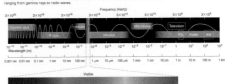

Annotated Figures throughout the book present clear, accurate science in visually compelling ways.

We believe the science must be portrayed accurately, even when it is a little complicated. Hence we kept our use of mathematical expressions within the chapter narrative, where it was in the First Edition. We rarely use more than basic algebra. And we make every effort to apply the equations with real world values, to make the math and the underlying concepts easier to understand.

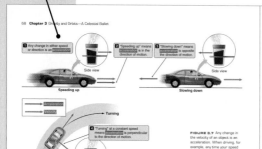

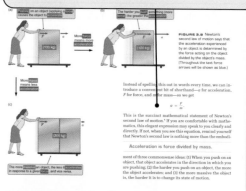

Summaries provide an outline of the key concepts as they were applied in the chapter.

Key Terms lists are ideal for quick review and reference.

Seeing the Forest through the Trees sections conclude each chapter with a thematic synthesis and a glimpse ahead to future chapters.

Applying the Concepts questions involve problem solving and critical thinking.

Thinking about the Concepts questions provide an opportunity for review and emphasize factual recall.

Each chapter ends with a reminder of additional resources available on the StudySpace website for further chapter review. See right for more details.

We retain four types of boxes sprinkled throughout the chapters. These boxes touch on material either because it is somewhat out of the mainstream of our journey or because it deserves to be highlighted.

StudySpace
wwnorton.com/astro21
provides a Study Plan for each chapter that includes a reading outline, animations, keyword flash cards, and gradebook-enabled multiple-choice quizzes. From StudySpace you can also access premium content in the ebook and SmartWork.

Foundations boxes discuss the basic science that is central to our physical understanding of the universe.

Connections boxes draw attention to recurring themes—bridges between different parts of our journey. Connections boxes have an "origins" theme in the Second Edition.

Excursions boxes are short but interesting field trips that highlight how scientists apply what they learn to solve problems.

Tools address the technology and techniques that astronomers and planetary scientists use.

So What's New in the Second Edition?

We were gratified so many instructors found the First Edition of *21st Century Astronomy* a useful teaching tool. This Second Edition is updated, revised, and expanded, but we have kept the "big story" approach to astronomy that was so well-received the first time around.

New Chapter 5 on Telescopes and Instruments The discussion of instrumentation was located principally in the First Edition's Chapter 4. We have consolidated and expanded this material to create a new self-standing Chapter 5, "The Tools of the Astronomer," to close Part I:

- Section 5.1 has expanded coverage of refractors and reflectors, resolution and diffraction, and the limitations imposed by Earth's atmosphere.

- Section 5.2 features new content about optical detectors and instruments, including expanded coverage of charge-coupled devices (CCD).

- Section 5.3 has expanded coverage of radio telescopes and new coverage of radio and optical interferometric arrays.

- Section 5.5 features new coverage of space observatories.

- Section 5.6 has new coverage of flybys, orbiters, rovers, and atmospheric probes.

- Section 5.7 discusses how high-energy colliders help astronomers understand the earliest moments of the universe.

- Section 5.8 has new coverage of high-speed computer use in astronomy.

Updated Science throughout the Second Edition The past few years have been exciting ones in astronomy, with new discoveries revealing themselves all the time. And as mentioned earlier, the IAU announced a new definition for planets in August 2006. We have infused new or updated research findings throughout the book, for example:

- Chapter 6: A new section on and expanded coverage of extrasolar planets.

- Chapter 7: New results from the *Mars Express* and *Mars Odyssey* orbiters and ground discoveries by the *Opportunity* and *Spirit* rovers.

- Chapter 8: An expanded discussion of convection and weather in a new Excursions box about thunderstorms, lightning, and tornadoes.

- Chapter 11: Exciting new discoveries about Saturn's large moon Titan and the geologically active Enceladus from the Cassini/Huygens mission, and a new section on Dwarf Planets, which includes Ceres, Pluto, and Eris.

- Chapter 12: New results from spacecraft visits to Comets Borrelly and Wild 2 and the impact on Comet Tempel 1, and new discoveries of Kuiper Belt objects.

- Chapter 17: New material on gravity waves.

- Chapter 20: New results from the Wilkinson Microwave Anisotropy Probe on cosmic background radiation.

***Origins of Life* Theme** Given the student interest in the origin of life and the growing importance of cosmology, we have added new Connections essay boxes throughout the Second Edition:

- Connections 1.1: Origins—An Introduction

- Connections 6.1: Origins—Choosing the Right Kind of Planet

- Connections 7.1: Origins—Where Have All the Dinosaurs Gone?
- Connections 8.1: Origins—On Atmospheres and Life
- Connections 10.1: Origins—Lunar Tides and Life
- Connections 11.1: Origins—Extreme Environments and an Organic Deep Freeze
- Connections 12.1: Origins—Comets, Asteroids, and Life
- Connections 16.1: Origins—Choosing the Right Kind of Star
- Connections 17.1: Origins—The Chemistry of Life
- Connections 21.2: Origins—Life, the Universe, and Everything

New Pedagogy New Summaries, Key Terms lists, and additional review questions for each chapter round out a robust program in support of the reading.

Many New Photographs and Line Art Figures To keep up with the terrific and pedagogically useful images coming out of the astronomical research community every day, we have added over 100 new photographs In addition, to help illustrate the newly added concepts we have revised almost 100 line art figures and added 30 new ones.

New Design One consequence of adding so much new art to the Second Edition is a new layout of the chapters, so readers can more readily follow the major story within chapters and continue to see the forest for the trees.

New Electronic Resources The StudySpace website is the free and open portal through which students access the new resources that accompany this text.

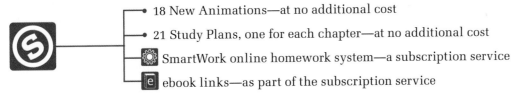

- 18 New Animations—at no additional cost
- 21 Study Plans, one for each chapter—at no additional cost
- SmartWork online homework system—a subscription service
- ebook links—as part of the subscription service

See below for more details.

Student Resources

StudySpace at wwnorton.com/astro21

Ann Schmiedekamp, *Pennsylvania State University–Abbington*

This free and open student Web site offers a sensible study plan that gives students practical assignments to integrate review and assessment resources for each chapter. It includes reading outlines, Keyword flash cards, and Gradebook-enabled concept tests and multiple-choice quizzes. It also features 18 brand new Animations by Science Technologies, which use interactivity to enhance students' understanding of core concepts. From StudySpace students can also access premium content in the ebook and SmartWork (see below for details).

Animations

Brand new for the Second Edition and developed specifically for use with *21st Century Astronomy,* these brief lessons use animation and interactivity to enhance students' understanding of core concepts:

A. The Celestial Sphere and the Ecliptic
B. The Earth Spins and Revolves
C. The Moon's Orbit: Eclipses and Phases
D. Kepler's Laws
E. Newton's Laws and Universal Gravitation
F. Light as a Wave, Light as a Photon
G. Doppler Effect
H. Atomic Energy Levels and the Bohr Model
I. Atomic Energy Levels and Light Emission and Absorption
J. Geometric Optics and Lenses
K. Solar System Formation
L. Processes That Shape the Planets
M. Tides and the Moon
N. Solar Spectrum
O. The H-R Diagram
P. The Solar Core
Q. Star Formation
R. Hubble's Law

Animations are available from the free StudySpace student web site, and are also integrated into assignable SmartWork exercises. Offline versions of the animations for classroom presentation are available from the Norton Media Library instructor's CD-ROM.

SmartWork Online Homework System

SmartWork—Norton's online homework management system—provides ready-made self-grading assignments, including guided problems, simple feedback questions, and animated tutorials—all specifically designed to extend the text's emphasis on critical thinking.

Developed in collaboration with Sapling Systems, SmartWork features an intuitive and easy-to-use interface that offers instructors flexible tools to manage assignments. Helpful and immediate feedback makes it easy for students to assess their understanding of basic concepts. Two types of questions expand on the exposition of concepts in the text:

- **Simple Feedback Problems** anticipate common misconceptions and offer prompts to help them discover the correct answer.

- **Guided Tutorial Problems** address more challenging topics. If a student answers a problem incorrectly, SmartWork guides the student through a series of discrete tutorial steps. Each step is a simple feedback question that the student answers—with hints, if necessary. After completing all of the tutorial steps, the student returns to the original question ready to apply this newly obtained knowledge.

Instructors can easily use these ready-made questions and assignments, customize them to address specific course objectives, or use SmartWork to create their own.

SmartWork and ebook integration SmartWork is available as a standalone purchase or with an integrated ebook version of *21st Century Astronomy*. Links to the ebook make it easy for students to consult the text while completing their homework assignments.

Starry Night Pro 5.0 CD-ROM and Workbook

Workbook by Donald Terndrup, *The Ohio State University*

The remarkably realistic and user-friendly Starry Night Pro 5.0 CD-ROM allows students to explore stars and objects in our cosmic neighborhood and beyond. The accompanying workbook includes 12 observation exercises that guide students' virtual explorations of the night sky and help them apply what they've learned from the text.

Ebook

21st Century Astronomy is also available in a Norton ebook format, a convenient alternative that retains all of the print book's content. The ebook offers a variety of tools for study and review, including sticky notes, highlighters, zoomable images, links to Animations, and a search function.

The ebook is available as a stand-alone item or packaged with SmartWork. The SmartWork/ebook package makes it easy for students to consult the text when completing their homework assignments. A downloadable PDF version of the ebook is also available from Powells.com. Go to norton**ebooks**.com for more information.

Instructor Resources

Instructor's Manual and Test Bank

Scott Miller, *Pennsylvania State University, University Park*
David Wood, *San Antonio College*

Thoroughly revised and expanded for the Second Edition, this resource includes brief chapter overviews, worked solutions to the end-of-chapter problems, instructor's notes for the Starry Night Workbook activities and approximately 1,000 true-false, multiple-choice, and short answer test questions. The Test Bank is also available in rich-text, ExamView® Assessment Suite, BlackBoard, and WebCT formats.

PowerPoint Lecture Outlines

Donald Terndrup, *The Ohio State University*

These ready-made lecture outlines include selected art from the text, "clicker" questions, and offline and lecture-ready versions of the Animations.

Norton Media Library CD-ROM

This helpful resource includes the PowerPoint lecture outlines with "clicker" questions, offline versions of the Animations, plus selected photographs and all drawn art from the text.

BlackBoard and WebCT Course Cartridges

Course cartridges for BlackBoard and WebCT include access to the Animations, a Study Plan for each chapter, multiple-choice tests, and links to ebook and SmartWork premium content.

Transparencies

Acetates for approximately 200 figures from the text are available to qualified instructors. Contact your local representative for details.

Video Library

A video library will be available to qualified instructors. Contact your Norton representative for details.

About the Authors

Jeff Hester received his Ph. D. from Rice University and is currently a professor of physics and astronomy at Arizona State University. His research interests are the interstellar medium in the Milky Way and external galaxies; structure of the diffuse ISM; interstellar shock waves; supernova remnants; pulsar wind interactions; Herbig-Haro objects; and H II region structure.

David Burstein was born in Englewood, New Jersey on May 19, 1947. After graduating as valedictorian from his high school, he went to Wesleyan University in Middletown, Connecticut for his undergraduate career and University of California–Santa Cruz for his graduate career. After getting his Ph.D. in astronomy and astrophysics, he went to the Department of Terrestrial Magnetism, Carnegie Institution of Washington for 2 years, then on to the National Radio Astronomy Observatory for 3 years. He is now at Arizona State University. Gail and Dave have been married for 35 years as of June 2006, and have two children—Jon who is a reporter for the *South Florida Sun-Sentinel,* and Elizabeth, who is a first grade teacher in the same class in which she was a student.

George Blumenthal is Acting Chancellor at the University of California–Santa Cruz, where he has been a Professor of Astronomy and Astrophysics since 1972. Chancellor Blumenthal received his B.S. degree from the University of Wisconsin–Milwaukee and his Ph.D. in physics from the University of California–San Diego. As a theoretical astrophysicist, Chancellor Blumenthal's research encompasses several broad areas, including the nature of the dark matter that constitutes most of the mass in the universe; the origin of galaxies and other large structures in the universe; the earliest moments in the universe; astrophysical radiation processes; and the structure of active galactic nuclei such as quasars. Besides his teaching and research, Chancellor Blumenthal has served as the chair of the UC Santa Cruz Astronomy and Astrophysics Department, has chaired the Academic Senate for both the UC Santa Cruz campus and the entire University of California system, and has served as the Faculty Representative to the UC Board of Regents.

Ronald Greeley is Regents' Professor in the School of Earth and Space Exploration at Arizona State University. After completing his Ph.D. in geology in 1966, he worked for Standard Oil for a year in exploration and then joined NASA-Ames Research Center, where he remained until 1978 when he went to ASU. While at Ames, he led a consortium of planetary scientists and engineers to design and implement the Mars Surface Wind Tunnel, which he continues to manage as a NASA faculty. Ron was a science team member on *Viking, Galileo-Jupiter, Magellan-Venus, Mars Pathfinder,* and the ill-fated *Soviet Mars 96* orbiter and is currently on the *Mars Exploration Rover* and *Mars Express* orbiter science teams. He has chaired numerous NASA and National Research Council committees charged with formulating plans for solar system exploration.

Bradford Smith has served as an Associate Professor of Astronomy at New Mexico State University, a Professor of Planetary Sciences and Astronomy at the University of Arizona, and a Research Astronomer at the University of Hawaii. Through his interest in Solar System astronomy, he has participated as a team member or imaging team leader on several US and international space missions, including *Mars Mariners* 6, *7* and *9, Viking, Voyager,* and the Soviet *Vega* and *Phobos* missions. More recently, Smith's interests have turned to other planetary systems, working as a team member of the HST NICMOS experiment. He has four times been awarded the NASA Medal for Exceptional Scientific Achievement. Smith is a member of the IAU Working Group for Planetary System Nomenclature and is Chair of the Task Group for Mars Nomenclature. He is now semi-retired but remains affiliated with the Institute for Astronomy at the University of Hawaii, Manoa.

Howard G. Voss is Professor Emeritus of Physics at Arizona State University, where he taught for over four decades, served as chair of the Department of Physics, and received the Distinguished Faculty Award and the Dean's Teaching Award. He was awarded the Melba Phillips Medal by the American Association of Physics Teachers, which he served as president, secretary, a member of the Executive Board, and in other offices. He also served the American Institute of Physics in several positions including as chair of the Publishing Policy Committee and as a member of the Governing Board.

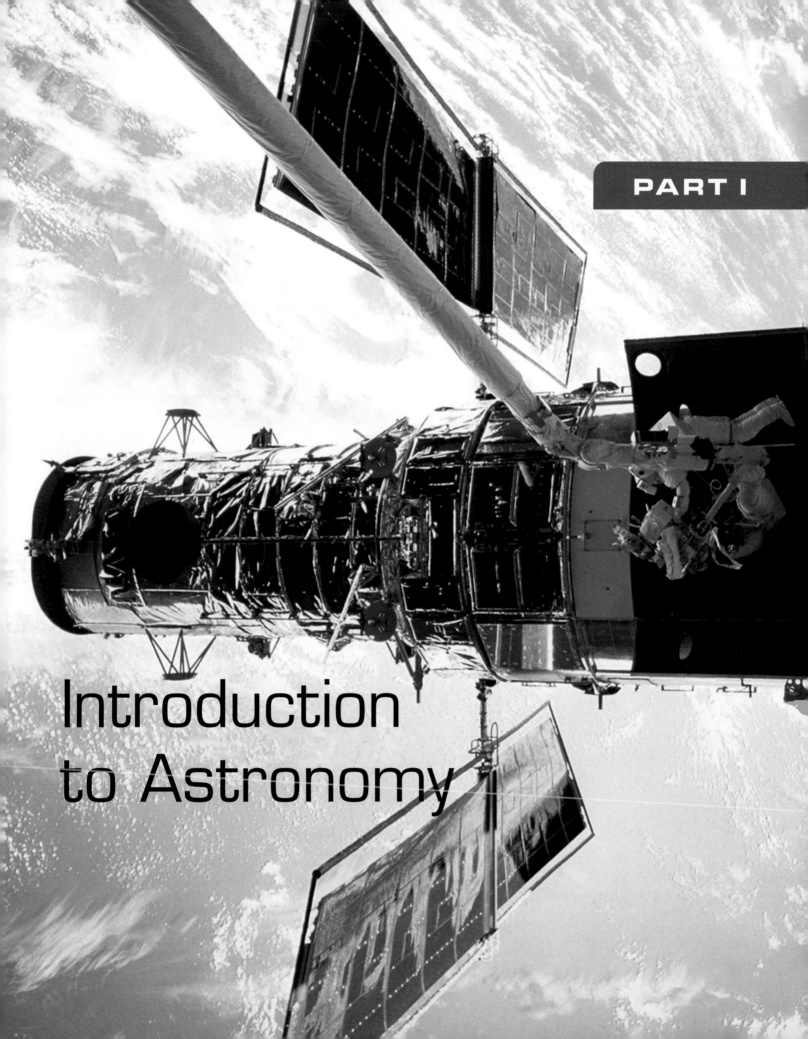

PART I

Introduction to Astronomy

The most beautiful thing we can experience is the mysterious.
It is the source of all true art and all science.
He to whom this emotion is a stranger,
who can no longer pause to wonder and stand rapt in awe,
is as good as dead: his eyes are closed.

ALBERT EINSTEIN (1879–1955)

The moon seen rising over the 2400 year old Greek Temple of Poseidon.

CHAPTER 1

Why Learn Astronomy?

1.1 Starting with a Spark of Interest

Not everyone is fascinated by science, but almost everyone harbors a spark of interest in astronomy. Because you are reading this book, you probably share this spark as well. The spark may have been struck when you were a child looking at the sky and found yourself wondering about what you saw there. What are the Sun and Moon made of? How far away are they? What are the stars? How do they work? Do they have anything to do with me? The prominence of the Sun, Moon, and stars in cave paintings and rock drawings (such as those in **Figure 1.1**) dating back thousands of years tells us that these questions have long occupied the human imagination. Your initial spark of interest in astronomy may have grown over the years as you saw or read news reports about spectacular discoveries made in your lifetime. Some of these discoveries may have sounded so amazing that it was difficult to draw the line between science fact and science fiction.

If you nurture your spark of interest in astronomy as you continue through this book, you may be surprised to find that spark growing into a flame. The title of this book—*21st Century Astronomy*—was chosen to emphasize that this is the most fascinating time in history to be studying this most ancient of sciences. This book will take you to places you never imagined going and will lead you to insights and understandings you never imagined having. To those of you who are reading this book for a course in astronomy at your college or university, we have a special note. The authors of this text have taught introductory astronomy many times over the years. We recognize that you may be

KEY CONCEPTS

Before traveling through unfamiliar terrain, it helps to have some idea of where you are going, what you might see along the way, and what you should pack for the journey. In *21st Century Astronomy*, we will learn not only about the wonders of the universe but also about what it means to look at the world through the eyes of science. In this chapter's overview of what is to come, we will find that

- The universe is vast beyond all human experience, yet it is governed by the same physical laws that shape our daily lives.

- We are a product of that universe; the very atoms of which we are made were formed in stars that died long before the Sun and Earth were formed.

- Science is a creative human activity like art, literature, and music, and it is also a remarkably powerful, successful, and aesthetically beautiful way of viewing the world.

- Understanding comes from thinking carefully and deeply about patterns in the world, not simply from memorization of facts.

- Like climbing a mountain, the journey we are about to make requires effort, but the view from the top is amazing to behold.

FIGURE 1.1 Ancient petroglyphs often include depictions of the Sun, Moon, and stars.

in this course primarily because you need a science credit to graduate. As you flipped through your course catalog, perhaps you were reminded of your interest in astronomy, and that led you to choose astronomy over your other options. (Or perhaps you simply considered astronomy to be the least of the available evils!) Whatever your expectations, the story in *21st Century Astronomy* can fascinate you if you open your mind to it.

The journey of discovery on which we are embarking is not always easy, but few worthwhile journeys are. A hike in the mountains can at times be an easy stroll and at other times a more strenuous climb; but when you arrive, the view from the top is hard to beat. In much the same way, this book will ask you to exercise your mental muscles in different, possibly unaccustomed ways. But as with the hike in the mountains, we feel certain that you will find the rewards worth the investment.

Getting a Feel for the Neighborhood

If you are like many people, your conception of astronomy may not go much beyond learning about the constellations and the names of the stars in them. Loosely translated, the word **astronomy** means "patterns among the stars." But modern astronomy—the astronomy we will talk about in this book—has become far more than looking at the sky and cataloging what is visible there. It may seem something of a contradiction, but a great deal of frontline astronomy is now carried out in physics laboratories like the one shown in **Figure 1.2**. Today astronomers work along with their colleagues in related fields such as **physics**, **chemistry**, **geology**, and planetary science to sharpen our understanding of the physical laws that govern the behavior of **matter** and

energy and to use this understanding to make sense of our observations of the cosmos.

We are confident in our answers to many of the questions that you may have asked yourself as a child when you looked at the sky. We all live on a planet called Earth, which is orbiting under the influence of gravity about a star called the Sun. The Sun is an ordinary, middle-aged star, more massive and luminous than some stars but less massive and

---

Earth exists in the context of the Universe.

---

luminous than others. The Sun is extraordinary only because of its importance within our own Solar System. The Sun is located about halfway out from the center in a flattened collection of approximately 100 billion stars referred to as the **Milky Way Galaxy**. The Milky Way in turn is a member of a small collection of a few dozen galaxies called the **Local Group**, which is part of a vastly larger collection of thousands of galaxies called a **supercluster**. But even this vast structure is part of the *local* universe. The part of the universe that we can see extends outward for the distance that **light** travels in 13.7 billion years, and in this volume we estimate that there are about 100 billion galaxies— roughly as many galaxies as there are stars in the Milky Way!

One of the first conceptual hurdles that we face as we begin to think about the universe is its sheer size. If a hill is big, then a mountain is really big. If a mountain is really big, then Earth is enormous. But where do we go from there? We quickly run out of superlatives as the scale of what we are talking about comes to dwarf our human experience. One technique that can help us develop a sense for the size

FIGURE 1.2 This laboratory, where physicists are studying the properties of atoms, might seem an unlikely place to be doing astronomy. But laboratory astrophysics, studying astronomically important physical processes in a laboratory, has become an important part of astronomy.

of things in the universe is to use a little sleight of hand and move from discussing distance to talking instead about time. If you are driving down the highway at 60 miles per hour, a mile is how far you go in a minute. Sixty miles is how far you go in an hour. Six hundred miles is how far you go in 10 hours. So to get a feeling for the difference in size between 600 miles and 1 mile, you can think about the difference between 10 hours and a single minute.

We can play this same game in astronomy, but the **speed** of a car on the highway is far too low to be useful. Instead we will use the greatest speed in the universe—the speed of light. Light travels at a speed of 300,000 kilometers per second. At that speed, light can circle Earth (a distance of 40,000 kilometers) in just under $\frac{1}{7}$ of a second—about the time it takes you to snap your fingers. Fix that comparison in your mind. The size of Earth is like—*snap!*—a snap of your fingers. Follow along in **Figure 1.3** as we move outward into the universe. We next encounter the Moon, 384,000 kilometers away, or a bit over $1\frac{1}{4}$ seconds when moving at the speed of light. So if the size of Earth is a snap of your fingers, the distance to the

---

Light travel time helps in understanding size.

Moon is about the time that it takes to turn a page in this book. Continuing on, we find that at this speed the Sun is $8\frac{1}{3}$ minutes away, or the length of a hurried lunch at the student union. Crossing from one side of the orbit of Eris, the outermost planetary body in our Solar System, to the other takes about 19 hours. Think about that for a minute. Let it sink in. Comparing the size of Eris's orbit to the circumference of Earth is like comparing the time of a long plane ride to Australia to a single snap of your fingers.

Yet in crossing Eris's orbit we have only just begun our journey. Many steps remain. It takes us a bit over four years to cover the distance from Earth to the nearest star (other than the Sun), or as much time as you spent in high school. At this point even our analogy using light travel time can no longer bring astronomical distance to a human scale. Light takes about 100,000 years to travel across our galaxy—about the time that modern humans (*Homo sapiens*) have walked the surface of Earth. To reach the nearest large galaxies beyond our own takes several million years, or the time since our australopithecine ancestors appeared on the scene. To reach the limits of the currently observable universe takes light 13.7 billion years—the age of the universe, or about three times the age of Earth.

Look at that comparison again. The size of Earth is to the vast expanse of the universe as a single snap of your fingers is to three times the amount of time that has passed since the Sun and Earth were formed! Here is something to ponder the next time you look up at a star-filled summer sky.

As you read this text, you will occasionally see material set aside in boxes. A topic has been boxed either because it is somewhat out of the mainstream of our journey or because we wish to highlight it. In particular,

- **Foundations** boxes address material that is central to our physical understanding of the universe.

- **Connections** boxes draw attention to recurring themes —bridges between different parts of our journey.

- **Excursions** boxes are short but interesting side trips.

- **Tools** boxes discuss the technology and techniques that astronomers and planetary scientists use.

Glimpsing Our Place in the Universe

While seeking knowledge about the universe and how it works, modern astronomy and physics have repeatedly come face-to-face with a number of age-old questions long thought to be solely within the domain of philosophers. Issues as seemingly metaphysical as the origin and fate of the universe and the nature of space and time have become the subjects of rigorous scientific investigation. The answers we are finding to these questions are often far more wondrous than our predecessors could have dreamed. They are changing not only our view of the cosmos, but our view of ourselves as well.

Figure 1.4 envisions a traveler raising the veil of the heavens to see what lies there. Throughout most of history, philosophers looked at the universe and saw it as remote and different from Earth—disconnected from our terrestrial existence. When modern astronomers look at the universe, they see instead a network of ongoing processes that we are a part of. Astronomy begins by looking out at the universe, but increasingly that outward gaze turns introspective as we come to appreciate that our very existence is a consequence of those same processes.

The study of the chemical evolution of the universe is such a case. As a result of both observation and theoretical work, we now understand that when the universe was young (13.7 billion years ago), the only chemical elements found in abundance were hydrogen and helium, plus tiny

---

We are stardust.

---

amounts of lithium, beryllium, and boron. Yet we are not made exclusively of these lightest elements. Our bodies are built of carbon, nitrogen, oxygen, sodium, phosphorus, and a host of other chemical elements. We live on a planet with a core consisting mostly of iron and nickel, surrounded by a mantle made up of rocks containing large amounts of silicon and other elements. If these more massive elements were not present in the early universe, where did they come from?

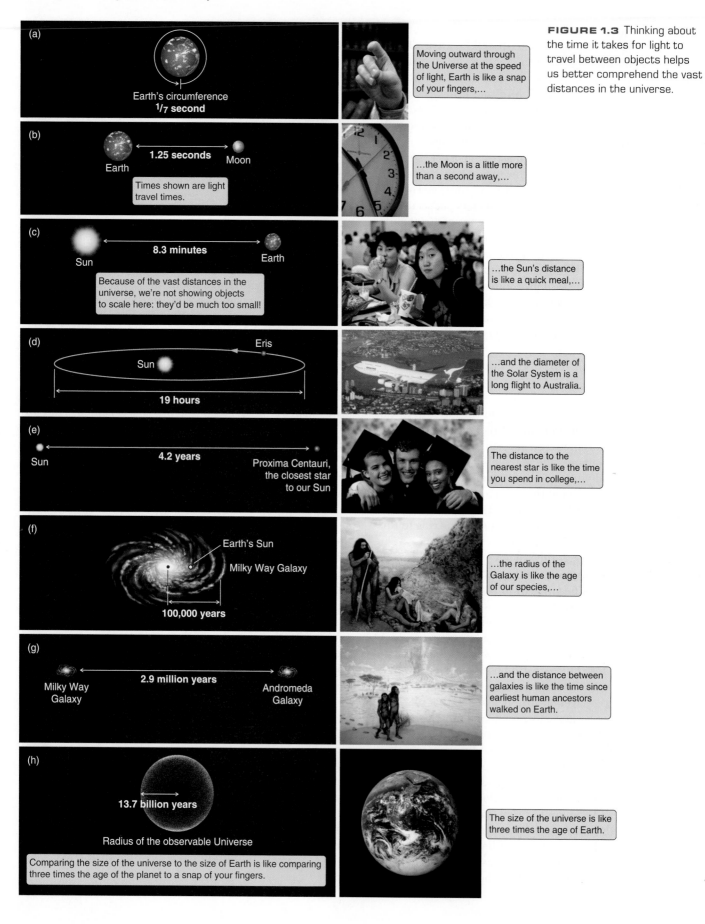

FIGURE 1.3 Thinking about the time it takes for light to travel between objects helps us better comprehend the vast distances in the universe.

(a) Earth's circumference ¹/₇ second

Moving outward through the Universe at the speed of light, Earth is like a snap of your fingers,…

(b) Earth — 1.25 seconds — Moon

Times shown are light travel times.

…the Moon is a little more than a second away,…

(c) Sun — 8.3 minutes — Earth

Because of the vast distances in the universe, we're not showing objects to scale here: they'd be much too small!

…the Sun's distance is like a quick meal,…

(d) Sun — Eris — 19 hours

…and the diameter of the Solar System is a long flight to Australia.

(e) Sun — 4.2 years — Proxima Centauri, the closest star to our Sun

The distance to the nearest star is like the time you spend in college,…

(f) Earth's Sun — Milky Way Galaxy — 100,000 years

…the radius of the Galaxy is like the age of our species,…

(g) Milky Way Galaxy — 2.9 million years — Andromeda Galaxy

…and the distance between galaxies is like the time since earliest human ancestors walked on Earth.

(h) 13.7 billion years — Radius of the observable Universe

Comparing the size of the universe to the size of Earth is like comparing three times the age of the planet to a snap of your fingers.

The size of the universe is like three times the age of Earth.

FIGURE 1.4 Throughout most of history humans conceived of the rest of the universe as a place apart from us. Here a traveler raises the curtain of the firmament to get a glimpse of what lies beyond.

To answer this question we have but to look at the lights in the night sky. The energy to power stars comes from nuclear fusion reactions that occur deep within their interiors. Fusion reactions in stars take less massive **atoms** like hydrogen and combine them, forming more massive atoms, accomplishing the alchemist's dream of transforming one element into another. When a star exhausts its nuclear fuel and nears the end of its life, it often loses much of its mass—including some of the new atoms formed in its interior—blasting it back into interstellar space. We will talk later about the life and death of stars. For now it is enough to note that our Sun and Solar System formed from a cloud of interstellar gas and dust that had been "polluted" by the chemical effluent from earlier generations of stars. This chemical legacy supplies the building blocks for the interesting chemical processes that go on around us—chemical processes such as life. **Figure 1.5** symbolizes this intimate relationship between the world around us and our heritage in the stars. Look around you. The atoms that make up everything you see were formed in the hearts of stars. Poets sometimes say that "we are stardust," but this is not poetry. It is literal truth.

As humans, we have long speculated about our beginnings. Who or what is responsible for our existence? How were Heaven and Earth created? In the modern world, primitive creation myths have largely given way to scientific theory.[1] And so the topic of *origins* is one that you will en-

---

[1] A few cultures still retain their creation myths. And even within "enlightened" society, certain factions have ignored science in favor of nonscientific "creationism" and "intelligent design."

counter frequently in our journey. We have developed this as a recurring theme that we call (what else?) "Origins." It begins with **Connections 1.1**.

We Live in an Age of Exploration and Discovery

Another reason this is a fascinating time to be learning about astronomy is that we live in an age of exploration. The 1957 launch of *Sputnik,* the first human-made satellite, occurred just one year before the birth of the youngest of the authors of this book. Five decades later, as we begin the 21st century, we have seen humans walk on the Moon (**Figure 1.6**) and have sent unmanned probes to visit all of the classical planets. (We will occasionally refer to eight "classical" planets—Mercury, Venus, Earth, Mars, Jupiter, Saturn, Uranus, and Neptune—to distinguish them from the new "dwarf" class of planets adopted by the International Astronomical Union

---

Space exploration has expanded our view of the universe.

(IAU).) Spacecraft have flown by asteroids, comets, and even the Sun. Our inventions have landed on Mars, Venus, and Titan (Saturn's largest moon) and have plunged into the atmosphere of Jupiter. Most of what we know of the Solar System has been learned over these past five decades as a result of this burst of exploration.

Satellite observatories in orbit around Earth have also given us many new perspectives on the universe. The same atmosphere that shields us from harmful solar radiation also

FIGURE 1.5 You and everything around you, including beautiful waterfalls, are composed of atoms that were forged in the interior of stars that lived and died before the Sun and Earth were formed. The left panel shows a cloud of chemically enriched material that has been ejected from the star Eta Carinae.

CONNECTIONS 1.1

Origins—An Introduction

How and when did the universe begin? What combination of events—some probable, others much less likely —have led to our existence as sentient beings living on a small rocky planet orbiting a typical middle-aged star? Was this a unique happening, or are there others like us scattered throughout the galaxy?

Throughout our journey we will encounter a recurring theme that addresses questions such as these. We call it **origins**. To get a good feeling for this subject we must cast a wide net because origins involves much more than how we humans came to be. We'll look into the genesis of terrestrial life, but we will also examine the possibilities of life elsewhere in our Solar System and beyond, a subject called **astrobiology.** For life to exist around other stars in the galaxy, there must be life-sustaining planets to support it. So our origins theme will include the discovery of extrasolar planets and how they compare with the planets of our own Solar System.

For example, we have learned in this chapter that the early universe was made up almost entirely of only two elements, hydrogen and helium, with just a touch of lithium, beryllium, and boron thrown in. Further along we'll discover why hydrogen and helium were the big winners in a minutes-old universe that had barely cooled down from the Big Bang, and how they were later transformed into more massive elements such as carbon, nitrogen, oxygen, sulfur, and phosphorus, the very atoms that make up the **molecules** of life. We will find that some elements were created in a relatively benign environment in the cores of stars like our Sun, whereas others had their origin in the violence of unimaginable cosmic explosions. Putting it all together, we'll see that we ourselves, the terrestrial life around us, our Earth, the other planets in our Solar System, and those beyond have a common origin: All are made of recycled stardust.

---

blinds us to much of what is going on around us. Space astronomy continues to show us vistas hidden from the gaze of groundbased **telescopes** by the protective but obscuring blanket of our atmosphere. Satellites capable of detecting radiation—ranging from gamma rays and X-rays, to ultraviolet radiation, to infrared radiation to microwaves—have brought surprising discovery after surprising discovery. Each has forever altered our perception of the universe, further expanding the domain of the human mind. At the same time, the closing years of the 20th century also witnessed a renewed vigor in astronomical observations from the surface of Earth. The view of the sky seen by radio telescopes shown in **Figure 1.7** illustrates the new perspectives that have been opened by our growing technological prowess.

Astronomy has also benefited enormously from the computer revolution. The 21st-century astronomer spends far more time peering at a computer screen than peering through the eyepiece of a telescope. Computers are used to do everything from collecting and analyzing data from telescopes, to calculating physical models of the conditions that exist in the hearts of stars, to preparing and disseminating the results of our work.

We truly live in a golden age of exploration and discovery. When we look back at the Renaissance (15th and 16th centuries), we recall few of the concerns that dominated the day-to-day existence of those alive at that time. Instead we

remember the Renaissance as a time of great art, literature, and music. We remember it especially as a time when the spirit of inquiry was reawakened and when much of what we think of as science was born. What might a historian 500 years in the future consider to be of lasting significance about *our* time? It is doubtful that our hypothetical historian of the future will care much about who won the Super Bowl, or which performer was at the top of the pop charts, or which brand of toothpaste tasted best. The media spectacles that often dominate public consciousness will merit little more than an occasional footnote.[2]

Much of what seems so significant to us today will be dust in the wind 500 years from now. However, we can be certain that our future historian *will* remember ours as the time when humankind first stepped beyond the world of our birth and began to reach out with our minds and our science to touch the fabric of the universe itself. It is probably safe to say that few things will have a more lasting impact on our culture than this revolution in our understanding of the universe and our place in it. No history book will ever again be complete without the headline in **Figure 1.8**. What has yet to be determined is whether our future historian

---

[2] That footnote will probably comment on a civilization whose technical ability to spread information had far outstripped its judgment about what information was worth spreading.

(a)

Apollo lunar rover (1971)

KEY	Space observatories	Lunar and planetary explorers	Historical

Galileo (1989–), first Jupiter orbiter, atmospheric probe

Chandra (1999–), Advanced X-ray Astronomy Observatory (AXAF)

Hubble Space Telescope (1990–), UV, visible, infrared astronomy

(b)

Mariner 4 (1964–1965), first images of Mars

Surveyor 1 (1966), lunar lander

Sputnik (1957–1958), USSR, first human-made satellite

Viking lander 1 (1975–1982), first of two Mars landers

FIGURE 1.6 (a) *Apollo 15* astronaut James B. Irwin stands by the lunar rover during an excursion to explore and collect samples from the Moon. (b) Artificial satellites and space probes have progressed a long way since the 1957 launch of *Sputnik I*. These spacecraft are all shown to the same scale. Some are astronomical observatories that view space from Earth's orbit. Others are interplanetary explorers sent to investigate other worlds within our Solar System.

will remember us for reaching out to touch the universe and embracing what we found, or whether we will instead be remembered for a loss of spirit—for stepping back and turning away from the frontier of exploration and discovery. The direction we take from here is a decision in which you will play a part.

1.2 Science Is a Way of Viewing the World

As we look at the universe through the eyes of astronomers, we will also learn something of how **science** works. It is almost impossible to overstate the importance of science in our civilization. One obvious manifestation of science is the fact that almost everything you use in your everyday life is a product of our scientific understanding of the world. The historical "simple life" is often romanticized, but it would be hard for any of us today even to begin to imagine what a truly pretechnological existence was like. Even a "primitive" trip into the backcountry is often accomplished today by eating freeze-dried food, sheltering under ultralight, ultrastrong synthetic fabrics, using a cellular phone to stay minutes

FIGURE 1.7 In the 20th century, new tools opened new windows on the universe. This is the sky as we would see it if our eyes were sensitive to radio waves, shown as a backdrop to the National Radio Astronomy Observatory site in Green Bank, West Virginia.

FIGURE 1.8 There is no doubt that history will remember ours as the time when humankind first stepped beyond our home world and reached out with our minds to embrace the universe.

away from emergency medical assistance, and following maps using a handheld global positioning system (GPS) receiver. Given the visible importance of technology in our lives, you might even be tempted to say that science *is* technology. It is true that science forms the basis of technology and that a mutually supportive relationship exists between science and technology, in which each enables advances in the other. Yet science is much more than technology. Science can no more be reduced to its practical technological application than the accomplishment of an Olympic athlete can be reduced to the utilization of the athlete's Olympic fame to market shoes and other products.

If science is more than technology, you might instead suggest that science is the **scientific method**. When you took science in high school you probably had the scientific method drilled into you—hypothesis and theory, followed by prediction, followed by experiments to test those predictions. There is good reason for placing emphasis on the sci-

entific method. For all practical purposes, it defines what we mean when we use the verb *to know*. It is sometimes said

The scientific method involves trying to falsify ideas.

that the scientific method is how scientists prove things to be true, but actually it is a way of proving things to be *false*. Before scientists accept something as true, they work hard to show that it is false. Only after repeated attempts to disprove an idea have failed do scientists begin to accept its likely validity. This is important! For a theory to be given serious scientific consideration, it *must* be *falsifiable*. It must be capable of being shown to be false. Scientific theories are accepted only as long as they are able to be tested and are not shown to be false.

But science can no more be said to *be* the scientific method than music can be said to *be* the rules for writing down a musical score. The scientific method provides the rules for asking **nature** whether an idea is false, but it offers no insight into where the idea came from in the first place, or how an experiment was designed. If you were to listen to a group of scientists discussing their work, you might be surprised to hear them using words such as *insight, intuition,* and *creativity.* Scientists speak of a beautiful theory in the same way that an artist speaks of a beautiful painting or a musician speaks of a beautiful performance. Yet science is not the same as art or music in one important re-

Nature is the arbiter of science.

spect. Whereas art and music are judged by a human jury alone, in science, it is nature (through the application of the scientific method) that provides the final decisions about which theories can be kept and which theories must be discarded. Nature is completely unconcerned about what we *want* to be true. In the history of science many a beautiful and beloved theory has been abandoned. At the same time, however, **Figure 1.9** makes the point that there is an aesthetic to science that is as human and as profound as any found in the arts.

It is also incorrect to say that science is a body of facts. We do not pretend to have all the answers, and we are constantly having to refine our ideas in response to new data and new insights. (This is, after all, what it means to learn.) The vulnerability of knowledge that is implicit in the sci-

All scientific knowledge is provisional.

entific method may seem like a weakness at first. "Gee, you really don't know anything," the cynical student might say. But this vulnerability is actually science's great strength. It is what keeps us honest. Once an idea is declared to be "truth," then all progress stops. In science even our most cherished ideas about nature remain fair game, subject to challenge by evidence according to the rules of the scien-

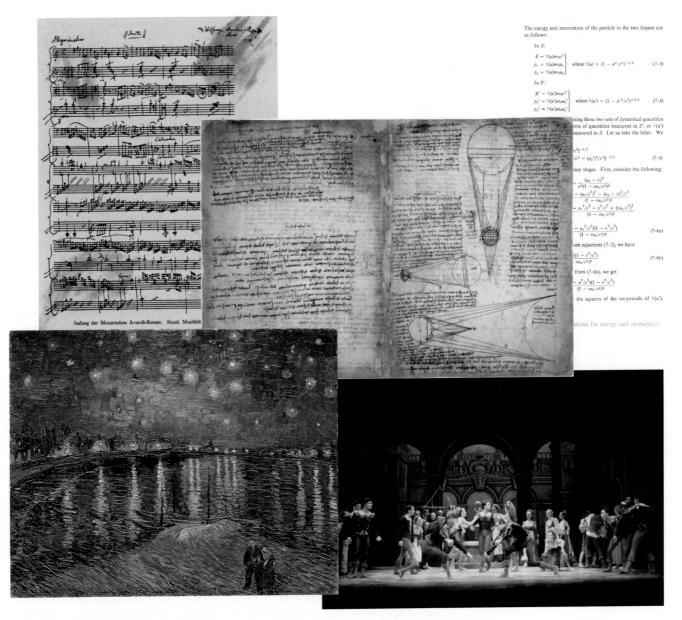

FIGURE 1.9 The scientific worldview is as aesthetically pleasing as that of art, music, or literature; but unlike the arts, nature has the final say about what scientific theories have lasting value.

tific method. Many of history's best scientists earned their place in the forward march of knowledge by successfully goring a sacred cow.

Scientists spend most of their time working within a framework of understanding, extending and refining that framework, and testing its boundaries. Occasionally, however, major shifts occur in the framework of some scientific field itself. Many books have been written about how science progresses, perhaps the most influential being *The Structure of Scientific Revolutions* by Thomas Kuhn. In this work Kuhn emphasizes the constant tension between the scientist's human need to construct a system of beliefs within which to interpret the world, and the occasional

(and likely painful) need to drastically overhaul that system of beliefs.

A scientific revolution is not a trivial thing. We cannot just wish that the universe were one way or another and then expect the universe to oblige. A new theory or way of viewing the world must be able to explain everything that the previous theory could while extending this understanding to new territory into which the earlier theory could not go. If one face can be said to symbolize modern science, it is that of Albert Einstein (**Figure 1.10**). Einstein's theories of special and general relativity replaced the 300-year-old edifice of Newtonian mechanics not by proving Newton wrong, but by showing that Newton's mechanics was a

FIGURE 1.10 Albert Einstein, perhaps the most famous scientist of the 20th century, and *Time* magazine's selection as Person of the Century. Einstein helped to usher in two different scientific revolutions, one of which he himself was never able to accept.

special case of a far more general and powerful set of physical laws. Einstein's new ideas unified our concepts of mass and energy and destroyed our conventional notion of space and time as separate things. Yet scientific revolutions are seldom comfortable for those who live through them, and even the greatest of scientists can be left behind. Einstein actually helped to start two scientific revolutions. He saw the first of these—relativity—through and embraced the world that it opened. Yet Einstein was unable to accept the implications of the second revolution he helped start —quantum mechanics—and went to his grave unwilling to embrace the view of the world it offered.

Science is not simply a body of facts. Science is not simply technology. Science is not simply the scientific method. Perhaps more than anything, science is a way of thinking about the world. It is a way of relating to nature. It is a search for the relationships that make our world what it is. It is a belief that nature is not capricious but instead operates by consistent, explicable, inviolate rules. It is a collection of ideas about how the universe works, coupled with an acceptance of the fact that what is known today may be superseded tomorrow. The scientist's faith is that there is an order in the universe and that the human mind is capable of grasping the essence of the rules underlying that order—or at least of inventing ever better approximations to those rules. The scientist's creed is that nature, through observation and experiment, is the final arbiter of the only thing worthy of the term *objective truth*. Science is an exquisite blend of aesthetics and practicality. And in the final analysis, science has found such a central place in our civilization because *science works.*

It sometimes may seem to you that science is arbitrary. For example, in August 2006 the International Astronomical Union (IAU) redefined what astronomers mean when they call something a planet. Under this new definition Pluto, known to all of us as the "ninth planet" since its discovery in 1930, was unceremoniously stripped of its "planet" status and demoted to "dwarf planet." Ceres, long regarded as the largest asteroid, was also redefined as a "dwarf planet." But it is important to keep in mind that this controversial decision, unsupported by many astronomers, was really more a matter of semantics than science. Pluto and Ceres are still the same scientifically important Solar System bodies that they always were. It is just that "science" has now put new labels on them.

It is beyond the scope of this book for us to provide you with a detailed justification for all that we will say. However, we will try to offer some explanation of where an idea comes from and why we believe it to be valid. We will not present something as fact unless there is a compelling reason to believe it. We will be honest when we are on uncertain, speculative ground, and we will admit it when the truth is that "we really do not know." This book is not a compendium of revealed truth or a font of accepted wisdom. Rather, it is an introduction to a body of knowledge and understanding that was painstakingly built (and sometimes torn down and rebuilt) brick by brick.

For those of us who grew up in a world transformed by science, the scientific worldview might seem anything but subversive. However, for much of history, knowledge was sought in the pronouncements of "authority" rather than through observation of nature. This authoritarian view slowed the advance of knowledge throughout Western Europe for the millennium prior to the European Renaissance, and it was largely the Chinese and Arab cultures that kept the spark of inquiry alive during this time. The greatest scientific revolution of all was the one that overthrew "authority" and replaced it with rational inquiry and the scientific method. Science is not *just* one of many possible worldviews. Science is the most successful worldview in the history of our species. It is worth noting that so far science itself has passed its own test. The foundations of the scientific worldview have withstood centuries of fine minds trying to prove them false.

The Cosmological Principle

The scientific revolution brought about a dramatic shift in our ideas about what knowledge is and how it is sought, but this change alone was not enough to open the universe to our probing gaze. It is likely that every civilization has had some system of beliefs about the relationship between the heavens and Earth. Before the Renaissance most of these included two fundamental tenets. The first was the belief that Earth occupies a special, unique place, usually at the center of the universe. The second was the belief that objects in the heavens are made of a different type of substance than

Earth and behave according to their own rules. The key to our understanding of the universe turned out to be the literal and total negation of both of these beliefs.

At the heart of modern astronomy is a fundamental idea called the **cosmological principle**. The cosmological principle asserts that there is nothing special or unique about Earth, either in our place in the universe or in the rules that govern the behavior of matter and energy here. The cosmological principle states that we are a part of the universe, rather than apart from the universe. The cosmological principle has two important aspects. The first is that

> There is nothing special about our place in the universe.

when we look out around us, what we see is representative of what the universe is generally like. Our location in the universe is what it happens to be by chance—nothing more, nothing less. In a deep image (an image showing very faint objects), even a piece of apparently empty sky is filled with distant galaxies. There are as many galaxies in the observable universe as there are stars in our own Milky Way. The cosmological principle says that nothing sets the Milky Way apart from this group of galaxies. The impression that we get of the universe from our vantage point is representative of the whole. The second aspect of the cosmological principle is the premise that matter and energy obey the same physical laws throughout space and time as they do today on Earth. This means that the same physical laws we learn about in terrestrial laboratories can be used to understand what goes on in the centers of stars or in the hearts of distant galaxies.

The cosmological principle is a theory, and like all scientific theories it is subject to whatever tests our ingenuity and the scientific method can bring to bear. We will not visit a distant star or galaxy in our lifetimes, nor relive the early days of the universe. Yet we can analyze the light that reaches us from events distant in time and space, and ask whether those events follow the same laws that apply today on Earth. Each new success that comes from applying the cosmological principle to observations of the universe around us—each new theory that succeeds in explaining or predicting patterns and relationships among celestial objects—adds to our confidence in the validity of this cornerstone of our worldview.

1.3 Patterns Make Our Lives and Science Possible

Imagine what life would be like if sometimes when you let go of an object it fell up instead of down. What if one day apples were essential nutrition, but when you bit into an apple the next day you discovered they were deadly poison? What if, unpredictably, one day the Sun rose at noon and set at 1:00 P.M., the next day it rose at 6:00 A.M. and set at 10:00 P.M., and the next day the Sun did not rise at all? In fact, objects do fall toward the ground. Our biochemistry remains stable. The Sun rises, sets, then rises again. Spring turns into summer, summer turns into autumn, autumn turns into winter, and winter turns into spring. The rhythms of nature produce patterns in our lives, and we count on these patterns for our very survival. If nature did not behave according to regular patterns, then our lives—indeed, life itself—would not be possible.

The same patterns that make our lives possible also make science possible. The goal of science is to identify and characterize these patterns and to use them to understand the world around us. Some of the most regular and easily identified patterns in nature are the patterns that we see in the sky. What in the sky will look different or the same a week from now? A month from now? A year from now? Most of us probably lead an indoor and in-town existence, removed from an everyday awareness of the patterns in the sky. However, away from the smog and glare of our cities, the patterns and rhythms of the sky are as easy to see today as they were in ancient times. Patterns in the sky mark the changing of the seasons (**Figure 1.11**), the coming of the rains, the movement of the herds, and the planting and harvesting of

> Patterns in our lives echo patterns in the sky.

the crops. Patterns in the sky share the rhythms of our lives. It is no surprise that astronomy, which is the expression of our human need to understand these patterns, is the oldest of all sciences.

Mathematics Is the Language of Science

There are many kinds of mathematics, most of which deal with more than just numbers. Arithmetic is about counting things. Algebra is about manipulating symbols and the relationships between things. Geometry is about shapes and special relationships. Calculus is about change. Other types of mathematics deal with topics such as topology,

> Mathematics is the science and language of patterns.

the properties of surfaces, or statistics, involving groups of objects and their relationships. What do they have in common? Why do we consider all of them to be part of a single discipline called "mathematics"? All share one thing —they deal with patterns. The best working definition of

mathematics is that "mathematics is the science and language of patterns."

We have seen that science is about patterns—patterns in relationships, patterns in behaviors, and patterns in characteristics. Astronomy is the part of science concerned with patterns related to celestial objects. If patterns are the heart of science, and mathematics is the language of patterns, it should come as no surprise that *mathematics is the language of science.* Trying to study science while avoiding mathematics is the practical equivalent of trying to study Shakespeare while avoiding the written or spoken word.

It quite simply cannot be done, or at least cannot be done meaningfully.

On the other hand, as the authors of this book we understand (and as is humorously pointed out by **Figure 1.12**) that for many of you, *math* is not *a* four-letter word—it is *the* four-letter word. Many people decide early in their education that they cannot "do" math, and from that day forward the mere mention of the word causes their eyes to glaze over and their palms to sweat. A distaste for mathematics is one of the most common obstacles standing between a nonscientist and an appreciation of the beauty and elegance of

FIGURE 1.11 Since ancient times our ancestors recognized that patterns in the sky, such as which stars are overhead at night, change together with the coming and going of the seasons and other patterns that shape our lives.

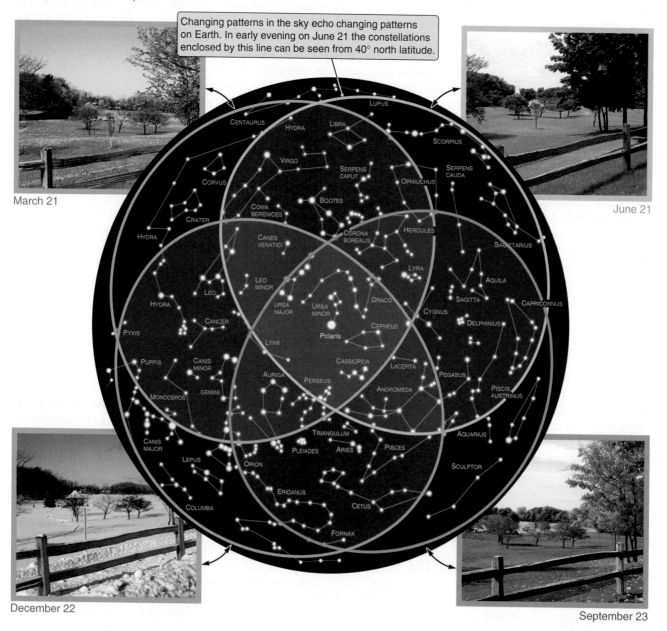

Changing patterns in the sky echo changing patterns on Earth. In early evening on June 21 the constellations enclosed by this line can be seen from 40° north latitude.

March 21

June 21

December 22

September 23

FIGURE 1.12 Mathematics is the science of patterns, which makes mathematics the language of science.

the world as seen through the eyes of a scientist. To move beyond this obstacle, both scientist and nonscientist need to find common ground.

Part of the responsibility for moving beyond this obstacle lies with us, the authors. It is our job to take on the role of translators, using words to express as many concepts as possible, even when these concepts are more concisely and accurately expressed mathematically. When we do use mathematics, we will explain in everyday language what the equations mean and try to show you how equations express concepts that you can connect to the world. We will also limit the mathematics to a few basic tools that all college students should have been exposed to. *Scientific notation* (see **Foundations 1.1**) is needed because of the vast range of sizes of the objects involved. Units are needed to distinguish between time, distance, mass, and energy. A bit of geometry is necessary for understanding the distances, sizes, shapes, and volumes of things. Finally, some algebra —mostly a few ratios and proportionalities—will provide a way of expressing the patterns that relate one physical quantity to another. "Basic" does not necessarily mean easy, but it does mean that we will use the most accessible tools that will make our journey of discovery as comfortable and informative as possible.

Your responsibility is to accept the challenge and make an honest effort to think through the mathematical concepts that we use. Do not concede defeat while still in the starting blocks. It is likely that you know what it means to square a number, or to take its square root, or to raise it to the third power. The mathematics in this book is on a par with what it takes to balance a checkbook, build a

bookshelf that stands up straight, check your gas mileage, estimate how long it will take you to drive to another city, figure your taxes, or buy enough food to feed an extra guest or two at dinner. Foundations 1.1 describes several of the basic mathematical tools we will use throughout this book. (Also see Appendix 1.)

1.4 Bending Your Brain into Shape

This book will likely ask you to think in ways that are different from the ways in which you are accustomed to thinking, and to learn to view the world from new and unfamiliar perspectives. Knowledge and understanding have nothing to do with shoving facts into short-term memory so they can be regurgitated on an exam, and that is certainly not what astronomy is about. Changing the way you think about things takes more effort. It also means studying in ways that may be different from your normal habits. Here are a few practical suggestions for how you might better study this text:

- **Read the text actively.** Think about each section after you have read it. What major concepts were discussed in the section? How are they related to what you have read so far? Have you run into similar concepts elsewhere? Why are the contents of that section important enough to

FOUNDATIONS 1.1

Mathematical Tools

Mathematics gives scientists many of the tools they need to understand the patterns they see and to communicate that understanding to others. As the authors of this text, we are aware that mathematics is not a friend to many of you taking this course, and so we have worked to keep the math in this text to a minimum. Even so, there are a few tools that we will need:

Scientific notation: Scientific notation is the way that scientists deal with numbers of vastly different sizes. Rather than writing out 7,540,000,000,000,000, 000,000, we write 7.54×10^{21}. Rather than writing out 0.000000000005, we write 5×10^{-12}.

Ratios: Ratios are the most common way that astronomers use to compare things. A star may be 10 times as massive as the Sun or 10,000 times as luminous as the Sun. These are ratios.

Geometry: To describe and understand objects in astronomy and physics, we use concepts such as distance, shape, area, and volume. Apparent separations between objects in the sky are expressed as *angles*. Earth's orbit is an *ellipse* with the Sun at one *focus*. The planets in the Solar System lie close to a *plane*. Geometry provides the tools for working with these concepts.

Algebra: Algebra provides a way of using and manipulating symbols that represent numbers or quantities. We will use algebra to express relationships that are valid not just for a single case, but for many cases. Algebra lets us conveniently express ideas such as "the distance that you travel is equal to the speed at which you are moving times the length of time you go that speed" (in other words, $d = s \times t$, where d is distance, s is speed, and t is time). Algebra also lets us combine these ideas with other ideas to arrive at new relationships.

Proportionality: Often understanding a concept amounts to understanding the *sense* of the relationships that it predicts or describes. "If you have twice as far to go, it will take you twice as long to get here." "If you have half as much money, you will be able to buy only half as much gas." These are examples of proportionality. If you are traveling at a constant speed, then time is proportional to distance. We write $t \propto d$, where $\propto$ means "is proportional to." Proportionalities often involve quantities raised to some power. A circle of radius r has an area A equal to πr^2, so we say that area is proportional to the square of the radius, and write $A \propto r^2$. This means that if you make the radius of a circle three times as large, its area will grow by a factor of 3^2, or 9.

be included in the book? Briefly summarize the section and your thoughts about it in your class notes.

- **Draw a picture.** Many physical and mathematical concepts are most easily understood if they are visualized. If you understand a concept well enough to draw a picture that expresses it, you probably understand the concept fairly well. Trying to sketch a picture will also help you better identify what things you understand and what things you do not.

- **Ask yourself "What if?"** when trying to understand a concept. What if Earth were more massive? How would that affect Earth's gravity? What if the Sun were hotter? How would that affect the color of the light from the Sun or the amount of energy that the Sun radiates?

If you cannot "what if" a concept, you probably do not really understand it yet.

- **Try to teach it.** It is often said that you do not really understand something until you try to explain it to someone else. At the end of a reading assignment, talk about the material with a friend or family member. Try to find a partner or a group in your class with whom you can meet, and *take turns teaching each other the material.* Each of you should read the assigned text, then divide up responsibility for who will present which sections. Ask each other lots of questions—the harder the better! (Make a game of playing "stump the instructor.") Even explaining a concept out loud to yourself can help.

- **Share your ideas and insights** with your discussion group. If a particular concept really "clicks" for you—if you really think it is neat—share both your understanding and your enthusiasm with your group.

- **Be honest with yourself** about what things you understand and what things you do not, and try not to avoid concepts you find difficult. Personal growth comes from real accomplishment. Getting the most from this journey will come from facing challenging concepts.

- **Focus your discussion on concepts, relationships, and connections.** You have to know the facts, but the facts are the starting point rather than the end. Use the key concepts, study questions, and other study aids to identify and concentrate your effort on the most important ideas.

- **Do not let discomfort with math keep you from succeeding.** Mathematical formulas are not magical incantations. Rather, they are expressions of logical ideas. We will always present a plain-English discussion of the idea behind any mathematics that we use. Begin by focusing on this discussion. After you grasp the idea, then look at the math. Try to see how the relationships between the quantities in the equation embody the idea. If you need help with basic math skills, work through the appendixes or explore Study Space (wwnorton.com/astro21) for other aids available to you. Your instructor is also there to help.

- **Above all, remember that building understanding is always an *active* process, never passive!**

Our best suggestion for a successful journey through *21st Century Astronomy* is to nurture your spark of interest in astronomy until it grows enough to draw you in. When, as Einstein said in the opening quote, you "pause to wonder and stand rapt in awe"—*then* you will have learned the secret!

1.5 Let the Journey Begin

The journey we are about to begin is sometimes entertaining, sometimes enlightening, often surprising, usually challenging, and always remarkable. We will explore the wonders of the universe, and along the way take a look at what science is, how it is done, and what it means to look at the world as a scientist does. It is a journey that will

- Teach us as much about ourselves as it will about what is "out there."

- Open our minds to new ways of thinking about and experiencing the world.

- Pay tribute to what the human mind and the human spirit are capable of achieving.

Reading the book and taking the journey with us are two different things. This is only a guidebook. It can lead you to the trailhead and tell you something of what you might find along the way, but *you* have to walk the path! If you become an active participant in this adventure, rather than a passive spectator, then what you gain from the journey will remain a part of you long after the final exam is forgotten. And if along the way you find yourself applying your understanding to new situations and new information, and if you learn to combine your understandings and arrive at new insights that are greater than the sum of their parts, then you will have learned something far more than just astronomy.

Summary

- The entire universe is governed by the same physical laws that shape our lives here on Earth.

- There is nothing special about our particular place in the universe.

- The scientific method is a way of trying to *falsify*, not prove, ideas.

- *All* scientific knowledge is provisional.

- Mathematics is the science and language of patterns, and thus it is the language of science.

Seeing the Forest through the Trees

At the end of each chapter of *21st Century Astronomy* you will find a brief narrative about the content of that chapter. The purpose of "Seeing the Forest through the Trees" is not to rehash the entire contents of the chapter, but rather to pick out a few of the high points and put them into a broader context. Building on the analogy of a hike in the mountains, we will spend a lot of time looking in detail at the rocks and the trees, but every so often we need to step back and look around at the forest as a whole.

We live in a world that has been profoundly shaped by the scientific revolution that took hold of Western thought during the Renaissance. That revolution fundamentally altered our way of thinking about the world, as well as our view of the relationship between ourselves and the universe of which we are a part. A new spirit of rational inquiry was turned on the heavens, dislodging Earth and humankind from the center of the cosmos. Observation, experiment, and rigorously applied reason came to replace dogma and authority as the arbiter of knowledge. The heavens became a realm not of mysticism and magic, but instead of physical law—the same physical law that governs the behavior of matter and energy in laboratories here on Earth.

The closing years of the 20th century saw our knowledge of the universe charge ahead at an ever-accelerating pace. This progress comes courtesy of a great many advances both in our technology and in the sophistication of our physical understanding of matter and energy and of space and time themselves. We have seen many fundamental questions about the origin and fate of the universe and the threads that tie our existence to the cosmos move from the realm of philosophical speculation into the realm of rigorous scientific inquiry. The insights that this age of exploration and discovery have brought are often far more profound and startling than dreamed of even a few decades ago. There can be little doubt that this time will be looked on as one of the more significant moments in the intellectual and cultural history of our species.

Like a hike through the mountains, *21st Century Astronomy* will not be an effortless journey. Many travelers will find that they have to flex a few mental muscles in ways they are not used to, and may even have to face an old adversary or two on the trail. But muscles that are sore after the first day of a hike grow comfortable and strong with time, and adversaries can become the best of friends.

In Chapter 2 we begin the journey in earnest, and as with most journeys our starting point is home. What patterns do we see in the skies of our planet Earth, and how do those patterns come to be? This is not an easy or gentle slope on which to begin our trek, but our vistas will change rapidly as we climb.

Key Terms

astronomy, p. 4
matter, p. 4
energy, p. 4
Milky Way Galaxy, p. 4
Local Group, p. 4
supercluster, p. 4

light, p. 4
speed, p. 5
origins, p. 8
astrobiology, p. 8
science, p. 9
scientific method, p. 10
nature, p. 10
cosmological principle, p. 13
mathematics, p. 14
scientific notation, p. 16
algebra, p. 16
proportionality, p. 16

Student Questions

THINKING ABOUT THE CONCEPTS

1. Imagine yourself on a planet orbiting a star in a faraway galaxy. What does the cosmological principle tell you about the way you would perceive the universe from this distant location?

2. It is said that we are made of stardust. Explain why this is a true statement.

3. List patterns in your own life that repeat regularly. How do these patterns affect you? Which patterns are of your own making, which are set by others, and which are determined by nature?

4. The scientific method states that scientific theories must be falsifiable. List some beliefs or views that you conclude are *not* falsifiable.

5. A textbook published in 1945 stated that it takes 800,000 years for light to reach us from the Andromeda Galaxy. In *21st Century Astronomy* we say that it takes 2,900,000 years. What does this tell you about a scientific "fact" and how our knowledge evolves with time?

6. Astrology makes testable predictions. For example, it predicts that the horoscope for your star sign on any day should fit you better than horoscopes for other star signs. Read each of the horoscopes in yesterday's paper without regard to your own sign. How many of them might fit the day that you had yesterday? Repeat the experiment every day for a week and keep records. Was your horoscope consistently the best description of your experiences?

7. A scientist on television states that it is a known fact that life does not exist beyond Earth. Would you consider this scientist reputable? Explain your answer.

8. Some astrologers use elaborate mathematical formulas and procedures to predict the future. Does this show that astrology is a science? Why or why not?

9. You run across an old newspaper with the headline "EINSTEIN PROVES NEWTON WRONG!" Did the newspaper get this story right? Explain your answer.

APPLYING THE CONCEPTS

10. If it takes about 8 minutes for light to travel from the Sun to Earth, and Pluto is 40 times this distance from us, how long does it take light to reach Earth from Pluto? Radio waves travel at the speed of light. What does this imply about the problems you would have if you tried to conduct a two-way conversation between Earth and a spacecraft orbiting Pluto?

11. Imagine the Sun to be the size of a grain of sand and Earth a speck of dust 83 mm away. (On this scale, each light-minute of distance equals 10 mm.) How far would it be from Earth to the Moon on this scale? From the Sun to Pluto? From Earth to the nearest star? To the nearest large galaxies? (Note that $1 \text{ m} = 10^3 \text{ mm}$ and that $1 \text{ km} = 10^3 \text{ m}$.) At what point do you lose your "feeling" for these distances?

12. The average distance from Earth to the Moon is 384,000 km. How many days would it take, traveling at 800 km/h (the typical speed of a jet airplane), to reach the Moon?

13. The surface area of a sphere is proportional to the square of its radius. If the Moon has a radius only one-quarter that of Earth, how does the surface area of the Moon compare with that of Earth?

14. Write 86,400 (the number of seconds in a day) and 0.0123 (the Moon's mass compared to Earth's) in scientific notation.

15. Write 1.60934×10^3 (the number of meters in a mile) and 9.154×10^{-3} (Earth's diameter compared to the Sun's) in standard notation.

16. The time (t) it takes for light to reach us from a distant galaxy is equal to the distance (d) of the galaxy divided by the speed of light (c). Use algebra to describe this relationship more simply.

17. If you understand proportionality, then you understand most of the math you need to follow this text. Make a list of five different proportionalities from your daily life. (For example, the price of a bag of apples is proportional to the weight of the bag of apples.) For each proportionality, identify the constant of proportionality (such as the price per pound of apples). How are these constants determined?

18. The circumference of a circle is given by $C = 2\pi r$.
    a. Calculate the approximate circumference of Earth's orbit around the Sun, assuming that the orbit is a circle with a radius of 1.5×10^8 km. You can approximate π as being about equal to 3.
    b. Noting that there are 8,766 hours in a year, how fast, in kilometers per hour, does Earth move in its orbit?
    c. How far along in its orbit does Earth move in one day?

StudySpace
wwnorton.com/astro21
provides a Study Plan for each chapter that includes a reading outline, animations, keyword flash cards, and gradebook-enabled multiple-choice quizzes. From StudySpace you can also access premium content in the ebook and SmartWork.

...marking the conclave of all the night's stars,
those potentates blazing in the heavens
that bring winter and summer to mortal men,
the constellations, when they wane, when they rise.

AESCHYLUS (525–456 B.C.)

The Moon seen rising among the ancient stones of Stonehenge.

CHAPTER 2

Patterns in the Sky— Motions of Earth

2.1 A View from Long Ago

The herds have reached the high meadows where they spend the warm season, and for a time the life of the tribe has settled in as well. The weather is comfortable, and the days are long, with none of the hardships that accompany the time of cold and snow and long, dark nights. It is a time of plenty and a time for telling the age-old stories of the tribe around a fire that guards against the chill of the gathering night. You feel a sense of contentment and are thankful to the gods for this time when life is good. As the embers die down, you gaze upward at the familiar canopy of stars overhead, and as you often do in such moments, you wonder about what you see.

To survive, you must learn the subtle patterns of your world. You must know the ways of the herds, and recognize the gathering of clouds that heralds a coming storm. When you turn your keen eye toward the heavens, you find subtle and changing patterns there as well—patterns that somehow echo those of your life. The spirits of the great animals dwell in the sky; as a child you learned to recognize their pictures there. Above you now are the stars that rule the time of the short nights. These are the stars that bring summer and lead the herds to this pleasant place.

Some of the spirits of the sky can be difficult to please. The mischievous planets wander from place to place, using their fearsome powers to sow chaos through the heavens. The Moon sometimes turns blood red, and the Sun is consumed by an ominous beast. But as long as the tribe remembers them, the gods and spirits of the sky will continue to bring the seasons and send the stars to guide the tribe. This is as your elders taught you when you were young, and this is as

KEY CONCEPTS

In this chapter we begin our journey in earnest, starting out as our ancestors did when they first gazed at the Sun, Moon, and stars, and tried to understand what they saw. With the benefit of knowledge hard-won over the centuries, we will look at patterns present both in the sky and on Earth, and then look beyond appearances to the underlying motions that cause those patterns. Here we will discover

- How the stars appear to move through the sky as Earth rotates on its axis, and how those motions differ when seen from different latitudes on Earth.

- The fundamental concept of a frame of reference, and how Earth's rotating frame of reference affects weather patterns and other terrestrial phenomena.

- How Earth's motion around the Sun and the tilt of Earth's axis relative to the plane of its orbit combine to determine which stars we see at night and the seasons we feel through the year.

- The motion of the Moon in its orbit about Earth, and how that motion, together with the motion of Earth and the Moon around the Sun, shapes the phases of the Moon and the spectacle of eclipses.

you teach the young ones today. So it has always been, and so it shall always be.

Our ancestors lived their lives attuned to the ebb and flow of nature, and the patterns in the sky were a part of that ebb and flow. The coming of night and day, the changing of the seasons, the rising and falling of the tides, the movement of the herds—all of these march in lockstep with the changes that we see in the sky. The repeating patterns of the Sun, Moon, and stars echo the rhythms that have defined the lives of humans since before the beginning of recorded history. By watching the patterns in the sky our ancestors found that they could predict when the seasons would change and the rains would come and the herds would move. Knowledge of the sky offered knowledge of the world, and knowledge of the world was power. It was a small step from here to thinking of the unreachable, untouchable stars as not only

Patterns in the sky have always been important to our species.

a reflection of the patterns in the world, but also as the *cause* of those patterns. The stars found a special place in legend and mythology as the realm of gods and goddesses, holding sway over the lives of humankind. As writing came to replace oral traditions and legends, mythologies of the sky became more elaborate as well. And as humans invented numbers and mathematics to describe and predict and account for things in the world, predictions of the motions of the stars and planets were among their greatest successes. Some of our ancestors came to look upon the orderly and predictable patterns of the sky as the *true* patterns of the world, and our own lives as imperfect reflections of this heavenly reality. They looked for ways to use their knowledge of the sky to find order in the seeming chaos of their everyday lives, and **astrology** was born.

Elements of this same basic history played themselves out many times over and in every part of our globe. From Africa to Asia, from Europe to Central America, from North America to the British Isles, the archeological record holds evidence of early humans who projected ideas from their own cultures onto what they saw in the sky (see **Excursions 2.1**). The connection between the patterns in the sky and the patterns in their lives was simply too compelling to be missed. The idea of the sky as a realm of mysticism and magic is deeply rooted in the traditions and beliefs and history of our species. There is no mystery about the currents of mind that led our ancestors to their belief in astrology and other celestial mythologies. At a time when the causes of things were unknown, and humans existed at the seeming whim of forces they could not comprehend, the sky seemed to offer a window into a mystical and powerful world of spirits, gods, devils, and angels.

What an unwelcome shock it must have been when a few remarkable individuals, with names like Copernicus

and Kepler and Galileo and Newton, tugged at the threads of this comfortable and familiar tapestry, *only to discover that it fell apart in their hands!* The stars and other heavenly bodies do not rotate about Earth each day, as humans have thought since they first took notice of the sky. Rather, it is *Earth* that spins on its axis, giving the stars, planets, Sun, and Moon the appearance of following daily paths through the heavens. Nor is Earth at the center of all existence, as

Science shattered ancient mystical views of the heavens.

befits the home of humankind, the pinnacle of all Creation. Earth is just one of eight planets orbiting the Sun. The subtle complexity of the changing patterns we see in the sky results from the motions of planets and moons as they step through their gravitational dance with the Sun. Even the Sun itself, whose radiant energy makes our world what it is, is only one of countless stars, adrift in a universe whose full extent is unknown even today.

The magic of astrology properly belongs to a time long dead, when in the minds of humans Earth rode on the back of a giant sea turtle, and with each passing month the Sun moved from one stellar "house" to the next. Today it is a matter of experimentally verifiable fact that the imaginary patterns seen in the stars hold no more influence over our lives than the random patterns of leaves blowing down the street on an autumn day. The astrologers' quest for deep connections between our lives and the patterns in the sky was both understandable and well placed, but knowledge of the true nature of those connections had to wait for the birth of modern science. Today the sky has become a window of knowledge on the *physical* world. This knowledge has proven worth the wait.

Today saw a long hard climb before you reached the meadow by the river where you now camp, and tomorrow's trek promises to be just as demanding. Even so, the evening is pleasant, and you are content. The embers of your campfire have almost died away when the distant sound of a jetliner interrupts your reverie. Without really thinking you look up to catch sight of the plane high overhead, and are caught off guard by the blazing spectacle of the summer Milky Way and the thousands of pinpoints of light, which seem so close that you can almost reach out and touch them. For a moment the thousands of years separating you from a long-dead tribal nomad vanish as you share the same sense of wonder and awe that has always defined humankind's experience of the universe.

It is here that we begin the journey of *21st Century Astronomy*—with the changing patterns in the sky that captured the attention and imagination of that long-ago nomad and that still beacon overhead on a dark, cloudless night. Yet unlike that nomad, we look on those changing patterns

EXCURSIONS 2.1

Where Are the Constellations?

Where are the **constellations**? The answer may seem obvious: "The constellations are overhead in the sky, for all to see." Yet if you look at the sky, no pictures of winged horses or dragons or chained maidens are painted there. Instead there is only the random pattern of stars —about 6,000 of them visible to the naked eye—spread out across the sky. Constellations exist only within the imagination of the human mind. Constellations are the ideas and pictures that humans imposed on the lights in the sky in an effort to connect our lives on Earth with the workings of the heavens.

As illustrated in **Figure 2.1**, there have been as many different sets of constellations and stories to go with them as there have been cultural traditions in our history. Modern constellations visible from the Northern Hemisphere draw heavily from the list compiled 2,000 years ago by the Alexandrian astronomer Ptolemy. Con-

stellations in the southern sky are drawn from the lists put together by European explorers visiting the Southern Hemisphere during the 17th and 18th centuries. Today astronomers use an officially sanctioned set of constellations as a kind of road map of the sky. The entire sky is broken into 88 different constellations, much as continental landmasses are divided into countries by invisible lines. Every star in the sky lies within the borders of a single constellation, and the names of constellations are used in naming the stars that lie within their boundaries. For example, Sirius, the brightest star in the sky, lies within the boundaries of the constellation Canis Major (meaning the "big dog"). Sirius's official name is therefore Alpha *Canis Majoris* (this is the Latin genitive or "possessive" form—see Appendix 6), indicating that it is the brightest star in that constellation and earning its nickname, the Dog Star.

FIGURE 2.1 The region around what we now call the Big Dipper (Ursa Major, or the "Great Bear") as viewed by three different civilizations. Constellations exist only in the human mind.

Egyptian—1275 B.C.

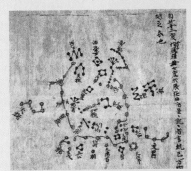
Chinese—A.D. 940

European—A.D. 1540

with the perspective of centuries of hard-won knowledge. We will find that patterns of change in the sky are often the understandable and even unavoidable consequences of the daily rotation of Earth about its axis and Earth's annual trip around the Sun. This is an example of science at its best —the discovery of wonderful variety arising from simple and elegant underlying causes. And just as happened over the history of our species, curiosity about the changing patterns in the sky will show us the way outward into a universe far more vast and awesome than our distant ancestor could have imagined.

2.2 Earth Spins on Its Axis

Despite the apocryphal stories you may have learned in grade school, Columbus did not discover that the world is round. Long before his famous (or possibly infamous) journey to the New World, anyone who had read **Aristotle** or the other Greek philosophers (as had Columbus) knew that Earth is a ball. Far more difficult to accept was the idea that the changes occurring in the sky from day to day and month to month are the result of the motion of Earth rather than the motion of the Sun and stars. The most apparent of these motions is Earth's rotation on its axis, which sets the very rhythm of life on Earth—the passage of day and night. When our remote ancestors first noticed the sky with something approaching human awareness, it was doubtless the daily motion of the Sun in the sky that drew their attention.

As viewed from above Earth's **North Pole**, Earth rotates in a counterclockwise direction (**Figure 2.2**), completing one rotation in a 24-hour period. As the rotating Earth carries us from west to east, objects in the sky *appear* to move in the other direction, from east to west. The path a celes-

---

Earth's rotation is counterclockwise when viewed from above the North Pole.

---

tial body makes across the sky as seen from Earth is called its *apparent daily motion*. The Sun is one such object. When we say "noon," we mean the time of day when our location on Earth faces most directly toward the Sun. By convention, astronomers divide the sky evenly into eastern and

FIGURE 2.2 The rotation of Earth and the Moon, the revolution of Earth and the planets about the Sun, and the orbit of the Moon about Earth are counterclockwise as viewed from above Earth's North Pole (not drawn to scale).

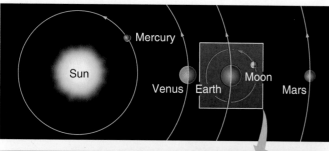

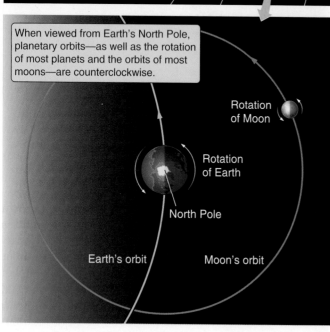

When viewed from Earth's North Pole, planetary orbits—as well as the rotation of most planets and the orbits of most moons—are counterclockwise.

Rotation of Moon

Rotation of Earth

North Pole

Earth's orbit   Moon's orbit

western halves along an imaginary north–south arc called the **meridian**. The meridian runs from due north to due south, passing through the point directly overhead in the sky, called the **zenith**. True *local noon* occurs when the Sun crosses the meridian at our location. Half a day later our spot on Earth comes closest to facing directly away from the Sun. This is *local midnight*.

The View from the Poles

The apparent daily motions of the stars and the Sun witnessed by ancient nomadic tribes would have depended on where on the surface of the planet they happened to live. The apparent daily motions of celestial objects in northern Europe, for example, are quite different from the apparent daily motions seen from a tropical island. (Differences in apparent motions of the Sun are responsible for many of the differences in culture between people living in these two regions.) The daily motions of the stars are easiest to understand when viewed from a place where humans did not set foot until 1909—Earth's North Pole.

Imagine you are standing on the North Pole watching the sky, as shown in **Figure 2.3**. (Ignore the Sun for the moment, and suppose that you can always see stars in the sky.) You are standing where Earth's axis of rotation intersects its surface, which is much the same as standing at the center of a rotating carousel. As Earth rotates, the spot directly above you remains fixed while everything else in the sky appears to revolve in a counterclockwise direction around this spot. (If you are having trouble visualizing this, find a globe and, as you spin it, imagine standing at the pole of the globe.) The direction in which Earth's axis of rotation points, and about which the stars appear to revolve as Earth turns, is called the **north celestial pole**. The greater the angular distance between a celestial object and the north celestial pole, the larger the circular path the object appears to follow. Objects close to the pole appear to follow small circles, whereas the largest circles are followed by objects nearest to the horizon.

You can never see more than half of the sky at any one time, regardless of where you are on the surface of our planet. The other half of the sky is blocked from view by Earth. The boundary between the part of the sky you

---

The same half of the sky is always visible from the North Pole.

---

can see and the part that is blocked by Earth is called the **horizon**. From most locations on Earth, the half of the sky that we can see above the horizon changes constantly as Earth rotates. (The direction in space in which our zenith points at the moment is different now from what it was 12 hours ago, or even 12 seconds ago.) In

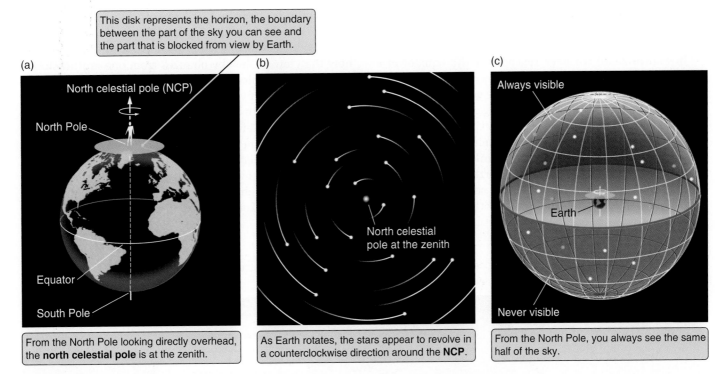

(a)

This disk represents the horizon, the boundary between the part of the sky you can see and the part that is blocked from view by Earth.

North celestial pole (NCP)

North Pole

Equator

South Pole

From the North Pole looking directly overhead, the **north celestial pole** is at the zenith.

(b)

North celestial pole at the zenith

As Earth rotates, the stars appear to revolve in a counterclockwise direction around the **NCP**.

(c)

Always visible

Earth

Never visible

From the North Pole, you always see the same half of the sky.

FIGURE 2.3 (a) As viewed from Earth's North Pole (b) stars move throughout the night on counterclockwise circular paths about the zenith. (c) The same half of the sky is always visible from the North Pole.

contrast, Earth's North Pole points in the *same* direction, hour after hour and day after day. From the North Pole we always see the *same* half of the sky. Nothing rises or sets as Earth turns beneath you. If you look off toward the horizon, you will see that the objects visible there follow circular paths that keep them always the same distance above the horizon.

The view from Earth's **South Pole** is much the same, but with two major differences. First, the South Pole is on the opposite side of Earth from the North Pole, so the half of the sky you see overhead is precisely the half that is hidden from the North Pole. The direction in space that is at the zenith at the South Pole, and about which everything ap-

> From the South Pole, the other half of the sky is visible, and stars circle in a clockwise direction.

pears to spin as Earth rotates, is the **south celestial pole**. The second difference is that instead of appearing to move counterclockwise around the sky, stars appear to move *clockwise* around the south celestial pole. (To see this, sit in a swivel chair and spin it around from right to left. As you look at the ceiling things appear to move in a counterclockwise direction, but as you look at the floor it will appear to be moving clockwise.)

Away from the Poles the Part of the Sky We See Is Constantly Changing

Latitude is a measure of how far north or south we are on the face of Earth. Imagine a line from the center of Earth to your location on the surface of the planet. Now imagine a second line from the center of Earth to the point on the **equator** closest to you. (Refer to **Figure 2.4** for help imagining these lines.) The angle between these two lines is your latitude. The latitude of any point on the equator is 0°. The latitude of the North Pole is 90° north latitude, whereas the South Pole is at 90° south latitude. The latitude of Phoenix, Arizona, is 33.5° north.

Imagine what we see as we leave the North Pole and travel south to lower latitudes. As we follow the curve of Earth, our horizon tilts and our zenith moves away from the north celestial pole. By the time we reach a latitude of 60° north (as shown in Figure 2.4(b)), the north celestial pole has dropped to 60° above the northern horizon. This equality between north latitude and the height of the north celestial pole above the northern horizon holds everywhere. When we reach a latitude of 30° north (as in Figure 2.4(c)), the direction of the north celestial pole has dropped to 30° above the horizon. (A more accurate way to say it is this: The north celestial pole lies in the same direction regardless of where we are on Earth. It is *our horizon* that is tilted at 30° from the north celestial pole when we are at a latitude of 30° north.)

In Figure 2.4(d) we have reached Earth's equator at a latitude of 0°. The north celestial pole is now sitting on the northern horizon. At the same time we get our first look at the south celestial pole, which is sitting opposite the north celestial pole on the southern horizon. Continuing on into the Southern Hemisphere, the south celestial pole is now visible above the southern horizon, while the north celestial pole is hidden from view by the northern horizon. At a latitude of 45° south (Figure 2.4(e)), the south celestial pole lies 45° above the southern horizon. At the South Pole (90°

south latitude—Figure 2.4(f)), the south celestial pole is at the zenith, 90° above the horizon.

Probably the best way to cement your understanding of how the view of the sky changes from one latitude to another is to draw pictures like those in **Figure 2.5(a)** for different latitudes. If you can draw a picture like this for any latitude—filling in the values for each of the angles in the drawing and imagining what the sky looks like from that location—then you will be well on your way to developing a working knowledge of the appearance of the sky. That knowledge will prove useful later when we discuss a variety of phenomena such as the changing of the seasons. When practicing your sketch, however, take care not to make the

FIGURE 2.4 Our perspective on the sky depends on our location on Earth. Here we see how the locations of the celestial poles and celestial equator depend on an observer's latitude.

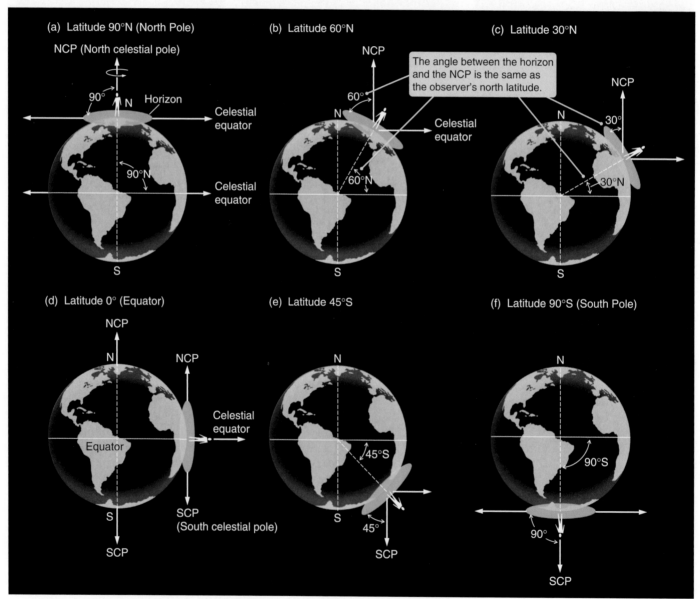

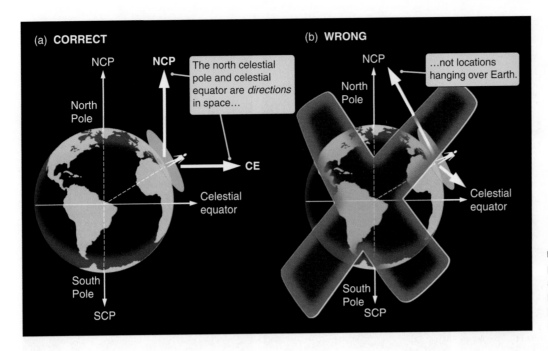

FIGURE 2.5 (a) The celestial poles and the celestial equator are directions in space, not locations hanging above Earth. Do not make the mistake shown in (b).

common mistake shown in **Figure 2.5(b)**. *The north celestial pole is not a location* in space, hovering over Earth's North Pole. Instead, *it is a direction* in space—the direction parallel to Earth's axis of rotation.

For many centuries travelers, including sailors at sea, have used the stars for navigation. Perhaps the simplest of the navigator's techniques is to use the equality between latitude and the altitude of the north (or south) celestial pole. The north or south celestial poles can be found by recognizing the stars that surround them. In the Northern Hemisphere it happens by chance that a moderately bright star is seen within about ¾° of the north celestial pole. This star is called **Polaris**, or more commonly the **north star**. If

---

The star Polaris marks the north celestial pole.

---

you can find Polaris in the sky and measure the angle between the north celestial pole and the horizon, then you know your latitude. If you are in Phoenix, Arizona, for example (latitude 33.5° north), you will find the north celestial pole 33.5° above your northern horizon. On the other hand, if you are studying astronomy in Fairbanks, Alaska (latitude 64.6° north), Polaris sits much higher overhead, 64.6° above the horizon in the north.

The location of the north celestial pole in the sky can be used to measure the size of Earth. Suppose we start out in Phoenix, Arizona, and head north. By the time we reach the Grand Canyon, about 290 km (kilometers) later, we notice that the north celestial pole has risen from 33.5° to about 36° above the horizon. This change—2.5°—is 1/144 of the way around a circle. (A circle is 360°, and 2.5°/360° = 1/144.)

This means that we must have traveled 1/144 of the way around the circumference of Earth, so the circumference of Earth must be about 144 × 290 km, or about 42,000 km. The actual circumference of Earth is just a shade over 40,000 km, so our simple measurement was not too bad, given our sloppy measurements of angles and distances. The radius of Earth is this circumference divided by 2π, or about 6,400 km. It was in much this way that the Greek astronomer Eratosthenes (276–194 B.C.) made the first accurate measurements of the size of Earth around 230 B.C. (well before Columbus's time).

The apparent motions of the stars about the celestial poles also differ from latitude to latitude. **Figure 2.6(a)** shows an observer at a point in the Northern Hemisphere other than the North Pole. As Earth rotates, the part of the sky visible to this observer is constantly changing. Of course, from the perspective of the observer it is the horizon that seems to remain fixed, while the stars appear to move past overhead. If we focus our attention on the north celestial pole, from this perspective we still see much the same thing we saw from Earth's North Pole. The north celestial pole remains fixed in the sky, and all of the stars appear to move in daily counterclockwise circular paths around that point. But because the north celestial pole is no longer directly overhead, the apparent circular paths of the stars are now tipped relative to the horizon. (More correctly, our horizon is now tipped relative to the apparent circular paths of the stars.)

Stars located close enough to the north celestial pole are above the horizon 24 hours a day (even if we can't see them in daytime) as they complete their apparent paths around the pole (see **Figures** 2.6(a) and **2.7**). This always-visible re-

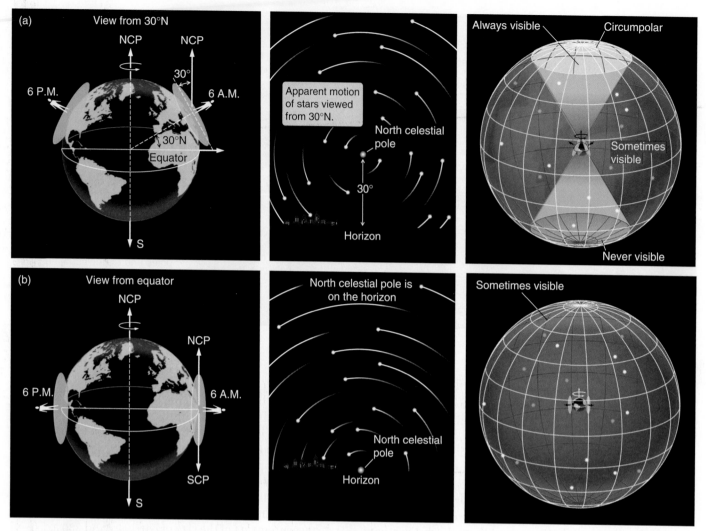

FIGURE 2.6 (a) As viewed from 30° north latitude the north celestial pole is 30° above the northern horizon. Stars appear to move on counterclockwise paths around this point. At this latitude some parts of the sky are always visible, while others are never visible. (b) From the equator the north and south celestial poles are seen on the horizon, and the entire sky is visible over 24 hours.

gion of our sky is referred to as being **circumpolar**, which means "around the pole." There also remains a part of the sky that can *never* be seen from this latitude. This is the part of the sky near the *south* celestial pole that never rises above your horizon. And between this region and the always-visible circumpolar region lies a portion of the sky

> Circumpolar stars are always above the horizon.

that can be seen for *part but not all* of each day. Stars in this intermediate region appear to rise above and set below Earth's shifting horizon as Earth turns. The only place on Earth where you can see the entire sky over the course of 24 hours is the equator. From the equator (**Figure 2.6(b)**) the north and south celestial poles sit on the northern and

southern horizons, respectively, and the whole of the heavens passes through the sky each day.

The Celestial Sphere Is a Useful Fiction It is sometimes useful to think about the sky as if it were a huge sphere with the stars on its surface and Earth at its center. (Our ancient tribesman probably thought this really was the case.) Astronomers refer to this imaginary sphere as the **celestial sphere**.[1] The celestial sphere is a useful concept because it is easy to draw and visualize, but never forget that it is imaginary! Each point on the celestial sphere actually corresponds to a *direction* in space. **Figure 2.8** shows

---

[1] You can find a description of celestial coordinates used with the celestial sphere in Appendix 6.

the celestial sphere as seen by observers at different places on Earth.

We divide the celestial sphere into a northern half and a southern half with an imaginary circle called the **celestial equator**. Just as the north celestial pole is the projection of the direction of Earth's North Pole into the sky, the celestial equator is the projection of the plane of Earth's equator into the sky. If you are on Earth's equator (Figure 2.8(c)), then the celestial equator runs east to west, passing directly overhead through the zenith. As you move north away from

---

The celestial equator is the projection of Earth's equator into space.

Earth's equator, the celestial equator tips toward the southern horizon by the same amount that the north celestial pole appears to rise above the northern horizon. Just as Earth's North Pole is 90° away from Earth's equator, the north celestial pole is always 90° away from the celestial equator. If you point one arm at a point on the celestial equator and one arm toward the north celestial pole, your arms will always form a right angle. If you are in the Southern Hemisphere, the same holds true there. The angle between the celestial equator and the south celestial pole is 90° as well.

Look at the location of the celestial equator in Figure 2.8. The points where the celestial equator intersects the horizon are always due east and due west. (The only exception to this is at the poles, where the celestial equator is coinci-

dent with the horizon.) An object on the celestial equator rises due east and sets due west. Objects that are north of the celestial equator rise north of east and set north of west. Objects that are south of the celestial equator rise south of east and set south of west.

Also note from Figure 2.8 that regardless of where you are on Earth, half of the celestial equator is always visible above the horizon (again with the exception of the poles). Because half of the celestial equator is always visible, it follows that you can see any object that lies in the direction of the celestial equator half of the time. An object that is in the direction of the celestial equator rises due east, is above the horizon for exactly 12 hours, and sets due west. This is not true for objects that are not on the celestial equator. A look at Figure 2.8(b) shows that from the Northern Hemisphere, you can see more than half of the apparent circular path of any star that is north of the celestial equator. And if you can see more than half of a star's path, then the star is above the horizon for more than half of the time.

As seen from the Northern Hemisphere, stars north of the celestial equator remain above the horizon for more than 12 hours each day. The farther north the star is, the longer it stays up. The circumpolar stars near the north celestial pole are the extreme example of this. They are up 24 hours a day. In contrast, objects south of the celestial equator are above the horizon for less than 12 hours a day, and the farther south you look, the less time a star is visible. Near the south celestial pole there are stars that never rise above our horizon.

From a location in the Canadian woods, the north celestial pole appears high in the sky…

…but at lower latitudes the north celestial pole appears closer to the horizon.

FIGURE 2.7 Time exposures of the sky showing the apparent motions of stars through the night. Note the difference in the circumpolar portion of the sky as seen from the two different latitudes.

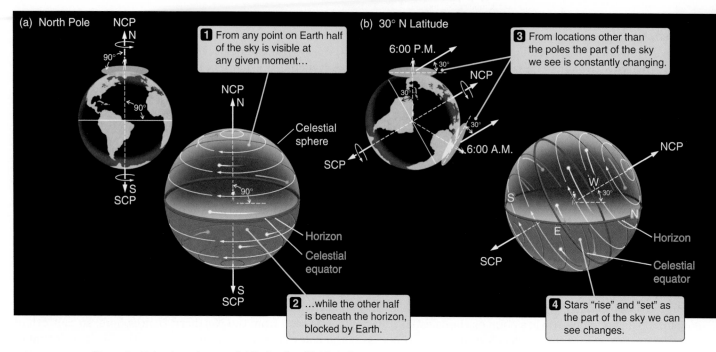

FIGURE 2.8 The celestial sphere is a useful fiction for thinking about the appearance and apparent motion of the stars in the sky. Here the celestial sphere is shown as viewed by observers at four different latitudes. At most latitudes, stars rise and set as the part of the celestial sphere that we see changes during the day.

If you were an observer in the Southern Hemisphere (Figure 2.8(d)), the reverse of the preceding discussion would be true: Objects on the celestial equator would still be up for 12 hours a day, but now objects south of the celestial equator would be up longer than 12 hours, and objects north of the celestial equator would be up less than 12 hours.

A Swinging Pendulum, a Flying Cannonball, and a Swirling Storm Feel the Effect of Earth's Rotation

One reason the ancients did not believe that Earth rotates is that they could not perceive the spinning motion of Earth. Put yourself in their place. As a result of Earth's rotation, the surface of Earth is moving along at a respectable speed—1,674 km/h (kilometers per hour) at the equator (calculated by dividing the circumference of Earth by the period of its rotation). Even so, we do not feel that motion any more than we would "feel" the speed of a car with a perfectly smooth ride cruising down a straight highway. However, Earth's rotation *does* have a number of measurable effects on objects riding along on its surface.

Jean-Bernard-Léon Foucault (pronounced "Foo-coe"; 1819–1868) was the first to carry out an experiment dem-

onstrating that Earth rotates. In 1851 Foucault constructed a **pendulum** by suspending a 28 kg (kilogram) weight at the end of a steel wire 67 meters long from the dome of the Panthéon in Paris, and he set the pendulum swinging. A pendulum swings back and forth because of the combined effects of Earth's gravity and the tension in the wire. A child swinging on a rope is a good example of a pendulum. He falls toward Earth, picking up speed, then slows to a stop as he climbs on the other side of his swing. He then repeats the processes in reverse, returning to his starting point. Left on its own, the laws of physics say that a pendulum will keep swinging back and forth, its motion remaining in the same plane. Yet when Foucault observed his pendulum over several hours, he found that the plane in which the pendulum swung gradually *changed,* rotating in a clockwise direction as viewed from above!

It is easiest to understand the behavior of a **Foucault pendulum** when it is placed at Earth's North or South Pole, as shown in **Figure 2.9**. The pendulum swings back and forth, staying in the same plane, *but Earth is rotating underneath it.* This is obvious to an observer watching from space (Figure 2.9(a)), yet to us riding on the rotating Earth (Figure 2.9(b)), the direction of the pendulum's swing appears to change. After 6 hours, Earth will have rotated 90° on its axis. The pendulum, which is still swinging in the same plane as it always was, appears to us to have changed the direction of its swing by 90° in the opposite direction. (We stress that

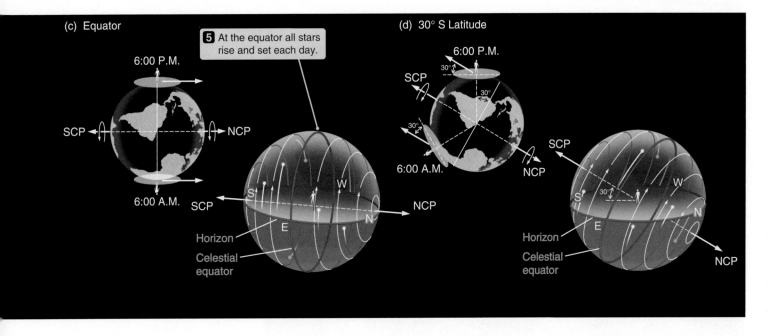

(c) Equator

5 At the equator all stars rise and set each day.

6:00 P.M.

SCP ← → NCP

6:00 A.M.   SCP ← → NCP

Horizon

Celestial equator

(d) 30° S Latitude

6:00 P.M.

30°

SCP

30°

6:00 A.M.   SCP

30°

NCP

Horizon

Celestial equator   NCP

Foucault's pendulum provided the first experimental demonstration of Earth's rotation.

the back-and-forth motion of any pendulum is due to Earth's gravity and the tension in the wire, and *not* to anything having to do with Earth's rotation. It is only the *apparent* change in the *direction* of the back-and-forth swing relative to the ground that is caused by Earth's rotation.)

The plane in which a Foucault pendulum at the North Pole swings appears to make one complete rotation in 24 hours. In contrast, a Foucault pendulum on Earth's equator appears to behave very differently. Imagine that the pendulum is set swinging back and forth along the equator (Figure 2.9(c)). As Earth rotates, the pendulum keeps swinging in the same plane; so from the standpoint of an observer riding along on Earth, the pendulum shows no change in direction. Earth is no longer spinning underneath the pendulum. Instead the pendulum is riding around on a big circle with the surface of Earth.

For latitudes between the pole and the equator, the situation is more difficult to understand in detail, but you can probably guess the answer. If it takes exactly one day for the plane of the swing of a pendulum at the pole to appear to rotate once, and if it takes forever for a pendulum on the equator to change the plane of its swing, then for latitudes

between the two it probably takes longer than a day but less than forever to seem to go around once. This guess would be correct. Foucault's original pendulum, located in Paris at a latitude of 49°, took about 32 hours to complete one rotation. (In science we often start by working out the "easy" or "limiting" cases—the pendulum at the pole or the equator, for example—and then use these to guide our thinking about what happens in more complicated situations.)

Differences in Speed Between Different Latitudes Cause the Coriolis Effect   Foucault may have been the first to conduct an experiment showing that Earth rotates, but he was not the first to experience effects of this rotation. Earth's rotation actually influences things as diverse as the motion of weather patterns on Earth and how an artillery gunner must aim at a distant target. Any object sitting on the surface of Earth follows a circle each day as Earth rotates on its axis. This circle is larger for objects near Earth's equator and smaller for objects nearer to one of Earth's poles; but because Earth is a solid body, all objects must complete their circular motion in exactly one day. Because an object nearer to the equator has farther to go each day than an object nearer a pole, the object nearer the equator must be moving *faster* than the object at a greater latitude. If an object starts out at one latitude and then moves to another, its apparent motion over the surface of Earth is influenced by this difference in speed.

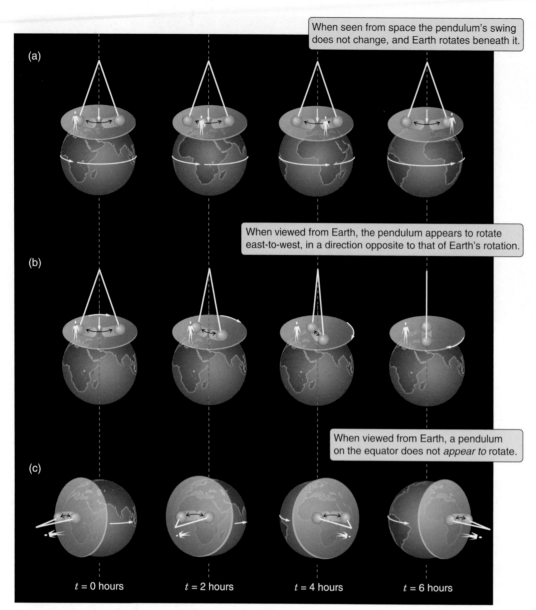

(a)

When seen from space the pendulum's swing does not change, and Earth rotates beneath it.

(b)

When viewed from Earth, the pendulum appears to rotate east-to-west, in a direction opposite to that of Earth's rotation.

(c)

When viewed from Earth, a pendulum on the equator does not *appear to* rotate.

| $t = 0$ hours | $t = 2$ hours | $t = 4$ hours | $t = 6$ hours |

FIGURE 2.9 (a) A Foucault pendulum at Earth's North Pole swings back and forth in the same plane while Earth rotates underneath it. (b) To an observer on the rotating Earth, however, the swing of the pendulum seems to change directions. (c) A Foucault pendulum on the equator does not appear to rotate.

Imagine that you are riding in a car traveling down a straight section of highway at a constant speed. The windows are blacked over so that you cannot see the scenery go by. How would you tell the difference between this and sitting completely still in your driveway? The fact is that, apart from the vibration due to the roughness of the road,

---

The motion of one object *relative* to another is important.

---

you could not. You might try throwing a ball around in the car. You toss it up and it falls right back in your lap. You can play catch with the person sitting next to you. Even though the car is speeding down the road, the relative mo-

tions between the objects in the car are small, and it is only the **relative motions** that count.

Now imagine that two cars are driving down the road at *different* speeds, as shown in **Figure 2.10**. Ignoring for the moment any real-world complications like wind resistance, if you were to throw a ball from the faster-moving car directly at the slower-moving car as the two cars pass, you would miss. The ball shares the forward motion of the faster car, so the ball outruns the forward motion of the slower car. From your perspective in the faster car (Figure 2.10(b)), the slower car lagged behind the ball. From the slower car's perspective (Figure 2.10(c)), your car and the ball sped on ahead.

Now do the same experiment; but instead of two cars, think about two locations at different latitudes. Suppose you

(a) Frame of reference: Viewer on the street

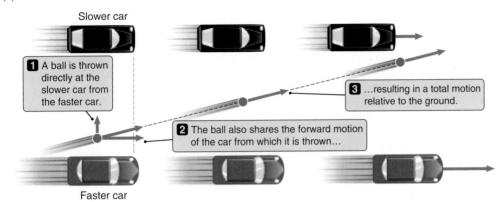

FIGURE 2.10 The motion of an object depends on the frame of reference of the observer.

(b) Frame of reference: Viewer in faster car

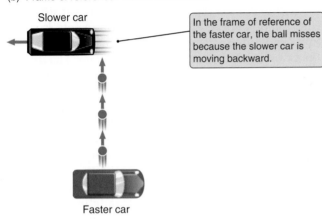

(c) Frame of reference: Viewer in slower car

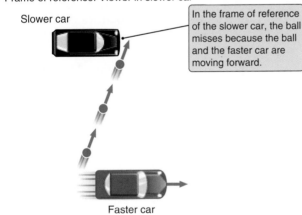

fire a cannon directly north from a point in the Northern Hemisphere, as shown in **Figure 2.11(b)**. Because the cannon is located nearer to the equator than its target is, the cannon itself is moving toward the east faster than its target. Even though the cannonball is fired toward the north, it shares in the eastward velocity of the cannon itself. This means that the cannonball is *also* moving toward the east

faster than its target! Recall how the ball thrown from the faster car outpaced the slower-moving car. Similarly, as the cannonball flies north, it finds itself moving toward the east faster than the ground underneath it is. To an observer on the ground, it looks like the cannonball curves toward the east as it outruns the eastward motion of the ground it is crossing. The farther north the cannonball flies, the greater the difference between its eastward velocity and the eastward velocity of the ground. As a result the cannonball follows a path that appears to curve more and more to the east the farther north it goes. If you are located in the Northern Hemisphere and fire a cannonball *south* toward the equator (**Figure 2.11(c)**), the opposite effect will occur. Now the cannon is moving toward the east more slowly than its target. As the cannonball flies toward the south, its eastward motion lags behind that of the ground underneath it, and the cannonball appears to curve toward the west.

This effect of Earth's rotation is called the **Coriolis effect**. In the Northern Hemisphere the Coriolis effect causes a cannonball fired north to drift to the east as seen from the surface of Earth. In other words, the cannonball appears to curve to the right. A cannonball fired south appears to curve to the west, which also gives it the appearance of curving

---

Deflection caused by the Coriolis effect is to the right in the Northern Hemisphere.

---

to the right. In the Northern Hemisphere the Coriolis effect seems to deflect things to the *right*. If you think through this example for the Southern Hemisphere, you will see that south of the equator the Coriolis effect seems to deflect things to the *left*. In between, at the equator itself, the Coriolis effect vanishes.

These differences in the Coriolis effect between the Northern and Southern Hemispheres are seen in the rotation of weather systems. As air is pushed from regions of higher pressure toward regions of lower pressure, these motions are influenced by the Coriolis effect. Think about a low-pressure region in the Northern Hemisphere. When air is pushed toward this region of low pressure from the south,

The Coriolis effect causes the counterclockwise rotation of northern hurricanes.

the Coriolis effect deflects this flow of air toward the east. Similarly, air moving toward the region of low pressure from the north is deflected to the west by the Coriolis effect. The net effect is that as air moves toward a region of low pressure in the Northern Hemisphere, the Coriolis effect deflects it into a counterclockwise circulation (**Figure 2.12**). (Think about this carefully. The Coriolis effect deflects objects toward the right as you face north in the Northern Hemisphere, which results in weather patterns that rotate toward the *left*.) The next time you see a television weather

map with a low-pressure region, look at the direction of the winds around the region and you will see this counterclockwise flow. The most spectacular example of this **cyclonic motion** is the swirl of wind and clouds around the deep low-pressure area at the eye of a hurricane or typhoon. As the air moves in closer and closer to the central region of low pressure, it rotates faster and faster, giving rise to the winds of hundreds of kilometers per hour that make hurricanes so destructive.

With what you now know about the Coriolis effect, you can show that hurricanes in the Southern Hemisphere rotate in the opposite direction from hurricanes in the Northern Hemisphere. Instead of curving to the right, air moving into

Southern Hemisphere hurricanes rotate clockwise.

a region of low pressure curves to the left, causing a clockwise rotation around the low-pressure region. In crossing from the Northern Hemisphere to the Southern Hemisphere, the Coriolis effect goes away at the equator. The difference in the direction of rotation between the hemispheres and the weakness of the Coriolis effect near the equator mean that a northern hurricane would literally collapse if it tried to cross into the Southern Hemisphere. (Imagine throwing your car into reverse while traveling down the road at 150 km/h.) This is why countries around Earth's equator do not experience the ravages of hurricanes.

As discussed in **Foundations 2.1**, the Coriolis effect is only one example of the type of effect associated with relative motions. On a humorous note, the Coriolis effect has nothing to do with the direction that water swirls in a toilet bowl, as many people imagine. The difference in the speed of Earth's motion between the two sides of a toilet bowl is not enough to matter much. Other effects, such as the direction in which the water flows into the bowl, are much more important. However, the Coriolis effect is enough to deflect a fly ball hit north or south into deep left field

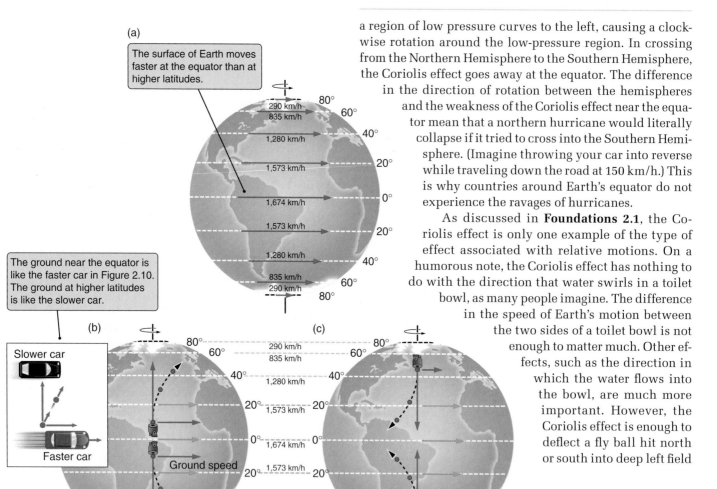

(a) The surface of Earth moves faster at the equator than at higher latitudes.

290 km/h 80°
835 km/h 60°
1,280 km/h 40°
1,573 km/h 20°
1,674 km/h 0°
1,573 km/h 20°
1,280 km/h 40°
835 km/h 60°
290 km/h 80°

The ground near the equator is like the faster car in Figure 2.10. The ground at higher latitudes is like the slower car.

(b) Slower car Faster car

A cannonball fired away from the equator outruns the ground it flies over, so its ground track curves to the east.

(c) Ground speed

A cannonball fired toward the equator lags behind the eastward motion of the ground, so its ground track curves to the west.

FIGURE 2.11 The Coriolis effect causes objects to appear to be deflected as they move across the surface of Earth.

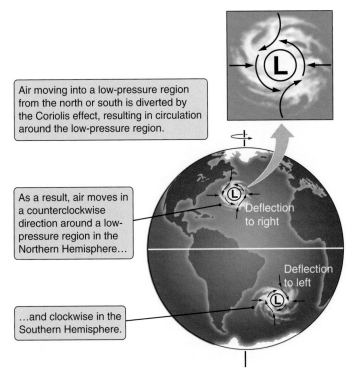

Air moving into a low-pressure region from the north or south is diverted by the Coriolis effect, resulting in circulation around the low-pressure region.

As a result, air moves in a counterclockwise direction around a low-pressure region in the Northern Hemisphere…

…and clockwise in the Southern Hemisphere.

Deflection to right

Deflection to left

FIGURE 2.12 As a result of the Coriolis effect, air circulates around regions of low pressure on the rotating Earth.

in a stadium in the northern United States by about a half a centimeter. At some time or other the Coriolis effect has probably determined the outcome of a ball game.

2.3 Revolution about the Sun Leads to Changes during the Year

The second motion we will discuss is the motion of Earth about the Sun. Earth revolves around the Sun in the same direction Earth spins about its axis—counterclockwise as viewed from above Earth's North Pole. A **year**, by definition, is the time it takes for Earth to complete one revolution around the Sun. The motion of Earth around the Sun is re-

> Earth's orbital motion is counterclockwise as viewed from above Earth's North Pole.

sponsible for many of the patterns of change we see in the sky and on Earth, including changes in which stars we see at night. When you look overhead at midnight, you are looking away from the Sun. As Earth moves around the Sun, this direction changes. Six months from now, Earth will be

on the other side of the Sun, and the stars that we see overhead at midnight will be in nearly the opposite direction from the stars we see near overhead at midnight tonight. The stars that were overhead at midnight six months ago are the stars that are overhead today at noon, but we cannot see those stars today because of the glare of the Sun.

If you could note the position of the Sun relative to the stars each day for a year, you would find it traces out a **great circle** against the background of the stars (**Figure 2.13**). On September 1, the Sun appears to be in the direction of the constellation of Leo. Six months later, on March 1, Earth is on the other side of the Sun, and the Sun appears to be in the direction of the constellation of Aquarius. The apparent path that the Sun follows against the background of the stars is called the **ecliptic**. The constellations that lie along

> The ecliptic is the Sun's apparent yearly path against the background of stars.

the ecliptic and through which the Sun appears to move are called the constellations of the **zodiac**. This is why ancient astrologers assigned special mystical significance to these stars. Actually, the constellations of the zodiac are nothing more than random patterns of distant stars that happen by chance to lie near the plane of Earth's orbit about the Sun.

Earth's Motion through Space Is Measured Using Aberration of Starlight

Just as it is difficult to "feel" the effects of Earth's rotation on its axis, it is even harder to sense the motion of Earth around the Sun. Through most of the history of our species, humans believed that Earth remains stationary while the Sun, the Moon, and the heavens revolve around us. The history of modern astronomy, and to some degree the story of the rise of modern science, can be told as the story of how this view was overthrown during the 17th and 18th centuries. However, the first direct measurement of the effect of Earth's motion was not made until the 18th century. To understand how this measurement was made, we return to our automotive example of a moving frame of reference.

Imagine you are sitting in a car in a windless rainstorm, as shown in **Figure 2.14**. If the car is sitting still and the rain is falling vertically, when you look out your side window you see raindrops falling straight down. That is, if you were to hold a vertical tube out the window, raindrops would fall straight through the tube. When the car is moving forward, however, the situation is different. Between the time a raindrop appears below the top of your window and the time it disappears beneath the bottom of your window, the car has moved forward. The raindrop disappears beneath the window *behind* the point at which it appeared, which means

FOUNDATIONS 2.1

Relative Motions and False Forces

Aside from looking out the window or feeling road vibrations, there is no experiment that you could easily do to tell the difference between riding in a car down a straight section of highway at constant speed and sitting in the car while it is parked in your driveway. Because everything in the car is moving together, the relative motions between objects in the car are all that count. In fact, the only reason you can feel the roughness of the road is that it slightly changes the motion of the car. You feel these brief accelerations as the car's vibration.

The idea that only relative motions count occurs again and again in astronomy and physics. There are numerous examples in this chapter alone. For example, even though Earth is spinning on its axis and flying through space in its orbit about the Sun, the resulting relative motions between objects that are near each other on Earth are small—so small that for most of history humans assumed that Earth was sitting still. Newton's realization that motions are meaningful only when tied to the **frame of reference** of some observer is also at the heart of Einstein's theories of relativity. These theories, which we will return to later, wound up changing the way we think about space and time.

The Coriolis effect, discussed in this chapter, is an example of what can happen if motion is viewed from a frame of reference with a changing velocity. The Coriolis effect is sometimes wrongly called the "Coriolis force" because to someone on the ground it looks like a force has acted on the cannonball, pushing it to the side. In fact, no additional force is acting on the cannonball. It is flying true. The ball *appears* to curve because of the shifting frame of reference of the ground over which it travels. The Coriolis effect and other false forces can be seen in many contexts. When you turn a sharp corner in a car, it seems as though you are thrown against the door by some force. Actually, no force is throwing you against the door at all. Instead, your body is trying to continue moving in a straight line while the frame of reference of the car (and you) is changing. For a playground example of the Coriolis effect and such false forces, return to the same carousel we used to understand the apparent motions of the stars. Sit near the center of the spinning carousel, and try to play catch with someone sitting near the edge. This exercise, if carried out physically and not just in your mind, will help you develop a feeling for how the Coriolis effect works. (It might be good for a few laughs or a queasy stomach as well.)

As we continue on our journey, keep your eyes open for more examples of relative motions. Relative motions among planets, among stars, among galaxies, among atoms—they will all be there.

---

the raindrop *looks as if* it falls at an angle, even though in reality it is falling straight down. For raindrops to fall directly through the tube you are holding out the window now, you would have to tilt the top of the tube forward. The faster you go, the greater the apparent front-to-back motion of the raindrops, and the more tilted their apparent paths become. An observer by the side of the road would say the raindrops are coming from directly overhead, but to you in the moving car they are coming from in front of the car.

This same phenomenon occurs with starlight, as shown in **Figure 2.15**. The light from a distant star arrives at Earth from the direction to this star. Because Earth moves, however, to an observer on Earth the starlight seems to be coming from a slightly different direction, just as the raindrops

appeared to be coming from in front of the car.[2] As the direction of Earth's motion around the Sun continuously changes during the year, the apparent position of a star in the sky moves in a small loop. This shift in apparent position is what we refer to as the **aberration of starlight**.

The speed and direction of Earth's motion and the part of the sky a star is in determine the exact shift in the apparent position of a star. The apparent deflection of a star located at a right angle to Earth's motion is about half of what the typical unaided human eye can detect. This much deflection

---

[2] Actually, in the frame of reference of Earth, the starlight *is* coming from a different direction, and in the frame of reference of the moving car, the raindrops *are* coming from in front of the car.

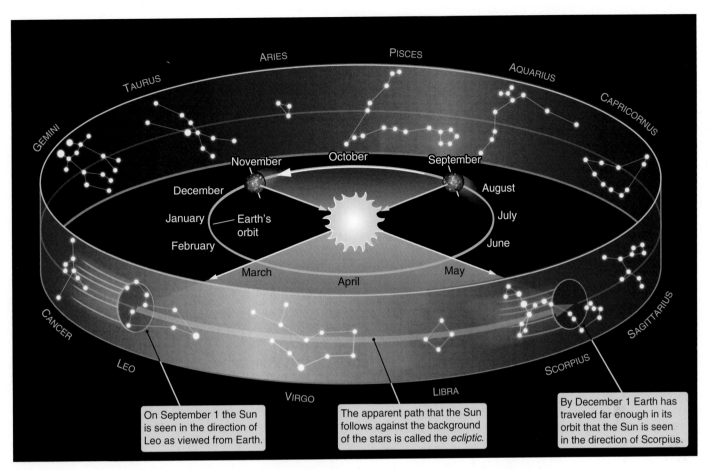

ARIES PISCES AQUARIUS CAPRICORNUS TAURUS GEMINI CANCER LEO VIRGO LIBRA SCORPIUS SAGITTARIUS

November October September December August January Earth's orbit July February June March May April

On September 1 the Sun is seen in the direction of Leo as viewed from Earth.

The apparent path that the Sun follows against the background of the stars is called the *ecliptic*.

By December 1 Earth has traveled far enough in its orbit that the Sun is seen in the direction of Scorpius.

FIGURE 2.13 As Earth orbits about the Sun, the Sun's apparent position against the background of stars changes. The imaginary circle apparently traced by the Sun is called the ecliptic. Constellations along the ecliptic form the zodiac.

Earth's orbital motion around the Sun was first measured from the aberration of starlight.

is easy to measure with a telescope. Aberration of starlight was first detected in the 1720s by two English astronomers, Samuel Molyneux and James Bradley. Measurement of the aberration of starlight shows that Earth is moving on a roughly (but not exactly) circular path about the Sun with an average speed of just under 30 km/s (kilometers per second). Because distance equals speed times time, the distance around this near-circle—its circumference—is just the speed of Earth (29.8 km/s) times the length of one year (3.16×10^7 s). The circumference of Earth's orbit is then

$$\text{Distance} = \text{Speed} \times \text{Time}$$

$$= \left(29.8 \, \tfrac{\text{km}}{\text{s}}\right) \times (3.16 \times 10^7 \, \text{s})$$

$$= 9.42 \times 10^8 \, \text{km}.$$

The radius of Earth's nearly circular orbit is this circumference divided by 2π, or 1.50×10^8 km, or 150 million kilometers. Astronomers refer to this distance—the average distance between the center of the Sun and the center of Earth[3]—as one **astronomical unit**, abbreviated AU. The astronomical unit is a good unit for measuring

The astronomical unit (AU) is the average distance between the Sun and Earth.

distances within the Solar System. Modern measurements of the size of the astronomical unit are made in very different ways, such as bouncing radar signals off Venus. However, the aberration of starlight provided a simple and compelling demonstration that Earth orbits about the Sun—and a pretty good value for the size of Earth's orbit as well.

[3] Distances between celestial objects are almost always taken to be between their centers.

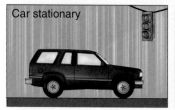
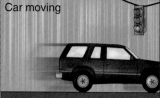

FIGURE 2.14 On a windless day the direction in which rain falls depends on the reference frame in which it is viewed. From a stationary car, rain is seen to fall vertically downward. From a moving car, rain is seen to fall at an angle determined by the speed and direction of the car's motion.

Seasons Are Due to the Tilt of Earth's Axis

If you ask most people why it is cold in the winter and warm in the summer, they are likely to tell you that it is because Earth is closer to the Sun in the summer and farther away in the winter. This is a common (and commonsense) idea that *does* have something to do with the seasons on Mars, but has virtually *nothing* to do with the seasons on Earth! Earth's orbit around the Sun is almost a perfect circle centered on the Sun,[4] and so the distance from the Sun changes little during the year. In fact, Earth is slightly closer to the Sun during the northern winter than it is during the northern summer. So far we have discussed the rotation of Earth on its axis and the revolution of Earth about the Sun, and the consequences of each. To understand the changing of the seasons, we need to consider the combined effects of these two motions.

If Earth's spin axis were exactly perpendicular to the plane of Earth's orbit (the ecliptic plane), then the Sun would always appear to lie on the celestial equator. Because the position of the celestial equator is fixed in our sky, the Sun

would follow the same path through the sky day after day, rising due east each morning and setting due west each evening. If the Sun were always on the celestial equator, it would be above the horizon for exactly half the time, and days and nights would always be exactly 12 hours long. In short, if Earth's axis were exactly perpendicular to the plane of Earth's orbit, each day would be just like the last, and there would be no seasons.

However, Earth's axis of rotation is *not* exactly perpendicular to the plane of the ecliptic. Instead it is tilted by 23.5° from the perpendicular. (Astronomers use the term **obliquity** to refer to the angle between a planet's equatorial and orbital planes.) As Earth moves around the Sun, its axis points in almost exactly the same direction through the

---

Seasons result from the 23.5° tilt of Earth's axis with respect to a line perpendicular to its orbital plane.

---

year and from one year to the next. As a result, sometimes Earth's North Pole is tilted more toward the Sun, and at other times it is pointed more away from the Sun. When Earth's North Pole is tilted toward the Sun, an observer on Earth sees the Sun as lying *north* of the celestial equator. Six months later, when Earth's axis is tilted away from the Sun, the Sun is seen as lying *south* of the celestial equator. If we look at the circle of the Sun's apparent path through the stars—the ecliptic—we see that it is tilted by 23.5° with respect to the celestial equator.

To understand the effect that this has on Earth, begin by looking at **Figure 2.16(a)**. This shows the situation on June 21, the day that Earth's North Pole is tilted most directly toward the Sun.[5] Note first that the Sun is north of the celestial equator. We found earlier in the chapter that from the perspective of an observer in the Northern Hemisphere, an object north of the celestial equator can be seen above the horizon for more than half the time. This is true for the Sun as well as for any other celestial object. Saying that the Sun is above the horizon for more than half the time is just another way of saying that the days are longer than 12 hours. You can see this directly in Figure 2.16(a) by noting that when Earth's North Pole is tilted toward the Sun, over half of Earth's Northern Hemisphere is illuminated by sunlight. These are the long days of the northern summer. Six months later, on December 22, the situation is very different. On December 22 (**Figure 2.16(b)**) Earth's North Pole is tilted away from the Sun, so the Sun appears in the sky south of the celestial equator. Someone in the Northern Hemisphere

---

[4] Planetary orbits are actually elliptical, as will be discussed in Chapter 3.

[5] We are a bit sloppy with language here. Earth's North Pole tilts in the same direction year-round. On the first day of the northern summer, Earth is on the side of the Sun where the tilt of the North Pole is toward the Sun. On the first day of the northern winter, Earth is on the opposite side of the Sun, so the tilt of the North Pole is away from the Sun.

will see the Sun for less than 12 hours each day. Less than half of the Northern Hemisphere is illuminated by the Sun. It is winter in the north.

Over the course of the year the length of the day changes, courtesy of the 23.5° tilt of Earth's axis relative to the perpendicular to the plane of its orbit. In the preceding paragraph we were careful to specify the length of the day in the *Northern* Hemisphere because things are very different in the Southern Hemisphere. In fact, things in the Southern

Seasons in the Southern Hemisphere are the reverse of those in the Northern Hemisphere.

Hemisphere are exactly reversed from what is going on in the north. Look again at Figure 2.16. On June 21, while the Northern Hemisphere is enjoying long days and short nights, Earth's South Pole is tilted in the direction away from the Sun. Less than half of the Southern Hemisphere is illuminated by the Sun, and days are shorter than 12 hours. Similarly, on December 22 Earth's South Pole is tilted toward the Sun, and the southern days are long.

The differing length of days through the year is part of the explanation for the changing seasons, but we need to

consider another important effect. The Sun appears higher in the sky during the summer than it does during the winter, so sunlight strikes the ground *more directly* during the summer than during the winter. To see why this is important, hold a piece of cardboard toward the Sun and look at the size of its shadow. If the cardboard is held so that it is

The angle of sunlight to the ground is closer to perpendicular in summer than in winter, so there is more heating per unit area in summer.

directly face-on to the Sun, then its shadow is large. However, as you turn the cardboard more edge-on, the size of its shadow shrinks. The size of the cardboard's shadow tells you that the cardboard catches less energy from the Sun each second when it is tilted relative to the Sun than it does when it is face-on to the Sun. This is exactly what happens with the changing seasons. During the summer Earth's surface is more nearly face-on to the incoming sunlight, so more energy falls on each square meter of ground each second. During the winter the surface of Earth is more inclined with respect to the sunlight, so less energy falls on each square meter of the ground each second. That is the main

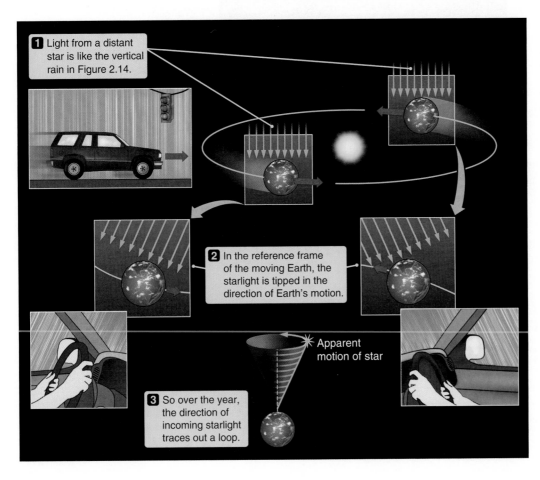

1 Light from a distant star is like the vertical rain in Figure 2.14.

2 In the reference frame of the moving Earth, the starlight is tipped in the direction of Earth's motion.

3 So over the year, the direction of incoming starlight traces out a loop.

✳ Apparent motion of star

FIGURE 2.15 The apparent positions of stars are deflected slightly toward the direction in which Earth is moving. As Earth orbits the Sun, stars appear to trace out small ellipses in the sky. This effect is called aberration of starlight.

reason why it is hotter in the summer and colder in the winter.

Look at **Figure 2.17**, which shows the direction of incoming sunlight striking Earth at the border between Kansas and Nebraska (40° north latitude). At noon on the first day of summer the Sun is high in the sky—73.5° above the horizon and only 16.5° away from the zenith. Sunlight strikes the ground almost face-on. In contrast, at noon on the first day of winter, the Sun is only 26.5° above the horizon, or 63.5° from the zenith. Sunlight strikes the ground at a rather shallow angle. The difference between the two cases is important. At the Kansas–Nebraska border, over twice as much solar energy falls on each square meter of ground per second at noon on June 21 as falls there at noon on December 22. Together these two effects —the directness of sunlight and the differing length of the day—mean that during the summer there is more heating from the Sun and during the winter there is less heating from the Sun.

We do not have to wait for the seasons to change to see the effect that the height of the Sun in the sky has on terrestrial climate. We need only compare the climates found at different latitudes on Earth. Near the equator the Sun passes high overhead every day, regardless of the season. As a result, the climate is warm throughout the year. At high latitudes, however, the Sun is *never* high in the sky, and the climate can be cold and harsh even during the summer.

Four Special Days Mark the Passage of the Seasons

The apparent path that the Sun follows through the stars each year (the ecliptic) is a great circle that is tilted 23.5° with respect to the celestial equator. Follow along in **Figure 2.18** as we note the four special points on this path that

FIGURE 2.16 (a) On the first day of the northern summer (June 21, the summer solstice), the northern end of Earth's axis is tilted most nearly toward the Sun, while the Southern Hemisphere is tipped away. Seasons are opposite in the Northern and Southern Hemispheres. (b) Six months later, on the first day of the northern winter, the situation is reversed.

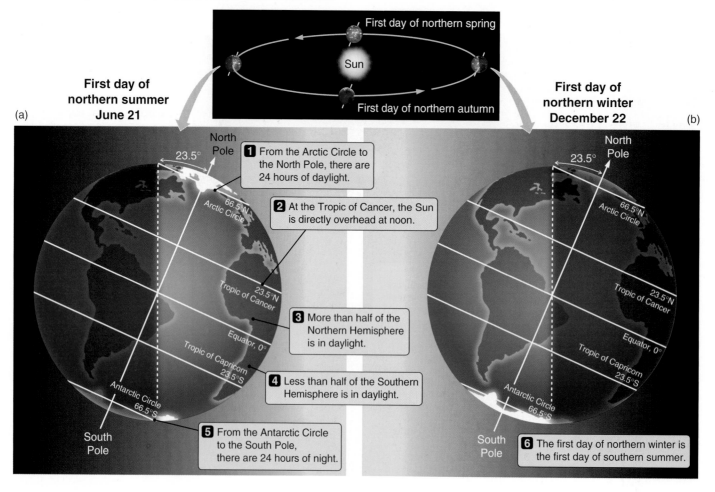

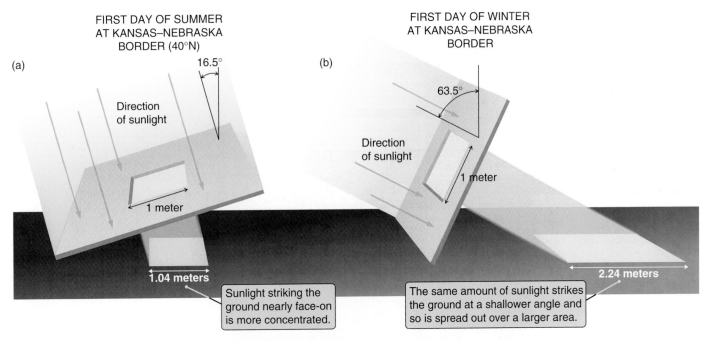

FIGURE 2.17 Local noon at the Kansas–Nebraska border (40° north latitude). (a) On the first day of northern summer, sunlight strikes the ground almost face-on. (b) At local noon on the first day of northern winter, sunlight strikes the ground more obliquely, and less than half as much sunlight falls on each square meter of ground each second.

mark the passage of the seasons. Begin with the point in March when Earth's axis is perpendicular to the direction to the Sun. This is the point where the Sun's apparent motion along the ecliptic crosses the celestial equator moving from the south to the north. That direction in the sky, located in the constellation Pisces, is called the **vernal equinox**. The term *vernal equinox* also refers to the day—about March 21—when the Sun appears at this location. When

---

The changing seasons are marked by equinoxes and solstices.

---

the Sun lies on the celestial equator on the vernal equinox, days are 12 hours long. (The term **equinox** means literally "equal night": Everywhere on Earth, night and day are the same length on the days of the equinoxes.) In the Northern Hemisphere, this is the first day of spring.

As Earth continues its journey around the Sun, the direction to the Sun moves into closer alignment with the tilt of the northern end of Earth's axis, and the Sun climbs higher into the northern sky. The Sun is lined up with the tilt of Earth's axis about three months after the vernal equinox. When this happens, the Sun reaches its northernmost point in the sky, located in the constellation Taurus, near its border with Gemini. This day, which marks the longest day of the year and the beginning of summer in the Northern Hemisphere, occurs around June 21. This is the **summer solstice**. (**Solstice** literally means "sun standing still": The

Sun's north–south motion stops as it reverses its direction.) Note that on this same day, the southern end of Earth's axis is tipped directly away from the Sun. This is the shortest day of the year in the Southern Hemisphere, marking the beginning of the southern winter.

Three months later, around September 23, the Sun is again crossing the celestial equator. This point on the Sun's apparent path, located in the constellation Virgo, is called the **autumnal equinox**. *Autumnal equinox* refers both to the location in the sky and the date when this happens. Because the Sun is on the celestial equator, days and nights are exactly 12 hours long. It is the first day of autumn in the Northern Hemisphere and the first day of spring in the Southern Hemisphere.

Around December 22 the Sun reaches its southernmost point in the sky as its apparent path takes it through the constellation Sagittarius. This day is called the **winter solstice**. Earth's North Pole is tipped most directly away from the Sun on this day. In the Northern Hemisphere this is the shortest day of the year—the first day of winter. As the Sun passes the winter solstice and moves on toward the vernal equinox, the northern days begin growing longer again. Almost all cultural traditions in the Northern Hemisphere include some major celebration in late December (**Figure 2.19**). Christmas, for example, is celebrated just three days after the winter solstice. These winter festivals have many different meanings to their various celebrants, but they all share one thing: They celebrate the return of the source of

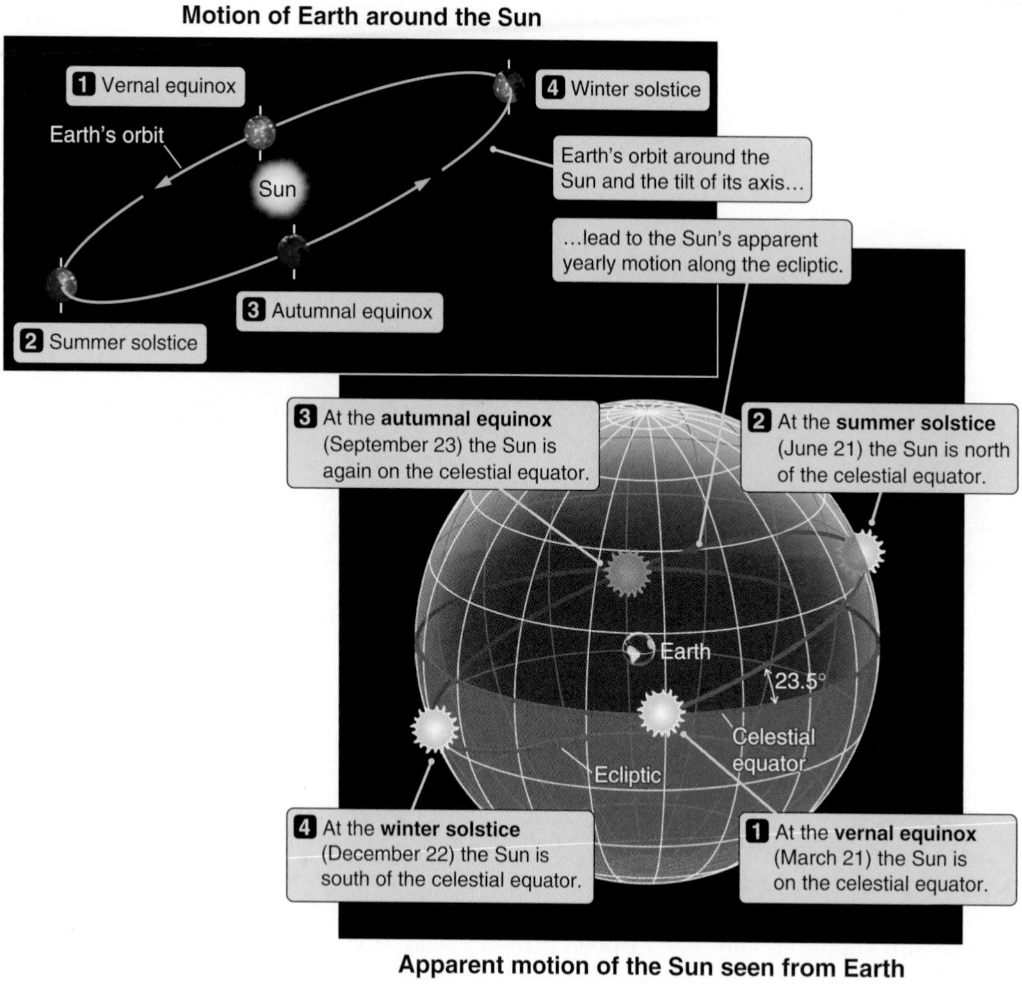

Motion of Earth around the Sun

1 Vernal equinox

Earth's orbit

Sun

4 Winter solstice

Earth's orbit around the Sun and the tilt of its axis…

…lead to the Sun's apparent yearly motion along the ecliptic.

2 Summer solstice

3 Autumnal equinox

3 At the **autumnal equinox** (September 23) the Sun is again on the celestial equator.

2 At the **summer solstice** (June 21) the Sun is north of the celestial equator.

Earth

23.5°

Celestial equator

Ecliptic

4 At the **winter solstice** (December 22) the Sun is south of the celestial equator.

1 At the **vernal equinox** (March 21) the Sun is on the celestial equator.

FIGURE 2.18 The motion of Earth about the Sun as seen from the frame of reference of the Sun (upper panel) and Earth (lower panel).

Apparent motion of the Sun seen from Earth

Earth's light and warmth. The days have stopped growing shorter and are beginning to get longer. It is a "new year," and once again the Sun will hold sway over night. Spring will come again.

As an aside, it is interesting to note that there is more to the seasons we feel than the amount of energy we are receiving from the Sun. Just as it takes a pot of water on a stove time to heat up when the burner is turned up and time to cool off when the burner is turned down, it takes Earth time to respond to changes in heating from the Sun. The

Seasonal temperatures lag behind changes in the directness of sunlight.

hottest months of the summer are usually July and August, which come *after* the summer solstice, when the days are growing shorter. Similarly, the coldest months of winter are usually January and February, which occur *after* the winter solstice, when the days are growing longer. The climatic seasons on Earth lag behind changes in the amount of heating we are receiving from the Sun.

Our picture of the seasons must be modified somewhat near Earth's poles. At latitudes north of 66.5° north latitude and south of 66.5° south latitude, the Sun is circumpolar for a part of the year surrounding the first day of summer. These lines of latitude are called the **Arctic Circle** and the **Antarctic Circle**, respectively. When the Sun is circumpolar, it is above the horizon 24 hours a day, earning the arctic regions the nickname "land of the midnight Sun." The Arctic and Antarctic regions pay for these long days, however, with an equally long period surrounding the first day of winter when the Sun never rises and the nights are 24 hours long. The Sun never rises high in the Arctic or Antarctic sky, which means that sunlight is never very direct. This is why even with the long days at the height of summer, the Arctic and Antarctic regions remain relatively cool.

The seasons are also different near the equator. Recall from Figure 2.8(c) that for an observer on the equator, *all* stars are above the horizon 12 hours a day, and the Sun is no exception. On the equator, days and nights are 12 hours long throughout the year. Changes in the directness of sunlight through the year are also different on the equator. Here the

Sun passes directly overhead on the first day of spring and the first day of autumn because these are the days when the Sun is on the celestial equator. Sunlight is most direct on the equator on these days. At the summer solstice the Sun is at its northernmost point along the ecliptic. It is on this day, and on the winter solstice, that the Sun is *farthest* from the zenith at noon, and therefore sunlight is *least* direct. Strictly speaking, the equator experiences only two seasons: summer, when the Sun passes directly overhead, and winter, when the Sun is at its northernmost and southernmost points on the ecliptic. However, summer and winter are not very different. The Sun is always up for 12 hours a day, and the Sun is always so close to being overhead at noon that the directness of sunlight changes by only 8 percent throughout the year.

If you live between the latitudes of 23.5° south and 23.5° north, twice during the year the Sun will be directly overhead at noon. This band is called the **Tropics**. The northern limit of this region is called the Tropic of Cancer. The southern limit is called the Tropic of Capricorn. (As a challenge, think about what the seasons are like at different locations within the Tropics.)

Earth's Axis Wobbles, and the Seasons Shift through the Year

After our discussion of the apparent motion of the Sun along the ecliptic, you might wonder about the names given to the Tropics. The summer solstice is located in the constellation Taurus, so why do we call the northern tropic the tropic of Cancer rather than the tropic of Taurus? Similarly, why is the southern tropic not referred to as the tropic of Sagittarius since the Sun is seen in that constellation on the winter solstice? The answer has to do with the fact that Earth's axis

FIGURE 2.19 Most cultural traditions in the Northern Hemisphere include a major celebration in late December, around the time when days begin to grow longer.

(a)

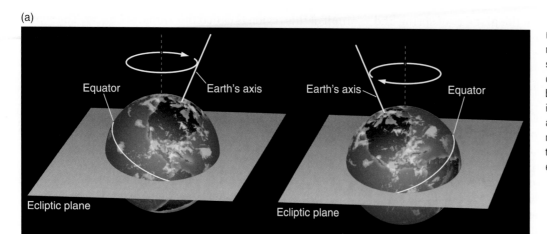

FIGURE 2.20 (a) Earth's rotation axis precesses like a spinning top. (b) Precession causes the projection of the Earth's rotation axis to move in a 47° diameter circle with a period of 25,800 years. The red cross shows the location of the projection of the axis in the early 21st century.

(b)

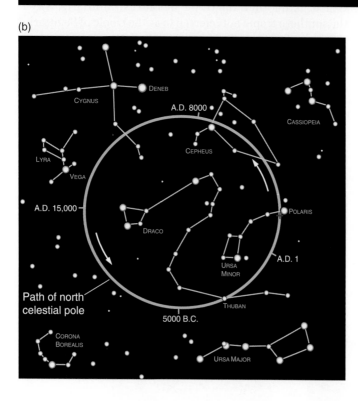

the celestial equator wobbles, the locations where it crosses the ecliptic—the equinoxes—change as well. During each 26,000-year wobble of Earth's axis, the locations of the equinoxes make one complete circuit around the celestial equator. This phenomenon is called the **precession of the equinoxes**. Ptolemy and his cohorts were formalizing their

---

A 26,000-year wobble causes the position of the equinoxes to slowly precess.

---

knowledge of the positions and motions of objects in the sky two thousand years ago. At that time the Sun *was* in the constellation of Cancer on the first day of northern summer and it *was* in the constellation of Capricorn on the first day of northern winter—hence the names of the Tropics.[6]

The tendency of the seasons to shift through the year from century to century has played havoc with human efforts to construct reliable calendars, and history is full of interesting anecdotes related to this difficulty. For example, for much of the 16th, 17th, and 18th centuries the calendars in Protestant Europe lagged behind the calendars in Catholic Europe by first 10, and later 11, days. It was not until 1752 that England and her colonies, including those in America, dropped 11 days from their calendars to bring them into line. This conciliatory step was met by riots among the people, who somehow felt that these 11 days had been stolen from them. Today's calendar (which was not adopted in Russia until the 1917 Bolshevik Revolution) is based on the **tropical year**, which is 365.242199 **solar days** long. The tropical year measures the time from one vernal equinox to the next—from the start of spring to the start of spring. Notice that the tropical year is not an integral number of days long. An elaborate system of **leap years** is used in our calendar

wobbles like the axis of a spinning top (**Figure 2.20**). The wobble is very slow, taking about 26,000 years to complete one cycle. During this time the north celestial pole makes one trip around a large circle through the stars. Polaris is

---

Earth's axis wobbles like the axis of a top.

---

a modern name for the star we see near the north celestial pole. If you could travel several thousand years into the past or future, you would find that the point about which the northern sky appears to rotate is no longer near Polaris.

Recall that the celestial equator is the set of directions in the sky that are perpendicular to Earth's axis. As Earth's axis wobbles, then, so must the celestial equator. And as

---

[6] In fact, due to precession, the location of the summer solstice just passed from Gemini into Taurus in 1990. In about 600 years precession will cause the vernal equinox to pass from Pisces into Aquarius, marking the beginning of the "Age of Aquarius."

Why Is It Surprising That A.D. 2000 Was a Leap Year?

Everyone knows that years that are divisible by 4 are leap years, and A.D. 2000 was no exception to that rule. Yet A.D. 2000 *was* a special case. To understand why, we need to take a look at the way our calendar is constructed, and the purpose that leap years serve.

It takes Earth 365.242199 days to travel from one vernal equinox to the next. This tropical year is *not* exactly equal to the 365 days we normally think of as making up a year. (And thanks to the precession of the equinoxes, is it also not equal to the time—365.256366 days—it takes for Earth to complete one orbit about the Sun!) For convenience we count years as starting at midnight on the morning of January 1, and ending 365 days later at midnight on the night of December 31. But what about that extra fraction of a day (0.24 plus a bit) that we have not accounted for? Here is where leap years come in. A true tropical year is 0.242199 days—or about ¼ day—longer than a 365-day year, which means that after 4 years these extra parts have added up to make about 1 extra day. So every 4 years we add the extra day back in, and February

gets a 29th day. If we did not add in this extra day, our calendar would slip by ¼ day each year, which means that the seasons would shift by not quite a month each century. If we did not correct for leap years, the first day of summer would come in July, then in August, then in September, and so on.

Even after we add a leap year, the calendar still has problems. A 365-day year is short by 0.242199 days, which is a bit *less* than ¼. If we keep adding an extra day *every* 4 years, then over time we would accumulate an average of 0.007801 extra days per year. In 400 years the calendar would have slipped by about 3 days. To fix this it is necessary to get rid of 3 days every 400 years. This is accomplished by making century years into common 365-day years, *except* for those century years that are divisible by 400—such as A.D. 2000—which remain leap years. With one slight further revision—making years divisible by 4,000 into common 365-day years—the modern **Gregorian calendar** now slips by only about one day in 20,000 years.

to make up for the extra fraction of a day, preventing the seasons from slowly "sliding through" the year (getting increasingly out of sync with the months): winter in December one year, in August in another. **Excursions 2.2** gives further interesting details on this topic.

2.4 The Motions and Phases of the Moon

The second most prominent object in the sky after the Sun is the Moon. Just as Earth orbits about the Sun, the Moon orbits around Earth. (Actually, Earth and the Moon orbit around each other and together they orbit the Sun, as we will see in Chapter 3.) In some respects the appearance of the Moon is constantly changing, but we begin our discus-

sion of the motion of the Moon by talking about an aspect of the Moon's appearance that does *not* change.

We Always See the Same Face of the Moon

The Moon constantly changes its lighted shape and position in the sky, but one thing that does *not* change is the face of the Moon that we see. If we were to go outside next week or next month or 20 years from now or 20,000 *centuries* from

The Moon rotates on its axis once each orbit around Earth.

now, we would still see the same side of the Moon as tonight. Because of this fact, there is a common misconception that the Moon does not rotate. The Moon *does* rotate on its axis—exactly once for each revolution that it makes about Earth.

Imagine walking around the Washington Monument while keeping your face toward the monument at all times (a reasonable thing to do—you want to get a good look at it). By the time you complete one circle around the monument, your head has turned completely around once. (When you were south of the monument, you were facing north; when you were west of the monument, you were facing east; and so on.) But someone looking at you from the monument would never have seen anything other than your face. The Moon does exactly the same thing, rotating on its axis once per revolution around Earth, always keeping the same face toward Earth, as shown in **Figure 2.21**. This phenomenon is referred to as the Moon's **synchronous rotation**. (The Moon's synchronous rotation is not an accident. In Chapter 10 we will find that its cause is related to why we have tides on Earth.)

The Changing Phases of the Moon

Sometimes the Moon appears as a circular disk in the sky. At other times it is nothing more than a thin sliver. At still others its face is dark. The Moon has no light source of its own. Like the planets, including Earth, it shines by reflected sunlight. Like Earth, half of the Moon is always in bright

---

The Moon shines by reflected sunlight.

---

daylight, and half of the Moon is always in darkness. The different **phases** of the Moon result from the fact that the illuminated portion of the Moon that we see is constantly changing. Sometimes (during a new Moon) the side facing away from us is illuminated, and sometimes (during a full

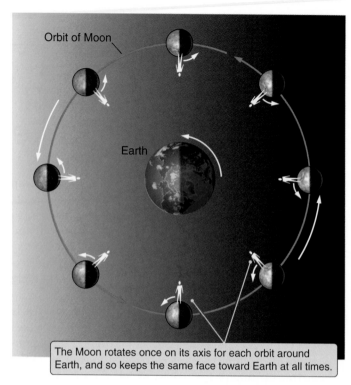

The Moon rotates once on its axis for each orbit around Earth, and so keeps the same face toward Earth at all times.

FIGURE 2.21 The Moon rotates once on its axis for each orbit around Earth, an effect called synchronous rotation.

Moon) the side facing toward us is illuminated. The rest of the time, only part of the illuminated portion can be seen from Earth (see **Excursions 2.3**).

To help you visualize the changing phases of the Moon, go outside at night and have a friend hold up a soccer ball so that it is illuminated from one side by a nearby streetlight (representing the Sun). Have your friend walk around you in a circle, and watch the changes in the ball's appearance.

EXCURSIONS 2.3

The Dark Side of the Moon

Popular culture often refers to the side of the Moon away from Earth as the "dark side of the Moon." This is even the title of a popular piece of 20th-century music. But in fact, there is no dark side of the Moon. At any given time, half of the Moon is in sunlight and half of the Moon is in darkness—just as at any given time, half of Earth is in sunlight and half of Earth is in darkness. The side of

the Moon that faces away from Earth, the *far side,* spends just as much time in sunlight as the side of the Moon that faces toward Earth.

On the other hand, "I'll see you on the backside of the Moon" might not have been as wildly successful as a song lyric.

The phase of the Moon is determined by how much of its bright side you can see.

When you are between the ball and the streetlight, the face of the ball that is toward you is fully illuminated. The ball appears to be a bright circular disk. As the ball moves around its circle, you will see a progression of lighted shapes, depending on how much of the bright side and how much of the dark side of the ball you can see. This progression of shapes exactly mimics the changing phases of the Moon.

Figure 2.22 shows the changing phases of the Moon. When the Moon is between Earth and the Sun, the illuminated side of the Moon faces away from us, and we see only its dark side. This is called a **new Moon**. Study Figure 2.22

to see that a new Moon can only be "seen" from the illuminated side of Earth. It appears close to the Sun in the sky, and so it rises in the east at sunrise, crosses the meridian near noon, and sets in the west near sunset. A new Moon is never visible in the nighttime sky.

As the Moon continues on its orbit around Earth, a small part of the portion being illuminated becomes visible. This shape is called a **crescent**. Because the Moon appears to be "filling out" from night to night at this time, the full name for this phase of the Moon is a waxing crescent Moon. (**Waxing** here means "growing in size and brilliance.") From our perspective the Moon has also moved away from the Sun in the sky. Because the Moon travels around Earth in the same direction in which Earth rotates, we now see the Moon located to the east of the Sun. A waxing crescent Moon is visible in the western sky in the evening, near the setting

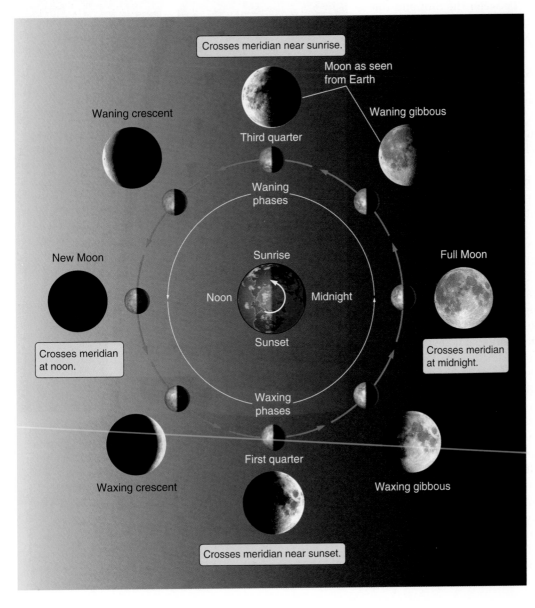

Crosses meridian near sunrise.

Moon as seen from Earth

Waning crescent

Waning gibbous

Third quarter

Waning phases

New Moon

Sunrise

Full Moon

Noon   Midnight

Sunset

Crosses meridian at noon.

Crosses meridian at midnight.

Waxing phases

First quarter

Waxing crescent

Waxing gibbous

Crosses meridian near sunset.

FIGURE 2.22 The inner circle of images shows the Moon as it orbits Earth, as seen by an observer far above Earth's North Pole. The outer ring of images shows the corresponding phases of the Moon as seen from Earth.

Sun but remaining above the horizon after the Sun sets. The "horns" of the crescent always point away from the Sun.

As the Moon moves farther along in its orbit, more and more of its illuminated side becomes visible each night, so the crescent continues to fill out. At the same time the angular separation in the sky between the Moon and the Sun grows. After about a week the Moon has moved a quarter of the way around Earth. We now see half the Moon illuminated in daylight and half the Moon as dark—a phase that we call **first quarter Moon**. A look at Figure 2.22 shows that the first quarter Moon rises at noon, crosses the meridian at sunset, and sets at midnight. Note that *first quarter* refers not to how much of the face of the Moon that we see illuminated, but rather to the fact that the Moon has completed the first quarter of its cycle from new Moon to new Moon.

As the Moon moves beyond first quarter, we are able to see more than half of its bright side. This phase is called a waxing **gibbous Moon**. The gibbous Moon continues nightly to "grow" until finally Earth is between the Sun and the Moon and we see the entire bright side of the Moon—a **full Moon**. The Sun and the Moon now appear opposite each other in the sky. The full Moon rises as the Sun sets, crosses the meridian at midnight, and sets in the morning as the Sun rises.

The second half of the Moon's orbit proceeds just like the first half but in reverse. The Moon continues in its orbit, again appearing gibbous but now becoming smaller each night. This phase is called a waning gibbous Moon. (**Waning** means "becoming smaller.") A **third quarter Moon** occurs when we once again see half of the illuminated part of the Moon and half of the dark part of the Moon. A third quarter Moon rises at midnight, crosses the meridian near sunrise, and sets at noon. The Moon continues on, visible now as a waning crescent Moon in the morning sky, until the Moon again appears as nothing but a dark circle rising and setting with the Sun, and the cycle begins again.

It takes the Moon 27.32 days to complete one revolution about Earth; this is called its **sidereal** period. However, because of the changing relationship between Earth, the Moon, and the Sun due to Earth's orbital motion, it takes 29.53 days to go from one full Moon to the next; this is called its **synodic** period. (**Figure 2.23** shows how this works.) You can always tell a waxing from a waning Moon because the side that is illuminated is always the side facing the Sun. When the Moon is waxing, it appears in the evening sky, so its western side is illuminated (the right side as viewed from the Northern Hemisphere). Conversely, when the Moon is waning, the eastern side (the left side as viewed from the Northern Hemisphere) appears bright.

Do not try to memorize all of the possible combinations of where the Moon is in the sky at what phase and at what time of day. You do not have to. Instead, work on *understanding* the motion and phases of the Moon, then use your understanding to figure out the specifics of any given case. As a way to study the phases of the Moon, draw a picture like

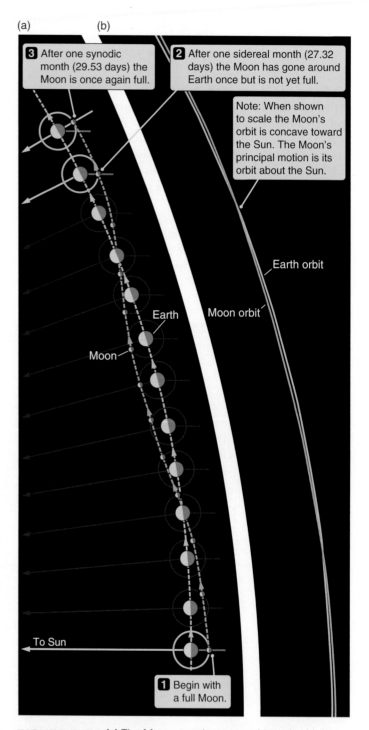

FIGURE 2.23 (a) The Moon completes one sidereal orbit in 27.32 days, but the synodic period (the period between phases seen from Earth) from one full Moon to the next is 29.53 days. (The orange line to the right of the Moon indicates a fixed direction in space.) (b) The orbits of Earth and the Moon are shown here to scale.

Figure 2.22, and use it to follow the Moon around its orbit. Figure out from the drawing what phase you would see and where it would appear in the sky at a given time of day. You might also enjoy thinking about what phase someone on the Moon would see when looking back at Earth.

FIGURE 2.24 Stonehenge is an ancient artifact in the English countryside, used 4,000 years ago to keep track of celestial events.

2.5 Eclipses: Passing through a Shadow

Put yourself in the place of our nomadic friend from the opening section of this chapter. You are finely attuned to the patterns of the sky, and you view these patterns not as the inexorable consequences of physical law but as visible signs from the gods. Can you imagine any celestial event that would strike more terror in your heart than to look up and see the Sun, giver of light and warmth, being eaten away as if by a giant dragon? There is archeological evidence that our ancestors put great effort into trying to find the pattern of eclipses and thereby bring them into the orderly scheme of the heavens. For example, the ancient and massive stone artifact in the English countryside called Stonehenge, pictured in **Figure 2.24**, may have allowed its builders to predict when eclipses might occur. The lives and motivations of the builders of Stonehenge, 4,000 years dead, may be lost in antiquity. Yet how can we doubt their desire to exert some control over the terror of eclipses by learning a few of their secrets—and in the process assuring themselves that an eclipse did not mean that all was lost?

Varieties of Eclipses

The type of eclipse just described, in which Earth moves through the shadow of the Moon, is called a **solar eclipse**.

Three different types of solar eclipses are possible: *total, annular,* and *partial.* To see why, begin by looking at the structure of the shadow of the Sun cast by a round object such as the Moon, as shown in **Figure 2.25**. An observer at point A could see no part of the surface of the Sun. This darkest, inner part of the shadow is called the **umbra**. If a point on Earth passes through the Moon's umbra, the Sun's light is

> There are three types of solar eclipses: partial, annular, and total.

totally blocked by the Moon. This is called a **total solar eclipse (Figure 2.26)**. Now look instead at points B and C in Figure 2.25. From these points an observer can see one side of the disk of the Sun but not the other. This outer region, which is only partially in shadow, is the **penumbra**. If a point on the surface of Earth passes through the Moon's penumbra, the result is a **partial solar eclipse**, in which the disk of the Moon blocks the light from a portion of the Sun's disk.

In the third type of eclipse, called an **annular solar eclipse**, the Sun appears as a bright ring surrounding the dark

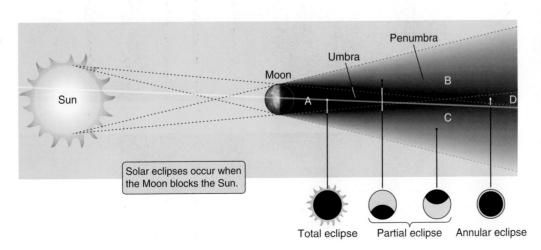

Sun

Moon

Penumbra

Umbra

B

A

D

C

Solar eclipses occur when the Moon blocks the Sun.

Total eclipse   Partial eclipse   Annular eclipse

FIGURE 2.25 Different parts of the Sun are blocked at different places within the Moon's shadow. An observer in the umbra (A) sees a total solar eclipse, observers in the penumbra (B and C) see a partial eclipse, while observers in region D see an annular eclipse.

FIGURE 2.26 The full spectacle of a total eclipse of the Sun.

FIGURE 2.27 An annular eclipse, in which the Moon does not quite cover the Sun.

disk of the Moon (**Figure 2.27**). An observer at point D is far enough from the Moon that the Moon appears to be smaller than the Sun. You may be wondering how one eclipse can be total and another annular. Two things make this possible. One is a fluke of nature: The diameter of the Sun is about 400 times the diameter of the Moon, and the Sun is about 400 times farther away from Earth than the Moon is. As a result, the Moon and Sun have almost exactly the same apparent size in the sky. The other factor is that the Moon's orbit is not a perfect circle. So when the Moon and Earth are a bit closer together than average, the Moon appears larger in the sky than the Sun. An eclipse occurring at that time will be

total. When the Moon and Earth are farther apart than average, the Moon appears smaller than the Sun, so eclipses occurring during this time will be annular. Pictures of total and annular solar eclipses are shown in **Figure 2.28**. Among solar eclipses, one-third are total, one-third are annular, and one-third are seen only as partial eclipses.

Figure 2.29(a) shows a drawing of the geometry of a solar eclipse, with the Moon's shadow falling on the surface of Earth. It is important to realize that figures like this (or like Figures 2.21 or 2.22) are seldom drawn to scale. Instead they show Earth and the Moon much closer together than they are in reality. The reason for distorting figures this way

FIGURE 2.28 Time sequences of images of the Sun taken (a) during a total solar eclipse and (b) during an annular solar eclipse.

(a)

(b)

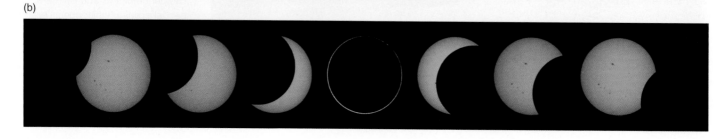

(a) Solar eclipse geometry (not to scale)

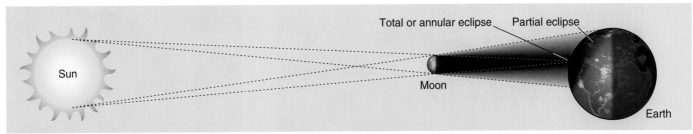

(b) Solar eclipse to scale

(c) Lunar eclipse geometry (not to scale)

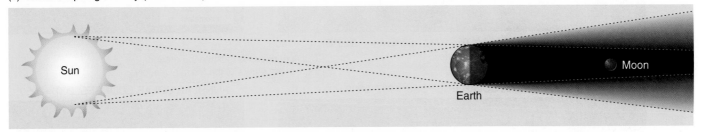

(d) Lunar eclipse to scale

FIGURE 2.29 A solar eclipse occurs when the shadow of the Moon falls on the surface of Earth. A lunar eclipse occurs when the Moon passes through Earth's shadow. Note that (b) and (d) are drawn to proper scale.

is simple: There is not enough room on the page to draw them correctly and still keep the smaller details visible. The 384,400 km distance between Earth and the Moon is over 60 times the radius of Earth and over 220 times the radius of the Moon. The relative sizes and distances between Earth and the Moon are roughly like a quarter and a dime held 2 meters apart. **Figure 2.29(b)** shows the geometry of a solar eclipse with Earth, the Moon, and the separation between them drawn to scale. Compare this to Figure 2.29(a), and you will understand why artistic license is normally taken in drawings of Earth and the Moon. If the Sun were drawn to scale in Figure 2.29(b), it would be 6⁄10 of a meter across and located almost 64 meters off the left side of the page.

The Moon's penumbra is quite large. In fact, with a bit of thought and a pencil and paper you can convince yourself that the Moon's penumbra, where it hits Earth, must have about twice the diameter of the Moon itself, or almost 7,000 km. This part of the shadow is large enough to cover

a substantial fraction of Earth, so partial solar eclipses are often seen from much of the planet. In contrast, the path along which a total solar eclipse can be seen (**Figure 2.30**) covers only a tiny fraction of the surface of Earth. Earth is so close to the tip of the Moon's umbra that even when the distance between Earth and the Moon is at a minimum, the umbra is only 269 km wide at the surface of Earth. As the Moon moves along in its orbit, this tiny shadow sweeps across the face of Earth at breakneck speed. The Moon moves in its orbit around Earth at a speed of about 3,400 km/h, and its shadow sweeps across the disk of Earth at the same rate. Earth is also rotating on its axis with a velocity of 1,670 km/h at the equator (and less than that at other latitudes). The situation is further complicated by the fact that the Moon's shadow falls on the curved surface of Earth. You may have noticed that the image projected by an overhead projector is distorted when the beam is not perpendicular to the screen. Similarly, the curvature of Earth often causes the region

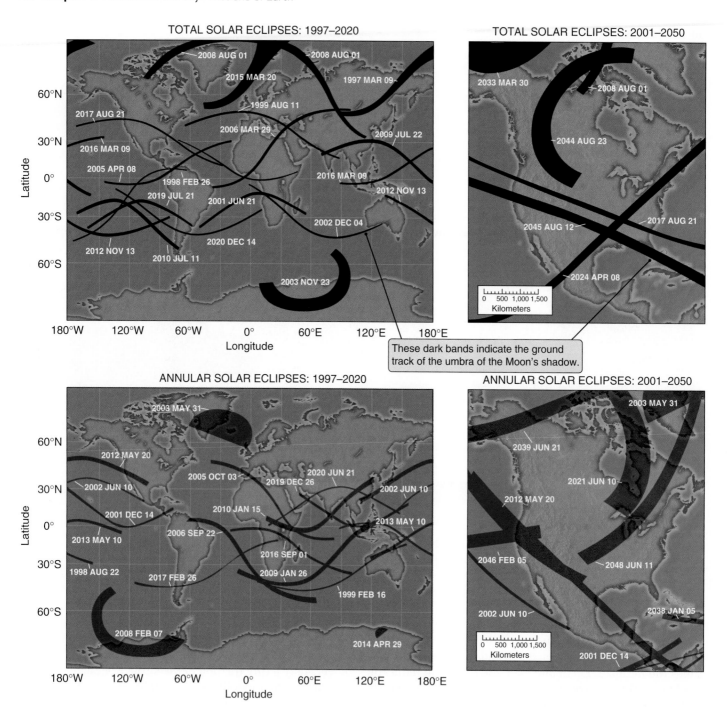

FIGURE 2.30 The paths of total and annular solar eclipses predicted for the late 20th and early 21st centuries.

shaded by the Moon during a solar eclipse to be elongated by differing amounts. The curvature can even cause an eclipse that started out as annular to become total.

When all of these effects are considered, the result is that a total solar eclipse can never last longer than 7½ minutes and is usually significantly shorter. Even so, it is one of the most amazing and awesome sights in nature. People flock from the world over to the most remote corners of Earth to witness the fleeting spectacle of the bright disk of the Sun blotted out of the daytime sky, leaving behind the eerie glow of the Sun's outer atmosphere.

Lunar eclipses are very different in character from solar eclipses. The geometry of a lunar eclipse is shown in **Figure 2.29(c)** (and is shown drawn to scale in **Figure 2.29(d)**). Because Earth is much larger than the Moon, the dark umbra of Earth's shadow at the distance of the Moon is about

Lunar eclipses last much longer than solar eclipses.

9,200 km in diameter, or over 2.5 times the diameter of the Moon. A **total lunar eclipse** is a much more leisurely affair than a total solar eclipse, with the Moon spending as long as 1 hour and 40 minutes in the umbra of Earth's shadow. A **penumbral lunar eclipse** occurs when the Moon passes through the penumbra of Earth's shadow. A penumbral eclipse can be unspectacular: Its appearance from Earth is nothing more than a fading in the brightness of the full Moon. Although the penumbra of Earth is 16,000 km across at the distance of the Moon—over four times the diameter of the Moon—a penumbral eclipse is noticeable only when the Moon passes within about 1,000 km of the umbra.

Eclipse Seasons Occur Roughly Twice Every 11 Months

If the Moon's orbit were in exactly the same plane as the orbit of Earth (imagine Earth, the Moon, and the Sun all sitting on the same flat tabletop), then the Moon would pass directly between Earth and the Sun at every new Moon. The Moon's shadow would pass across the face of Earth, and we would see a solar eclipse. Similarly, Earth would pass directly between the Sun and the Moon every synodic month, and each full Moon would be marked by a lunar eclipse.

Solar and lunar eclipses do *not* happen every month because the Moon's orbit does not lie in exactly the same plane as the orbit of Earth. Look at **Figure 2.31** to see how this works. The plane of the Moon's orbit about Earth is inclined by about 5.2° with respect to the plane of Earth's orbit about the Sun. The line along which the two orbital planes intersect is called the **line of nodes**. For part of the year, the line of nodes passes close to the Sun. During these times, called **eclipse seasons**, a new Moon passes between the Sun and Earth, casting its shadow on Earth's surface and causing a solar eclipse. Similarly, a full Moon occurring during an eclipse season passes through Earth's shadow, and a lunar eclipse results. An eclipse season lasts for only 38 days. That is how long the Sun is close enough to the line of nodes for eclipses to occur. Most of the time, the line of nodes points farther away from the Sun, and Earth, Moon, and Sun cannot line up closely enough for an eclipse to occur. A solar eclipse cannot be seen because the shadow

FIGURE 2.31 Eclipses are only possible when the Sun, Moon, and Earth lie along a line. When the Sun does not lie along the line of nodes, Earth passes under or over the shadow of a new Moon, and a full Moon passes under or over the shadow of Earth.

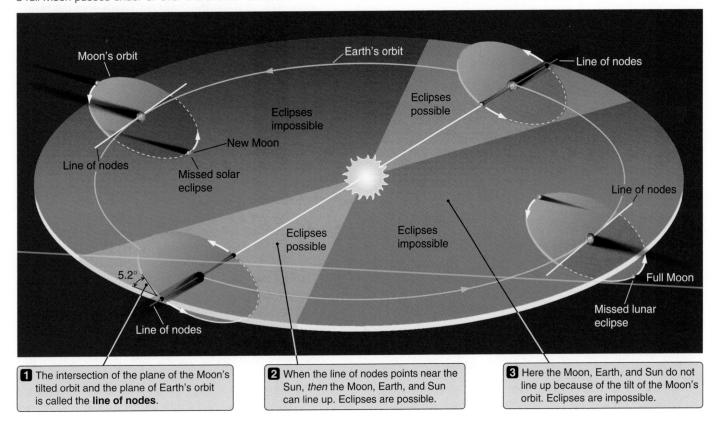

1 The intersection of the plane of the Moon's tilted orbit and the plane of Earth's orbit is called the **line of nodes**.

2 When the line of nodes points near the Sun, *then* the Moon, Earth, and Sun can line up. Eclipses are possible.

3 Here the Moon, Earth, and Sun do not line up because of the tilt of the Moon's orbit. Eclipses are impossible.

of a new Moon passes "above" or "below" Earth. Similarly, no lunar eclipse can be seen because a full Moon passes "above" or "below" the shadow of Earth.

If the plane of the Moon's orbit always had the same orientation, then eclipse seasons would occur twice a year, as suggested by the drawing in Figure 2.31. In actuality, eclipse seasons occur about every 5 months and 20 days. The roughly 10-day difference is due to the fact that the plane of the Moon's orbit slowly wobbles, much like the wobble of a spinning plate balanced on the end of a circus performer's stick. As it does so, the line of nodes changes direction. This wobble rotates in the direction opposite the direction of the motion of the Moon in its orbit. (That is, the line of nodes moves clockwise as viewed from the north.) It takes the Moon's orbit 18.6 years to complete one "wobble" of 360°, so we say that the line of nodes *regresses* by about 360°/18.6 years, or 19.4° per year. This amounts to about a 20-day regression each year. If January 1 marks the middle of an eclipse season, the next eclipse season would be centered around June 20, and the one after that around December 10.

We have come far since our nomadic ancestor looked at the sky and saw there a mystical reflection of the patterns and events that marked the life of the tribe. Yet when we look at the sky today our sense of awe and majesty is no less than that experienced by that long-ago tribesman. Our distant ancestors had to look for patterns in the world to survive. This same human impulse to seek patterns led astrologers to look for connections between ourselves and the heavens, and later led people to seek patterns that gave birth to science. We now know that the patterns of the sky are connected to us much more directly than any mystical link invented by an astrologer. The patterns and changes that we see in the sky are caused by the same forces of nature that bind us to our planet and that cause the wind to blow and the rain to fall. They are the same forces that push the blood through our veins and carry the electric impulses of thought through our brains. So far in our look at the changing patterns of the sky, we have only glimpsed these connections, so it is to these underlying causes that we now turn our attention.

Summary

- Stars appear to move through the sky as Earth rotates on its axis.

- The specific stars we see at night depend on where Earth is in its orbit around the Sun.

- The Coriolis effect causes hurricanes to rotate.

- The tilt of Earth's axis determines the seasons.

- The motion of the Moon in its orbit around Earth shapes the phases of the Moon.

- The phase of the Moon is determined by how much of its bright side you see.

- Special alignments of the Sun, Earth, and Moon result in solar and lunar eclipses.

Seeing the Forest through the Trees

Patterns in the sky change in lockstep with the cycles of life on Earth. It is no surprise that our ancestors sought some link between the stars and the events in their lives, but early astrologers were ultimately limited by their misconceptions about the universe. Earth does not reside at the center of Creation, nor does the Sun move from "house" to "house" along the ecliptic. The eerie spectacle of a solar eclipse blotting out the Sun is no more mystical than the shadow you cast on the sidewalk on a sunny afternoon. The heavens are a place not of magic but of physical law, knowable though observation and experiment and subject to test by the scientific method. The signs of the zodiac that grace the entertainment section of your local newspaper and the covers of supermarket tabloids are anachronisms—pictures born from the human imagination and painted onto the random splash of stars across the night sky.

Yet while astrologers were off the mark in their conclusions, their quest was well motivated. The motions of Earth and the properties of the Sun do, indeed, give rise to the most basic of all of the patterns faced by life on Earth. Earth's rotation is responsible for the coming of night and day. Earth's axial tilt and its passage around the Sun bring the changing of the seasons. Changes in the direction in which sunlight falls on Earth cause dramatic differences in climate from the equator to the poles. These patterns set the stage for the evolution of life on Earth, and they remain with us today, buried deep within the genetic code of our species. Even as humans begin to venture beyond Earth and into space, we carry these patterns with us in everything from the length of our cycle of waking and sleeping, to the temperature we prefer, to the amount of light we need in order to see. Much of human culture also has its roots in the apparent motions of celestial objects. Many of our legends and traditions arose in the ancient view of the sky as a place of gods and spirits. At the same time, patterns of objects in the sky spurred the development of mathematics and, as we will see in Chapter 3, the development of a physical understanding of the world around us. Much of who and what we are as a thinking species has its origins in our experience of the larger universe.

The errors in perception that shaped our views of the universe throughout most of human history are both understandable and forgivable. It is remarkably difficult to directly sense the effects of Earth's motion. As you read this, you are probably (depending on your latitude) moving at more than 1,000 km/h on a circular path around Earth's axis, while Earth itself is moving at over 100,000 km/h in its orbit about the Sun. (And the Sun itself is moving through our galaxy at almost 1 million km/h.) Yet you feel none of this motion. Objects around you share your motion, and so to you are motionless. The telltale signs of your motion are far too subtle to perceive directly. As you watch the Sun and stars cross the sky, it certainly seems as if it is you about which the cosmos revolves. In fact, without the benefit of discussions such as ours, it would be difficult to avoid that impression.

When we replace simple perception with careful experiment and reason, however, this commonsense notion evaporates before our eyes. Minute changes in the direction of starlight through the year provide proof of Earth's motion in its orbit, and they allow us to measure our path around the Sun. The behavior of a simple pendulum hung from the ceiling of a dome in Paris, when carefully considered, provides direct evidence of Earth's rotation about its axis. And as our physical understanding of Earth's motion improves, even the blowing of the wind and the need for a fire on a cold, dark winter night become side effects of Earth's motions through space.

In our journey we will again encounter many of the motions discussed in this chapter, but from a rather different perspective. So far we have concentrated on *describing* the motion of Earth about the Sun and the Moon about Earth. From here we will take the step that separates post-Renaissance science from all that came before by taking up the question of *why* these motions are as they are. The search for understanding will lead us to a discovery that changed our perception of the universe and our place in it—that the same natural forces that dictate the path of a well-hit baseball also govern the clockwork motions of Earth, the Moon, and planets.

Key Terms

constellations, p. 23
meridian, p. 24
zenith, p. 24
celestial poles, north and south, pp. 24, 25
latitude, p. 25
equator, p. 25
Polaris (north star), p. 27
circumpolar, p. 28
celestial sphere, p. 28
celestial equator, p. 29
Foucault pendulum, p. 30
relative motions, p. 32
Coriolis effect, p. 33
cyclonic motion, p. 34
year, p. 35
great circle, p. 35
ecliptic, p. 35
zodiac, p. 35
frame of reference, p. 36
aberration of starlight, p. 36
astronomical unit, p. 37
obliquity, p. 38
equinoxes, vernal and autumnal, p. 41
solstices, summer and winter, p. 41
precession of the equinoxes, p. 44
tropical year, p. 44
solar day, p. 44
leap year, p. 44
synchronous rotation, p. 46
phases, p. 46
sidereal, p. 48
synodic, p. 48
solar eclipse, p. 49
umbra, p. 49
total solar eclipse, p. 49
penumbra, p. 49
partial solar eclipse, p. 49
annular solar eclipse, p. 49
total lunar eclipse, p. 53
penumbral lunar eclipse, p. 53
line of nodes, p. 53
eclipse seasons, p. 53

Student Questions

THINKING ABOUT THE CONCEPTS

1. What is the approximate time of day when you see the full Moon near the meridian? At what time is the first quarter (waxing) Moon on the eastern horizon? Use a sketch to help explain your answers.

2. The Coriolis *effect* is sometimes wrongly called the Coriolis *force* because it can make a cannonball appear to deviate from its projected path. Yet the cannonball truly follows its projected path. Explain the illusion and how it appears to make the projected path deviate.

3. A soldier fires a cannon directly at a distant target toward the east and makes a perfect hit. She then fires a shot directly at another distant target toward the north and finds that her shot has hit to the west of the target. Was the soldier in Australia or Canada? Explain.

4. We tend to associate certain constellations with certain times of year. For example, we see the zodiacal constellation Gemini in the Northern Hemisphere's winter (Southern Hemisphere's summer) and the zodiacal constellation Sagittarius in the Northern Hemisphere's summer. Why do we not see Sagittarius in the Northern Hemisphere's winter (Southern Hemisphere's summer) or Gemini in the Northern Hemisphere's summer?

5. The tilt of Jupiter's rotational axis is 3°. If Earth's axis had this tilt, explain how it would affect our seasons.

6. Why do we not see a lunar eclipse each time the Moon is full or witness a solar eclipse each time the Moon is new?

7. Assume that the Moon's orbit is circular. Suppose you are standing on the side of the Moon that faces Earth. How would Earth appear to move in the sky as the Moon made one revolution around Earth? How would the "phases of Earth" appear to you, as compared to the phases of the Moon as seen from Earth?

8. Sometimes artists paint the horns of the crescent moon pointing toward the horizon. Is this realistic? Explain.

9. The true length of a year is not 365 days, but is actually about 365¼ days. How do we handle this extra quarter day to keep our calendars from getting out of sync?

10. Many cities have main streets laid out in east–west, north–south alignments. Why are there frequent traffic jams on east–west streets during both morning and evening rush hours within a few weeks of the equinoxes? Considering this, if you work in the city during the day, would you rather live east or west of the city?

11. What is the advantage of launching satellites from spaceports located near the equator? Why are satellites never launched in a westerly direction?

12. Does the occurrence of solar and lunar eclipses disprove the notion that the Sun and the Moon both orbit around Earth? Explain your reasoning.

APPLYING THE CONCEPTS

13. The Moon's orbit is tilted by about 5° relative to Earth's orbit around the Sun. What is the highest altitude in the sky that the Moon can reach as seen in Philadelphia (latitude 40°)?

14. Assume that you and your sister can both throw a baseball at a speed of 100 km/h. If you are in an airplane traveling at 800 km/h and play catch with your sister who is near the front of the plane (you being toward the rear), how fast would the ball be traveling as seen by an observer on the ground when thrown (a) by you and (b) by your sister? How fast does it appear to move as seen by you and your sister?

15. You are out in open ocean sailing from the Carolinas to Bermuda (that is, heading due east) and you know there is a hurricane nearby. A strong wind is blowing straight out of the south. Do you continue on to Bermuda or head back? To arrive at your answer, draw a diagram of the Coriolis effect acting on air circulating around the center of a hurricane.

16. Using an atlas, determine the latitude where you live. Draw and label a diagram showing that your latitude is the same as (a) the altitude of the north celestial pole and (b) the angle (along the meridian) between the celestial equator and your local zenith. What is the noontime altitude of the Sun as seen from your home at the times of winter solstice and summer solstice?

17. Solar (and lunar) eclipses can occur only when the Moon is near one of its nodes and is new (or full) during its orbit around Earth. The time between successive passages of the Moon through the same node is 27.2122 days. The period between successive new Moons is 29.5306 days. Show that 223 successive new Moons (223 lunar months) is the same interval (to within one part in 250,000) as 242 successive crossings of the same node, and that this is very close to 18 years in length. It seems likely that the builders of Stonehenge knew about this cycle some 4,000 to 5,000 years ago.

18. Assume that rain is falling at a speed of 5 meters per second (m/s) and you are driving along in the rain at a leisurely 5 m/s (or 18 km/h). Estimate the angle from the vertical at which the rain appears to be falling.

19. Suppose the tilt of Earth's equator relative to its orbit were 10° instead of 23.5°. At what latitudes would the Arctic and Antarctic Circles and the two Tropics be located?

20. Assume Earth is a perfect sphere with a radius of 6,400 km. What is the distance at Earth's equator that corresponds to 1° of longitude? What would be the distance corresponding to 1° of latitude? Estimate the dis-

tance corresponding to 1° of longitude at latitudes of 30° and 60°. What would be the distance of 1° of latitude at these same latitudes?

21. The vernal equinox is now in the zodiacal constellation of Pisces. Wobbling of Earth's axis will eventually cause the vernal equinox to move into Aquarius, beginning the legendary, long-awaited "Age of Aquarius." How long, on average, does the vernal equinox spend in each of the 12 zodiacal constellations?

StudySpace
wwnorton.com/astro21
provides a Study Plan for each chapter that includes a reading outline, animations, keyword flash cards, and gradebook-enabled multiple-choice quizzes. From StudySpace you can also access premium content in the ebook and SmartWork.

The Newtonian principle of gravitation is now more firmly established, on the basis of reason, than it would be were the government to step in, and to make it an article of necessary faith. Reason and experiment have been indulged, and error has fled before them.

THOMAS JEFFERSON (1743–1826)

Sir William Herschel, Sir Isaac Newton, and Johannes Kepler.

CHAPTER 3

Gravity and Orbits—
A Celestial Ballet

3.1 Gravity!

Today it is a rare person who does not know that the planets, including Earth, orbit around the Sun. Yet this was not always so. Only 500 years or so have passed since a soft-spoken Polish monk named **Nicholaus Copernicus** (1473–1543) started a revolution when he revived the idea, discarded by the Greeks two thousand years earlier, that the Sun rather than Earth lies at the center of creation. At the time this suggestion seemed outlandish. To think that Copernicus wanted us to believe that humankind—the pinnacle of creation—resides anywhere but at the center of all things! To imagine that we occupy but one of several planets circling the Sun's central fire, that we are nothing but a pebble among all the pebbles on the beach—absurd!

It would be easy from our "modern" perspective to chuckle at the naiveté of our ancestors and their Earth-centered view of the universe, but to do so would be unfair. The previous chapter showed that hard evidence of the motions of Earth was remarkably difficult to come by. To an ancient scholar, educated in Greek and Roman philosophy, Copernicus's view of the universe simply made no sense. Copernicus knew quite well that his ideas flew in the face of authority and would not be welcomed by the powers that be. Wishing to avoid the controversy his theory would certainly cause, Copernicus chose not to publish his ideas until late in his life. His great work *De revolutionibus orbium coelestrium* (On the Revolution of the Celestial Spheres) did not appear until the year of his death. This work pointed the way toward our modern cosmological principle. Copernicus knew his ideas would be unpopular, but he had no way of guessing the consequences of what he had started. In retrospect we see that he knocked the bottom out from

KEY CONCEPTS

In this chapter we follow the story of the birth of modern science as humans discovered regular patterns in the motions of the planets, then went on to explain those patterns with fundamental physical laws. Along the way we will explore

- Empirical rules discovered by Kepler that describe the elliptical orbits of planets around the Sun.
- Theoretical physical laws discovered by Newton and Galileo that govern the motion of all objects.
- The use of proportionality to describe patterns and relationships in nature.
- The nature of scientific theories, the roles of empirical and theoretical science, and the difference between science and pseudoscience.
- How Newton's laws of motion and an inverse square law of gravitation combine to explain an orbit as one body falling freely around another.
- The cosmological principle that grew from the startling realization that the heavens and Earth are governed by the same physical laws.
- How theories lead to new knowledge, such as the way Newton's derivation of Kepler's third law is used to measure the masses of objects from observations of orbital motions.

under the house of cards representing humankind's view of the world around them, and the following centuries would see that house of cards slowly fall apart. The repercussions of Copernicus's insight not only would shape our understanding of the universe around us but would change the direction of the progress of human civilization itself.

Even today we need to avoid too much complacency about what we think we know. People often confuse knowing the name of something with actually *understanding* something. For example, we happily talk about spacecraft in orbit about Earth, or planets in orbit about the Sun, but remarkably few people understand what those words mean. When asked why astronauts float about the cabin of a spacecraft, most educated adults answer, "The spacecraft has escaped Earth's gravity." Yet the gravitational force acting on a space shuttle orbiting Earth is only slightly weaker than when the shuttle is sitting on the launchpad. In fact, were it not for Earth's gravity, the spacecraft would not orbit Earth at all!

Copernicus knew nothing of gravity, but his ideas inevitably led to it. As physicists and astronomers have come to better understand gravity, they have realized that in most respects it is gravity that holds the universe together. Our Solar System is a gravitational symphony. The Sun's gravity shapes the motions of the planets and every other object in its vicinity. These motions range from the almost circular orbits of some planets to the extremely elongated orbits of comets. A comet's orbit may carry it from tens of thousands of astronomical units[1] from the Sun to somewhere inside the orbit of Mercury and back out again. Within this grand symphony, subthemes arise. A chorus of particles orbiting the giant planets give rise to majestic systems of rings, which, in turn, play counterpoint to the gravitational ballet of the planets and their moons. The analogy between the motions of objects in the Solar System and the patterns of music is not new. **Johannes Kepler** (1571–1630), who will figure prominently in our story, titled his great work of 1619 *Harmonice mundi,* or "Harmonies of the World."

As our grasp of the universe has expanded, we have come to realize that our Solar System is but one gravitational opus in a far larger opera. Gravity binds stars into the colossal groups we call galaxies and slows the expansion of the universe. It is gravity that holds the planets and stars together and keeps the thin blanket of air we breathe close to the surface of the planet we live on. It is gravity that caused a vast interstellar cloud of gas and dust to collapse 4.5 billion years ago to form our Sun and Solar System. It is gravity that gives space and time their very shape. As we continue our journey outward through the cosmos, we will come to each of these ideas in turn, and time and time again we will find gravity at the center of our growing understanding.

3.2 An Empirical Beginning: Kepler's Laws Describe the Observed Motions of the Planets

Now that we have extolled the wonders of gravity and the role it plays in the universe, you might expect us immediately to discuss what gravity is and how it works. But the great minds that brought us to our modern understanding did not have the benefit of our 20–20 hindsight. All they could do was watch the motions of the planets over the course of months and years and puzzle over what they saw. With no way even to judge the distances to the planets, it is no wonder that it took humans thousands of years to begin to see the reality behind the celestial patterns before their eyes.

In Chapter 1 we painted a picture of science as a worldview in which nature is governed by physical laws and in which mathematical descriptions of these physical laws are used to explain natural phenomena. But how do scientists go about *discovering* these physical laws? When facing phenomena as complex and puzzling as the motions of the planets in the heavens, where can we find a toehold? Just as the wise sailor settles for any port in a storm, the wise scientist knows that when facing a complex and poorly understood phenomenon, there may be little choice but to turn directly to the information our senses provide. We carefully observe the phenomenon under study, systematically recording as much information as we can as accurately as possible. As our observations start piling up, we look for patterns in those observations and start trying to formulate rules that seem to describe what we have seen.

Imagine that you are a scientist from another world, setting foot for the first time on Earth. You notice right away that many interesting structures are sticking out of the ground on this unexplored planet. As you record your observations, you find that some of these structures are large, some are small, some spread out over the ground, some stick up in the air, and so on. But after a time you realize that almost all these structures are covered with some kind of appendage, and those appendages are green. So you form a descriptive rule about these objects (call them "plants"): Most have green appendages. You decide that green appendages must be fundamental to the nature of plants, so you begin to study what makes these appendages green. After a time you discover that the green appearance always comes

---

[1] Recall from Chapter 2 that an astronomical unit (abbreviated AU) is the average distance between the Sun and Earth.

from the same chemical substance, and when you study that chemical substance you find that it can absorb light and turn water and carbon dioxide into more complex organic molecules. You have discovered photosynthesis, the process responsible for powering the majority of life on Earth. *But you did not start out to discover photosynthesis. You started out noting that plants have green leaves.*

The quest to first note and then accurately describe patterns in nature is called **empirical science**. Empirical science often involves a great deal of creativity and not a small amount of pure (but educated) guesswork. Copernicus's theory that Earth and the planets move in circular orbits about the Sun is an example of empirical science. Copernicus did not understand *why* the planets move about the Sun, but he did realize that his Sun-centered picture provided a much *simpler* description of the observed motions of planets than a model with Earth at its center did. Copernicus's work was made great by the fact that he was able to see beyond the prejudice of his time and to think the unthinkable—that perhaps Earth is "merely" one planet among many.

Copernicus's work paved the way for another great empiricist, Johannes Kepler. Science has often benefited from unlikely and chancy collaborations, and Kepler's is one such story. Kepler, a mathematician who had studied the ideas of Copernicus, worked in 1600 as an assistant to **Tycho Brahe** (1546–1601), a firm believer in an Earth-centered universe. Although Tycho Brahe is described as anything but a pleasant individual, he was also one of the greatest observational astronomers of all time. Toiling away through long nights with primitive equipment, Tycho Brahe amassed a wealth of remarkably accurate observations of the positions of the planets over the course of decades. Kepler, using Tycho's Brahe's data, took the next major step toward understanding the motions of the planets. Working first primarily with Brahe's observations of Mars, Kepler was able to deduce three empirical rules that elegantly and accurately describe the motions of the planets. These three rules are now almost universally referred to as **Kepler's laws**.

Kepler's First Law: Planets Move on Elliptical Orbits with the Sun at One Focus

When Kepler used Copernicus's model to calculate where in the sky a planet should be at a particular time, he found quite a lot of disagreement between these predictions and

---

Planetary orbits are ellipses.

Brahe's data. He was not the first to notice such discrepancies. Rather than discarding Copernicus's ideas, however, Kepler played with Copernicus's Sun-centered model. Kep-

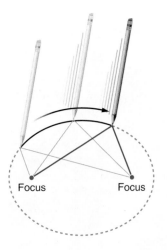

FIGURE 3.1 We can draw an ellipse by attaching a length of string to a piece of paper at two points (called foci), then pulling the string around as shown.

ler discovered that if he replaced Copernicus's circular orbits with elongated elliptical orbits and displaced these orbits to one side so that the Sun was no longer at the center, his calculations and Brahe's observations agreed almost perfectly.

You probably think of an ellipse as an oval shape, but to make sense of Kepler's discovery we need to be a bit more precise about what an ellipse is. The most concrete way to define an **ellipse** is to call it the shape that results when you attach the two ends of a piece of string to a piece of paper, stretch the string tight with the tip of a pencil, and then draw around those two points keeping the string taut (see **Figure 3.1**). Each of the points at which the string is attached is called a **focus** of the ellipse. The closer the two *foci* (plural of focus) are to each other, the more nearly circular the ellipse is. In fact, a circle is just an ellipse with the two foci at the same place. (To see this, just think about the shape you would draw if the two ends of the string were attached at the same spot. In this case, each half of the string would become a radius of the circle.) As the two foci are moved

---

The Sun is at one focus of a planet's elliptical orbit.

---

farther apart, however, the ellipse becomes more and more elongated. Kepler found that the orbit of each planet is an ellipse with the Sun located at one focus. This result is now known as **Kepler's first law** of planetary motion. (You might ask, "If the Sun is located at *one* focus, what is at the other focus?" The answer is "nothing but empty space.")

Figure 3.2 shows a drawing of an ellipse. Half of the length of the long axis of the ellipse is called the **semimajor axis** of the ellipse, often denoted by the letter A. The semimajor axis of an orbit turns out to be a handy way to describe the orbit because, apart from being half the longer dimension of the ellipse, it is also the average distance between one focus and the ellipse itself. The average distance between the Sun and Earth, for example, equals the

Kepler's First Law

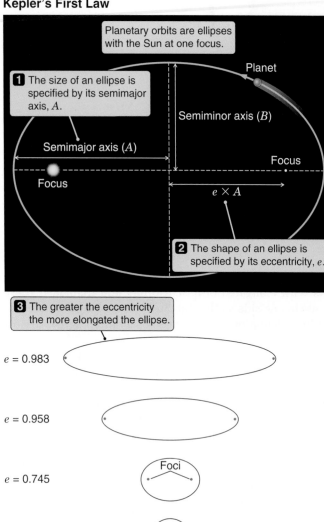

FIGURE 3.2 Planets move on elliptical orbits with the Sun at one focus. Ellipses range from circles to elongated eccentric shapes.

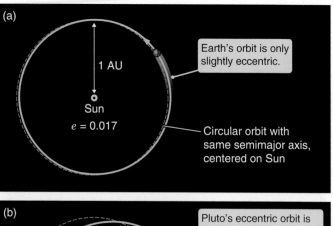

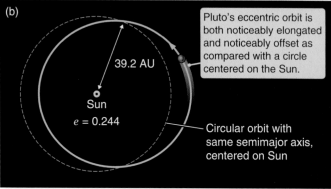

FIGURE 3.3 The shapes of the orbits of Earth and Pluto compared with circles centered on the Sun.

semimajor axis of the Earth's orbit. The same is true for the orbits of all the planets.

In the case of a circular orbit, the semimajor axis is just the radius of the circle. Some ellipses, on the other hand, are very elongated. When describing the shape of an ellipse, we speak of its **eccentricity**. The eccentricity of an ellipse is defined as the separation between the two foci divided by the length of the long axis of the ellipse.[2] A circle has an eccentricity of 0. The more elongated the ellipse becomes,

the closer its eccentricity gets to 1 (see Figure 3.2). Most planets have nearly circular orbits with eccentricities close to 0. The eccentricity of Earth's orbit, for example, is 0.017, which means that the distance between the Sun and Earth departs from its average value by only 1.7 percent. The distance between the two bodies varies from about 0.983 AU to 1.017 AU. It is hard to tell the difference between the orbit of Earth and a circle centered on the Sun (**Figure 3.3(a)**). One of the characteristics that distinguishes the dwarf planet Pluto from its classical cousins is its highly eccentric orbit. With an eccentricity of 0.244, the distance between the Sun and Pluto varies by 24.4 percent from its average value, ranging from 75.6 percent of average to 124.4 percent of average. Pluto's orbit is noticeably oblong, and its center is noticeably displaced from the Sun (**Figure 3.3(b)**).

Kepler's Second Law: Planets Sweep Out Equal Areas in Equal Times

The next empirical rule that Kepler found has to do with how fast planets move at different places on their orbits. A

---

[2] This book tends to use operational definitions. An ellipse, for example, is defined according to how it is drawn. Eccentricity is defined according to how you would calculate its value. Definitions do not mean much in science unless they are connected to how something is actually measured or calculated.

Planets move fastest when they are closest to the sun.

planet moves most rapidly when it is closest to the Sun, and is at its slowest when it is farthest from the Sun. The average speed of Earth in its orbit about the Sun is 29.8 km/s. When Earth is closest to the Sun, it travels at 30.3 km/s. When it is farthest from the Sun, it travels at 29.3 km/s.

Kepler found an elegant way to describe the changing speed of a planet in its orbit about the Sun. Look at **Figure 3.4**, which shows a planet at six different points in its orbit. Imagine a straight line connecting the Sun with this planet. We can think of this line as "sweeping out" an area as it moves with the planet from one point to another. Area A (in red) is swept out between times t_1 and t_2, area B (in blue) is swept out between times t_3 and t_4, and area C (in green) is swept out between times t_5 and t_6. When the planet is closest to the Sun it is moving rapidly, but the distance between the planet and the Sun is small (area A in Figure 3.4). Kepler realized that changes in the distance between the Sun and a planet and changes in the speed of a planet work together to produce a surprising result: The area swept out by a planet in the same amount of time is always the same regardless of the location of the planet in its orbit. In Figure 3.4, this means that if the three time intervals are equal ($t_1 \rightarrow t_2 = t_3 \rightarrow t_4 = t_5 \rightarrow t_6$) then the three areas A, B, and C will be equal as well.

This is **Kepler's second law**, which is also referred to as Kepler's **Law of Equal Areas**. It states that the imaginary

A planet "sweeps out" equal areas in equal times.

line connecting a planet to the Sun sweeps out equal areas in equal times, regardless of where the planet is in its orbit. Note that this law applies to only one planet at a time. The area swept out by Earth in a given time is always the same. Likewise, the area swept out by Mars in a given time is always the same. But the area swept out by Earth and the area swept out by Mars in a given time are *not* the same.

Kepler's Third Law: The Harmony of the Worlds

Kepler's first law describes the shapes of planetary orbits, and Kepler's second law describes how the speed of a planet changes as it goes around its orbit. But neither of these laws tells us how long it takes a planet to complete one orbit about the Sun (referred to as the **period** of the orbit). Nor do these laws tell us how this time depends on the distance between the Sun and a planet.

Planets that are closer to the Sun do not have as far to go to complete one orbit as do planets that are farther from the Sun. Jupiter, for example, has an average distance of 5.2 AU from the Sun—5.2 times as far from the Sun as Earth is. That means Jupiter has 5.2 times farther to travel in its orbit about the Sun than Earth does. We might guess, then, that if the

Outer planets have farther to go and move more slowly in their orbits around the sun.

two planets travel at the same speed, Jupiter would complete one orbit in 5.2 years. But such a guess would be wrong. Jupiter takes almost 12 years to complete one orbit. Clearly Jupiter not only has farther to go in its orbit but must be *moving more slowly than Earth* as well. This trend holds true for all the planets. As we go farther out from the Sun, the circumferences of planetary orbits become longer while the speeds at which the planets travel become less. Mercury, at an average distance of 0.387 AU from the Sun, whizzes around its short orbit at an average speed of 47.9 km/s, completing one revolution in only 88 days. At a distance of 30.1 AU from the Sun, Neptune lumbers along at an average speed of 5.48 km/s, taking 163.7 years to make it once around the Sun.

Kepler discovered a simple mathematical relationship between the period of a planet's orbit and its distance from

The square of a planet's orbital period equals the cube of the orbit's semimajor axis.

the Sun. **Kepler's third law** states that the square of the period of a planet's orbit, measured in years, is equal to the cube

FIGURE 3.4 An imaginary line between a planet and the Sun sweeps out an area as the planet orbits. Kepler's second law states that if the three intervals of time shown are equal, then the three areas A, B, and C will be the same.

Kepler's Second Law

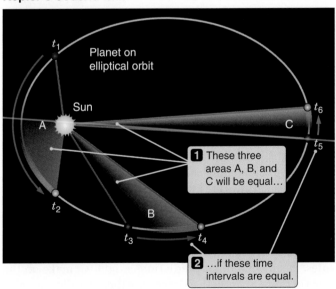

t_1

Planet on elliptical orbit

Sun

A

C

t_6

t_5

1 These three areas A, B, and C will be equal…

t_2

B

t_3

t_4

2 …if these time intervals are equal.

of the semimajor axis of the planet's orbit, measured in astronomical units. Written as an equation, this says that

$$(P_{\text{years}})^2 = (A_{\text{AU}})^3,$$

where P_{years} represents the period of the orbit divided by 1 year, and A_{AU} is the semimajor axis of the orbit divided by 1 AU.

This is a case where astronomers use nonstandard units as a matter of convenience. Years are handy units for measuring the periods of orbits, and AU are handy units for measuring the sizes of orbits. When we use years and AU as our units, we get the simple relationship just shown. However, it is important to realize that *our choice of units in no way changes the physical relationship* we are studying. If we instead stayed with standard metric units, this relationship would read $(P_{\text{seconds}})^2 = 3 \times 10^{19} (A_{\text{meters}})^3$. **Table 3.1** lists the periods and semimajor axes of the orbits of the classical and dwarf planets, together with the values of the ratio P^2 divided by A^3.

Judge for yourself how well Kepler's third law works. These data are also plotted in **Figure 3.5**. This relationship was so beautiful to Kepler that he referred to it as his **harmonic law** or, more poetically, as the "Harmony of the Worlds."

3.3 The Rise of Scientific Theory: Newton's Laws Govern the Motion of All Objects

When investigating a newly discovered or poorly understood phenomenon, an empirical approach is often the only available way to proceed. This was certainly the case with Kepler. The development of Kepler's laws of planetary motion was an intellectual accomplishment with few peers. Yet to a modern scientist the empirical rules that Kepler spent his life pursuing are only the first step in the study of a phenomenon. Such empirical laws *describe* some phenomenon and are even useful in predicting what will happen in the future, but they do little to *explain* that behavior. Taken at face value, empirical rules offer little insight into the more fundamental laws describing nature. Kepler was able to characterize the orbits of planets as ellipses, but he did not understand *why* they should be so.

Once the empirical rules that describe some phenomenon have been discovered, a modern scientist will next try to understand those empirical rules in terms of more general physical principles or laws. Beginning with basic physical principles and using the tools of mathematics, the scientist works to *derive* the empirically determined rules. At other times a scientist may start with physical laws and predict relationships, which are then verified empirically. This technique is sometimes referred to as the **theoretical** approach to science. In practice, if the relevant physical laws are already understood, this process is often short-circuited. A scientist may make a theoretical prediction about the behavior of a system, then compare the prediction with experimental data directly to see how well they fit.

Today a great deal of science is done without ever trying to invent an empirical rule. This works only if the relevant physical laws are known ahead of time. If the relevant physical laws are *not* known—as was the case for planetary motion—the empirical rules become a way of *discovering* the physical laws themselves. Can we invent **hypothetical** physical laws that will allow us to derive the empirical rules? If so, what other predictions might we make on the basis of these hypothetical laws? Are these predictions also

TABLE 3.1

Kepler's Third Law: $P^2 = A^3$
The Orbital Properties of the Classical and Dwarf Planets

Planet	Period P years	Semimajor axis A (AU)	$\frac{P^2}{A^3}$
Mercury	0.241	0.387	$\frac{0.241^2}{0.387^3} = 1.00$
Venus	0.615	0.723	$\frac{0.615^2}{0.723^3} = 1.00$
Earth	1.000	1.000	$\frac{1.000^2}{1.000^3} = 1.00$
Mars	1.881	1.524	$\frac{1.881^2}{1.524^3} = 1.00$
Ceres	4.599	2.765	$\frac{4.559^2}{2.765^3} = 1.00$
Jupiter	11.86	5.204	$\frac{11.86^2}{5.204^3} = 1.00$
Saturn	29.46	9.582	$\frac{29.46^2}{9.582^3} = 0.99*$
Uranus	84.01	19.201	$\frac{84.01^2}{19.201^3} = 1.00$
Neptune	164.79	30.047	$\frac{164.79^2}{30.047^3} = 1.00$
Pluto	247.68	39.236	$\frac{247.68^2}{39.236^3} = 1.02*$
Eris	557.00	67.696	$\frac{557.00^2}{67.696^3} = 1.00$

*These ratios are not exactly 1.00, due to slight perturbations from the gravity of other planets.

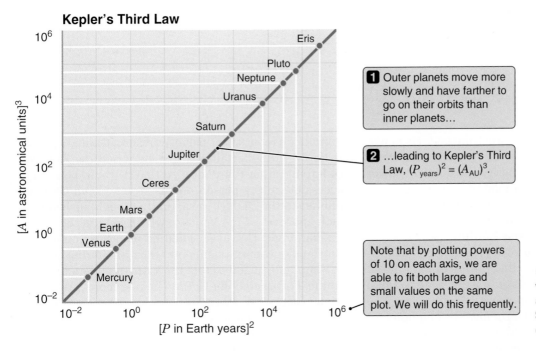

Kepler's Third Law

❶ Outer planets move more slowly and have farther to go on their orbits than inner planets...

❷ ...leading to Kepler's Third Law, $(P_{years})^2 = (A_{AU})^3$.

Note that by plotting powers of 10 on each axis, we are able to fit both large and small values on the same plot. We will do this frequently.

FIGURE 3.5 A plot of A^3 versus P^2 for the eight classical and three dwarf planets in our Solar System shows that they obey Kepler's third law.

borne out by experiment and observation? If so, then we may have discovered something more fundamental about the way the universe works. This is how physical laws are discovered and tested.

One of the earliest great advances in theoretical science was also arguably one of the greatest intellectual accomplishments in the history of our species. In many ways the work of **Sir Isaac Newton** (1642–1727) on the nature of motion set the standard for what we now refer to as *scientific theory* and *physical law.* Building on the work of Kepler and others, Newton proposed three laws that he believed to govern the motions of all objects in the heavens and on Earth.

Newton's laws of motion are the basis of classical mechanics.

Today **Newton's laws** remain the basis for all of what is known as **classical mechanics**. By the time physicists and astronomers complete their formal education, they have spent many hours studying the wondrously subtle and complex consequences of Newton's three laws of motion. (Physics professors pride themselves on their ability to invent truly nasty problems to challenge their graduate students' understanding of Newton's laws.) Even so, Newton's laws themselves are beautifully elegant, and the relationships they describe between such everyday concepts as force, velocity, acceleration, and mass are accessible to all.

However fascinating Newton's laws of motion may be, you might reasonably ask why they are an essential stop on our journey through *21st Century Astronomy.* "After all," you might say, "this is a book about astronomy, not physics." Yet in a very real sense it is with Newton's laws, pub-

lished in 1687, that truly modern astronomy got its start. It was these laws that allowed Newton to look at the motion of a shot fired from a cannon and see instead the motions of the planets on their orbits around the Sun. It was with Newton's laws that the chasm between our thinking about the heavens and about Earth was banished once and for all, and Earth took its true place in the universe.

Newton's First Law: Objects at Rest Stay at Rest; Objects in Motion Stay in Motion

In a strange quirk of history, the physical law almost invariably referred to today as *Newton's first law of motion* did not originate with Newton at all. It was the brainchild of a contemporary of Kepler's by the name of **Galileo Galilei** (1564–1642). Galileo is probably best known to the general public as the first person to use a telescope to make significant discoveries about the heavens and to report those discoveries. In the history of science, however, Galileo's work on the motion of objects is at least as fundamental a contribution as his astronomical observations.

By Galileo's day, Copernicus and others had begun to turn toward the view that knowledge comes from observing nature rather than only from reading the works of classical Greek and Roman philosophers. Yet even in the 16th and 17th centuries the works of one of the greatest of these philosophers, **Aristotle** (384–322 B.C.), who lived almost 2,000 years earlier, still carried the weight of authority. Aristotle

Challenging Authority

It is seldom safe to challenge the entrenched wisdom of your day, and that was especially true at a time when intellectual and religious authority and political power resided in the same hands. In 1600 Giordano Bruno fell victim to the Inquisition and was burned at the stake for his beliefs. These included his support of Copernicus, his belief that the universe is infinite, and his suggestion that Earth is but one of many habitable planets. In 1632 Galileo actually invited the ire of the powers that be when he published his great work, *Dialogo sopra i due massimi sistemi del mondo* (*Dialogue on the Two Great World Systems*). In the *Dialogo* the champion of the Copernican (Sun-centered) view of the universe is a brilliant, witty, and erudite philosopher named Salviati. The *Dialogo*'s defender of Aristotelian authority is named Simplicio and is as much an ignorant buffoon as the name might imply. In Galileo's story the "neutral" but intelligent moderator, Sagredo, is quick to see the truth in Salviati's arguments and dismisses

Simplicio's rebuttals as patently absurd. Galileo's prose is lively and entertaining, and he wrote the *Dialogo* in Italian rather than Latin so that it would be easily accessible to the person on the street. When Galileo published the *Dialogo,* he actually thought he had the tacit approval of the Vatican, which held to the Aristotelian view. However, when he placed a number of the Pope's own arguments into the unflattering mouth of Simplicio, he found that the Vatican's tolerance had limits. Fortunately for Galileo he had more friends in high places than Bruno did, so he spent the closing years of his life under house arrest rather than ending up tied to the stake. To escape a harsher sentence Galileo was forced to publicly recant the Copernican theory that he had supported with such fervor. In one of the great apocryphal stories of the history of astronomy, it is said that as he left the courtroom following his sentencing, he stamped his foot on the ground and muttered, "But it moves!"

believed that the natural state of all objects was to be at rest and that an object in motion would tend toward this natural state. This seemed to be a good empirical rule about how objects in the world around us behaved, and Aristotle had elevated this observation into a fundamental tenet of his philosophy of nature. Aristotle was an extremely sharp individual, and this idea was not easy to refute. A cart rolling down the street coasts to a stop when it is no longer being pulled. A bouncing ball eventually settles to the ground. Even an arrow shot from a bow loses much of its speed before striking its target.

As discussed in **Excursions 3.1**, Galileo was no stranger to controversy. In his writings on motion, Galileo challenged Aristotle's authority by proposing that this apparent tendency of objects to come to rest was a mirage. Galileo argued that in all of the cases just mentioned—indeed in *every* such case—there are hidden reasons why objects come to rest. There is friction as the axle of the cart rubs against its bearing, resisting the motion and eventually bringing it to a halt. Every time a ball bounces, its shape is distorted, and what we might think of as "internal friction" within the ball causes it to bounce less high each time. The resistance of air, which the arrow must push out of the way and which drags against the arrow's shaft, slows the arrow's progress.

Galileo agreed with Aristotle that an object at rest remains at rest unless something causes it to move. But Galileo disagreed with Aristotle by asserting that, *left on its own, an object in motion will remain in motion.* Specifically, Galileo said that *an object in motion will continue moving along a straight line with a constant velocity until an* **unbalanced force** *acts on it to change its state of motion.* Galileo

Galileo found that an object left in motion remains in motion.

referred to the resistance of an object to changes in its state of motion as **inertia**. Galileo's great insight formed the starting point for Newton's tour de force that was to come. The work of great scientists is always built on the foundation of the great scientists who came before.[3] It is a tribute to Galileo that his law of inertia became the cornerstone of physics as **Newton's first law of motion**.

The idea of inertia has come a long way since the days of Galileo and Newton. In fact, we have already seen a number of very sophisticated applications of this idea. What Galileo

[3]Newton himself is credited with the famous quote, "If I have seen further [than you] it is by standing upon the shoulders of giants."

and Newton called *inertia* can actually be viewed as a consequence of what we discovered in Chapter 2's discussions of relative motion and frames of reference. To say that only *relative* motions between objects have meaning is the same as saying that there is *no difference* between an object at rest and an object in uniform motion. What objects are at rest and what objects are in motion, anyway? The object at rest beside you on the front seat of your car as you drive down the highway is moving at 60 miles per hour according to a bystander along the side of the road—but it is moving at 120 mph according to a car in oncoming traffic. All of these perspectives are equally valid.

The connection between inertia and the relative nature of motion is so fundamental that a reference frame moving in a straight line at a constant speed is referred to as an **inertial frame of reference**. Motion is meaningful only when measured relative to an inertial frame of reference, and *all* inertial frames of reference are as good as any other. The realization that the laws of physics are the same in *any* inertial frame of reference is one of the deepest insights ever made into the nature of the universe. When thought of in this way, *of course* an object moving in a straight line at a constant speed remains in motion. As illustrated in **Figure 3.6**, in the frame of reference of that object, *it is already at rest*.

Newton's Second Law: Motion Is Changed by Unbalanced Forces

Newton often gets credit for Galileo's insight about inertia because it was Newton who took the crucial next step. Newton's first law says that in the absence of an unbalanced

---

Unbalanced forces cause changes in motion.

---

force an object's motion does not change; **Newton's second law of motion** goes on to say that *if there is an unbalanced force acting on an object, then the object's motion does*

change. Even more, Newton's second law tells us *how* the object's motion changes in response to that force.

Before going any further, it would be wise to pause to be sure we are all together on this. In the preceding paragraphs we spoke of changes in an object's motion, but what does that phrase really mean? When you are in the driver's seat of a car, a number of controls are at your disposal. On the floor of the car are a gas pedal and a brake. You use these to make the car speed up or slow down. A *change in speed* is one way the motion of an object can change. But also remember the steering wheel beneath your hands. When you are moving down the road and you turn the wheel, your speed does not necessarily change, but the direction of your motion does. A *change in direction* is also a kind of change in motion.

Together the speed and direction of an object's motion are called the object's **velocity**. A change in velocity is called an **acceleration**. Acceleration actually refers to how rapidly the change in velocity happens. If you go from 0 to 60 mph in 4 seconds, you feel the back of your seat shoving your

---

Acceleration measures how quickly a change in motion takes place.

---

body forward, causing you to accelerate along with the car. If you take 2 minutes to get from 0 to 60 mph, on the other hand, the acceleration is so slight that you hardly notice it. To formalize this a bit, your acceleration is determined by how much your velocity changes divided by how long it takes for that change to happen:

$$\text{Acceleration} = \frac{\text{How much velocity changes}}{\text{How long the change takes}}$$

For example, if an object's speed goes from 5 m/s (meters per second) to 15 m/s, then the change in velocity is 10 m/s. If that change happens over the course of 2 seconds, then the acceleration is 10 m/s divided by 2 seconds, which equals 5 meters per second per second. This is the same as saying "5 meters per second squared," which is written 5 m/s^2 or 5 m s^{-2}.

Because the gas pedal on a car is often called the accelerator, some people think *acceleration* means that an object is speeding up. But we need to stress that *any* change in motion is an acceleration. **Figure 3.7** illustrates the point. Slamming on your brakes and going from 60 to 0 mph in 4 seconds is just as much acceleration as going from 0 to 60 mph in 4 seconds. Similarly, the acceleration you experience as you go through a fast, tight turn at a constant speed is every bit as real as the acceleration you feel when you slam your foot on the gas pedal or brake. Faster, slower, turn left, turn right—*if you are not moving in a straight line at a constant speed, you are experiencing an acceleration.*

Newton's second law of motion says that changes in motion—accelerations—are caused by unbalanced forces. The acceleration that an object experiences depends on two

FIGURE 3.6 An object moving in a straight line at a constant speed is at rest in its own inertial reference frame.

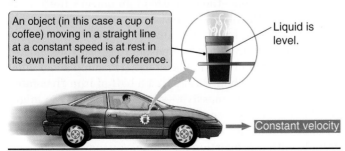

An object (in this case a cup of coffee) moving in a straight line at a constant speed is at rest in its own inertial frame of reference.

Liquid is level.

Constant velocity

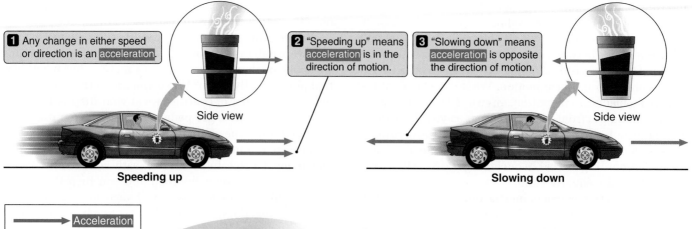

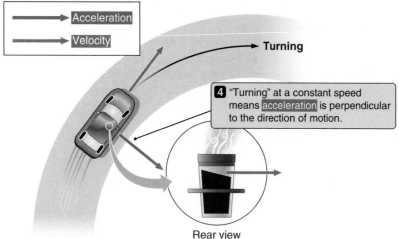

1 Any change in either speed or direction is an acceleration.

2 "Speeding up" means acceleration is in the direction of motion.

3 "Slowing down" means acceleration is opposite the direction of motion.

Side view

Side view

Speeding up

Slowing down

Acceleration

Velocity

Turning

4 "Turning" at a constant speed means acceleration is perpendicular to the direction of motion.

Rear view

FIGURE 3.7 Any change in the velocity of an object is an acceleration. When driving, for example, any time your speed changes or you follow a curve in the road, you are experiencing an acceleration. (Throughout the text velocity arrows will be shown as red and acceleration arrows will be shown as green.)

things, as shown in **Figure 3.8**. First, it depends on the unbalanced force acting on the object to change its motion. This is a pretty commonsense idea. The stronger the unbalanced force, the greater the acceleration. In fact, the accel-

Greater force means greater acceleration.

eration of an object is *proportional* to the unbalanced force applied. Push on something twice as hard (Figure 3.8(b)) and it experiences twice as much acceleration. Push on something three times as hard and its acceleration is three times as great. (The idea of proportionality, discussed in **Foundations 3.1**, will be used over and over again throughout our journey.) The resulting change in motion occurs in the direction in which the unbalanced force is imposed. Push something forward and it speeds up. Push it to the left and it veers in that direction.

The acceleration that an object experiences also depends on the degree to which the object resists changes in motion (Figure 3.8(c)). Some objects—say a baseball—are easily shoved around by humans. A baseball is thrown at great velocity by the pitcher, only to be hit with a bat and have its motion abruptly changed again. The hard-hit line drive comes suddenly to a stop in the glove of the second base

player. A baseball resists changes in its motion; that is, it has inertia—but not *too* much inertia. Other objects are less obliging. A piece of solid iron the size of a baseball would make a poor substitute in the game. A pitcher would be *very* hard pressed to throw such an iron ball hard enough to get it over home plate; and if he did, being catcher would become an even more dangerous job. A baseball and a ball of

Mass is the property of matter that resists changes in motion.

iron may be the same size, but they are quite different in the degree to which they resist changes in their motion. The property of an object that determines its resistance to changes in motion—the measure of an object's inertia—is referred to as the object's **mass**.

You probably knew this answer intuitively. The iron ball is "heftier" and therefore harder to throw than a baseball. However, you may not have thought much about what we really mean when we say that a ball of iron "has more mass"—or "is more massive"—than a baseball. You might say that the ball of iron is made up of "more stuff" than the baseball; but again, what is meant by "more stuff"? If you grapple with this question for a time, you may find yourself

Newton's Second Law:

$$\text{Acceleration } (a) = \frac{\text{Force } (F)}{\text{Mass } (m)}$$

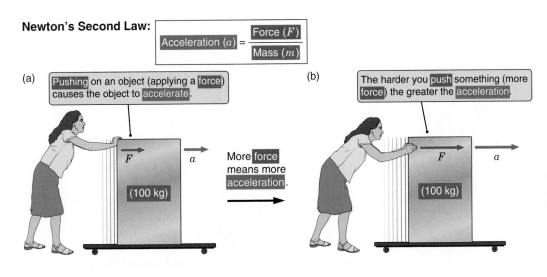

(a) Pushing on an object (applying a force) causes the object to accelerate.

More force means more acceleration.

(b) The harder you push something (more force) the greater the acceleration.

FIGURE 3.8 Newton's second law of motion says that the acceleration experienced by an object is determined by the force acting on the object divided by the object's mass. (Throughout the text force arrows will be shown as blue.)

More mass means less acceleration.

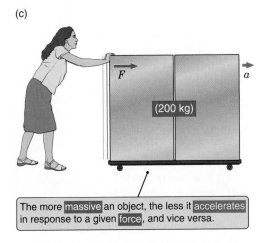

(c) The more massive an object, the less it accelerates in response to a given force, and vice versa.

Instead of spelling this out in words every time, we can introduce a convenient bit of shorthand—a for acceleration, F for force, and m for mass—so we get

$$a = \frac{F}{m}.$$

This is the succinct mathematical statement of Newton's second law of motion.[4] If you are comfortable with mathematics, this elegant expression may speak to you clearly and directly. If not, when you see this equation, remind yourself that Newton's second law is nothing more than the embodi-

---

Acceleration is force divided by mass.

---

ment of three commonsense ideas: (1) When you push on an object, that object accelerates in the direction in which you are pushing; (2) the harder you push on an object, the more the object accelerates; and (3) the more massive the object is, the harder it is to change its state of motion.

chasing the question around in circles. When it comes right down to it, *the property of matter that we refer to as "mass" is nothing more and nothing less than the degree to which an object resists changes in its motion.* (Mass is measured in units of kilograms. An object with a mass of 2 kg is twice as hard to accelerate as an object with a mass of 1 kg. An object with a mass of 9 kg is three times as hard to accelerate as an object with a mass of 3 kg.)

So if we want to know how an object's motion is changing, we need to know two things: What unbalanced force is acting on the object, and what is the resistance of the object to that force? We can put this into equation form as follows:

$$\frac{\text{The acceleration experienced by an object}}{} = \frac{\text{The force acting to change the object's motion}}{\text{The object's resistance to that change}} = \frac{\text{Force}}{\text{Mass}}$$

Newton's Third Law: Whatever Is Pushed, Pushes Back

Imagine you are a child again, sitting in a wagon or standing on a skateboard and pushing yourself along with your foot. Each shove of your foot against the ground sends you faster along your way. But why does this happen? Your muscles flex and your foot exerts a force on the ground. (Earth does not respond much to that force because its great mass gives it great inertia.) Yet this does not explain why *you* experience an acceleration. The fact that you accelerate at

---

[4] Newton's second law is often written as $F = ma$, giving force as units of mass times units of acceleration, or kg m/s². These units are aptly named "newtons," abbreviated N.

Proportionality

Often in this text we will say that one quantity is *proportional* to another. Proportionality is a way of getting the gist of how something works—understanding the relationships between things—without having to actually calculate the details of one case after another.

PROPORTIONALITY

If two quantities are **proportional** to each other, then making one of them larger means making the other quantity larger by the same factor. In other words, the ratio between the two remains constant. For example, think about the weight of a bag of apples and how much the bag costs. Double the weight of the bag of apples and you double the cost. Increase the weight of the bag of apples by a factor of 5, and the cost goes up by a factor of 5 as well. *The cost of a bag of apples is proportional to the weight of the bag of apples.* We write this relationship as

$$\text{Cost} \propto \text{Weight},$$

where the symbol $\propto$ means "is proportional to." This expression captures the essence of the relationship between the cost and the weight of apples. It tells us that the more apples we buy, the more they will cost us.

CONSTANTS OF PROPORTIONALITY

Sometimes it is enough to know that two quantities are proportional to each other, but sometimes it is not. What if you need to know how much one of those bags of apples will actually set you back? We know that the full relationship between the cost and weight of a bag of apples is that the cost is equal to the price per pound of apples times the weight of the bag. We write

$$\text{Cost} = \text{Price per pound} \times \text{Weight}.$$

Compare this expression with the previous one. When we say that two quantities are proportional to each other, what we mean is that one quantity equals some number *times* the other quantity. The number by which one quantity is multiplied to get the other number is called the **constant of proportionality**. In our example, the constant of proportionality is just the price per pound of apples.

Look at the difference between the two expressions. The fact that the cost of a bag of apples is proportional to the weight of the apples is a statement about the *relationship* between things. It is a statement about how things work. More apples do not cost *less* than fewer apples. More apples cost *more* than fewer apples. The constant of proportionality—here the price per pound of apples—means something very different. Hidden within the price per pound of apples is a great deal of information, such as the cost of growing apples, the cost of transporting them from the orchard, and the profit margin the grocer needs to stay in business. The constant of proportionality carries information about this aspect of the world.

Very often physical laws work in this same fashion. Proportionalities tell us about *relationships*—how two things vary with one another. They let us get a feeling for the "how" in how something works. In this chapter, for example, we find that gravitational force is proportional to an object's mass. Constants of proportionality more precisely tell us about the way the universe is. The universal gravitational constant G is a constant of proportionality that tells us about the intrinsic strength of gravitational interactions and allows calculation of the numerical value of this force. Constants of proportionality are needed if we are to turn an understanding of relationships into hard numbers.

the same time means that as you push on the ground, *the ground must be pushing back on you.*

Part of Newton's genius was his ability to see sublime patterns in such mundane events. Newton realized that *every* time one object exerts a force on another, a matching force is exerted by the second object on the first. That second force is exactly as strong as the first force but is in

exactly the *opposite* direction. The child pushes back on Earth, and Earth pushes the child forward. A canoeist's paddle pushes backward through the water, and the water pushes forward on the paddle, sending the canoe along its way. A rocket engine pushes hot gases out of its nozzle, and those hot gases push back on the rocket, propelling it into space.

For every force there is an equal and opposite force.

All of these are examples of **Newton's third law of motion**, which says that *forces always come in pairs, and those pairs are always equal in magnitude but opposite in direction.* The forces in these action–reaction pairs always act on two different objects. Your weight pushes down on the floor, and the floor pushes back on you with the same amount of force. For every force there is *always* an equal and opposite force. This is one of the few times when we can say "always" and really mean it. **Figure 3.9** gives a few examples. There is a great game hiding in Newton's third law. It is called "find the force." Look around you at all the forces at work in the world, and for each force find its mate. It will *always* be there!

To see how Newton's three laws of motion work together, think about the situation shown in **Figure 3.10**. An astronaut is adrift in space, motionless with respect to the nearby space shuttle. With no tether to pull on, how can the astronaut get back to the ship? The answer? Throw something. Suppose the 100 kg astronaut throws a 1 kg wrench directly away from the shuttle at a speed of 10 m/s. Newton's second law says that in order to cause the motion of the wrench to change, the astronaut has to apply a force to it in the direction away from the shuttle. Newton's third law says that the wrench must therefore push back on the astronaut with as much force but in the opposite direction. The force of the wrench on the astronaut causes the astronaut to begin drifting toward the shuttle. How fast will the astronaut move? Turn to Newton's second law again. A force that causes the 1 kg wrench to accelerate to 10 m/s will not have much effect on the 100 kg astronaut. Because acceleration equals force divided by mass, the 100 kg astronaut will experience only 1/100 as much acceleration as the 1 kg wrench. The astronaut will drift toward the shuttle at the leisurely rate of 1/100 × 10 m/s, or 0.1 m/s.

3.4 Gravity Is a Force between Any Two Objects Due to Their Masses

Drop a ball and the ball falls toward the ground, picking up speed as it falls. It accelerates toward Earth. Newton's second law says that where there is acceleration, there is

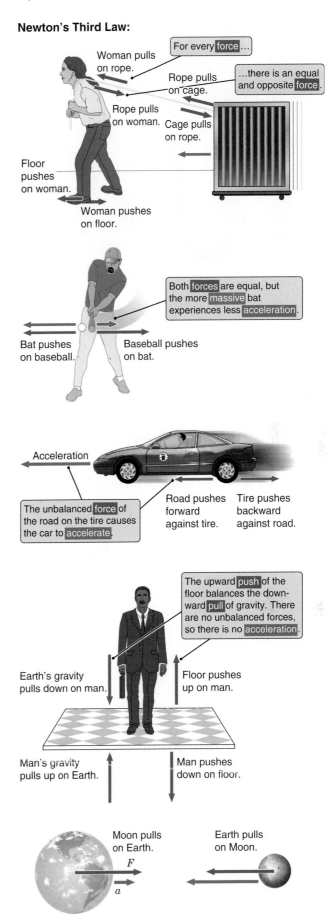

Newton's Third Law:

FIGURE 3.9 Newton's third law states that for every force there is always an equal and opposite force. These opposing forces always act on the two different objects in the same pair.

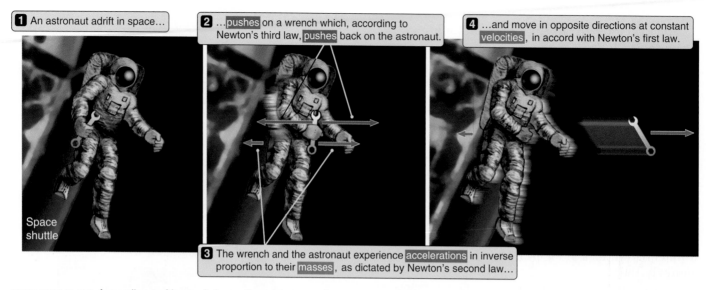

FIGURE 3.10 According to Newton's laws, if an astronaut throws a wrench the two will move in opposite directions at speeds that are inversely proportional to their masses. (Acceleration and velocity arrows are not drawn to scale.)

force. But where is the force that causes the ball to accelerate? Many forces that we see in everyday life involve "direct contact" between objects.[5] The cue ball slams into the eight ball, knocking it into the pocket. The shoe of the child in the wagon shoves directly onto the surface of the pavement. In cases where there is physical contact between two objects, the source of the forces between them is easy

---

Gravity is "force at a distance."

---

to see. But the ball falling toward Earth is an example of a different kind of force, one that acts at a distance across the intervening void of space. The ball falling toward Earth is accelerating in response to the force of **gravity**. We began this chapter with a qualitative discussion of the fundamental role that gravity plays in the universe. Having explored both Kepler's empirical description of the motions of planets about the Sun and Newton's laws of motion, it is time to return to gravity, for it is gravity that unites these two pillars of empirical and theoretical science.

You probably will not be surprised to learn that once again it is Newton we turn to for a **law of gravitation**. At this point in an introductory textbook it is customary to simply present Newton's law of gravitation as a "done deal" and go straight to its application, but such a leap misses one of the most interesting aspects of this stretch of our journey. A common misconception about how science works is the no-

tion that new theories just spring fully formed into the mind of a scientist as if by magic. This idea is certainly supported by the grade school story of the apple falling on Newton's head, literally knocking the idea of gravity into his brain. One could almost get the idea that scientific theories are arbitrary—that Newton could have invented some *other* law of gravity that would have worked just as well. (The idea that scientific knowledge is a cultural construct is discussed in **Excursions 3.2**.) Although it might seem at first glance that his work was arbitrary, nothing could be further from the truth. Where did Newton get his ideas about gravity? What guided him in his development of those ideas, and how did he turn them into a theory with testable predictions? How did he confront that theory in the crucible of experiment and observation? By answering these questions, rather than simply stating Newton's law of gravitation, we will gain some insight into what science is.

Where Do Theories Come From? Newton Reasons His Way to a Law of Gravity

As with inertia, the story of gravity begins with the insight and observation of Galileo. Galileo discovered that all freely falling objects accelerate toward Earth at the same rate, regardless of their mass. Drop a marble and a cannonball, at the same time and from the same height, and they will hit the ground together. If proof were needed, it was provided by astronaut David Scott on the lunar surface (**Figure 3.11**). The gravitational acceleration near the surface of Earth is

---

[5] Actually the "direct contact" between billiard balls or other "solid" objects is also force at a distance—electric force acting at a distance between the electrons and protons in the atoms of which the objects are made.

EXCURSIONS 3.2

Science and Culture

In recent years some critics of science have drawn attention to how science is influenced by culture. It is hard to avoid the conclusion that political and cultural considerations strongly influence which scientific research projects are funded. This choice of funding channels the directions in which scientific knowledge advances and can lead to serious ethical issues. For example, moral judgments about nontraditional lifestyles greatly restricted the funding available for AIDS research during the decade or so after its discovery.

Some critics even carry this view a step further, arguing that scientific knowledge itself is an arbitrary cultural construct. Yet, as illustrated by the discussion of Newton's law of gravity, successful scientific theories are *never* arbitrary. Scientific theories must be consistent with all that we know of how nature works, and turning a clever idea into a real theory with testable predictions is a matter of careful thought and effort. One of the most remarkable aspects of scientific knowledge is its *independence* from culture. Scientists are people, and politics and culture enter into the day-to-day practice of science. But in the end, *scientific theories are judged not by cultural norms but by whether their predictions are borne out by observation and experiment.* As long as the results of experiments are repeatable and do not depend on the culture of the experimenter—that is, as long as there is such a thing as objective physical reality—scientific knowledge cannot be called a cultural construct.

Nor does it seem that the path to knowledge embodied in science is any more arbitrary than the logic it is built on. It is significant that no philosopher critical of science has ever offered a viable alternative for obtaining reliable knowledge of the workings of nature. Had science not arisen when and where it did, something much like it would have arisen at some time and in some location. Furthermore, no other category of human knowledge is subject to standards as rigorous and unforgiving as those of science. For this reason, scientific knowledge is reliable in a way that no other form of knowledge can claim. Whether you want to design a building that will not fall over, choose the best treatment for a disease, or calculate the orbit of a spacecraft on its way to the Moon, you had better consult a scientist rather than a psychic —regardless of your cultural heritage.

Finally, we must admit that some disreputable scientists purposefully try to influence results by inventing or ignoring data, often when claiming "a major breakthrough" or challenging a well-established scientific principle. Fortunately, attempts by others to repeat the experiment will eventually expose such scientific misconduct.

---

usually written as g and has a value of 9.8 m/s^2. Whether you drop a marble or a cannonball, after 1 second it will be falling at a speed of 9.8 m/s, after 2 seconds at 19.6 m/s, and

---

All objects on Earth fall with the same acceleration, *g*.

---

after 3 seconds at 29.4 m/s. (These numbers assume that we can neglect air resistance, which is reasonably negligible for relatively dense objects.)

Having worked out the laws governing the motion of objects, Newton saw something deeper in Galileo's findings. Newton realized that if all objects fall with the same acceleration, then the gravitational *force* on an object must be determined by the object's *mass*. To see why, look back at Newton's second law (acceleration equals force divided by mass). The only way gravitational acceleration can be the same for all objects is if the value of the force divided by the mass is the same for all objects. A greater mass *must*, therefore, be accompanied by a stronger gravitational force. In other words, the gravitational force on an object on Earth is, according to Newton's second law, the object's mass times the acceleration due to gravity, or $F_{grav} = mg$. Make an object twice as massive, and you double the gravitational force acting on it. Make an object three times as massive, and you triple the gravitational force acting on it.

The gravitational force acting on an object is commonly referred to as the object's **weight**. It is easy to see why people often confuse mass and weight. On the surface of Earth, weight is just mass times the constant g. The situation is not helped by the sloppy way we use language. We often say that an object with a mass of 2 kg "weighs 2 kg," but

FIGURE 3.11 Astronaut Alan Bean's portrait of fellow astronaut David Scott standing on the Moon and dropping a hammer and falcon feather together. Both reached the lunar surface simultaneously. (Their lunar module was nicknamed "Falcon.")

it is more correct to express a weight in terms of **newtons** (N): Thus an object with a *mass* of 2 kg has a *weight* of 2 kg × 9.8 m/s² or 19.6 N.

Newton's next great insight came from applying his third law of motion to gravity. For every force there is an equal and opposite force. If Earth exerts a force of 19.6 newtons on a 2 kg mass sitting on its surface, then that 2 kg mass must exert a force of 19.6 newtons on Earth as well. Drop a 20 kg cannonball, and it falls toward Earth, but at the same time Earth falls toward the 20 kg cannonball! The reason we do not notice the motion of Earth is because Earth is very massive. It has a lot of resistance to a change in its motion. In the time it takes a 20 kg cannonball to fall to the ground from a height of 1 km, Earth has "fallen" toward the cannonball by about 3.4×10^{-21} meters, which is only about 1/300 of the diameter of a hydrogen atom!

Newton reasoned that this should work both ways. If doubling the mass of an object doubles the gravitational force between the object and Earth, then doubling the mass of Earth ought to do the same. In short, the gravitational

force between Earth and an object must be equal to the product of the two masses times something:

$$\text{Gravitational force} = \text{Something} \times \text{Mass of Earth} \\ \times \text{Mass of object}.$$

If the mass of the object is three times greater, then the force of gravity will be three times greater. Likewise, if the mass of Earth were three times what it is, the force of gravity would have to be three times greater as well. If *both* the mass of Earth *and* the mass of the object were three times greater, the gravitational force would increase by a factor of 3 × 3, or 9 times. Because objects fall toward the center of Earth, we know that this force is an attractive force acting along a line between the two masses.

"And by the way," reasoned Newton, "why are we restricting our attention to Earth's gravity?" If gravity is a force that depends on mass, then there should be a gravitational

---

The force of gravity is proportional to the product of two masses.

---

force between *any* two masses. Say we have two masses— call them mass 1 and mass 2, or m_1 and m_2 for short. The gravitational force between them is something times the product of the masses:

$$\text{Gravitational force between two objects} = \\ \text{Something} \times m_1 \times m_2.$$

Realize that we have gotten this far just by combining Galileo's observations of falling objects with (1) Newton's laws of motion and (2) Newton's belief that Earth is a mass just like any other mass. There has been no wiggle room —there is nothing arbitrary in what we have done. But what about that "something" in the previous expression? Today we have sensitive enough instruments to allow us to put two masses close to each other in a laboratory, measure the force between them, and determine that something directly. Yet Newton had no such instruments. He had to look elsewhere to go further with his exploration of gravity.

It turns out that Kepler had already thought about this question. He reasoned that because the Sun is the focal point for planetary orbits, the Sun must be responsible for exerting an influence over the motions of the planets. Kepler speculated that whatever this influence is, it must grow weaker with distance from the Sun. (After all, it must surely require a stronger influence to keep Mercury whipping around in its tight, fast orbit than it does to keep the outer planets lumbering along their paths around the Sun.) Kepler's speculation went even further. Although he did not know about forces or inertia or gravity, he did know quite a lot about geometry, and geometry alone suggested how this solar "influence" might change for planets progressively farther from the Sun.

Imagine you have a certain amount of plaster to spread over the surface of a sphere. If the sphere is small, then when you spread the plaster over the sphere you get a thick coat. But if the sphere is larger, the plaster has to spread farther, and so you get a thinner coat. The surface area of a sphere depends on the square of the sphere's radius. Double the radius of a sphere, and the sphere's surface becomes four times what it was. If you plaster this new, larger sphere, the plaster must cover four times as much area, and so the thickness of the plaster will only be a fourth of what it was on the smaller sphere. Triple the radius of the sphere, and the sphere's surface is nine times as large, and the thickness of the coat of plaster will be only a ninth as thick.

Kepler thought that the influence that the Sun exerts over the planets might be like the plaster in this example. As the influence of the Sun extends farther and farther into space, it would have to spread out to cover the surface of a larger and larger imaginary sphere centered on the Sun. (We will learn later that light works in exactly this way.) If so, then, like the thickness of the plaster, the influence of

Gravity is an inverse square law.

the Sun should be proportional to 1 divided by the square of the distance between the Sun and a planet. Double the distance between the Sun and a planet, and this influence declines by a factor of $2 \times 2 = 4$, to $\frac{1}{4}$ of its original strength. Triple the distance, and this influence declines by a factor of $3^2 = 9$, becoming $\frac{1}{9}$ of its initial strength. When something (like the thickness of the plaster or the strength of Kepler's solar "influence") changes in proportion to 1 divided by the square of the distance, we refer to this relationship as an **inverse square law** (see **Connections 3.1**).

Kepler had an interesting idea, but not a scientific theory with testable predictions. What he lacked was a good idea of the true source of this influence and the mathematical tools to calculate how an object would move under such an influence. Newton had both. If gravity is a force between *any* two objects, then there should be a gravitational force between the Sun and each of the planets. Might this gravitational force be the same as Kepler's "influence"? If so, then the something in Newton's expression for gravity might be a term that diminishes according to the square of the distance between two objects. Gravity might behave according to an inverse square law. Newton's expression for gravity now came to look like this:

$$\text{Gravitational force between two objects} =$$
$$\text{Something} \times \frac{m_1 \times m_2}{(\text{Distance between objects})^2}.$$

There is still a "something" left in this expression. Newton guessed that it was a measure of the intrinsic strength of gravitation interactions and that it would turn out to be the same for all objects. He named this something the **universal gravitational constant**. This quantity is now written as G.

Putting the Pieces Together: A Universal Law for Gravitation

Newton had good reasons every step of the way in his thinking about gravity—reasons directly tied to observations of how things in the world behave. Newton's chain of logic and reason brought him to what has come to be known as

---

CONNECTIONS 3.1

Inverse Square Laws

In this chapter we discover that gravity obeys what is called an *inverse square law*. This means that the force of gravity is proportional to 1 divided by the square of the distance between two objects, or

$$F_{\text{grav}} \propto \frac{1}{r^2}.$$

If two objects are moved so that they are twice as far apart as they were originally, the force of gravity between them becomes only $\frac{1}{4}$ of what it was. If two objects are

moved three times as far apart, the force of gravity drops to $\frac{1}{9}$ of its original value.

Gravity is only one of several inverse square laws found in nature. The other important one that we will deal with in this book involves radiation. The intensity of radiation from an object is also proportional to 1 over the square of the distance between the objects. Our discussion of radiation in Chapter 4 will present a clear picture of why an inverse square law applies in that case.

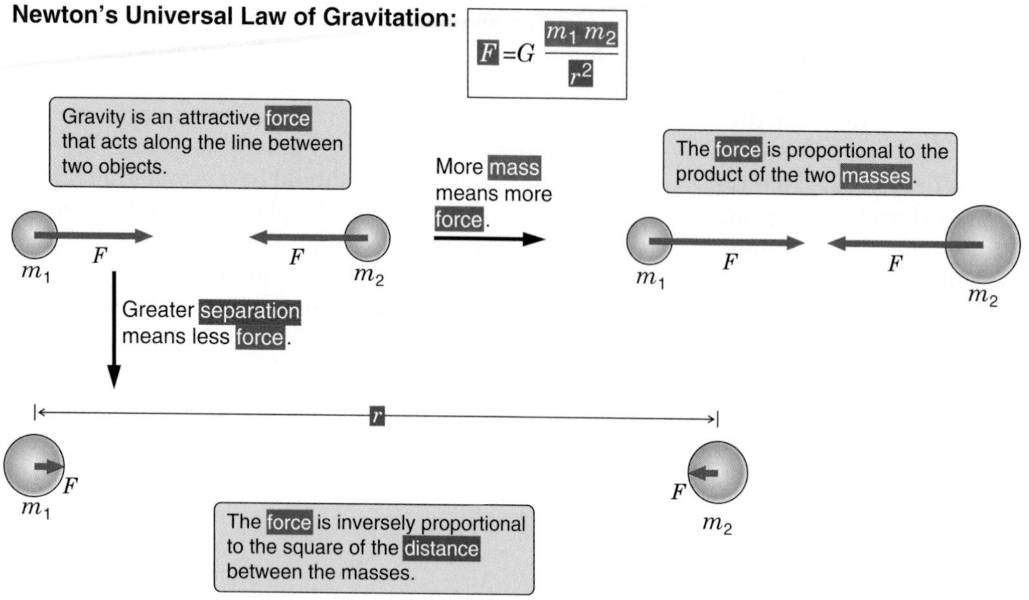

Newton's Universal Law of Gravitation:

$$F = G \frac{m_1 m_2}{r^2}$$

Gravity is an attractive force that acts along the line between two objects.

More mass means more force.

The force is proportional to the product of the two masses.

F m_1 F m_2

F m_1 F m_2

Greater separation means less force.

r

F m_1

F m_2

The force is inversely proportional to the square of the distance between the masses.

FIGURE 3.12 Gravity is an attractive force between two objects. The force of gravity depends on the masses of the objects and the distance between their centers of mass.

Newton's **universal law of gravitation**. This law, illustrated in **Figure 3.12**, states that gravity is a force between any two objects and has these properties:

1. It is an attractive force acting along a straight line between the two objects.

2. It is proportional to the mass of one object times the mass of the other object:

$$F_{grav} \propto m_1 m_2.$$

3. It decreases in proportion to 1 divided by the square of the distance between the two objects:

$$F_{grav} \propto \frac{1}{r^2}.$$

Written as a mathematical formula, the universal law of gravitation states that

$$F_{grav} = G \times \frac{m_1 \times m_2}{r^2},$$

where F is the force of gravity between two objects, m_1 and m_2 are the masses of objects 1 and 2, r is the distance between the centers of mass of the two objects, and G is the universal gravitational constant.

Any time you run across a statement like this, get into the habit of pulling it apart to be sure that it makes sense. First, gravity is an attractive force between two masses that acts along the straight line between the two masses.[6] Regardless

of where you stand on the surface of Earth, Earth's gravity pulls you *toward the center of Earth*. Second, the force of gravity depends on the product of the two masses. If you make m_1 twice as large, then the gravitational force between m_1 and m_2 becomes twice as large. Again, this should make sense. Doubling the mass of an object also doubles its weight. A subtle—but in retrospect very important—point lurks in this statement. The mass that appears in the universal law of gravitation is the *same* mass that appears in Newton's laws of motion. *The same property of an object that gives it inertia is the property of the object that makes it interact gravitationally.* (This equivalence between the effect of gravitation and the effect of inertia later became the basis for Einstein's general theory of relativity, in which mass literally warps space and time. We will return to this idea later in our journey when we discuss *spacetime* in Chapter 17.)

The third part of Newton's universal law of gravitation tells us that the force of gravity is inversely proportional to the *square* of the distance between two objects. Doubling the distance between two objects reduces the strength of gravity to $\frac{1}{2^2}$, or $\frac{1}{4}$ of its original value. Tripling the distance between two objects reduces the strength of gravity to $\frac{1}{3^2}$, or $\frac{1}{9}$ of its original value (see **Figure 3.13**). Gravity is only one of several laws that we will see in which the strength of some effect diminishes in proportion to the square of the distance.

Playing with Newton's Laws of Motion and Gravitation

If you have ever watched a child play with blocks, you have probably noticed how the child will put the blocks

---

[6] This is not obvious from the equation alone. More properly, this equation should be written using *vector notation,* which would include information about the direction of the force.

together in different ways, seeing what she can build. Perhaps if you are an artist, you play with colors and patterns of light and dark as you create new works. If you are a musician, you might play with tones and rhythms as you compose. Writers play with combinations of words. All of these uses of the term *play* mean the same thing. Play is very serious business because it is by playing that we explore the world. And in exactly the same sense, scientists often play with the equations describing natural laws as they seek new insights into the world around them. We can play a bit with Newton's laws of gravitation and motion and see what interesting things turn up. (If you have difficulty following the math here, do not worry too much. Pay attention instead to the sense of what we are doing.)

The universal gravitational constant G has a value of $6.67 \times 10^{-11}\,\text{Nm}^2/\text{kg}^2$. This makes gravity a *very* weak force. The gravitational force between two 16-pound (7.26 kg) bowling balls sitting 0.3 m (about a foot) apart is only

$$F_{\text{grav}} = 6.67 \times 10^{-11}\,\frac{\text{Nm}^2}{\text{kg}^2} \times \frac{7.26\,\text{kg} \times 7.26\,\text{kg}}{(0.3\,\text{m}^2)}$$

$$= 3.9 \times 10^{-8}\,\text{N},$$

or 0.000000039 newtons. This is about equal to the weight on Earth of a single bacterium! Gravity is such an important force in our everyday lives only because Earth is so very massive.

There are two different ways to think about the gravitational force that Earth exerts on an object with mass m. The first is to look at gravitational force from the perspective of Newton's second law of motion: Gravitational force equals mass times gravitational acceleration, or

$$F_{\text{grav}} = mg.$$

The other way to think about the force is from the perspective of the universal law of gravitation, which says that

$$F_{\text{grav}} = G\frac{M_\oplus m}{(R_\oplus)^2}.$$

(Here $M_\oplus$ is the mass of Earth and $R_\oplus$ is the radius of Earth.) Because the force of gravity on an object is what it is, the two expressions describing this force must be equal to each other. $F_{\text{grav}} = F_{\text{grav}}$, so

$$mg = G\frac{M_\oplus m}{(R_\oplus)^2}.$$

The mass m is there on both sides of the equation, so we can divide it out. The equation then becomes

$$g = G\frac{M_\oplus}{(R_\oplus)^2}.$$

This is interesting. We started out to calculate the gravitational acceleration experienced by an object of mass m on the surface of Earth. The expression that we arrived at says

FIGURE 3.13 The gravitational force between two objects falls off as 1 divided by the square of the distance between them.

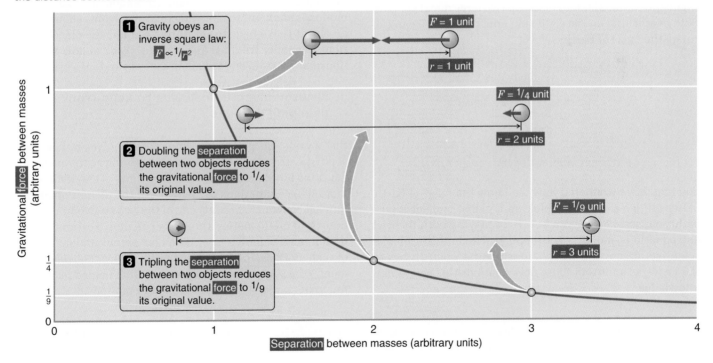

that this acceleration (g) is determined by the mass of Earth ($M_\oplus$) and by the radius of Earth ($R_\oplus$). But the mass of the object itself (m) appears nowhere in this expression. So according to this equation, changing m has no effect on the gravitational acceleration experienced by an object on Earth.

Galileo's discoveries are contained within Newton's laws.

In other words, our play with Newton's laws has shown us that all objects experience the same gravitational acceleration, regardless of their mass. *This is just what Galileo found in his experiments with falling objects!* We already saw that Galileo's work *shaped* Newton's thinking about gravity. Here we find that Galileo's discoveries about gravity are *contained within* Newton's laws of motion and gravitation.

What else can we discover? If we rearrange that last equation a bit so that the mass of Earth is on the left and everything else is on the right, we get

$$M_\oplus = \frac{g(R_\oplus)^2}{G}.$$

Everything on the right side of this equation is known. Galileo measured a value for g, the acceleration due to gravity on the surface of Earth, almost 400 years ago, and in about 235 B.C. Erastosthenes measured the radius of Earth

Newton's laws provide us with the tools to calculate the mass of Earth.

in the manner described in Chapter 2. The universal gravitational constant G is a bit tougher, but it too can be measured in the laboratory—for example, by measuring the slight gravitational forces between two large metal spheres. With everything on the right side now known, we can calculate the mass of Earth:

$$M_\oplus = \frac{gR_\oplus^2}{G}$$

$$= \frac{\left(9.80\,\frac{\text{m}}{\text{s}^2}\right) \times (6.38 \times 10^6\,\text{m})^2}{6.67 \times 10^{-11}\,\frac{\text{m}^3}{\text{kg}\,\text{s}^2}}$$

$$= 5.98 \times 10^{24}\,\text{kg}$$

You may have wondered how we know the mass of Earth. (After all, we cannot just pick up a planet and set it on a bathroom scale.) Now you know. By playing with theories and equations, much as a child plays with building blocks, scientists discover new relationships between things in the universe and from those new relationships comes new knowledge.[7]

3.5 Orbits Are One Body "Falling Around" Another

If you have been following our discussion closely, you may be about ready to take us to task. Kepler may have speculated about the dependence of the solar "influence" that holds the planets in their orbits, and Newton may have speculated that this influence is gravity, but physical law is *not* a matter of speculation! Newton could not measure the gravitational force between two objects in the laboratory directly, so how did he test his universal law of gravitation? Again it was Kepler who provided what Newton lacked. Newton used his laws of motion and his proposed law of gravity to *calculate* the paths that planets should follow as they move around the Sun. When he did so, his calculations predicted that planetary orbits should be ellipses with the Sun at one focus, that equal areas should be swept out during equal times, and that the square of the period of a planet's orbit should vary as the cube of the semimajor axis of that ellipse. In short, Newton's universal law of gravitation *predicted* that planets should orbit the Sun in just the way that Kepler's empirical laws described. This was the moment when it all came together. By *explaining* Kepler's laws, Newton found important corroboration for his law of gravitation. And in the process he moved the cosmological principle out of the realm of interesting ideas and into the realm of testable scientific theories. To see how this happened, we need to look below the surface of how scientists go about connecting their theoretical ideas with events in the real world.

Newton's laws tell us how an object's motion changes in response to forces and how objects interact with each other through gravity. To go from statements about how an object's motion is *changing* to more practical statements about where an object *is*, we have to carefully "add up" the object's motion over time. We must keep careful track of how the motion changes from one instant to the next and then ask where

Newton invented calculus to keep track of changing motions.

that motion has gotten us. Doing this requires rather more sophisticated play than we have time for at the moment. Learning how to make the jump from laws of gravitation and motion to calculations of the paths of the planets about the Sun led Newton to become one of the two coinventors of the branch of mathematics known as *calculus*. Fortunately we do not need calculus to build a conceptual understanding of such motions. Instead we can begin with a *thought experiment*—the same thought experiment that helped lead Newton to his understanding of planetary motions.

---

[7] Newton actually turned this around. He *guessed* at the mass of Earth by assuming it had about the same density as typical rocks. Then he

used this mass and the previous equation to get a rough idea of the value of G.

Newton Fires a Shot around the World

Drop a cannonball and it falls directly to the ground, just as any mass does. However, if instead we fire the cannonball out of a cannon that is level with the ground, as shown in **Figure 3.14(a)**, it behaves differently. The ball still falls to the ground in the same time as before, but while it is falling, it is also traveling *over* the ground, following a curved path that carries it some horizontal distance before it finally lands. The faster the cannonball is fired from the cannon (**Figure 3.14(b)**), the farther it will go before hitting the ground.

In the real world this experiment reaches a natural limit. To travel through air the cannonball must push the air out of its way—an effect we normally refer to as *air resistance*—which slows it down. *Air resistance* increases very rapidly with increasing speed. (For example, doubling the speed increases air resistance by much more than two times.) But because this is only a thought experiment, we can ignore such real-world complications. Instead imagine that, having inertia, the cannonball continues along its course until it runs into something. As the cannonball is fired faster and faster, it goes farther and farther before hitting the ground. If the cannonball flies far enough, the curvature of Earth starts to matter. As the cannonball falls toward Earth, Earth's surface "curves out from under it" (**Figure 3.14(c)**). Eventually we reach a point where the cannonball is flying so fast that the surface of Earth curves away from the cannonball at exactly the same rate at which the cannonball is falling toward Earth. This is the case shown in **Figure 3.14(d)**. At this point the cannonball, which always falls *toward the center of Earth,* is literally "falling around the world."

In 1957 the Soviet Union used a rocket to lift an object about the size of a basketball high enough above Earth's blanket of air that wind resistance ceased to be a concern, and Newton's thought experiment became a matter of great practical importance.[8] This object, called *Sputnik I,* was moving so fast that it fell around Earth, just as the cannonball did in Newton's mind. *Sputnik I* was the first human-made object to **orbit** Earth. This is what an orbit is: one object freely falling around another.

In Section 3.1 we asked why astronauts float freely about the cabin of a spacecraft. We now see that it is *not* because they have escaped Earth's gravity. It is Earth's gravity that holds them in their orbit. Instead the answer lies in Galileo's early observation that any object falls in just the same way, regardless of its mass. The astronauts and the spacecraft are both moving in the same direction, at the same speed, and are experiencing the same gravitational acceleration,

[8] Actually, wind resistance did not totally cease to be a concern. Objects in orbit within a few hundred kilometers of Earth are moving through the thin outer part of Earth's atmosphere. Friction caused by this thin atmosphere will oppose the object's motion and cause its orbit to *decay.*

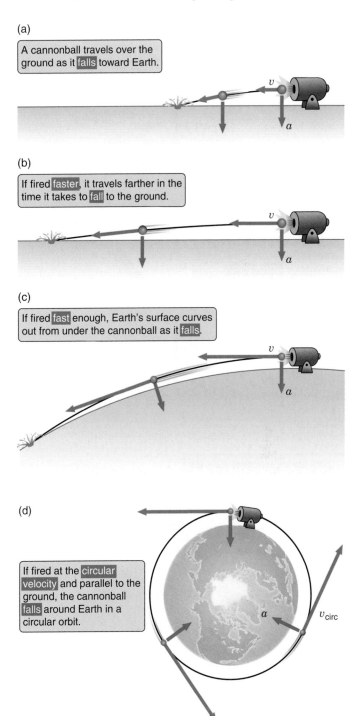

(a) A cannonball travels over the ground as it falls toward Earth.

(b) If fired faster, it travels farther in the time it takes to fall to the ground.

(c) If fired fast enough, Earth's surface curves out from under the cannonball as it falls.

(d) If fired at the circular velocity and parallel to the ground, the cannonball falls around Earth in a circular orbit.

FIGURE 3.14 Newton realized that a cannonball fired at the right speed would fall around Earth in a circle.

so they fall around Earth together. **Figure 3.15** demonstrates the point. The astronaut is orbiting Earth just as the spacecraft is orbiting Earth. On the surface of Earth our bodies try to fall toward the center of Earth, but the ground gets in the way. We experience our weight when we are standing on Earth because the ground pushes on us hard enough to

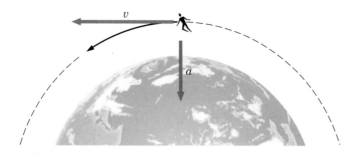

Like Newton's cannonball, an astronaut falls freely around Earth.

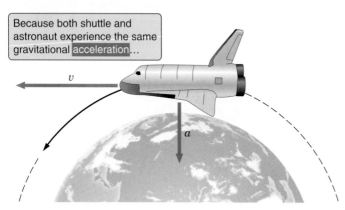

Because both shuttle and astronaut experience the same gravitational acceleration...

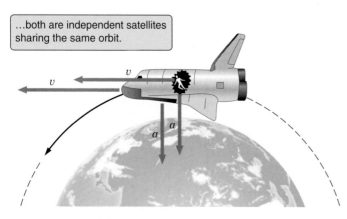

...both are independent satellites sharing the same orbit.

FIGURE 3.15 A "weightless" astronaut has not escaped Earth's gravity. Rather, an astronaut and a spacecraft share the same orbit as they fall around Earth together.

counteract the force of gravity, which is trying to pull us down. In the spacecraft, however, nothing interrupts the astronaut's fall because the spacecraft is falling around

Orbiting means falling around the world.

Earth in just the same orbit. The astronaut is not truly weightless. Instead the astronaut is in **free fall**.

When one object is falling around another, much more massive object, we say that the less massive object is a **satellite** of the more massive object. Planets are satellites of the

Sun, and **moons** are natural satellites of planets. Newton's imaginary cannonball is a satellite. *Sputnik I,* the first artificial satellite (*sputnik* means "satellite" in Russian), was the early forerunner to the spacecraft and the astronauts, which are independent satellites of Earth that conveniently happen to share the same orbit.

How Fast Must Newton's Cannonball Fly?

If fired fast enough, Newton's cannonball falls around the world; but just how fast is "fast enough"? Newton's orbiting cannonball moves along a circular path at constant speed. This type of motion, referred to as **uniform circular motion**, is discussed in more depth in Appendix 7. You are probably familiar with other examples of uniform circular motion. For example, think about a ball whirling around your head on a string, as shown in **Figure 3.16(a)**. If you were to let go of the string, the ball would fly off in a straight line in what-

Centripetal forces maintain circular motion.

ever direction it was traveling at the time, just as Newton's first law says. It is the string that keeps this from happening. The string exerts a steady force on the ball, causing it constantly to change the direction of its motion, always bending its flight toward the center of the circle. This central force is called a **centripetal force**. Using a more massive ball, speeding up its motion, or making the circle smaller so the turn is tighter all increase the force needed to keep the ball from being carried off in a straight line by its inertia.

In the case of Newton's cannonball, there is no string to hold the ball in its circular motion. Instead the centripetal force is provided by gravity, as illustrated in **Figure 3.16(b)**.

Gravity provides the centripetal force that holds a satellite in its orbit.

For Newton's thought experiment to work, the force of gravity must be just right to keep the cannonball moving on its circular path. In this case we can say that

$$\text{The force needed for uniform circular motion} = \text{Force provided by gravity} .$$

In Appendix 7 we derive an expression for the centripetal force needed to keep an object moving in a circle at a steady speed. If we put that expression on the left side of this equation and the universal law of gravitation on the right side (and then do some algebra), we arrive at

$$v_{\text{circ}} = \sqrt{\frac{GM}{r}} ,$$

(a)

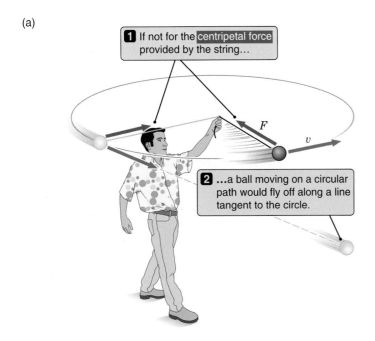

1 If not for the centripetal force provided by the string...

2 ...a ball moving on a circular path would fly off along a line tangent to the circle.

F

v

(b)

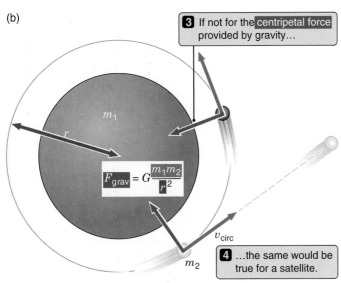

3 If not for the centripetal force provided by gravity...

m_1

r

$F_{\text{grav}} = G\dfrac{m_1 m_2}{r^2}$

v_{circ}

m_2

4 ...the same would be true for a satellite.

FIGURE 3.16 (a) A string provides the centripetal force that keeps a ball moving in a circle. (We have ignored the smaller force of gravity that also acts on the ball.) (b) Similarly, gravity provides the centripetal force that holds a satellite in a circular orbit.

where M is the mass of the orbiting object and r is the radius of the circular orbit. This value is called the **circular velocity**.

Here is the result we were looking for. If a satellite is in a stable circular orbit, then it *must* be moving at a velocity v_{circ}, where v_{circ} is given by this expression. *If the satellite were moving at any other velocity, it would not be moving in a circular orbit.* Remember the cannonball. If the cannonball were moving too slowly, it would drop below the circular path and hit the ground. Similarly, if the cannonball

were moving too fast, its motion would carry it above the circular orbit. Only a cannonball moving at just the right velocity—the circular velocity— will fall around Earth on a circular path. The circular velocity at Earth's surface is about 8 km/s (see **Foundations 3.2**).

Planets Are Just Like Newton's Cannonball

We can apply this same idea to the motion of Earth around the Sun. In the case of Earth's orbit, we already know from our discussion of the aberration of starlight that Earth travels at a speed of 2.98×10^4 m/s or (29.8 km/s) on its orbit

---

We use Earth's orbit to calculate the Sun's mass.

---

about the Sun. We also know that the radius of Earth's orbit is 1.50×10^{11} m. So we know everything about the circular orbit except for the mass of the Sun. If we can skip a step or two, a little algebra applied to the equation for v_{circ} gives the mass of the Sun ($M_\odot$) as

$$M_\odot = \frac{(v_{\text{circ}})^2 \times r}{G}$$

$$= \frac{\left(2.98 \times 10^4 \, \frac{\text{m}}{\text{s}}\right)^2 \times (1.50 \times 10^{11} \, \text{m})}{6.67 \times 10^{-11} \, \frac{\text{m}^3}{\text{kg s}^2}}$$

$$= 1.99 \times 10^{30} \, \text{kg}.$$

Once again a bit of play has taken us places we might not have imagined. We began with Newton's thought experiment about a cannonball fired around the world, and ended up knowing the mass of the star our planet orbits.

As long as we are at it, we can carry our game one step further. Kepler's third law talks about the time for a planet to complete one orbit about the Sun, known as the period P of a planet's orbit. The time it takes an object to make one trip around a circle is just the circumference of the circle ($2\pi r$) divided by the object's speed. (Time equals distance divided by speed.) If the object is a planet in a circular orbit about the Sun, then its speed must be equal to the circular velocity that we calculated. Bringing this together, we get

$$\text{Period } P = \frac{\text{Circumference of orbit}}{\text{Circular velocity}} = \frac{2\pi r}{\sqrt{\frac{GM}{r}}}.$$

Some algebra gives us

$$P^2 = \frac{4\pi^2}{GM_\odot} \times r^3.$$

Once again our play has really gotten us somewhere interesting. The square of the period of an orbit is equal to a

FOUNDATIONS 3.2

Circular Velocity

It is interesting to put some values into the equation for circular velocity to see how fast Newton's cannonball would really have to travel. The radius of Earth is 6.38×10^6 m, the mass of Earth is 5.98×10^{24} kg, and the gravitational constant is 6.67×10^{11} m³/kg s². (We have seen how each of these values is measured.) Putting these values into the expression for v_{circ} gives

Newton's cannonball would have to be traveling about 8 km/s—over 28,000 km/h—to stay in its circular orbit. That's well beyond the reach of a typical cannon but just what we routinely accomplish with rockets.

$$v_{circ} = \sqrt{\frac{\left(6.67 \times 10^{-11} \, \frac{m^3}{kg \, s^2}\right) \times (5.98 \times 10^{24} \, kg)}{6.38 \times 10^6 \, m}}$$

$$= 7.9 \times 10^3 \text{ m/s.}$$

constant $(4\pi^2/GM_\odot)$ times the cube of the radius of the orbit. *This is just Kepler's third law applied to circular orbits.* This is how Newton showed, at least in the special case of a circular orbit, that Kepler's third law—his beautiful "harmony of the worlds"—is a direct consequence of the way objects move under the force of gravity. A more complete treatment of the problem—the problem for which Newton invented calculus—shows that Newton's laws of motion and gravitation predict *all* of Kepler's empirical laws of planetary motion—for elliptical as well as circular orbits.

You may well be wondering why we are taking such a long and sometimes strenuous excursion through the work of Galileo, Kepler, and Newton. Sometimes when we are hiking in the mountains it is hard to see the summit from the perspective of the trail. Now that we have arrived at the top of this particular pass, we can look around and appreciate what we have gained. When Newton carried out his calculations, he found that his laws of motion and

---

Kepler's laws provided an empirical test of Newton's laws.

---

gravitation predicted elliptical orbits that agree exactly with Kepler's empirical laws. *This is how Newton tested his theory that the planets obey the same laws of motion as cannonballs and how he confirmed that his law of gravitation is correct.* Had Newton used *any* other rule for gravity, he would have predicted something different for the way planets orbit the Sun, and these predictions would have failed Kepler's observational test. If Newton's laws

had failed to predict motions that agreed with Kepler—if Newton's predictions had not been borne out by observation—then Newton would have had to throw them out and go back to the drawing board! All of his work would have fallen onto that large heap of beautiful ideas that do not pass nature's test.

We promised we would show you *how* science works. Well, this is how science works. Kepler's empirical rules for planetary motion pointed the way for Newton and provided the crucial observational test for Newton's laws of motion and gravitation. At the same time, Newton's laws of motion and gravitation provided a powerful new understanding of why planets and satellites move as they do. Theory and empirical observation work together hand in hand, and our understanding of the universe strides forward.

Rather than hiding from challenges that might prove their theories wrong, scientists actively seek and confront nature's judgment. And even if one scientist fails to find and probe the potential weaknesses of her favorite theory, she can rest assured that another scientist will. As discussed in **Excursions 3.3**, a well-tested scientific theory is a far cry from the "theory" that Elvis was abducted by aliens. A well-tested scientific theory is about as close to certain knowledge as we humans can come. This is the main difference between science and the pseudosciences—astrology, creationism, intelligent design, quack medicine, homeopathy, parapsychology, numerology, and a host of others—which hide slipshod thinking and untested, untestable, or even disproven speculation behind scientific-sounding jargon, and which actively ignore the wealth of evidence against them. Here is the key to the difference between science and what *pretends* to be science. In our discussion of Newton we have focused on the basis of scientific knowledge and on how science is done. If you run across activities that are *not* done this way—if they do not make testable predictions and do not search for and embrace every observation or experiment that might in principle prove them wrong—then they are *not* science, no matter how many "authorities" would have you believe otherwise.

Real-World Orbits Are Not Circles

So far we have concentrated on circular orbits and simply asserted that everything works out for elliptical orbits as well. As we pointed out before, it is much more difficult to carry out these calculations for ellipses than for circles, so we will leave that particular peak unscaled on this journey. Even so, from our current vantage point we can get an idea of what that peak looks like—of how elliptical orbits differ

---

EXCURSIONS 3.3

"After All, It's Only a Theory"

Science is sometimes misunderstood because of the special ways that scientists use everyday words. An example is the word *theory*. In everyday language a theory may mean something that is little more than a conjecture or a guess: "Have you any theory about who might have done it?" "My theory is that a third party could win the next election." In everyday parlance a theory is something worthy of little serious regard. "After all," we say, "it is only a theory."

In stark contrast, a *scientific* theory is a carefully constructed proposition that takes into account all the relevant data and all our understanding of how the world works, and makes testable predictions about the outcome of future observations and experiments. A theory is a well-developed idea that is ready to be confronted by nature. A well-corroborated theory is a theory that has survived many such tests. Rather than being simple speculation, scientific theories represent and summarize bodies of knowledge and understanding that provide our fundamental insights into the world around us. A successful and well-corroborated theory is the pinnacle of human knowledge about the world.

Theories fill a place in a loosely defined hierarchy of scientific knowledge. In science the word *idea* has its everyday use. An idea is just a notion about how something might be. A **hypothesis** is an idea that leads to testable predictions. A hypothesis may be the forerunner of a scientific theory, or it may be based on an existing theory, or both. When an idea has been thought about carefully enough, has been tied solidly to existing theoretical and experimental knowledge, and makes testable predictions, then that idea has become a **theory**. Scientists build **theoretical models** that are used to connect theories with the behavior of complex systems. Competing theories are ultimately decided among on the basis of the success of their predictions. Some theories become so well-tested and are of such fundamental importance that we come to refer to them as **physical laws**. A scientific **principle** is a general idea or sense about how the universe is that guides our construction of new theories. **Occam's razor**, for example, is a guiding principle in science that says that when faced with two hypotheses that explain some phenomenon equally well, we should adopt the simpler of the two.

Unfortunately, there is room for confusion here because our terminology does not always follow these categories. For example, Newton's laws are physical laws, but Kepler's laws are really empirical rules (based on data and observation, but not on more fundamental principles or laws). Furthermore, science is a living enterprise, and the status of ideas changes in time. For example, principles can themselves become theories. The cosmological principle has been instrumental in shaping countless theories about the universe. In turn, the success of those theories has effectively become the experimental test of the cosmological principle itself.

Newton's theory of motion illustrates the point of our digression. This theory is the basis of our worldwide technological civilization and comes about as close to certain knowledge as humankind can hope for. Yet to scientists this knowledge remains a theory, subject to observational and experimental tests via the formal and rigorous application of the scientific method.

So think twice the next time you hear someone casually dismiss some body of scientific knowledge as being "only a theory."

from circular orbits, and why Newton's calculations work out the way they do.

Begin again with a satellite in a circular orbit about Earth. The satellite is traveling at the circular velocity, so it remains the same distance from Earth at all times, neither speeding up nor slowing down in its orbit. But now change the rules a bit. What if the satellite were in the same place in its orbit and moving in the same direction, but traveling *faster* than the circular velocity? The pull of Earth is as strong as ever, but because the satellite has a greater speed, its path is not bent by Earth's gravity sharply enough to hold it in a circle. So the satellite begins to climb above a circular orbit.

As the distance between Earth and the satellite begins to increase, an interesting thing starts to happen. Think about a ball thrown into the air, as shown in **Figure 3.17(a)**. As the ball climbs higher, the pull of Earth's gravity opposes its motion, slowing the ball down. The ball climbs more and more slowly until its vertical motion stops for an instant, then is reversed; the ball begins to fall back toward Earth, picking up speed along the way. Our satellite does exactly the same thing as the ball. As the satellite climbs above a circular orbit and begins to move away from Earth, Earth's gravity opposes the satellite's outward motion, slowing the satellite down. The farther the satellite pulls away from Earth, the more slowly the satellite moves—just as happened with the ball thrown into the air. And just like the ball, the satellite reaches a maximum height on its curving path, then begins falling back toward Earth. Now as the satellite falls back in toward Earth, Earth's gravity is pull-

ing it along, causing it to pick up more and more speed as it gets closer and closer to Earth.

What is true for a satellite orbiting Earth in an elliptical orbit is also true for any object in an elliptical orbit, including a planet orbiting the Sun. As we saw earlier, Kepler's Law of Equal Areas says that a planet moves fastest when it is closest to the Sun and slowest when it is farthest from the Sun. Now we know why. As shown in **Figure 3.17(b)**, planets lose speed as they pull away from the Sun, then gain that speed back as they fall inward toward the Sun.

Newton's laws do more than explain Kepler's laws. Newton's laws also predict different types of orbits that are beyond Kepler's empirical experience. **Figure 3.18** shows a whole series of satellites, each with the same point of closest approach to Earth but with different velocities at that point. A look at the figure shows that the greater the speed a satellite has at its closest approach to Earth, the farther the satellite is able to pull away from Earth, and the more eccentric its orbit becomes. Yet no matter how eccentric it becomes, as long as it remains elliptical, an orbit will eventually bring a satellite back to the planet that it orbits or bring a planet back to the Sun.

You might imagine, though, that somewhere in this sequence of faster and faster satellites there comes a point of no return—a point when the satellite is moving so fast that gravity is unable to reverse its outward motion, so the satellite coasts away from Earth, never to return. This indeed is possible. The lowest speed at which this happens is called the **escape velocity**. If we were to work through the calculation, we would find that the escape velocity is a factor

FIGURE 3.17 (a) A ball thrown into the air slows as it climbs away from Earth, then speeds up as it heads back toward Earth. (b) A planet on an elliptical orbit around the Sun does the same thing. (Although no planet has an orbit as eccentric as that shown, the orbits of comets can be far more eccentric.)

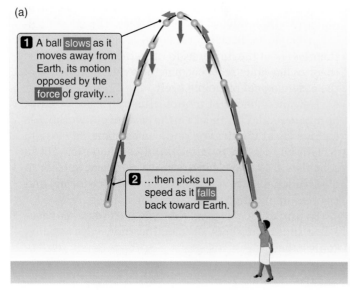

(a)

1 A ball slows as it moves away from Earth, its motion opposed by the force of gravity…

2 …then picks up speed as it falls back toward Earth.

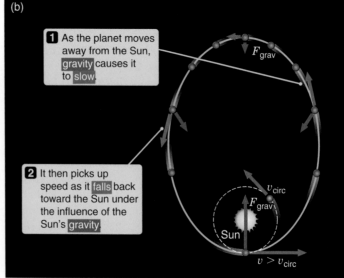

(b)

1 As the planet moves away from the Sun, gravity causes it to slow.

2 It then picks up speed as it falls back toward the Sun under the influence of the Sun's gravity.

F_{grav}

v_{circ}

F_{grav}

Sun

$v > v_{circ}$

(a) Representative orbits

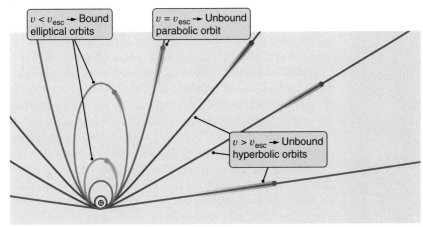

(b) Velocity at closest approach

FIGURE 3.18 (a) A range of different orbits that share the same perigee but differ in velocity at that point. (b) Perigee velocities for the orbits in (a). An object's velocity determines the orbit shape and whether the orbit is bound.

A satellite moving fast enough will escape a planet's gravity.

of $\sqrt{2}$ or 1.414..., larger than the circular velocity. This can be expressed as

$$v_{esc} = \sqrt{\frac{2GM}{R}} = \sqrt{2} \; v_{circ}.$$

Look at this equation for a minute to be sure it makes sense to you. The larger the mass (M) of a planet, the stronger its gravity, so it stands to reason that a more massive planet would be harder to escape from than a less massive planet. Indeed, the equation says that the more massive the planet, the greater the required escape velocity. Also, it stands to reason that the closer we are to the planet, the harder it will be to escape from its gravitational attraction. Again, the equation confirms our intuition. As the distance R becomes larger (that is, as we get farther from the planet), v_{esc} becomes smaller (in other words, it is easier to escape from the planet's gravitational pull). For objects at the surface of Earth, the escape velocity is 11.2 km/s (about 40,000 km/h). Once an object reaches escape velocity, the shape of its orbit is no longer an ellipse. But when Newton solved his equations of motion, he found that ellipses are not the only possible shape an orbit can have.

Unbound orbits are hyperbolas or parabolas. If a satellite's velocity is *less* than the escape velocity (v_{esc}), its orbit

Bound orbits are ellipses.

will have an elliptical shape. Elliptical orbits close on themselves. Thus an object traveling in an elliptical orbit is des-

tined to follow the same path over and over again.[9] For this reason *elliptical orbits* are also called **bound orbits**: The satellite is bound to the object it is orbiting about. If a satellite has a velocity *greater* than the escape velocity, it is not bound to the object it is orbiting. Such orbits are called **unbound orbits** and have a hyperbolic shape. A hyperbola does not close like an ellipse but instead keeps opening up forever. A space probe traveling on a *hyperbolic orbit* makes only a single pass around a planet, then is back off into deep

Unbound orbits are hyperbolas or parabolas.

space, never to return. The third type of orbit is the borderline case where the orbiting object moves at exactly the escape velocity. If it had any less velocity it would be traveling in a bound elliptical orbit; any more and it would be moving on an unbound hyperbolic orbit. A body moving with a velocity *equal* to the escape velocity follows a *parabolic orbit*. Like the hyperbolic orbit, the parabolic orbit involves only a single pass by the planet. As an object traveling in a parabolic orbit moves away from a planet, its velocity relative to the planet gets closer and closer to zero. An object traveling in a hyperbolic orbit always has excess velocity relative to the planet, even when it has moved infinitely far away.

[9] Strictly speaking, this is true only for a single body orbiting a second body. The presence of other bodies can cause the ellipse to not quite close, causing the orientation of the ellipse to slowly swing around in the orbital plane. This is referred to as *orbital precession*. Relativistic effects can also cause orbital precession, as we will see in the case of Mercury's orbit in Chapter 17.

Newton's Theory Is a Powerful Tool for Measuring Mass

As mentioned earlier, Kepler's empirical laws describe the motion of the planets but do not explain them. On the basis of Kepler's laws alone we might imagine that angels carry the planets around in their orbits, just as many people believed during the 16th century! Newton's derivation of Kepler's laws changed all of that. Newton showed that the *same* physical laws that describe the flight of a cannonball on Earth—or the fall of the apocryphal apple on his head—also describe the motions of the planets through the heavens. In this way Newton shattered the prevailing concept of the heavens and Earth, and at the same time opened up an entirely new way of investigating the universe. Copernicus may have dislodged Earth from the center of the universe and started us on the way toward the cosmological principle, but it was Newton who moved the cosmological principle out of the realm of philosophy and into the realm of testable scientific theory. And it was through Newton's work that **astrophysics** was born.

Not only is Newton's method more philosophically satisfying than simple empiricism—it is far more powerful as well. We have already seen, for example, how Newton's laws can be used to measure the mass of the Sun and Earth. This could never be done with Kepler's empirical rules. This is especially important when we remember that Newton's laws apply to *all* objects, not just the Sun and Earth. This fact will prove handy as we continue our journey.

Astronomers often rearrange Newton's form of Kepler's third law to read

$$ M = \frac{4\pi^2}{G} \times \frac{A^3}{P^2}. $$

Everything on the right side of this equation is either a constant (such as 4, π, and G) or a quantity we can measure (like the semimajor axis A and period P of an orbit). The left side of the equation is the mass of the object at the focus of the ellipse.

It is important to note that we cut a couple of corners to get to this point. For one thing, we arrived at this relationship by thinking about circular orbits, then simply asserting that it holds for elliptical orbits as well. Another corner we cut was assuming that a low-mass object such as a cannonball is orbiting a more massive object such as Earth. Earth's gravity has a strong influence on the cannonball; but as we have seen, the cannonball's gravity has little effect on Earth. For this reason we can imagine that Earth remains motion-

less while the cannonball follows its elliptical orbit. In the same way it is a good approximation to say that the Sun remains motionless as the planets orbit about it.

This picture changes when two objects are closer to having the same mass. In this case *both* objects experience significant accelerations in response to their mutual gravitational attraction. We now must think of the two objects as falling around *each other,* with each mass moving on its own elliptical orbit around a point (the **center of mass**) located between the two. Yet even in this most general case the equation remains valid. The mass M now refers to the *sum* of the masses of the two objects. So if we can measure the size and period of an orbit—*any* orbit—then we can use this equation to calculate the mass of the orbiting objects. This is true not only for the masses of Earth and the Sun, but also for the masses of other planets, distant stars, our galaxy and distant galaxies, and vast clusters of galaxies. In fact, it turns out that *almost all of our knowledge about the masses of astronomical objects comes directly from the application of this one equation.* In a sense, this single equation even allows us to tackle the question "What is the mass of the universe itself?"

We have come a long way in this chapter. We have gone from Copernicus's simple picture of planets moving on circles around the Sun to the pinnacle of Newton's comprehensive laws of motion and gravitation. The work of Copernicus, Galileo, Kepler, and Newton was not just *a* scientific revolution, it was *the* scientific revolution, which changed forever not only our view of the universe but also our very notion of what it means "to know."

Before getting too carried away, however, we need to put our accomplishment in perspective. Newton's grand theoretical edifice leaves us with the feeling that we understand the "how come" of Kepler's laws and a great deal more—as if we have lifted up the hood of the universe and peeked underneath. And indeed we have. But remember that we have not talked about *why* acceleration equals force divided by mass, or *why* objects have inertia, or *why* one inertial reference frame is as good as any other. These are good, fundamental laws—so fundamental that they form the basis of our understanding of the world we live in and the foundation of our entire technological civilization. Yet the law equating acceleration with force divided by mass, the inverse square law of gravity, and the law of inertia are still basically empirical facts about how objects are observed to move. These are not articles of faith, but rather scientific hypotheses, the validity of which remains to this day subject to test through experimentation and observation.

Summary

- Proportionality describes patterns and relationships in nature.

- The scientific method and nature provide the distinction between science and pseudoscience.

- Newton's three laws govern the motion of all objects.

- Unbalanced forces cause changes in motion.

- Mass is the property of matter that gives it resistance to changes in motion.

- Gravity is a force between any two objects due to their masses.

- *All* objects near Earth's surface fall with the same acceleration, *g*.

- Planets orbit the Sun in elliptical orbits.

- Unbound orbits are parabolas or hyperbolas.

Seeing the Forest through the Trees

The story of planetary motions is also the story of how we as a species learned to do science. No one taught Copernicus or Galileo or Kepler or Newton about the differences between empiricism and theory or about how the scientific method could be used to test their ideas. They had to figure these lessons out on their own as they went along. Yet the accomplishments they made remain near the top of the all-time intellectual feats of humankind, and the trail they blazed pointed the way for all that was to come. Galileo offered powerful insights into the nature of matter and gravity. Kepler built on Copernicus's revolutionary ideas about planetary motions and uncovered three empirical rules that showed the orbits of all the planets to be reflections of the same underlying patterns. Newton combined the two sets of ideas: He built Galileo's insights into a powerful theoretical edifice describing the motions of all objects and then tested this edifice against Kepler's empirical reality. The culmination of this intellectual campaign was far greater than the sum of its parts. The walls separating the heavens and Earth came down once and for all, and the science of astrophysics was born.

The model for science that emerged during this era remains the template for how science is done to this day. Careful empiricism uncovers patterns in nature in need of explanation. Scientific theory seeks to discover the fundamental truths underlying all things. And in the meeting of the two—the test of theory against the unforgiving and stalwart challenge of empirical fact—new knowledge and understanding emerge.

Newton's work became the cornerstone of what is often referred to as classical mechanics. All objects have inertia. An object will continue to move in a straight line at a constant speed unless an unbalanced force acts to change its motion. Mass is the property of matter that resists changes in motion. Every force is matched by another force that is equal in magnitude but opposite in direction. Gravity is a force between any two masses, proportional to the product of the two masses and inversely proportional to the square of the distance between them. Putting all of this together, we find that objects "fall around" the Sun and Earth on elliptical, parabolic, or hyperbolic paths. Orbits are ultimately given their shape by the gravitational attraction of the objects involved, which in turn is a reflection of the mass of these objects. We now understand that it is gravity that holds the universe together, giving planets, stars, and galaxies their very shapes as well as controlling their motions through space. Using Newton's theoretical insight we look backward along this chain, turning observations of the motions of objects throughout the universe into measurements of the masses of objects that no human has ever, or in most cases, will ever visit.

This brings us to the next stage of our journey. If the ancients could have stepped off Earth and touched the planets, they would never have fallen into the conceptual errors that muddled our thinking for millennia —but they had no such luxury. Today we have sent robotic surrogates to all of the planets in the Solar System except one, and humans have walked on the surface of the Moon. Even so, we have made only the most cursory visits to our immediate neighborhood. Even the nearest stars remain thousands of times more distant than the most far-flung of our robotic planetary explorers. For the most part we, like the ancients, are left with nothing on which to base our knowledge of the universe but the signals reaching us from across space. Far and away, the most common of these signals is electromagnetic radiation, which includes light (such as the light by which you are reading this book). Our ability to interpret these signals depends on what we know about light. What is it? How does it originate? How does it interact with matter? What changes does it experience during its journey? In Chapter 4 we turn to these questions.

Key Terms

empirical science, p. 61
Kepler's laws, p. 61

Student Questions

THINKING ABOUT THE CONCEPTS

1. The orbits of the planets around the Sun and the orbits of satellites around these planets are always ellipses rather than perfect circles. Why?

2. When riding in a car, we can sense changes in speed or direction through the forces the car applies on us. Do we wear seat belts in cars and airplanes to pro-tect us from speed or from acceleration? Explain your answer.

3. Weight on Earth is proportional to mass. On the Moon weight is also proportional to mass, but the constant of proportionality is different on the Moon than it is on Earth. Why?

4. Erich von Daniken proposed the theory that Earth was visited by extraterrestrials in the remote past. Would you regard this as scientific or pseudoscientific theory? Is the theory falsifiable? Can you think of any tests that could support or refute the theory?

5. Picture a swinging pendulum. During a single swing, the bob of the pendulum first falls toward Earth and then moves away until the gravitational force between it and Earth finally stops its motion. Would a pendulum swing if it were in orbit? Explain your reasoning.

6. Had Kepler lived on one of a group of planets orbiting a star three times as massive as our Sun, would he have deduced the same empirical laws? Explain your answer.

7. Kepler's and Newton's laws all tell us something about the motion of the planets, but there is a fundamental difference between them. What is the difference?

8. In 1920 a *New York Times* editor refused to publish an article based on rocket pioneer Robert Goddard's paper that predicted space flight, saying that "rockets could not work in outer space because they have nothing to push against" (a statement the *Times* did not retract until July 20, 1969, the date of the *Apollo 11* Moon landing). You, of course, know better. What was wrong with the editor's logic?

9. Aristotle taught that the natural state of all objects is to be at rest. Even though this seems consistent with what we observe around us, explain why Aristotle's conclusion was wrong.

10. Imagine a planet moving in a perfectly circular orbit around the Sun. Because the orbit is circular, the planet is moving at a constant speed. Is this planet experiencing acceleration? Explain your answer.

11. An astronaut standing on Earth could easily lift a wrench having a mass of 1 kg, but not a scientific instrument with a mass of 100 kg. In the International Space Station she is quite capable of manipulating both, although the scientific instrument moves more slowly than the wrench. Explain why.

12. Two comets are leaving the vicinity of the Sun, one traveling in an elliptical orbit and the other in a hyperbolic orbit. What can you say about the future of these two comets? Would you expect either of them to eventually return?

APPLYING THE CONCEPTS

13. During the latter half of the 19th century a few astronomers thought there might be a planet circling the Sun inside Mercury's orbit. They even gave it a name, Vulcan. We now know that Vulcan does not exist. If there were such a planet with an orbit a fourth the size of Mercury's, what would be its orbital period relative to that of Mercury?

14. Earth speeds along at 29.8 km/s in its orbit. Neptune's nearly circular orbit has a radius of 4.5×10^9 km, and the planet takes 164.8 years to make one trip around the Sun. Calculate the rate at which Neptune plods along in its orbit.

15. Venus's circular velocity is 35.03 km/s, and its orbital radius is 1.082×10^8 km. Calculate the mass of the Sun.

16. At the surface of Earth, the escape velocity is 11.2 km/s. What would be the escape velocity at the surface of a very small asteroid having a radius 10^{-4} of Earth's and a mass 10^{-12} of Earth's? If you were standing on the asteroid and threw a baseball with a strong pitch, what would happen to it?

17. How long does it take Newton's mythical cannonball, moving at 7.9 km/s just above Earth's surface, to complete one orbit around Earth?

18. Using values given in the appendixes for G and for Earth's radius and mass, show that the acceleration of gravity at the surface of Earth is 9.80 m/s^2.

19. What does an acceleration of 9.80 m/s^2 mean? If you jumped from a stationary balloon, after 1 second you would be falling at a speed of 9.80 m/s. After 2 seconds your speed would be $2 \times 9.80 = 19.6$ m/s or slightly more than 70 km/h! In the absence of any air resistance, how fast would you be falling after 20 seconds?

20. Weight refers to the force of gravity acting on a mass. We often calculate the weight of an object by multiplying its mass by the local acceleration due to gravity. The value of gravitational acceleration on the surface of Mars is 0.39 times that on Earth. Assume your mass is 85 kg. Then your weight on Earth is 833 N (833 N $= m \times g =$ 85 kg $\times$ 9.8 m/s^2). What would be your mass and weight on Mars?

StudySpace
wwnorton.com/astro21
provides a Study Plan for each chapter that includes a reading outline, animations, keyword flash cards, and gradebook-enabled multiple-choice quizzes. From StudySpace you can also access premium content in the ebook and SmartWork.

Then God said, "Let there be light,"
and there was light.
And God saw that the light was good;
and God separated the light from
the darkness.
And God called the light day, and the
darkness He called night.
And there was evening and there was
morning, one day.

GENESIS 1:3–5

The setting sun colors clouds over the Australian bush.

CHAPTER 4

Light

4.1 "Let There Be Light"

Light is a fundamental part of our experience of the world because it is through light that we most directly perceive the world beyond our physical grasp. The symbolic nature of light is everywhere in our language. A close companion may be the "light of our lives." To understand something is to "see it." A "bright idea" is symbolized by a lightbulb going off over our heads. As we leave our ignorance behind, we become "enlightened." A symbol of hope is a "light at the end of the tunnel." Throughout our language and culture, light is a metaphor for knowledge and information. Those who have lost their sight face a challenge greater than most of us can imagine. At the same time, light plays another, even more important role in our lives. Light from the Sun warms Earth, drives the wind and rain, and powers photosynthesis in plants, which lie at the bottom of the terrestrial food chain. The energy you expend as you move through your day arrived on the planet in the form of light.

The role of light in astronomy closely parallels the role of light in our everyday existence. Your mental picture of an astronomer is probably of a denizen of the night with head bent to the eyepiece of a telescope, peering at the distant universe. Although this picture is a bit quaint in this modern era of electronic cameras and telescopes orbiting Earth, in many ways it remains on the mark. Whether the telescope is on a mountaintop, in orbit about Earth, or part of a spacecraft hurtling toward a comet makes little difference. Our knowledge of the universe beyond Earth comes overwhelmingly from light given off or reflected by astronomical objects. Fortunately, light is a *very* informative messenger. It carries with it information about the temperatures

KEY CONCEPTS

Unlike the physicist or the chemist, who has control over the conditions in a laboratory, the astronomer must try to glean the secrets of the universe from the light and other particles such as *neutrinos* that reach us from distant objects. On this leg of our journey we turn our attention to light, a most informative messenger, and find that

- Light is an electromagnetic wave with a spectrum extending far beyond the colors of the rainbow.
- Light is *also* a stream of particles called photons.
- Reconciling the wave and particle nature of light and matter points beyond Newton's physics and challenges our everyday ideas about what is "real."
- Measurements of the speed of light also require that we think beyond classical physics and reassess our understanding of time and space.
- The wave/particle nature of light and matter gives different types of atoms unique spectral "fingerprints" that we can use to measure the composition and properties of distant objects.
- Temperature measures the thermal energy of an object and determines the amount and spectrum of light that a dense object emits.
- Light is not only a messenger but is also a way in which energy is carried throughout the universe.

of objects, what they are made of, their speed, and even the nature of the material that the light passed through on its way to Earth.

Yet light plays a far larger role in astronomy than just being a messenger. Light is one of the main ways in which energy is transported throughout the universe. Light carries energy generated in the heart of a star outward through the star and off into space. From stars to planets to vast clouds of gas and dust filling interstellar space—absorption of light heats objects up while emission of light cools them off.

Although light allows us to see the world, we cannot actually "see" light. That is, we do not see light in the same way that we see a tree or we see the stars at night. Light is the messenger but is itself invisible. It is hard to study something that cannot be seen, so an understanding of light was a long time coming. The property of light that is easiest to *try* to measure is the speed at which it travels. This might seem straightforward, yet the answer led 19th- and 20th-century physicists to change how we think about the very fabric of space and time. As we continue our journey we will find that light sets the standard for what we mean by "when," "where," or "how fast" because nothing can travel faster than light.

4.2 Our Picture of Light Evolved with Time

Suppose you are a scientist living in the 16th or 17th century. How might you go about measuring the speed of light? One obvious way would be to have a friend stand on a hilltop far away. You uncover a lantern, and the instant your friend sees the light from your lantern, your friend uncovers her own lantern. The time it takes from when you uncover your lantern to when you see your friend's light will be the light's round-trip travel time—plus, of course, your friend's reaction time. Galileo measured the speed of sound in just this way, but when he tried this method on light, he failed. He could not measure any delay. Galileo concluded that the speed of light must be very great indeed, possibly even infinite.

If light travels so rapidly, then to measure its speed we will need either very large distances over which to measure its flight or very good clocks. Galileo had neither at his disposal, but by the end of the 18th century astronomers had both. The great distances were the distances between the planets, whereas the good clock was provided courtesy of Kepler and Newton. According to Newton's derivation of Kepler's laws, orbital periods should be completely constant, with each orbit taking exactly as much time as the orbit before. This applies to moons orbiting planets just as it applies to planets orbiting the Sun.

In the 1670s **Ole Rømer** (1644–1710) was studying the moons of Jupiter, taking measurements of the times when each moon disappeared behind the planet. Much to his amazement Rømer found that rather than maintaining a regular schedule, the observed times of these events would slowly drift in comparison with predictions. Sometimes the moons disappeared behind Jupiter too soon, and at other times they went behind Jupiter later than expected. Rømer realized that the difference depended on where Earth was in its orbit. If he began tracking the moons when Earth was closest to Jupiter, then by the time Earth was farthest from Jupiter, the moons were a bit over 16½ minutes "late." But if he waited until Earth was once again closest to Jupiter, the moons "made up" the lost time and once again passed behind Jupiter at the predicted times.

It is often the case in science that a difference between theoretical predictions and experimental results points the way to new knowledge, and Rømer's work was no exception. Rømer correctly surmised that rather than a failure of Kepler's laws, he was seeing the first clear evidence that light travels at a finite speed. As shown in **Figure 4.1**, the moons appeared "late" when Earth was farther from Jupiter because of the time needed for light to travel the extra distance between the two planets. Over the course of Earth's yearly trip around the Sun, the distance between Earth and

---

Rømer used Jupiter's moons to measure the speed of light.

Jupiter changes by 2 AU (astronomical units), which is about 3×10^{11} m. The speed of light equals this distance divided by Rømer's 16.7-minute delay, or about 3×10^8 m/s. The value Rømer actually announced in 1676 was a bit on the low side—2.25×10^8 m/s—because the length of 1 astronomical unit was not well known. But Rømer's result was more than adequate to make the point. The speed of light is very great indeed! The orbiting space shuttle moves around Earth at a dazzling speed of about 28,000 km/h (almost 8,000 m/s). Light travels almost 40,000 times faster than this. It could circle Earth in only ⅐ of a second. No wonder Galileo's attempts to measure the speed of light failed!

A good deal of work has been done to improve on Rømer's original result. Modern measurements of the speed of light made with the benefit of high-speed electronics give a value of 2.99792458×10^8 m/s in a vacuum. As of October 1983 the length of a meter is now *defined* as the distance traveled by light in a vacuum in 1/299,792,458 of a second.

The speed of light is 300,000 km/s in a vacuum.

The speed of light in a vacuum, about 300,000 km/s, is one of nature's fundamental constants, usually written as *c*. Keep in mind, however, that this is true *only* in a vacuum.

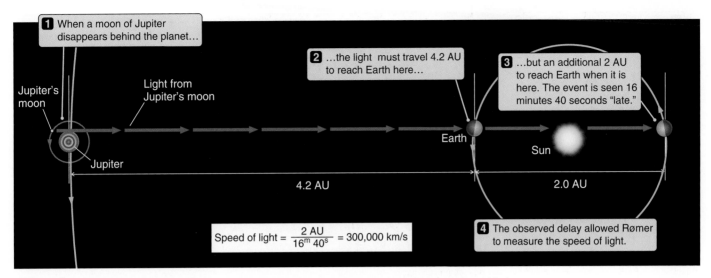

1 When a moon of Jupiter disappears behind the planet…

2 …the light must travel 4.2 AU to reach Earth here…

3 …but an additional 2 AU to reach Earth when it is here. The event is seen 16 minutes 40 seconds "late."

Jupiter's moon

Light from Jupiter's moon

Earth

Jupiter

Sun

4.2 AU

2.0 AU

$$\text{Speed of light} = \frac{2\ \text{AU}}{16^m\ 40^s} = 300{,}000\ \text{km/s}$$

4 The observed delay allowed Rømer to measure the speed of light.

FIGURE 4.1 Ole Rømer measured the speed of light by noting that apparent delays in the orbital motions of Jupiter's moons depend on the distance between Earth and Jupiter.

The speed of light through any medium, such as air or glass, is *always* less than c. We refer to the ratio of light's speed in a vacuum to its speed, *v*, in a medium as the medium's **index of refraction**:

$$n = \frac{c}{v}.$$

For typical glass *n* is approximately 1.5. Rearranging this equation we find the speed of light in glass is

$$v = \frac{c}{n} = \frac{300{,}000\ \text{km/s}}{1.5} = 200{,}000\ \text{km/s}.$$

We will come back to the index of refraction in Chapter 5 when we discuss refraction and refracting telescopes.

Recall that in the opening chapter we spoke of distances expressed not in kilometers or miles, but in units of time. For example, the Moon's distance is such that it takes light 1¼ seconds to travel between Earth and Moon. In other words, we can say the Moon is 1¼ light-*seconds* from Earth.

A light year is the distance light travels in one year.

The Sun is 8⅓ light-*minutes* away, and the next nearest star is 4⅓ light-*years* distant. Light travel time is a convenient way of expressing cosmic distances, and the basic unit is the **light-year**.[1] A light-year is defined as the distance traveled by light in one year, or about 9.5 trillion kilometers. Remember, *a light-year is a measure of distance. It is* **not** *a*

---

[1] As we will learn in Chapter 13, professional astronomers frequently use another yardstick to describe stellar and galactic distances, the *parsec*. One parsec is equal to 3.26 light-years.

measure of time. ("My goodness, it's been simply light-years since we last saw them.")

Light Is an Electromagnetic Wave

Since the earliest investigations of light, there has been a good deal of controversy over the question of whether light is composed of particles, as Newton believed, or is instead a wave. (A **wave** is a disturbance that travels from one point to another.) This controversy was seemingly put to rest once and for all in 1873 by the Scottish physicist **James Clerk Maxwell** (1831–1879). One of Maxwell's many accomplishments was the discovery of the fundamental laws that describe electricity and magnetism. The electric force and the magnetic force are actually two aspects of the same electromagnetic phenomenon. The **electric force** is the push and pull between electrically charged particles. Opposite charges attract and like charges repel. The **magnetic force**, on the other hand, is a force between electrically charged particles arising from their motion.

To describe the electric and magnetic forces Maxwell introduced the concepts of the **electric field** and the **magnetic field**. A charged particle creates an electric field that points away from it if the charge is positive, as shown in **Figure 4.2(a)**, or toward it if the charge is negative. To find out how much force is exerted on a charged particle, we multiply the charge of the particle by the strength of the electric field at its location. Because the electric field points directly away from a positively charged particle (or directly toward a negatively charged particle), the force that a second charge feels is either directly toward or directly away from the first charged particle.

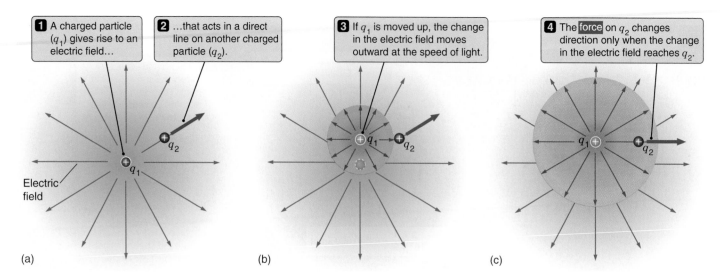

1 A charged particle (q_1) gives rise to an electric field...

2 ...that acts in a direct line on another charged particle (q_2).

3 If q_1 is moved up, the change in the electric field moves outward at the speed of light.

4 The force on q_2 changes direction only when the change in the electric field reaches q_2.

Electric field

(a)   (b)   (c)

FIGURE 4.2 When a charged particle accelerates, changes in the electric field move outward at the speed of light. (In (b) the charge is shown moving instantly from one place to another for clarity. In reality this could not happen.)

The picture gets a bit more interesting if we quickly move the first charged particle (q_1) by some amount, as shown in **Figure 4.2(b)**. We might expect the force on the second particle (q_2) to change immediately so that it points away from the new position of the first particle. Yet experiments show that it does not. Immediately after the first charge moves, there is *no* change in the force felt by the second charge. Only later does the second particle feel the change in location of the first (**Figure 4.2(c)**). The situation is something like what happens if you are holding onto one end of a long piece of rope and a friend is holding the other end. When you yank your end of the rope up and down, your friend does not feel the result immediately. Instead your yank starts a pulse—a wave—that travels down the rope. Your friend notices the yank only when this wave arrives at his end. Similarly, when you move a charged particle, information about the change travels outward through space as a wave in the electric field. Other charged particles do not know that the first particle has moved until the wave reaches them.

Maxwell summarized the behavior of electric and magnetic fields in four elegant equations. Among other things, these equations say that a changing electric field causes a magnetic field, and that a changing magnetic field causes

Changing electric and magnetic fields lead to a self-sustaining electromagnetic wave.

an electric field. These changes "feed" on themselves. A change in the motion of a charged particle causes a changing electric field, which causes a changing magnetic field, which causes a changing electric field.... Once the process starts, a self-sustaining procession of oscillating electric

and magnetic fields moves out in all directions through space. Instead of a purely electric wave, an accelerating charged particle gives rise to an **electromagnetic wave**.

In addition to predicting that electromagnetic waves should exist, Maxwell's equations also predict how rapidly the disturbance in the electric and magnetic fields should move. In short, Maxwell's equations *predict* the speed at which an electromagnetic wave should travel. When Maxwell carried out this calculation, he discovered that electromagnetic waves should travel at 3×10^8 m/s—which is the speed of light! This agreement could not be simple coincidence. Maxwell had shown that light is an electromagnetic wave.

Maxwell's wave description of light also gives us an idea of how light originates and how it interacts with matter. Imagine a cork floating on a lake on a perfectly calm day. The surface of the lake is as smooth and flat as a mirror until a fish tugs on the hook and line dangling beneath the cork. The motion of the cork causes a disturbance that moves

Accelerating charges cause electromagnetic waves.

outward as a ripple on the surface of the lake (see **Figure 4.3(a)**). In much the same way, an accelerating electric charge (**Figure 4.3(b)**) causes a disturbance that moves outward through space as an electromagnetic wave. (An electric charge that is moving at a constant velocity is stationary in its inertial frame of reference and so does not radiate.) According to Maxwell's equations, *any* time an electrically charged particle is accelerated, the result is an electromagnetic wave. *Accelerating charges are the sources of electromagnetic radiation.*

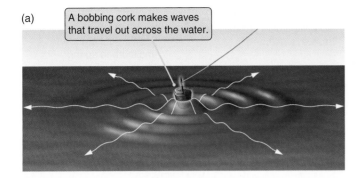

(a) A bobbing cork makes waves that travel out across the water.

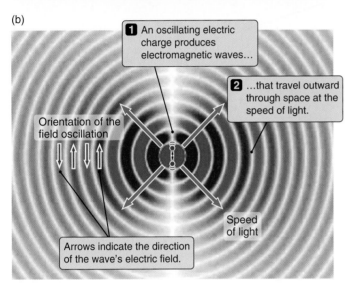

(b)

1 An oscillating electric charge produces electromagnetic waves...

2 ...that travel outward through space at the speed of light.

Orientation of the field oscillation

Arrows indicate the direction of the wave's electric field.

Speed of light

FIGURE 4.3 (a) A fish pulling downward on a cork generates waves that move outward across the water's surface. (b) In similar fashion, an accelerated electric charge generates electromagnetic waves that move away at the speed of light.

Waves Are Characterized by Wavelength, Frequency, Speed, and Amplitude

Along our journey we will encounter waves of different kinds, ranging from electromagnetic waves crossing the vast expanse of the universe to seismic waves traveling through Earth. In the most general sense a wave is a disturbance that travels away from its source. If you drop a pebble in a pond, ripples spread out over the surface of the water. This kind of wave is called a **transverse wave** because the wave's displacement is perpendicular, or "transverse," to its direction of travel (see **Figure 4.4 (a)**). The waves on a plucked guitar string are also transverse waves. Such waves travel because when the material is disturbed, forces try to even out that disturbance. A portion of the material moves back toward its undisturbed position. But because it is still moving and has inertia, when it reaches that position, it overshoots, distorting the material in the opposite way from

A wave is a disturbance that travels away from a source.

how it was originally distorted. Forces now try to push the material back in the other direction, but again there is an overshoot. The material oscillates back and forth from one side of its undisturbed position to the other, and the wave moves along. In the guitar string, the force responsible for the wave is the tension that tries to keep the string straight. In the case of water, the forces responsible for creating the wave are gravity, which tries to pull down the crest of the wave, and water pressure (also caused by gravity), which tries to push the trough of the wave up.

Another kind of wave is called a **longitudinal wave.** Sound waves are an example of longitudinal waves, and so are the waves in a spring (**Figure 4.4(b)**). In a spring, compressed regions try to push into the stretched-out regions to even the spring out. But when a portion of the spring reaches its undisturbed position, it is still moving, and it overshoots. Parts of the spring that were originally compressed are now stretched, and the parts that were originally stretched are now compressed. Regions in which the spring is alternately compressed and stretched move along the length of the spring as the cycle at each point repeats itself. In the case of sound waves, air pressure provides the forces that keep the wave moving.

In these examples the waves result from *mechanical* distortions of the medium the wave travels through (for example, distortion of the surface of the pond or the coils in the spring). Mechanical waves involve distortions measured as distances, and media that have mass. Maxwell showed that light waves are a fundamentally different type of wave. Light waves involve no mechanical distortion of a medium, but instead involve periodic changes in the strength of the electric and magnetic fields. Even so, light waves are generally thought of as transverse waves because the directions of the electric and magnetic fields are perpendicular to the direction in which the wave travels (see **Figure 4.5**).

Waves are generally characterized by four quantities, which are shown in **Figure 4.6**. The **amplitude** of a wave is the maximum excursion from its undisturbed or relaxed position. The wave travels at some speed, which is usually written as v (except in the case of light, where the speed of

Waves are characterized by their wavelength, frequency, speed, and amplitude.

light is written as c). The number of wave crests passing a point in space each second is called the wave's **frequency**, denoted by f. The unit of frequency is cycles per second, which is generally referred to as **hertz** (abbreviated Hz) after the 19th-century physicist **Heinrich Hertz** (1857–1894), who

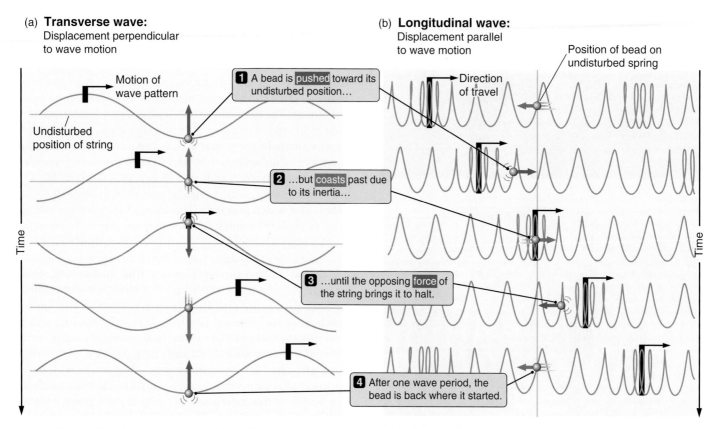

(a) **Transverse wave:**
Displacement perpendicular to wave motion

Motion of wave pattern

Undisturbed position of string

Time

1 A bead is pushed toward its undisturbed position…

2 …but coasts past due to its inertia…

3 …until the opposing force of the string brings it to halt.

4 After one wave period, the bead is back where it started.

(b) **Longitudinal wave:**
Displacement parallel to wave motion

Position of bead on undisturbed spring

Direction of travel

Time

FIGURE 4.4 Mechanical waves result from forces that try to even out disturbances. (a) A transverse wave involves oscillations that are perpendicular to the direction in which the wave travels. (b) A longitudinal wave involves oscillations along the direction of travel of the wave.

was the first to experimentally confirm Maxwell's predictions about **electromagnetic radiation**. The time taken for one complete cycle is called the *period, P,* which is measured in seconds.

The distance a wave travels during one complete oscillation is called the **wavelength**. This is just the distance from one wave crest to the next, or the distance from one wave trough to the next. The wavelength is usually denoted by the Greek letter λ (pronounced "lambda"). There is a clear

relationship between the frequency of a wave and its wavelength. If the period of a wave is ½ second—that is, if it takes ½ second for one wave to pass by, crest to crest—then two waves will go by in 1 second. So a wave with a period of ½ second per cycle has a frequency of 2 cycles per second. Similarly, if a wave has a period of 1/100 second per cycle, then 100 waves will pass by each second. This wave has a frequency of 100 Hz. More generally, the frequency of a wave is just 1 divided by its period:

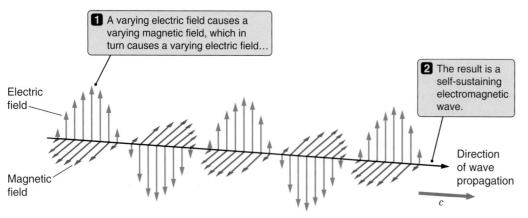

1 A varying electric field causes a varying magnetic field, which in turn causes a varying electric field…

2 The result is a self-sustaining electromagnetic wave.

Electric field

Magnetic field

Direction of wave propagation

c

FIGURE 4.5 Far from its source, an electromagnetic wave consists of oscillating electric and magnetic fields that are perpendicular both to each other and to the direction in which the wave travels.

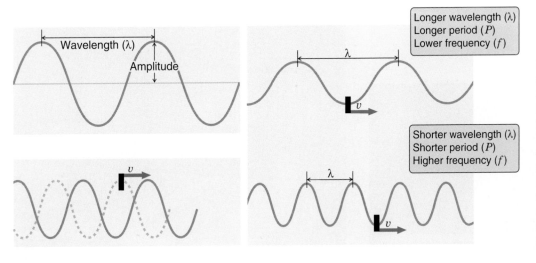

FIGURE 4.6 A wave is characterized by the distance over which the wave repeats itself (called the wavelength, λ), the maximum excursion from its undisturbed state (called the amplitude), and the speed (v) at which the wave pattern travels. In an electromagnetic wave, the amplitude is the maximum strength of the electric field, and the speed of light is written as c.

$$\text{Frequency} = \frac{1}{\text{period}} \quad \text{or} \quad f = \frac{1}{P}$$

There is also a relationship between the period of a wave and its wavelength. The period of a wave is the time between the arrival of one wave crest and the next. During this time the wave travels a distance equal to the separation between the two wave crests, or one wavelength. So far, so good. Now add to the picture the fact that distance traveled equals speed times time taken. Change "distance traveled" to one wavelength and change "time taken" to one period, and we find that the wavelength of a wave equals the speed at which the wave is traveling times the period of the wave:

$$\text{Wavelength} = \text{Speed} \times \text{Period}$$

Physicists use the letter c to represent the speed of light, so we can say that

$$\lambda = c \times P.$$

Using the relationship between period and frequency just given, we can also write

$$\text{Wavelength} = \frac{\text{Speed}}{\text{Frequency}}, \quad \text{or} \quad \lambda = \frac{c}{f}.$$

So if we know the speed of a wave, then knowing one of the three properties—its wavelength, period, or frequency—tells us the other two.

Look at this relationship more closely. The longer the length of a wave, the longer you have to wait between wave crests, so the frequency of the wave will be lower. A shorter wavelength means less distance between wave crests, which means a shorter wait until the next wave comes along. Therefore, a shorter wavelength means a higher frequency. A tremendous amount of information can be carried by waves—intelligible speech, for example, or complex and beautiful music. As we continue our study of the universe,

A long wavelength means low frequency and short wavelengths mean high frequency.

time and again we will find that the information we receive, whether about the interior of Earth or a distant star or galaxy, rides in on a wave.

Maxwell's equations also describe how electromagnetic waves interact with the matter they encounter. Returning to the analogy of the lake in Figure 4.3, imagine now that a second cork is afloat on the lake some distance from the first, as in **Figure 4.7(a)**. The second cork remains stationary until the ripple from the first cork reaches it. As the ripple passes by, the rising and falling of the water causes the cork to rise and fall as well. Similarly, the oscillating electric field of an electromagnetic wave causes an oscillating force on any charged particle that the wave encounters, and this force causes the particle to move about as well **(Figure 4.7(b))**. It takes energy to produce an electromagnetic wave, and that energy is carried through space by the wave. Matter far from the source of the wave can absorb this energy. In this way some of the energy lost by the particles generating the electromagnetic wave is transferred to other charged particles. The emission and absorption of light by matter are the result of the interaction of electric and magnetic fields with electrically charged particles.

Electromagnetic Waves of Different Wavelengths Make Up the Electromagnetic Spectrum

You have almost certainly seen a rainbow like the one in **Figure 4.8** spread out across the sky, or sunlight split into many different colors by a prism. This sorting of light by colors is really a sorting by wavelength. When we talk about light spread out according to wavelength, we refer to the

(a)

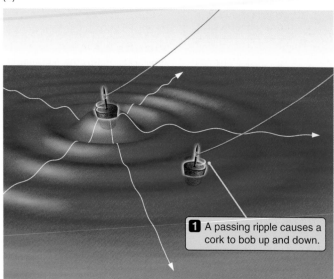

(b)

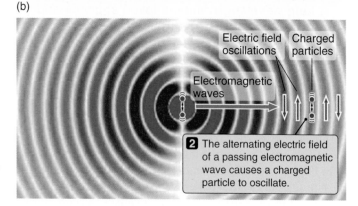

Electric field oscillations | Charged particles

Electromagnetic waves

1 A passing ripple causes a cork to bob up and down.

2 The alternating electric field of a passing electromagnetic wave causes a charged particle to oscillate.

FIGURE 4.7 (a) When waves moving across the surface of water reach a cork, they cause the cork to bob up and down. (b) Similarly, a passing electromagnetic wave causes an electric charge to wiggle in response to the wave.

FIGURE 4.8 The visible part of the electromagnetic spectrum is laid out in all its glory in the colors of this rainbow.

spectrum of the light. On the long-wavelength (and therefore low-frequency) end of the visible spectrum is red light. A **micrometer**, or **micron**, is the unit often used for measuring the wavelength of visible light. A micrometer is a millionth ($1/10^6$) of a meter. The abbreviation used for a micron

> The spectrum of visible light is seen as the colors of the rainbow.

is μm (where μ is the Greek letter "mu"). Another commonly used unit is the nanometer, abbreviated nm. A nanometer is one billionth ($1/10^9$) of a meter.[2] The wavelengths of the light we perceive as red fall between about 600 and 700 nm.

[2] Astronomers conventionally use nanometers (nm) when referring to wavelengths at visible and shorter wavelengths, micrometers or "microns" (μm) in the infrared, and centimeters (cm) and meters (m) in the microwave and radio regions of the electromagnetic spectrum.

At the other end of the visible spectrum is violet light, which is the bluest of blue light. The shortest-wavelength violet light that our eyes can see has a wavelength of around 350 nm. Stretched out between the two, literally in a rainbow, is the rest of the visible spectrum. The colors in the visible spectrum in order of decreasing wavelength can be remembered as a name: "Roy G. Biv," which stands for

Red Orange Yellow Green Blue Indigo Violet.

The human eye is most sensitive to light in the green to yellow part of the spectrum. This light has a wavelength of around 500 to 550 nm. Green light with a wavelength of 520 nm has a frequency of

$$f = \frac{c}{\lambda} = \frac{3.00 \times 10^8 \frac{m}{s}}{520 \times 10^{-9}\,m} = \frac{5.8 \times 10^{14}}{s}$$

or 5.8×10^{14} Hz

That frequency corresponds to 580 *trillion* wave crests passing by each second!

When we say "visible light," what we mean is "the light that the light-sensitive cells in our eyes respond to." But this is not the whole range of possible wavelengths for elec-

Visible light is only one small segment of the electromagnetic spectrum.

tromagnetic radiation. Radiation can have wavelengths that are much shorter or much longer than our eyes can perceive. The whole range of different wavelengths of light is collectively referred to as the **electromagnetic spectrum**.

Follow along in **Figure 4.9** as we take a tour of the electromagnetic spectrum, beginning with visible light and working our way to shorter and longer wavelengths. We start at the blue end of the visible spectrum. Beyond this short-wavelength, high-frequency side of the visible spectrum, there is light that is "bluer than blue" or actually "more violet than violet." This light, with wavelengths between 40 and 350 nm, is called **ultraviolet (UV) radiation**. You can remember what ultraviolet light is just by looking at the name. The prefix *ultra-* means "extreme," so *ultra*violet light is light that is more "extremely" violet than violet. It is important to remember that ultraviolet light is fundamentally no different from visible light, any more than high C on a piano is fundamentally different from middle C.

As we go to shorter wavelengths of light (and so to higher frequencies), we pass through the ultraviolet part of the spectrum. At a wavelength shorter than 40 nm, or 4×10^{-8} m, we stop calling radiation ultraviolet light and instead start calling it **X-rays**. This distinction comes for historical reasons. When X-rays were discovered in the last part of the 19th century, they were given the name "X" by their discoverer, **Wilhelm Conrad Roentgen** (1845–1923), to indicate they were "a new kind of **ray**." As we continue to even shorter wavelengths, we come to another somewhat arbitrary break. Electromagnetic radiation with the very shortest wavelengths (less than about 10^{-10} m) is referred to as **gamma rays**. Again the reasons are historical. Gamma rays (or γ-rays) were first discovered as a type of radiation given off by radioactive material. It was only later that their true nature became known.

So far we have been considering ever shorter wavelengths and higher frequencies. In principle, there is no limit to this process. We can conceive of gamma rays of arbitrarily short wavelengths and arbitrarily high frequencies (even though practical considerations eventually come into play). We can also go in the other direction. Just as there is light that is more violet than violet, there is also light that is "redder than red." Such light, covering wavelengths longer than about 700 nm and shorter than 500 μm (5×10^{-4} m), is referred to as **infrared (IR) radiation**. Again the key to remembering what infrared light is comes from looking at the word itself. *Infra-* is a prefix that means "below." *Infra*red light is light that has a frequency that is lower than (below) that of red light. When the wavelength of light gets longer than this, we start calling it **microwave radiation**. The longest-wavelength (and therefore lowest-frequency) electromagnetic

FIGURE 4.9 By convention the electromagnetic spectrum is broken into loosely defined regions ranging from gamma rays to radio waves.

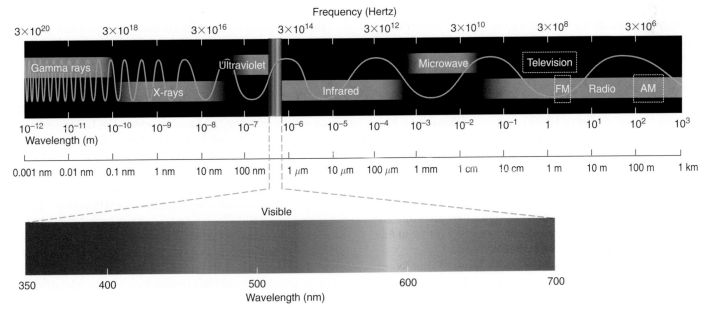

radiation, with wavelengths longer than a few centimeters and ranging up to arbitrarily long wavelengths, is called **radio waves**. Chapter 5 discusses the various kinds of telescopes used by astronomers to capture and analyze the wide range of electromagnetic radiation.

4.3 The Speed of Light Is a Very Special Value

Maxwell's description of light as a wave was a great success, but without realizing it he had also found a flaw in Newtonian physics. To understand this flaw, we need to think back to Chapter 2, where we used a moving car as an example of a moving frame of reference. Imagine that you are sitting in a moving car and there is a ball sitting on the seat beside you. In your frame of reference the ball is at rest. But if the car is moving at 50 mph down the highway, someone standing by the road will say that the ball is also moving at 50 mph. To someone in oncoming traffic moving at 50 mph, the relative speed of both your car and the ball would be 100 mph. There really is no difference between these three perspectives. The laws of physics are the same in *any* inertial frame of reference.

As a variant of the "ball in the car" experiment, imagine that as your car moves down the highway at 50 mph you pitch a fastball forward at 100 mph (see **Figure 4.10(a)**). In your frame of reference the ball is moving at 100 mph, but to an observer standing by the road the ball is moving at 150 mph. (The ball has the original 50 mph speed of the car plus the additional 100 mph that you gave it with your throw.) In the frame of reference of a car in oncoming traffic traveling at 50 mph, the ball is moving at 200 mph. (This is the 150 mph that the ball is moving relative to the ground plus the 50 mph motion of the oncoming car.) In our everyday experience velocities simply add. This is also how Newton's laws say the universe should behave.

Now do exactly the same thought experiment, but with two changes. Instead of a car traveling at 50 mph, imagine you are in a spaceship traveling at half the speed of light, or $0.5c$, as shown in **Figure 4.10(b)**. Instead of throwing a baseball, you shine a beam of light forward. To you the light is moving at the speed of light, c. If we replace "100 mph" with "c" in the previous paragraph, we think we know what to expect for other observers. To an observer on a nearby planet, the light should travel by at a speed equal to the speed of your spacecraft plus the speed of light, or $1.5c$. Similarly, to an observer in an oncoming spacecraft traveling at $0.5c$, the light should appear to travel at a speed of $2c$.

We do not have the luxury of performing this experiment while traveling through space at half the speed of light, but physicists are ingenious folk. During the closing years of the 19th century and the early years of the 20th century, physicists were conducting laboratory experiments that were the functional equivalent of our thought experiment. What they found puzzled them greatly. Rather than the speed of the beam of light differing from one observer

> Surprisingly, experiments showed that the speed of light is the same for all observers.

to the next, as expected on the basis of Newton's physics and "common sense," they found instead that *all observers measure exactly the same value for the speed of the beam of light, regardless of their motion!*

As you ride in your spaceship you measure the speed of the beam of light to be c, or 3×10^8 m/s. That is as expected because you are holding the source of the light. But the observer on the planet *also* measures the speed of the passing beam of light to be 3×10^8 m/s. Even the passenger in the oncoming spacecraft finds that the beam from your light is traveling at exactly c in her own frame of reference. In fact, it turns out that *every observer always finds that light in a vacuum travels at exactly the same speed* c, *regardless of his or her own motion or the motion of the source of the light.*

If at this point you are feeling uneasy and saying to yourself, "This is very bizarre," then you probably have followed what the last few paragraphs have stated. And if this discussion bothers you, imagine the reaction of those physicists! Newton's laws of motion had been the bedrock of science for 200 years, facing every experimental challenge that came their way. Now suddenly that bedrock seemed to turn to sand. How could light have the same speed for *all* observers, regardless of their own velocity? Preposterous! And yet that was the inescapable experimental result. Despite all its spectacular successes, Newtonian physics seemed to be in serious trouble.

Enter a young German-born Swiss patent clerk named **Albert Einstein** (1879–1955). As a 16-year-old schoolboy Einstein had already realized that there was trouble afoot. Light travels in a straight line at a constant speed. Einstein reasoned that according to Newton's laws of motion, there should be a perfectly good inertial frame of reference that moves along with the light and in which the light is stationary. That is, you should be able to "keep up" with light so that you are moving right along with it. But if you could do that, the light would be an oscillating electric and magnetic wave *that does not move*. This was impossible according to Maxwell's equations for electromagnetic waves. There was a contradiction here. Either Maxwell was wrong in his understanding of electricity and magnetism, or Newtonian physics did not apply at very large velocities. As the experimental results rolled in on measurements of the speed of light, it became clear that it was Newtonian physics that needed revision.

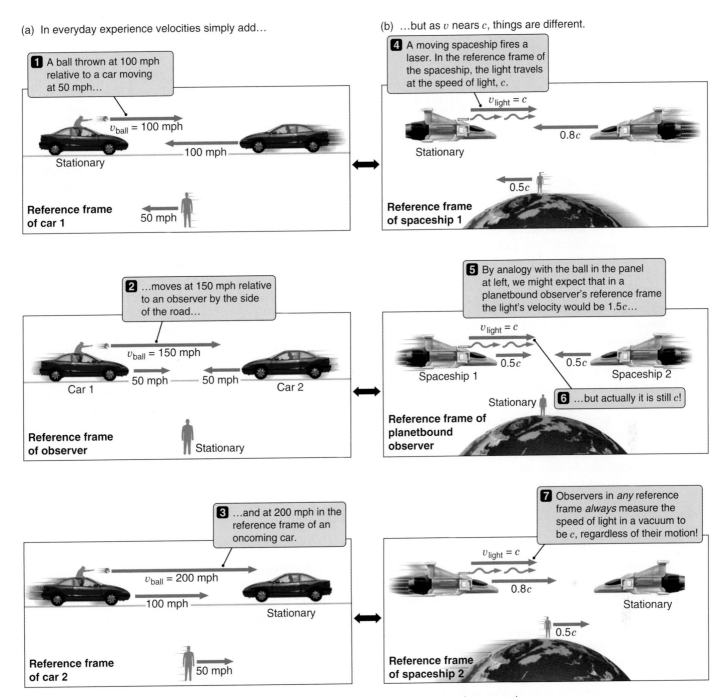

(a) In everyday experience velocities simply add...

1 A ball thrown at 100 mph relative to a car moving at 50 mph...

v_{ball} = 100 mph

100 mph

Stationary

Reference frame of car 1

50 mph

2 ...moves at 150 mph relative to an observer by the side of the road...

v_{ball} = 150 mph

50 mph   50 mph

Car 1   Car 2

Reference frame of observer

Stationary

3 ...and at 200 mph in the reference frame of an oncoming car.

v_{ball} = 200 mph

100 mph

Stationary

Reference frame of car 2

50 mph

(b) ...but as v nears c, things are different.

4 A moving spaceship fires a laser. In the reference frame of the spaceship, the light travels at the speed of light, c.

v_{light} = c

0.8c

Stationary

0.5c

Reference frame of spaceship 1

5 By analogy with the ball in the panel at left, we might expect that in a planetbound observer's reference frame the light's velocity would be 1.5c...

v_{light} = c

0.5c 0.5c

Spaceship 1   Spaceship 2

Stationary   **6** ...but actually it is still c!

Reference frame of planetbound observer

7 Observers in *any* reference frame *always* measure the speed of light in a vacuum to be c, regardless of their motion!

v_{light} = c

0.8c

Stationary

0.5c

Reference frame of spaceship 2

FIGURE 4.10 The rules of motion that apply in our daily lives break down when speeds approach the speed of light. The fact that light itself always travels at the same speed for any observer is the basis of special relativity. (Note that relativity also affects the relative speeds of the two spacecraft.)

Time Is a Relative Thing

Einstein resolved the contradiction between Maxwell and Newton and ushered in a scientific revolution (see **Connections 4.1**) with his theory of **special relativity**, which was published in 1905. Special relativity was Einstein's answer to the question, "What must the universe be like if every observer always measures the same value for the speed of light in a vacuum?" Einstein focused his thinking on pairs of *events*. In relativity, an **event** is something that happens at a particular location in space at a particular time. When you snap your fingers, that is an event. From everyday experience we know that the distance between any two events depends on the frame of reference of the person observing

A Scientific Revolution

Throughout the first three chapters of this book we interlaced our story of the motions of the sky and the discovery of Newton's laws of motion and gravitation with a discussion of the nature of scientific knowledge and the way that science progresses. We stressed that scientific knowledge differs from all other forms of knowledge in that even our most cherished and fundamental knowledge is open to challenge by new observations and experiments. In this chapter we will see this drama play itself out several times over.

By the middle of the 19th century many physicists felt that our fundamental understanding of physical law was more or less complete. For over a century Newtonian physics had withstood the scrutiny of scientists the world over. It seemed that little remained but cleanup work —filling in the details. Some even went so far as to pronounce this period the "end of science." Yet during the late 19th and early 20th centuries, physics was rocked by a series of scientific revolutions that shook the very foundations of our understanding of the nature of reality. In this chapter we have come across several of these revolutions. Einstein's theory of relativity erased the classical distinction between space and time and united our concepts of matter and energy. Quantum mechanics forced us to abandon our everyday understanding of "substance" and even to part with the notion that we live in a universe in which effect follows cause in lockstep. Together these revolutions led to the birth of what has come to be known as **modern physics**. Although modern physics *contains* Newtonian physics, the understanding of the universe offered by modern physics is far more sublime and powerful than the earlier understanding that it subsumed.

As we continue on our journey, we will encounter many other discoveries and successful ideas that forced scientists either to abandon their treasured notions or be left behind, hopelessly locked into a worldview that had ultimately failed the test of observation and experiment. The point is this: In physical science, we are not just paying lip service to a hollow ideal when we say that the rigorous standards of scientific knowledge respect no authorities. No theory, no matter how central or how strongly held, is immune from the rules.

them. Suppose you are sitting in a car that is traveling down the highway in a straight line at a constant 60 mph. You snap your fingers (event 1), and a minute later you snap your fingers again (event 2). In your frame of reference *you* are stationary and the two events happened at exactly the same place. They are separated by a minute in *time,* but there is no separation between the two events in *space.* This is very

---

Special relativity concerns the relationship between events in space and time.

---

different from what happens in the frame of reference of an observer sitting by the road. This observer agrees that the second snap of your fingers (event 2) occurred a minute after the first snap of your fingers (event 1), but to this observer the two events were separated from each other in space by a mile. In this everyday, "Newtonian" view, the *distance* between two events depends on the motion of the observer, but the *time* between the two events does not.

Einstein questioned why there was such a distinction between the way Newton treated space and the way New-ton treated time. Einstein realized that the *only* way the speed of light can be the same for all observers is if *the passage of time is different from one observer to the next!* This is a *very* counterintuitive idea, but it is so central to our modern understanding of the universe that it is worth wrestling with a bit. Hang onto your hat while we reconstruct some of the reasoning that led Einstein to this remarkable conclusion.

To measure time, the first thing we need is a clock. The best way to build a clock is to base it on a value that everyone can agree on—such as the speed of light. **Figure 4.11(a)** shows just such a clock as seen by observer 1, who is stationary with respect to the clock. At time t_1 a flashlamp gives off a pulse of light. Call this event 1. The light bounces off a mirror a distance l meters away, then heads back toward its source. At time t_2 the light arrives and is recorded by a photodetector. Call this event 2. The time between events 1 and 2 is just the distance the light travels ($2l$ meters), divided by the speed of light, or $t_2 - t_1 = 2l/c$.

So far so good, but now look at the clock from the perspective of observer 2 in a frame of reference that is moving relative to the clock. In *this* observer's frame of reference he is stationary, and it is the *clock* that is moving at speed *v*, as shown in **Figure 4.11(b)**. (Recall that because any inertial frame of reference is as good as any other, this observer's perspective is as valid as the first observer's perspective.) We see the same two events as before: event 1 when the light leaves the flashlamp and event 2 when the light arrives at the detector. There is a difference, however. In this frame of reference the clock *moves* between the two events, so the light has *farther to go*. (If you do not see this right away, use a ruler to measure the total length of the light path in Figure 4.11(b) and compare it with the total length of the light path in Figure 4.11(a).) The time between the two events is still the distance traveled divided by the speed of light, but now that distance is *longer* than 2*l* meters. Because the speed of light is the same for all observers, the time between the two events must be longer as well!

Go over that again. The two events are the *same two events,* regardless of the frame of reference from which they are observed. The question is, how much time passed between the two events? Because the speed of light is the same for all observers, there *must* be more time between the two events when they are viewed from a frame of reference in which the clock is moving. It takes a moving clock more time than a stationary clock to complete one "tick." Moving clocks *must* run slow, and the passage of time *must* depend on an observer's frame of reference.

To Newton, and to us in our everyday lives, the march of time seems immutable and constant. But in reality the only thing that is truly constant is the speed of light, and even time itself flows differently for different observers.

Here, in a nutshell, is the heart of Einstein's theory of special relativity. In our everyday Newtonian view of the world, we live in a three-dimensional space through which time marches steadily onward. Events occur in space at a certain

---

Space and time together form a four-dimensional "spacetime."

---

time. By the time Einstein finished working out the implications of his insight, he had reshaped this three-dimensional universe into a four-dimensional **spacetime**. Events occur at specific locations within this four-dimensional spacetime, but how this spacetime translates into what we perceive as "space" and what we perceive as "time" depends on our frame of reference.

It is very important to state that Einstein did not throw out Newtonian physics. We were not wasting our time in Chapter 3 when we studied Newton's laws of motion. Instead Einstein found that Newtonian physics is *contained within* special relativity. In our everyday experience we never encounter speeds that approach that of light. Even the breakneck speed of the space shuttle is only about 0.000025*c*. When Einstein's special relativity is restricted

FIGURE 4.11 The "tick" of a light clock as seen in two different reference frames. As Einstein's thought experiment demonstrates, if the speed of light is the same for every observer, then moving clocks *must* run slow.

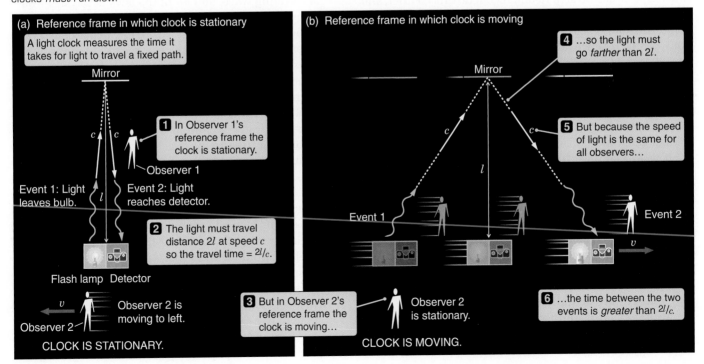

Newtonian physics is contained within special relativity.

to cases where velocities are much less than the speed of light, then Einstein's equations become the very equations that describe Newtonian physics! In our everyday lives we experience a Newtonian world. Only when relative velocities approach that of light do things begin to depart from the predictions of Newtonian physics. When great velocities cause something to turn out differently than we would expect based on Newtonian physics, this is referred to as a **relativistic** effect.

The Implications of Relativity Are Far-Ranging

The story of special relativity is another case study of how science works. Newton's laws had proven for a long time to be an extraordinarily powerful way of viewing the world. But as science turned its attention to a different phenomenon —the phenomenon of light—difficulties arose. Newton's theory of motion, Maxwell's theory of electromagnetic radiation, and empirical measurements of the speed of light met head on. Such conflicts are what scientists live for: They point the way to new knowledge and new understanding.

Conflicts between theory and experiment point the way to new knowledge.

Einstein was able to step in and reconcile this conflict, and in the process he changed the way we think about the universe. Einstein's ideas remained controversial well into the 20th century. However, as one experiment after another confirmed the strange and counterintuitive predictions of relativity, scientists came to accept its validity. Today special relativity is an integral and indispensable part of all of physics, shaping our thinking about the motions of the tiniest subatomic particles as well as the motions of the most distant galaxies.

It would be great fun to linger here for a time and explore. But our journey has hardly begun, and there is so much more to see. We hope you will find the time at some point to come back and explore the wonders of the relativistic world in which we live. Puzzling out relativity is time well spent. In the meantime, here are a few of the interesting insights that come from Einstein's work:

1. **What we think of as "mass" and what we think of as "energy" are actually two manifestations of the same thing.** Usually we think of the energy of an object as depending on its speed. The faster it moves, the more energy it has. But Einstein's famous equation $E = mc^2$

says that even a *stationary* object has an intrinsic "rest" energy that equals the mass m of the object times the speed of light, c, squared. The speed of light is a very large number. This relationship between mass and energy says that a single tablespoon of water has a rest energy equal to the energy released in the explosion of over 300,000 tons of TNT! All reactions that produce energy do so by converting some of the mass of the reactants into other forms of energy. But even the most efficient chemical or nuclear reactions release only tiny fractions of the total energy available. Exploding TNT, for example, converts less than a trillionth of its mass into energy. Even the explosion of a hydrogen bomb releases far less than 1 percent of the energy contained in the mass of the bomb.

The equivalence between mass and energy points both ways. In Chapter 3 we defined *mass* as the property of matter that resists changes in motion. Does the energy of an object really increase its resistance to changes in motion? Yes. Even adding to the energy of motion of an object increases its inertia. For example, a proton in a high-energy particle accelerator may approach the speed of light so closely that its total energy is 1,000 times greater than its rest energy. Such an energetic proton is, indeed, harder to "push around" (in other words, it has more inertia) than a proton at rest.

2. **The speed of light is the ultimate speed limit.** There are several ways to think about this. We already discussed the insight that led Einstein to relativity in the first place. Were it possible to travel at the speed of light, then in that frame of reference light would cease to be a traveling wave, and all of the laws of physics would come tumbling down around our ears. We can also think about this limit in terms of the equivalence of mass and energy just discussed. As the speed of an object gets closer and closer to the speed of light, its energy, and therefore its mass, become greater and greater, so it becomes increasingly resistant to further changes in its motion. We can continue to push on it all we like, making it go faster, but we face diminishing returns. The situation is like trying to get from 0 to 1 by halving the remainder again and again. The resulting sequence— $0, \frac{1}{2}, \frac{3}{4}, \frac{7}{8}, \frac{15}{16}, \frac{31}{32}, \frac{63}{64}, \ldots$ —gets arbitrarily close to 1 but never actually reaches it. In the same way, a continuous force applied to an object will cause its velocity to get closer and closer to the speed of light, but it will never actually reach the speed of light. You just cannot get there. It would take an *infinite* amount of energy to accelerate an object with a nonzero rest mass to the speed of light. In short, all the energy in the entire universe is inadequate to accelerate a single electron to the speed of light. We can get the electron arbitrarily close to that number—0.999999999999999999999999... $\times c$ is no problem, at least in principle—but there is no getting over

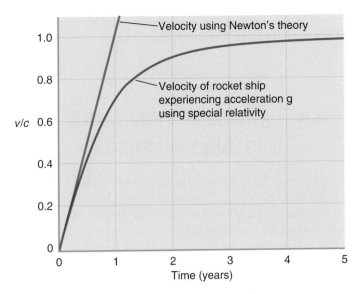

FIGURE 4.12 The speed of a rocket ship experiencing an acceleration equal to Earth's acceleration of gravity. The rocket ship approaches the speed of light but never gets there.

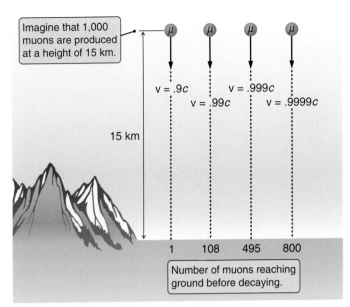

FIGURE 4.13 Plot of the muon lifetime (the time during which half of all muons decay) versus energy, or the velocity at which the muons travel. Also shown is the distance these muons can travel compared to the 15-km height at which they originate.

the hump. In **Figure 4.12** we show how a rocket ship, which experiences a constant acceleration equal to that of gravity on Earth (so its occupants will feel at home), moves faster and faster but never reaches the speed of light. Faster-than-light travel may be a mainstay of science fiction, but "Sorry, Jim, I just cannot make her go faster than *c*!" (We leave the Scottish accent up to your imagination.) Here is one of those cases where wishing that something is physically possible does not necessarily "make it so."

3. **Time passes more slowly in a moving reference frame.** This phenomenon is referred to as **time dilation** because time is "spread out" in the moving reference frame. Were you to compare clocks with an observer moving at 9/10 the speed of light (0.9*c*), you would find that the other observer's clock was running less than half as fast as your clock (about 0.44 times as fast).[3] You might guess that to the other observer, your clock would be fast, but actually the other observer would find instead that it is *your* clock that was running slow! A bit of thought shows why it must be this way. To you, the other observer may be moving at 0.9*c*, but to the other observer, *you* are moving. Either frame of reference is equally valid, so it stands to reason that if a clock in a moving reference frame runs slow, then you would each find the other's clock to be slow. An everyday scientific experience illustrates this effect. Fast particles

called cosmic ray muons provide an example of time dilation, as illustrated in **Figure 4.13**. Cosmic ray muons are produced at about the 15-km level in Earth's atmosphere when high-energy primary cosmic rays strike atmospheric atoms or molecules. Muons at rest decay very rapidly into other particles. Within 2.2 microseconds, half of all muons will have changed their identity. This happens so quickly that, even traveling at the speed of light, virtually all muons would have decayed long before traveling the 15 km to reach Earth's surface. However, time dilation causes the muon clock to run slower, so they live longer and can travel farther. That is why we are able to detect cosmic ray muons on the ground.

4. **"At the same time" is a relative concept.** Two events that occur at the *same* time for one observer may occur at *different* times for a different observer. Hold out your arms and snap the fingers on both hands at the same time. For you, the two snaps were simultaneous. But to an observer moving by you from right to left at nearly the speed of light, you snapped the fingers of your left hand first and the fingers of your right hand later.

5. **An object in motion is shorter than it is at rest.** More specifically, moving objects are compressed in the direction of their motion. A meter stick moving at 0.9*c* is only 43.6 cm (centimeters) long.[4]

---

[3] The factor by which time is dilated and space is contracted is given by $\frac{1}{\sqrt{1-\frac{v^2}{c^2}}}$. This factor is often referred to as γ.

[4] See footnote 3.

These different consequences of relativity can be combined in what is often called the *twin paradox.* You head off on a trip to the center of the Milky Way Galaxy, roughly 25,000 light-years distant. Your spectacularly powerful star drive accelerates your ship up to 0.9999999992c. To you, the galaxy is moving by at this speed, so the 25,000 light-year

---

The twin paradox illustrates many aspects of relativity.

---

distance to the center of the galaxy is compressed by a factor of 25,000 to a distance of a single light-year (see number 5 in the previous list). At your speed, you cross this distance in a single year. You snap a picture of the gas swirling around the black hole at the center of the galaxy, then turn around to head home and show it to your twin. Again, the return trip takes only a year. So in two years you have traveled to the center of the galaxy and back again. (Who says interstellar travel is such a big deal?) But when you return, you find that your twin died 50,000 years ago. In the reference frame of Earth, your spacecraft crossed the 25,000 light-year distance to the center of the galaxy moving at just under the speed of light. The only reason you survived the journey, according to an Earth-bound observer, is because in your moving frames of reference time ran extremely slow (number 3 in our list). Each leg of the two-way journey took 25,000 years to observers on Earth, and your twin just could not wait that long.

You might puzzle over the twin paradox a bit. Both on the way out and on the way back, in your reference frame it is the clocks on Earth that are running slowly, so you are aging *faster* than your twin. Yet when you return, more time has passed for your twin than for you. How can this be? The answer is that, unlike your twin, you *changed reference frames* during your trip. Event 1 is when you left Earth, and event 2 is when you returned to Earth. Your twin went from one event to the other, riding along in Earth's frame of reference. You, on the other hand, changed reference frames when you left Earth, changed again when you stopped at the center of the galaxy, changed a third time when you left the galactic center to return home, and changed reference frames one final time when you arrived back at Earth. It happens that the path through spacetime that you followed between the two events involved the passage of only two years of what you experienced as time, while your twin's path involved 50,000 years of what your twin experienced as time.

Another way to view this is that the key difference between you and your twin is that you experienced acceleration during your trip, while your twin did not. When two observers are in *uniform motion* relative to one another, *neither* of them can lay claim to being in a unique frame of reference. However, *acceleration is a real phenomenon.* You *feel* acceleration when you are riding in a car, and you would surely *feel* the acceleration of the spaceship in this

example. It is the fact that you experienced an acceleration that allowed you to "outlive" your twin.

4.4 Light Is a Wave, but It Is Also a Particle

Maxwell may have achieved great success with his theory of electromagnetic waves, and he may have opened the crack in Newtonian physics that led to the theory of relativity; but Maxwell's accomplishments themselves would soon need serious revision. As mentioned earlier, from early on scientists disagreed over whether light consists of waves or particles. Maxwell's work seemed to put the issue to rest by showing that light is an electromagnetic wave; but before too many years had gone by, the particle description raised its head again. Although the electromagnetic wave theory of light has had many successes in describing phenomena, there are also many phenomena that it does not describe well. These range from the presence of sharp bright and dark "lines" at specific wavelengths in the light from some objects, to the shape of the continuous spectrum of light emitted by a lightbulb. Many of these difficulties with the wave model of light have to do with the way in which light interacts with atoms and molecules.

Scientists working in the late 19th and early 20th centuries discovered that many of the puzzling aspects of light could be better understood if light energy came in discrete packages. In 1905 Einstein published a paper in which he argued that light consists of particles. He based his argument on the **photoelectric effect**, the emission of electrons from surfaces illuminated by electromagnetic radiation above a certain frequency. Einstein showed that the *rate* at which electrons are ejected depends only on the *intensity* of the incident radiation, and that the electron *velocity* depends only on the *frequency* of the incident radiation.[5] Effectively, scientists were reintroducing the particle picture of light. In some ways the particle description of light is easier to think about than its wave description. In this model we think about light as being made up of particles called **photons** (*phot-* means "light," as in *photograph,* and *-on* signifies a particle, as in *electron, neutron,* and *proton*). Photons always travel at the speed of light, and they carry energy. (After our discussion of relativity in Section 4.3, you may wonder how a particle can travel at the speed of light. The answer is that a photon has no mass. A massless particle can travel *only* at the speed of light.)

---

[5] It is interesting to note that it was for his work on the photoelectric effect, not special or general relativity, that Einstein received the Nobel Prize in 1921.

The particle description of light is tied to the wave description of light by a relationship between the energy of a

The energy of a photon is proportional to its frequency.

photon and the frequency or wavelength of the wave. The higher the frequency of the electromagnetic wave, the greater the energy carried by each photon. Specifically we write

$$E = hf \quad \text{or} \quad E = \frac{hc}{\lambda}.$$

The h in this equation is called **Planck's constant** and has the value $h = 6.63 \times 10^{-34}$ joule-second. (Planck's constant is named after the German physicist **Max Planck**, 1858–1947.) According to the particle description of light, the electromagnetic spectrum is a spectrum of photon energies. Photons of shorter wavelength (higher frequency) carry more energy than photons of longer wavelength (lower frequency). For example, photons of blue light carry more energy than photons of longer-wavelength red light. Ultraviolet photons carry more energy than photons of visible light, and X-ray photons carry more energy than ultraviolet photons. The lowest-energy photons are radio wave photons.

The **intensity** of light measures the *total* amount of energy that a beam of the light carries. A beam of red light can be just as intense as a beam of blue light—that is, it can carry just as much energy—but because the energy of a red photon is less than the energy of a blue photon, it will take

Photons are the *quantum mechanical* description of light.

more red photons to reach that intensity than it would take blue photons. This relationship is a lot like money. A hundred dollars is a hundred dollars, but it takes a lot more pennies (low-energy photons) to make up a hundred dollars than it takes 50-cent pieces (high-energy photons).

When physicists speak of the energy of light as broken into discrete packets called photons, they say that the light energy is **quantized**. The word *quantized,* which has the same root as the word *quantity,* means that something is subdivided into discrete units. A photon is referred to as a **quantum of light**. The branch of physics that deals with the quantization of energy and of other properties of matter is called **quantum mechanics**.

Quantum mechanics, like special relativity, is counterintuitive for humans, but its predictions have been confirmed over and over again by experiment. The conflict between everyday, commonsense ideas about the world and the world as revealed through modern science is discussed in **Connections 4.2**.

Atoms Can Occupy Only Certain Discrete Energy States

If we want to understand better how light interacts with matter, we need to start by pinning down exactly what we mean by *matter* in the first place. To a physicist, matter is anything that occupies space and has mass. Virtually *all* of the matter we have direct experience with is composed

Virtually all matter we encounter is composed of atoms.

of **atoms**. The computer keyboard this book is being typed on is made of atoms, and the neurons in your brain that are changing their structure as you read are made of atoms. Atoms are incredibly tiny—so tiny that a single teaspoon of water contains about 10^{23} atoms. (There are more atoms in a single teaspoon of water than there are stars in the observable universe.) When we talk about the interaction of light with matter, what we are really talking about is the interaction of light with atoms, and the things atoms themselves are composed of. So the next question is, "What are atoms?"

Atoms are built from three types of **elementary particles** as illustrated in **Figure 4.14(a)**. Sitting in the center of the atom is the **nucleus**, which is composed of positively charged **protons** and electrically neutral **neutrons**. An atom may have many protons and neutrons in its nucleus. Surrounding the nucleus of the atom are negatively charged **electrons**. For an atom to be electrically neutral, it must have the same number of electrons as protons. Electrons have much less mass than protons or neutrons, so almost all the mass of an atom is found in its nucleus. This naturally leads to a mental picture of an atom as a "tiny solar system," with the massive nucleus sitting in the center and the smaller electrons orbiting about much as planets orbit about the Sun (**Figure 4.14(b)**). We refer to this as the **Bohr model** after the Danish physicist **Niels Bohr** (1885–1962), who proposed it in 1913.

Unless you have thought about atoms a great deal, this is probably your concept of the structure of an atom. It is much the same picture that scientists in the early 20th century held as well. But it has a fatal problem. In this view, an electron whizzing about in an atom is constantly undergoing an acceleration—the direction of its motion is constantly changing. The wave description of electromagnetic radiation says that *any* electrically charged particle that is accelerating must also be giving off electromagnetic radiation. This electromagnetic radiation should be carrying away the orbital energy of the electron. (Imagine that electron as the wiggling electric charge in Figure 3.3(b).) If you calculate how much energy should be carried off by radiation from the electron, you find that only a tiny fraction of a second

Thinking Outside the Box

As you read about the combined wave and particle description of light, you will likely find yourself scratching your head in confusion over exactly what light really is. If you do, consider yourself in good company. The scientists who invented the seemingly bizarre quantum description of nature had a great deal of trouble thinking about light as well. The wave model of light is clearly the correct description to use in many instances, just as Maxwell has shown. At the same time the particle description of light is also clearly the correct description to use in other cases, as scientists like Planck and Einstein demonstrated. But how can the same thing—light—be both a wave *and* a particle? It is hard for us to imagine a single thing sharing the properties of a wave on the ocean *and* a beach ball, yet light does just that.

Our trouble with thinking of light as both a wave and a particle only hints at the puzzling and philosophically troublesome world of quantum mechanics. As we go further, things only get worse. Light is not the only thing that shares wave and particle properties. In fact, *all* matter shares wave and particle properties. Sometimes a "particle" such as an electron behaves as if it were a wave, while at other times a "wave" of light clearly exhibits the properties of a discrete particle. Early quantum physicists would sometimes joke that on Monday, Wednesday, and Friday, light and matter were particles, whereas on Tuesday, Thursday, and Saturday, light and matter were waves. (And on Sunday it was best just not to think about them at all!)

Light is what light is, and an electron is what an electron is. The trouble with quantum mechanics lies not with the nature of reality, but with what our brains can easily think about. This chapter earlier provided another

example of the limitations of our genetic programming. Our brains deal well with objects that are sitting still or even moving as fast as a hard-hit fly ball. Basically our brains cope best with things moving at the speeds of things in nature that we might want to eat or that might want to eat us. (Animals whose brains could not deal with such speeds tended not to survive long enough to pass their genes for those brains on to future generations.) But the brains of our ancestors did not have to deal with things moving at nearly the speed of light, so we should not be surprised that special relativity seems to defy our intuition. Likewise, there was no evolutionary pressure for our ancestors to be able to think easily about the wave/particle duality of light and matter.

Quantum mechanics and special relativity are not the only places where our ease in thinking about nature breaks down. Quantum mechanics deals with the very smallest scales in nature. At the other extreme, our brains did not evolve to think about things as large or as massive as stars and galaxies and the universe. When we move on to these larger scales later in the book, we will find our ideas about the nature of space and time themselves further challenged as we seek ways of visualizing curved spacetime or understanding why the question "What came before the beginning of the universe?" is in some ways much like asking, "What was to the left of last Thursday?"

Our brains exist in a box that is defined by the experiences and circumstances that we and our ancestors had to cope with. One exciting thing about modern physics and astronomy is that they force us to break down the walls of that conceptual box and find tools for understanding what lies beyond its boundaries.

should be needed for the electrons in an atom to lose all their energy and fall into the atom's nucleus! Fortunately for us this does not happen. Atoms exist for very long periods of time, and electrons never "fall into" the nuclei of atoms. So something must be wrong with this concept of an atom. A way out of this difficulty came when scientists realized that, just as waves of light have particlelike properties, so too do particles of matter have wavelike properties. With this realization, the *miniature solar system* model of the atom was

modified so that a positively charged nucleus is surrounded *not* by planetlike electrons moving in their orbits, but by electron "clouds" or electron "waves" as illustrated in **Figure 4.14(c)** and discussed further in **Foundations 4.1**.

The strings on a guitar can vibrate only at certain discrete frequencies, giving rise to the discrete notes we hear. In much the same way, the electron waves in an atom can assume only certain specific forms. So instead of being able to take on *any* arbitrary energy, atoms can absorb or emit

(a) Parts of an atom

(b) "Solar system" model

(c) Quantum mechanical model

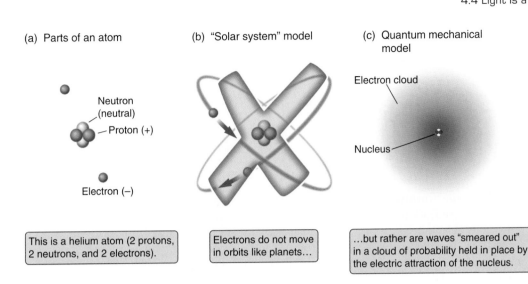

This is a helium atom (2 protons, 2 neutrons, and 2 electrons).

Electrons do not move in orbits like planets...

...but rather are waves "smeared out" in a cloud of probability held in place by the electric attraction of the nucleus.

FIGURE 4.14 (a) An atom is made up of a nucleus consisting of positively charged protons and electrically neutral neutrons, surrounded by less massive negatively charged electrons. (b) Atoms are often drawn as miniature "solar systems," but this model is incorrect. (c) Electrons are actually smeared out around the nucleus in quantum mechanical clouds of probability.

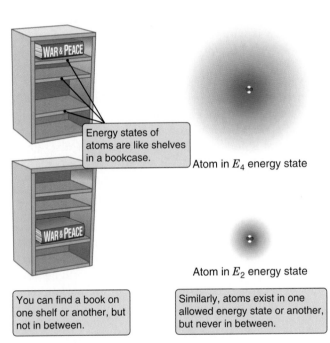

Energy states of atoms are like shelves in a bookcase.

Atom in E_4 energy state

Atom in E_2 energy state

You can find a book on one shelf or another, but not in between.

Similarly, atoms exist in one allowed energy state or another, but never in between.

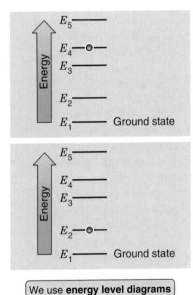

We use **energy level diagrams** to represent the allowed states of an atom.

FIGURE 4.15 Atoms can have only certain discrete energies.

only certain specific energies corresponding to the allowed waveforms of their electron clouds. A given atom may have a tremendous number of different energy states available to it, but these states are *discrete.* An atom might have the en-

Atoms can have only certain discrete energies, much as guitar strings can play only certain notes.

ergy of one of these allowed states, or it might have the energy of the next allowed state, *but it cannot have an energy somewhere in between.* We can imagine the energy states of atoms as being a bookcase with a series of shelves as

shown in **Figure 4.15**. The energy of an atom might correspond to the energy of one shelf or to the energy of the next shelf; but the energy of the atom will *never* be found *between* the two shelves.

The lowest possible energy state of an atom—the "floor" —is called the **ground state** of the atom. Allowed states with energies lying above the ground state are called **excited states** of the atom. When the atom is in its ground state, it has nowhere to go. An electron cannot "fall" into the nucleus because there is no allowed state there with less energy for it to occupy. It cannot move up to a higher-energy state without getting some extra energy from somewhere. For this reason an atom will remain in its ground

Uncertainty Is Ordinary in the Quantum World

In the world of the very small, nothing seems intuitive. We have learned that all electromagnetic radiation behaves as both waves and particles. Possibly more surprising was the discovery that things like electrons and protons, which you may have visualized as "solid" particles, also have wave characteristics. There is, of course, a nice symmetry here. Waves have particlelike characteristics and particles have wavelike characteristics. This is not just a curious observation. It has huge implications in both science and technology. For instance, the wave–particle property is the principle by which electron microscopes work, as you will see in the next chapter.

Possibly the most significant implication and important outcome of wave–particle duality is the famous **Heisenberg uncertainty principle**, named for the German physicist **Werner Heisenberg** (1901–1976). If particles have wave characteristics, you cannot simultaneously pin down both their exact location and their **momentum**. There will *always* be some uncertainty in one or the other. Momentum (p) is defined as the product of mass and velocity ($p = m \times v$). Keep in mind that *velocity* includes both *speed* and *direction*. We can be more quantitative about this. The product of the uncertainty in a particle's position (Δx) and the uncertainty in its momentum (Δp) is always equal to or greater than a particular constant, which is of the order of Planck's constant, h. We can express this as a simple equation: $\Delta x \times \Delta p \sim h$. In other words, the more you know about *where* something is (Δx approaching zero), the less you can know about *how fast* and *in what direction* it is moving (Δp approaching infinity). Conversely, the better you know the momentum of something, the less you know about its location. This is not a matter of scientists making inferior measurements. You simply *cannot* do better, no matter how precisely you measure!

How does the uncertainly principle work in the real world? You should not be surprised to find that the best examples occur in the realm of subatomic particles. Earlier in this chapter we talked about the Bohr model of the hydrogen atom and imagined electrons as particles sailing around the proton in well-behaved orbits. The Bohr model says the angular momentum of the electron in an orbit has to be given *exactly* by an integer times a constant. There is no room for any uncertainty here. Yet the uncertainty principle must be obeyed. It tells us that if the angular *momentum* has no uncertainty, the angular *position* of the electron must be *completely uncertain*! That is right—you cannot know where the electron is in its orbit. This is why we use a featureless cloud to represent electrons in orbit around an atomic nucleus, as shown in Figure 4.14(c).

Another example involves a different form of the Heisenberg uncertainty principle: $\Delta E \times \Delta t \sim h$, where ΔE is the uncertainty of energy and Δt is the time over which the energy is measured. In Section 4.4 we make the point that atoms can occupy only certain discrete energy states, implying that their electrons are restricted to specific well-defined excited states. But can there really be *no* uncertainty whatsoever in these energy states? The answer becomes apparent if we rewrite the previous equation as $\Delta t \sim h/\Delta E$. If we say that $\Delta E = 0$ (no uncertainty in the energy state), then Δt becomes infinite.[6] If an electron were forced to stay at some excited state forever, it could never drop to a lower level, and we would never see narrow spectral emission lines. We do, of course, see emission lines, so there *must* be a certain amount of uncertainty in the electron's energy. Remember that wavelength is related to energy. A narrow range in energy therefore represents a narrow range in wavelength. So the longer an electron resides in an elevated energy state (Δt is large), the narrower will be the spectral emission line when it finally drops to a lower level (ΔE, and therefore $\Delta \lambda$, is small).

The bottom line here is that we can be absolutely *certain* that there is *uncertainty* at the root of *everything* physical. If you are bothered by this, you are in good company. It *really* bothered Einstein too.

[6]If you divide any number by zero, the result will be infinite.

state forever unless something happens to knock it into an excited state. A book sitting on the floor has nowhere left to fall, and it cannot jump to one of the higher shelves of its own accord.

An atom in an excited state is a very different matter, however. Just as a book on an upper shelf might fall to a lower shelf, an atom in an excited state might **decay** down to a lower state by getting rid of some of its extra energy. An important difference between the atom and the book on the shelf, however, is that whereas a snapshot might catch the book between the two shelves, the atom will never be caught between two energy states. When the transition from one state to another occurs, the energy difference between the two states must be carried off all at once. A common way for an atom to do this is to give off a photon. But not just any photon will do. The photon emitted by the atom must carry away exactly the amount of energy lost by that atom as it goes from the higher-energy state to the lower-energy state.

The Energy Levels of an Atom Determine the Wavelengths of Light It Can Emit and Absorb

To better understand the relationship between the energy levels of an atom and the radiation it can emit or absorb, imagine a hypothetical atom that has only two available energy states. Call the energy of the lower-energy state (the ground state) E_1 and the energy of the higher-energy state (the excited state) E_2. The energy levels of this atom can be represented in an energy level diagram like those in Figure 4.15, but with only two levels (see **Figure 4.16(a)**).

To understand the process of *emission*, imagine that the atom begins in the upper state (E_2) and then spontaneously drops down to the lower-energy state (E_1). This is shown in **Figure 4.16(b)**, where the downward arrow indicates that the atom went from the upper state to the lower state. The atom just lost an amount of energy equal to the difference between the two states, or $E_2 - E_1$. However, energy is never truly lost or created, so the energy lost by the atom has to show up somewhere. In this case, the energy

> When an atom drops to a lower energy state, the lost energy is carried away as a photon.

shows up in the form of a photon that is emitted by the atom. The energy of the photon emitted must just match the energy lost by the atom, so the energy of the photon must be $E_{photon} = E_2 - E_1$.

We have already seen the relationship between the energy of a photon and the frequency or wavelength of electromagnetic radiation. Using this relationship we can say that the frequency of the photon emitted by a transition from E_2 to E_1, which we will denote as $f_{2 \to 1}$, is just the energy difference divided by Planck's constant (h):

$$f_{2 \to 1} = \frac{E_{photon}}{h} = \frac{E_2 - E_1}{h}.$$

FIGURE 4.16 (a) The energy levels of a hypothetical two-level atom. (b) A photon with energy $hf = E_2 - E_1$ is emitted when an atom in the more energetic state decays to the lower-energy state.

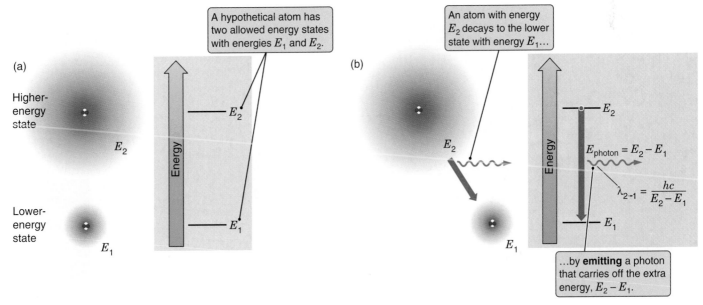

Similarly, the wavelength of the photon is just $\lambda = c/f$, or

$$\lambda_{2 \to 1} = \frac{c}{f_{2 \to 1}} = \frac{hc}{E_2 - E_1}.$$

This shows that the wavelengths of photons emitted by an atom—the color of the light that the atom gives off—are determined by the energy level structure of the atom. An atom can emit photons with energies corresponding only to the difference between two of its allowed energy states.

Imagine what the light coming from a cloud of gas consisting of our hypothetical two-state atoms would be like. This case is illustrated in **Figure 4.17**, which shows a collection of our two-state atoms. Any atom that finds itself in the upper energy state (E_2) will quickly decay and emit a photon in some random direction. A cubic meter of the air around you contains about 10^{25} atoms. Even if only a tiny fraction of these atoms emit a photon each second, an enormous number of photons would still come pouring out of the cloud of gas. But instead of containing photons of all different energies (that is, light of all different colors), like sunlight, this light would instead contain only photons with the specific energy $E_2 - E_1$ and wavelength $\lambda_{1 \to 2}$. In other words, all of the light coming from the cloud would be the same color.

We have all seen what happens to sunlight when it passes through a prism. Sunlight contains photons of all different colors, so when sunlight passes through a prism, it spreads out into all colors of a rainbow. But if we were to pass the light from our cloud of gas through a slit and a prism, as in Figure 4.17, the results would be very different. This time there would be no rainbow. Instead all of the light from the cloud of gas would show up on the screen as a single bright line. The process we have just described—the production of a photon when an atom decays to a lower-energy state—is

The spectrum of a cloud of glowing gas contains emission lines.

referred to as **emission**. The bright, single-colored feature in the spectrum of the cloud of gas is referred to as an **emission line**.

So far in this discussion we have ignored an important question: "How did the atom get to be in the excited state E_2 in the first place?" An atom sitting in its ground state will remain in the ground state unless it is somehow given just the right amount of energy to kick it up to an excited state. Most of the time this extra energy comes in one of two forms: (1) The atom absorbs the energy of a photon (we will talk about this possibility shortly); or (2) the atom collides with another atom, or perhaps an unattached electron, and the collision knocks the atom into an excited state. This is how a neon sign works. When a neon sign is turned on, an alternating electric field is set up inside the glass tube that pushes electrons in the gas back and forth through the neon gas inside the tube. Some of these electrons crash into atoms of the gas, knocking them into excited states. The atoms then drop back down to their ground states by emitting photons, causing the gas inside the tube to glow. (In like fashion, an electron beam in a television picture tube collides with atoms in the screen, knocking them into excited states. When those atoms decay they emit the photons, producing what we perceive as the picture on the television.)

FIGURE 4.17 A cloud of gas containing atoms with two energy states, E_1 and E_2, emits photons with an energy $E = hf = E_2 - E_1$, which appear in the spectrogram (right) as an emission line.

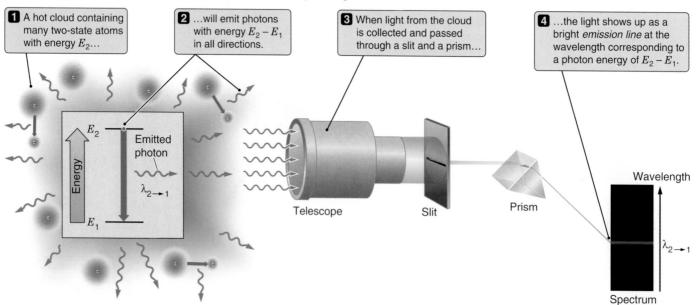

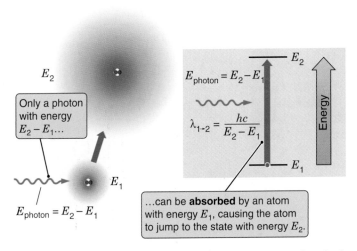

FIGURE 4.18 A photon of energy $hf = E_2 - E_1$ may be absorbed by an atom in the lower-energy state, leaving the atom in the higher-energy state.

So far we have focused on the emission of photons by atoms in an excited state, but what about the opposite process? An atom in a low-energy state can absorb the energy of a passing photon and jump up to a higher-energy state, as shown in **Figure 4.18**, but not just any photon can be absorbed by the atom. As before, the energy that it takes to get from E_1 to E_2 is the difference in energy between the two states, or $E_2 - E_1$. For a photon to cause an atom to jump from E_1 to E_2, it must provide just this much energy. Using the relationship that $E_{photon} = hf$ or $f = E_{photon}/h$, we find that the *only* photons capable of exciting atoms from E_1 to E_2 are photons whose frequency and wavelength are, respectively,

$$f_{1\to2} = \frac{E_{photon}}{h} = \frac{E_2 - E_1}{h}.$$

and

$$\lambda_{1\to2} = \frac{c}{f_{1\to2}} = \frac{hc}{E_2 - E_1}.$$

This is exactly the same energy photon—the same color of light—that is emitted by the atoms when they decay from E_2 to E_1. This is not a coincidence. The energy difference

Atoms can absorb photons with the same energies as the photons they emit.

between the two levels is the same whether the atom is emitting a photon or absorbing one, so the energy of the photon involved will be the same in either case.

What might the spectrum of light look like when viewed through a cloud composed of our hypothetical gas of two-state atoms? If we shine photons of all different wavelengths (that is, light of all different colors) through the gas from

one side, almost all of these photons will pass through the cloud of gas unscathed. So counting how many photons come *out* the other side of the cloud of gas should give us the number of photons that we shined *into* the gas. There is only one exception. Rather than passing through the gas, some of the photons with just the right energy ($E_2 - E_1$) might instead be absorbed by atoms.

If we shine light from a lightbulb directly though a glass prism, we will see that a rainbow of colors comes out as shown in **Figure 4.19(a)**. If we instead shine the light through a cloud of our two-state atoms before putting the light through a prism, the rainbow of colors will be unchanged

When viewed through a cloud of gas, the spectrum of a lightbulb contains absorption lines.

except for one detail. Because some of the photons with energies equal to $E_2 - E_1$ will have been absorbed by the gas, these photons will be missing in the light passing through the prism. If we look at the screen, we will see a sharp, dark line at the color corresponding to these photons (**Figure 4.19(b)**). The process of atoms capturing the energy of passing photons is referred to as **absorption**, and the dark feature seen in the spectrum is called an **absorption line**.

There is one final point worth making before leaving the subject of emission and absorption of radiation. When an atom absorbs a photon and jumps up to an excited energy state, there is a good chance that the atom will quickly decay back down to the lower-energy state by emitting a photon with the same energy as the photon it just absorbed. If the atom reemits a photon just like the one it absorbed, you might reasonably ask why the absorption really matters. After all, the photon that was taken out of the passing light was replaced, was it not? The answer is yes and no. The photon was replaced, true enough, but while all of the photons that were absorbed were originally traveling in the *same direction*, the photons that are reemitted travel off in *random directions*. In other words, some of the photons with energies equal to $E_2 - E_1$ are in effect diverted from their original paths by their interaction with atoms. If you look at a lightbulb *through* the cloud, you will notice an absorption line at a wavelength of $\lambda_{1\to2}$; but if you look at the cloud from the side (looking perpendicular to the original beam), you will see it as a glowing light with an emission line at this wavelength.

Emission and Absorption Lines Are the Spectral Fingerprints of Types of Atoms

In the previous section we used a hypothetical atom with only two allowed energy states to help us think about emission and absorption of photons. Real atoms have many more than just two possible energy states that they might occupy,

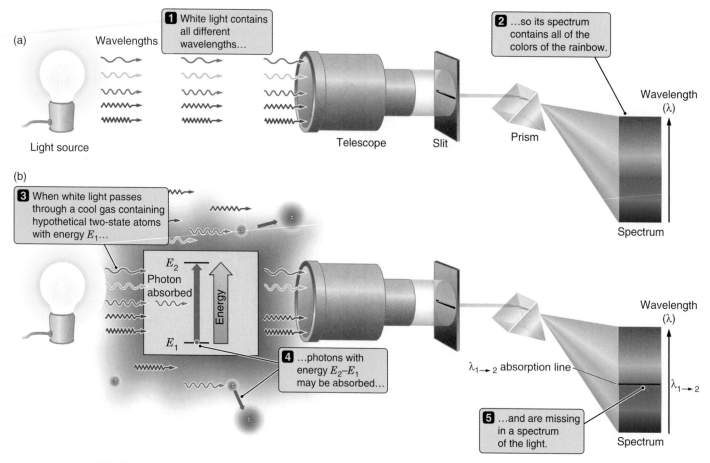

FIGURE 4.19 (a) When passed through a prism, white light produces a spectrum containing all colors. (b) When light of all colors passes through a cloud of hypothetical two-state atoms, photons with energy $hf = E_2 - E_1$ may be absorbed, leading to the dark absorption line in the spectrogram.

so a given type of atom will be capable of emitting and absorbing photons at many different wavelengths. An atom with three energy states, for example, might jump from state 3 to state 2, or from state 3 to state 1, or from state 2 to state 1. The emission lines from such a gas might have wavelengths of $hc/(E_3 - E_2)$, $hc/(E_3 - E_1)$, and $hc/(E_2 - E_1)$.

The allowed energy states of an atom are determined by the complex quantum mechanical interactions among the electrons and the nucleus that comprise the atom. Every hydrogen atom consists of a nucleus containing one proton, plus a single electron in a cloud surrounding the nucleus. As a result, every hydrogen atom has the same energy states available to it. It follows that all hydrogen atoms are capable of emitting and absorbing photons with the same wavelengths. **Figure 4.20(a)** shows the energy level diagram of hydrogen, along with the spectrum of emission lines for hydrogen in the visible part of the spectrum (**Figures 4.20(b)** and **(c)**).

Every hydrogen atom has the *same* energy states available to it, so all hydrogen atoms are in principle capable of producing the same spectral lines. But the energy states of a hydrogen atom are *different* from the energy states avail-

able to a helium atom, a lithium atom, or a boron atom, just as the energy states of these kinds of atoms differ from each other. Each different type of atom has a unique set of available energy states and therefore a unique set of wavelengths

> The wavelengths at which atoms emit and absorb radiation form unique spectral fingerprints for each type of atom.

at which it can emit or absorb radiation. **Figure 4.20(d)** shows the set of emission lines that are given off by discharge tubes (like those in a neon sign) containing different kinds of atoms. These unique sets of wavelengths serve as unmistakable spectral fingerprints for each type of atom.

Spectral fingerprints are of crucial importance to astronomers. They let us figure out what types of atoms (or molecules) are present in distant objects by doing nothing more than looking at the spectrum of light from those objects. If we see the spectral lines of hydrogen, or helium, or carbon, or oxygen, or any other element in the light from a distant object, then we know that some of that element is present in that object. The strength of a line is determined in part by how many atoms of that type are present in the source. By measuring the strength of the lines from different types of atoms in the spectrum of a distant object, astronomers can often infer the relative amounts of different types of atoms of which the object is composed. But it gets even better. The fraction of atoms of a given kind that are in some particular energy state (as opposed to some other energy state) is often determined by factors such as the temperature or the **density** of the gas. By looking at the relative strength of different lines from the same kind of atom, it is often possible to determine the temperature, density, and pressure of the material as well.

How Are Atoms Excited, and Why Do They Decay?

In the last section we sidestepped an aspect of the emission process that has troubled physicists and philosophers alike since the earliest days of quantum mechanics. To appreciate this question, return to the analogy between emission of a photon and a book falling off a shelf. If we place a book on a level shelf and do not disturb it, the book will sit there forever. Once the book is resting on the upper shelf, something must *cause* the book to fall off the shelf. So what about the atom? Once an atom is in an excited state, what causes it to jump down to a lower-energy state and emit a photon? What triggers the event? Sometimes an atom in an upper-energy state can be "tickled" into emitting a photon —a process called *stimulated emission*—but under most circumstances the answer is that *nothing causes the atom to jump to the lower-energy state.* There is no trigger. Instead the atom decays *spontaneously.* And while we can say *about* how long the atom is *likely* to remain in the excited state, the rules of quantum mechanics say (and experiment shows) that we cannot know exactly when a given atom will decay until *after* the decay has happened. The atom decays at some *random* time that is not influenced by anything in the universe and cannot be known ahead of time.

FIGURE 4.20 (a) The energy states of a hydrogen atom are shown. Decays to level E_2 emit photons in the visible part of the spectrum. (b) This is what you might see if you looked at the light from a hydrogen lamp projected through a prism onto a screen. (c) This graph of the brightness of lines versus their wavelength is an example of how spectra are traditionally plotted. (d) Emission lines from several other types of gases: helium, argon, neon, and sodium.

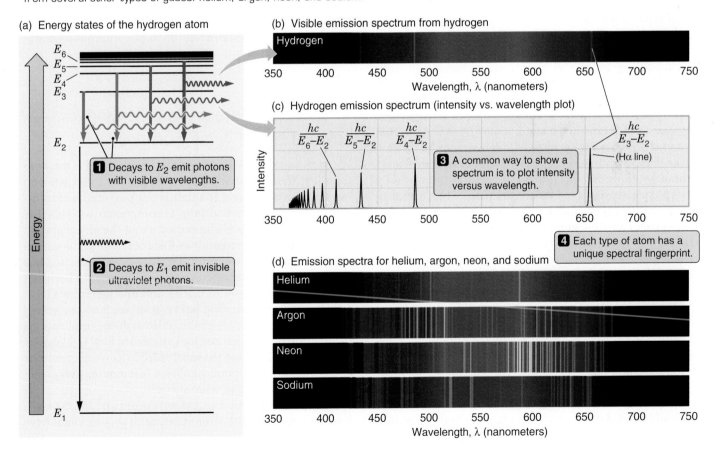

You have seen many examples of this rather amazing phenomenon. For example, you have probably owned a "glow in the dark" toy or a watch with a "glow in the dark" face. Photons in sunlight or from a lightbulb are absorbed by the atoms in the watch face, knocking those atoms into excited energy states. Unlike many excited energy states of atoms that tend to decay in a small fraction of a second, the excited states of the atoms in the watch face instead tend to live for many seconds before they decay. Suppose, for example, that on average these atoms tend to remain in their excited state for 1 minute before decaying and emitting a photon. In other words, suppose that if we wait for 60 seconds, there is a 50–50 chance that any particular atom will have decayed and a 50–50 chance that the atom will remain in its excited state. There are trillions upon trillions of such atoms in the watch face. Although it is impossible to say exactly which atoms in the watch face will decay after a minute, we can say with certainty that *about* half of them *will* decay within 60 seconds. From the standpoint of the glow that we see from the watch, it makes little practical difference which half of the atoms decay and which half do not. All we need to know is that if we wait 1 minute, half of the atoms will have decayed, and the brightness of the glow from the watch will have dropped to half of what it was. If we wait another minute, half of the remaining excited atoms will decay, and the brightness of the glow will be cut in half again. Each 60 seconds, half of the remaining excited atoms decay, and the glow from the watch drops to half of what it was 60 seconds earlier. The glow from the watch slowly fades away.

We have now come upon one of the most philosophically troubling aspects of quantum mechanics. In deep space, where atoms can remain undisturbed for long periods of time, there are certain excited states of atoms that, on average, live for tens of million years or even longer. Envision an atom in such a state. It may have been in that excited state for a few seconds, a few hours, or 50 million years when in an instant it decays to the lower-level state *without anything causing it to do so.* Newton and virtually every physicist who lived before the turn of the 20th century envisioned a clockwork universe in which every effect had a cause. They imagined that if we knew the exact properties of every bit of the universe today, it was just a matter

---

Quantum mechanics undermines the orderly, causal universe of Newtonian physics.

---

of turning the crank on the laws of physics to predict what the state of the universe would be tomorrow. Then quantum mechanics came along and turned this view on its head. Instead of dealing with strict cause-and-effect relationships, physicists found themselves calculating the *probabilities* of certain events taking place and facing fun-

damental limitations on what can ever be known about the state of the universe. We have mentioned that while Einstein helped start the scientific revolution of quantum mechanics, in the end that revolution left him behind. He could never shake his firm belief in Newton's clockwork, causal universe. "God does not play dice with the universe!" he insisted emphatically. As more of the predictions of quantum mechanics were borne out by experiment, most physicists came to accept the implications of the strange new theory. Einstein, on the other hand, went to his grave looking unsuccessfully for a way to save his notion of order in the universe.

It is interesting to note that although Einstein refused to accept quantum mechanics, our understanding of the quantum mechanical nature of reality owes him a great debt. His was one of the greatest minds of all time, and as he searched tirelessly for flaws in quantum mechanics, he presented challenge after challenge to those who were trying to work out the details of the new theory. It was in responding to Einstein's objections that physicists were forced to confront the full implications of their own work. As an epilogue to this story, this struggle continues to this day. At the dawn of the 21st century a few theoretical physicists are still pursuing Einstein's dream, trying to recast quantum mechanics in a way that recaptures the strict causality that seemed irretrievably lost shortly after the beginning of the 20th century. So far they have had little success, and most physicists doubt that they ever will. However, like Einstein before them, their healthy skepticism has led to ever deeper understanding of the implications and limitations of the theory.

The Doppler Effect—Is It Moving Toward Us or Away from Us?

We have begun to see that to an astronomer light is far more than just the stuff that bounces off the page and lets you read these words. Light is a tightly packed bundle of information that when spread into its component wavelengths can reveal a wealth of information about the physical state of material located tremendous distances away. The nature of light shapes how we think about space and time and has forced physicists to abandon many of their most cherished ideas about the nature of matter and energy. Yet we have only begun to explore what light can tell us. It is time to step back from the precipice of the philosophical implications of quantum mechanics and look instead at how light can be used to measure one of the most straightforward questions about a distant astronomical object: Is it moving away from us or toward us, and at what speed?

Have you ever stood on a street corner and listened as a fire truck sped by with sirens blaring? If so, you might have noticed something funny about the way the siren sounded. As the fire truck came toward you, its siren had a certain high pitch; but as it passed by, the pitch of the siren dropped noticeably. If you were to close your eyes and listen, you would have no trouble knowing when the fire truck passed, just from the change in pitch of its siren. You do not even need a fire engine to hear this effect. The sound of normal traffic behaves in the same way. As a car drives past, the pitch of the sound that it makes suddenly drops.

The pitch of a sound is like the color of light. It is determined by the wavelength or frequency of the wave. What we perceive as higher pitch corresponds to sound waves with higher frequencies and shorter wavelengths. Sounds that we perceive as lower pitch are waves with lower frequencies and longer wavelengths. When an object is moving toward us, the waves that it gives off "crowd together" in front of the object. You can see how this works by looking

The motion of a source toward or away from us changes the wavelength of the waves reaching us.

at **Figure 4.21**, which shows the locations of successive wave crests given off by a moving object. If you are standing in front of an object moving toward you, the waves that reach you have a shorter wavelength and therefore a higher frequency than the waves given off by the object when it is not moving. In the case of sound waves, the sound reaching you from the object has a higher pitch than the sound given off by the object if it were stationary. Conversely, if an object

is moving away from you, the waves reaching you from the object are spread out. (Again, refer to Figure 4.21.) In the case of sound, this means that the pitch of the sound drops, in line with our experience with the fire truck.

This same phenomenon, which is referred to as the **Doppler effect**, occurs with light as well as with sound. If an object is moving toward you, the light reaching you from

Light from approaching objects is blueshifted. Receding objects are redshifted.

the object has a shorter wavelength than the light emitted by the object. In other words, the light that you see is bluer than if the source were not moving toward you. We say that the light from an object moving toward us is **blueshifted**. On the other hand, light from a source that is moving away from you is shifted to longer wavelengths. The light that you see is redder in color than if the source were not moving toward you, so we say that the light is **redshifted**.

As long as the speed of an object is much less than the speed of light (which means we can ignore relativistic effects), the observed wavelength of the Doppler-shifted light, λ_1, is given by the equation

$$\lambda_1 = \left(1 + \frac{v_r}{c}\right)\lambda_0.$$

In this expression λ_0 is called the **rest wavelength** of the light. It is the wavelength of the light as measured in the frame of reference of the source of the light. The velocity v_r is the velocity at which the object is moving *away* from you. To be more precise, v_r is the rate at which the distance between you and the object is changing. If v_r is positive, the object is getting farther from you. If v_r is negative, the object is getting closer to you.

Take a moment to be sure this expression makes sense to you. Our argument says that if an object is moving away from you, the wavelength of the light coming to you from that object should be greater than it would be if the object were "at rest" (not moving away from you). If the object is moving away, v_r will be greater than 0. If v_r is greater than 0, then $1 + v_r/c$ is greater than 1, and therefore $\lambda_1 = (1 + v_r/c)\lambda_0$ will be larger than λ_0. This is as we expected. What about the opposite case? If an object is moving toward us, v_r is less than 0, so $(1 + v_r/c)\lambda$ will be less than 1. Now we find that λ_1 is shorter than λ_0. The observed wavelength is shorter than the rest wavelength, so we see a blueshift. This is again in line with our expectations.

Remember that the Doppler shift provides information only about whether an object is moving toward you or away from you. That is what the subscript r in v_r signifies. This stands for **radial velocity**, which is the rate at which the distance between you and the object is changing. At the moment that the fire truck is passing you, it is neither getting

FIGURE 4.21 Motion of a light source relative to an observer may cause waves to be spread out (redshifted) or squeezed together (blueshifted). This change in wavelength is called a Doppler shift.

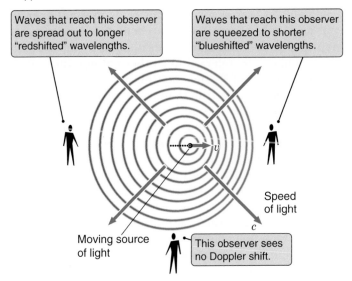

closer to you nor getting farther away from you, so the pitch you hear is the same as the pitch heard by the crew riding on the truck. You can see this directly by looking at Figure 4.21 or by referring to the previous equation. If an object is moving perpendicular to your line of sight, then $v_r = 0$ and $\lambda_1 = \lambda_0$. The observed wavelength equals the rest wavelength. The light is neither redshifted nor blueshifted.

In the case of light we need to stress again the caveat that this holds true only if the object's velocity is much less than the speed of light. If an object is moving at nearly the speed of light, special relativity says that time will be running slowly for that object. In this case the frequency of light from the object will be reduced even if the object is moving perpendicular to our line of sight. (The time between ticks of the object's clock will be spread out, which is the same as saying that the frequency of those ticks will be reduced.) The relativistic correction to the Doppler effect can be important for certain astronomical sources.

The amount by which the wavelength of light is shifted by the Doppler effect is called the **Doppler shift** of the light.

Radial velocity is measured from Doppler shifts of emission or absorption lines.

Doppler shifts become especially useful if we are looking at an object that has emission or absorption lines in its spectrum. If we can identify a spectral line as being from a certain type of atom, then we know the rest wavelength λ_0 of the line from measurements in terrestrial laboratories. If we measure that this line has a wavelength of λ_1 in the spectrum of a distant object, then we know all we need to know to calculate how rapidly the object is moving toward or away from us. Turning the above expression around a bit, we can write

$$v_r = \frac{\lambda_1 - \lambda_0}{\lambda_0} c.$$

If you know λ_0, just measure λ_1, plug both values into this expression, and *voilà!*—you know what v_r is.

It is probably useful to illustrate this with an example. A prominent spectral line of hydrogen atoms is seen at a rest wavelength of $\lambda_0 = 656.3$ nm (Figure 4.20). Suppose at a telescope you measure the wavelength of this line in the spectrum of a distant object and find that instead of seeing the line at 656.3 nm you instead see the line at a wavelength of $\lambda_1 = 659.0$ nm. You could then infer that the object is moving at a velocity of

$$v_r = \frac{\lambda_1 - \lambda_0}{\lambda_0} c$$

$$= \frac{659.0 \text{ nm} - 656.3 \text{ nm}}{656.3 \text{ nm}} \times (3 \times 10^8 \text{ m/s})$$

$$= 1.2 \times 10^6 \text{ m/s}.$$

The object is moving away from you with a radial velocity of 1.2×10^6 m/s, or 1,200 km/s.

4.5 Why Mercury Is Hot and Pluto Is Not

If you ask a child in elementary school why Mercury is hot and Pluto is cold, the child will probably look at you like you are hopelessly uninformed, then patiently explain to you that Mercury is hot because it is close to the Sun, whereas Pluto is cold because it is much farther away. This explanation is fine as far as it goes, but we should push on it a bit harder. Why are the planets as hot as they actually are? Why does the surface of Mercury reach temperatures that are hot enough to melt lead while the surface of Pluto remains so cold that even substances such as methane and ammonia remain permanently frozen? And closer to home, why is the surface of Earth hot enough for water to melt over most of the planet, but cold enough for that water to remain liquid?

The temperature of any object is determined by two things: what is trying to heat up the object and what is trying to cool it off. If an object's temperature is constant, then these two must be in balance with each other. Your body, for instance, is heated by the release of chemical energy from inside. It is also heated by energy from your surroundings. If you are standing in the sunshine on a hot day, the hot air around you and the sunlight falling on you both are working to heat you up. In response to this heating, there must be some way that your body manages to cool itself off. This is where perspiration comes in. When you perspire, water seeps from the pores in your skin and evaporates. It takes energy to evaporate water, and much of this energy comes from your body. Thus as the perspiration

---

If an object's temperature is constant, then heating must balance cooling.

---

evaporates, it carries away your body's **thermal energy,** cooling your body down. For your body temperature to remain stable, the heating must be balanced by the cooling. Such a balance is referred to as **thermal equilibrium.** If your body is out of thermal equilibrium in one direction —if there is more heating than cooling—then your body temperature climbs. If your body is out of thermal equilibrium in the other direction—if there is more cooling than heating—then your body temperature falls. (When your body is in thermal equilibrium, we can write this down by putting the amount of cooling on one side of an equation, the amount of heating on the other side, and an equal sign in between.)

Planets have a thermal equilibrium as well, and electromagnetic radiation plays a crucial role in maintaining that equilibrium. The energy from sunlight heats the surface of Earth, driving its temperature up. This is one side of the equilibrium. The other side of Earth's thermal equilibrium is also controlled by light energy. Earth radiates energy back into space, cooling Earth. Our eyes are not sensitive to the wavelengths of light that Earth radiates, but it is there nonetheless. Overall Earth must radiate away just as much energy into space as it absorbs from the Sun. If there were

> Earth is heated by sunlight and cooled by energy radiating back into space.

more heating than cooling, the temperature of Earth would climb. (It would absorb more energy from the Sun than it got rid of, and this imbalance would show up in increasing temperature.) If there were more cooling than heating, the temperature would fall. For Earth to remain at the same average temperature over time—which it has done for many centuries—the energy that Earth radiates into space must exactly balance the energy absorbed from the Sun. Thermal equilibrium must be maintained.

Equilibrium is an important concept in science. There are many kinds of equilibrium besides thermal equilibrium, some of which we will encounter later in the book. See **Foundations 4.2** for an explanation of equilibrium's basic properties.

We now have a qualitative understanding that is interesting but still is not yet useful. We would like to turn this intuitive idea of thermal equilibrium into a real prediction for the temperatures of the planets. To do this we need to find out more about light and temperature and the relationship between the two. Before going too far down this path, however, we should start by better understanding what we mean by *temperature*.

Temperature Is a Measure of How Energetically Particles Are Moving About

When we say that something is hot or cold, we know exactly what we mean. In everyday life *hot* and *cold* are defined in terms of our subjective experiences. Something is hot when it *feels* hot or cold when it *feels* cold. But our perceptions of hot and cold are a layer of subjective experience that comes between us and a definable, quantifiable concept called **temperature**. When we talk about temperature, we speak of degrees on a thermometer. The way we define a *degree* is arbitrary—a matter of convention. If you grew up in the United States, for example, you probably think of temperatures in "degrees **Fahrenheit**," whereas if you grew up anywhere else in the world, you think of temperatures in "degrees **Celsius**." Both of these are perfectly reasonable scales for measuring temperatures. But what does the thermometer actually measure?

What we refer to as *temperature* is actually a measurement of how energetically the atoms that make up an object

are moving about. The air around you is composed of vast numbers of atoms and molecules. Those molecules are moving about every which way. Some move slowly, while some move more rapidly. Similarly the atoms that make up the chair you are sitting in or the floor that you are walking on (or the anatomical parts of your body that are involved in those two activities) are constantly in motion. We can characterize these motions by talking about the average **kinetic energy,** the energy of motion:

$$E_K = \tfrac{1}{2}mv^2$$

where m is the mass of a particle and v its velocity.

The more energetically the atoms or molecules in something are bouncing about, the higher its temperature. In fact, the random motions of atoms and molecules are often referred to as their **thermal motions** to emphasize the connection between these motions and temperature.

This definition of temperature should make some intuitive sense. If something is hotter than we are, our experience says that thermal energy flows from that object into us. At the atomic level that means the object's atoms are bouncing more energetically than the atoms in our bodies, so if we touch the object, its atoms collide with our atoms, causing the atoms in our body to move faster. Our body gets hotter as thermal energy flows from the object to us. (At the same time these collisions rob the particles in the object of some of their energy. Their motions slow down. The hotter object becomes cooler.) In general, when we talk about heating, we mean processes that increase the average thermal energy of the particles something is composed of, and when we talk about cooling, we mean a process that decreases the average thermal energy of those particles. The connection between the temperature and the motion of particles is illustrated in **Figure 4.24**.

The change in thermal energy associated with a change of one unit, or degree, is arbitrary on any temperature scale. On the Fahrenheit scale there are 180 degrees between the melting point (32°) and the boiling point (212°) of water at sea level. On the Celsius scale there are 100 degrees between those two temperatures. For these two scales, the temperature

> Absolute zero, the temperature at which thermal motions stop, is zero on the Kelvin scale.

corresponding to "zero degrees" is also arbitrary. On the Celsius scale, 0°C is chosen to be the temperature at which water freezes, whereas on the Fahrenheit scale 0°F is placed at a temperature corresponding to –17.78°C. However, there is a lowest possible physical temperature below which no object can fall. As the motions of the atoms in an object slow down, the temperature drops lower and lower. When the motions of the particles finally stop, things have gotten as cold as they can get. This lowest possible temperature, where thermal

FOUNDATIONS 4.2

Equilibrium Means Balance

Equilibrium is the term we use to refer to systems that are in balance. Imagine two well-matched teams struggling in a contest of tug-of-war. Each team pulls steadfastly on the rope, but the force of their pull is only enough to match but not overcome the force exerted by their opponents. Muscles flex, but the scene does not change. A picture taken now and another taken five minutes from now would not differ in any significant way. In this book we will frequently encounter this kind of **static equilibrium**, represented by a tug-of-war where opposing forces just balance each other. The equilibrium between the downward force of gravity and the pressure that opposes it will play a central role in the stories of planetary interiors, planetary atmospheres, and the interiors of stars. Static equilibrium can be stable, unstable, or neutral as shown in **Figure 4.22**. Consider the center image in Figure 4.22 or a book standing on its edge, unsupported on either side by other objects or books. If you nudge the book, it will fall over rather than settling back into its original position. This is an example of an

unstable equilibrium. When an unstable equilibrium is perturbed, it will move further away from equilibrium rather than back toward it.

However, not all types of equilibrium are static. Equilibrium can also be dynamic, in which the system is constantly changing. Here one source of change is exactly balanced by another source of change so that the configuration of the system remains the same. **Figure 4.23** shows a simple example of **dynamic equilibrium**. A can with a hole cut in the bottom has been placed under an open water faucet. The depth of the water in the can determines how fast water pours out through the hole in the bottom of the can. Once the water reaches just the right depth, as in Figure 4.23(a), water continues to pour out of the bottom of the can at exactly the same rate it pours from the faucet into the top of the can. The water leaving the can balances the water entering the can, and equilibrium is established. If you were to take a picture now and another picture a few minutes from now, little of the water in the can would be the same. Even so, the pictures would be indistinguishable.

If a system is not in equilibrium, its configuration will change. Look at Figure 4.23(b). Here the level of the water in the can is too low, so water will not flow out of the bottom of the can fast enough to balance the water flowing into the can, causing the water level to rise. A picture taken now and another a short time later will not look the same. The configuration of the system is changing. Likewise, if the water level in the can is too high, as in Figure 4.23(c), water will flow out of the can faster than it flows into the can, and the water level will fall. Once again, if the system is not in equilibrium, its configuration will change.

FIGURE 4.22 Examples of stable (a), unstable (b), and neutral (c) equilibrium. Imagine what happens to the ball if, in each case, you give it a small nudge.

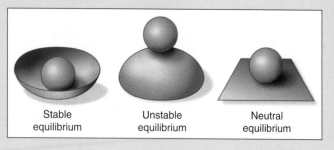

| Stable equilibrium | Unstable equilibrium | Neutral equilibrium |

motions have come to a standstill, is called **absolute zero**. Absolute zero corresponds to –273.16°C or –459.16°F.

The preferred temperature scale for most scientists is the **Kelvin scale**. For convenience, the size of one unit on the Kelvin scale, called a **kelvin** (abbreviated K) is the same as the Celsius degree. What makes the Kelvin scale special is that 0 kelvins is set equal to that absolute lowest temperature where thermal motions stop—absolute zero.

There are no negative temperatures on the Kelvin scale. The importance of the Kelvin scale is that *when temperatures are measured in kelvins, we know that the average thermal energy of particles is proportional to the measured temperature.* The average thermal energy of the atoms in an object with a temperature of 200 K is twice the average thermal energy of the atoms in an object with a temperature of 100 K.

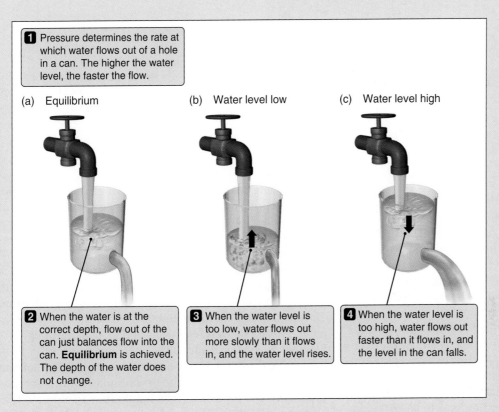

1 Pressure determines the rate at which water flows out of a hole in a can. The higher the water level, the faster the flow.

(a) Equilibrium

(b) Water level low

(c) Water level high

2 When the water is at the correct depth, flow out of the can just balances flow into the can. **Equilibrium** is achieved. The depth of the water does not change.

3 When the water level is too low, water flows out more slowly than it flows in, and the water level rises.

4 When the water level is too high, water flows out faster than it flows in, and the level in the can falls.

FIGURE 4.23 Water flowing into and out of a can determines the water level in the can. This is an example of dynamic equilibrium.

Water passing through a can is an example of a **stable equilibrium**. When a stable equilibrium is *perturbed* (forced away from its equilibrium configuration), it will tend to return to its equilibrium state. If the water level is too high or too low, it will move back toward its equilibrium level. The equilibrium that we discuss in this chapter, between sunlight falling on a planet and thermal energy radiated away into space, is a stable equilibrium. If you take Figure 4.23 and replace "water in" with "sunlight absorbed," "water out" with "energy radiated by the planet," and "water level" with "the planet's temperature," then the stable equilibrium that sets the level of the water in the can becomes the stable equilibrium that sets the temperature of a planet.

Hotter Means More Luminous and Bluer

So far we have focused our attention on the way discrete atoms emit and absorb radiation. This led us to a useful understanding of emission lines and absorption lines and how we might use these lines to learn about the physical state and motion of distant objects. But not all objects have spectra that are dominated by discrete spectral lines. For instance, if you pass the light from an incandescent lightbulb through a prism, as we saw in Figure 4.19(a), then instead of discrete bright and dark bands you will see light spread out smoothly from the blue end of the spectrum to the red. Similarly, if you look closely at the spectrum of the Sun, you will see absorption lines, but mostly you will see light smoothly spread out across all colors of the spectrum—"Roy G. Biv." What is the

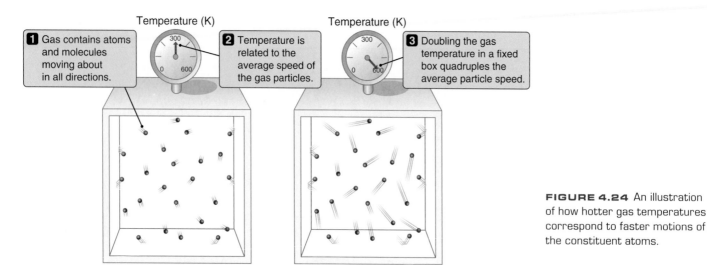

1 Gas contains atoms and molecules moving about in all directions.

Temperature (K)

2 Temperature is related to the average speed of the gas particles.

Temperature (K)

3 Doubling the gas temperature in a fixed box quadruples the average particle speed.

FIGURE 4.24 An illustration of how hotter gas temperatures correspond to faster motions of the constituent atoms.

Objects like lightbulbs emit continuous radiation at all wavelengths.

origin of such **continuous radiation**, and what clues might this kind of radiation carry about the objects that emit it?

We can think of a dense material as being composed of a collection of charged particles that are being jostled as their thermal motions cause them to run into their neighbors. The hotter the material is, the more violently its particles are being jostled. Any time a charged particle is subjected to an acceleration it radiates, and so the jostling of particles due to their thermal motions causes them to give off a continuous spectrum of electromagnetic radiation. This is why *any* material that is sufficiently dense for its atoms to be jostled by their neighbors will emit light *simply because of its temperature.* Radiation of this sort is called **thermal radiation**.

We can guess how the radiation from an object changes as the object heats up or cools off. Start with the question of the **power** (energy per second, measured in watts) radiated, which is referred to as the object's **luminosity**. The hotter the object, the more energetically the charged particles within it wiggle. (Again, this is what it *means* to be hot.) The more energetically the charged particles move, the more energy they emit in the form of electromagnetic radiation. So as an object gets hotter, we expect the light that it emits to get more intense. Here is our first intuition about thermal radiation—*hotter means more luminous.*

Now we move to the question of what color light an object emits. Again, as the object gets hotter, the thermal mo-

Making an object hotter also makes its thermal radiation bluer and more luminous.

tions of its particles become more energetic. These more energetic motions are capable of producing more energetic photons. So as an object gets hotter, we might expect the

average energy of the photons that it emits to become greater. In other words, we might expect the average wavelength of the emitted photons to get shorter. The light from the object gets bluer. Here is our second intuition about thermal radiation—*hotter means bluer.*

Both of these intuitive predictions are borne out by a simple experiment you can do. An incandescent lightbulb is a good example of an object that emits thermal radiation. The electric current in the lightbulb filament heats the filament. (More precisely, electrons being pushed through the filament by electric fields collide with atoms in the filament, increasing the thermal motions of those atoms.) The hot filament then glows. Somewhere in your home, dorm, or classrooms you can probably find a lightbulb with a rheostat, or a "dimmer." When you turn the knob on the dimmer, this changes the amount of electric current in the filament in the lightbulb. Turning the dimmer up increases the current, which increases the number and strength of collisions between electrons and atoms, which in turn increases the temperature of the filament. The hotter filament is more luminous. This confirms the first of our expectations: *hotter means more luminous.*

But what about the color of the emitted light? Look again at the lightbulb as you turn up the dimmer. When the bulb first comes on, it glows a dull red; but as the current through the bulb increases, driving up the temperature of the filament, the perceived color of the light changes. When the dimmer is turned all the way up, the light from the bulb has lost its red tint. The hotter the lightbulb gets, the more energetic blue photons become mixed with the less energetic red photons, and the light becomes whiter. The color of the light shifts from red toward blue, confirming our second intuitive expectation—*hotter means bluer.*

These observations offer an intuitive grasp of how the light given off by an object depends on the temperature of the object, but to be really useful we need to get quantitative. We need to know *how much* more luminous and *how much* bluer. The detailed answers to these questions were worked out around 1900 by Max Planck. Planck was thinking about a special situation—a hollow, totally enclosed cavity of material at a specific temperature, *T*. The crucial point here is that inside the cavity all the radiation emitted

---

A blackbody emits thermal radiation that has a Planck spectrum.

---

by the cavity walls is also absorbed by the walls of the cavity. In this situation a balance is set up, with each bit of the wall emitting just as much thermal radiation as it is absorbing from its surroundings. Physicists refer to such a special situation as a **blackbody**. Planck used this balance to calculate the spectrum of the light inside such a cavity. The result of his calculation, which beautifully matches the results of experiments, is called a **Planck spectrum** or a **blackbody spectrum**.

You might reasonably ask what this hypothetical cavity has to do with the light from the filament of a lightbulb. Surely the filament of a lightbulb is not a cavity of this sort! But in a certain sense it is. The light emitted by charged particles within the filament is mostly absorbed by other charged particles within the filament. This is exactly the assumption that Planck made when calculating the shape of the spectrum of a blackbody. As a result, we expect that the radiation existing *inside the filament of the lightbulb* will have a Planck spectrum. This radiation "leaks out" of the filament, just as light might leak out of a small hole in the side of Planck's cavity. As a result, the radiation from the filament of a lightbulb is very close to a Planck spectrum. The light from stars such as the Sun and the thermal radiation from a planet also often come close to having a blackbody spectrum.

Stefan's Law Says That Hotter Means Much More Luminous

Figure 4.25 shows plots of the Planck spectra for objects at several different temperatures. We now ask the question that scientists must always ask: Does the theoretical prediction agree with observation and experiment? Do these spectra agree with our intuitive ideas and with our experiment with the lightbulb and the dimmer? Begin with luminosity. As the temperature of an object increases,

---

The luminosity of a blackbody is proportional to T^4.

Planck's theory says that the object gives off more radiation at every wavelength, so the luminosity of the object should increase. In fact, it increases in a hurry. Adding up all of the energy in a Planck spectrum shows that the

increase in luminosity is proportional to the *fourth power* of the temperature: Luminosity $\propto T^4$. This result is known as **Stefan's Law** because it was discovered in the laboratory by **Josef Stefan** (1835–1893) before Planck's theory came along to explain it.

What Stefan's Law actually says is that the amount of energy radiated *by each square meter* of the surface of an object is given by the equation

$$\mathcal{F} = \sigma T^4.$$

In this equation $\mathcal{F}$ is called the **flux**. It is a measurement of the total amount of energy coming through each square meter of the surface each second. The constant σ (pronounced "sigma") is called the **Stefan-Boltzmann constant**, named after the discoverer of this relationship. The value of σ is the same for all cases and is given by 5.67×10^{-8} W/(m^2 K^4) (where W stands for watts). To find the total amount of energy emitted by an object in the form of electromagnetic radiation, multiply $\mathcal{F}$ by the surface area of the object.

Returning to our lightbulb example, we can actually use Stefan's Law to figure out what the surface area of the filament in a lightbulb must be. Suppose the filament in an incandescent bulb operates at a temperature of about 2,500 K. The amount of energy radiated by the bulb is stamped right

FIGURE 4.25 Planck spectra emitted by sources with temperatures of 2,000 K, 3,000 K, 4,000 K, 5,000 K, and 6,000 K. At higher temperatures the peak of the spectrum shifts toward shorter wavelengths, and the amount of energy radiated per second from each square meter of the source increases.

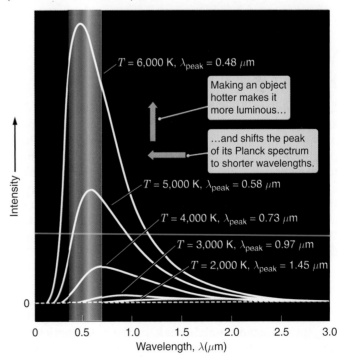

on the face of the bulb—a 100 W bulb has a luminosity of 100 W, which means that it radiates away 100 joules each second. (A **joule** is a unit of energy, abbreviated J.) If the total amount of light from the filament is equal to the flux ($\mathcal{F}$) times the surface area of the filament (A), we can write

$$100\ \text{W} = A \times \sigma T^4.$$

Solving this equation for the area, we get

$$A = \frac{100\ \text{W}}{\sigma T^4} = \frac{100\ \text{W}}{\left(5.67 \times 10^{-8} \frac{\text{W}}{\text{m}^2\text{K}^4}\right) \times (2{,}500\ \text{K})^4}$$

$$= 4.5 \times 10^{-5}\ \text{m}^2.$$

Note that not much filament surface area is needed to provide the light that turns night into day in our homes and cities.

Stefan's Law says that an object rapidly becomes more luminous as the temperature increases. If the temperature of an object goes up by a factor of 2, the amount of energy being radiated each second increases by a factor of 2^4 or 16. If the temperature of an object goes up by a factor of 3, then

---

Slight changes in temperature mean large changes in brightness.

---

the energy being radiated by the object each second goes up by a factor of $3^4 = 81$! A lightbulb with a filament temperature of, say, 3,000 K radiates 16 times as much light as it would if the filament temperature were 1,500 K. Even modest changes in temperature can result in large changes in the amount of power radiated by an object.

Wien's Law Says That Hotter Means Bluer

Look again at Figure 4.25, but this time instead of paying attention to how high each curve is, notice where the peak of each curve falls along the horizontal axis. As the temperature increases, the *peak* of the Planck spectrum shifts toward shorter wavelengths, which means the average energy of the photons becomes greater. Just as we surmised, increasing the temperature causes the light from the object to get bluer. The shift in the location of the peak of the Planck spectrum with increasing temperature is given by the equation

$$\lambda_{\text{peak}} = \frac{2{,}900\ \mu\text{m K}}{T}.$$

This result is referred to as **Wien's Law**. In this equation λ_{peak} (pronounced "**lambda peak**") is the wavelength where the Planck spectrum is at its peak. It is the wavelength where

the electromagnetic radiation from an object is greatest. Wien's Law says that the location of the peak in the spectrum is inversely proportional to the temperature of the

---

The peak wavelength of a blackbody is inversely proportional to its temperature.

---

object. If you increase the temperature by a factor of 2, the peak wavelength becomes half of what it was. If you increase the temperature by a factor of 3, the peak wavelength becomes a third of what it was.

It is useful to put a few numbers into Wien's Law. The surface of the Sun, for example, has a temperature of about 5,800 K. Wien's Law says that the peak in the light from the Sun occurs at a wavelength of

$$\lambda_{\text{peak},\odot} = 0.5\ \mu\text{m}.$$

The light given off by the Sun is concentrated at a wavelength of about 0.5 μm or 500 nm, which is in the middle of what we refer to as the visible part of the spectrum.

Wien's Law will prove handy as we continue our study of the universe. If we can measure the spectrum of an object emitting thermal radiation and find where the peak in the spectrum is, we can use Wien's Law to calculate the temperature of the object. Turning our previous example around, we have no way of dropping a thermometer into the Sun and directly measuring its temperature, but we *can* observe the spectrum of the light coming from the Sun. When we do so, we find that the peak in the spectrum occurs at a wavelength of about 0.5 μm. Wien's Law can be rewritten as

$$T = \frac{2{,}900\ \mu\text{m K}}{\lambda_{\text{peak}}}.$$

If we plug the observed peak of the spectrum of the Sun ($\lambda_{\text{peak}} \approx 0.5\ \mu$m) into this equation, we get

$$T = \frac{2{,}900\ \mu\text{m K}}{0.5\ \mu\text{m}} = 5{,}800\ \text{K}.$$

This is how we know the temperature of the Sun.

4.6 Twice as Far Means One-Fourth as Bright

You might have noticed that we have consistently spoken of the "luminosity" of objects, where in everyday language we probably would have just said that one object is "brighter" than another. This is a case where everyday language is too sloppy for science. To a physicist or an astronomer, the **brightness** of electromagnetic radiation refers to the amount of light that is *arriving* at some location, such as the page of the book you are reading or the pupil of your eye. Luminosity refers to the amount of light *leaving* a source. The concept of brightness is certainly related to the concept of luminosity. For example, replacing a lightbulb with a luminosity of 50 W with a 100 W bulb succeeds in making a room twice as bright because it doubles the light reaching any point in the room. But brightness also depends on the distance from a source of electromagnetic radiation. If you needed more light to read this book by, you could replace the bulb in your lamp with a more luminous one, but it would probably be easier to just move the book closer to the light. Conversely, if a light were too bright for you, you would move away from it. Our everyday experience says that as we move away from a light, its brightness decreases.

The particle description of light provides a convenient way to think about the brightness of radiation and how brightness depends on distance. Suppose you had a piece

Brightness measures how much light falls per square meter per second.

of cardboard that was 1 meter on a side. Intuitively you might imagine that making the light that falls on the cardboard twice as bright would mean doubling the number of photons that hit the cardboard each second. Tripling the brightness of the light would mean increasing the number of photons hitting the cardboard each second by a factor of 3, and so on. Here is a beginning point for understanding brightness. Brightness depends on the number of photons falling on each square meter of a surface each second.

Working with this idea of brightness, now imagine a lightbulb sitting at the center of a spherical shell, as shown in **Figure 4.26**. Photons from the bulb travel in all directions and land on the inside of the shell. To find the number of photons landing on each square meter of the shell during each second (that is, to find the brightness of the light), take the *total* number of photons given off by the lightbulb each second and divide by the number of square meters those photons have to be spread over. The surface area of a sphere is given by the formula $A = 4\pi r^2$, where r is the distance between the bulb and the surface of the sphere (thus r = the radius of the sphere). When this is written as a formula, we find that

$$\text{Number of photons striking one square meter each second} = \frac{\text{Total number of photons emitted per second}}{\text{Number of square meters the photons are spread over}}$$

$$= \frac{\text{Total number of photons emitted each second}}{4\pi r^2}.$$

The next step in building an understanding of brightness is to change the size of the spherical shell while keeping the total number of photons given off by the lightbulb

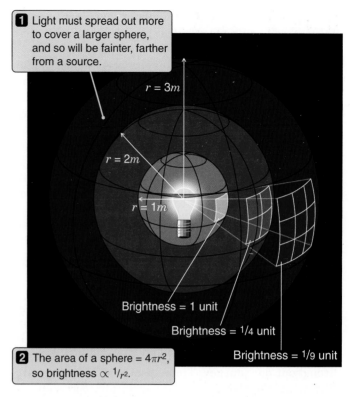

1 Light must spread out more to cover a larger sphere, and so will be fainter, farther from a source.

$r = 3m$
$r = 2m$
$r = 1m$

Brightness = 1 unit
Brightness = 1/4 unit
Brightness = 1/9 unit

2 The area of a sphere = $4\pi r^2$, so brightness $\propto 1/r^2$.

FIGURE 4.26 Light obeys an inverse square law as it spreads out away from a source. Twice as far means one-fourth as bright.

the same. As the shell becomes larger, the photons from the lightbulb must spread out to cover a larger surface area. Each square meter of the shell receives fewer photons each second, so the brightness of the light falls. If the shell's surface is moved twice as far from the light, the area over which the light must spread increases by a factor of $2^2 = 2 \times 2 = 4$.

Like gravity, light obeys an inverse square law.

The photons from the bulb spread out over four times as much area, so the number of photons falling on each square meter each second becomes ¼ of what it was. If the surface of the sphere is three times as far from the light as illustrated in Figure 4.26, the area over which the light must spread increases by a factor of $3 \times 3 = 3^2 = 9$, and the number of photons per second falling on each square meter becomes ⅑ of what it was originally. We encountered just this kind of relationship earlier when we talked about gravity (Chapter 3). Just like gravity, light obeys an inverse square law. The brightness of the light from an object is inversely proportional to the square of the distance from the object. *Twice as far means one-fourth as bright.*

It is nice to think of brightness in terms of photons streaming onto a surface from a light because this gives us a nice mental picture of the physical nature of brightness and why brightness follows an inverse square law. In practice, however, it is usually more convenient to speak of the *energy* coming to a surface each second, rather than the number of photons received.

The luminosity of an object is the total number of photons given off by the object times the energy of each photon. Instead of talking about how the number of photons must spread out to cover the surface of a sphere (brightness), we now talk about how the *energy* carried by the photons must spread out to cover the surface of a sphere. When speaking of brightness in this way, we mean the amount of energy falling on a square meter in a second. If L is the luminosity of the bulb, then the brightness of the light at a distance r from the bulb is given by

$$\text{Brightness} = \frac{\text{Energy radiated per second}}{\text{Area over which energy is spread}}$$

$$= \frac{L}{4\pi r^2}.$$

Before moving on we offer the following aside. Usually the only information that astronomers have to work with is the light from a distant object. For this reason we will use our understanding of radiation over and over again throughout our journey. Time spent now thinking carefully about the electromagnetic spectrum, emission and absorption of photons, Planck radiation, and the inverse square law for brightness will be a *very* good investment for what is to come.

4.7 Radiation Laws Allow Us to Calculate the Equilibrium Temperatures of the Planets

We began our discussion of thermal radiation by asking a straightforward question: "Why does a planet have the temperature that it does?" In a qualitative way we said that the temperature of a planet is determined by a balance between the amount of sunlight being absorbed and the amount of energy being radiated back into space. We now have the tools we need to turn this qualitative idea into a real prediction of the temperatures of the planets.

Begin with the amount of sunlight being absorbed. The amount of energy absorbed by a planet is just the area of the planet that is absorbing the energy times the brightness of sunlight at the planet's distance from the Sun. When we look at a planet, we see a circular disk with a radius equal to the radius of the planet. The area of this circular disk is

πR^2, where R is the radius of the planet. We found in our discussion in Section 4.6 that the brightness of sunlight at a distance d from the Sun is equal to the luminosity of the Sun ($L_\odot$ in watts) divided by $4\pi d^2$. (This d is the same as the r in the previous section. We use d here to avoid confusion with the planet's radius, R.) We must consider one additional factor. Not all of the sunlight falling on a planet is absorbed by the planet. The fraction of the sunlight that is reflected from a planet is called the **albedo**, a, of the planet. The corresponding fraction of the sunlight that is absorbed by the planet is 1 minus the albedo. A planet with an albedo of 1 reflects all the light falling on it. A planet that absorbs 100 percent of the sunlight falling on it has an albedo of 0.

Writing this as an equation, we say that

$$\begin{pmatrix}\text{Energy absorbed}\\\text{by the planet}\\\text{each second}\end{pmatrix} = \begin{pmatrix}\text{Absorbing}\\\text{area of}\\\text{planet}\end{pmatrix} \times \begin{pmatrix}b = \text{Brightness}\\\text{of sunlight}\end{pmatrix} \times \begin{pmatrix}\text{Fraction}\\\text{of sunlight}\\\text{absorbed}\end{pmatrix}$$

$$= \quad \pi R^2 \quad \times \quad \frac{L_\odot}{4\pi d^2} \quad \times \quad (1-a)$$

where a is the albedo of the planet.

Moving to the other piece of the equilibrium, the amount of energy that the planet radiates away into space each second is just the number of square meters of surface area that the planet has times the power radiated by each square meter. The surface area for the planet is given by $4\pi R^2$. Stefan's Law tells us that the power radiated by each square meter is given by σT^4. So we can say that

$$\begin{pmatrix}\text{Energy radiated by}\\\text{planet per second}\end{pmatrix} = \begin{pmatrix}\text{Surface area}\\\text{of planet}\end{pmatrix} \times \begin{pmatrix}\text{Energy radiated by}\\\text{each m}^2\text{ each second}\end{pmatrix}$$

$$= \quad 4\pi R^2 \quad \times \quad \sigma T^4.$$

If the planet's temperature is to remain stable—if it is to keep from heating up or cooling off—then it must be radiating away just as much energy into space as it is absorbing in the form of sunlight, as indicated in **Figure 4.27**. That means that we can equate these two expressions. We can set the quantity "Energy radiated by planet" equal to the quantity "Energy absorbed by planet." When we do this, we arrive at the expression

$$\begin{array}{c}\text{Energy radiated by}\\\text{the planet each second}\end{array} = \begin{array}{c}\text{Energy absorbed by}\\\text{the planet each second}\end{array}$$

or

$$4\pi R^2 \sigma T^4 = \pi R^2 \frac{L_\odot}{4\pi d^2}(1-a).$$

Look at this equation for a moment. It may seem rather complex, but when broken into pieces it becomes more digestible. On the left side of the equation, $4\pi R^2$ tells how many square meters of the planet's surface are radiating energy back into space, while σT^4 tells how much energy

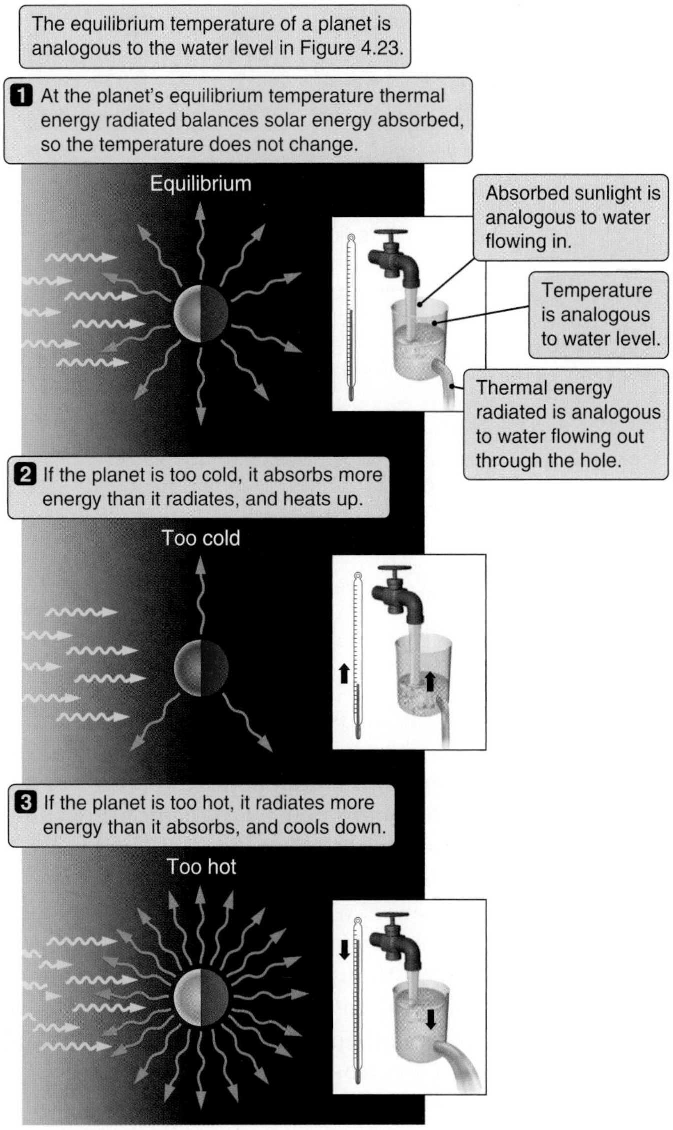

The equilibrium temperature of a planet is analogous to the water level in Figure 4.23.

1 At the planet's equilibrium temperature thermal energy radiated balances solar energy absorbed, so the temperature does not change.

Equilibrium

Absorbed sunlight is analogous to water flowing in.

Temperature is analogous to water level.

Thermal energy radiated is analogous to water flowing out through the hole.

2 If the planet is too cold, it absorbs more energy than it radiates, and heats up.

Too cold

3 If the planet is too hot, it radiates more energy than it absorbs, and cools down.

Too hot

FIGURE 4.27 Planets are heated by sunlight and cooled by emitting thermal radiation into space. If there are no other sources of heating or means of cooling, then the equilibrium between these two processes determines the temperature of the planet.

each one of those square meters radiates each second. Put them together, and you get the total amount of energy radiated away by the planet each second. On the right side of the equation, πR^2 is the area of the planet as seen from the Sun. That amount times the brightness of the sunlight reaching the planet, $L_\odot/4\pi d^2$, tells how much energy is falling on the planet each second. The final $1 - a$ tells how much of that energy the planet actually absorbs. Put everything on the right side of the equation together, and you get the amount of energy absorbed by the planet each second. The equal sign says that the energy radiated away needs to balance the sunlight absorbed. There is no magic here. In

fact, when broken down, this formidable equation embodies little more than a few straightforward ideas such as "hotter means more luminous," "twice as far means one-fourth as bright," and "heating and cooling must balance each other." The math just gives us a convenient way to work with these concepts.

We started down this path hoping to find a way to predict the temperatures of the planets, and a bit of algebra gets us the rest of the way there. Rearranging the previous equation to put T on one side and everything else on the other gives

$$T^4 = \frac{L_\odot(1 - a)}{16\sigma\pi d^2}.$$

If we take the fourth root of each side, we wind up with

$$T = \left[\frac{L_\odot(1 - a)}{16\sigma\pi}\right]^{\frac{1}{4}} \times \frac{1}{\sqrt{d}}.$$

We have now produced a full-fledged physical model for why the temperatures of planets are what they are. Restating the meaning in words, T is the temperature at which the energy radiated by a planet exactly balances the energy absorbed by the planet. If the planet were hotter than this equilibrium temperature, it would radiate energy away faster than the planet absorbed sunlight, and the temperature would fall. If the planet were cooler than this temperature, it would radi-

---

Balancing cooling and heating sets an equilibrium temperature.

---

ate away less energy than was falling on it in the form of sunlight, and the temperature of the planet would rise. Only at this equilibrium temperature do the two balance.

The equation tells us that as the distance from the Sun increases—in other words, as d gets bigger—the temperature of the planet decreases. No surprise there. But now we know how much the temperature should decrease. It should be inversely proportional to the square root of the distance. Using this formula can turn our intuition about why planets that are close to the Sun are hot into a prediction of just how hot they should be.

Figure 4.28 shows a graph of the predicted temperatures of the planets. The vertical bars show the range of temperatures found on the surfaces of each planet (or, in the case of the giant planets, at the tops of their clouds). The black dots show our predictions using the equation above. From the figure, you can see that overall we are not too far off. That should give us a sense of accomplishment: It says that our basic understanding of *why* planets have the temperatures that they do is probably not too far off. Mercury, Mars, and Pluto agree particularly well. (The agreement for Mercury would improve if we took into account the huge difference in temperature between the daytime and nighttime sides of the planet and recomputed our equilibrium accordingly.)

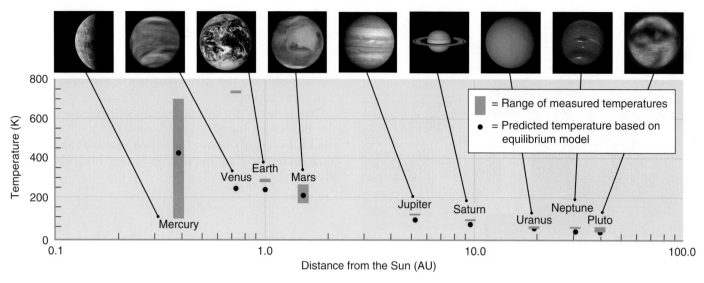

FIGURE 4.28 Predicted temperatures for the classical planets and Pluto, based on the equilibrium between absorbed sunlight and thermal radiation into space, are compared with ranges of observed surface temperatures. Some predictions are correct. Interestingly, others are not.

In other cases, however, our predictions are wrong. For Earth and the giant planets the actual temperatures are a bit higher than the predicted temperatures. In the case of Venus the actual surface temperature is wildly higher than our prediction. Rather than cause for despair, these discrepancies between theory and observation are cause for excitement. As we built our physical model for the equilibrium temperatures of planets, we made a number of assumptions. For example, we assumed that the temperature of the planet was the same everywhere. This is clearly not true: We might expect planets to be hotter on the day side than on the night side. We also assumed that a planet's only source of energy is the sunlight falling on it. Finally, we assumed that a planet is able to radiate energy into space freely as a blackbody. The discrepancies between our theory and the measured temperatures of some of the planets tell us that for these planets, some or all of these assumptions must be incorrect. In other words, the places where the predictions of our theory are not confirmed by observation point to areas where there is something still to be discovered and understood. The question of *why* these planets are hotter than the prediction will lead us to a number of new and interesting insights into how these planets work. Scientific theories sometimes succeed and sometimes fail, but even when they fail they can teach us a lot about the universe.

Summary

- From gamma rays to visible light to radio waves, all radiation is an electromagnetic wave.

- Light is also a stream of particles called photons.

- The speed of light in a vacuum is 300,000 km/s, and nothing can travel faster.

- Like gravity, light obeys the inverse square law.

- Light from receding objects is redshifted. Light from approaching objects is blueshifted.

- Special relativity concerns the relationship between events in space and time.

- Space and time together form a four-dimensional spacetime.

- Nearly all matter is composed of atoms.

- Atoms absorb and emit radiation at unique wavelengths like spectral fingerprints.

- Temperature is a measure of the thermal energy of an object.

Seeing the Forest through the Trees

In our daily lives, light is the ideal messenger, faithfully telling us about the world around us. As with any good courier, we usually concentrate on the information that light carries, taking the messenger itself for granted. It is easy to forget that our seemingly immediate visual perception of reality is actually a derived experience—the result of a sophisticated interplay between our eyes, our brains, and the flood of electromagnetic radiation that is emitted, absorbed, transmitted, and reflected by objects in the world around us.

The light that our eyes see is only a tiny portion of the full span of the electromagnetic spectrum. Electromagnetic radiation carries with it a wealth of information about the temperature, density, composition, state of motion, and other physical characteristics of the place of its origin and the material it interacts with en route. In place of the carefully controlled laboratory experiments of many other sciences, the astronomer uses a combination of ingenuity and technology to "slice up" the light reaching us and tease out the information it carries about conditions throughout the universe. The role of electromagnetic radiation in astronomy is far more than that of messenger. Radiation is also a participant in the processes that we study. For example, light carries energy from the Sun outward through the Solar System, heating the planets, and light carries energy away from each planet, allowing it to cool. The balance between these two processes establishes the conditions of our existence.

Light may be both an informative messenger and an important player in the ebb and flow of the universe, but the very nature of light itself plays havoc with our commonsense ideas about the world. When we explored the motions of Earth, the Moon, and the planets, we relied heavily on our intuition about the world. The pull or shove of one object on another and the force of gravity that holds us tightly to the surface of Earth are well within the realm of our everyday experience. But when we consider the properties of light, the boundaries of our experience and intuition are shattered. We are forced beyond the confines of the "box" within which our brains evolved. Answering a simple question like "How fast does light travel?" demands that we abandon our most cherished ideas about the nature of space, time, matter, and energy. We ask, "What is light?" and confront abstract concepts like electric and magnetic fields while running headlong into the seemingly impossible question of how something can be both a wave *and* a particle. We delve into the interaction between light and matter and find that at the scale of atoms and photons, Newton's clockwork universe crumbles. In its place we discover a world of random chance and uncertainty, governed not by the strict march of cause and effect but by laws of probability and statistics. We ask about the nature of light—and we collide squarely with the shortcomings of our intuitive ideas about the nature of reality itself.

If electromagnetic radiation is the messenger, how do we hear and interpret the message? In the next chapter we'll learn about the many and varied tools that astronomers use to detect and decipher the meaning of these communications that come to us from the Solar System and beyond.

Key Terms

index of refraction, p. 93
light-year, p. 93
wave, p. 93
electric field, p. 93
magnetic field, p. 93
electromagnetic wave, p. 94
transverse wave, p. 95
longitudinal wave, p. 95
amplitude, p. 95
frequency, p. 95
hertz, p. 95
wavelength, p. 96
spectrum, p. 98
ultraviolet (UV) radiation, p. 99
X-rays, p. 99
gamma rays, p. 99
infrared (IR) radiation, p. 99
microwave radiation, p. 99
radio waves, p. 100
special relativity, p. 101
spacetime, p. 103
relativistic, p. 104
time dilation, p. 105
photoelectric effect, p. 106
photon, p. 106
Planck's constant, p. 107
intensity, p. 107
quantum of light, p. 107
quantum mechanics, p. 107
atom, p. 107
nucleus, p. 107
proton, p. 107
neutron, p. 107
electron, p. 107
ground state, p. 109

4. Imagine a future cosmonaut traveling in a spaceship at 0.866 times the speed of light. Special relativity says that the length of his spaceship along the direction of flight is only half of what it was when it was at rest on Earth. He checks this with a meter stick that he brought along with him. Would his measurement confirm the contracted length of his spaceship? Explain your answer.

5. Patterns of emission or absorption lines in spectra can uniquely identify individual atomic elements, just as DNA testing uniquely identifies individual human beings. Explain how positive identification of atomic elements can be used as one way of testing the validity of the cosmological principle.

6. Many physical properties are proportional to the temperature of an object, raised to some power. For example, as we have seen in this chapter, the luminosity of a radiating body is proportional to T^4. Why must the temperature T be expressed in kelvins rather than degrees Celsius or Fahrenheit?

7. Consider two hypothetical planets with no atmospheres. One orbits the Sun at an average distance of 5.0 AU and the other at an average distance of 10.0 AU, yet both have the same average surface temperature. Explain how this could be possible.

8. During a popular art exhibition, the museum staff finds that to protect the artwork they must limit the total number of viewers in the museum at any time. Therefore, new viewers are admitted at the same rate that others leave. Is this an example of static or dynamic equilibrium? Explain.

9. The difference between brightness and luminosity can confuse many people. How would you explain the difference to a family member or a friend who is not taking this class?

Student Questions

THINKING ABOUT THE CONCEPTS

1. Our eyes are not sensitive to electromagnetic radiation with wavelengths much shorter than 400 nm, which lies in the ultraviolet portion of the electromagnetic spectrum. Why is this the case?

2. Einstein's theory of special relativity tells us that no object can travel faster than, or even at, the speed of light. We know that light is an electromagnetic wave, but we know that it is also a particle called a photon. If it acts as a particle, how can a photon travel at the speed of light?

3. The Sun's mass declines by more than 4 million tons of mass each second. What happens to this mass?

APPLYING THE CONCEPTS

10. You are tuned to 790 on AM radio. This station is broadcasting at a frequency of 790 kilohertz (7.9×10^5 Hz). What is the wavelength of the radio signal? You switch to 98.3 on FM radio. This station is broadcasting at a frequency of 98.3 megahertz (9.83×10^7 Hz). What is the wavelength of this radio signal?

11. If a spaceship approaching us at 0.9 times the speed of light shines a laser beam at Earth, how fast will the photons in the beam be moving when they arrive at Earth?

12. We are given a bar of tungsten, a metal with a melting point of 3,640 K and a boiling point of 6,170 K. We heat

the bar until it starts to melt and plot a graph of its emitted Planck energy distribution. We then continue to heat the tungsten until it begins to boil and plot a similar graph.

a. Sketch and label these graphs on the same set of axes. At what wavelength does the spectrum of each peak?

b. How many times more luminous is the boiling tungsten than the melting tungsten?

13. Imagine that you have been transported to Neptune, 30 AU from the Sun. How bright would the Sun appear compared to its brightness as seen from Earth?

14. A planet with no atmosphere at 1 AU from the Sun would have an average blackbody surface temperature of 279 K if it absorbed all the Sun's electromagnetic energy falling on it (albedo = 0).

a. What would be the average temperature on this planet if its albedo were 0.1, typical of a rock-covered surface?

b. What would be the average temperature if its albedo were 0.9, typical of a snow-covered surface?

15. On a dark night you notice that a distant lightbulb happens to be the same brightness as a firefly that is 5 m away from you. If the lightbulb is a million times more luminous than the firefly, how far away is the lightbulb?

16. Consider a hypothetical planet named Vulcan. If such a planet were in an orbit ¼ the size of Mercury's and had the same albedo as Mercury, what would be the average temperature on Vulcan's surface? Assume that the average temperature on Mercury's surface is 450 K.

17. The average temperature of Earth's surface is approximately 290 K. If the Sun were to become 5 percent more luminous (1.05 times its present luminosity), how much would Earth's temperature increase? (You might consider these to be only modest increases in solar luminosity and Earth's average temperature, but they would nevertheless have a drastic effect on our planet's climate.)

18. Your body, at a temperature of about 37°C (98.6°F), emits radiation in the infrared region of the spectrum.

a. What is the peak wavelength, in μm, of your emitted radiation?

b. Assuming an exposed body surface area of 0.25 m², how many watts of power do you radiate to your environment?

c. In general, the radiating surface of your body will have a temperature that is different from your internal body temperature. Why is this so?

All truths are easy to understand
once they are discovered. The point
is to discover them.

GALILEO GALILEI (1564–1642)

Robotic rovers *Spirit* and *Opportunity* have roamed the surface of Mars.

CHAPTER 5

The Tools of the Astronomer

5.1 The Optical Telescope— An Extension of Our Eyes

Do you remember the first time you saw the Moon up close? Perhaps your family or a neighbor had a backyard telescope like the one shown in **Figure 5.1**. Or the Moon might have been the featured attraction during a visit to your local planetarium. With that first view came recognition of what the Moon really is—a nearby planetary world covered with craters and vast lava-flooded basins. You might also have been treated to a breathtaking look at the Orion Nebula,

FIGURE 5.1 Amateur astronomers with their telescope. There are several hundred thousand amateur astronomers in the United States alone.

KEY CONCEPTS

In the previous chapter we learned how our understanding of the physical and chemical properties of distant planets, stars, and galaxies comes to us in the form of electromagnetic radiation. But this information must first be collected and processed before it can be analyzed and converted to useful knowledge. Here we will learn about the tools astronomers use to capture and scrutinize that information. We will find that

- Telescopes of various types collect radiation over the entire range of the electromagnetic spectrum—from gamma rays to radio signals.

- Optical telescopes come in two basic types, refractors and reflectors; but all of the larger astronomical telescopes are reflectors.

- Telescope resolution increases with aperture; image size is proportional to a telescope's focal length.

- Earth's atmosphere distorts telescopic images and prevents large parts of the electromagnetic spectrum from reaching the ground. Telescopes in orbit overcome this problem.

- The most effective way to study the planets and moons of our Solar System is to go there.

A Brief History of the Telescope

As long ago as 1350, craftsmen in Venice were making small disks of glass that could be mounted in frames and worn over the eyes to improve vision. The glass disks were convex on both sides, shaped something like lentils. And so they became known as *lenses*, from *lens*, which is Latin for "lentil." Looking back, it's rather remarkable that more than 250 years would pass before these lenses would be employed for something other than spectacles!

Hans Lippershey (1570–1690) was a German-born spectacle maker living in the Netherlands around the turn of the 17th century. In 1608 legend has it that children, playing with his lenses, put two of them together and saw a distant object magnified. Lippershey looked for himself and mounted the lenses together in a tube to produce a *kijker* (or "looker"), as he called it. As you might imagine, news of his invention spread rapidly, eventually reaching the Italian instrument maker Galileo Galilei. Galileo at once saw the potential of the "looker" for studying the heavens and constructed one of his own, as seen in **Figure 5.2**. By 1610 Galileo became the first to see craters on the Moon, the phases of Venus, and the moons of Jupiter. He was also the first to realize that the Milky Way is made up of countless numbers of individual stars. As the story goes, Galileo was demonstrating

FIGURE 5.3 Newton's reflecting telescope.

his instrument to guests when one of them christened it the "telescope," from the Greek meaning "farseeing," and the name stuck. With its ability to see far beyond the range of the human eye, the refracting telescope quickly revolutionized the science of astronomy.

Unfortunately, all simple-lens telescopes suffer from a serious problem called *chromatic aberration* (see Foundations 5.1). Realizing this, Sir Isaac Newton in 1671 designed a telescope using mirrors instead of lenses. He cast a 2-inch mirror made of speculum (basically copper and tin) and polished it to spherical curvature. He then placed the primary mirror at the bottom of a tube with a secondary flat mirror mounted above it at a 45° angle, which directed the focus to an eyepiece on the outside of the tube (see **Figure 5.3**). As it turned out, others did not share Newton's talent for instrument making, and the reflecting telescope remained a curiosity for decades. The spherical mirror surface used by Newton works only for small mirrors. Larger mirrors require a *parabolic* surface to produce a sharply focused image, and parabolic surfaces are much more difficult to fabricate. It was not until the latter half of the 18th century that large reflecting telescopes came into their own.

Throughout the 19th century both refracting and reflecting telescopes continued to grow in size. By 1897,

FIGURE 5.2 A replica of Galileo's refracting telescope.

Yerkes: 1-meter diameter lens
World's largest refracting telescope

Keck: 10-meter diameter mirrors
World's largest reflecting telescopes

FIGURE 5.4 (a) The Yerkes 1-m refractor uses a lens to collect light. (b) The twin Keck 10-m reflectors are more compact and use mirrors to collect light.

though, refracting telescopes had reached their limit with the completion of the Yerkes 1-m (40-inch) refractor (see **Figure 5.4(a)**). Gravitational distortion of a massive lens and a long telescope tube severely limits the size of refractors. The Yerkes telescope was destined to become the world's largest. Reflecting telescopes, on the other hand, seem to have no such size limits. Today the world's largest reflecting telescopes are the 10-m twin Keck telescopes located on 4-km high Mauna Kea in Hawaii (see **Figure 5.4(b)**). Each of the two Keck telescopes has a mirror with a diameter of 10 meters, giving it 4,000,000 times the light-gathering power of the human eye! And even larger reflecting telescopes are in the works. Several organizations are considering telescopes with apertures in the 30-m range, and ESO's 100-m OWL (OverWhelmingly Large) telescope may be in operation by 2020, at an estimated cost of more than €1.0 billion (see **Figure 5.5**). It seems that the only limitation on the size of reflecting telescopes is the cost of fabricating them.

FIGURE 5.5 The proposed OWL 100-m reflecting telescope.

a giant assemblage of gas and dust 1,500 light-years away, or the larger and more remote Andromeda Galaxy. These are but a few of many celestial wonders that come alive in the eyepiece of even a small telescope. As it has for so many before you, the telescope can change the way you view the heavens. From such experiences you might understand why the **telescope** is the astronomer's most important instrument (see **Excursions 5.1**). Yet it is only within the past century and a half that its capabilities have been fully exploited.

Turn a telescope or binoculars on a field of stars and what do you see? As you might expect, the stars appear both closer and brighter. Now look at a distant landscape. The scene seems closer, but its surface is no brighter. What is going on here? It turns out that only *point sources* such as stars appear brighter in a telescope. Like the distant landscape, the Orion Nebula and other extended astronomical

---

The telescope is the astronomer's most important tool.

---

objects look bigger in the eyepiece, but their surfaces are no brighter than they appear to the unaided eye. A telescope gives you a closer view of the Moon, but it does not increase the Moon's surface brightness. For more than two centuries after the invention of the telescope, astronomers struggled with this **surface brightness** problem. No matter how big they built their telescopes, nebulae and galaxies might appear larger, but their faint detail remained elusive. The problem, of course, was not with their telescopes but with the limitations of optics and the human eye. Only with the discovery of photography and the later development of electronic cameras were astronomers finally able to discern the faint but intricate fabric of the cosmos.

Here is a professional secret we can share with you. Today's working astronomers never—well, hardly ever—look through the eyepiece of a telescope. Why ignore such an opportunity? The reason is that they cannot afford to waste valuable observing time. Telescope time can be very expensive: Operating a large telescope for just a single night can cost tens of thousands of dollars! So although it might be exhilarating to glimpse Saturn through the eyepiece of a really big telescope, astronomers can learn much more and make better use of precious observing time by permanently recording the planet's image at a variety of wavelengths or seeing its light spread out into a revealing spectrum. Long after the observing session is over, the wistful peek through the eyepiece would be just a distant memory, but the recorded data remain as a permanent quantitative record for subsequent analysis.

As important as telescopes are, they are not the only instruments in the astronomer's bag of gear. The tools used by modern astronomers are many and remarkably diverse, ranging from physics laboratories to powerful supercom-

puters to robotic probes sent to cruise the Solar System. Astronomy's tools are not only diverse; they are also changing rapidly. It seems that every few months we are likely to see the commissioning of a new mountaintop telescope, the launch of a satellite observatory, or the arrival of a spacecraft at some remote planetary destination. In fact, in the time it takes between the final changes made by the authors of this volume and its appearance on your bookshelf, much of what we might say about the latest astronomy tools will already have become yesterday's news. Rather than give in to instant obsolescence, we will instead make use of the technology of the 21st century to bring you a discussion of (what else?) the technology of the 21st century!

Refractors and Reflectors

When most people think of astronomy, the mental image that comes to mind is a toy store telescope pointed at the night sky. The image is fitting. Some form of telescope forms the heart of almost every tool that astronomers have used to directly observe the heavens. In fact, humanity's use of telescopes dates back to far earlier than the night Galileo first turned his telescope skyward. You were born with two telescopes of your own—your eyes. The human eye is a refracting telescope, which means that it uses a lens to bend the light passing through it, bringing that light to a sharp **focus** on the retina, as shown in **Figure 5.6**. Eyes are such amazingly useful things that biologists believe they have evolved *independently* as many as 60 different times during the history of terrestrial life! Although the human eye is wonderfully evolved to meet our daily needs, it is not so well suited for doing astronomy. For that we need large, powerful telescopes.

Astronomical telescopes come in two basic types, *refracting* and *reflecting*, illustrated here in Figure 5.4. As we noted in our discussion of the eye, a **refracting telescope** forms an image in the **focal plane** when light from a distant

---

There are two types of optical telescopes: refractors and reflectors.

---

object is refracted by the objective lens (see **Figure 5.7**). The diameter of the objective lens defines the telescope's **aperture**, which in turn defines its light-gathering power. (Keep in mind that the light-gathering power of a telescope is proportional to the *area* of its aperture—that is, to the *square* of its diameter.) The distance between the telescope lens and the images formed is referred to as the **focal length** of the telescope. Now we know the two most important parameters of a telescope. The aperture determines a telescope's light-collecting power, and the focal length establishes the size of the image. But herein lies a major

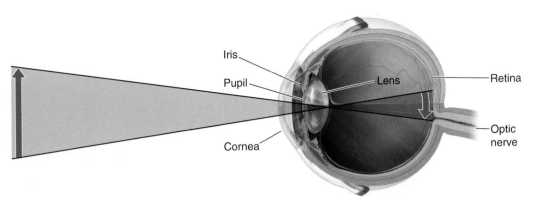

FIGURE 5.6 A schematic view of the human eye.

problem with refractors. To get the most light-gathering power and produce the largest images, a refractor must suspend a massive piece of glass (the objective lens) at the end

A telescope's aperture determines its light-gathering power and resolution.

of a very long tube without sagging unduly under the force of gravity. Take a look at the 1-m Yerkes telescope, shown in Figure 5.4(a). It carries a 450-kg objective lens mounted at the end of a 19.2-m tube. This is as big as refractors get.

The structural limitation of size is not the only problem with refractors. **Refraction** depends on the wavelength of light. (See **Foundations 5.1**.) This means that the focal length of a simple lens is different for red light than it is for blue light, an effect called **chromatic aberration** (see Figure 5.12(a). Refracting telescopes typically use **compound lenses**, such as the one shown in Figure 5.12(b), which partially

correct for chromatic aberration, although some residual effects always remain.

A **reflecting telescope** forms an image in its focal plane when light is reflected from a specially curved mirror, as seen in Figure 5.9. We call this the **primary mirror** because, as we will see, modern reflecting telescopes usually have two or more mirrors. Reflectors have a number of important advantages over refractors. Because the direction of a reflected ray does not depend on the wavelength of light,

All of the world's largest telescopes are reflecting telescopes.

chromatic aberration is no longer a problem. Primary mirrors can be made thinner and therefore less massive than objective lenses. This becomes a distinct advantage when dealing with astronomers' insatiable appetite for ever larger telescopes. Finally, the light path from the primary mirror

FIGURE 5.7 (a) A refracting telescope uses a lens to collect and focus light from two stars, forming images of the stars in its focal plane. (b) Longer focal length telescopes produce larger, more widely separated images.

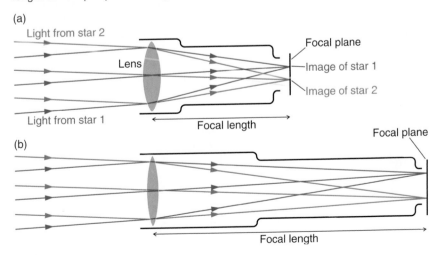

When Light Doesn't Go Straight

REFLECTION

Picture a beam of light striking a piece of glass. When light encounters a different medium, in this case going from air to glass, there will always[1] be a certain amount of **reflection** from the surface of the new medium. In other words, some of the light will change its direction of travel. If the medium is smooth and shiny, reflection is easier to understand. The most common example occurs when light encounters an ordinary flat mirror. In **Figure 5.8** an incoming or incident **ray**, *AB*, reflects from the surface, becoming the reflected ray, *BC*. The angle between *AB* and *PB*, the perpendicular to the surface, is called the *angle of incidence* (*i*). The angle between *BC* and *PB*, is called the *angle of reflection* (*r*). In the case of a flat mirror, the angles of incidence and reflection are always equal. What reflects *from* the mirror is a good representation of what falls *on* it, although left and right are interchanged. That's what makes a flat mirror so convenient for admiring our appearance.

Curved mirrors can also be very useful, especially in astronomical telescopes. The same rules of incidence and reflection hold here for each ray, but in this case the reflected rays do not maintain the same angle with respect to each another as they do with a flat mirror. A mirror that is concave toward the incoming light and has a parabolic surface (see **Figure 5.9**) will reflect the rays so that they converge to form an image. If the incoming rays are parallel, as from a distant source—think "star" —the reflected rays cross at a distance from the mirror called the focal length of the mirror. Rays from a distant source on the axis of the mirror will cross on the axis at a point called the *focus*. The surface at which all parallel rays cross is called the *focal plane* (see Figure 5.9).

REFRACTION

Returning to our light beam and the piece of glass, what can we say about the light that is not reflected? When a light wave enters a new medium its speed changes. Remember from Chapter 4 that the speed of light is always

[1] We have to be careful here. In those rare cases in which the index of refraction (see Chapter 4) is exactly the same in both media, there will be no reflection or refraction at the surface.

(a)

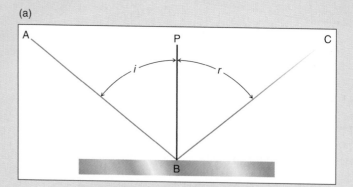

(b)

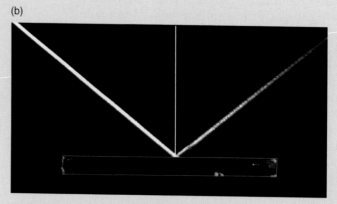

FIGURE 5.8 (a) Light incident on and reflecting from a flat surface. The angle of incidence (*i*) equals the angle of reflection (*r*). (b) Light from a laser beam is reflected from a flat glass surface.

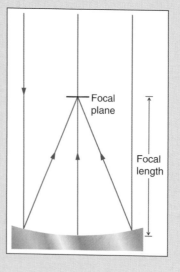

FIGURE 5.9 Parallel rays of light incident on a concave parabolic mirror are brought to a focus in the mirror's focal plane.

less in any material medium than it is in a vacuum. If the incident light is perpendicular to the surface, the speed changes but the direction of travel does not. If the incident light encounters the surface at some other angle, the direction of travel also changes. This change in the direction of the incident light is called *refraction*. The amount of refraction depends both on the incidence angle and the relative speeds in the two media, which are defined by their respective indices of refraction (see Chapter 4). In **Figure 5.10(a)** light waves are depicted as coming in from the upper left and hitting the surface of the new medium at a certain angle to the perpendicular. The part of each wave that enters the new medium first is slowed before the other part of the wave enters. This causes the wave to bend, taking up a new direction of travel. If the speed of light in the new medium is less than that in the initial medium (a higher refractive index), the light bends toward the perpendicular. If the speed of light in the new medium is greater than that in the initial medium (a lower refractive index), the light bends away from the perpendicular. **Figure 5.10(b)** shows a green laser beam being refracted by a plastic block.

A convex lens uses its curved surface and refraction to form images by making the rays cross at the focus of the lens. Figure 5.7 illustrates the path of light rays from a distant source passing through a simple convex lens and coming to a focus at the focal plane. Our eyes employ simple convex lenses.

(a)

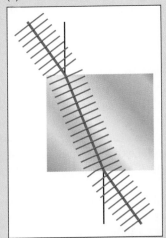

FIGURE 5.10 (a) Light waves are refracted (bent) when entering a medium with a higher index of refraction. They are refracted again as they reenter the medium with a lower index of refraction. (b) Light from a green laser beam is refracted as it enters and exits a plastic block.

(b)

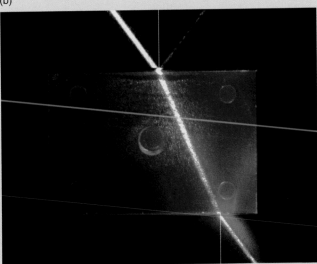

DISPERSION

Shine white light through a glass prism, as shown in **Figure 5.11**, and you'll get a rainbowlike spectrum. This demonstrates that the refraction, and therefore the

FIGURE 5.11 White light is dispersed into its component colors as it passes through a glass prism.

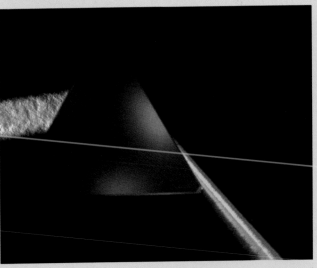

(continued on next page)

FOUNDATIONS 5.1

speed of light in glass, depends on wavelength. Glass, like most transparent materials, has a refractive index (see Chapter 4) that increases with decreasing wavelength. This means that shorter wavelengths (those toward the blue) are refracted more strongly than longer wavelengths (those toward the red). This wavelength-dependent difference in refraction, which spreads the white light out into its spectral colors, is what we call **dispersion**. Although dispersion is helpful in creating prism spectra, it creates a serious problem called *chromatic aberration* in refractive optics, as seem in **Figure 5.12(a)**. Chromatic aberration causes blue light to come to a shorter focus than the longer visible wavelengths. You can see this effect when you look at a bright object such as a distant streetlight through an inexpensive telescope. (Low-priced telescopes usually have only a simple convex objective lens.) The streetlight will appear to be surrounded by a blue halo, which is caused by the blue component of the light being out of focus. Manufacturers of quality cameras and telescopes avoid the use of simple convex lenses in favor of the *compound lens*, similar to that seen **Figure 5.12(b)**. By using two types of glass, a compound lens corrects for chromatic aberration.

FIGURE 5.12 (a) Light of different wavelengths (different colors) comes to different foci along the optical axis of a simple lens, causing chromatic aberration. (b) A compound lens using two types of glass (crown and flint) with different indices of refraction can compensate for much of the chromatic aberration.

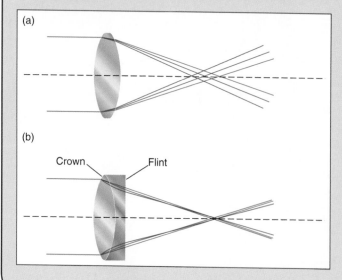

(a)

(b)

Crown   Flint

INTERFERENCE

The intersection of two sets of electromagnetic waves can produce patterns of high and low intensity called **interference**. Say we have a pair of slits and an opaque screen. **Figure 5.13(a)** illustrates monochromatic light (light having a single wavelength or a very narrow range of wavelengths) going through the slits. Each slit now becomes a source of wavefronts. Notice the regular pattern on the screen where the wavefronts from the two slits intersect. If the intersection point occurs where the amplitudes of both waves are at their maximum positive or maximum negative value, the two add and the light will be bright.[2] We call this **constructive interference**. If, on the other hand, one wave is at its maximum *positive* value and the other is at its maximum *negative* value, the sum is zero and the result will be darkness. We call this **destructive interference**. When we replace the two slits with a large number of very narrow, very closely spaced parallel slits, we call it a **grating**. The same effect can be produced by engraving closely spaced lines on a mirror.

If we now substitute a multiwavelength source of light, we get a similar pattern for each and every wavelength, as shown in **Figure 5.13(b)** for a reflective grating. For each wavelength, there will be a different point on the screen where constructive interference takes place. In other words, the grating produces a spectrum. Modern spectrographs use a grating to disperse incoming light into its constituent wavelengths. You can see this effect for yourself. Look at light reflected from a CD or DVD. The closely spaced tracks act as a grating and create a respectable spectrum **(Figure 5.13(c))**.

DIFFRACTION

When light passes through a small opening (or near an opaque edge), the effects of interference come into play. This is called **diffraction. Figure 5.14(a)** shows what happens when monochromatic light from a distant source passes through a lens and is brought to a focus. The image pattern is not a single point as we might expect, but rather is smeared out into a series of bright and dark concentric rings (as seen in **Figure 5.14(b)**) representing constructive and destructive interference from the edge of the aperture. If we make the aperture smaller, the diffracted image grows larger, causing more blurring and limiting how close two

[2]Don't be concerned about the negative value of the wave's amplitude. The intensity of light is actually equal to the *square* of the amplitude, so the negative sign goes away.

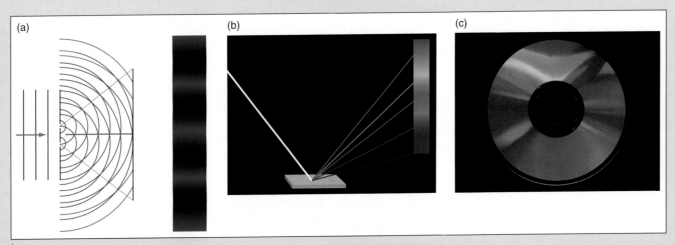

FIGURE 5.13 (a) Constructive and destructive interference patterns are created when monochromatic light passes through a pair of narrow slits. (b) Spectral dispersion is produced by interference when multiwavelength (white) light reflects from a grating. (c) A spectrum is created by the reflection of light from the closely spaced tracks of a CD.

images can be and still be resolved. The angular size of the diffraction pattern depends on the ratio (λ/D) between the wavelength of light (λ) and the aperture of the hole (D). At any given wavelength, larger apertures produce smaller diffraction patterns and therefore higher resolution. And at any given aperture, shorter wavelengths produce smaller diffraction patterns and higher resolution. Electron microscopes take advantage of this property of diffraction to achieve very high resolution by using (what else?) electrons instead of photons to illuminate the target. Recall the dual wave–particle nature of electromagnetic radiation, which we discussed in Chapter 4. Electrons can behave both as particles and as waves, with wavelengths shorter than 0.1 nm. This means that electron microscopes have more than five *thousand* times better resolution than conventional microscopes, which use visible light ($\lambda \sim 550$ nm).

FIGURE 5.14 (a) Light waves from a star are diffracted by the edges of a telescope's lens or mirror. (b) This diffraction causes the stellar image to be blurred, limiting a telescope's ability to resolve objects. (c) Diffraction from a circular aperture illuminated by green laser light.

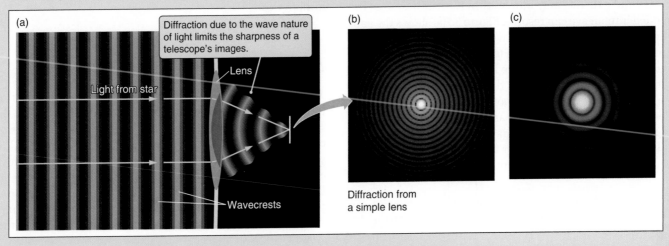

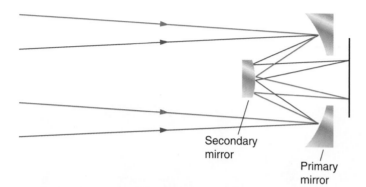

FIGURE 5.15 Reflecting telescopes use mirrors to collect and focus light. Large telescopes typically use a secondary mirror that directs the light back through a hole in the primary mirror to an accessible focal plane behind the primary mirror.

to the focal plane can be folded by introducing a **secondary mirror** (see **Figure 5.15**). This significantly reduces the length and weight of the telescope. For example, the focal length of each of the 10-m Keck telescopes, currently the world's largest reflectors, can be as long as 250 meters, even though the overall length of each telescope is a mere 25 meters. **Table 5.1** lists the world's largest optical telescopes. As you can see, all are reflecting telescopes.

Resolution

One of the eye's major limitations as an astronomical telescope is **resolution.** When we speak of resolution, we are referring to how close two points of light can be to each other before a telescope is no longer able to split the light into two separate images. Unaided, the human eye can resolve objects separated by an angular distance of 1 arcminute,[3] or a 30th the diameter of the full Moon.[4] This may seem small, and in our daily lives it is; yet when we look at the sky, thousands of stars and galaxies may hide within the smallest area the unaided human eye can resolve. Figure 5.7(a) shows the path followed by rays of light from two distant stars as they pass through the lens of a refracting telescope. Comparison with Figure 5.7(b) illustrates that the longer the focal length, the greater the separation between the images. The focal length of a human eye is typically about 20 mm. In comparison, telescopes used by professional astronomers often have focal lengths of tens or even hundreds of meters. Such telescopes make images that are far larger than those formed by your eye, and consequently they contain far more detail.

[3] A description of angular units—radians, degrees, arcminutes, and arcseconds—can be found in Chapter 13, Section 13.2, and in Appendix 1.

[4] Only the sharpest of human eyes achieve 1 arcminute resolution. Typical resolution for many of us would be more like 2 arcminutes.

Focal length explains only one difference between the resolution of telescopes and the unaided eye. The other results from the wave nature of light. As waves of light pass through the lens of a telescope, they spread out from the edges of the lens, as illustrated in Figure 5.14. The distortion of the wavefront as it passes the edge of an opaque object is called **diffraction** (see Foundations 5.1). Diffraction

> **Diffraction, or blurring of an image, depends on the ratio of wavelength to telescope aperture.**

"diverts" some of the light from its path, slightly blurring the image made by the telescope. The degree of blurring depends on the wavelength of the light in comparison with the diameter of the telescope lens. The larger the lens relative to the wavelength of the light it is focusing, the less of a problem is posed by diffraction. The ultimate limit on the angular resolution of a telescope, called the **diffraction limit**, is determined by the ratio of the wavelength of light passing through it to the diameter of the lens:

$$\theta = 2.06 \times 10^5 \left(\frac{\lambda}{D}\right),$$

where θ is the diffraction-limited angular resolution in arcseconds,[5] λ is the wavelength of light, and D is the diameter of the telescope. Both λ and D are expressed in the same units, usually meters. As we can see, the smaller the ratio of λ/D, the better will be the resolution of the telescope. We can apply this relationship to the **Hubble Space Telescope (HST)** operating in the visible part of the spectrum. The space telescope's primary mirror has a diameter (D) of 2.4 m. Visible (green) light has a wavelength (λ) of 550 nm or 5.5×10^{-7} m. Substituting into the previous equation, we have

$$\theta = 2.06 \times 10^5 \left(\frac{5.5 \times 10^{-7}}{2.4}\right) = 0.047 \text{ arcseconds}$$

or about 1,000 times better than the resolving power of the human eye.

Atmospheric Distortions—Seeing

The previous equation for the diffraction limit tells us that larger telescopes get better resolution. Theoretically the 10-m Keck telescopes have a diffraction-limited resolution of 0.0113 arcseconds in visible light, which would allow you to read newspaper headlines 60 kilometers away. However, for telescopes with apertures larger than about a meter, Earth's atmosphere stands in the way of better resolution. If you have ever looked out across the desert on a summer day, you have seen the distant horizon shimmer as light from that

[5] See footnote 3.

TABLE 5.1

The World's Largest Optical Telescopes

Mirror Diameter	Telescope	Sponsor	Location	Operational Date
10.4 m	Gran Telescopio Canarias	Spain, Mexico, University of Florida	Canary Islands	2006
10 m	Keck I	Caltech, University of California, NASA	Mauna Kea, Hawaii	1993
10 m	Keck II	Caltech, University of California, NASA	Mauna Kea, Hawaii	1996
10 m	Southern African Large Telescope	11 international partners	Southerland, South Africa	2005
9.2 m	Hobby-Eberly	University of Texas, Penn State, Stanford, Germany	Mount Lock, Texas	1997
2 × 8.4 m	Large Binocular Telescope	University of Arizona, Ohio State, Italy, Germany	Mt. Graham, Arizona	2006
4 × 8.2 m	Very Large Telescope	European Southern Observatory	Cerro Paranal, Chile	2000
8.3 m	Subaru	Japan	Mauna Kea, Hawaii	1999
8 1 m	Gemini North	USA, UK, Canada, Chile, Brazil, Argentina	Mauna Kea, Hawaii	1999
8.1 m	Gemini South	USA, UK, Canada, Chile, Brazil, Argentina	Cerro Panchon, Chile	2000
6.5 m	Magellan I	Carnegie Institute, University of Arizona, Harvard, University of Michigan, MIT	Las Campanas, Chile	2000
6.5 m	Magellan II	Carnegie Institute, University of Arizona, Harvard, University of Michigan, MIT	Las Campanas, Chile	2002

horizon is constantly bent this way and that by turbulent bubbles of warm air rising off the hot desert floor.

The problem is less pronounced when we look overhead, but the twinkling of stars in the night sky tells us the phenomenon is still there. As telescopes magnify the angular diameter of a planet, they also magnify the shimmering effects of the atmosphere. The limit on the resolution of a telescope on the surface of Earth caused by this atmospheric distortion is called **astronomical seeing**. One advantage of launching telescopes such as the Hubble Space Telescope

Earth's atmosphere distorts images.

into orbit around Earth is that from their vantage point above the atmosphere, telescopes get a much clearer view of the universe, unhampered by seeing. Does that mean that groundbased telescopes are becoming obsolete? Not at all. Modern technology has come to their rescue with computer-controlled **adaptive optics**, which compensate for much of the atmosphere's distortion.

To better understand how adaptive optics work, we need to look more closely at how Earth's atmosphere smears out an otherwise perfect stellar image. Look again at Figure 5.14(a). Light from a distant star arrives at the top of Earth's atmosphere as a series of flat, parallel waves called a **wavefront**. If Earth's atmosphere were perfectly homogeneous, the wavefront would remain flat as it reached the objective lens or primary mirror of a groundbased telescope. After making its way through the telescope's optical system, the wavefront would produce a tiny diffraction disk in the focal plane, as shown in Figure 5.14. But Earth's atmosphere is not homogeneous. It is filled with small bubbles of air that have slightly different temperatures than their surroundings. Different temperatures mean different densities, and different densities mean different refractive properties. The air bubbles act as weak lenses, and by the time the wavefront reaches the telescope it is far from flat, as shown in **Figure 5.16**. Instead of a tiny diffraction disk, the image in the telescope's focal plane is distorted and swollen, degrading the resolution. Now suppose we could measure the

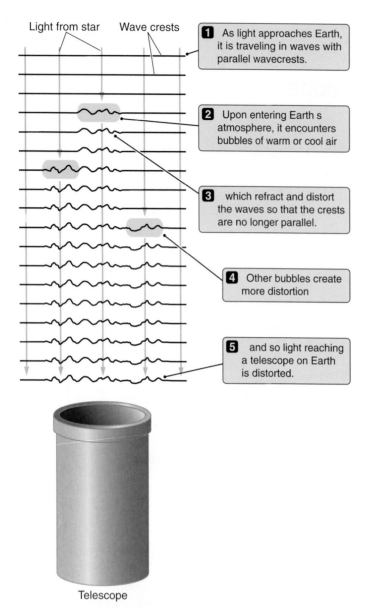

Light from star   Wave crests

1 As light approaches Earth, it is traveling in waves with parallel wavecrests.

2 Upon entering Earth s atmosphere, it encounters bubbles of warm or cool air

3 which refract and distort the waves so that the crests are no longer parallel.

4 Other bubbles create more distortion

5 and so light reaching a telescope on Earth is distorted.

Telescope

FIGURE 5.16 Distortion of the wavefront from a distant object after passing through bubbles of warmer or cooler air in Earth's atmosphere.

amount of distortion in the wavefront and somehow flatten it out. This is how adaptive optics work. First an optical device within the telescope constantly samples the wavefront, measuring its departure from flatness. Then, before

Adaptive optics can correct for atmospheric distortion of telescopic images.

reaching the telescope's focal plane, light is reflected from yet another mirror that has a deformable surface. (Astronomers sometimes call this a "rubber" mirror, although it is actually made of glass.) A computer analyzes the wavefront distortion and sends a signal to mechanisms that bend the

deformable mirror's surface so that it accurately corrects for the distortion of the wavefront. An example of an image corrected by adaptive optics is shown in **Figure 5.17**. The widespread use of adaptive optics has now made the image quality of groundbased telescopes competitive with those of the Hubble Space Telescope. But image distortion is not the only problem caused by Earth's atmosphere. Large regions of the electromagnetic spectrum are partially or completely absorbed by various atmospheric molecules.

Atmospheric Transmission— Windows and Blinds

The final limitation on the human eye is that it is sensitive only to light in the visible part of the electromagnetic spectrum. (That is, after all, why we call it the "visible" part of the spectrum!) Even though visible light is only a small part of the electromagnetic spectrum, it is anything but happenstance that our eyes work in this range of wavelengths. Our atmosphere is transparent in the visible part of the spectrum, but for most of the spectrum outside this restricted wavelength band, trying to see through our atmosphere is like trying to see through a brick wall. Almost all of the X-ray, ultraviolet, and infrared light arriving at Earth is blocked

Earth's atmosphere blocks much of the electromagnetic spectrum.

before it reaches the ground by the layer of atmosphere that surrounds our planet. The visible part of the spectrum, which is not blocked by Earth's atmosphere, is a fairly narrow range of wavelengths, or window, through which we can look at the universe. There are a few other **atmospheric windows** in the spectrum as well, as shown in **Figure 5.18**. Do not make the mistake of thinking that light that fails to reach the surface of Earth is uninteresting. There are many things that we can learn only by observing the universe outside the visible window. Although radio observations are also possible from the ground, we owe a large fraction of what we know about the universe to a host of ultraviolet, X-ray, gamma ray, and infrared telescopes that, beginning in the 1960s, were carried above Earth's atmosphere by rockets. We'll discuss space telescopes in more detail later in the chapter.

5.2 Optical Detectors and Instruments

Detectors are devices placed in a telescope's focal plane to transform images into something that we can see and record. The detector in the human eye is the retina (see Fig-

(a)

(b)

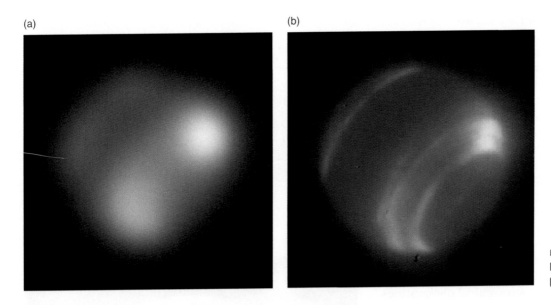

FIGURE 5.17 An image of Neptune taken (a) without and (b) with adaptive optics.

ure 5.6), and the individual receptor cells that respond to light falling on the retina are called *rods* and *cones*. Cones are located near the eye's optical axis at the center of our vision. They provide the highest resolution and allow us to recognize color. The size and spacing of cones determine the 1-arcminute resolution of the human eye, not its 7-mm pupil. Rods, located in our peripheral vision, provide the highest sensitivity to low light levels, but they have poorer resolution and cannot distinguish color. The photons to which the human eye is sensitive have wavelengths ranging from about 400 nm (deep violet) to 700 nm (far red).

So what is it that limits the faintest stars we can see with our unaided eyes, assuming a clear dark night and good eyesight? This limit is determined in part by two factors that are characteristic of all detectors: *integration time* and *quantum efficiency*. As photons from a star enter the aperture or pupil of our eyes, they fall on and excite cones at the center of our vision. The cones then send a signal to our brains, which interpret this message as "I see a star." We might now ask, "How many photons does it take to send that signal to the brain?" It turns out that your retina is hindered by short-term memory. The eye can add up photons for only a limited interval called the **integration time**. For the human eye, the integration time is about 100 ms. If two images on a television or computer screen appear 30 ms apart, you will see them as a single image because

FIGURE 5.18 Earth's atmosphere blocks most electromagnetic radiation.

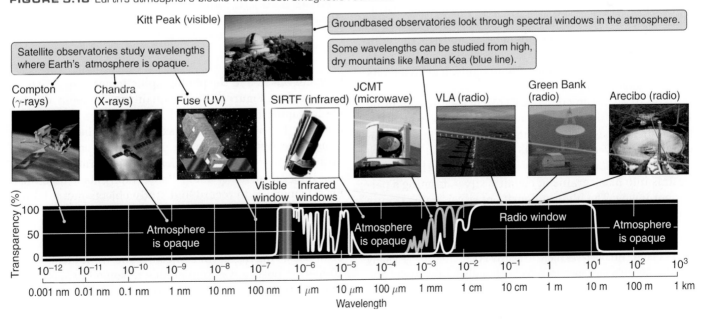

your eyes will sum up whatever they see over an interval of 100 ms. If the images occur, say, 200 ms apart, you will see them as separate images. So the signals the cones send to your brain include only those photons that arrive within an interval of 100 ms. This relatively brief integration time is the biggest factor limiting our nighttime vision. There is another effect called **quantum efficiency** that also restricts our nighttime vision. As the name implies, quantum efficiency is the likelihood that a particular photon landing on the retina will, in fact, produce a response. For the human eye, it takes about 10 photons landing on a cone to activate a single response. In other words, the quantum efficiency of our eyes is about 10 percent. Together integration time and quantum efficiency determine the rate at which photons must arrive on the retina before your brain says, "Aha, I see something." For more than two centuries after the invention of the telescope, the retina of the human eye was the only detector. Permanent records of astronomical observations were limited to what an experienced observer could sketch on paper while working at the eyepiece of a telescope, as illustrated in **Figure 5.19(a)**. Photography would eventually change all that.

Photographic Plates

In 1839 John W. Draper, a New York chemistry professor, created the earliest known astronomical photograph. His subject was the Moon, shown here in **Figure 5.19(b)**. Photography was not quick to catch on among astronomers, though, because this early *daguerreotype* process was slow and very messy. The relatively simple dry emulsion process

---

Photography opened the door to modern astronomy.

---

finally came along in the late 1870s, and with that, astronomical photography took off. Astronomers could now create permanent images of planets, nebulae, and galaxies with ease. Thousands of photographic plates soon filled the "plate vaults" of major observatories. Photography had created its own astronomical revolution.

In the dry emulsion process, a layer of gelatin containing tiny crystals of silver halide is coated onto glass plates or film.[6] During an exposure, photons landing on the emulsion energize the silver halide crystals, creating what is called a *latent* image. "Developing" the emulsion turns these small crystals into black grains of metallic silver, forming a permanent image. The highest density of silver grains occurs where the telescope's image was the brightest. So bright

---

[6]Glass plates, although far more expensive than film, are generally used for imaging and spectroscopy because they have greater geometric stability.

(a)

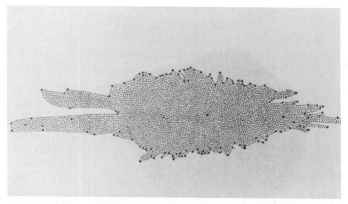

(b)

FIGURE 5.19 (a) Drawing of the Milky Way Galaxy made by William Herschel in the early 19th century. (b) Photograph of the Moon taken by J. W. Draper in 1839.

becomes black, and the photographic image is negative, as illustrated in **Figure 5.20**. The quantum efficiency of most photographic emulsions used in astronomy is very low: typically 1–3 percent, which is even poorer than that of the human eye. But unlike the eye, photographic emulsions can overcome poor quantum efficiency by integrating photons over intervals of many hours. Photography made it possible for astronomers to record and study objects much fainter than the human eye can see.

Photography is not without its own problems. Very faint objects often require long exposures that can take up much of an observing night. (Imagine how you would feel if your 10-hour exposure was spoiled due to some mishap.) Also, the spectral range of photographic emulsions is hardly

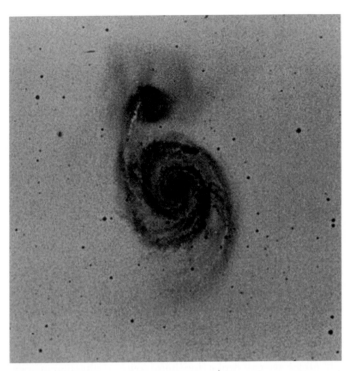

FIGURE 5.20 An image of Galaxy M 51 on a photographic plate.

broader than that of the human eye. In fact, for many years photographic plates were sensitive only to violet and blue light. Another problem is their nonlinear response to light, meaning that the optical density of the processed emulsion is not proportional to the intensity of light falling on it. Finally, there is the nontrivial matter of economics. Each photographic plate can be used only once, and they are expensive. By the middle of the 20th century the search was on for electronic detectors that would overcome many of the deficiencies of photographic plates.

Charge-Coupled Devices (CCD)

Throughout the latter half of 20th century, astronomers employed various electronic detectors to overcome the sensitivity, spectral range, and nonlinearity problems of photography. Some, such as *photoelectric photometers*, are nonimaging devices. They work extremely well for precision stellar photometry because they have excellent linearity, which means that their electronic output is directly proportional to the intensity of light falling on the photometric detector. But they can measure only one stellar image at a time. This resulted in truly labor-intensive observing. Other detectors, such as *vidicons*, are electronic imaging devices with sensitivity far superior to photographic emulsions. Vidicons unfortunately suffer from an electronic instability that causes small geometric distortions of the image. You

may have seen pictures taken by early spacecraft that had little + shaped marks superimposed on them. These *fiducial* marks were engraved on the faceplates of the vidicons to help remove geometric distortions in the image. The search for a better detector continued.

In 1969 scientists at Bell Laboratories were developing "picture phones"—telephones containing a small camera and viewing screen that could display an image of the person at the other end of the conversation. As it turned out, public opinion declared Bell's picture phones an invasion of personal privacy and they were never commercially produced, but the research led to the invention of a remarkable detector called a **charge-coupled device** or **CCD**. Astronomers soon realized that this was the detector they had been looking for. Shortly after it was first applied to astronomical imaging in the mid-1970s, the CCD became the detector of choice in almost all astronomical imaging applications. Gone were the problems associated with photographic emulsions, photoelectric photometers, and vidicon-type imagers. The CCD is a photometrically linear imaging device, able to perform precise photometry over large regions of sky. It responds over a wide spectral range from 200 nm to 1,200 nm and has a high quantum efficiency, typically 80 percent or greater. The output from a CCD is a computer-ready, digital signal that can be sent directly from the telescope to image-processing software or stored on disk for later analysis.

CCDs consist of an ultrathin wafer of silicon, less than the thickness of a human hair, which is divided into a two-dimensional array of picture elements, or **pixels**, as seen in **Figure 5.21(a)**. When a photon strikes a pixel, it creates a small electrical charge within the silicon. As each CCD pixel is "read out," the digital signal that flows to the computer is almost precisely proportional to the accumulated charge. This is what we mean when we say the CCD is a very linear device. Like many electronic detectors, CCDs are subject to thermal noise, but this can be minimized by cooling them down to liquid nitrogen temperatures (~80 K). The first astronomical CCDs were small arrays containing no more than a few hundred thousand pixels. The larger CCDs used in astronomy today may contain as many as 100 million pixels, like the one seen in **Figure 5.21(b)**.

The impact that CCDs have made on astronomy cannot be understated. Conventional photography is now a distant second for imaging at the telescope. As you surf the Internet, nearly every spectacular astronomical image that pops

> The CCD is the astronomer's detector of choice.

up on your screen was made with a CCD, whether from groundbased telescopes or from those in space. And professional astronomers are not the only ones making spectacular photos with CCDs. Amateur astronomers are now using commercially available, thermoelectrically cooled CCD

(a)

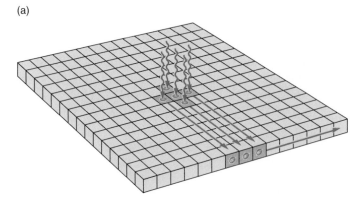

FIGURE 5.21 (a) A simplified diagram of a CCD. Photons from a star land on pixels (gray squares) and create electrons within the silicon. The electron charges are electronically moved sequentially to the collecting register at the bottom. Each row is then moved out to the right to an electronic amplifier, which converts the electrical charge of each pixel into a digital signal. (b) A very large charge-coupled device (CCD).

(b)

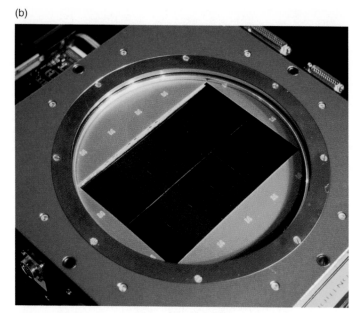

imagers with impressive results. **Figure 5.22(a)** is an image of Saturn taken by an amateur astronomer with a 36-cm telescope. This image shows more detail than the best professional photographs of Saturn taken before CCDs became available to astronomers, an example of which is shown in **Figure 5.22(b)**.

You may never have seen a CCD, but in recent years they have found their way into many devices that we now take for granted—digital cameras, video cameras, and the ubiquitous picture phones, just to name a few.

Spectrographs

Spectrographs, or **spectrometers** as they are often called, are another of the astronomer's essential tools. As we learned in Chapter 4, we can probe the chemistry and physical properties of distant objects by studying their spectra. Early spectrographs used glass prisms to disperse the incoming light into its component wavelengths (see **Figure 5.23**), creating a spectrum like the ones shown in Figure 4.19 and Figure 5.11. A photographic plate recorded the spectrum for precise measurement of the wavelengths of its spectral lines. One disadvantage of glass prisms is that **dispersion** is not uniform with wavelength. Prism spectrographs produce spectra with more dispersion at the shorter-wavelength (violet) end of the spectrum than at the longer-wavelength (red) end. Another drawback is that glass is opaque to ultraviolet and long-wavelength infrared light. Prism spectrographs are more or less limited to the visible part of the spectrum. Most modern spectrographs use a *diffraction grating* to disperse the light (see Figure 5.13(b)) and a CCD to record

FIGURE 5.22 (a) A CCD image of Saturn taken in 2005 by an amateur astronomer with a 36-cm telescope. (b) A pre-CCD photograph of Saturn taken in 1974 with a 1.5-m telescope at the Catalina Observatory of the University of Arizona.

(a)

(b)

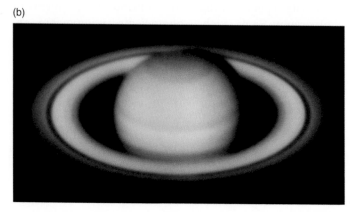

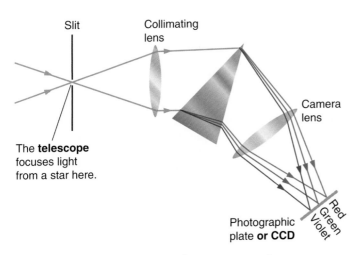

FIGURE 5.23 Diagram of a prism spectrograph.

the spectrum. Refer to Foundations 5.1 for a more detailed discussion of dispersion and diffraction.

Spectrographs may be designed for either low or high dispersion. Low-dispersion spectrographs are most often used to identify the chemical components of feeble light sources such as nebulae or to measure the reflected spectral energy distribution of faint Solar System objects such as small or distant asteroids. Measurements of temperature and radial velocity, on the other hand, usually require very high dispersion. Today's high-dispersion spectrographs have evolved into substantial scientific instruments, hardly the sort of thing that one would transport from one telescope to another. The high-dispersion spectrographs now associated with most major telescopes tend to be huge and weigh several metric tons—think SUV. For obvious reasons, they are not attached directly to the telescope! A system of mirrors feeds light from the telescope's focal plane into the spectrograph, which is located nearby.

We'll encounter many applications of **spectroscopy** throughout the chapters to come as we continue in our journey through the Solar System and beyond.

5.3 Radio Telescopes

Karl Jansky (1905–1950) was a young physicist working for Bell Telephone Laboratories in the early 1930s when he was assigned the job of identifying sources of static in transatlantic radiotelephone service. He built a pointable antenna and soon identified the major sources of static as nearby and distant thunderstorms; but one source was mysterious. A faint steady hiss rose and fell once every 23 hours and 56 minutes, a characteristic of celestial objects far beyond our Solar System. In 1932 Jansky identified the mysterious source. It was in the Milky Way in the direction of Sagittarius, the galactic center. Excited by his discov-

ery, he submitted a request to build a large dish antenna, a **radio telescope**, to study these signals in more detail. Bell Labs turned down the request. After all, Jansky had already given them the information they needed. Nevertheless, Jansky's discovery marked the birth of radio astronomy. In his honor, the basic unit for the strength of a radio source is called the **jansky**.

In 1937 Grote Reber, a radio engineer and ham radio operator, decided to build his own radio telescope. It consisted of a parabolic sheet of metal, 9 meters in diameter, with a radio receiver mounted at the focus. With this instrument he conducted the first survey of the sky at radio frequencies, and he published the first radio frequency map in 1941. Reber was largely responsible for the rapid advancement in radio astronomy that blossomed in the post–World War II era.

Radio telescopes are yet another of astronomy's indispensable tools. From our Solar System to the most distant galaxies, the penetrating power of radio waves unlocks secrets not possible with shorter-wavelength optical or infrared telescopes. Look back at Figure 5.18 and notice the wide radio window in Earth's atmosphere, covering wavelengths ranging all the way from a centimeter to 10 meters.[7] This ability of radio waves to pass unattenuated through our atmosphere is also the property that allows us to peer through

---

Radio telescopes allow astronomers to "see" through obscuring gas and dust.

---

the vast amounts of gas and dust found in many galaxies. Most radio telescopes are large steerable parabolic dishes, typically tens of meters in diameter such as the one shown in **Figure 5.24(a)**. The world's largest radio telescope is the 305-m Arecibo dish built into a natural bowl-shaped depression in Puerto Rico, as seen in **Figure 5.24(b)**. But there can be a price to pay for size. As you might guess from looking at the picture, this huge structure is too big to steer. Instead it must point by moving its radio receiver, suspended in the focal plane above the dish. Arecibo's targets are therefore limited to those celestial sources that pass within 20° of the zenith as Earth's rotation carries them overhead.

As large as radio telescopes are, they have relatively poor angular resolution. Recall our earlier discussion about diffraction. A telescope's angular resolution is determined by the ratio λ/D, where λ is the wavelength of electromagnetic radiation and D is the telescope's aperture. (Keep in mind that a larger ratio means poorer resolution.) Radio telescopes have diameters much larger than the apertures of most optical telescopes, and that helps. But the wavelengths of radio waves are typically several hundred times greater than the

---

[7] Microwave astronomy is considered a branch of radio astronomy. Microwaves are very high-frequency radio waves with wavelengths ranging from about 1 mm to 10 cm at the short-wavelength end of the radio spectrum. As seen in Figure 5.18, Earth's atmosphere is only partially transparent to microwaves.

(a)

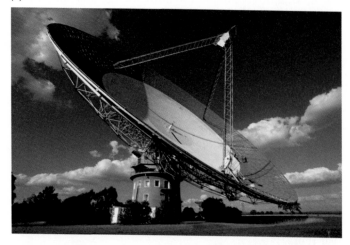

(b)

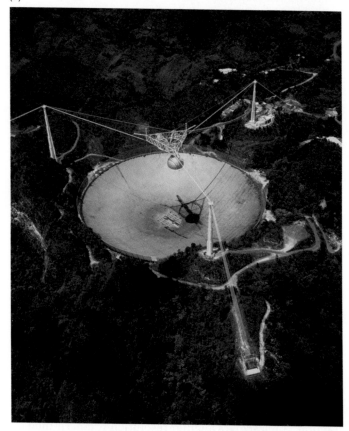

FIGURE 5.24 (a) A large radio telescope in Australia. (b) The Arecibo radio telescope is the world's largest. The steerable receiver suspended above the dish permits limited pointing toward celestial targets as they pass close to the zenith.

wavelengths of visible light, and that hurts. Radio telescopes are thus hampered by the very long wavelengths they are designed to receive. Consider the huge Arecibo dish. Its resolution is typically about 1 arcminute, no better than the unaided human eye! So radio astronomers have had to

develop their own bag of tricks, and one of the cleverest is the interferometer.

Single radio telescopes have relatively poor resolution...

When we combine the signals from two radio telescopes in a certain way, the separation between them—not the diameters of the individual telescopes—determines the angular resolution. For example, if two 10-m telescopes are located 1,000 meters apart, the D in λ/D is 1,000, not 10. Such an arrangement is called an **interferometer** because it makes use of the wavelike properties of electromagnetic radiation, in which signals from the individual telescopes *interfere* with one another (see Foundations 5.1). Usually several telescopes are employed, an arrangement called an **interferometric array**. Through the use of very large arrays, radio astronomers can attain and exceed the angular resolution enjoyed by their optical colleagues. One of the larger radio interferometric arrays is the Very Large Array (VLA) in New Mexico, shown in **Figure 5.25**. The VLA is made up of 27 individual movable dishes spread out in a Y-shaped configuration 30 km across. At a

...but interferometric arrays overcome this problem.

wavelength of 10 cm, this array can achieve resolutions of less than 1 arcsecond. Not satisfied, radio astronomers have sought still larger arrays—and no one can accuse them of having limited imagination. The Very Long Baseline Array (VLBA) employs 10 radio telescopes spread out over more than 8,000 km from the Virgin Islands in the Caribbean to Hawaii in the Pacific. At a wavelength of 10 cm, this array can reach resolutions better than 0.003 arcseconds. It might seem that Earth's diameter would set the ultimate limit on resolution for radio astronomers, but plans are under way to build the Very Long Baseline Interferometer (VLBI) in which one of the radio telescopes is put into near-Earth space. The combined Earth and space-based array would extend over 30,000 km, yielding resolutions far exceeding those of any existing optical telescope.

Before leaving our discussion of interferometers, we should point out that radio astronomers are not the only ones using the interferometer's greater resolving power. Optical telescopes can also be arrayed to yield resolutions greater than those of single telescopes, although for technical reasons the individual units cannot be spread as far apart as radio telescopes. The Very Large Telescope (VLT), operated by the European Southern Observatory (ESO) in Chile, consists of the four VLT 8-m telescopes (see **Figure 5.26**) and four movable 1.8-m auxiliary telescopes. When fully operational it will have a baseline of up to 200 m, yielding angular resolution in the milli-arcsecond range.

FIGURE 5.25 The Very Large Array (VLA) in New Mexico.

5.4 Neutrino and Gravity Wave Detectors

FIGURE 5.26 The Very Large Telescope (VLT) operated by the European Southern Observatory in Chile. Movable auxiliary telescopes allow the four large telescopes to operate as an optical interferometer.

In learning about the Sun in Chapter 14 you will be introduced to the **neutrino**, an elusive particle that plays a major role is the physics of stellar interiors. Of course we can't see beneath the Sun's surface, but observations of neutrinos can give us important insight into what is happening deep within. There's only one problem, and it's a big one: Neutrinos are extremely difficult to detect. To study a neutrino you first have to grab one, and they are nearly impossible to catch. In a sense this is fortunate for us. In less time than it takes you to read this sentence, a thousand trillion (10^{15}) solar neutrinos from the Sun are passing through your body. It doesn't matter a bit if you are reading this at night. Neutrinos are so nonreactive with matter that they can pass right through Earth (and you) as though it (or you) weren't there at all. In fact, half of the neutrinos produced by the Sun would make it through a slab of lead one light-year thick. For us to detect a neutrino, it has to interact with a detector. Several neutrino detectors are in operation today. All are buried deep underground to avoid false detection of cosmic background radiation. Neutrino detectors typically record only one out of every 10^{22} (10 billion trillion) neutrinos passing though them, but that's enough to reveal processes deep within the Sun or witness the violent death of a star 160,000 light-years away.

Another elusive phenomenon is the gravity wave. In fact, **gravity waves** are so elusive we've never actually observed them. You might even say the several facilities constructed to detect gravity waves have been built on faith. But there is strong, although indirect, observational evidence for their existence, as we will see later in Chapter 17 when we discuss *binary pulsars*. Gravity waves are disturbances in a gravitational field, similar to the waves that spread out from the disturbance you create when you toss a pebble into the quiet surface of a pond. Scientists are eager to detect gravity waves, not so much to confirm their existence but to study the physical phenomena they are likely to reveal. To understand the importance of gravity waves and what they might tell us about disturbances in the fabric of spacetime (refer back to the discussion of special relativity in Chapter 4), we will have to wait until Chapter 17 for a discussion of **general relativity**.

5.5 Getting above Earth's Atmosphere: Airborne and Orbiting Observatories

Imagine strolling through a Hawaiian rainforest while taking in the fragrance of the warm, humid tropical air. In your bliss you might be unaware that 4 km above you on the summit of Mauna Kea astronomers are at war with the same atmospheric water vapor that has so stimulated your senses. Water vapor is the enemy of the infrared astronomer. We have already seen that Earth's atmosphere distorts telescopic images and that certain molecules in Earth's atmosphere, including water, block large parts of the electromagnetic spectrum from getting through to the ground. It shouldn't surprise us then to find that astronomers have put considerable effort into getting their instruments above as much of the atmosphere as possible. Look for an astronomical observatory and you'll be looking at the summit of a tall mountain. Most of the world's larger astronomical telescopes are located 2,000 m and more above sea level. Mauna Kea, a dormant volcano and home of the Mauna Kea Observatory, rises 4,200 m above the Pacific Ocean. At this altitude the MKO telescopes sit above 40 percent of Earth's atmosphere; but more important, 90 percent of Earth's atmospheric water vapor lies below. Still, for the infrared astronomer, the remaining 10 percent is troublesome.

One way to solve the water vapor problem is to make use of high-flying aircraft. NASA's Kuiper Airborne Observatory (KAO), a modified C-141 cargo aircraft, carried a 90-cm telescope and was among the first of these flying observatories. It could cruise at an altitude of 14 km, above 98 percent of Earth's water vapor. NASA retired KAO in 1995 and will replace it with the Stratospheric Observa-tory for Infrared Astronomy (SOFIA), expected to become operational in 2007. SOFIA will carry a 2.5-m telescope and work in the far infrared region of the spectrum, from 30 μm to 350 μm.

Having full access to the complete electromagnetic spectrum is yet another matter. This means getting completely above Earth's atmosphere.[8] In the late 1940s scientists put ultraviolet and cosmic ray instruments in the nose cones of captured German V-2 rockets and launched them from the White Sands Proving Grounds in New Mexico to altitudes greater than 100 km. Of course, such observations had to be brief because, thanks to gravity, the rockets and their scientific instruments invariably came back down. The next step was to put astronomical instruments into orbit. The first astronomical satellite was the British Ariel 1, launched in 1962 to study solar ultraviolet and X-ray radiation and the energy spectrum of primary cosmic rays. Today we have a multitude of orbiting astronomical telescopes covering the electromagnetic spectrum from gamma rays to microwaves, with many more in the planning stage (see **Table 5.2**).

Optical telescopes, such as the Hubble Space Telescope (HST), can operate successfully at modest altitudes in what is called low earth orbit (LEO), 600 km above Earth's surface. This is also the region where the International Space Station (ISS) and many scientific satellites orbit. For others 600 km is not nearly high enough. Chandra, an X-ray telescope, cannot tolerate even the tiniest traces of atmosphere and so flies

> Orbiting observatories explore regions of the spectrum inaccessible from the ground.

in an orbit that keeps it more than 16,000 km above Earth's surface. And even this is not distant enough for some telescopes. Spitzer, an infrared telescope, is so sensitive it needs to be completely free from Earth's own infrared radiation. The solution was to put it into a *solar* orbit, trailing tens of millions of kilometers behind Earth. Many future space telescopes, including NASA's replacement for HST, will orbit free of Earth, bound only to the Sun.

5.6 Getting Up Close with Planetary Spacecraft

As we have stressed, we live in a remarkable time of discovery, when our newfound technological prowess has allowed us to begin the process of exploring our local corner of space.

---

[8] In a sense, there is no definable upper limit to Earth's atmosphere. As we will learn in Chapter 8, our atmosphere simply blends into outer space at an altitude of about 10,000 kilometers.

TABLE 5.2

Selected Present and Future Space Observatories

Telescope	Space Agency	Description	Launch Year
Hubble Space Telescope	NASA	Optical, infrared, ultraviolet observations	1990
Far Ultraviolet Spectroscopic Explorer	NASA	Ultraviolet spectroscopy	1999
Chandra X-Ray Observatory	NASA	X-ray imaging and spectroscopy	1999
Wilkenson Microwave Anisotropy Probe	NASA	Cosmic background radiation	2001
Spitzer	NASA	Infrared observations	2004
Swift	NASA	Gamma ray bursts	2006
Herschel	ESA	Far-infrared and submillimeter observations	2007
Kepler	NASA	Planet finder	2008
James Webb Telescope	NASA	Replacement for HST	2013

The general strategy for exploring our Solar System begins with a reconnaissance phase, using spacecraft that fly by or orbit a planet or other body. At the opening of the 21st century, we have conducted preliminary reconnaissance of much of the Solar System. We have sent spacecraft flying by all of the classical planets, giving humanity its first ever close-up views of these distant worlds and their moons. We have even seen comets and asteroids at close range. As they sped by, instruments aboard these spacecraft briefly probed the physical and chemical properties of their targets and their environments.

Reconnaissance spacecraft use **remote sensing** instrumentation much like the remote sensing techniques used by Earth-orbiting satellites to study our own planet. These include tools such as cameras capable of taking images in different wavelength ranges, radar for mapping surfaces hidden beneath obscuring layers of clouds, and spectrom-eters that spread out the target's light into a diagnosable spectrum. Remote sensing allows planetary scientists to map other worlds, measure the heights of mountains, identify geological features, learn about types of rocks present, watch weather patterns develop, measure the composition

Planetary spacecraft take our instruments directly to the planets.

of atmospheres, and in general get a feeling for the "lay of the land." Still other instruments make *in situ* measurements of the extended atmospheres and space environment through which they travel.

The study of our Solar System from space is a truly international collaboration involving NASA, the European Space Agency (ESA), and the Japanese Space Agency. Other countries, including China and India, may soon join the endeavor.

Flybys and Orbiters

Since the dawn of history no human had ever seen the far side of the Moon. This is because, as we learned in Chapter 2, the orbital and rotational periods of the Moon are equal to one another. This keeps one side of the Moon permanently facing Earth and the other side forever hidden. Hidden, that is, until October 18, 1959. On that date the Soviet **flyby** probe *Luna 3* sent back humanity's first view of the far side of our nearest celestial neighbor (see **Figure 5.27**). No matter how powerful we make our groundbased or Earth orbiting telescopes, *Luna 3* showed us there is nothing quite like going there.

Flyby missions have several distinct advantages in the reconnaissance phase of exploration. First, they are relatively inexpensive and the easiest missions to design and execute. Second, flyby spacecraft such as *Voyager*, shown in **Figure 5.28(a)**, may be able to visit several different worlds during their travels. The downside of flyby missions is that, thanks to the physics of orbits, these spacecraft must move by very swiftly. They are limited to just a few hours or at most a few days in which to conduct close-up studies of their targets. Yet flyby spacecraft give us our first intimate views of our planetary neighbors and provide the details we need to plan follow-up studies.

More detailed reconnaissance work uses spacecraft that orbit around planets. These are intrinsically more difficult missions than flyby missions; but **orbiters** can linger, looking in detail at more of the surface of the object they are orbiting and studying things that change with time, like planetary weather. Spacecraft have orbited the Moon, Venus, Mars, Jupiter, Saturn, and even an asteroid. **Figure 5.28(b)** shows the *Cassini* spacecraft, which, as this book goes to print, is still sending us data from its orbit around Saturn.

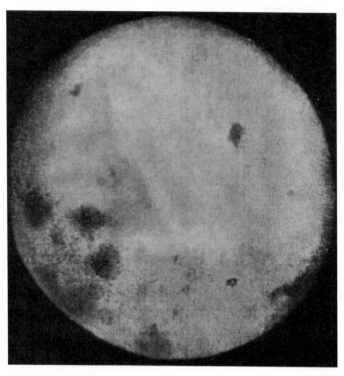

FIGURE 5.27 Humanity's first view of the far side of the Moon seen in this image sent back by the Soviet probe *Lunar-3* in 1959.

Landers, Rovers, and Atmospheric Probes

Reconnaissance spacecraft provide a wealth of information about a planet, but there is no better way of obtaining "ground truth" than to put our instruments where they can get right to the heart of it—within a planet's atmosphere or on solid ground. We have landed spacecraft on the Moon, Mars, Venus, Saturn's large moon Titan, and the asteroid Eros. One spacecraft even shot a massive bullet into a comet nucleus to observe the splash. These spacecraft have returned pictures of the surfaces, measured surface chemistry, and conducted experiments to determine the physical properties of the surface rocks and soils.

One disadvantage of using landed spacecraft is that only a few landings in limited areas are practical because of the expense, and the results may apply only to the small area around the landing site. Imagine, for example, what a different picture of Earth we might get from a spacecraft that landed in Antarctica, as opposed to a spacecraft that landed in the caldera (the summit crater) of a volcano or the floor of a dry riverbed. Sites to be explored with landed spacecraft must be very carefully chosen on the basis of reconnaissance data if we are to know what to make of the information they provide. We can mitigate some of the limitations of **landers** by putting their instruments on wheels and sending them from place to place, exploring the vicinity of the

landing site. Such remote-controlled vehicles, called **rovers**, were used first by the Soviet Union on the Moon more than a quarter century ago, and more recently by the United States on Mars. The opening photo for this chapter shows an artist's view of one of two rovers still roaming about the Martian landscape.

We have also sent probes into the atmospheres of Venus, Jupiter, and Titan. As they descend, **atmospheric probes** continuously measure and send back physical properties such as temperature, pressure, and wind speed along with other properties, such as chemical composition. Meteorologists take measures of our terrestrial atmosphere from the surface up by sending their instruments aloft in balloons. Planetary scientists must work from the top down by suspending their instruments from parachutes. The end result is much the same. Atmospheric probes have survived all the way to the solid surfaces of Venus and Titan, sending back streams of data during their descent. An atmospheric probe sent into Jupiter's atmosphere never reached that planet's surface because, as we will learn later, Jupiter does not have a solid surface in the same sense that terrestrial planets and moons do. After sending back its data, the Jupiter probe eventually melted and vaporized as it dropped into the hotter layers of the planet's atmosphere.

Sample Returns

If you pick up a rock from a road cut, there is a lot you might learn from the rock using the tools that you could easily carry in your pocket. On the other hand, the sophistication of the tools you could carry with you would be limited. It would be much better to pick up a few samples and carry them back to a laboratory equipped with a full range of state-of-the-art instruments capable of measuring chemical compositions, mineral types, radiometric ages (see Foundations 7.1), and other information needed to reconstruct the story of their origin and evolution. So, too, is the case in Solar System exploration. One of the most powerful methods for investigating remote objects is to collect samples of the objects and bring them back to Earth for detailed study. So far, only samples of the Moon, a comet, and the **solar wind** (a stream of charged particles from the Sun) have been collected and returned to Earth.

As we will learn in Chapter 12, we do have meteorites that are considered to be parts of Mars, but there is a problem in putting them in their proper geological perspective.

Sample returns provide "ground truth."

We just talked about picking up a rock from a road cut. This is quite different from simply grabbing any old rock by the side of the road because that rock could have come from anywhere. In geological sampling it is important to have

(a)

(b)

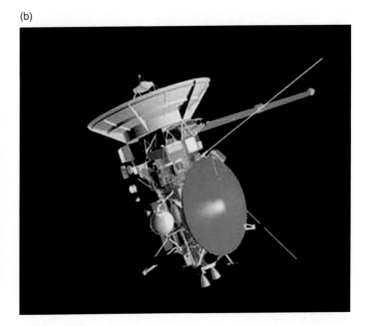

FIGURE 5.28 Explorers of the planets. (a) *Voyager* flew past Jupiter, Saturn, Uranus, and Neptune. (b) *Cassini* orbits Saturn.

samples from a source of known context. Some have claimed that we do not need samples returned from Mars because we already have the Martian meteorites. The problem is that we do not know where on Mars they came from and, therefore, how they fit into the planet's global geology. Plans are currently under way for unmanned "sample and return" missions to Mars.

Of course, we could not collect specimens in a national park without permission and a scientifically valid reason. Similarly, the return of extraterrestrial samples to Earth is governed by international treaties and standards to ensure that contamination of Earth does not occur. For example, before the lunar samples brought back by the Apollo missions could be studied, they (and the astronauts) were placed in quarantine and tested for alien life forms. The same international standards apply to spacecraft landing on planets. The goal of these standards is to avoid *forward contamination,* or transporting life forms from Earth to another planet. If there is life on other planets, then not only is there concern about introducing potential harm, but from a scientific perspective we do not want to "discover" life that we, in fact, have introduced.

With numerous missions under way and others on the horizon, unmanned exploration of the Solar System is an ongoing, dynamic activity. In our journey we will frequently refer to space missions and the information they return, but today's hot results may be tomorrow's old news in light of other, even more exciting discoveries. We hope that you will make use of the *21st Century Astronomy* website as a gateway to the wealth of exciting results that the future holds.

5.7 High-Energy Colliders

Ever since the early years of the 20th century, physicists have been peering into the structure of the atom by observing what happens when small particles collide. In 1906 **Ernest Rutherford** (1871–1937) found that positively charged **alpha particles**[9] are deflected when they pass through a thin sheet of mica, a shiny mineral that readily splits into thin layers. This discovery proved for the first time that atoms must contain heavy, centrally concentrated nuclei. By the 1930s physicists had developed the means to accelerate charged particles such as protons to very high speeds and then observe what happens when they slam into a target. From such experiments (which are continuing even today) physicists have discovered many kinds of **elementary particles**[10] and learned about their physical properties. High-energy particle colliders have proven to be an essential tool for physicists.

Why then do we regard the high-energy particle collider as a tool of the astronomer? How does the astronomer's interest in the largest objects in the universe relate to what happens on the very smallest scales? As we will see in

---

[9]Alpha particles are positively charged He^4 nuclei emitted in certain types of radioactive decay.

[10]In Chapter 4 we learned about three elementary particles: the proton, neutron, and electron. In Section 5.4 we were introduced to another, the neutrino, and in Foundations 14.1 we will meet still another, the positron. However, many other elementary particles are now known to physicists—in fact too many to list here.

FIGURE 5.29 The Fermi National Accelerator Laboratory in Batavia, Illinois.

facility, when it becomes operational in 2007, will produce collisions with nearly 20 times more energy than the Fermi Collider—about 3×10^{-6} joules.

5.8 High-Speed Computers

Imagine your life without computers. From laptops to desktops to larger servers and computers, we depend on computers to surf the Internet, process our data, and organize our daily lives. Tiny computer processors control every modern convenience from automobiles to cameras to washing machines. And as you might assume, computers are essential in the world of science. Data gathering, analysis, and interpretation are entirely dependent on computers—and the more powerful, the better. Consider, for example, analyzing a night's worth of astronomical images recorded by a very large CCD. A single image may contain as many

Chapter 21, to understand the very largest structures we see in the universe—indeed the large-scale universe itself—we need to understand the physics that took place during the earliest moments in the universe, when everything was unbelievably hot and dense. Although we have not yet reached that level of comprehension, the high-energy particle colliders that physicists use today are designed to lead

> Colliders teach us the physics we need to understand the early universe and the formation of structure.

us there. As we will see in Chapters 20 and 21, this knowledge may help us to understand such issues as the nature of the dominant matter and energy in the universe, why the universe consists of matter rather than **antimatter**, and whether there really is a beginning or an end to our universe.

Two factors determine the effectiveness of particle accelerators: the energy they can achieve and the number of particles they can accelerate. Whereas the first particle accelerator, the *cyclotron*, attained an energy of about 10^{-13} joules, modern particle colliders now reach much higher energies. For example, the accelerator at the Fermi National Accelerator Laboratory (see **Figure 5.29**) can accelerate protons up to 1.6×10^{-7} joules. This may seem like a small number, but it corresponds to more than a thousand times the rest mass energy (mc^2) of the proton. To put it in perspective, this is the energy of a flying mosquito all concentrated into a single tiny proton. Yet this energy will be dwarfed by the new Large Hadron Collider at CERN, the European Organization for Nuclear Research, shown in **Figure 5.30**. This

FIGURE 5.30 The tunnels and magnets for the Large Hadron Collider at CERN.

as 100 million pixels, with each pixel displaying roughly 10,000 levels of brightness. That adds up to a *trillion* pieces of information in each image! And that's only one image. To analyze their data, astronomers typically do calculations on *every single pixel* of an image in order to remove unwanted contributions from Earth's atmosphere or correct for instrumental effects. From the astronomer's point of view, without high-speed computers, the CCD would be just another electronic curiosity.

High-speed computers also play an essential role in generating and testing theoretical models of astronomical objects. Even when we completely understand the underlying physical laws that govern the behavior of some particular object, it is frequently the case that the object is so complex

that it would be impossible to calculate its properties and behavior without the assistance of high-speed computers. For example, as we learned in Chapter 3, we can use Newton's laws to easily compute the orbits of two stars that are gravitationally bound to one another because their orbits take the form of simple ellipses. However, it is not so easy to understand the orbits of the hundred billion (10^{11}) stars that comprise our Milky Way Galaxy, even though *the underlying physical laws remain the same*. If that were not complicated enough, consider the problem involving the collision of *two* such galaxies—and, for good measure, throw in some gas in addition to the stars. We can see the result in **Figure 5.31**. When we apply even the fastest computers available to this problem, the sequence shown in Figure 5.31 is only an

FIGURE 5.31 Numerical simulation of the merging of two galaxies, including the gravitational attraction of all forms of matter. (a) Just before the merge. (b) After passing through one another. (c) As the cores orbit each other and merge. (d) When the central core begins to settle down.

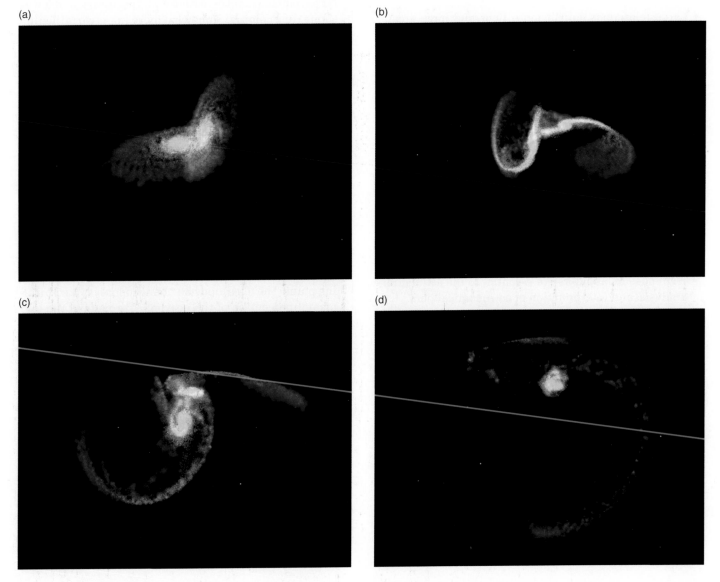

approximation of what we believe really happens. Modern computers have enough speed and memory to handle the behavior of a few million stars at best, so we are forced to assume that a single star in the computer simulation really represents hundreds of thousands of real stars.

Similar modeling procedures have worked well in determining the interior properties of stars and planets, including our own Earth. Although we cannot "see" beneath their surfaces, we have a surprisingly good understanding of their interiors, as we will learn in later chapters. We begin a model by assigning well-understood physical properties to tiny volumes within a planet or star. The computer assembles an enormous number of these individual elements into an overall representation of the complete body. When it is all put together, we have a rather good picture of what the interior of the star or planet is like.

Summary

- Optical telescopes come in two basic types, refractors and reflectors.

- All large astronomical telescopes are reflectors.

- Large telescopes collect more light and have greater resolution.

- The CCD is today's astronomical detector of choice.

- Earth's atmosphere blocks many spectral regions and distorts telescopic images.

- Putting telescopes in space solves problems created by Earth's atmosphere.

- Most of what we know about the planets comes from spacecraft we have sent there.

- Infrared and radio telescopes can see through vast clouds of cosmic gas and dust.

- Radio and optical telescopes can be arrayed to greatly increase angular resolution.

- High-speed computers are essential to the acquisition, analysis, and interpretation of astronomical data.

Seeing the Forest through the Trees

From our perspective in the 21st century, we can only imagine what went through the minds of our distant ancestors as they gazed upward toward the heavens. We might guess that they felt completely comfortable with the daily movements of the Sun and the Moon, yet experienced fear when confronted with an eclipse or the appearance of an occasional comet. They probably took for granted the tiny points of light that filled the nighttime sky, but must have wondered about those mysterious few that seemed to move freely among the others. With nothing more than their eyes to fulfill their curiosity, our early ancestors could only observe and wonder. But wondering alone does not give rise to comprehension. Insight far beyond personal experience was necessary, and it was a long time coming. Somewhat more than two millennia ago enlightenment of a sort came to the classical world. Greek philosophers concluded that these "wandering stars" or planets were unlike the distant canopy of fixed stars—that they were much closer—but stopped short of claiming to know *what* they were. And so these and other heavenly mysteries lingered on until the turn of the 17th century, when two historic events took place.

It all started when an obscure Flemish spectacle maker put a pair of lenses together and saw that distant objects appeared closer. Although Lippershey may have regarded his "looker" as an amusing toy, it was a Tuscan instrument maker who first recognized its potential for studying the heavens. It didn't take Galileo Galilei long to discover that the Moon was a nearby world covered with craters, and that those "mysterious moving points of light" were tiny disks of planets. This was the turning point. The telescope had changed forever our fundamental perception of the heavens. Throughout the four centuries since Galileo's historic discoveries, astronomers have continued to perfect the astronomical telescope. Reaching far outside the visible, telescopes now cover the entire electromagnetic spectrum from gamma rays to radio waves. Some telescopes reside on mountaintops, while others make their home in Earth orbit or nearby space. Still others take the long and arduous journey through interplanetary space to observe Earth's neighbors up close. But telescopes alone cannot provide all the answers. Tools of a different type have recently joined the astronomer's repertoire of gear. Some are buried deep underground quietly observing one of nature's most elusive particles. Others sit on desktops crunching numbers. All are devoted to decoding the cryptic messages sent to us from the cosmos.

Having stopped briefly to examine the many tools used by astronomers, we are now ready to resume our outward journey, armed with a new appreciation of how important it is to proceed carefully. Common sense is not enough. We must instead rely on our growing understanding of physical law and on rigorously tested predictions of carefully constructed theories. These also are useful tools that will allow us to look beyond the surface of spectacular vistas that otherwise would be devoid of sense or meaning or connection. Kepler's laws, Newton's laws, relative motion, Doppler shifts, Wien's Law, Stefan's Law, energy states, spectral lines, and the rest—these are the keys we will use to build an understanding of planets, stars, galaxies, and ultimately the universe itself. The first place in which we will bring these tools to bear will be in our immediate neighborhood as we consider the nature and origin of our Solar System. And to put our own Solar System in perspective, we'll take a look at some of the many other planetary systems that lie far beyond.

Key Terms

surface brightness, p. 136
refracting telescope, p. 136
aperture, p. 136
focal length, p. 137
reflecting telescope, p. 137
primary mirror, p. 137
reflection, p. 138
ray, p. 138
dispersion, p. 140
interference, p. 140
constructive interference, p. 140
destructive interference, p. 140
grating, p. 140
secondary mirror, p. 142
resolution, p. 142
diffraction, p. 142
diffraction limit, p. 142
astronomical seeing, p. 143
adaptive optics, p. 143
wavefront, p. 143
atmospheric window, p. 144
integration time, p. 145
quantum efficiency, p. 146
charge-coupled device (CCD), p. 147
pixel, p. 147
spectrograph, p. 148
radio telescope, p. 149

jansky, p. 149
interferometer, p. 150
interferometric array, p. 150
flyby, p. 153
orbiter, p. 153
lander, p. 154
rover, p. 154
atmospheric probe, p. 154

Student Questions

THINKING ABOUT THE CONCEPTS

1. Optical telescopes reveal much about the nature of astronomical objects. Why do astronomers also need information provided by gamma ray, X-ray, infrared, and radio telescopes?

2. The largest astronomical refractor has an aperture of 1 m. List several reasons why it would be impractical to build a still larger refractor with, say, twice the aperture.

3. Your camera may have a zoom lens, ranging between wide angle (short focal length) and telephoto (long focal length). How would the size of an object in the camera's focal plane differ between wide angle and telephoto?

4. CCDs are now the most commonly used astronomical detector. List three advantages that CCDs have over photographic emulsions.

5. Why do we not have groundbased gamma ray and X-ray telescopes?

6. Some people believe that we put astronomical telescopes in space because it gets them closer to the objects they are observing. As an enlightened student taking this class, you know better. Explain what is wrong with this popular misconception.

7. We have now sent various kinds of spacecraft—including flybys, orbiters, and landers—to all of the classical planets. Explain the advantages and disadvantages of each of these types of spacecraft.

8. Why are the world's largest telescopes located on high mountains?

9. "Twinkle, twinkle, little star. How I wonder what you are." Explain *why* stars twinkle. (In a later chapter we will find out *what* the stars are.)

10. Humans had our first look at the far side of the Moon only as recently as 1959. Why had we not been able to see it earlier—say, when Galileo first observed the Moon with his telescope in 1610?

APPLYING THE CONCEPTS

11. Compare the light-gathering power of a large astronomical telescope (aperture = 10 m) with that of the dark-adapted human eye (aperture = 7 mm).

12. Compare the angular resolution of the Hubble Space Telescope (aperture = 2.4 m) with that of a typical amateur telescope (aperture = 20 cm).

13. Assume that the maximum aperture of the human eye, D, is approximately 7 mm and the average wavelength of visible light, λ, is 5.5×10^{-4} mm.
    a. Calculate the diffraction limit of the human eye in visible light.
    b. How does this compare with its actual resolution of 1 to 2 arcminutes (60 to 120 arcseconds)?
    c. To what do you attribute the difference?

14. The diameter of the full Moon's image in the focal plane of an average amateur's telescope (focal length = 1.5 m) is 13.8 mm. How big would the Moon's image be in the focal plane of a very large astronomical telescope (focal length = 250 m)?

15. One of the earliest astronomical CCDs had 160,000 pixels, each recording 8 bits (256 levels of brightness). Today's CCDs may contain 100 million pixels, each recording 15 bits (32,768 levels of brightness). Compare the number of bits of data that each produces in single image.

16. The VLBA employs an array of radio telescopes ranging across 8,000 km of Earth's surface from the Virgin Islands to Hawaii.
    a. Calculate the angular resolution of the array when radio astronomers are observing interstellar water molecules at a microwave wavelength of 1.35 cm.
    b. How does this compare with the angular resolution of two large optical telescopes separated by 100 m and operating as an interferometer at a visible wavelength of 550 nm?

17. Rovers *Spirit* and *Opportunity* can move across the Martian landscape at speeds of up to 5 cm/s. In contrast, our typical walking speed is about 4 km/h.
    a. How quickly could *Spirit* move the length of a football field (91.44 m)?
    b. Compare this with the time it would take to walk the same distance.

18. *Voyager 1* is now about 100 AU from Earth, continuing to record its environment as it approaches the limits of our Solar System.
    a. What is the distance of *Voyager 1* expressed in kilometers?
    b. How long does it take observational data to get back to us from *Voyager 1*?
    c. How does its distance compare with that of the nearest star (other than the Sun)?

StudySpace
wwnorton.com/astro21
provides a Study Plan for each chapter that includes a reading outline, animations, keyword flash cards, and gradebook-enabled multiple-choice quizzes. From StudySpace you can also access premium content in the ebook and SmartWork.

PART II

The Solar System

You are a child of the universe no less than the trees and the stars; *you have a right to be here,* and whether or not it is clear to you, no doubt the universe is unfolding as it should.

DESIDERADA (AUTHOR UNKNOWN)

From clouds of gas and dust, solar systems such as ours are born.

CHAPTER 6

A Brief History of the Solar System

6.1 Forming Stars and Evolving Planets

In thinking about how it all began, a good place to start is right here in our own small corner of the universe. We will begin this chapter by learning how the Solar System formed, and we will then see that it is really only one of an enormous number of other **planetary systems** scattered throughout the galaxy. Later, in Part IV, we will learn about the birth of the much vaster universe. Do not confuse the two. Planetary systems are infinitesimal compared to the universe as a whole. On average, light takes a mere 4 hours and 9 minutes to travel here from Neptune, the outermost classical planet in our Solar System. Light from the most distant galaxies takes nearly *14 billion years* to reach us!

Until the last part of the 20th century, every discussion of the Solar System would necessarily start with an accounting of its pieces. "There are planets with such and such properties. There are **moons**, there are **asteroids**, and there are **comets**. And so forth." Any mention of the origin of the Solar System would wait until the end of the discussion—the part of a work to which speculation is generally relegated. Yet the last few decades have seen this ordering turned on its head. Today astronomers studying the formation of stars and planetary scientists studying clues about the history of the Solar System find themselves arriving at the same picture of our early Solar System but from two very different directions. This unified understanding provides the foundation for the way we now think about the Sun and the myriad objects that orbit about it. On the other hand, other issues may be less clear. How astronomers defined planets in the past was called into question. In August

KEY CONCEPTS

In these early years of the 21st century we have come to see our Solar System as an unmistakable by-product of the birth of the Sun, rather than a random collection of planets and moons. It is this understanding that brings order to what we see around us today. Here we begin our investigation of the Sun's family by recounting something of the story of the formation of the Solar System—a story in which we will learn that

- A star forms when a cloud of interstellar gas and dust collapses under its own weight.
- While a star forms it is surrounded by a flat, rotating disk that provides the raw material from which a planetary system might form.
- Dust grains in the disk around a young star stick together to form larger and larger solid objects.
- Differences in temperature from place to place within the disk determine the kinds of materials from which solid objects can form.
- Giant planets form when solid planet-sized bodies capture gas from the surrounding disk.
- The atmospheres of smaller terrestrial planets are gases released by volcanism and volatile materials that arrived on comets.
- Planetary systems around other stars are common.

International Astronomical Union (IAU) issued a
inition of a planet (see Appendix 8). Under this new
ition, Pluto is no longer considered a classical planet
but is instead defined as a "dwarf planet." Ceres, once a
planet, then an asteroid, is now a dwarf planet, as is Eris,
previously known as a large KBO (see Chapter 12).

Later in our journey we will extend our attention beyond
the confines of our Solar System and learn about the incomprehensibly vast and tenuous clouds of gas that fill the expanse of interstellar space. We will discuss how, when

Young stars are surrounded by rotating disks.

conditions are right, this gas collapses under the force of
its own gravity to form stars. In this chapter we will skip
the details of this process, except for one important fact.
For various reasons, some of which have to do with why a
spinning ball of pizza dough spreads out to form a flat crust,

the cloud that eventually produced the Sun collapsed first
not into a ball but instead into a rotating disk. Most of the
material in the disk eventually either traveled inward onto
the forming star at its center or was thrown back into interstellar space. However, a small fraction of the material in
the disk was left behind. As astronomers' tools improved,
this scenario of how stars form was confirmed by discovery
after discovery of disks of gas and dust surrounding young
stellar objects like the ones shown in **Figure 6.1**.

During the same years that astronomers were beginning
to ferret out the secrets of star formation, other groups of
scientists with different backgrounds—mainly geochemists and geologists—were piecing together the history of
our Solar System. Some of the characteristics that the early
Solar System must have had are fairly obvious. The orbits
of all of the planets in the Solar System lie very close to a
single plane, which says that the early Solar System must
have been flat. The fact that all the planets orbit the Sun in
the same direction says that the material from which the

(a)

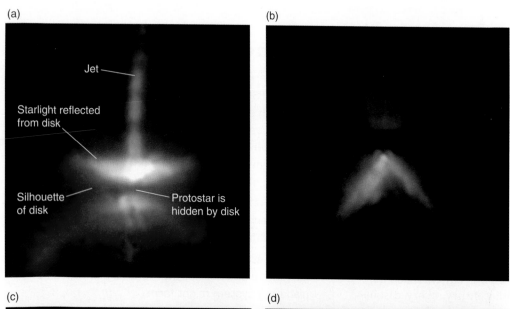

(b)

(c)

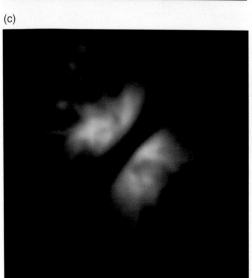

(d)

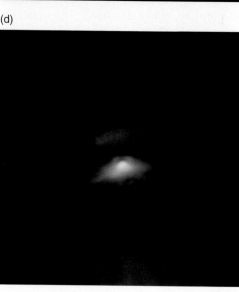

FIGURE 6.1 Hubble Space Telescope images showing accretion disks around newly formed stars. The dark bands are the shadows of the disks seen more or less edge on. Bright regions are dust illuminated by starlight. Some disk material may be expelled perpendicular to the plane of the disk in the form of violent jets.

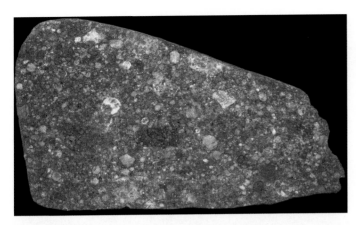

FIGURE 6.2 Meteorites are fragments of the young Solar System that have fallen to Earth. It is clear from this cross section that this meteorite formed from many smaller pieces that stuck together.

planets formed must have been swirling about the Sun in the same direction as well. Other clues about what the early Solar System was like were harder to puzzle out. *Meteorites*, for example, include bits and pieces of material that are left over from the Solar System's youth. These fragments of the early Solar System can be captured by Earth's gravity and fall to the ground, where they can be picked up and studied. Many meteorites, such as the one in **Figure 6.2**, look something like a piece of concrete in which pebbles and sand are mixed in with a much finer filler. This structure is surely telling us *something* about how these pieces of interplanetary debris formed, but *what*?

Beginning in the 1960s a flood of information about Earth and other objects in the Solar System poured in from a host of sources including space probes, groundbased telescopes, laboratory analysis of meteorites, and theoretical calculations. Scientists working with this wealth of information began to see a pattern. What they were learning made sense

The Solar System formed from a rotating disk of gas and dust.

only if they assumed that the larger bodies in the Solar System had grown from the aggregation of smaller bodies. Following this chain of thought back in time, they came to envision an early Solar System in which the young Sun was surrounded by a flattened disk of both gaseous and solid material (see the opening paragraph for this chapter). This swirling disk of gas and dust provided the raw material from which the objects in our Solar System would later form.

While reading the previous few paragraphs, you may have noticed a remarkable similarity between the disks that astronomers find around young stars and the disk that planetary scientists hypothesize as the cradle of the Solar System. This similarity is not happenstance. As astronomers and planetary scientists compared notes, they realized they had arrived at the *same* picture of the early Solar Sys-

tem from two completely different directions. The rotating disk from which the planets formed was none other than the remains of the disk that accompanied the formation of the Sun. The planet we live on, along with all of the other orbiting bodies that make up our **Solar System**, evolved from the remnants of the interstellar cloud that collapsed to form our local star, the Sun.

The connection between the formation of stars and the origin and subsequent evolution of the Solar System has become one of the cornerstones of both astronomy and planetary science—a central theme around which a great deal of our understanding of our Solar System revolves. As we begin the 21st century, the story of the Sun's formation and the history of the material in the surrounding disk bring order to our understanding of our Solar System.

6.2 In the Beginning Was a Disk

Later in the book we will turn our attention beyond the boundaries of our Solar System to the process of star formation itself. For now it is enough to jump into this story of star formation, midstream. Begin by holding the picture shown in **Figure 6.3** firmly in your mind. Roughly 5 billion years ago the newly formed Sun was adrift in interstellar space. The Sun was not yet a star in the true sense of the word because the nuclear fires that power the Sun today had yet to ignite. It was still a **protostar**—a large hot ball of gas that shone due to gravitational energy being turned into thermal energy and radiation as the protostar collapsed from a cloud of interstellar gas. (The often-used prefix *proto-* means "early form" or "in the process of formation.") Surrounding the protostellar Sun was a flat, rotating disk of gas and dust called the **protostellar disk**. "Orbiting" is perhaps a better word than "rotating" because each bit of the material in this thin disk was orbiting around the Sun according to the same laws of motion and gravitation that govern the orbits of the planets today. The disk around the Sun was much like the disks that astronomers see today surrounding protostars and newly formed stars elsewhere in our galaxy. This disk is referred to as a **protoplanetary disk**, which is a protostellar disk capable of producing planets. It probably contained less than 1 percent as much mass as the nascent star at its center, but this amount was more than enough to account for the bodies that make up the Solar System today.

Okay, so we know the Solar System formed from a protoplanetary disk and that disks are seen around newly formed stars—but *why* is this the case? What is it about the process of star formation that leads not only to a star itself, but to a flat orbiting collection of gas and dust as well? The answer to this question lies with a skater spinning on the ice and something called *angular momentum*.

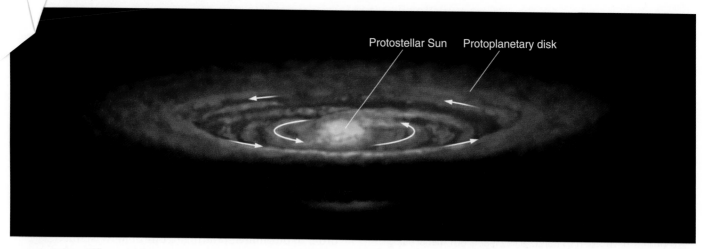

Protostellar Sun   Protoplanetary disk

FIGURE 6.3 When you think of the young Sun, think of it as being surrounded by a flat but flared rotating disk of gas and dust.

The Collapsing Cloud Rotates

We have all seen an Olympic skater spinning on the ice (see **Figure 6.4**). Like any rotating object or isolated group of objects, the spinning ice skater has some amount of **angular momentum**. The amount of angular momentum (L) that an object possesses depends on three things. First, it depends on how fast (v) the object is rotating: The faster an object is rotating, the more angular momentum it has. A top that is spinning rapidly has more angular momentum than the same top does when it is spinning slowly. Second, an object's angular momentum depends on the mass (m) of the object. Suppose you compare two spinning tops. Both tops have the same size, shape, and rate of spin. They are the same except for the fact that one top is made of lead while the other top is made of balsa wood. The lead top has more angular momentum. The third thing that angular momentum depends on is how the mass of the object is distributed —how "spread out" (r) the object is. For an object of a given mass and rate of rotation, the more spread out the object is, the more angular momentum it has. An object that is rotating slowly but is very spread out might have more angular momentum than a more rapidly rotating but more compact object. Putting this together, we have

$$L = m \times v \times r.$$

In the case of *orbital* angular momentum, such as the orbital angular momentum of the Moon about Earth, v is the Moon's orbital speed, m is the Moon's mass, and r is the Moon's orbital radius. It is a bit more complicated to determine the *rotational* or *spin* angular momentum of a single object, such as the spinning top just referred to or a rotating planet or interstellar gas cloud. Here we must add up the individual angular momentum of every tiny mass element within the body. For example, the spin angular momentum of a uni-

form sphere is proportional to the square of its radius. See Student Question 13 at the end of the chapter for details.

What makes angular momentum such an important and useful idea in physics and astronomy is that *the amount of angular momentum possessed by an object or isolated group*

FIGURE 6.4 A figure skater can change the speed with which she spins simply by invoking the conservation of angular momentum.

of objects does not change unless those objects are affected in just the right way by something other than themselves. This statement is referred to as the law of **conservation of angular momentum**. In the parlance of physics, if something is "conserved," it means that the amount of that quantity does not change by itself. This idea might remind you of Newton's first law of motion, which says that in the absence

Angular momentum is conserved.

of some external force, an object continues to move in a straight line at a constant speed. Indeed, both Newton's first law and the conservation of angular momentum are examples of **conservation laws**. There are many other conservation laws in physics, including the law of conservation of momentum, the law of conservation of energy (discussed later in this chapter), and the law of conservation of electric charge.

Conservation of angular momentum brings us back to the ice skater we opened the section with and to the collapsing interstellar cloud. You have probably noticed that an ice skater can control how rapidly she spins by doing nothing other than pulling in or extending her arms or legs. A compact object must spin more rapidly to have the same amount of angular momentum as a more extended object with the same mass. As our skater spins, her angular momentum does not change much. (The slow decrease is due to friction, an external force.) When her arms and leg are fully extended, she spins slowly; but as she pulls her arms and leg in she spins faster and faster. With the skater's arms held tightly in front of her and one leg wrapped around the other, her spin becomes a blur. She finishes with a flourish by throwing her arms and leg out, which abruptly slows her spin. Despite the dramatic effect, her angular momentum remains the same throughout. This impressive athletic spectacle comes courtesy of the law of conservation of angular momentum, and from the difference between an extended object and a compact object.

Now we turn our attention to how conservation of angular momentum affects a forming star. We begin with a cloud of interstellar gas that is collapsing under the force of its own gravity. It might seem most natural for the cloud to collapse directly into a ball—and so it would, but for the cloud's own angular momentum. *Interstellar clouds* are truly vast objects, light-years in size. (Recall that a light-year is the distance traveled by light in one year, or about 9.5 trillion kilometers.) As interstellar clouds orbit about the galaxy's center, they are constantly being pushed around by stellar explosions or by collisions with other interstellar clouds. This constant "stirring" guarantees that all interstellar clouds will have *some* amount of rotation. As spread out as an interstellar cloud is, even a tiny amount of rotation corresponds to a huge amount of angular momentum. Imagine our ice skater now with arms that reach from here to the other side of Earth. Even if she were rotating very slowly at

Interstellar clouds have far more angular momentum than the stars they form.

first, think how fast she would be spinning by the time she pulled those long arms to her sides! Just as the ice skater speeds up when she pulls in her arms, the cloud rotates faster and faster as it collapses. Suppose, for example, that we start with a cloud that is about a light-year across—say 10^{16} meters —and is rotating so slowly that it takes a million years to complete one rotation. By the time such a cloud collapsed to the size of our Sun—a mere 1.4×10^9 meters across, or only one 10-millionth the size of the original cloud—it would be spinning 50 trillion times faster, completing a rotation in only 0.6 of a second! This is over 3 million times faster than our Sun is actually spinning. At this rate of rotation, the Sun's self-gravity would have to be almost 200 million times stronger to hold the Sun together!

An Accretion Disk Forms

There is a puzzle here. Conservation of angular momentum would seem to say that stars cannot form from collapsing interstellar clouds, yet there is no other way for stars to form. When scientists find what appears to be a contradiction—like the apparent contradiction between the principle of conservation of angular momentum and the fact that a star has far less angular momentum than the cloud that formed it—it is cause for great excitement. Such seeming contradictions do not mean that nature is breaking the rules. (Nature *never* breaks its own rules!) Instead it means that our understanding of what is going on is incomplete or that we have the wrong rule. It means that we have found a place where there are new things to be learned.

The key to solving the riddle of angular momentum in a collapsing interstellar cloud lies in realizing that the *direction* of the collapse is important. The cloud's rotation may thwart the collapse of the cloud toward its axis of rotation, but there is nothing to prevent collapse *parallel* to the axis of rotation (see **Figure 6.5**). Instead of collapsing into a ball, the interstellar cloud becomes flattened as it collapses. As the cloud collapses more and more, the strength of gravity causing the collapse gets greater and greater. Eventually the flattening cloud reaches a point where the

The cloud collapses into a disk rather than directly into a star.

inner parts of the cloud begin to fall freely inward, raining down on the growing object at the center. As this happens, the outer portions of the cloud lose the support of the collapsed inner portion of the cloud, and they start falling inward too. The whole cloud collapses inward, much like a house of cards with the bottom layer knocked out. As this

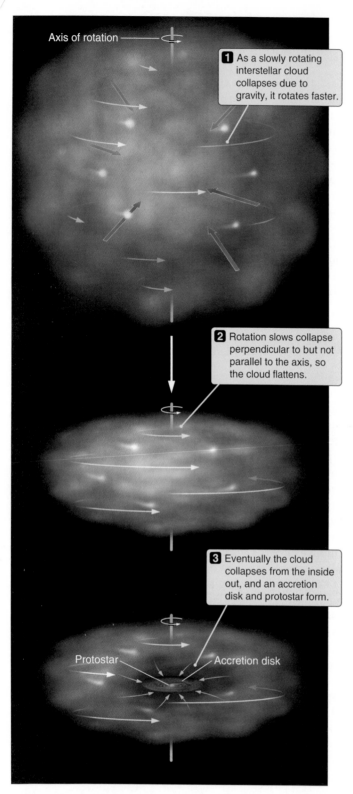

Axis of rotation

1 As a slowly rotating interstellar cloud collapses due to gravity, it rotates faster.

2 Rotation slows collapse perpendicular to but not parallel to the axis, so the cloud flattens.

3 Eventually the cloud collapses from the inside out, and an accretion disk and protostar form.

Protostar   Accretion disk

FIGURE 6.5 A rotating interstellar cloud is free to collapse parallel to its axis of rotation, but not perpendicular to that axis. As a result, the cloud collapses into a disk.

material makes its final inward plunge, it lands on a thin, rotating, pizza-dough-like structure called an **accretion disk**. The accretion disk serves as a way station for material on its way to becoming part of the star that is forming at its center.

Formation of accretion disks occurs in many situations in astronomy, so it is worth taking a moment to think carefully about this process. We can use what we learned about orbits in Chapter 3 to better understand what happens during this final stage of the collapse of an interstellar cloud. As the material falls toward the forming star, it travels on curved, almost always elliptical paths, just as Kepler's laws say that it should. These orbits would carry the material around the forming star and back into interstellar space except for one problem: The path inward toward the forming star is a one-way street. When material nears the center of the cloud, it runs headlong into material that is falling in from the *other* side. Imagine a huge rotary, or traffic circle with lots of entrances but no exits (**Figure 6.6**). As traffic flows into the traffic circle, it has nowhere else to go, resulting in a continuous growing line of traffic driving around and around. Eventually, as more and more cars try to pack in, the traffic piles up. This situation is analogous to an accretion disk. As material falls onto the disk, its motion perpendicular to the disk stops abruptly, but its motion parallel to the surface of the disk adds to the disk's angular momentum. The angular momentum of the infalling material has been transferred to the disk.

Traffic in a traffic circle moves on a flat surface, but the accretion disk around a protostar forms from material coming in from all directions in three-dimensional space. Where the disk forms—the *plane* of the accretion disk—is determined by a balance between the amounts of material falling onto the disk from each side of the disk. There is only one plane that satisfies this requirement: the plane perpendicular to the cloud's axis of rotation. The gas settles down into a rotating accretion disk that has a radius of hundreds of astronomical units—and thousands of times greater than the radius of the star that will eventually form at its center —making it large enough to accommodate the angular momentum of the infalling material.

The next obvious question concerns how material falling onto the accretion disk finds its way inward onto the growing protostar. We will pick up this question in Chapter 15, when we return to follow the story of the growth and evolution of the star at the center of the disk. For now it is enough to know that most of the matter that lands on the accretion disk either ends up as part of the star or is ejected back into interstellar space, sometimes in the form of violent jets as seen Figure 6.1(a). However, a small amount of material is left behind in the disk. It is this leftover disk —the dregs of the process of star formation—to which we next turn our attention.

Before going on with our story, however, we should stress that theoretical calculations by astronomers have long pre-

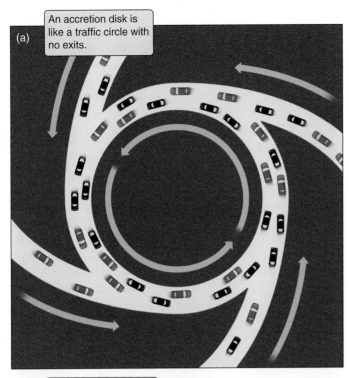

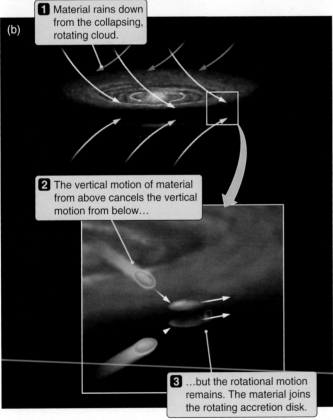

FIGURE 6.6 (a) Traffic piles up in a traffic circle with entrances but no exits. (b) Similarly, gas from a rotating cloud falls inward from above and below, piling up onto a rotating disk.

dicted that accretion disks should be found around young stars, and illustrations like Figure 6.3 have been a mainstay in textbooks for years. In the last decade or so of the 20th century, however, these cartoon drawings had to make room for images of the real thing. Figure 6.1 shows Hubble Space Telescope images of edge-on accretion disks around young stars. The dark bands are the shadows of the edge-on disks, the top and the bottom of which are illuminated by light from the forming star. It would be nice if we could go back 5 billion years and watch as our own Sun formed from a cloud of interstellar gas, but we do not have to. All we have to do is look at objects like the ones in Figure 6.1 to know what we would have seen.

6.3 Small Objects Stick Together to Become Large Objects

The chain of events that connects the accretion disk around a young star to a planetary system such as our own begins with the same basic physics that leaves you with dust in your eye on a windy day. A stiff breeze picks up dust and sand and blows them about but leaves the pebbles and rocks

---

Gas motions push small particles into larger particles.

---

behind. In like fashion, the random motions of the gas within the protoplanetary disk push the smaller grains of solid material back and forth past the larger grains; and as this happens, the smaller grains stick to the larger grains.

In matters of social dynamics we sometimes hear that "the rich get richer at the expense of the poor." Although we can always debate this social axiom, the principle certainly holds true when it comes to the dynamics within a protoplanetary disk, as seen in **Figure 6.7**. The larger dust grains get larger at the expense of the smaller grains. Starting out at only a few microns (micrometers) across, the slightly larger bits of dust grow to the size of pebbles, then to the size of boulders. The rate at which objects grow in this way is thought to decrease when clumps of boulders are about 100 meters across. Such objects are so few and far between in the disk that chance collisions become less and less frequent. Even so, the process of growth continues at a slower pace as 100-meter clumps join together to produce still larger bodies.

In order for two such clumps to stick together, they must bump into each other gently, very gently. Otherwise the energy of collision would cause the two colliding bodies to fragment into many smaller pieces instead of forming a

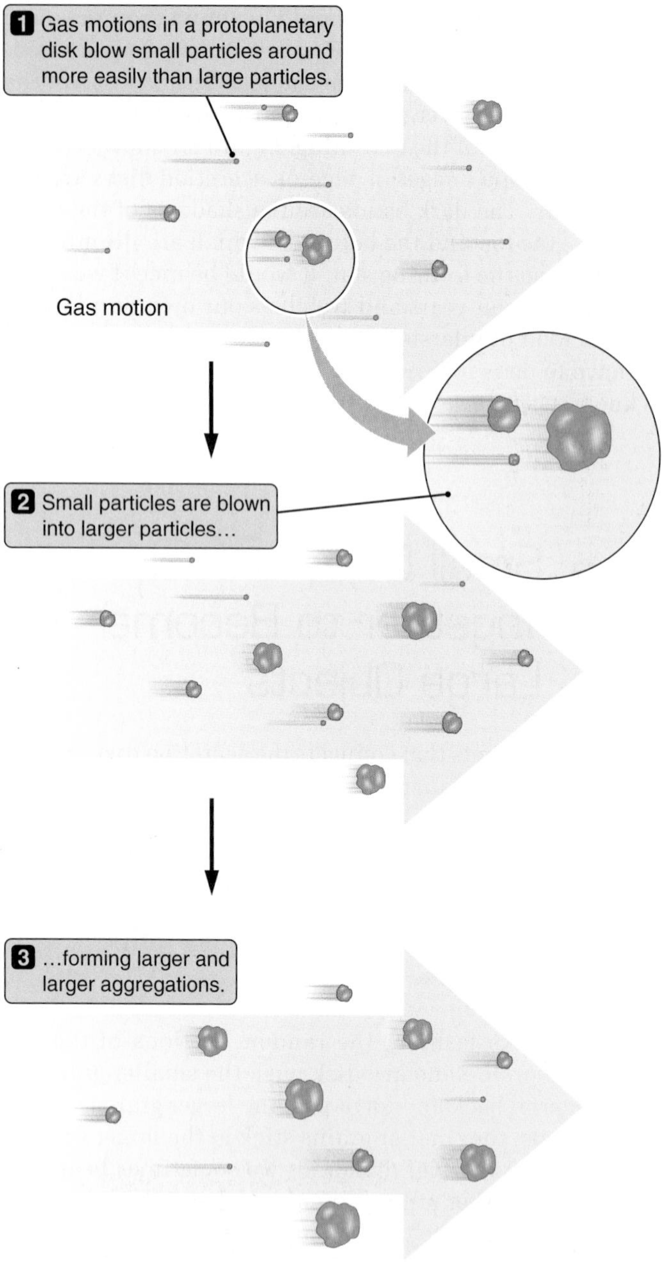

1 Gas motions in a protoplanetary disk blow small particles around more easily than large particles.

Gas motion

2 Small particles are blown into larger particles…

3 …forming larger and larger aggregations.

FIGURE 6.7 Motions of gas in a protoplanetary disk blow smaller particles of dust into larger particles, making the larger particles larger still. This process continues, eventually making objects many meters in size.

single larger one. Typical collision speeds cannot be much greater than 0.1 m/s if colliding boulders are to stick together. If you were to walk that slowly, it would take you 15 minutes to travel the length of a football field. In a real accretion disk, collisions more violent than this certainly happen on occasion, breaking these clumps back into smaller pieces. Likely there are many reversals in this growth of clumps of boulders.

Up to this point, larger objects have grown mainly by "sweeping up" smaller objects that run into them or that get in their way. As the clumps reach a size of about a kilometer, a different process becomes important (see **Figure 6.8**). These kilometer-sized objects, now called **planetesimals** (literally "tiny planets"), are massive enough that their

Gravity helps planetesimals grow into planets.

gravity starts to be important, exerting a significant attraction on nearby bodies. No longer is growth of the planetesimal fed only by chance collisions with other objects: The planetesimal's gravity can now pull in and capture other smaller planetesimals that lie outside its direct path. The growth of planetesimals speeds up, with the larger planetesimals quickly consuming most of the remaining bodies in the vicinity of their orbits. The final survivors of this process are now large enough to be called **planets**. As with the major bodies in orbit about the Sun, some of the planets may be small and others quite large.

6.4 The Inner Disk Is Hot, but the Outer Disk Is Cold

The accretion disks surrounding young stars form from interstellar material that may have a temperature of only a few kelvins, but the disks themselves reach temperatures of hundreds of kelvins or more. What is it that heats up the disk around a forming star? The answer lies with our old friend gravity. Material from the collapsing interstellar cloud falls inward toward the protostar, but because of its angular momentum it "misses," falling instead onto the surface of the disk. When the material that is raining onto the disk hits the disk, its bulk motion comes to an abrupt halt, and the velocity that the atoms and molecules in the gas had before hitting the disk is suddenly converted into random *thermal* velocities instead. That is to say, the cold gas that was falling toward the disk gets very hot when it lands on the disk.

To help you visualize this, imagine dumping a box of marbles from the top of a tall ladder onto a rough, hard floor below (**Figure 6.9**). The marbles fall, picking up speed as they go. Even though the marbles are speeding up, however, they are all speeding up *together*. As far as one marble is concerned, the other marbles are not moving very fast at all. (If you were riding on one of the marbles, the other marbles would not appear to you to be moving very much; it would be the rest of the room that was whizzing by.) The atoms and molecules in the gas falling toward the protostar are like these marbles. They are picking up speed as they fall as a group toward the protostar, but the gas is still *cold* because the random *thermal* velocities of atoms and molecules with respect to each other are still low. Now imagine what happens when the marbles hit the rough floor. They bounce every which way. They are still moving rapidly, but they are no longer moving together. A change has taken place from the ordered motion of marbles falling together to the random motions of marbles traveling in all directions. The atoms and molecules in the gas falling toward the central star behave in the same fashion when they hit the disk. They are no longer moving as a group, but their random "thermal" velocities are now very large. The gas is now *hot*.

Another way to think about why the gas that falls on the disk makes the disk hot is to apply the ever-useful concept of conservation of energy. The law of **conservation of energy** means that unless energy is added to or taken away from a system from the outside, the total amount of energy in the system must remain constant. But the *form* the energy takes *can* change.

Imagine lifting a heavy object—say a brick. It is hard to do because you are working against gravity. It takes energy to lift the brick, and conservation of energy says that energy is never lost. But where does that energy go? It is changed into a form called **gravitational potential energy**. In a sense

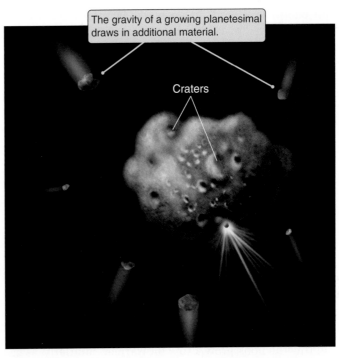

FIGURE 6.8 The gravity of a planetesimal is strong enough to attract surrounding material, causing the planetesimal to grow.

FIGURE 6.9 (a) Marbles dropped as a group fall together until they hit a rough floor, at which point their motions become randomized. (b) Similarly, atoms in a gas fall together until they hit the accretion disk, at which point their motions become randomized, which raises the temperature of the gas.

(a)

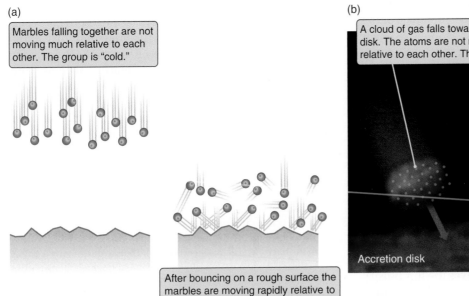

(b)

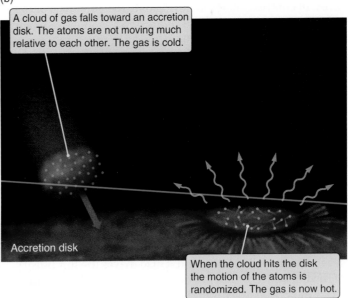

this energy is "stored" in a way that is reminiscent of how energy is stored in a battery. Potential energy is energy that "has potential"—it is waiting to show up in some more obvious form. If you drop the brick, it falls, and as it falls, it speeds up. The gravitational potential energy that was stored is being converted instead to energy of motion, which, as you may recall from Chapter 4, is called *kinetic energy.* When the brick hits the floor, it stops suddenly. The brick loses its energy of motion, so what form does this energy take now? If the brick cracks, part of the energy goes into breaking the chemical bonds that hold it together. Some of the energy is converted into the sound the brick makes when

---

The gravitational energy of infalling material turns into thermal energy.

---

it hits the floor. Some goes into heating and distorting the floor. But *most* of the energy is converted into thermal energy. The atoms and molecules that make up the brick are moving about within the brick a bit faster than they were before the brick hit; so the brick and its surroundings, including the floor, grow a tiny bit warmer. Similarly, as gas falls toward the disk surrounding a protostar, gravitational potential energy is converted first to kinetic energy, so the gas picks up speed. When the gas hits the disk and stops suddenly, that kinetic energy is turned into thermal energy. (**Foundations 6.1** discusses why it can be useful to think about the same thing in different ways.)

In this way, as material falls onto the accretion disk around a forming star, the disk is made hot. Material hitting the inner part of the disk (which we will call the *inner disk*) has fallen far in the gravitational field of the forming star. Like a rock dropped from a tall building, material hitting the inner part of the disk is moving quite rapidly when it hits the disk, so it heats the inner disk to high temperatures. In contrast, material falling onto the outer part of the disk (which we will call the *outer disk*) is moving much more slowly (envision a rock dropped from only a foot or

---

The inner disk is hotter than the outer disk.

---

so), leaving the outermost parts of the disk at temperatures that may be little higher than the original interstellar cloud. Stated another way, material falling onto the inner disk converts more gravitational energy into thermal energy than does the material falling onto the outer disk.

The energy released as material falls onto the disk is not the only source of thermal energy in the disk. Even before the nuclear fires that will one day power the new star have ignited, conversion of gravitational energy into thermal energy drives the temperature at the surface of the protostar to several thousand kelvins, and it also drives the luminosity of the huge ball of glowing gas to many times the luminosity of the present-day Sun. For the same reasons that Mercury is hot while Pluto is not (see Section 4.5), the radiation

streaming outward from the protostar at the center of the disk drives the temperature in the inner parts of the disk even higher, increasing the difference in temperature between the inner and outer disk.

Rock, Metal, and Ice

We have all noticed that temperature affects which materials can and cannot exist in a solid form. On a hot summer day, ice melts and water quickly evaporates, whereas on a cold winter night even the water in our breath freezes into tiny ice crystals before our eyes. Some materials, such as iron, **silicates**, and carbon—rocky materials and metals—remain solid even at quite high temperatures. Such materials, which are capable of withstanding high temperatures

---

Refractory materials remain solid even at high temperatures.

---

without melting or being vaporized, are referred to as **refractory materials**. Other materials, such as water, ammonia, and methane, can remain in a solid form only if their temperature is quite low. These less refractory substances are called **volatile materials**. The solid form of a volatile material is generally referred to as an **ice**.[1]

Differences in temperature from place to place within the disk have a significant effect on the makeup of the dust grains in the disk (**Figure 6.10**). In the hottest parts of the

---

Volatile ices survive in the outer disk, but only refractory solids survive in the inner disk.

---

disk (closest to the protostar), only the most refractory substances can exist in solid form. In the inner disk, dust grains are composed of refractory materials only.[2] Somewhat farther out in the disk, some hardier volatiles such as water ice and certain **organic** substances can survive in solid form, adding to the materials that dust grains are made of.[3] Highly volatile components such as methane, ammonia, and carbon monoxide ices and some organic molecules survive in solid form only in the coldest, outermost parts of the accretion disk, far from the central protostar. The differences in

---

[1] In scientific usage, the term *ice* is used to describe the volatile itself, whether it is in solid, liquid, or gaseous form.
[2] Some geochemists believe that certain volatiles, such as water, did survive in the hot inner disk because they were bound chemically to refractory materials and were thus able to withstand the high temperatures.
[3] The term *organic* does not mean "life" but refers instead to a large class of chemical compounds containing the element carbon. All terrestrial life is organic, but not all organic compounds come from living organisms.

Thinking about Energy in Different Ways

In the text we discuss why the gas falling onto a proto-stellar disk heats the disk. First we give an example of marbles falling on a floor, noting how the motion of marbles is jumbled up when they hit, like the atoms in the gas hitting the disk. Then we give a completely different explanation, talking about how energy is conserved but changes its form from gravitational energy to kinetic energy and finally to thermal energy. These may be two different explanations of why the disk gets hot, but they are *not* two different *reasons* why the disk gets hot. The protostar gets hot as a result of the same physical process—the "reason" is the same—regardless of the words we use to describe it. Instead these two explanations offer two different ways to *think about* why the disk gets hot.

Both ways of thinking about the process are correct. Both are included here because sometimes students—or scientists—need to look at the same thing from several different directions before they understand it. In this case, even though both ways of thinking about the process are correct, most scientists would agree that the second way of thinking about the problem is far more *powerful* than the first. The idea of energy changing forms connects how disks are heated with a wide variety of other phenomena. Properly stated, the heating of proto-stellar disks is an example of one of the most far-reaching patterns in nature: the conservation of energy.

Once you understand the different forms of energy and how energy is conserved, you begin to see this pattern of nature everywhere. For example, when water falls through the turbines in a hydroelectric generator,

it turns the generator and produces electric energy that is carried out over power lines and lights up the lights in your home. Where did the energy to heat the water that pours from your tap come from? If you have an electric water heater and get your power from a hydroelectric plant, it comes from the gravitational potential energy of the water in a reservoir near you—just as the energy to heat the newly formed Sun came from the gravitational potential energy of the reservoir of gas from which the Sun formed.

Conservation of energy is also a more powerful way to think about the heating of disks around young stars because it allows astronomers to *calculate* how hot these disks get. If we know the mass of the protostar and the surrounding disk, and how far away from the protostar the gas is falling from, we can calculate how much gravitational energy the gas started out with. According to the conservation of energy, this gravitational energy must eventually be converted to thermal energy, so we can calculate the expected temperature of the protostar. So without calculating any details about how the disk formed, we can say right off the bat how much thermal energy will be deposited onto the disk.

Scientists spend much of their time trying to come up with new ways of "thinking about" problems, looking for particularly powerful ways that point to new insights and discoveries. The most powerful means of thinking about a problem usually tie the problem to ever grander patterns in nature. Conservation of energy is one of the grandest and most useful patterns around.

composition of dust grains within the disk are reflected in the composition of the planetesimals formed from that dust. Planets that form closest to the central star tend to be made up mostly of refractory materials such as rock and metals. Those that form farthest from the central star also contain refractory materials, but they contain in addition large quantities of ices and organic materials.

As we turn to a study of our own Solar System, we will find that the trend in composition expected in a protoplane-

tary disk is closely echoed in the makeup of the solid bodies orbiting the Sun. The inner planets are composed of rocky material surrounding metallic cores of iron and nickel. In contrast, objects in the outer Solar System, including moons, giant planets, KBOs, and comets, are composed largely of ices of various types. In the years to come, as astronomers learn more about planetary systems around other stars, it seems likely that this trend will prove to be quite common. In fact, based on our understanding of the way stars and

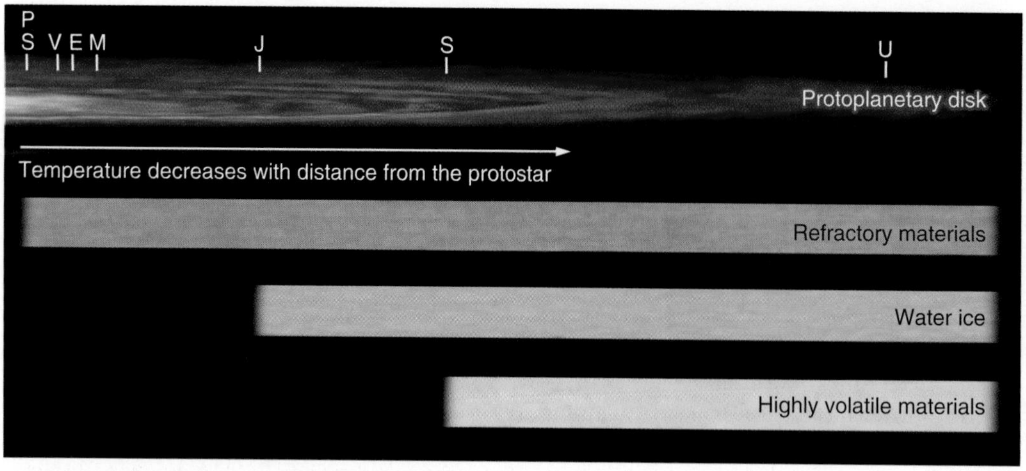

FIGURE 6.10 Differences in temperature within a protoplanetary disk determine the composition of dust grains that then evolve into planetesimals and planets. Shown here are the protostar (PS) and the orbits of Venus (V), Earth (E), Mars (M), Jupiter (J), Saturn (S), and Uranus (U).

planetary systems form, this change in composition with distance from the central star would seem to be almost unavoidable. Even so, chaotic encounters like those we will discuss in Chapter 10 can shuffle the deck, adding diversity to the organization of planetary systems.

Solid Planets Gather Atmospheres

Once a solid planet has formed, it may have a chance to continue growing by capturing gas from the protoplanetary disk. However, if it is going to do so, it must act quickly. Young stars and protostars are known to be sources of strong "winds" and intense radiation that can quickly disperse the gaseous remains of the accretion disk. Gaseous planets such as Jupiter probably have only about 10 million years or so to form and to grab whatever gas they can. Tremendous mass is a great advantage in a planet's ability to accumulate and hold onto the hydrogen and helium gas that makes up the bulk of the disk. Because of their strong gravitational fields, more massive young planets are thought to create their own mini-accretion disks as gas from their surroundings falls toward them. What follows is much like the formation of a star and protoplanetary disk, but on a smaller scale. Just as happened in the accretion disk around the star, gas from a mini-accretion disk moves inward and falls onto the solid planet.

The gas that is captured by a planet at the time of its formation is referred to as the planet's **primary atmosphere**. The primary atmosphere of a large planet can become massive enough to dominate the mass of the planet, as in the case of giant planets such as Jupiter. Some of the solid material in the mini-accretion disk might stay behind to coalesce into larger bodies in much the same way that dust in the protoplanetary disk came together to form planets. The result is a "mini-solar system"—a group of moons that orbit about the planet.

A less massive planet may also capture some gas from the protoplanetary disk, only to lose its prize. Here again, more massive planets have the advantage. As we will learn in Chapter 8, the gravity of small planets may not be strong enough to prevent less massive atoms and molecules such as hydrogen or helium from escaping back into space. Even if a small planet is able to gather an amount of hydrogen and helium from its surroundings, this temporary primary atmosphere will be short-lived. The atmosphere that remains around small planets like our Earth is a **secondary atmosphere**. A secondary atmosphere forms later in the life of a planet. Carbon dioxide and other gases released from

---

Less massive planets lose their primary atmospheres, then form secondary atmospheres.

---

the planet's interior by widespread volcanism are probably one important source of a planet's secondary atmosphere. Also, as we will see later, volatile-rich comets that formed in the outer parts of the disk continue to fall inward toward the new star long after its planets have formed, and will sometimes collide with planets. Comets possibly provide a significant source of water, organic compounds, and other volatile materials on planets close to the central star.

6.5 A Tale of Eight Planets

We are now at a point in our discussion where we can take our general ideas about the evolution of the material in a protoplanetary disk and apply them to our own Solar System. Only in the closing years of the 20th century did our knowledge progress to the point where this story could be told. It is still sketchy in places and doubtless wrong in some ways. But error and uncertainty are part of the advance of science. In this section we bring together a tremendous wealth of information taken both from what we know of our local star and planetary neighbors and from what we have learned about stars forming around us today. It is a synthesis of the painstaking efforts of hundreds of astronomers and planetary scientists over the course of decades. Many more scientists will devote their lives to unraveling this story before its details are fully known. There are good reasons to believe that the explanations we give in this section are basically correct, although we know they are not yet complete. Returning to our early discussion of what science is and how science works, the story we tell here is one that many people have "tried to prove wrong" but that has withstood all such tests … so far.

Nearly 5 billion years ago, our Sun was still a protostar surrounded by a protoplanetary disk of gas and dust. Over the course of a few hundred thousand years, much of the dust in the disk had collected into planetesimals—clumps of rock and metal near the emerging Sun and aggregates of rock, metal, ice, and organic materials in the more distant parts of the disk. Within the inner few astronomical units (AU) of the disk, several rock and metal planetesimals, probably fewer than a half dozen, quickly grew in size to become the dominant masses at their respective distances from the Sun. With their ever-strengthening gravitational fields, they

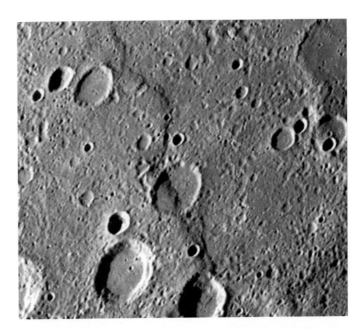

FIGURE 6.11 Large impact craters on Mercury (and on solid bodies throughout the Solar System) record the final days of the Solar System's youth, when planets and planetesimals grew as smaller planetesimals rained down on their surfaces.

Rocky terrestrial planets formed in the inner Solar System.

either captured most of the remaining planetesimals or ejected them from the inner part of the disk. These dominant planetesimals had now become planet-sized bodies with masses ranging between that of Earth and about one-20th of that value. They were to become the *terrestrial planets.* Mercury, Venus, Earth, and Mars are the surviving terrestrial planets. One or two others are thought to have formed in the young Solar System but were later destroyed. For several hundred million years following the formation of the four surviving terrestrial planets, leftover pieces of debris still in orbit around the Sun continued to rain down on their surfaces. Much of this barrage may have originated in the outer Solar System where the gravitational tug of the massive, newly formed outer planets acted like slingshots shooting debris inward. Today we can still see the scars of these postformation impacts on the cratered surfaces of all the terrestrial planets, such as the surface of Mercury shown in **Figure 6.11**. This rain of debris continues today, albeit at a much lower rate.

Before the proto-Sun emerged as a true star, gas in the inner part of the protoplanetary disk was still plentiful. During this early period the two larger terrestrial planets, Earth and Venus, may have held onto weak primary atmospheres of hydrogen and helium. If so, these thin atmospheres were soon lost to space. For the most part the terrestrial planets were all born devoid of thick atmospheres and remained so until the formation of the secondary atmospheres that now surround Venus, Earth, and Mars. Mercury's proximity to the Sun and the Moon's small mass must have prevented these bodies from retaining significant secondary atmospheres. They remain nearly airless today.

Farther out in the nascent Solar System, 5 AU from the Sun and beyond, planetesimals coalesced to form a number of bodies with masses around 5 to 10 times that of Earth. Why such large bodies formed in the region beyond the terrestrial planets remains an unanswered question. Located in a much colder part of the accretion disk, these planet-sized objects formed from planetesimals containing volatile ices and organic compounds in addition to rock and metal. Four such massive bodies, found between 5 AU and 30 AU from the Sun, would later become the cores of the *giant*

The giant planets formed cores from planetesimals and captured gaseous hydrogen and helium onto their cores.

planets, which we know today as Jupiter, Saturn, Uranus, and Neptune. Some planetologists believe that all the giant planets may have formed closer to where Jupiter is now and that their mutual gravitational interactions caused them to migrate to their present orbits. Mini-accretion disks formed around these planetary cores, capturing massive amounts of hydrogen and helium and funneling this material onto the planets. Jupiter's solid core was able to capture and retain the most gas—a quantity roughly 300 times the mass of Earth, or 300 $M_\oplus$. (The symbol $\oplus$ signifies "Earth.") The other planetary cores captured much smaller amounts of hydrogen and helium, perhaps because they were located in more remote parts of the protoplanetary disk where the gas density was lower. Saturn ended up with less than 100 $M_\oplus$ of gas, whereas Uranus and Neptune were able to grab only a few $M_\oplus$ of gas.

Before moving on, we should mention that some planetary scientists do not believe our protoplanetary disk could have survived long enough to form gas giants such as Jupiter through the general process of **core accretion**. The core accretion model indicates that it could take up to 10 million years for a Jupiter-like planet to accumulate. The problem here is that all the gas in the protoplanetary disk may have been blown away in a little more than half that time, abruptly cutting off Jupiter's supply of hydrogen and helium. This is not just a Solar System predicament. The available time dilemma would apply as well to other protoplanetary disks and to the formation of their massive planets (see Section 6.6). To get around this problem several scientists have proposed a process called *disk instability*, in which the protoplanetary disk suddenly and quickly fragments into massive clumps equivalent to that of a large planet. Although core accretion and disk instability appear to be competing processes, they may not be mutually exclusive. It is possible that both played a role in the formation of our own and other planetary systems.

For the same reasons that a forming protostar is hot—namely, conversion of gravitational into thermal energy—the gas surrounding the cores of the giant planets compressed under the force of gravity and became hotter. Proto-Jupiter and proto-Saturn probably became so hot that they actually glowed a deep red color, similar to the heating element on an electric stove. Their internal temperatures may have reached as high as 50,000 K. In a sense, the more massive protoplanets were trying to become stars. But for them this was an unreachable goal. In Chapter 15 we will find that a ball of gas must have a mass at least 0.08 times that of the Sun if it is to become a star. This minimum mass is some 80 times the mass of Jupiter. Science fiction notwithstanding, Jupiter never had a chance.

Some of the material remaining in the mini-accretion disks surrounding the giant planets coalesced into small

Moons formed from the mini-accretion disks around the giant planets.

bodies, which became moons. (A *moon* is any natural satellite in orbit about a planet or asteroid.) The composition of these moons of the giant planets followed the same trend as the planets that formed around the Sun: The innermost moons formed under the hottest conditions and therefore contained the smallest amounts of volatile material. As very young moons, Io and Europa may have experienced nearby

Jupiter glowing so intensely that it rivaled the distant Sun. The high temperatures created by the glowing planet would have evaporated most of the volatile substances in the inner part of its mini-accretion disk. Io today contains no water at all. However, water is relatively plentiful on Europa, Ganymede, and Callisto because these moons formed farther from hot Jupiter.

Not all planetesimals in the protoplanetary disk went on to become planets. Jupiter is a true giant of a planet. Its gravity kept the region of space between it and Mars so "stirred up" that most planetesimals there never coalesced into a single planet. (The one exception is Ceres, once considered to be the largest asteroid but is now redefined as a **dwarf planet** under the new IAU planet definition; see Appendix 8.) This region, now referred to as the **asteroid belt**, contains many planetesimals that remain from this early time. In the outermost part of the

Asteroids and comets are planetesimals that survive to this day.

Solar System planetesimals also persist to this day. Born in a "deep freeze," these objects retained most of the highly volatile materials found in the grains present at the formation of the protoplanetary disk. Unlike conditions in the crowded inner part of the disk, planetesimals in the outermost parts of the disk were too sparsely distributed for large planets to grow. Icy planetesimals in the outer Solar System remain today as **comet nuclei**—relatively pristine samples of the material from which our planetary system formed. Frozen Pluto and the distant dwarf planet Eris are especially large examples of these denizens of the outer Solar System.

The early Solar System must have been a remarkably violent and chaotic place. Many objects in the Solar System show evidence of cataclysmic impacts that reshaped worlds. A dramatic difference in the terrain of the northern and southern hemispheres on Mars, for example, has been interpreted as the result of one or more colossal collisions. Mercury has a crater on its surface from an impact so devastating that it caused the crust to buckle on the opposite side of the planet. In the outer Solar System, one of Saturn's moons, Mimas, sports a crater roughly one-third the diameter of the moon itself. Uranus suffered a collision that was violent enough to literally knock the planet on its side. Today its axis of rotation is tilted at an almost right angle to its orbital plane.

Not even our own Earth escaped devastation by these cataclysmic events. In addition to the four terrestrial planets that remain, there was at least one other terrestrial planet in the early Solar System—a planet about the same size and mass as Mars. As the newly formed planets were settling down into their present-day orbits, this fifth planet suffered a grazing collision with Earth and was completely destroyed. The remains of the planet, together with material

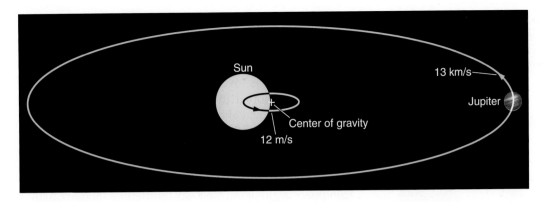

FIGURE 6.12 Both the Sun and Jupiter orbit around a common center of gravity, which lies just outside the Sun's surface. Spectroscopic measurements made by an extrasolar astronomer would reveal the Sun's radial velocity varying by ±12 m/s over an interval of 11.86 years, Jupiter's orbital period.

knocked from Earth's outer layers, formed a huge cloud of debris encircling Earth. For a brief period Earth may have displayed a magnificent group of rings like those of Saturn. In time this debris coalesced into the single body we know as our Moon.

6.6 There Is Nothing Special about Our Solar System

We began this chapter by discussing the formation of a generic planetary system around a generic star. Only later did we turn to the specifics of our own Solar System. We did this to make a very important point: According to our current understanding, *there is absolutely nothing special about the conditions that led to the formation of our Sun and our Solar System*. When astronomers turn their telescopes on Sun-like stars forming around us today, they see just the sort of disks from which our own Solar System formed. Based on what we know, the physical processes that led to the formation of our Solar System should be commonplace wherever new stars are being born.

In 1994 astronomers found their first **extrasolar planet**, a Jupiter-sized body orbiting surprisingly close to 51 Pegasi, a solar-type star. Today the number of known extrasolar planets has grown to more than 170, with new discoveries occurring almost daily.

The Search for Extrasolar Planets

How do astronomers search for extrasolar planets? Why were such planets not found earlier? The "why" is easy to answer. Until the closing years of the 20th century, astronomers lacked the sophisticated instruments needed to detect these elusive objects. We turn now to "how." The most successful technique employed so far has been the spectro-

Several techniques are being used to find extrasolar planets.

scopic radial velocity method. As a planet orbits about a star, the planet's gravity causes the position of the star to wobble back and forth ever so slightly. By observing the Doppler shift of the star, we can sometimes detect this wobble and infer the properties of the planet—its mass and distance from the star. (We will use an almost identical method in Chapter 13 to measure the masses of pairs of stars that orbit each other.) We can see how this would work using our own Solar System as an example. As a simplification, we will consider the Solar System as consisting of only the Sun and Jupiter. This makes sense because Jupiter's mass is greater than the mass of all the other planets, asteroids, and comets combined. Imagine an alien astronomer pointing her spectrograph toward the Sun. Both the Sun and Jupiter orbit a common center of gravity that lies just outside the surface of the Sun, as shown in **Figure 6.12**. Our alien astronomer would find the Sun's radial velocity to vary by ± 12 m/s, with a period equal to Jupiter's orbital period of 11.86 years. From this she would rightly conclude that the Sun has at least one Jupiter-mass planet, but without greater precision, she would be unaware of the other eight planets.[4] The technology of today limits the precision of our own radial velocity instruments to about 3 m/s. This enables astronomers to detect giant planets around solar-type stars, but we are still far from being able to reveal any smaller bodies with masses similar to our own terrestrial planets.

Another technique for finding extrasolar planets is the transit method. If a planet should pass in front of its parent star, we would see the light from the star diminished by a tiny amount, as illustrated in **Figure 6.13**. Current technology limits the sensitivity of the transit technique to about 0.1 percent of a star's brightness, but that has been good enough for this method to find more than a dozen extrasolar

---

[4] If our alien astronomer could measure radial velocities as small as 2.7 m/s, she could detect Saturn. If she could measure radial velocities as small as 0.09 m/s, she could detect Earth.

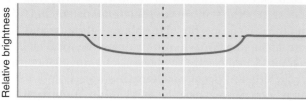

FIGURE 6.13 As a dark planet passes in front of a star it blocks some of the light coming from the star's surface, causing the brightness of the star to decrease slightly.

planets. (In fact, amateur astronomers have confirmed the existence of two extrasolar planets by observing transits with telescope apertures as small as 20 cm!) Our alien astronomer could infer the existence of Earth if she were located somewhere in the plane of Earth's orbit (that is the only way she could see Earth pass in front of the Sun) and could detect a 0.009 percent drop in the Sun's brightness. Our own astronomers will soon have that capability. In 2008 NASA plans to launch an orbiting spacecraft called *Kepler*, with photometric instruments able to detect transits of Earth-size planets.

Still another procedure involves a relativistic effect called gravitational lensing[5] (see Chapter 17), whereby the gravitational field of an unseen planet bends light from a distant star in such a way that it causes the star to brighten temporarily while the planet is passing in front of it. So far, three extrasolar planets have been found using this tech-

---

[5] *Gravitational lensing* generally refers to the relativistic lensing of distant quasars by intervening galaxies. When referring to the lensing of stars by small bodies such as planets, astronomers use the term *microlensing*.

nique. Like the transit method, lensing is also capable of detecting Earth-size planets.

A fourth method is by direct imaging. This is the most difficult technique because it involves finding a relatively faint planet in the overpowering glare of a bright star—a challenge far more difficult than looking for a firefly in the dazzling brilliance of a searchlight. As of this writing, astronomers have confirmed the first planet spotted by direct imaging, although it seems to be orbiting not a star, but instead a *brown dwarf* (see the following subsection). More discoveries may be in store a few years from now when astronomers begin using the Keck Interferometer (see Chapter 5) for direct imaging searches. However, the most exciting discoveries are sure to come in about a decade when NASA launches its Terrestrial Planet Finder (TPF). As designed, TPF will not only detect Earth-like planets around nearby stars, it will also measure the planets' physical and chemical characteristics. As we will learn later in Connections 8.1, certain telltale gases in a terrestrial planet's atmosphere would be a strong indicator of life.

Planetary Systems Seem to Be Commonplace

Our searches for extrasolar planets have been remarkably successful. Since the first was found in 1994, astronomers have added at least another 170 planets to their inventory, bringing the total number of planetary systems to 146, includ-

---

At least 170 extrasolar planets have been discovered so far.

---

ing 18 with multiple planets. A representative few are shown in **Figure 6.14**. Keep in mind that none are terrestrial-type planets. The least massive found so far is estimated to have a mass about equal to ⅓ that of Neptune, or about 6 $M_\oplus$.

One of the first things astronomers noticed is that most of these planetary systems *do not look like our own.* Many contain **hot Jupiters**: massive gaseous planets orbiting solar-type stars in tight circular orbits that are closer to their parent stars than Mercury is to our own Sun. Others have

---

Most planetary systems found to date do not resemble our own.

---

planets in highly eccentric orbits, unlike the relatively circular planetary orbits in our Solar System. Does this mean that the Solar System is an oddball? Not necessarily, but maybe. A massive planet orbiting close to its parent star is the easiest kind to detect, and so those that we do see may *not* be representative of *most* planetary systems. Scientists refer to this bias as the *selection effect*. The selection effect does *not* explain the large number of extrasolar planets hav-

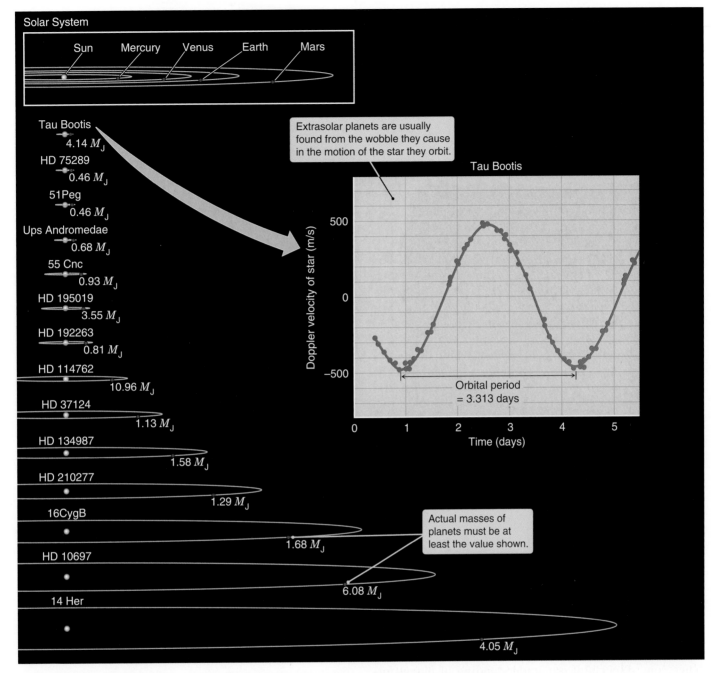

FIGURE 6.14 Planetary systems have been discovered around scores of stars other than the Sun, confirming what astronomers have long suspected—that planets are a natural and common by-product of star formation. A few of these systems are represented here. (Masses of planets are given in units of Jupiter masses, M_J)

ing highly eccentric orbits. Perhaps chaotic gravitational interactions take place within most planetary systems, knocking their planets out of the circular orbits in which they formed. Why, then, should our Solar System seem so stable? This is a question we cannot yet answer.

The discovery of massive planets around other stars raises an interesting question. When is a "massive Jupiter"

too large to be labeled a "planet"? In Chapter 15 we will introduce you to a curious object called a *brown dwarf*, an in-between body not massive enough to be considered a star, yet too massive to be called a planet. In their searches for planets, astronomers have come across many brown dwarfs —some orbiting stars, some alone in space. Although the distinction between massive planets and brown dwarfs is

Origins—Choosing the Right Kind of Planet

In this chapter we learn that planetary systems abound in the universe, and among those detected so far, we see only massive Neptune to Jupiter-like planets. This is understandable because we do not yet have the technology to detect small Earth-like planets. Hardly any of these planetary systems, though, are like our own. Some contain massive planets closer to their parent star than Mercury is to the Sun. Others have planets in highly eccentric orbits that carry them far from the warmth of their star. It is clear that such planets would not be included as candidates in any search for extraterrestrial life.

Given a choice, what kind of planet would life choose for its home? For starters, liquid water is essential for the formation and evolution of any life that we are familiar with. The planet must be at just the right distance from its parent star for water to remain in a liquid state. If too close, water would exist only as a vapor. Too far and it would be forever frozen as ice. The average planetary

temperature therefore should be in the comfort range, say somewhere around 300K.

The planet should have formed at a distance from the central star where refractory grains were the dominant material within the protoplanetary nebula. This would provide a solid rocky surface for supporting oceans and perhaps dry land, although dry land may not be a necessary condition for the formation of life, as we will see later in Connections 11.1.

Finally, the planetary system must be stable, meaning that its planets can remain in well-behaved circular orbits which preserve stable climatological and oceanic environments. Noncircular orbits would lead to wild temperature swings, which more than likely would be detrimental to the survival of life.

In summary, any life as we know it would choose a planet much like Earth.

somewhat arbitrary, the International Astronomical Union has placed the boundary at 13 Jupiter masses (13 M_J).

Many astronomers were surprised at the "hot Jupiters" because it is not yet clear how such massive planets can form so close to stars. It seems likely that these planets actually formed farther out from their parent stars and later moved inward. Yet even though these findings challenge some aspects of our understanding of planet formation, and even though discovery of *Earth-like planets* —planets similar to Earth in size and distance from their stars—around other stars is still years away, the message conveyed by these discoveries is clear: The formation of planets frequently, and perhaps always, accompanies the formation of stars.

The implications of this conclusion are profound. Planets are a common by-product of star formation. So in a galaxy of a hundred billion stars, and a universe of hundreds of billions of galaxies, how many Earth-like planets (or Earth-like moons) might exist? And with all of these Earth-like worlds in the universe, how many might play host to the particular category of chemical reactions that we refer to

as "life"? **Connections 6.1** tells us what kind of planet life might choose for its home.

In learning about the formation of stars and then our Solar System, our journey has come full circle. We began our journey by looking outward in wonder at the lights in the night sky. Now that outward exploration becomes instead a look inward as we discover the processes and events

---

Modern discoveries about the formation of the Sun and Solar System "complete" the Copernican revolution.

---

that set the stage for our own existence. In a sense, the insights we have gained in this chapter "complete" the revolution Copernicus started long ago when he had the audacity to challenge authority and suggest that Earth was not the center of all things. Not that long ago, our ancestors viewed Earth and the heavens as fundamentally different regimes —forever apart, each with its own rules and reality. Today we know there is nothing unusual about the conditions of our own existence. All we have to do to see our own history is to look at the stars that continue to form around us today.

It is ironic that at a time when science has turned such a dazzling spotlight on our place in the universe, many otherwise educated individuals continue to cling to outdated and fanciful notions about the heavens and Earth.[6]

---

[6] We cannot resist the temptation to quote the famous science fiction author Arthur C. Clarke at this juncture. "In one sense, of course, every age renews itself, as indeed it should. But the nitwits currently parroting this slogan [New Age] seem unable to understand that their

For someone interested in truly mind-boggling insights, the speculations and "revelations" of the mystic pale beside the discoveries of the scientist. Through the sometimes stodgy, often painstaking, and always uncompromising standards of science, we have come to appreciate that we are not *apart from* the universe. Rather we are *a part of* the universe. And the processes and events that link us to this larger universe are fascinating and wondrous indeed.

'New Age' is exactly the opposite, being about a thousand years past its sale date."

Summary

- Stars and their planetary systems form from collapsing interstellar clouds of gas and dust.

- Planets are a common by-product of star formation, and many stars are surrounded by planetary systems.

- Our Solar System formed about 4.6 billion years ago, nearly 10 billion years after the birth of the universe.

- Planets grew from a protoplanetary disk of gas and dust that surrounded the forming Sun.

- Solid terrestrial planets formed in the inner disk where temperatures were high.

- Giant gaseous planets formed in the outer disk where temperatures were low.

- Dwarf planets formed in the asteroid belt and in the region beyond the orbit of Neptune.

- Asteroids and comet nuclei remain today as leftover debris.

Seeing the Forest through the Trees

In this chapter we have seen two great lines of investigation merge into a single picture that shapes our understanding of the context of our own existence. Working first from the perspective of the stellar astronomer, basic physical principles—in particular the conservation of angular momentum—demand that when a star like our Sun forms, it will be surrounded by a thin, orbiting disk of gas and dust. This conclusion went from hypothesis to fact with the discovery of such disks around numerous young stars. In the meantime, as stellar astronomers were trying to understand star formation, planetary scientists were scrutinizing the worlds that make up our Solar System. The pieces of the planetary scientists' puzzle range from meteorites collected in Antarctica, to samples of the Moon brought back by *Apollo* astronauts, to data sent back by spacecraft visiting remote worlds. Those pieces fit together to form a clear picture of a flat, swirling cloud of gas and dust from which Earth and its neighbors coalesced 4.6 billion years ago. When the stellar astronomers and the planetary scientists compared notes, they realized that they had arrived at exactly the same description. The disks from which our Solar System and the many other planetary systems formed were none other than the stellar accretion disks that surrounded the young suns.

The joining of these two great rivers of thought, observation, and exploration—the link between the accretion disks that surround young stars and the local collection of planets that we call home—is the starting point for a modern study of the Solar System. It also ties our journey of understanding within the Solar System to the course we will later chart outward into the larger universe. Stated more bluntly, all that we know about the Sun, planets, and the world of our birth makes sense only when viewed within the context of the evolving universe as a whole. We have seen a first glimpse of the power of this insight in the current chapter. As we move on to look at the planets, moons, asteroids, and comets that orbit the Sun, we will time and again see the fingerprints left behind by our birth among the stars.

The journey that we are taking in *21st Century Astronomy* is one of scientific discovery, but the philosophical implications of the connection between star formation and the formation of the Solar System should not be overlooked. Our very existence is set within the context of the larger universe. We are the legacy of the processes we see at work all around us, even today.

Key Terms

Solar System, p. 165
protostar, p. 165
protostellar disk, p. 165
protoplanetary disk, p. 165
angular momentum, p. 166
conservation of angular momentum, p. 167
conservation laws, p. 167
accretion disk, p. 168
planetesimal, p. 170
planet, p. 170
conservation of energy, p. 171
gravitational potential energy, p. 171
refractory materials, p. 172
volatile materials, p. 172
ice, p. 172
organic, p. 172
primary atmosphere, p. 174
secondary atmosphere, p. 174
comet nucleus, p. 176
asteroid belt, p. 176
extrasolar planet, p. 177
hot Jupiter, p. 178

Student Questions

THINKING ABOUT THE CONCEPTS

1. What is the source of the material that now makes up the Sun and the rest of the Solar System?

2. In a poetic sense, planets are sometimes referred to as "children of the Sun." A more accurate description, though, would be "siblings of the Sun." Explain why.

3. Explain the process by which tiny grains of dust grow to become massive planets.

4. Why do we find rocky material everywhere in the Solar System, but large amounts of volatile material only in the outer regions? Would you expect the same to be true of other solar systems? Explain your answer.

5. Why were the four giant planets able to collect massive gaseous atmospheres, whereas the terrestrial planets could not?

6. Explain the source of the atmospheres now surrounding three of the terrestrial planets.

7. What happened to all the leftover Solar System debris after the last of the planets had formed?

8. Physicists describe certain properties, such as angular momentum and energy, as being "conserved." What does this mean? Do these conservation laws imply that an individual object can never lose or gain angular momentum or energy? Explain your reasoning.

9. Nearly all the total angular momentum in the Earth–Moon system resides in Earth's spin angular momentum and the Moon's orbital angular momentum in its orbital motion around Earth. (Note that Earth's orbital angular momentum is not included here.) Friction caused by ocean tides is very gradually slowing down Earth's rate of rotation (causing our days to become longer).
   a. Explain why, as Earth's rotation rate slows, our planet loses spin angular momentum.
   b. Explain why Earth's lost spin angular momentum *must* be transferred to the Moon's orbital angular momentum.
   c. What effect do you suppose this has on the Moon's orbit around Earth?

10. Step outside and look at the nighttime sky. Depending on the darkness of the sky, you may see dozens or hundreds of stars. Would you expect many or very few of those stars to be orbited by planets? Explain your answer.

11. Nearly all the planets astronomers have found orbiting other stars have been giant planets with masses more like that of Jupiter than Earth, and with orbits located very close to their parent stars. Does this prove that our Solar System is unusual? Explain your answer.

APPLYING THE CONCEPTS

12. Use information about the planets given in the appendixes to answer the following:
    a. What is the total mass of all the planets in our Solar System, expressed in Earth masses $(M_\oplus)$?
    b. What fraction of this total planetary mass does Jupiter represent?
    c. What fraction does Earth represent?

13. Orbital angular momentum is defined as $Lo = mvr$, where m is the mass, v is the speed, and r is the orbital radius of the orbiting body. Spin angular momentum (for a uniform sphere) is defined as $Ls = (4\pi mR^2)/5P$, where m is the mass of the sphere, R is the radius of the sphere, and P is the period of rotation. Compare Earth's orbital angular momentum with its spin angular momentum using the following values: $m = 6 \times 10^{24}$ kg, $v = 2.98 \times 10^4$ m/s, $r = 1.5 \times 10^{11}$ m, $R = 6.4 \times 10^6$ m, and $P = 8.64 \times 10^4$ s. What fraction does each contribute to Earth's total angular momentum?

14. Assume the Sun is a uniform sphere with a radius of 700,000 km and a rotation period of 26 days. Near the

end of its life, the remnant of the Sun will be a white dwarf with a radius of only 5,000 km. Assuming the mass remains unchanged, what will be the Sun's new rotation period as a white dwarf?

15. Jupiter has a radius 11.2 times that of Earth and a mass 318 times that of Earth. Its rotation period is 9.9 hours. What is the ratio of Jupiter's spin angular momentum to that of Earth? Here we have to make the unrealistic assumption that the densities of both bodies are uniform throughout.

16. Jupiter has an orbital radius of 5.2 AU and an orbital velocity of 13.1 km/s. Earth's orbital velocity is 29.8 km/s. What is the ratio of Jupiter's orbital angular momentum to that of Earth?

17. The asteroid Vesta has a diameter of 530 km and a mass of 2.7×10^{20} kg.
   a. Calculate the density of Vesta.
   b. The density of water is 1,000 kg/m^3 and that of rock is about 2,500 kg/m^3. What does this tell you about the composition of this primitive body?

18. A "hot Jupiter" nicknamed "Osiris" has been found around a solar-mass star, HD 209458. It orbits the star in only 3.524 days.
   a. What is the orbital radius of this extrasolar planet?
   b. Compare its orbit with that of Mercury around our own Sun. What environmental conditions must Osiris experience?

19. The extrasolar planet Osiris passes directly in front of its solar-type parent star, HD 209458 (diameter = 1.4×10^6 km), every 3.524 days, decreasing the brightness of the star by about 1.8 percent (0.018).
   a. What is the diameter of Osiris?
   b. Compare the diameter of this extrasolar planet with that of Jupiter (mean diameter = 139,800 km).

StudySpace
wwnorton.com/astro21

provides a Study Plan for each chapter that includes a reading outline, animations, keyword flash cards, and gradebook-enabled multiple-choice quizzes. From StudySpace you can also access premium content in the ebook and SmartWork.

We've sent a man to the Moon, and that's 240,000 miles away. The center of the Earth is only 4,000 miles away. You could drive that in a week, but for some reason nobody's ever done it.

ANDY ROONEY (1919–)

Planet Earth and the Moon seen from the Space Shuttle *Discovery*.

CHAPTER 7

The Terrestrial Planets and Earth's Moon

7.1 How Are Planets the Same, and How Are They Different?

Most of what we know about our planetary neighbors we have learned only in the recent few decades. The second half of the 20th century was a wonderfully exciting time of exploration and discovery about Earth and its sibling worlds. It was a time that saw unmanned probes visit every classical planet and watched men walk on the surface of the Moon. The variety of techniques used to explore the Solar System is discussed in Chapter 5. The information returned from these missions has revolutionized our understanding of our planetary system, offering us insights into the current state of each of our neighbors and clues about their histories.

The four innermost planets in our Solar System are Mercury, Venus, Earth, and Mars, collectively known as the **terrestrial planets** (**Table 7.1**). Although technically the Moon is known as Earth's lone natural satellite, we include it here in this chapter because of its close similarity to the terrestrial planets.

The vast quantity of information about the planets returned by space probes can be hard to digest. What information is truly fundamental and exciting, and what information is flashy but less important? The key to sorting through this information has proven to be comparison among the different planets. The ways the planets are alike and how they differ draw our attention to the

KEY CONCEPTS

The objects that formed in the inner part of the protoplanetary disk around the Sun were relatively small rocky worlds, one of which we call home. Comparing those worlds with one another teaches us lessons about what shapes a planet's fate. Among the lessons we will learn are

- That each terrestrial planet is shaped by impacts, tectonism, volcanism, and gradation.
- How impacts scarred planets early in the history of the Solar System, and still occur on occasion today.
- Why the concentration of craters on a planetary surface tells us how old the surface is.
- How radioactive dating of lunar rocks is used to calibrate the cratering clock.
- That larger worlds remain geologically active longer because smaller worlds cool off sooner.
- How predictions of models of Earth's interior are tested using seismic waves from earthquakes.
- That tectonism takes different forms on different planets, but plate tectonics is unique to Earth.
- That among the volcanoes found on Earth, Venus, and Mars, the most colossal are on Mars.
- The many ways that gradation wears down what other processes form.

TABLE 7.1

Comparison of Physical Properties of the Terrestrial Planets and the Moon

	Mercury	Venus	Earth	Mars	Moon*
Orbital radius (AU)	0.387	0.723	1.000	1.524	384,000 km
Orbital period (years)	0.241	0.615	1.000	1.881	$27^d.32$
Orbital velocity (km/s)	47.9	35.0	29.8	24.1	1.02
Mass ($M_\oplus = 1$)	0.055	0.815	1.000	0.107	0.012
Equatorial diameter (km)	4,880	12,102	12,756	6,794	3,476
Equatorial diameter ($D_\oplus = 1$)	0.351	0.949	1.000	0.533	0.272
Density (water = 1)	5.43	5.24	5.52	3.93	3.34
Rotation period	$87^d.97$	$224^d.7$	$23^h 56^m$	$24^h 37^m$	$27^d.32$
Obliquity (degrees)[†]	0.5	177.4	23.4	25.2	6.7
Surface gravity (m/s²)	3.70	8.87	9.78	3.71	1.62
Escape speed (km/s)	4.25	10.36	11.18	5.03	2.38

*The Moon's orbital radius and orbital period are given in kilometers and days, respectively.
[†]An obliquity greater than 90° indicates that the planet rotates in a retrograde, or backward, direction.

most fundamental issues, helping us to ask and answer the right questions. The correct explanation for some aspect of one planet must work together with what we know about the other planets. For example, when we explain why the Moon is covered with craters, our explanation must also allow for an understanding of why preserved craters on Earth are rare. An explanation for why Venus has such a massive atmosphere should point

---

We learn about planets by comparing them with each other.

---

to reasons why Earth and Mars do not. Such comparisons —an approach called **comparative planetology**—have provided the guideposts to planetary scientists. When making comparisons we need a place to start. Earth is the planet that we know best, so it is here at home that we begin our appraisal of the worlds of the inner Solar System.

7.2 Four Main Processes Shape Our Planet

For most of human history we have looked upon Earth as a vast, almost limitless expanse. This view of our planet changed forever with a single snapshot taken by *Apollo* astronauts looking back at Earth from space. We no longer have an excuse to view the world as anything other than what it is: a small blue ball, a tiny and fragile lifeboat adrift in the vastness of space, our home.

The psychological impact of these images is rivaled only by the change in scientific perspective of which they are a part. Seen from space (**Figure 7.1**), Earth is awash with color. White clouds drift in our atmosphere, and white snow and ice cover the planet's frozen poles. The blue of oceans, seas, lakes, and rivers of liquid water—Earth's **hydrosphere**—covers most of the planet. Brown shows us the outer rocky shell of Earth, referred to as Earth's **litho-**

FIGURE 7.1 Seen from space, the colors of Earth tell of the diversity of our planet.

sphere. And green is the telltale sign of vegetation, part of Earth's biosphere—the most extraordinary among Earth's many distinctions.

Earth is a place of change: Geological processes constantly work to reshape our planet. Some geological processes originate in the interior of Earth, powered by the energy generated there. Earthquakes are sudden reminders of the ongoing deformation of Earth's lithosphere, referred to as **tectonism**. **Volcanism** is a form of **igneous activity**, the formation and action of gas and molten rock, or **magma**.

Collisions involving planetary objects—a process referred to as **impact cratering**—are extremely important in our planet's history. Most of us have seen meteors, the bright streaks that flash across the sky when chunks of material from outer space hit our atmosphere; but for a few the experience with celestial debris is more intimate. In 1954 a meteorite crashed through the roof of a woman's home in Alabama, striking her hip. (Fortunately the roof of the house slowed the meteorite, so she was left with only a bad bruise.) You should not lose much sleep worrying about meteorites landing in your lap. Unlike being struck by lightning, being hit by a meteorite is even less likely than winning the state lottery. However, when a very large object hits a planet, as still happens occasionally in the Solar System, the resulting devastation can be global. At times in our past, such catastrophic events have altered Earth's climate and have changed the history of life itself.

These three processes—tectonism, volcanism, and impacts—affect Earth's surface in their own characteristic ways. Tectonism folds and breaks Earth's crust, forming mountain ranges, valleys, and deep ocean trenches. Volcanic eruptions, like the one shown in **Figure 7.2**, can spill sheets of lava and ash over vast areas, forming mountains or plains in the process. Distinctive scars in Earth's crust tell of impacts in our past. But as these processes work to form **topographic relief**, a fourth process called **gradation** works slowly

> Tectonism, volcanism, and impacts rough up planetary surfaces; gradation smooths them.

but persistently to level Earth's surface. Erosion by running water, wind, and other agents wears down the hills, mountains, and continents. The eroded debris collects downslope, filling in valleys, lakes, and ocean basins. Left on its own, gradation would eventually leave the surface of our planet smooth and featureless. Coupled with biological processes, including the actions of humans, the surface of our planet is an ever-changing battleground between processes that build up topography and those that tear it down.

Each of the four types of geological processes at work shaping the surface of Earth leaves its own distinctive signature. Planetary scientists have learned to read these signatures on the surfaces of other planets as well.

FIGURE 7.2 Volcanism, such as this eruption in Hawaii, spills molten rock and other materials onto planetary surfaces.

7.3 Impacts Help Shape the Evolution of the Planets

Objects orbiting the Sun and the planets in the Solar System move at very high speeds. Earth moves at about 30 km/s in its orbit around the Sun, and meteors can enter Earth's atmosphere at relative speeds in excess of 70 km/s. Because the *kinetic energy* of an object is proportional to the square of its speed ($KE = \frac{1}{2}\, mv^2$), collisions between such objects release huge amounts of energy. In fact, large impacts are by far the most concentrated and sudden release of energy of any geological process, including earthquakes and volcanic eruptions.

When an object hits a planet, its kinetic energy goes into heating and compressing the surface that it strikes and throwing material far from the resulting **impact crater** (see **Figure 7.3**). Sometimes material thrown from the crater, called **ejecta**, falls back to the surface of the planet with enough energy to cause **secondary craters**. In impacts the rebound of heated and compressed material can also lead to the formation of a central peak or ring of mountains in the floor of the crater, in ways similar to what happens when a drop lands in a glass of milk, as in **Figure 7.4**.

Impacts can melt and vaporize rock.

The energy of an impact can be great enough to melt or even vaporize rock. The floors of some craters are the cooled surfaces of pools of rock melted by the impact. The energy released in an impact can also lead to the formation of new minerals. In fact, some minerals, such as shock-modified

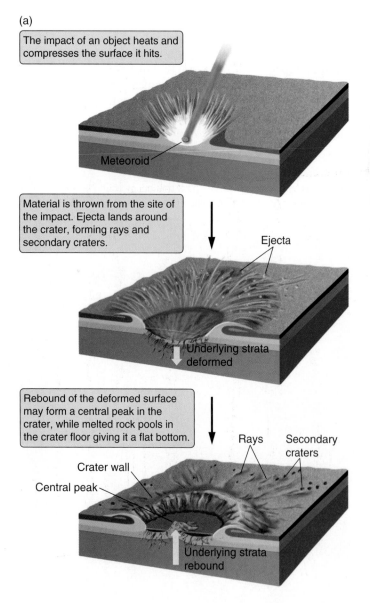

(a)

The impact of an object heats and compresses the surface it hits.

Meteoroid

Material is thrown from the site of the impact. Ejecta lands around the crater, forming rays and secondary craters.

Ejecta

Underlying strata deformed

Rebound of the deformed surface may form a central peak in the crater, while melted rock pools in the crater floor giving it a flat bottom.

Rays   Secondary craters

Crater wall

Central peak

Underlying strata rebound

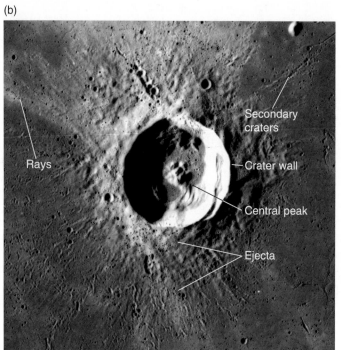

(b)

Secondary craters

Rays

Crater wall

Central peak

Ejecta

FIGURE 7.3 (a) Stages in the formation of an impact crater. (b) A lunar crater photographed by *Apollo* astronauts, showing rays, secondary craters, the crater wall and surrounding material, and a central peak—all typical features associated with impact craters.

(a)

(b)

(c)

(d)

(e)

(f)

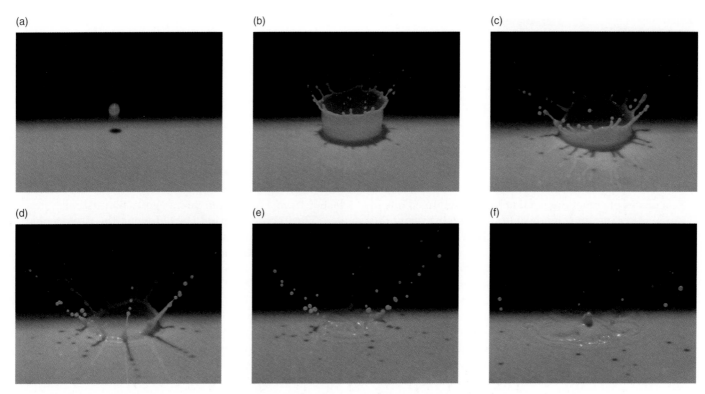

FIGURE 7.4 A drop hitting a pool of milk illustrates the formation of features in an impact crater, including crater walls (b and c), secondary craters (d and e), and a central peak (f).

quartz, are known to form *only* during the momentary fury of an impact. Geologists look for these distinctive minerals as evidence of ancient impacts on Earth's surface.

Meteor Crater in Arizona is the best-preserved impact structure on our planet (**Figure 7.5**). It is thought to be the result of the impact of an iron meteorite about 40 meters in diameter with a mass of about 10^7 kg that hit Earth traveling at 12 km/s about 40,000 years ago. Such a collision would have released about 1,000 times as much energy as the first atomic bomb, detonated in New Mexico in 1945. Yet at only 1 km in diameter, Meteor Crater is tiny compared with impact craters seen on the Moon or ancient impact scars on Earth. See **Connections 7.1** to read about the consequences of an especially violent impact that took place 65 million years ago.

One of the most obvious differences among the terrestrial planets is that whereas on some planets the surfaces are covered by

impact craters, on others (especially Earth) impact craters seem to be rare. Although all terrestrial planets are subject to impacts, tectonism, volcanism, and gradation, the relative intensity of these processes varies among the planets. For example, when we look at the Moon, we find

FIGURE 7.5 Meteor Crater (also known as Barringer Crater), located in northern Arizona, is a 1.2-km diameter impact crater formed some 40,000 years ago by the impact of an iron meteoroid.

CONNECTIONS 7.1

Origins—Where Have All the Dinosaurs Gone?

In 1994 a host of astronomical instruments watched as the pieces of comet Shoemaker-Levy 9 slammed one after another into the clouds of Jupiter, leaving temporary scars that were visible through backyard telescopes. The Jupiter comet crash provided graphic evidence that although impacts are not as frequent today as they were when the Solar System was young, they still do occur. The giant planets are not the only targets for such cosmic bombardment either. In 2004 a very small asteroid, about 30 m across, whizzed past Earth at nearly 30,000 km/h, missing our planet by only 43,000 km. A close call to be sure, but it was after all a relatively small body that probably would have broken up or exploded in our atmosphere, causing only local damage had we been hit. That same year University of Hawaii astronomers discovered an asteroid that was given the rather forgettable name 2004 MN_4. Forgettable, that is, until further observations and refined calculations showed that 2004 MN_4 (now named Apophis), a 320-m diameter mass of rock, will come within 30,000 km of Earth's surface on April 13, 2029. Although the probability of a collision with Earth is small, the energy of impact would be 880 megatons —15 times more powerful that the most powerful hydrogen bomb ever detonated. When large impacts happen on Earth, they can have far-reaching consequences for Earth's climate and for terrestrial life.

Earth's fossil record shows that on occasion large numbers of species vanish from the face of the planet in a geological blink of an eye. The most famous of these extinctions occurred 65 million years ago, when over 70 percent of *all living species*, including the dinosaurs, became extinct. This mass extinction is marked in Earth's fossil record by the **Cretaceous–Tertiary boundary**, or simply the K–T boundary.[1] (The **Cretaceous Period** lasted from 146 million years ago to 65 million years ago. The **Tertiary Period** started when the Cretaceous period ended and lasted until 1.8 million years ago. We live in the **Quaternary Period** of Earth's history, which started at the end of the Tertiary Period.) In older layers found below the K–T boundary, fossils of dinosaurs and other now-extinct life forms abound. In the newer rocks above the K–T boundary, more than half of all previous species are absent, and in their place are found fossils of newly evolving species. Big winners in the new order were the mammals—our own distant ancestors—which moved into ecological niches vacated by extinct species.

The K–T boundary is marked in the fossil record in many areas by a layer of clay. Studies at more than 80 locations around the world have found that this layer contains large amounts of the element iridium as well as traces of soot. Iridium is very rare in Earth's crust but is common in meteorites. The soot at the K–T boundary tells of a time when widespread fires burned the world over. The thickness of the layer of clay at the K–T boundary and the concentration of iridium increase as we move toward what is today the Yucatán Peninsula in Mexico. Although the original crater has been erased by gradation, geophysical surveys and rocks from drill holes in this area show a highly deformed subsurface rock structure, similar to that seen at known impact sites. Together these results provide compelling evidence that 65 million years ago a 10-km diameter asteroid struck the area, throwing great clouds of red-hot dust and other debris into the at-

[1] From *Kreide,* German for Cretaceous.

millions of craters of all different sizes, one on top of another (**Figure 7.7**). Nearly all of these craters are the result of impacts. By comparison, only about 150 impact craters, or scars from impacts, have been identified on Earth. The primary culprit behind Earth's crater shortage is erosion. Most craters have simply been obliterated by wind and water. There is another reason for the shortage of small craters on Earth. Whereas the surface of the Moon is directly exposed to this cosmic bombardment, the surface of Earth is partly protected by the blanket of Earth's atmosphere. For example, rock samples from the Moon show craters smaller than a pinhead, formed by micrometeoroids. In contrast, most meteoroids smaller than 100 m are either burned up or broken up by friction in Earth's atmosphere

FIGURE 7.6 This artist's rendition depicts a comet or asteroid, perhaps 10 km across, striking Earth 65 million years ago in what is now the Yucatán Peninsula in Mexico. The effects of the impact killed off most forms of terrestrial life, including the dinosaurs.

mosphere (**Figure 7.6**) and igniting a worldwide conflagration. The energy of the impact is estimated to have been more than that released by 5 *billion* nuclear bombs.

An impact of this magnitude clearly would have had a devastating effect on terrestrial life. Could this cosmic impact have been responsible for the sudden disappearance of forms of life that had ruled Earth for 150 million years? Many scientists believe so. In addition to igniting a nearly global firestorm, dust thrown into Earth's upper atmosphere by the impact would have remained there for years, blocking out sunlight and plunging Earth into decades of cold and darkness. The firestorms, temperature changes, and decreased food supplies could have led to mass starvation that would have been especially hard on large animals such as the dinosaurs.

Not all paleontologists believe that this mass extinction was the result of an impact. They point out that the evolution of species is a complex process and that simple answers are seldom complete. However, the evidence is compelling that a great impact *did occur* at the end of the Cretaceous Period. The rock record also shows other instances of mass extinctions associated with colossal impacts, suggesting that impacts have played a central role in the saga of life on Earth.

before they reach the surface. Venus is even better protected. With an atmosphere far more massive than that of Earth, still larger objects may be burned up or broken apart by its atmosphere before reaching the surface, creating clusters of craters in a pattern similar to a shotgun blast, or sometimes leaving only a dark "splotch" on the surface with no crater at all.

The characteristics of a crater also depend on the properties of the planetary surface. An impact in a deep ocean on Earth might create an impressive wave but leave no lasting crater. In contrast, an impact scar formed in granite can be preserved for billions of years. Planetary scientists can tell a lot about the surface of a planet by studying its craters. For example, craters on the Moon are often surrounded by

FIGURE 7.7 Terrain on the Moon, photographed by the *Apollo 17* astronauts, has been heavily cratered by impacts.

strings of *secondary craters* formed from material thrown out by the impact, like those in Figure 7.3(b). Some craters on Mars have a very different appearance. These craters are surrounded by what appears to be flows of material, much

---

The forms of impact scars give clues about the surfaces they are on.

---

like the pattern you might see if you threw a rock into mud (**Figure 7.8**). The apparent flows seem to indicate that unlike the lunar case, the Martian surface rocks contained water or ice at the time of the impact. Not all Martian crater ejecta deposits look like this, so the water or ice must have been concentrated in only some areas, and the locations might have changed with time.

It is possible that these craters were formed at a time in the past when there was liquid water on the surface of Mars. Dry riverbeds on Mars seen today attest to this possibility. But there is another intriguing possibility. Today the surface of Mars is dry and mostly frozen. This suggests that water that might once have been on the surface has soaked into the ground, much like the water frozen in the ground in Earth's polar regions. Recently the *Mars Odyssey* space-

---

Mars was once wetter than it is today.

---

craft's neutron spectrometer has detected a hydrogen signature,[2] implying subsurface water ice not only in the polar areas as expected, but also in some lower latitudes. The energy released by the impact of a meteoroid would melt this ice, possibly turning the surface material into a slurry with a consistency much like wet concrete. When thrown from the crater by the force of the impact, this slurry would have hit the surrounding terrain and slid out across the surface, forming the craters we see today.

Calibrating a Cosmic Clock

Although many factors affect the formation of craters, the biggest difference among planets is in the rate at which craters are destroyed. Earth experienced an impact history similar to that of the Moon and the other terrestrial planets, and yet preserved impact craters are rare on Earth. Obliteration of impact scars on Earth is due to a combination of gradation, tectonic processes, and volcanism acting throughout geological time. Geologically speaking, the Moon has been nearly dead for more than a billion years, and its surface still preserves the scars of craters dating from the early days of the Solar System. In contrast, geologically active planets such as Earth bear the scars of only more recent impacts.

The number of craters on a surface indicates the age of the surface.

Planetary scientists use this simple relationship—more craters means an older planetary surface—to estimate the ages and geological histories of planetary surfaces. Although we have seen only half of Mercury's surface, its preserved craters suggest that Mercury, like the Moon, has been geologically inactive for a long time. This contrasts to the much less heavily cratered surfaces of Mars, Venus, and Earth.

Some impact craters are young, while most are old. But how old is "old," and how young is "young"? We can use the amount of cratering as a clock to measure the ages of surfaces, but first we need to know how fast that clock runs. We need to be able to say that a surface with *this many* craters is *this* old, but a surface with *that many* craters is *that* old. We need a way of "calibrating the cratering clock."

The key to calibrating the cratering clock came mostly from our exploration of the Moon. The surface of the Moon is not uniform. Some parts of the Moon are heavily cratered,

---

[2] The neutron spectrometer does not actually detect the water molecule (H_2O), but it does reveal the presence of one of its component elements, hydrogen (H). Water, scientists agree, is the only likely source of elemental hydrogen.

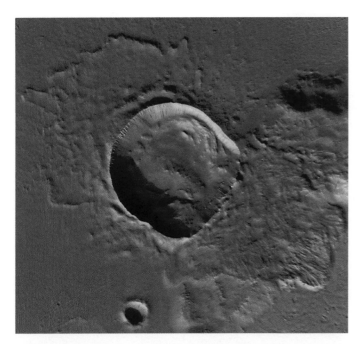

FIGURE 7.8 Some craters on Mars look like those formed by rocks thrown in mud, suggesting that material ejected from the crater contained large amounts of water. This crater is about 20 km across.

with craters overlapping their neighbors. Other parts of the Moon are much smoother, telling of more recent geological activity. Between 1969 and 1976, *Apollo* astronauts and Soviet machines visited the Moon and brought back samples taken from nine different locations on the lunar surface. By measuring relative amounts of various radioactive **elements** and the elements into which they decay, scientists were able to assign ages to these different lunar regions (see **Foundations 7.1**). The results of that work were surprising. Although smooth areas on the Moon were indeed found to be younger

> Most of the surface of the Moon is older than 3.4 billion years.

than heavily cratered areas, the differences in age were not great. The oldest, most heavily cratered regions on the Moon date back to about 4.4 billion years ago, whereas most of the smoother parts of the lunar surface are typically 3.1 to 3.9 billion years old.[3] The clear implication is that the vast preponderance of the cratering in the Solar System took place within the first billion years of the Solar System's formation (**Figure 7.10**). Heavily cratered surfaces such as those of Mercury and the Moon are ancient indeed.

---

[3] Although they have not been radiometrically dated, some of the youngest flows show very few impact craters and are therefore believed to be no more than 1 to 2 billion years old.

7.4 The Interiors of the Terrestrial Planets Tell Their Own Tale

To understand the processes responsible for all but wiping away the visible record of impacts in Earth's past and for continually remaking the surface of our planet, we begin by looking below its surface. What lies hundreds and thousands of kilometers below our feet, and how do we know? These are the questions to which we now turn.

We Can Probe the Interior of Earth in Many Ways

For all the effect that human activity has had on Earth, we have literally only scratched the surface of our planet. The deepest holes ever drilled have gone only about 12 km down, and Earth's center is 6,350 km deeper! Even so, scientists are confident about what the interior of Earth is like. Information about Earth's interior comes to us in many ways. For example, the size of Earth and the strength of Earth's gravity, together with Newton's universal law of gravitation, tell us the mass and hence the density of Earth. In this way we know that the average density of Earth is 5,500 kg/m³, or five and a half times the density of water (1 g/cm³ or 1,000

> The density of matter inside terrestrial planets increases with depth.

kg/m³). Even though we have no direct samples of material from deep within Earth's interior, this number alone tells us that the composition of the interior of our planet has a density higher than does the surface, which averages about 2,900 kg/m³. Other clues about Earth's interior come from studies of meteorites. Because these fragments are left over from a time when the Solar System was young and Earth was forming from similar materials, the overall composition of Earth should resemble the composition of meteorite material, which includes minerals with abundant iron.

By far the most important source of information about Earth's interior comes from monitoring the vibrations from earthquakes. When an earthquake occurs, vibrations spread

> Seismic waves provide information about the interior configuration of Earth.

out through and across the planet as **seismic waves**. There are two different classes of seismic waves. As their name implies, **surface waves** travel across the surface of a planet, much like waves on the ocean. If conditions are right, surface

FOUNDATIONS 7.1

Determining the Ages of Rocks

Look at a picture of the Grand Canyon. The rock layers tell the story of the canyon's geological history (**Figure 7.9**). Most of the layers were laid down by a process called *sedimentation,* in which material carried by water or wind buries what lies below. Volcanism also contributes to the layering as lava flows over Earth's surface. At the top of the stack are the latest deposits, such as those found on the rim of the Grand Canyon. Moving down through these layers takes us progressively further and further back into Earth's past.

In order to assign real dates to these different layers—or to rocks from any location, including the Moon—scientists use a technique called **radiometric dating**. Radiometric dating makes use of the steady decay of radioactive **parent elements** into more stable **daughter products**. Some minerals can contain radioactive isotopes as part of their chemical structure. (Different **isotopes** of an *element* have the same number of protons in their nuclei but different numbers of neutrons.) Chemical analysis of such a mineral immediately after its formation would reveal the presence of the radioactive parents, but the daughter products of the radioactive decay would be absent. Over time, as radioactive atoms decay, however, the amount of parent elements decreases and the amount of the trapped daughter products builds up. Chemical analysis of an old sample of such a mineral would reveal both remaining radioactive parent atoms as well as daughter products trapped within the structure of the mineral.

By comparing the relative amounts of radioactive parent and daughter products, scientists can determine when the mineral was formed and hence the age of the rock. For example, the most abundant isotope of the element uranium (the parent) decays through a series of intermediate daughters to an isotope of the element lead (its final daughter). The **half-life** of uranium is 4.5 billion

FIGURE 7.9 Layers of rock within the Grand Canyon reveal its geological history.

years. This means that in 4.5 billion years, a sample that originally contained uranium (the parent) but no lead (its final daughter) would be found instead to contain equal amounts of uranium and lead. If we were to find such a mineral, we would know that half the uranium atoms have turned to lead and that the mineral formed 4.5 billion years ago.

waves from earthquakes can be seen rolling across the countryside like ripples on water. These waves are responsible for much of the heaving of Earth's surface during an earthquake, causing damage such as the buckling of roadways.

The other type of wave travels through Earth, rather than along its surface, probing the interior of our planet. These include primary waves and secondary waves. **Primary waves** are longitudinal pressure waves (Figure 4.4(b)) that result from alternating compression and decompression of material. These are much like sound waves traveling through air or water, or a compression wave that moves along the length of a spring. **Secondary waves** are more like the motion that occurs when we pluck a guitar string. They are transverse waves (Figure 4.4(a)) that result from sideways motion of

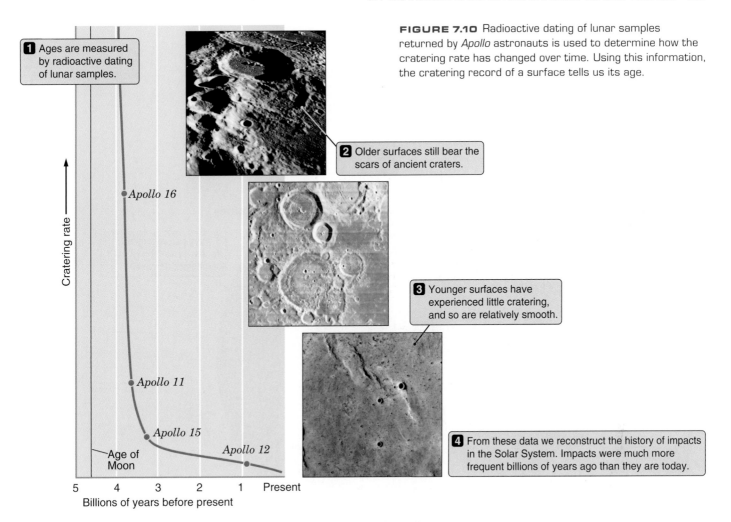

1 Ages are measured by radioactive dating of lunar samples.

FIGURE 7.10 Radioactive dating of lunar samples returned by *Apollo* astronauts is used to determine how the cratering rate has changed over time. Using this information, the cratering record of a surface tells us its age.

2 Older surfaces still bear the scars of ancient craters.

3 Younger surfaces have experienced little cratering, and so are relatively smooth.

4 From these data we reconstruct the history of impacts in the Solar System. Impacts were much more frequent billions of years ago than they are today.

Cratering rate

Apollo 16

Apollo 11

Apollo 15

Apollo 12

Age of Moon

5   4   3   2   1   Present
Billions of years before present

material. Unlike primary waves, which travel because rock rebounds after being compressed, secondary waves travel because rock springs back after being bent.

The progress of seismic waves through Earth's interior depends upon the characteristics of the material they are moving through (**Figure 7.11**). For example, an important difference between primary and secondary waves in studying Earth's interior is that whereas primary waves can travel through either solids or liquids, secondary waves cannot travel through liquids because liquids do not "spring back" when they are "bent." In addition, the speed at which seismic waves travel depends on the density and composition of rock. As a result, seismic waves moving through rocks of varying densities or composition are bent in much the same way that waves of light are bent when they enter or leave glass.[4] In fact, when a wave comes to a place where

the density of rock layers varies abruptly, the wave can be refracted (bent) or even reflected just as light is refracted or reflected by a pane of glass (see Foundations 5.1).

For nearly a hundred years, thousands of **seismometers** scattered around the globe have measured the vibrations from countless earthquakes and other seismic events, such as volcanic eruptions and nuclear explosions. When seismic waves arrive at a seismometer station, geologists ask many questions, including these: What types of waves were measured? How strong were they? When were they received at the station? Alone, a single seismometer can record ground motion at only one place on Earth. But when we combine such measurements with those of many seismometers placed all over Earth, we can use the data to get a comprehensive picture of the interior of our planet.

Building a Model of Earth

How geologists go from raw seismic data to an understanding of Earth's interior is a very good example of the interplay

---

[4]Light does not take a curved path through glass because the index of refraction remains constant throughout. Seismic waves will take a curved path through Earth's interior where the density is continuously changing.

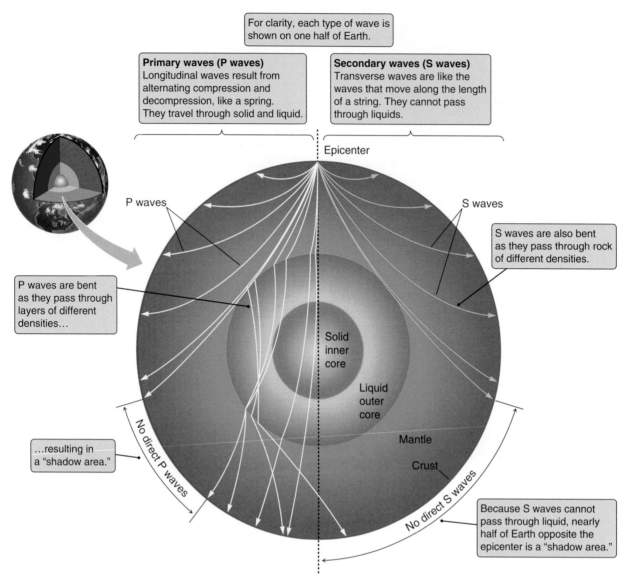

For clarity, each type of wave is shown on one half of Earth.

Primary waves (P waves) Longitudinal waves result from alternating compression and decompression, like a spring. They travel through solid and liquid.

Secondary waves (S waves) Transverse waves are like the waves that move along the length of a string. They cannot pass through liquids.

Epicenter

P waves

S waves

S waves are also bent as they pass through rock of different densities.

P waves are bent as they pass through layers of different densities…

Solid inner core

Liquid outer core

Mantle

Crust

No direct P waves

No direct S waves

…resulting in a "shadow area."

Because S waves cannot pass through liquid, nearly half of Earth opposite the epicenter is a "shadow area."

FIGURE 7.11 Different types of seismic waves propagate through the interior of Earth in different ways. Measurements of when and where different types of seismic waves arrive after an earthquake allow us to test predictions of detailed models of Earth's interior.

between theory and observation in modern science. To construct a model of Earth's interior, geologists begin with some obvious clues from the seismic data. Are there liquid regions where secondary waves cannot penetrate? Are there jumps in the density of the rock from which waves are reflected? How do waves bend, and what does that say about the density profile of Earth? The model must also be consistent with Earth's average density of 5,500 kg/m³.

To go beyond this basic sketch, geologists turn to the laws of physics, combined with knowledge of the properties of materials and how they behave at different temperatures and pressures. Scientists then construct a model of Earth's interior that follows the rough outline of their basic sketch but that is also physically consistent. (For example, the pressure at any point in Earth's interior must be just high enough to balance the weight of all the material above it, as discussed in **Foundations 7.2**.) They next "set off" earthquakes in their model, calculate how seismic waves would propagate through a model Earth with that structure, and predict what those seismic waves would look like at seismometer stations around the globe. They then test their model by comparing these predictions with actual observations of seismic waves from real earthquakes. The extent to which the predictions agree with observations points out both strengths and weaknesses of the model. The structure of the model is adjusted (always remaining consistent with the known physical properties of materials) until a good match is found between prediction and observation.

That is how geologists arrived at our current picture of the interior of Earth. Changing any single part of this picture would so change the way that waves travel through Earth's interior that the model would no longer agree with seismic observations.

The first thing you should notice about the interior structure of Earth is that its composition is far from uniform. Based on physical models and seismograph data, we find that the major subdivisions of Earth's interior, shown in Figure 7.11, include a **core**, surrounded by a thick **mantle**, on top of which is the **lithosphere**. At the center, Earth's core consists primarily of iron, nickel, and other dense metals. In contrast, the outer parts of Earth are made of materials that are of lower density. The **crust**, which is the outermost part of the lithosphere, comes in two forms: low-density continental crust that is rich in silica (SiO_2) and higher-density oceanic crust. Common continental rocks include **granite**, whereas most oceanic crust is **basalt**, a heavy, dark volcanic rock that is rich in iron and magnesium.

Geologists speak of Earth's interior as being "differentiated" and the process of separating materials by density as **differentiation**. Differentiation of the interiors of Earth, other terrestrial planets, and the Moon is a result of the fact that these planetary interiors were once molten. If rocks of different types are mixed together, they tend to stay mixed.

Differentiation of planets shows that they once were molten.

However, once this rock is melted, the denser materials are free to sink to the bottom, and the less dense materials are free to float to the top (just as less dense oil floats on denser vinegar in a bottle of salad dressing). Today little of Earth's interior is molten, but the differentiated structure of the planet tells of a time when Earth was much hotter, and its interior was liquid throughout.

Figure 7.12 shows the differentiated structure of each of the terrestrial planets and Earth's Moon. As we continue our study of the composition of objects in the Solar System, differentiation will be an important concept. For example, in Chapter 12 we will find that by analyzing the chemical composition of meteorite material, we can often determine whether it was once part of a body that was chemically differentiated.

The Moon Was Born from Earth

Models of the interior of the Moon show that it has only a tiny core and is composed mostly of material that is similar to that found in Earth's mantle. The best model explaining the Moon's composition is that when Earth was very young a Mars-sized protoplanet collided with Earth, blasting off

The Moon was probably formed from a collision between a protoplanet and the young Earth.

and vaporizing parts of Earth's partly differentiated crust and mantle. This debris condensed into orbit around Earth, evolving into our Moon. This model explains the similarities in composition between the Moon and Earth's mantle, and the model can account for the Moon's general lack of volatiles while Earth and its closest neighbors—Mars and Venus—are volatile-rich. According to this model, during the vaporization stage of the collision, most gases were lost to space, leaving only the nonvolatiles to condense as the Moon. Earth, however, was large enough to keep its volatiles,

FOUNDATIONS 7.2

Pressure and Weight

The pressure at any point within a planet's interior is determined by the weight of the material above it. To see why this is so, think about what would happen if it were not. If the outward pressure at some point within a planet were less than the weight per unit area of the overlying material, then that material would fall inward, crushing what was underneath it; if the pressure at some point within a planet were greater than the weight per unit area of the overlying material, then the material would

not be confined. The material would be able to expand, lifting the overlying material. The only stable situation is when the weight of matter above is just balanced by the pressure within the whole interior of the planet.

This balance between pressure and weight is a general and useful result that scientists refer to as **hydrostatic equilibrium**. We will see this balance repeatedly as we talk about the structure of planetary interiors, planetary atmospheres, and the structure and evolution of stars.

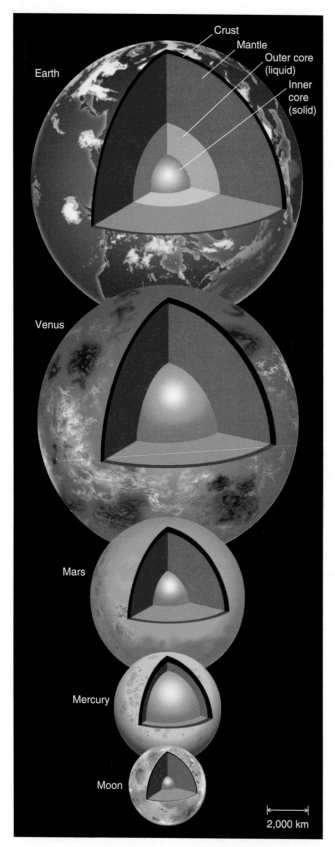

FIGURE 7.12 A comparison of the interiors of the terrestrial planets and Earth's Moon. Some fractions of the cores of Mercury, Venus, and Mars are probably liquid.

which continued to be released from the interior. Because of the stronger gravity of Earth, these gases were retained as part of our atmosphere.

The Evolution of Planetary Interiors Depends on Heating and Cooling

A general feature of planetary interiors is that the deeper we go within the planet, the higher the temperature climbs. A moment's thought about what happens to thermal energy in the interior of a planet shows why this must be so. A planet cools by radiating energy from its surface into space, so we would expect the surface of the planet to be the coolest part. Because thermal energy flows from hotter to cooler regions, the temperature must increase as we move inward from the surface if thermal energy is to flow from the interior toward the surface. The pressure in the interior of a planet also increases as we go deeper because the pressure at any given point is determined by the weight of all of the material above it.

The structure of the core of Earth results from an interesting interplay between the effects of increasing pressure and increasing temperature. We normally think about freezing and melting points—that is, whether a material is solid or liquid—as being related to temperature. When something gets hot enough, it melts and becomes a liquid. When something gets cold enough, it freezes and becomes a solid. The center of Earth is the hottest location in Earth's interior. With a temperature of perhaps 6,000 K, it is hotter than the surface of the Sun. Yet the center of Earth is solid! It is the *outer* core of Earth that is molten, despite the fact that it is hundreds of kelvins cooler than the inner core.

Whether a material is solid or liquid depends on pressure as well as on temperature. With most materials (water ice being a rare exception), the solid form of the material is more compact than the liquid form. Putting the material under higher pressure forces atoms and molecules closer together and makes the material more likely to become a solid. Moving toward the center of Earth, the effects of temperature and pressure oppose each other: The higher temperature makes it more likely that material would melt, but the higher pressure favors a solid form. It is only in the outer core of Earth that the high temperature wins, allowing the material to exist in a molten state. At the center of Earth, even though the temperature is higher, the pressure is so great that the inner core of Earth is solid.

Part of the thermal energy in the interior of Earth is left over from when Earth formed. We know that the tremendous energy liberated by the collisions responsible for the formation of Earth, together with energy from short-lived radioactive elements, was enough to melt the planet. The differentiated structure of Earth is evidence of this fact. The surface of Earth then cooled rather rapidly by radiating

Collisions and radioactive heating made the forming Earth molten, but as it ages, Earth is cooling off.

energy away into space, forming a solid crust on top of a molten interior. Because a solid crust does not conduct thermal energy well, it served as an insulator—a "wool sweater," if you will. (Anyone who has watched a lava flow "crust over" and then walked across its surface has experienced this fact. Molten rock is still there, possibly only centimeters beneath the adventurous walker's feet.) But the crust is not a perfect insulator. Over the eons, energy from the interior of the planet continued to leak through the crust and radiate into space. The interior of the planet slowly cooled, and the mantle and the inner core solidified.

But there must be more to the story than leftover energy from the time of Earth's formation. If this were the only source of heating in Earth's interior, calculations show that the interior of Earth should be much cooler than it is today, and Earth should have long ago solidified completely. There must be additional sources of energy continuing to heat the interior of Earth if we are to account for the high temperatures that persist there today.

One source of heating Earth's interior is friction generated by tidal effects of the Moon and Sun. When we discuss tides in Chapter 10, we will find that tidal heating is responsible for keeping the interior of Jupiter's moon Io molten. However, tidal heating fails by a wide margin to explain the elevated temperature of Earth's interior. Instead, most of the extra energy in Earth's interior comes from long-lived radioactive elements. As these radioactive elements trapped in the interior of Earth decay, they liberate energy, heating the planet's interior. Today the temperature of Earth's interior is determined by dynamic equilibrium (Foundations 4.2) between heating of the interior and the loss of energy that is radiated away into space. As radioactive "fuel" in Earth's interior is consumed by decay, the amount of thermal energy generated declines, and Earth's interior becomes cooler as it ages.

Because of this equilibrium, the internal temperature of a planet depends on the planet's size. The amount of heating produced in a planet is determined by the planet's *volume* because it is the volume of the planet that determines how much radioactive material ("fuel") there is. On the other hand, the planet's ability to get rid of the thermal energy in its interior depends on the planet's *surface area* because thermal energy has to escape through the planet's surface. (If you want to keep warm, you huddle up in a ball, reducing the amount of exposed skin through which thermal energy can escape. But if you want to cool off, you spread out your arms and legs, exposing as much skin as possible and allowing it to get rid of thermal energy.)

The energy-producing volume of a planet is proportional to the cube of the planet's radius ($\propto R^3$), whereas the cool-

ing surface area of the planet is proportional to only the square of the radius ($\propto R^2$). The ratio of the two—the amount of energy there is to lose divided by the surface area through which thermal energy can escape—is proportional to R^3/R^2, or R. Smaller planets therefore have less energy to lose in relation to their surface areas, and so should be cooler. Larger planets have more energy to lose per square meter

Generally, smaller terrestrial planets cool faster than larger terrestrial planets.

of surface, and so remain hotter. Indeed, in Section 7.5 we will find that the geological activity of planets is driven by the thermal energy in their interiors. It should not be surprising to learn that the smaller objects—Mercury and the Moon—are geologically inactive in comparison to the larger terrestrial planets—Venus, Earth, and Mars.

Most Planets Generate Their Own Magnetic Fields

As most schoolchildren know, a magnetic compass consists of a small bar magnet that is allowed to swing about freely. The compass needle lines up with Earth's magnetic field and points "north" and "south." But if we were to map the orientation of compass needles at every place on Earth, the north arrows would all converge not at the North Pole but at a location in northern Canada. Earth behaves as if it contains a giant bar magnet that is slightly tilted with respect to Earth's rotation axis (**Figure 7.13**). The location in Canada that compasses point to is one of Earth's magnetic poles. An

Earth's magnetic field behaves like a giant bar magnet, but the orientation of the poles changes with time.

opposite magnetic pole exists near Earth's South Pole. You may logically ask why Earth's hypothetical bar magnet, shown in Figure 7.13, has its *south* pole coincident with Earth's *north* magnetic pole. The answer is quite simple when you think about it. Bar magnets are attracted to each other's *opposite* poles. We have *defined* the north-pointing end of a compass's bar magnet as its *north* pole. So by this definition alone, Earth's magnetic field must be defined as we have shown it in Figure 7.13.

Earth's magnetic field actually originates deep within the interior of the planet, and the processes responsible for generating Earth's magnetism are not understood in detail. However, one thing we are certain of is that Earth's magnetic field is *not* due to *permanent magnets* (naturally occurring magnetic materials whose individual atoms are magnetically aligned) buried within the planet. Even though naturally occurring magnetic materials do exist, permanent

(a)

(b)

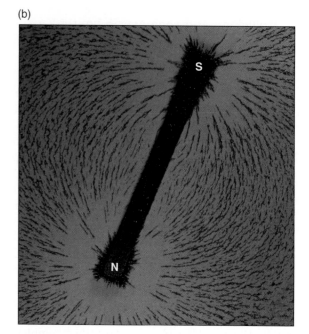

FIGURE 7.13 (a) Earth's magnetic field is similar to that of a giant bar magnet tilted relative to Earth's axis of rotation. Compass needles line up along magnetic field lines and point toward Earth's north magnetic pole. Note that the magnetic pole near Earth's North Pole is a south magnetic pole. (b) Iron filings sprinkled around a bar magnet help us visualize such a magnetic field.

magnets cannot explain the fact that Earth's magnetic field is constantly changing. Some changes in Earth's magnetic field, such as shifting in the exact location of the magnetic north pole, can occur over periods that are much shorter than a human lifetime.[5] In addition, the geological record shows that much more dramatic changes have occurred over the history of our planet.

The study of **paleomagnetism**—the fossil record of Earth's changing magnetic field—is an important part of geology. If a magnetic material such as iron gets hot enough, it loses its "memory" of its previous magnetization. As the material cools, it again becomes magnetized, but along the direction of any externally imposed magnetic field. In this way a memory of that magnetic field becomes "frozen" into the material. (This basic piece of physics is used in many ways, including in some computer data storage technologies. This is also another reason why permanent magnets cannot be responsible for Earth's magnetic field. At the temperatures in the interior of Earth, permanent magnets lose their magnetization.) Lava extruded from a volcano carries a record of Earth's magnetic field at the moment when it cooled. Using the radiometric techniques discussed in Foundations 7.1 to date these materials gives us a record of

how Earth's magnetic field has changed over time. Although Earth's magnetic field has probably existed for billions of years, the north–south *polarity* reverses from time to time. On average, these reversals in Earth's magnetic field take place about every 500,000 years.

Although the details are not known, we have a general idea of how Earth's magnetic field originates in the motions of material in Earth's interior. Magnetic fields result from electric currents, which are moving electric charges. Earth's magnetic field is thought to be a side effect of three factors: Earth's rotation about its axis; an electrically conducting, liquid outer core; and fluid motions including convection within the outer core. The interior of Earth acts as if it were a giant **dynamo**, converting mechanical energy into magnetic energy. (Other objects in the Solar System, including the Sun and the giant planets, are also thought to act like dynamos.)

The Moon is the most magnetically surveyed astronomical object other than Earth. During the *Apollo* program astronauts used surface *magnetometers* (devices for measuring magnetic fields) to make local measurements; and on two missions small satellites were placed into orbit to search for global magnetism. Results show that the Moon either lacks a magnetic field today or, at the most, has a very weak field. The lack of a lunar magnetic field can probably be understood because of the small size of the Moon and its correspondingly cooler interior. It also has a very small core. Overall, it would be difficult to make a lunar dynamo work

---

[5] At the moment, the north magnetic pole is on the move, traveling several tens of kilometers per year toward the northwest. If this rate continues, the north magnetic pole could leave Canada and be in Siberia by the end of the century.

The Moon had a magnetic field long ago, but it lacks one today.

today. However, remnant magnetism is preserved in lunar rocks from an earlier time when there likely was a magnetic field.

When we study the other terrestrial planets, however, there are a few surprises in store. Other than Earth, Mercury is the only terrestrial planet with a significant magnetic field today. Mercury's having a field is understandable. This planet has a large iron core, parts of which may be molten, so the ingredients for a dynamo are present. Like the Moon, Mars has a weak magnetic field, presumably frozen in place when its dynamo stopped working many billions of years ago. But the lack of a detected magnetic field

The lack of magnetic fields on Venus and Mars is a puzzle.

from Venus presents a puzzle. Venus should, like Earth, have an iron-rich core and partly molten interior. Its lack of a magnetic field might be attributed to its extremely slow rotation, which could make its dynamo ineffective. On the other hand, Mercury also rotates very slowly (once every 58.6 Earth days) but still has a planetary magnetic field.

The lack of a magnetic field today on Mars might be the result of its small core; but given that it is expected to have a partly molten interior and rotates rapidly, the lack of a field is still surprising. However, similar to the Moon, Mars has a pronounced remnant magnetic field, as discovered by the *Mars Global Surveyor* orbiter in the late 1990s. The magnetic signature occurs only in the ancient crustal rocks, showing that early in Mars's history some sort of a magnetic field existed. Geologically younger rocks lack this residual magnetism, so the planet's magnetic field has long since disappeared.

7.5 Tectonism— How Planetary Surfaces Evolve

As you drive through mountainous or hilly terrain, take a look at places like that shown in **Figure 7.14**, where the roadway has been cut through rock. These cuts show layers of rock that have been bent, broken, or fractured into pieces. Sometimes the force responsible for tectonism—the deformation of Earth's crust—is just gravity. For example, for more than 60 million years, rivers have dumped trillions of metric tons of rock and sediment into the Gulf of

Mexico. Layers of sediment over 25 km thick have built up, and some have solidified into rock. This enormous mass has been pulled downward by gravity, causing the rock layers to bend or to break along **faults**. Faults and **folds** in these rocks form traps for the accumulation of petroleum, creating some of the richest oil fields in the world.

Although the weight of the crust is responsible for some of the deformation of Earth's crust, most of the faulting and buckling that we see at Earth's surface originates instead from forces deep within Earth's interior. Early in this century, some scientists recognized that Earth's continents could be fit together like pieces of a giant jigsaw puzzle. The fit was particularly striking between the Americas and Africa-plus-Europe. Other evidence also suggested that this fit was more than coincidence. For example, the layers in the rock on the east coast of South America and the fossil records they hold match those on the west coast of Africa. Based on evidence such as this, in the 1920s the German scientist **Alfred Wegener** (1880–1930) proposed that over millions of years the continents had shifted their positions. This theory is popularly referred to as **continental drift**. Wegener proposed that the continents were originally joined in one large landmass that subsequently broke apart as the continents began to "drift" away from each other.

Originally the idea of continental drift met with great skepticism among geologists. However, in subsequent decades the evidence supporting the once highly controversial

FIGURE 7.14 Tectonic processes fold and warp Earth's crust, as seen in these rocks along a roadside in Israel.

Paleomagnetism: A Ticker-Tape Record of Plate Tectonics

The discovery that plates and continents are moving did not by itself confirm the hypothesis of plate tectonics because it did not demonstrate that these motions continue over geological time spans. An important clue to this puzzle came from studies of Earth's paleomagnetism.

In this chapter we note that ocean floor rifts such as the mid-Atlantic rift are spreading centers. These are locations where hot material rises toward Earth's surface and fills the gap between tectonic plates to become new ocean floor. In our discussion of paleomagnetism we found that as hot material cools, it becomes magnetized along the direction of the local magnetic field. Ocean floor "remembers" the direction of Earth's magnetic field at the time it cooled as it moves away from the spreading center. In this way the spreading ocean floor carries with it a record of the changes in Earth's magnetic field over time. The farther away from a spreading center, the older the ocean floor, and the earlier the time that its magnetization reflects. The ocean floor acts much like a ticker-tape recorder for Earth's magnetic history.

The discovery of this magnetic record was made during surveys of the ocean floor in the 1960s, and this record became one of the most important pieces of evidence supporting the theory of plate tectonics (**Figure 7.15**). These surveys showed a banded magnetic structure surrounding ocean rifts. Material near the rifts is magnetized in the same sense as Earth's current magnetic field, but the magnetization is different farther away from a rift. This banded magnetic structure is often symmetric about rifts. If a change in the magnetization of the ocean floor is seen 100 km on one side of a rift, then the same change will usually be seen about 100 km on the other side of the rift.

Combined with radiometric dates for the rocks, this magnetic record proved that the spreading of the seafloor and the motions of the plates have continued over long geological time spans.

FIGURE 7.15 (a) As new seafloor is formed at a spreading center, the cooling rock becomes magnetized. The magnetized rock is then carried away by tectonic motions. (b) Maps like this one of banded magnetic structure in the seafloor near Iceland provide support for the theory of plate tectonics.

theory of continental drift became impossible to ignore. Paleomagnetism of the ocean floor provided early evidence of continental drift, as discussed in **Excursions 7.1**. Today precise surveying techniques and satellite positioning techniques allow locations on Earth to be determined to within a few centimeters. These measurements confirm that Earth's lithosphere is indeed moving. Some areas are being pulled apart by more than 15 centimeters (about the length of a pencil) each year. This rate is slower than a snail's pace, but over millions of years of geological time such motions add up. Over 10 million years—not a long time by geological standards—15 cm/year becomes 1,500 km, by which time the maps definitely need to be redrawn!

Today geologists recognize that Earth's outer shell is composed of a number of segments, or **lithospheric plates**, and that motion of these plates is constantly changing the surface of Earth. This theory, which is perhaps the greatest

> The theory explaining motions of Earth's lithosphere is called *plate tectonics*.

advance in 20th-century geology, is referred to as **plate tectonics**. Plate tectonics is ultimately responsible for a wide variety of geological features on our planet.

Plate Tectonics Is Driven by Convection

The forces required to set lithospheric plates into motion are immense. We now understand that these forces are the result of thermal energy escaping from the interior of Earth. The motions of lithospheric plates are caused by **convection** in Earth's mantle. If you have ever noticed how water moves about in a pot on a stovetop, then you have seen convection

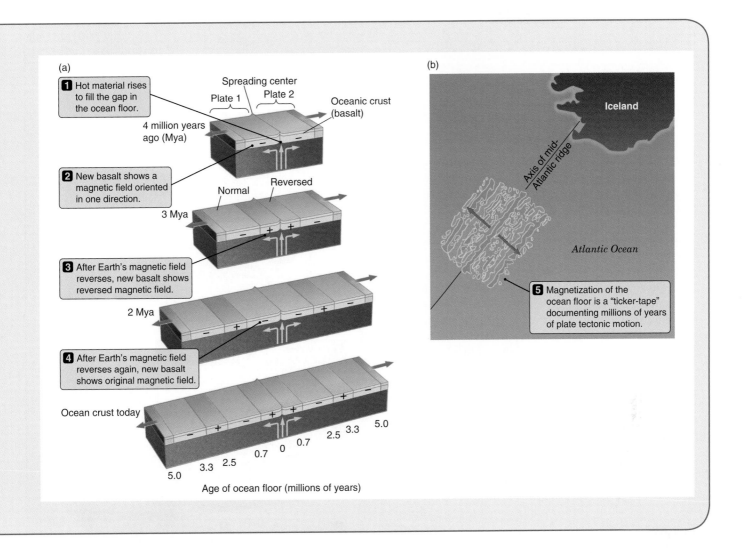

(a)

1 Hot material rises to fill the gap in the ocean floor.

Spreading center
Plate 1 Plate 2
Oceanic crust (basalt)

4 million years ago (Mya)

2 New basalt shows a magnetic field oriented in one direction.

Normal   Reversed

3 Mya

3 After Earth's magnetic field reverses, new basalt shows reversed magnetic field.

2 Mya

4 After Earth's magnetic field reverses again, new basalt shows original magnetic field.

Ocean crust today

5.0 3.3 2.5 0.7 0 0.7 2.5 3.3 5.0

Age of ocean floor (millions of years)

(b)

Iceland

Axis of mid-Atlantic ridge

Atlantic Ocean

5 Magnetization of the ocean floor is a "ticker-tape" documenting millions of years of plate tectonic motion.

at work. Thermal energy from the stove warms water at the bottom of the pot. The warm water expands slightly, becoming less dense than the cooler water above it, and the cooler water with higher density sinks, displacing the warmer water upward. When the lower-density water reaches the surface, it gives up part of its energy to the air, and in so doing cools, becomes denser, and sinks back toward the bottom of the pot. If you watch the water on the stove carefully, you can learn a lot about this process. Water cannot rise and fall in the same place, so the convection becomes organized into a circulating pattern in which warm water rises in some locations and cool water sinks in others. As discussed in **Connections 7.2**, convection is an important process in many contexts.

In the case of Earth, thermal energy generated by radioactive decay in the interior of the planet causes convection to take place in the mantle. Earth's mantle is not molten (if

it were, secondary seismic waves could not travel through it), but it is somewhat mobile. You can think of the mantle as having the plastic consistency of hot glass. This allows convection to take place, albeit very slowly.

Careful mapping shows that Earth's lithosphere is divided into about seven major plates and about a half dozen smaller ones. These plates are driven by convection cells

Earth's outer rocky shell is divided into seven major plates and about six smaller ones.

in Earth's mantle, carrying both continents and ocean crust along with them. In some places mantle material rises up and cools to form new crust, which slowly spreads out. The ocean floor around these **spreading centers** is the youngest part of Earth's crust.

Convection: From the Stovetop to Interiors of Stars

Convection can occur any time energy is introduced at the *bottom* of a gas, fluid, or even a deformable solid. The heated material expands and becomes less dense than the material above it; so it floats upward, being displaced by the denser, cooler surrounding material. At a higher level the material may give up some of its energy to its surroundings, cool and become denser, and so sink back toward the source of the energy. This sets up a circulating pattern or *cell* in which some material is rising while other material is sinking (**Figure 7.16**). In this chapter we see how convection in Earth's mantle can explain how Earth's continents move, and how convection in Earth's core can play a role in the formation of Earth's magnetic field. In Chapters 8 and 9 we will see that convection is also important in the atmospheres of planets, where it helps to explain everything from thunderstorms in the desert and the layer of smog that hangs over Los Angeles to the colorful bands of Jupiter. Convection will appear again when we turn our attention to the structure of the Sun and stars in Part III. All of these phenomena follow from a single physical process—one you can watch in action in a pan of water on the stove.

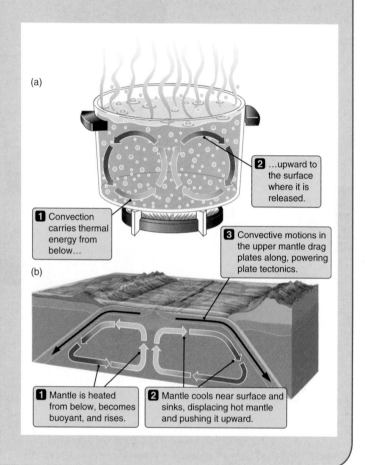

(a)

2 ...upward to the surface where it is released.

1 Convection carries thermal energy from below...

3 Convective motions in the upper mantle drag plates along, powering plate tectonics.

(b)

1 Mantle is heated from below, becomes buoyant, and rises.

2 Mantle cools near surface and sinks, displacing hot mantle and pushing it upward.

FIGURE 7.16 (a) Convection occurs when a fluid is heated from below. (b) Convection in Earth's mantle drives plate tectonics.

Figure 7.17 illustrates the process of plate tectonics and some of its consequences. If you think about convection, you will understand that if material is rising and spreading out in one location, it must be colliding and converging in another. Locations where plates converge and convection currents turn downward are called **subduction zones**. In a subduction zone one plate slides beneath the other, and convection drags the submerged lithospheric material back down into the mantle. The Mariana Trench—the deepest part of Earth's ocean floor—is such a subduction zone. Much of the ocean floor lies between spreading centers and subduction zones, and as a result the ocean floor tends to remain the youngest portion of Earth's crust. In fact, the *oldest*

seafloor rocks are less than 200 million years old. (This is one reason we do not see evidence for very large impact craters beneath the oceans.) In some places, however, the plates are not subducted but collide with each other and are shoved upward. For example, the highest mountains on

> Plates separate, or spread apart, in some regions and collide in other regions.

Earth, the Himalayas, result from the collision of the Indian–Australian subcontinental plate as it pushes northward into the Eurasian plate. In still other places, lithospheric plates meet at oblique angles and slide along past each other. A

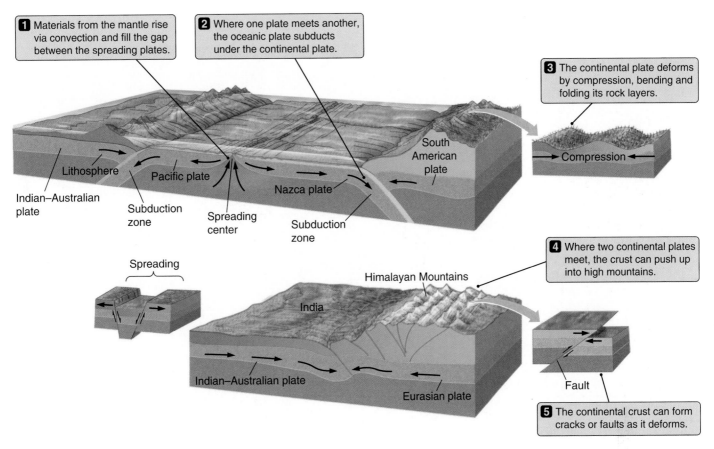

1 Materials from the mantle rise via convection and fill the gap between the spreading plates.

2 Where one plate meets another, the oceanic plate subducts under the continental plate.

3 The continental plate deforms by compression, bending and folding its rock layers.

Lithosphere

Indian–Australian plate

Pacific plate

Subduction zone

Spreading center

Nazca plate

South American plate

Compression

Subduction zone

Spreading

Himalayan Mountains

India

Indian–Australian plate

Eurasian plate

4 Where two continental plates meet, the crust can push up into high mountains.

Fault

5 The continental crust can form cracks or faults as it deforms.

FIGURE 7.17 Divergence and collisions of tectonic plates create a wealth of geological features.

type of fault called a **transform fault** marks the actively slipping fracture zone between plate boundaries. The San Andreas Fault in California is one such shear zone.

Locations where plates meet tend to be very active geologically. In fact, one of the best ways to see the outline of Earth's plates is to look at a map of where earthquakes and volcanism occur, like that in **Figure 7.18**. At locations where plates run into each other, enormous stresses build up. Earthquakes result as the friction between the two plates

---

Most volcanoes and earthquakes occur along plate boundaries.

---

finally gives way and the plates slip past each other, relieving the stress. Volcanoes occur when friction between plates melts rock that is then pushed up through cracks to the surface. Lithospheric plates can be thousands of kilometers across and range in thickness from about 5 to 100 km. As they shift, some parts move more rapidly than others, causing the plates to stretch, buckle, or fracture. These effects are readily seen on the surface as folded and faulted rocks. Mountain chains also are common near converging plate boundaries, where plates buckle and break.

Tectonism on Other Planets Is Different from on Earth

Somewhat surprisingly, evidence of plate tectonics is found only on Earth. But while spreading centers and subduction zones appear to be unique to our planet, *all* of the terrestrial planets show evidence of tectonic disruptions. Fractures have cut the crust of the Moon in many areas, leaving fault valleys (**Figure 7.19**). Most of these features are the result of large impacts that crack and distort the lunar crust.

Mercury also has fractures and faults similar to those on the Moon. In addition, there are numerous cliffs on Mercury that are hundreds of kilometers long. These appear to be the result of compression of Mercury's crust. Like the

---

Mercury's surface shrank after it cooled from a molten state.

---

other terrestrial planets, Mercury was once molten. After the surface of the planet cooled and the crust formed, the interior of the planet continued to cool and shrink. As the planet shrank, Mercury's lithosphere cracked and buckled in much the same way that a grape skin wrinkles as it

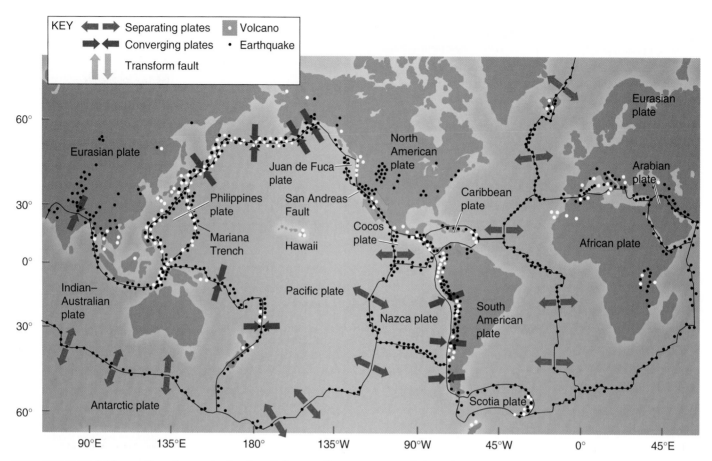

FIGURE 7.18 Major earthquakes and volcanic activity are often concentrated along the boundaries of Earth's principal tectonic plates.

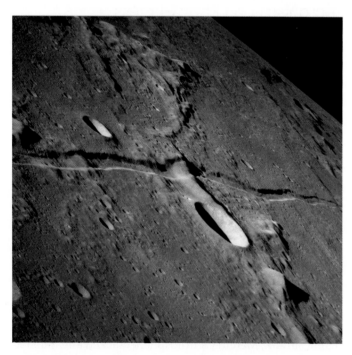

FIGURE 7.19 An *Apollo 10* photograph of Rima Ariadaeus, a 2-km wide valley between two tectonic faults on the Moon.

shrinks to become a raisin. Planetary scientists estimate that the volume of Mercury must have shrunk by about 5 percent after the formation of the planet's crust in order to explain the faults seen on the planet's surface.

On Mars the most impressive tectonic feature, and possibly the most impressive tectonic feature in the Solar System, is Valles Marineris (**Figure 7.20**). Stretching along the equatorial zone for nearly 4,000 km, this system, if occurring on Earth, would link San Francisco with New York.

Mars has experienced extensive tectonism.

Earth's Grand Canyon would be little more than a minor spur on the side of this chasm. Valles Marineris includes a series of massive cracks in Mars's lithosphere that are thought to have formed as local forces, perhaps related to mantle convection, pushed it upward from below. Once formed, the cracks were eroded by wind, water, and landslides, resulting in the massive chasm that we see today. Other parts of Mars show faults similar to those on the Moon, but the cliffs seen on Mercury are absent.

Venus is similar to Earth in many respects. Venus has a mass about 0.81 times that of Earth, a radius 0.95 times the

FIGURE 7.20 A mosaic of *Viking Orbiter* images shows Valles Marineris, the major tectonic feature on Mars stretching across the center of the image from left to right. This canyon system is more than 4,000 km long. The dark spots on the left are huge shield volcanoes in the Tharsis region.

radius of Earth, and a surface gravity 0.91 times that of Earth. As a result, many scientists speculated that Venus might also show evidence of plate tectonics. However, these speculations were not borne out by the *Magellan* mission, which used radar to peer through the thick dense clouds that enshroud the planet and map about 98 percent of the

> Most of the surface of Venus is less than 1 billion years old.

surface of Venus. *Magellan* provided the first high-resolution views of the surface of the planet. *Magellan's* view of one face of Venus is shown in **Figure 7.21**. Although Venus shows a wealth of volcanic features and tectonic fractures, there is no evidence of lithospheric plates or plate motion of the sort seen on Earth. Yet the relative scarcity of impact craters on Venus suggests that most of its surface is less than 1 billion years old.

The absence of plate tectonics on Venus is a puzzle. The interior of Venus should be very much like the interior of Earth, and convection should be occurring in its mantle. This presents a problem because it is not clear how the thermal energy from the interior of Venus is able to escape from

the planet. On Earth, mantle convection and plate tectonism provide a means for thermal energy to escape from the interior of the planet. Earth also has a few **hot spots**, where upwellings of hot mantle material rise, releasing thermal energy. The Hawaiian Islands are the result of one such hot spot. On Venus, hot spots may be the principal way that thermal energy escapes from the planet's interior. Circular fractures on the surface of Venus, ranging from a few hundred kilometers to more than 1,000 km across, may be the result of upwelling plumes of hot mantle that have fractured Venus's lithosphere.

Some planetary scientists have suggested that a radically different form of tectonism may be at work on Venus. They believe that hot spots are not adequate to allow the thermal energy generated within the planet to escape, and that as a result energy may continue to build up in the interior until large chunks of the lithosphere melt and overturn. This would suddenly release an enormous amount of energy, after which the surface of the planet would cool and resolidify. This idea remains highly controversial, but it could help explain the relatively young surface. It also drives home the point of how different the geological histories of the various planets appear to be.

FIGURE 7.21 Venus's atmosphere blocks our view of the surface in visible light. This false-color view of Venus is a radar image made by the *Magellan* spacecraft. Bright yellow and white areas are mostly fractures and ridges in the crust. Some circular features seen in the image may be regions of mantle upwelling, or hot spots. Most of the surface is formed by lava flows, shown in orange.

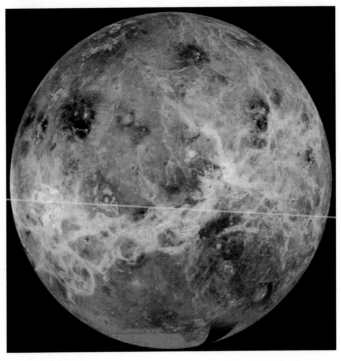

Earth's tectonic plates are unique in the Solar System.

Why Venus and Earth should have such different styles of tectonism remains an unsolved puzzle. The segmentation of Earth's lithosphere into moving plates seems to be unique among the planets and moons of the Solar System.

7.6 Igneous Activity: A Sign of a Geologically Active Planet

In December 2004 a strong earthquake off the coast of Sumatra triggered an enormous *tsunami*, which swept across the Indian Ocean. As many as 300,000 people may have died in this disaster, one of the deadliest in modern history. As with earthquakes, some of the most tragic natural disasters result from volcanism. The eruption of Mt. Vesuvius in A.D. 79 that buried Pompeii, the 1883 explosion of Krakatoa in the western Pacific that led to the loss of 36,000 lives, the hot ash flows from Mt. Pelée that demolished the Caribbean port of Saint Pierre in 1902 killing all but one of its 28,000 inhabitants—these are only a few of a long list of examples of death and destruction brought on by volcanic eruptions.

Terrestrial Volcanism Is Related to Tectonism

How do volcanoes form, and why are they found in some regions and not in others? Is there evidence of volcanoes on other planets? To answer these questions we must first understand how and where magma—the main component in volcanism—originates.

Magma does not come to Earth's surface from its molten core, as is sometimes imagined. We know from seismic signals that magma originates in the lower crust and upper mantle, where sources of thermal energy combine. These thermal energy sources include rising convection cells in the mantle, frictional heating generated by movement in the lithosphere and between tectonic plates, and concentrations of radioactive elements that produce energy from radioactive decay.

Because the thermal energy sources are not uniformly distributed inside our planet, volcanoes tend to be located only in specific areas, most notably (but not exclusively) over hot spots and along plate boundaries. Maps of geologi-

cal activity, such as Figure 7.17, leave little doubt that most terrestrial volcanism is ultimately linked to the same forces responsible for plate motions. For example, there is a tremendous amount of friction as plates slide under each other at a subduction zone. This friction generates a great deal of

Friction between moving plates generates thermal energy and leads to volcanism.

thermal energy, raising the temperature and pushing rock toward its melting point.

When we studied Earth's core, we found the counterintuitive result that even though the inner core of the planet is hotter than its surroundings, it is *solid* while its surroundings are liquid. This is because the inner core is under more pressure than its surroundings, and this higher pressure forces the material to stay solid. A similar effect occurs near Earth's surface. Material at the base of a lithospheric plate is under a great deal of pressure because of the weight of the plate pushing down on it. This pressure drives up the melting point of the material, forcing it to remain solid even though its temperature is above its normal melting point on Earth's surface. But as this material is forced up through the crust, its pressure drops, and as its pressure drops, so does its melting point. Because of this declining pressure, material that started out solid at the base of a plate may become molten as it nears the surface.

An obvious place to look for volcanic activity is along spreading centers, where convection carries hot mantle material toward the surface. As mentioned in the previous paragraph, the decrease in pressure as the material nears the surface may allow it to become molten. Spreading centers are indeed found to be frequent sites of eruptions. Iceland, which is one of the most volcanically active regions in the world, sits astride one such spreading center—the Mid-Atlantic Ridge (see Figures 7.15 and 7.18).

Once lava reaches the surface of Earth, it can form many types of structures (see **Figure 7.22**). Flows from spreading centers often form vast sheets, especially if the eruptions come from long fractures called **fissures**. If very fluid lava flows from a single "point source," it can spread out over the surrounding terrain or ocean floor, forming what is known as a **shield volcano**, so named because it resembles an upside-down warrior's shield. Pasty lava can form a steep-sided structure called a **composite volcano** or thick masses called **volcanic domes**.

A third setting for terrestrial volcanism is found where convective plumes rise toward the surface in the interiors of lithospheric plates, creating local *hot spots* (Figure 7.22(c)). Volcanism over hot spots works much like volcanism at a spreading center, except that the convective upwelling occurs at a single spot rather than along the length of a spreading center (Figure 7.17). These hot spots can melt mantle and lithospheric material and force it toward the surface of Earth.

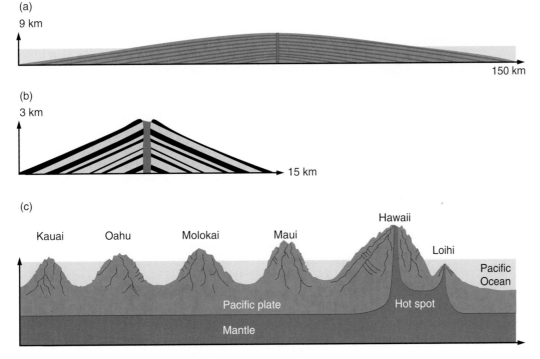

FIGURE 7.22 Magma reaching Earth's surface forms (a) shield volcanoes, such as Mauna Loa, (b) composite volcanoes, such as Vesuvius, and (c) hot spots, such as Mauna Loa and Loihi.

There are numerous hot spots on Earth, including the region around Yellowstone Park as well as the Hawaiian Islands. The Hawaiian Islands are a chain of shield volcanoes that formed as the lithospheric plate on which they ride was dragged across a relatively stationary hot spot. Volcanoes erupt over the hot spot, building an island. The island ceases to grow as the plate motion carries the island away from the hot spot, which is the source for the volcanic activity. The slower process of erosion, going on since the island's inception, continues to wear the island away. In the meantime, a new island grows over the hot spot. Today the Hawaiian hot spot is located off the southeast coast of the big island of Hawaii, where it continues to power the active volcanoes. On top of the hot spot the newest Hawaiian volcano is already forming. This volcano, called Loihi, remains submerged under the surface of the Pacific Ocean. However, viewed another way, Loihi is already a massive shield volcano, rising more than 5 km above the ocean floor. Geologists expect that it will eventually break the surface of the ocean and merge with the big island of Hawaii—but not anytime soon. Loihi is not expected to show itself above sea level for another 200,000 years!

Volcanism Also Occurs Elsewhere in the Solar System

Although Earth is the only planet on which plate tectonics is an important process, evidence of volcanism is found throughout the Solar System. Even before the *Apollo* astronauts set foot on the lunar surface, photographs showed it to have what appeared to be flowlike features in the dark regions. Early observers thought these looked like seas—thus the name *maria*, plural of **mare**, Latin for "seas." Their

---

The dark areas on the Moon's surface are ancient lava flows.

---

appearance suggested to planetary scientists that these are vast lava flows similar to basalts on Earth. Because the maria contain relatively few craters, we know that these volcanic flows occurred after the period of heavy bombardment ceased.

When the *Apollo* astronauts returned rock samples from the lunar maria, the rocks were indeed found to be basalts. Many of these Moon rocks contained gas bubbles typical of volcanic materials (**Figure 7.23**). Experiments show that when this lava flowed across the lunar surface, it must have been extremely fluid—something like the consistency of motor oil at room temperature. The fluidity of the lava, due partly to its iron- and titanium-rich chemical composition, explains why lunar basalts form vast sheets filling low-lying areas such as impact basins. It also explains the Moon's lack of classic volcanoes like Mt. Rainier: Motor oil poured from a can does not pile up; it spreads out (**Figure 7.24**).

The samples also showed that most of the lunar lava flows are older than 3 billion years! Only in a few limited areas of the Moon are younger lavas thought to exist; most of these have not been sampled directly. Samples from the

FIGURE 7.23 This sample of the Moon, collected by the *Apollo 15* astronauts from a lunar lava flow, shows gas bubbles typical of gas-rich volcanic materials. This rock is about 6 by 12 cm.

heavily cratered terrain of the Moon also originated from magma, which shows that the young Moon went through a molten stage. These rocks cooled from a "magma ocean" and are more than 4 billion years old, preserving the early history of the Solar System. Thus most of the sources of heating and volcanic activity on the Moon must have shut down some 3 billion years ago—unlike on Earth, where volcanism

continues. This conclusion is certainly consistent with our earlier argument that smaller planets should cool more efficiently and thus be less active than larger planets.

Mercury shows no conclusive evidence for past or present volcanism. There are, however, smooth plains that are similar in appearance to the lunar maria. These sparsely cratered plains are the youngest areas on Mercury and are thought to be volcanic in origin. Only half of Mercury has been explored, and the other half of the planet is mostly unknown. Moreover, most spacecraft images of Mercury have a resolution no better than that of groundbased telescope images of the Moon, so we still have much to learn about this innermost planet of the Solar System. This will all change soon. Launched in 2004, NASA's *Messenger* spacecraft will begin orbiting Mercury in 2010 and should return high-resolution images and other remote sensing data that will address many of the outstanding questions.

More than half the surface of Mars is covered with volcanic rocks. Plain-forming lavas covered huge regions of Mars, flooding the older, cratered terrain. Few of the vents

The largest mountains in the Solar System are Martian volcanoes.

or fissures for these flows are seen, suggesting that most were buried under the lava that poured forth from them (**Figure 7.25**). Among the most impressive features on Mars

FIGURE 7.24 The lava flowing across the surface of Mare Imbrium on the Moon must have been extremely fluid to spread out for hundreds of kilometers in sheets that are only tens of meters thick.

FIGURE 7.25 This mosaic of *Viking Orbiter* images shows a series of lava flows (shaded red) on Mars extending from the north *(left)* to the south *(right)* into cratered terrain. Note the crater rims that are partly "flooded" by the lava flows. The area shown is about 180 by 150 km.

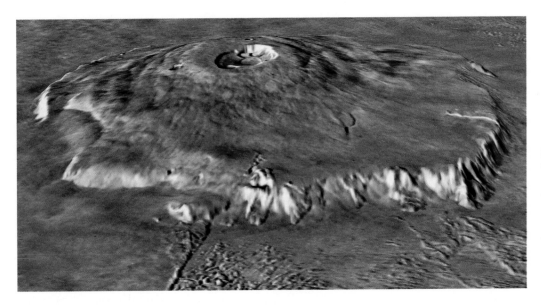

FIGURE 7.26 The largest volcano on Mars, Olympus Mons, is a 25-km high shield-type volcano, similar to but much larger than Hawaii's Mauna Loa. The oblique view was created using an overhead *Viking* image and topographic data provided by the *Mars Orbiter* Laser Altimeter.

are the enormous shield volcanoes. These volcanoes are the largest mountains in the Solar System. Olympus Mons, standing 25 km high at its peak and nearly 700 km wide at its base (**Figure 7.26**), would tower over Mount Everest and dwarf Hawaii's Mauna Loa. Standing a "mere" 9 km above the floor of the Pacific Ocean and spreading out to cover an area 120 km across, Mauna Loa is the largest mountain on Earth.

Despite the difference in size, the volcanoes of Mars are shield volcanoes, just like their Hawaiian cousins. Olympus Mons and its neighbors grew as the result of hundreds of thousands of individual eruptions that sent lava running down their flanks. The difference in size between Olympus Mons and Mauna Loa could be a result of the absence of plate tectonics on Mars. As discussed earlier, the motion of the plate that Hawaii rides on carries the Hawaiian volcanoes away from their hot spot after only a few million years. The Martian volcanoes, on the other hand, have remained over their respective hot spots for billions of years, growing ever taller and broader with each successive eruption.

Although no samples have been returned directly from Mars, analysis by instruments on landed spacecraft, remote sensing data, and the shapes of the lava flows and volcanoes all suggest that Martian lavas are basalts much like those found on Earth and the Moon. However, chemical analyses at the *Pathfinder* lander site suggest that the rocks contain slightly more silica than typical basalts, which could mean that the magma was partly differentiated chemically before it erupted. On the other hand, chemical analyses obtained in Gusev crater by the Mars exploration rover *Spirit* reveal basalts that are rich in iron, more like those of the Moon. In Chapter 12 we will discuss evidence that certain meteorites found on Earth were probably blasted from the surface of Mars by impacts. Chemical analysis of these meteorites supports the view that volcanism on Mars involved basaltic lavas.

Of all the terrestrial planets, Venus has the largest population of volcanoes. Radar images reveal a wide variety of volcanic landforms. These include highly fluid flood lavas covering thousands of square kilometers, shield volcanoes

Venus has the largest population of volcanoes among the terrestrial planets.

approaching those of Mars in size and complexity, dome volcanoes, and lava channels thousands of kilometers long. These lavas must have been extremely hot and fluid to flow for such long distances. Some of the volcanic eruptions on Venus are thought to have been associated with deformation of Venus's lithosphere above hot spots such as the circular fractures mentioned earlier.

Although we know little about the compositions of the volcanic rocks on Venus, the Soviet *Venera* landers measured some surface compositions; for the most part the results suggest that lavas on Venus are basalts, much like the lavas on Earth, the Moon, and Mars. It is presumed that the possible lavas on Mercury are basalts as well.

In summary, what can we say about the volcanic histories of the terrestrial planets and the Moon? After the Moon went though a molten state—a sort of "magma ocean" phase—shortly after its birth, the Moon developed an ever-thickening lithosphere overlying a mantle. Depending on locations of radioactive materials, local reservoirs of magma were generated. Some large impacts were able to penetrate these reservoirs or otherwise trigger the release of magma to the surface through fractures. Most of this volcanism ceased about 3 billion years ago, although some minor eruptions could have continued sporadically for another billion years. In all, less than 18 percent of the lunar surface is covered with volcanic rocks (excluding those cooled from the "magma ocean"). Volcanism on Mercury—if indeed volcanism occurred at all—probably mimicked that of the Moon. Many of the inferred volcanic plains on Mercury are also associated with impact scars. The ages of these plains are not known, but from superposed impact craters we can conclude that the plains are probably billions of years old. Until more and better data are available for Mercury, ideas about the formation and eruption of magma on that planet remain speculative.

Lava flows and other volcanic landforms span nearly the entire history of Mars, estimated to extend from the formation of crust some 4.4 billion years ago to geologically recent times, and cover more than half of the red planet's surface. But remember, "recent" in this sense could still be more than 100 million years ago, or back to our own dinosaur age. Although some "fresh"-appearing lava flows were identified on Mars, until rock samples are radiometrically dated, we will not know the age of these latest eruptions. Mars could, in principle, experience eruptions today.

Most of Venus is covered with volcanic materials or tectonically disrupted rocks of presumed volcanic origin. A geological time scale for Venus has not yet been devised, but from its relative lack of impact craters, most of the surface is considered to be less than 1 billion years old. When volcanism began on Venus and whether volcanoes are still active today remain unanswered questions.

Earth remains the champion for the diversity of volcanism. Volcanic rocks are found throughout the rock record, while compositions of magma span the spectrum of silica-rich to super-iron-rich materials erupted directly from the mantle.

7.7 Gradation: Wearing Down the High Spots and Filling in the Low

Gradation is the "great leveler" of planetary surfaces. The term *gradation* covers a wide variety of processes that together serve to smooth out planetary terrain, wearing down the high spots and filling in the low. The first step in the process of gradation is called *weathering,* in which rocks are broken into smaller pieces and may be chemically altered. For example, rocks on Earth are physically weathered along shorelines, where they are broken into beach sand by pounding waves, and along streambeds, where they are slammed together. Other weathering processes include chemical reactions, such as when oxygen in the air combines with iron in rocks to form a type of rust. One of the most efficient forms of weathering involves freeze–thaw cycles, during which liquid water runs into crevices and then freezes, expanding and shattering the rock.

After weathering, the resulting debris can be carried away by flowing water, glacial ice, or blowing wind and deposited in other areas as sediment. Where material is eroded, we can see features such as river valleys, wind-sculpted hills, or mountains carved by glaciers. Where eroded material is deposited, we see features such as river

The actions of water and wind produce the greatest amount of gradation on Earth.

deltas, sand dunes, or piles of rock at the bases of mountains and cliffs. It is not surprising to find that gradation is most efficient on planets where water and wind are present. On Earth, where water and wind are so dominant, most impact craters are worn down and filled in even before they are turned under by tectonic activity. If other processes were not at work to form mountains, valleys, and other topographic relief, gradation would eventually wear planets like Earth as smooth as billiard balls.

Yet even on the Moon and Mercury, which have no atmospheres or running water, a type of gradation (albeit *very* slow) is still at work. Radiation from the Sun and from deep space slowly decomposes some types of minerals, effectively weathering the rock. Such effects are only "skin deep"—usually a few millimeters at most—as a kind of rock "sunburn." Impacts of micrometeoroids also chip away at rocks. In addition, landslides can occur wherever gravity and differences in elevation are present. Although landslide activity is enhanced by the lubricating effects of water, landslides are also seen on dry planets like Mercury and the Moon. Debris from landslides has even been seen on the tiny moons of Mars and on asteroids (see Chapter 12).

Earth, Mars, and Venus, on the other hand, do have atmospheres, and all three planets show the effects of windstorms. Images of the surfaces of Mars and Venus returned by spacecraft landers show surfaces that have clearly been subject to the forces of wind (see **Figure 7.27**). Likewise,

The surfaces of Mars and Venus are also modified by winds.

orbiting spacecraft have returned pictures showing sand dunes, wind-eroded hills, and surface patterns called *wind streaks.* Planet-encompassing dust storms have been seen on Mars. These storms have been known to blot out the visibility of the surface of the planet for months on end.

Sand dunes are common on Earth and Mars, and some are identified on Venus. They occur frequently wherever moderately strong winds blow and there is a supply of loose grains. The largest field of windblown sand on Mars (**Figure 7.28**) is in the north polar area. Its size, covering more than 700,000 square kilometers, is comparable to the largest fields of sand dunes on Earth. The most common wind-related features on Mars and Venus are *wind streaks* (**Figure 7.29**). These are surface patterns that appear, disap-

FIGURE 7.27 A view of the surface of Mars taken by the Mars Rover, *Spirit*, showing the rock-littered surface of Gusev crater. Fine-grained material between and partly covering the rocks is wind-transported sand and dust.

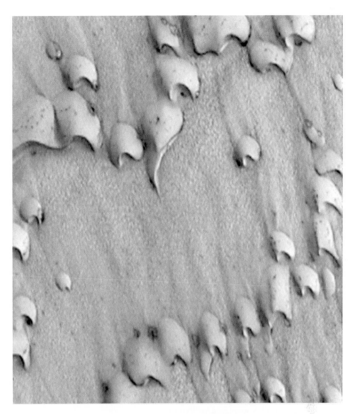

FIGURE 7.28 This high-resolution image taken by a camera aboard the *Mars Global Surveyor* shows windblown sand dunes in the north polar region of Mars.

pear, and change in response to winds blowing sediments around hills, craters, and cliffs. They serve as local "wind vanes," telling planetary scientists about the direction of local prevailing surface winds.

Today Earth is the only planet where the temperature and atmospheric conditions allow extensive liquid surface water to exist. Water is an extremely powerful agent of gradation and dominates erosion on Earth. Every year rivers and streams on Earth deliver about 10 billion metric tons of sediment into the oceans. Even though today there is no liquid water on the surface of Mars, at one time water flowed across its surface in such vast quantities as to make the Amazon River look like a backyard irrigation ditch. Huge dry riverbeds such as those shown in **Figure 7.30** attest to tremendous floods that poured across the Martian surface. In addition, many regions on Mars show small networks of valleys that are thought to have been carved by flowing water. Some parts of Mars may even have once contained oceans and glaciers.

The search for water is a primary goal in the exploration of Mars. In 2004 NASA sent two instrument-equipped roving vehicles, *Opportunity* and *Spirit*, to search for evidence of water on Mars (see the opening photograph for Chapter 5). The choice for *Opportunity*'s landing site was based on orbital remote sensing data that suggested the presence of hematite in an area near the Martian equator. *Hematite* is an iron-rich mineral that commonly forms in the presence of water, and the goal for this mission was to test the validity

of the remote sensing information. *Opportunity* hit a hole-in-one—literally and figuratively. Both Mars rovers were surrounded by inflatable airbags to cushion their impact on the Martian surface. Upon landing, the whole *Opportunity* package bounced across the Martian surface, finally coming to rest inside a small crater like a well-driven golf ball (see **Figure 7.31(a)**). The real payoff came when the walls of the crater revealed a treasure trove of information. For the first time, outcrops of Martian rocks were observed and available for study. *Outcrops* are places were exposed rocks remain in the original order in which they were laid down. Previously the only rocks that landers and rovers had come across were those that had been dislodged from their original settings by either impacts or river floods.

The layered rocks at the *Opportunity* site (shown in **Figure 7.31(b)**) revealed several lines of evidence indicating that they were once soaked in water. First of all, the form of the layers was typical of what we see here on Earth in sandy sediments that have been laid down by gentle currents of water.[6] Then rover instruments detected the presence of a

---

[6] Some geologists believe this layering was caused by deposits of volcanic ash rather than aqueous sediments. We find here an example of honest disagreement among scientists, all of whom are looking at the same data.

(a)

(b)

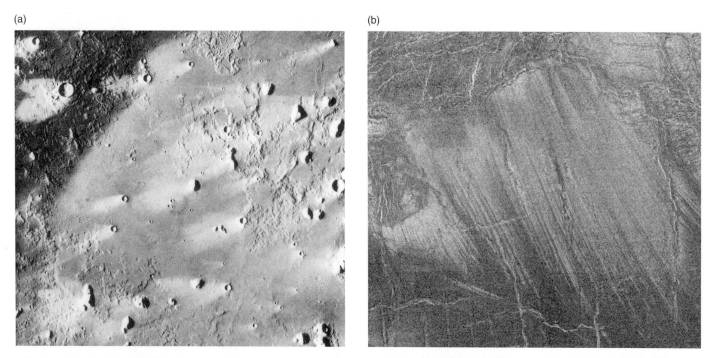

FIGURE 7.29 (a) The bright patterns associated with these craters and hills on Mars resulted from winds that redistribute sand and dust on the surface. The winds forming these streaks blew from right to left. The area shown is about 160 by 185 km. (b) Bright patches of windblown dust streak across the surface of Venus.

Water channels

250 m

FIGURE 7.30 *Viking Orbiter* images showing channels on Mars carved long ago by flowing water. Liquid water cannot exist on the surface of Mars today. The area shown is about 100 km wide. Inset: High-resolution *Mars Global Surveyor* image showing geologically recent "gullies" that may also have been carved by water.

(a)

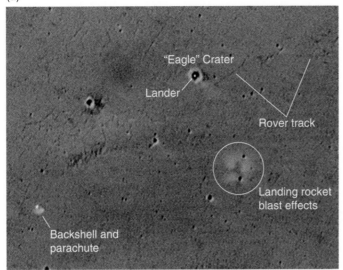

(b)

FIGURE 7.31 (a) An orbital view of *Opportunity's* landing site. After landing, the protected package bounced along the surface and ended up in "Eagle" crater, a hole-in-one. The white dot in the middle of the crater is the lander. Tracks leading out of the crater were made by *Opportunity Rover* as it began its exploration of Mars. (b) Layered rock seen near the *Opportunity* landing site may represent sediments deposited by water.

mineral commonly found in acid lakes and thermal springs, with so much sulfur content that it almost certainly formed by precipitation from water. Finally, magnified images of the rocks showed "blueberries": small spherical grains a few millimeters across that appear to have formed in place among the layered rocks in a manner similar to terrestrial features that form by the percolation of water through sediments. Analysis of the "blueberries" revealed abundant hematite, confirming the interpretations of the remote sensing data. Recent observations by the European Space Agency's *Mars Express* orbiter and NASA's *Mars Odyssey* show the hematite signature and the presence of sulfur-rich compounds in a vast area surrounding the *Opportunity* landing site. These observations suggest the existence of an ancient

Martian sea, larger than the combined area of the Great Lakes and as much as 500 meters deep.

The floor of Gusev, a 160-km impact crater, was chosen for *Spirit* because it showed signs of ancient flooding by a now dry river, and it was hoped that surface deposits would provide evidence for that past liquid water. Surprise, and perhaps some disappointment, followed when *Spirit* revealed that the flat floor of Gusev consisted primarily of basaltic rock. Only when the rover ventured cross-country to some low hills located 2.5 km from the landing site was evidence for water found. Here the basaltic rocks showed clear signs of chemical alteration by liquid water. As of this writing, *Opportunity* and *Spirit* continue to explore the Martian surface.

Where are Mars's water and ice today? At least some ice is locked in the polar regions, just as the ice caps on Earth hold much of our planet's water. (But most of the material in the Martian polar caps is frozen carbon dioxide rather

Water ice could exist on the Moon and Mercury, as well as on Mars.

than frozen water.) Some water can be found on the surface, as seen in **Figure 7.32**. However, much if not most of the water on Mars appears to be trapped below the surface of the planet.

It is also intriguing that while Earth and Mars are the only terrestrial planets that show evidence for liquid water at any time in their histories, water ice could exist on

FIGURE 7.32 Water ice is seen in this *Mars Express* image of a crater in the north polar region of Mars. The crater is about 35 km in diameter, and the ice is estimated to be about 200 m thick. The 300-m high crater wall blocks much of the sunlight at this high latitude, keeping the ice from vaporizing.

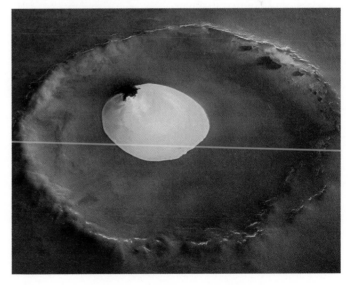

Mercury and the Moon today. Some deep craters in the polar regions of both Mercury and the Moon have floors that are in perpetual shadow. Because these planets lack atmospheres, temperatures in these permanently shadowed areas remain below 180 K. For many years planetary scientists speculated that ice—perhaps implanted by impacting comets—could be found in these craters. In the early 1990s Earth-based radar measurements of Mercury's north pole and infrared measurements of the Moon's polar areas by the joint Department of Defense/NASA *Clementine* mission returned information that seemed to support this possibility. In 1998 NASA sent another spacecraft, *Lunar Prospector*, to take a new look. Again the results suggested subsurface water ice in the polar regions. When its primary mission was completed, NASA crashed *Lunar Prospector* into a crater near the Moon's south pole while groundbased telescopes searched for evidence of water vapor above the impact site. None was seen. Perhaps it was the wrong place to look or the amount of water was too small to be detected.

Although these observations do not prove one way or another whether ice exists in these regions, the possibility is exciting and important. The existence of ice on the Moon could make any future lunar colonization much more practical than if all water had to be either brought up from Earth or synthesized from hydrogen and oxygen extracted from the lunar soil. NASA will try once again to find subsurface water ice when it launches *Lunar Reconnaissance Orbiter* in 2008.

Summary

- Comparative planetology is the key to understanding the planets.

- The Moon was created when a large planetesimal collided with Earth.

- Dark areas on the Moon are ancient lava flows.

- Impacts, volcanism, and tectonism create topography on the terrestrial planets.

- Gradation wears down the surfaces of Venus and Mars by wind and Earth by wind and water.

- Relative concentrations of impact craters divulge the ages of planetary surfaces.

- Radiometric dating tells us the age of rocks.

- Seismic waves reveal Earth's interior structure.

- Earth has a strong magnetic field, but Venus and Mars do not.

- The largest mountains in the Solar System are volcanoes on Mars.

- Mars today has subsurface water ice and once had liquid water on its surface.

Seeing the Forest through the Trees

In Chapter 6 we placed the formation of the Solar System and our Earth into the broader context of the ongoing process of star formation that we see occurring around us in the universe today. Earth is but one of several rocky planets that coalesced from the dust surviving in the hot inner parts of the disk that surrounded the young Sun. Mercury, Venus, Earth, and Mars share this common heritage and have been further shaped by the same fundamental processes of impacts, tectonism, igneous activity, and gradation over the ensuing 4.5 billion years. Earth is our benchmark for understanding the nature and effects of these processes. The cosmological principle not only suggests that we can apply physical laws discovered in terrestrial laboratories throughout the universe; it also guides us in our exploration of the Solar System. We live in a remarkable age of exploration and discovery, when our species has launched probes to visit almost all of the major bodies in the Solar System and many of the minor bodies as well. The resulting flood of information is a testament to 20th-century technology; but it is our understanding of Earth that has guided scientists as they have pieced together the story told by these data.

The impression we have gleaned from the results from the last four decades of planetary exploration is one of remarkable diversity. The surfaces of the four terrestrial planets and Earth's moon have all been stressed and fractured over the eons. One of the most startling discoveries of the last century was the fact that the outer shell of Earth itself consists of multiple plates that are in constant motion as they ride slow but inexorable convection currents within Earth's mantle. Exploration of the inner Solar System has provided new perspectives on the drifting motion of our planet's lithosphere by showing us that other possibilities exist. While Mercury's surface is cracked and wrinkled, tectonic activity on Mars fractured the planet, forming a vast canyon system. Even our sister planet, Venus, shows hints of a particularly violent form of tectonism in which the entire surface of the planet may have overturned and released bursts of pent-up thermal energy from the planet's interior. Likewise, igneous activity of several kinds has played an important role in the history of our planet, building such monuments to the power of volcanism as the Hawaiian Islands and the Cascade Range. Yet when looking at our neighbors we find that Earth's volcanoes shrink into relative insignificance beside the towering heights of Mars's Olympus Mons.

Along the way we have again seen the process of science at work. The realm of our species encompasses only the tiniest fraction of our planet. As alluded to in the humorous quote that opened the chapter, we are creatures of the *surface* of Earth, as far from direct exploration of the inner reaches of our own world as we are from direct exploration of the heart of the Sun. Even so, on this leg of our journey we have seen how the methods of science have allowed us to construct a detailed picture of the interior of Earth. We have built up that picture by applying physical laws to construct models of Earth, then tested the predictions of those models by analyzing the echoes from earthquakes. At first glance this might seem an insecure foundation for knowledge, but that is far from the case. We go through our lives building knowledge of the world around us through light waves that reach our eyes and sound waves that reach our ears, both interpreted by brains shaped by evolution to accomplish the task. Is it so different to say that we know about the interior of Earth through the seismic waves that reach our instruments, interpreted through computer models that are shaped for the job by our understanding of physical law? So it will be throughout the rest of our journey as we use the techniques and tools of science (and a carefully thought-out notion of what it means "to know") to "sense" the nature of the universe.

We have seen several ways in which the study of terrestrial planets has forced us to change the way we view our planet. Earth is but one of several rocky planets, and our world is only one small corner of the range of the possible. Yet no result of Solar System exploration has more dramatically changed our perspective on our own history—and perhaps our own future—than our understanding of the role of impact cratering in the inner So-

lar System. The active Earth is very effective at erasing its own memory. Plate tectonics, igneous activity, and gradation are continuously at work, wiping the record of our history clean. It took the ancient surfaces of Mercury and the Moon, and even the less ancient surfaces of Mars and Venus, to show us that impacts of objects from space have played a significant role in shaping Earth. With this realization, the cosmological principle truly comes full circle. Earth may provide many of the clues that we need to understand the Solar System, but it is the Solar System that provides the immediate context for the existence of Earth. Life on our planet has had its course turned time and again by sudden and cataclysmic events when asteroids and comets have slammed into Earth. It seems very likely that we owe our existence to the luck of our remote ancestors—small rodentlike mammals—that could live amid the destruction following such an impact 65 million years ago. The time since that event is but a brief moment in the history of our planet. The time until the next such event will almost certainly be no more than another short span of geological time.

Key Terms

terrestrial planet, p. 185
comparative planetology, p. 186
hydrosphere, p. 186
lithosphere, p. 186
tectonism, p. 187
volcanism, p. 187
igneous activity, p. 187
magma, p. 187
topographic relief, p. 187
gradation, p. 187
impact crater, p. 188
ejecta, p. 188
secondary crater, p. 188
Cretaceous–Tertiary boundary, p. 190
seismic wave, p. 193
surface wave, p. 193
radiometric dating, p. 194
parent element, p. 194
daughter product, p. 194
isotope, p. 194
half-life, p. 194
primary wave, p. 194
secondary wave, p. 194
seismometer, p. 195
core, p. 197
mantle, p. 197
lithosphere, p. 197

Student Questions

THINKING ABOUT THE CONCEPTS

1. Compare and contrast tectonism on Venus, Earth, and Mercury.

2. One region on the Moon is covered with craters, while another is a smooth volcanic plain. Which is older? How much older? How do we know?

3. Can all rocks be dated using radiometric methods? Explain.

4. Explain how we know that rocks at the bottom of Arizona's Grand Canyon are older than those found on the rim.

5. Describe the sources of heating that are responsible for the generation of Earth's magma. Explain why the Moon's core is cooler than Earth's.

6. How do we know that Earth's interior includes a liquid zone?

7. Explain plate tectonics, and identify the only planet on which this process has been observed.

8. Volcanoes can be found on most, perhaps all, of the terrestrial planets. Where are the largest volcanoes in the inner Solar System?

9. What is meant by *gradation*? What processes contribute to gradation?

10. Describe and explain the evidence for reversals in the polarity of Earth's magnetic field.

11. A current theory suggests that a mass extinction occurred as a consequence of an enormous impact on Earth 65 million years ago. What is the evidence for this theory?

12. What are the differences between a spreading center and a subduction zone?

13. Given images with adequate resolution, describe and explain the criteria you would apply in distinguishing between a crater formed by an impact and one formed by a volcanic eruption.

APPLYING THE CONCEPTS

14. Assume that Earth and Mars are perfect spheres with radii of 6,371 km and 3,390 km, respectively.
    a. Calculate the surface area of Earth.
    b. Calculate the surface area of Mars.
    c. If 0.72 (72 percent) of Earth's surface is covered with water, compare the amount of Earth's land area with the surface area of Mars.

15. Compare the kinetic energy of a 1-g piece of ice (about half the mass of a dime) entering Earth's atmosphere at a speed of 50 km/s with that of a 2–metric ton SUV (mass $= 2 \times 10^3$ kg) speeding down the highway at 90 km/h.

16. Using the information in Table 7.1, determine the relative rates of internal energy loss experienced by Earth and the Moon.

17. Although oceans cover 72 percent of Earth's surface, they represent but a tiny fraction of our planet's mass. Earth's mass is 6.0×10^{24} kg, and its oceans have a total mass of 1.5×10^{21} kg.
    a. What fraction of Earth's total mass do our oceans represent?
    b. Does surface and atmospheric water represent Earth's total aqueous inventory? Explain.

18. Earth's mean radius is 6,371 km, and its mass is 6.0×10^{24} kg.
    a. Calculate Earth's average density. Show your work. Do not look this value up.
    b. The average density of Earth's crust is 2,600 kg/m^3. What does this tell you about Earth's interior?

19. Say you find a piece of ancient pottery and take it to the laboratory of a physicist friend. He finds that the glaze contains radium, a radioactive element that decays to radon and has a half-life of 1,620 years. He tells you that there could not be any radon in the glaze when the pottery was being fired, but that it now contains three

atoms of radon for each atom of radium. How old is the pottery?

20. Assume that the east coast of South America and the west coast of Africa are separated by an average distance of 4,500 km. Assume also that Global Positioning System (GPS) measurements indicate that these continents are now moving apart at a rate of 3.75 cm/year. If you could assume that this rate has been constant over geological time, how long ago were these two continents joined together as part of a supercontinent?

21. Shield volcanoes are shaped something like flattened cones. The volume of a cone is equal to the area of its base times ⅓ its height. The largest volcano on Mars, Olympus Mons, is 25 km high with a base diameter of 700 km. Compare its volume with that of Earth's largest volcano, Mauna Loa, which is 9 km high and has a base diameter of 120 km.

StudySpace
wwnorton.com/astro21
provides a Study Plan for each chapter that includes a reading outline, animations, keyword flash cards, and gradebook-enabled multiple-choice quizzes. From StudySpace you can also access premium content in the ebook and SmartWork.

I come to carry you to yon shore, and lead
Into the eternal darkness, heat, and cold.

DANTE ALIGHIERI (1265–1321)
THE DIVINE COMEDY, INFERNO, CANTO III

Lightening strikes the ground near Kitt Peak National Observatory in Arizona.

CHAPTER 8

Atmospheres of the Terrestrial Planets

8.1 Atmospheres Are Oceans of Air

Earth's atmosphere surrounds us like an ocean of air. We see it in the blueness of the sky and feel it in the breezes that tousle our hair. People who live in large cities can sometimes even smell it. It can bring joy into our lives in the form of a spectacular sunset, or apprehension with the approach of a hurricane or tornado. It is responsible for all of our weather, be it pleasant or stormy. Without our atmosphere there would be neither clouds nor rain; no streams, lakes, or oceans. There would be no living creatures. Without an atmosphere Earth would look somewhat like the Moon. And we, quite simply, would not exist.

Among the five terrestrial bodies that we discussed in Chapter 7, only Venus and Earth have dense atmospheres (**Figure 8.1**). Mars has a very low-density atmosphere, while the atmospheres of Mercury and the Moon are so sparse they can hardly be detected. Why should some of the terrestrial planets have dense atmospheres while others have little or essentially none? Are atmospheres created right along with the planets they envelop, or do they appear at some later time?

Some Atmospheres Appeared Very Early

For these answers we need to look back nearly 5 billion years to the story told in Chapter 6—to a time when the planets were just completing their growth. The phases in

KEY CONCEPTS

Blankets of atmosphere warm and sustain Earth's temperate climates, but they also push the surface of Venus beyond Dante's worst nightmares of hell while leaving Mars frozen. As was the case with their surfaces, comparing these worlds is the key to understanding their atmospheres. Among many interesting insights, on this leg of our journey we will discover

- That terrestrial planets owe their atmospheres to volcanism and to volatiles captured from comets.

- Why some planets hold onto their atmospheres while others do not.

- That differences among Earth, Venus, and Mars are largely due to the atmospheric greenhouse effect.

- How Earth's atmosphere has been reshaped by life.

- That atmospheres are layered by convection and differences in how they are heated, while pressure steadily falls at higher and higher altitudes.

- How the Coriolis effect redirects convective flows into patterns of winds.

- That Earth's magnetic field interrupts the flow of the solar wind and traps charged particles in a huge magnetosphere responsible for auroras.

- The unsettling fact that we are living through an uncontrolled experiment in climate modification.

FIGURE 8.1 Global views of the atmospheres of (left to right) Venus, Earth, and Mars. Mercury and the Moon have no atmospheres to speak of.

the formation of planetary atmospheres are illustrated in **Figure 8.2**. The young planets at that time were still enveloped by the remaining hydrogen and helium that filled the protoplanetary disk surrounding the Sun, and they were able to capture some of this surrounding gas. Gas capture must have continued until the gaseous disk ultimately dissipated (soon after the formation of the planets) and the

---

Primary atmospheres consist of captured gas.

---

supply of gas ran out. The gaseous envelope collected by a newly formed planet is called its *primary atmosphere*. Although the giant planets still retain most of their original primary atmospheres, the terrestrial planets probably lost theirs soon after the protoplanetary disk was blown away by the emerging Sun. Why did only the terrestrial planets lose their primary atmospheres? The answer lies in their relatively small masses.

The terrestrial planets, with their weak gravity, lack the ability to hold light gases such as hydrogen and helium. As soon as the supply of gas in the disk ran out, their primary atmospheres began leaking back into space.[1] How can gas molecules escape from a planet? As discussed in **Excursions 8.1**, all it takes to escape a planet is a speed greater than the escape velocity (see Chapter 3), pointed in the right

---

A planet's atmosphere can sometimes escape into space.

---

direction. Intense radiation from the Sun, especially in the inner parts of the Solar System, raises the kinetic energy of atmospheric atoms and molecules so they move about more rapidly. The less massive the molecules, the faster they

move at any temperature. The very least massive, hydrogen and helium, can have their speeds raised so high that they actually escape from the uppermost levels of a planet's atmosphere and drift off into space. Heated by the Sun and lacking a strong gravitational grasp, the terrestrial planets soon lost the hydrogen and helium they had temporarily acquired from the protoplanetary disk. Born naked, they were naked once more.

Some Atmospheres Developed Later

If Earth's primary atmosphere was lost early in its history, what is the source of the air we breathe today? There are probably two principal sources. During the **accretion** process, minerals containing water, carbon dioxide, and other

---

Secondary atmospheres are a result of volcanism and comet impacts.

---

volatile matter collected in the interiors of the terrestrial planets. Later, as the interiors heated up, the higher temperatures released these gases from the minerals that had held them. Volcanism then brought the various gases to the surface, where they accumulated and created what we call a *secondary atmosphere*. Many planetary scientists now believe that there was another important source of gas that formed the secondary atmospheres of the terrestrial planets: impacts by huge numbers of comets, which had formed in the outer parts of the Solar System and were therefore rich in volatiles (see Chapter 6). Why did these icy bodies come into the inner Solar System? Their orbits were disrupted by the growth of the giant planets.

As the giant planets of the outer Solar System grew to maturity, their gravitational perturbations must have stirred up the entire population of icy planetesimals (*comet nuclei*) that orbited within their domain. Many of these icy bodies were flung outward by the giant planets to form the parts

---

[1] Not all atmospheric loss comes from slow leakage. Impacts by large planetesimals may have blasted away substantial amounts of the terrestrial planets' primary atmospheres.

of our Solar System we call the Kuiper Belt and Oort Cloud. We will discuss these remote regions later in Chapter 12. Others were scattered into the inner parts of the Solar System, where they could rain down on the surfaces of the terrestrial planets. These comet nuclei brought with them ices such as water, carbon monoxide, methane, and ammonia. Cometary water mixed together with the local water that was released into the atmosphere by volcanism. On Earth, at least, most of the water vapor then condensed as rain and flowed into the lower areas to form the earliest oceans. Some of the other cometary gases were not able to survive in their original form. Ultraviolet (UV) light from the Sun easily fragments cometary molecules such as ammonia and methane. Ammonia, for example, is broken down into hydrogen and nitrogen. When this happened, the lighter hydrogen atoms quickly escaped to space, leaving behind the much heavier nitrogen. Nitrogen atoms then combined to form even more massive nitrogen molecules, making it even less likely for these molecules to escape into space. This decomposition of ammonia by sunlight is likely the primary source of molecular nitrogen in the atmospheres of the terrestrial planets and on one of Saturn's moons, Titan. This molecular nitrogen makes up the bulk of Earth's atmosphere.

Today among the terrestrial planets only Earth, Venus, and Mars have significant secondary atmospheres. What happened in the case of Mercury and the Moon? Even if these two bodies had experienced less volcanism than the

Venus, Earth, and Mars have significant secondary atmospheres.

other terrestrial planets, they could hardly have escaped the early bombardment of comet nuclei from the outer Solar System. Some carbon dioxide and water must have accumulated during volcanic eruptions and comet impacts. Where are these gases now?

It appears that because of Mercury's relatively small mass and its proximity to the Sun, it lost its secondary atmosphere to space, just as it had earlier lost its primary atmosphere. More massive molecules such as carbon dioxide can escape from a small planet if the temperature is high

Mercury and the Moon are basically airless.

enough. Furthermore, intense ultraviolet radiation from the Sun can break molecules into less massive fragments, which are lost to space even more quickly. The Moon is much farther from the Sun than Mercury and is therefore cooler, but its mass is so small that even at relatively low temperatures molecules can easily escape. Because of their small mass and their proximity to the Sun, both Mercury and the Moon were doomed from the beginning to remain almost totally airless.

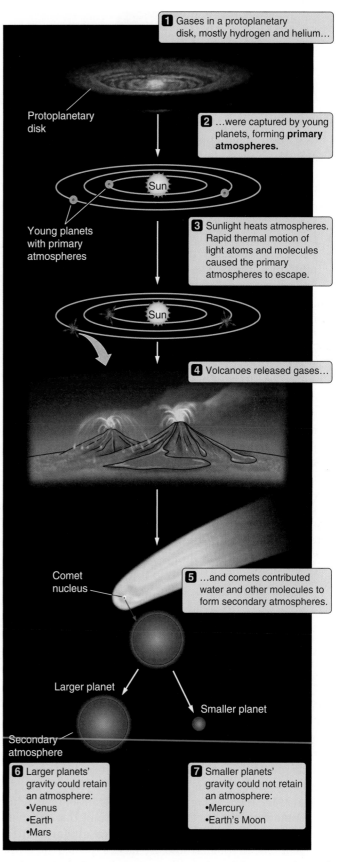

1 Gases in a protoplanetary disk, mostly hydrogen and helium…

Protoplanetary disk

2 …were captured by young planets, forming **primary atmospheres.**

Sun

Young planets with primary atmospheres

3 Sunlight heats atmospheres. Rapid thermal motion of light atoms and molecules caused the primary atmospheres to escape.

Sun

4 Volcanoes released gases…

Comet nucleus

5 …and comets contributed water and other molecules to form secondary atmospheres.

Larger planet

Smaller planet

Secondary atmosphere

6 Larger planets' gravity could retain an atmosphere:
•Venus
•Earth
•Mars

7 Smaller planets' gravity could not retain an atmosphere:
•Mercury
•Earth's Moon

FIGURE 8.2 The phases in the formation of the atmosphere of a terrestrial planet.

Escape of Planetary Atmospheres

In Chapter 4 we discussed the meaning of temperature and the equilibrium that determines the temperature of a planet. Sunlight is the primary source of thermal or kinetic energy in the atmospheres of the terrestrial planets. Atmospheric temperatures of the terrestrial planets, therefore, depend largely on the distance of the planet from the Sun. What we call the *temperature* of a gas is really just a measure of the average kinetic energy of its molecules. The kinetic energy of any object is determined by its mass and its speed. Hotter molecules have higher kinetic energies than cooler molecules and therefore move faster.

When a volume of air contains different kinds of molecules having different masses, there is a tendency for the average kinetic energy to be distributed equally among the different types. In other words, all of the types of molecules, from light to massive, will have the same average kinetic energy. But if all types have the same average energy, it means that the less massive molecules must be moving faster than the more massive ones. For example, in a mixture of hydrogen and oxygen at room temperature, hydrogen molecules will be rushing around at about 2,000 m/s on average, while the much more massive oxygen molecules will be poking along at a sluggish 500 m/s. Remember, though, that these are the *average* speeds. Some small fraction of the molecules will always be moving much slower than the average, while

a few will be moving much faster than average. A very few may be moving so fast that they exceed the planet's escape velocity.

Deep within a planet's atmosphere, these high-speed molecules will almost certainly collide with other molecules before they have a chance to escape. During the collision process, there is usually an exchange of energy. The high-speed molecule usually emerges with a lower speed, whereas the slower one tends to move faster. After a collision, then, both are likely to move with speeds that are closer to the average. Near the top of the atmosphere, where there are fewer surrounding molecules, a high-speed molecule has a good chance of escaping, if it is heading more or less upward, without first colliding with another molecule. Less massive molecules and atoms, such as hydrogen and helium, move faster and will be more quickly lost to space than more massive molecules such as nitrogen or carbon dioxide. Thus the primary atmospheres, composed of hydrogen and helium, were lost early in the evolutionary development of the terrestrial planets. The giant planets, on the other hand, are far more massive than their terrestrial cousins and are situated in the cooler environment of the outer Solar System. Stronger gravity and lower temperatures have allowed them to retain nearly all of their massive primary atmospheres.

8.2 A Tale of Three Planets

Two of the terrestrial planets—Venus and Earth—are similar in both mass and composition, and they have adjacent orbits that are less than 0.3 AU apart. Because of these similarities, we might think of them as rather close twins. The third—Mars—is also similar in composition but has a mass only about a tenth that of Earth or Venus. We might look at Mars, then, as being related to Venus and Earth but as a somewhat distant cousin. All three of these planets have secondary atmospheres, yet they are all quite different from one another. Would we have expected such differences? To understand this, we need to see how each of them got started.

All three planets are either volcanically active today or have been in their geological past, and all must have shared the intense cometary showers of the distant past. This suggests that their early secondary atmospheres might have been quite similar. In **Table 8.1** we see that the atmospheres of Venus and Mars today are nearly identical in composition: mostly carbon dioxide with much smaller amounts of nitrogen. This is what we would expect. But Table 8.1 also shows us that their surface pressures are very different. Mars and Venus have vastly different *amounts* of atmosphere. The atmospheric pressure on the surface of Venus is nearly a hundred times greater than Earth's, whereas the average surface pressure on Mars is less than a hundredth of our own. Earth differs in another important respect in that, alone among the planets, its atmosphere is made up primarily of nitrogen

TABLE 8.1

Atmospheres of the Terrestrial Planets

Physical Properties and Composition

	Planet		
	Venus	**Earth**	**Mars**
Surface pressure (bars)	92	1.0	0.006
Surface temperature (K)	737	288	210
Carbon dioxide (%)	96.5	0.039	95.3
Nitrogen (%)	3.5	78.1	2.7
Oxygen (%)	0.00	20.9	0.13
Water (%)	0.002	0.1 to 3	0.02
Argon (%)	0.007	0.93	1.6
Sulfur dioxide (%)	0.015	0.02	0.00

and oxygen, containing much less than 1 percent carbon dioxide. Although all of these planets must have started out with atmospheres of similar composition and comparable quantity, they ended up being very different from one another. Why did they evolve so differently? As with human beings, environment played the major role.

We can most easily see the similarities and differences between the atmospheres of Venus and Mars. Both experienced widespread volcanism at some time during their history. There is evidence that Venus might still be volcanically active, and Mars has certainly been volcanically active in the recent past. Carbon dioxide and water vapor must have poured out as volcanic gases into the emerging secondary atmospheres of both Venus and Mars. Decomposed cometary

Venus retained its atmosphere more effectively than Mars.

ammonia would be the likely source of nitrogen on both planets. As mentioned earlier, Venus and Mars have similar atmospheric compositions—except for water abundance, which we will discuss later. The major difference is that, even after allowing for differences in size and surface gravity, Venus today has nearly 2,500 times more atmospheric mass than Mars. Why such a large difference? We can find the answer by considering the relative masses of the two planets. Venus has nearly eight times as much mass as Mars.

This means that Venus had more bulk material with which to produce an atmosphere in the first place and, equally important, it has the gravitational pull necessary to hang onto its atmosphere. Mars had less material with which to produce an atmosphere and has less gravitational attraction to keep it. Furthermore, when a planet such as Mars loses much of its atmosphere to space, the process takes on a runaway behavior. With less atmosphere, there are fewer intervening molecules to keep breakaway molecules from escaping (see Excursions 8.1), and the rate of escape increases. This in turn leads to even less atmosphere and increased escape rates.

In this scenario we can understand the present-day differences between Venus and Mars. But why is the composition of Earth's atmosphere so different from that of the other two? We may find the answer in Earth's special location in the Solar System. Consider the early Earth and Venus, each having about the same mass, but with Venus orbiting just a little closer to the Sun. Volcanism must have poured out large amounts of carbon dioxide and water

The atmospheres of Earth and Venus have evolved very differently.

vapor to form early secondary atmospheres on both planets. Most of Earth's water quickly rained out of the atmosphere to fill vast ocean basins. However, Venus was closer to the Sun, and its surface temperatures were higher than Earth's. Most of the rainwater on Venus immediately reevaporated, much as it does today in Earth's desert regions. Venus was left with a planetwide surface containing very little liquid water but with an atmosphere filled with water vapor. The continuing buildup of both water vapor and carbon dioxide in the Venus atmosphere then led to a runaway **atmospheric greenhouse effect** that drove up the surface temperature of the planet, as discussed in **Foundations 8.1**. The surface of the planet became so hot that no water could survive there.

This early difference between a watery Earth and an arid Venus forever changed the ways their atmospheres and surfaces would evolve. On Earth water erosion caused by rain and rivers continuously exposed fresh minerals, which then reacted chemically with atmospheric carbon dioxide to form solid carbonates. This removed some of the atmospheric carbon dioxide, burying it within Earth's crust as a component of a rock called *limestone*. Later, as life developed in Earth's oceans, it accelerated the process of removing atmospheric carbon dioxide. Tiny sea creatures built their protective shells of carbonates, and as they died they built up massive beds of limestone on the ocean floors. As a result of water erosion and the chemistry of life, all but a trace of Earth's total inventory of carbon dioxide is now tied up in limestone beds. It seems that Earth's particular location in the Solar System spared it from the runaway atmospheric greenhouse effect. What if Earth had formed

FOUNDATIONS 8.1

The Atmospheric Greenhouse Effect

In Chapter 4 we calculated the expected temperatures of the planets, balancing absorbed solar radiation against emitted thermal radiation, and found that Earth is somewhat warmer than expected while Venus is very much hotter than our simple model predicts. When the predictions of a model fail, it means that we are leaving something out of the model. In this case the "something" is the atmospheric greenhouse effect.

What we call the atmospheric greenhouse effect in planetary atmospheres is in some ways a misnomer. The **greenhouse effect** traps the Sun's energy in a building somewhat differently from the way planetary atmospheres trap solar energy. A good example of what is commonly called the greenhouse effect happens in a car on a sunny day when you leave the windows closed. The consequences can be severe, especially in midsummer desert environments. Sunlight pours through the windows, heating the interior. With the hot air unable to escape, temperatures of the car's interior can approach 180°F. At this temperature the interior strongly radiates in the infrared (IR). But this radiation cannot escape from the car by the same path through which the sunlight entered. Glass is nearly opaque to IR radiation. Most of the energy ends up heating the trapped air within the car. Heating by solar radiation is most efficient if the enclosure is transparent, which is why the walls and roofs of real greenhouses are made mostly of glass.

In the case of planetary atmospheres, it is not hot air that is trapped, but rather the electromagnetic energy received from the Sun. The atmospheric greenhouse effect is illustrated in **Figure 8.3**. Atmospheric gases such as nitrogen, oxygen, carbon dioxide, and water vapor freely transmit visible solar energy, allowing the Sun to warm the planet's surface. The warmed surface now tries to radiate the excess energy back into space accord-ing to the temperature of that surface, which is much lower than that of the Sun. At the typical temperatures of planet surfaces, this energy is reradiated as infrared energy. But carbon dioxide and water vapor, among other kinds of molecules in our atmosphere, strongly absorb IR radiation, converting it to thermal energy. Some of this thermal energy is subsequently reradiated into space by the same molecules, but much goes back to the ground. The surface is now receiving thermal energy from both the Sun and from the atmosphere. Molecules such as water vapor and carbon dioxide that transmit visible radiation but absorb IR radiation are called **greenhouse molecules**. Methane and chlorofluorocarbons (CFCs) are other greenhouse molecules that are found in our atmosphere. As a result of greenhouse molecules in its atmosphere, the surface temperature of a planet will rise. This rise in temperature continues until the surface is hot enough, and therefore radiating enough energy, that even the fraction of infrared radiation leaking out through the atmosphere is enough to balance the absorbed sunlight. Convection will also transport thermal energy to the top of the atmosphere, where radiation to space will help balance absorbed sunlight. In short, the temperature rises until equilibrium between absorbed sunlight and thermal energy radiated away by the planet is reached, just as our earlier discussion in Chapter 4 said it must be.

Even though the mechanisms are different, both the greenhouse effect and the atmospheric greenhouse effect produce the same net result—the local environment is heated by trapped solar radiation.

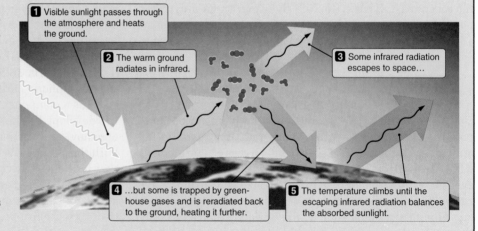

1 Visible sunlight passes through the atmosphere and heats the ground.

2 The warm ground radiates in infrared.

3 Some infrared radiation escapes to space…

4 …but some is trapped by greenhouse gases and is reradiated back to the ground, heating it further.

5 The temperature climbs until the escaping infrared radiation balances the absorbed sunlight.

FIGURE 8.3 Greenhouse gases such as water and carbon dioxide trap infrared radiation, increasing a planet's temperature.

a bit closer to the Sun? Look now at **Table 8.2**. If all the carbon dioxide locked up in limestone beds had remained in Earth's atmosphere, it would have a composition not much different from that of Venus and Mars.

Differences in the amount of water on Venus, Earth, and Mars are not so well understood. Geological evidence tells us that liquid water was once plentiful on the surface of Mars (see Chapter 7), and the *Mars Odyssey* spacecraft has found evidence that significant amounts of water still exist in the form of subsurface ice—far more than the atmospheric abundance indicated in Table 8.1. Earth's liquid and solid water supply is even greater: about 10^{21} kg or 0.02 percent of its total mass. More than 97 percent of Earth's water is in the oceans, which have an average depth of about 4 km. If all of that water were to go into the atmosphere, it would produce a crushing surface pressure 300 times greater than what we now experience. Earth today has 100,000 times more water than Venus. What happened to all the water on Venus? One possibility is that water molecules high in the Venus atmosphere were broken apart into hydrogen and oxygen by solar ultraviolet radiation. Hydrogen atoms, being very low-mass, were quickly lost to space. Oxygen, however, would eventually have migrated downward to the planet's surface, where it would remove itself from the atmosphere by oxidizing surface minerals.

Differences in the present-day masses of the atmospheres of Venus, Earth, and Mars have a large effect on their surface temperatures. Solar radiation can be trapped by the atmospheric greenhouse effect. What matters here is the actual numbers of greenhouse molecules in the atmosphere, not the percentage they represent. For example, even though the atmosphere of Mars is composed almost entirely of carbon

---

Earth would freeze without the greenhouse effect.

---

dioxide, a good greenhouse molecule, its tenuous atmosphere contains few carbon dioxide molecules compared with Venus or Earth. As a result, the atmospheric greenhouse effect on Mars raises the mean surface temperature by only about 5 K. Earth's more massive atmosphere is more efficient. Temperatures on Earth are 35 K warmer than they would be in the absence of an atmospheric greenhouse effect produced mainly by water vapor and carbon dioxide. Without this greenhouse warming, the mean global temperature of Earth would be well below the freezing point, leaving us with a world of frozen oceans and ice-covered continents!

Yet nowhere in the Solar System is the atmospheric greenhouse effect more dramatic than on Venus. Its massive atmosphere of carbon dioxide and sulfur compounds raises its surface temperature by more than 400 K, to about 737 K. At such high temperatures, any remaining water and most carbon dioxide locked up in surface rocks would long ago have been driven into the atmosphere, further enhancing the atmospheric greenhouse effect.

TABLE 8.2

Terrestrial Planet Atmospheres

If All Available Carbon Dioxide Were Included

	Planet		
	Venus	**Earth**	**Mars**
Carbon dioxide (%)	96.5	98.	96.
Nitrogen (%)	3.5	1.6	2.7
Oxygen (%)	0.0	0.4	0.1
All other constituents (%)	0.0	0.0	1.2

The conditions existing on Venus today could be created on an Earth-like planet by a runaway atmospheric greenhouse effect. Imagine a hypothetical situation in which large quantities of carbon dioxide or some other greenhouse gas suddenly appeared in Earth's atmosphere. The increased warming would raise surface temperatures, driving more water vapor into the atmosphere, which would in turn increase the strength of the atmospheric greenhouse effect. This would cause even greater warming and still larger amounts of atmospheric water vapor, ultimately creating a surface pressure about 300 times as great as it is now and a surface temperature that might exceed 800 K. Long before reaching this stage, our planet would have become devoid of all life.

In reality, the process is more complicated. Increased cloudiness caused by increased water in the atmosphere might decrease the amount of sunlight reaching Earth's surface to a point where the runaway effect would be turned off. The ocean currents are important in transporting energy from one part of Earth to another, and how they would

---

Humans are now conducting an uncontrolled experiment on Earth's atmosphere.

---

be affected by increased warming is something we really do not know. In fact, the process is so complicated by small changes leading to very large results that we still cannot predict the long-term outcome of the small changes that humans are now making in the composition of Earth's atmosphere. In a real sense, we are experimenting with the atmosphere of Earth. We are asking the question (whether we know it or not), "What happens to Earth's climate if we steadily increase the number of greenhouse molecules in its atmosphere?" We do not yet know the answer to this question, but we will eventually know the answer by seeing the results. Whether we will be happy with the results is another matter. We will explore this issue again later in the chapter.

8.3 Earth's Atmosphere— The One We Know Best

In simplest terms, we can describe Earth's atmosphere as a blanket of gas several hundred kilometers deep, with a total mass of approximately 5×10^{18} kg. As enormous as this may seem, it represents less than one-millionth of Earth's total mass. In Foundations 7.2 we learned about **hydrostatic equilibrium**, which tells us that the pressure at any point within a planet must be great enough to balance the weight of the overlying layers. The same thing must hold true in a planetary atmosphere. The atmospheric pressure on a planet is what it needs to be to hold up the weight of the overlying atmosphere. A difference between the two cases is what provides the pressure. In the interior of a solid planet it is

---

Earth's atmosphere has a surface pressure equal to a 10-meter layer of water.

---

the resistance of solid material to being compressed. In our atmosphere it is the motions of molecules in a gas. The relationship among density, temperature, and pressure of a gas is discussed in **Foundations 8.2**. The weight of Earth's atmosphere creates a force of approximately 100,000 N acting on each square meter of surface. This amount of pressure is expressed as a unit called a **bar** (from the Greek *baros*, meaning "weight" or "heavy"). Earth's average atmospheric pressure at sea level is approximately 1 bar. (Meteorologists frequently quote atmospheric pressures in *millibars* (mb), which are thousandths of a bar.) We can perhaps get a better feeling for how strongly our atmosphere presses on us if we realize that its pressure is equivalent to that of a layer of water 10 meters deep. (Imagine the extra pressure you experience at the bottom of a 33-foot-deep swimming pool.) Yet in air we seem to be completely unaware of atmospheric pressure. How can this be? It is because the very same pressure exists both within and outside of our bodies. The two precisely balance one another—another example of equilibrium.

Two principal gases make up Earth's atmosphere. About four-fifths of our atmosphere is nitrogen and one-fifth is oxygen, although there are many important minor constituents, such as water vapor and carbon dioxide. Atmospheric temperatures near Earth's surface can range from as high as

almost 60°C in the deserts to as low as –90°C in the polar regions, with a mean global temperature of about 15°C.

The Composition of Earth's Atmosphere Is Controlled by Life

As Table 8.1 shows, the composition of Earth's atmosphere is very different from that of Venus and Mars. We have already discussed the differences in carbon dioxide content, but what truly sets Earth apart from all other known plan-

---

Only Earth's atmosphere contains abundant oxygen.

---

ets is its oxygen. Our atmosphere contains abundant amounts of oxygen and the other planets do not. Why should this be so? Oxygen, it turns out, is a highly reactive gas. It chemically combines with, or *oxidizes,* almost any material it touches. Witness the rust that forms on metals, for example. For a planet to retain significant amounts of this reactive gas in its atmosphere, there would need to be some process to replace what is lost through oxidation. Such a process exists on Earth. We call it *plant life* (see **Connections 8.1**).

Figure 8.5 shows the change in oxygen concentration in Earth's atmosphere over the history of the planet. When Earth's secondary atmosphere first appeared about 4 billion years ago, it was almost totally free of oxygen. About 2.8 billion years ago, an ancestral form of cyanobacteria, single-celled organisms that contain chlorophyll, began releasing oxygen into Earth's atmosphere as a waste product of their metabolism. At first this biologically generated oxygen combined readily with exposed metals and minerals in surface rocks and soils and so was removed from the atmosphere

---

Life is responsible for the oxygen in Earth's atmosphere.

---

as quickly as it formed. In this way emerging life dramatically changed the very composition and appearance of Earth's surface, the first of many such widespread modifications imposed on our planet by living organisms. Ultimately the explosive growth of plant life accelerated the production of oxygen, building up atmospheric concentrations that approached today's levels only about 500 million years ago. All true plants, from tiny green algae to giant redwoods, use the energy of sunlight to build carbon compounds out of carbon dioxide and produce oxygen as a metabolic waste product in a process called *photosynthesis*. Earth's atmospheric oxygen content is held in a delicate balance primarily by plants. If plant life on our planet were to disappear, so too would nearly all of Earth's atmospheric oxygen.

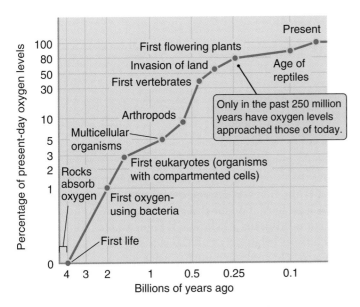

FIGURE 8.5 The amount of oxygen in Earth's atmosphere built up over time as a result of plant life on the planet.

light from the Sun breaks molecular oxygen (O_2) into its individual atoms (O), which can then recombine with other oxygen molecules to form ozone ($O_2 + O \rightarrow O_3$). Most of Earth's natural ozone is concentrated in the upper atmosphere at altitudes between 20 and 50 km. There it acts as

High-altitude ozone protects life. Our continuing survival depends on it.

a very strong absorber of ultraviolet sunlight. Without the ozone layer, this lethal radiation would reach all the way to Earth's surface, making it completely uninhabitable for nearly all forms of life (see **Excursions 8.2**). Ozone protects life. However, ozone is not always beneficial. Ozone in the lower atmosphere is found primarily in urban environments. It is mostly a human-made pollutant and is a health hazard.

Several minor gases in Earth's atmosphere also affect our daily lives. Over the current range of terrestrial temperatures, water is a volatile substance, and so its atmospheric abundance varies from time to time and from place to place. In warm, moist climates, water vapor may account for as much as 3 percent of the total atmospheric composition. In cold, arid climates, it may be less than 0.1 percent. The continuous process of condensation and evaporation of water involves the exchange of thermal and other forms of energy, making water vapor a major contributor to Earth's weather.

Carbon dioxide is another variable component of Earth's atmosphere, and a complex pattern of *sources* (places where it originates) and *sinks* (places where it goes) determines how much of it will be present at any one time. Plants consume carbon dioxide in great quantities as part of their metabolic process. Coral reefs are colonies of tiny ocean organisms that build their protective shells with carbonates produced from carbon dioxide. Fires, decaying vegetation, and human burning of fossil fuels all release carbon dioxide back into the atmosphere. This balance between sources and sinks can and does change with time. Meteorological records show that the amount of carbon dioxide in our atmosphere has been increasing for almost two centuries—since the Industrial Revolution. As noted earlier, this in turn has had a direct impact on the environment because carbon dioxide is also a powerful greenhouse gas. It should come as no surprise to learn that Earth's mean global temperature seems to be increasing with the growing abundance of this gas.

Another minor constituent in our atmosphere is *ozone* (O_3). This important molecule is formed when ultraviolet

Earth's Atmosphere Is Layered like an Onion

Earth's lowermost atmospheric layer, the one in which we live and breathe, is called the **troposphere** (**Figure 8.6**). It contains 90 percent of Earth's atmospheric mass and is the source of all of our weather. At Earth's surface, usually referred to as *sea level,* the troposphere has an average temperature of 15°C (288 K) and an average pressure of 1.013 bars. Within the troposphere, atmospheric pressure, density, and temperature all decrease with increasing altitude.

Pressure and temperature decrease with altitude in the troposphere.

For example, a few thousand feet below the summit of Mt. Denali in Alaska at an altitude of 5.5 km, the atmospheric pressure and density are only 50 percent of their sea level values and the average temperature has dropped to –20°C. Still higher, at an altitude of 12 km, where commercial jets cruise, the temperature is a frigid –60°C and the density and pressure are less than one-fifth what they are at sea level. Mountain climbers and astronomers are very much aware of this behavior of Earth's troposphere. At the Mauna Kea Observatory in Hawaii, even the most dedicated astronomers, surrounded by thin air and subfreezing temperatures, have been known to gaze longingly at the sunny beaches some 4 km below, where it is a pleasant 30°C warmer.

But why does the atmosphere get colder as we climb to higher elevations? From what we have already learned we might guess correctly that it is warmer near Earth's surface because the air is in contact with the sunlight-heated ground, which also warms the air by its infrared radiation. It also makes sense that the atmosphere would be cooler at very

What Is a Gas?

As we continue our journey of discovery through the universe, we will find that it is the gaseous state of matter that is most commonplace. More than those of any other form of matter, it is the properties of gases that we will turn to as we seek to understand the workings of planets, stars, and galaxies.

Matter is composed of atoms and molecules, and different forms of matter result from differences in how those atoms and molecules interact with each other. In a *solid,* molecules are packed tightly, held in place like bricks in a wall by the presence of their neighbors. In a *liquid,* molecules are able to move about but are constantly being jostled. Molecules in a liquid are like people in a crowded subway station. The crowd is free to flow, but the individuals that make it up are still limited in their movement by the people around them. The molecules in a *gas,* on the other hand, go their own way, traveling relatively long distances without interacting with other molecules. When you think of a gas, picture a swarm of tiny atoms and molecules flying about, each with its own speed and direction.

The temperature of a gas is a measure of the average kinetic energy of the individual molecules as they fly about. Two things determine a molecule's kinetic energy: the mass of the molecule and the speed at which it is moving. A massive, slow-moving molecule might have the same kinetic energy as a less massive, faster-moving molecule. Because of this, when a gas is composed of different types of molecules (such as the air around you), the average speed of less massive molecules will be higher than the average speed of more massive molecules. (They have to be if their average kinetic energies are to be the same.) The average speed of a molecule in a gas is inversely proportional to the square root of its mass. Oxygen molecules, for example, are 16 times more massive than hydrogen molecules. In a gas containing both hydrogen and oxygen, the hydrogen molecules will therefore be moving four times ($= \sqrt{16}$) as fast as the oxygen molecules. As we have seen, this difference explains why Earth can hold onto the oxygen in its atmosphere but loses the hydrogen to space.

As we study gases in different astronomical settings, we will find that one of the most important properties of a gas is how hard it pushes on its surroundings. Imagine a box containing gas, as shown in **Figure 8.4(a)**. Molecules are constantly bouncing off the walls of the box, pushing outward. This outward push, measured in units of force per square meter of the surface of the box, is referred to as **pressure**. (Any time that you think about the pressure of a gas, this is the mental picture that you should bring to mind: atoms and molecules slamming against the walls of a box, pushing outward.)

Armed with nothing more than this mental picture, we can draw some interesting conclusions about the pressure of a gas. First, if we increase the number of molecules in the box, more molecules will hit the walls of the box each second. The pressure will increase. Double the density of the gas but keep the temperature the same (**Figure 8.4(b)**) and you double its pressure.

Increasing the temperature of the gas increases the average speed at which molecules move, which also increases the pressure of the gas. There are two reasons for this. First, if the molecules in the box are moving faster, they will hit the walls *more frequently* (**Figure 8.4(c)**). More molecules hitting the walls of the box each second means that the pressure is higher. Second, faster-moving molecules hit the wall *harder,* exerting more force. Together these two effects mean that doubling the temperature of a gas doubles the pressure of the gas.

Pressure is proportional to density, and pressure is also proportional to temperature. We can combine these two relationships to get

$$\text{Pressure} \propto \text{Density} \times \text{Temperature}.$$

This relationship between density, temperature, and pressure is called the **ideal gas law**. The ideal gas law has a special place in the history of science. The empiri-

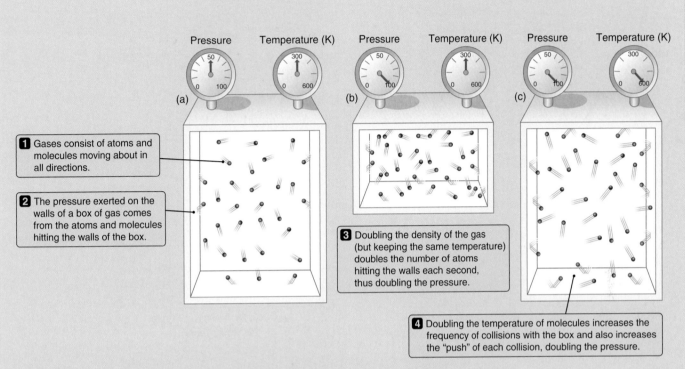

1 Gases consist of atoms and molecules moving about in all directions.

2 The pressure exerted on the walls of a box of gas comes from the atoms and molecules hitting the walls of the box.

3 Doubling the density of the gas (but keeping the same temperature) doubles the number of atoms hitting the walls each second, thus doubling the pressure.

4 Doubling the temperature of molecules increases the frequency of collisions with the box and also increases the "push" of each collision, doubling the pressure.

FIGURE 8.4 (a) The pressure of a gas comes from the motions of atoms and molecules. Making a gas (b) denser or (c) hotter (that is, increasing the speed of atoms or molecules) increases the pressure of the gas.

cally discovered ideal gas law provided strong support for the atomic theory of matter, in much the same way that Kepler's empirical laws of planetary orbits provided support for Newton's theories of motion and gravitation. The fact that laboratory gases come very close to obeying the ideal gas law provided compelling early evidence that atoms and molecules are real.

We need to know about one other property of gases as our journey continues. If you compress a gas, you not only increase its density, but you also increase its temperature. Conversely, when a gas is allowed to expand, it cools off. There are many everyday examples of this behavior. Pump up a bicycle tire, and the pump becomes hot because the air is being compressed. Hold down the nozzle on an aerosol can, and it feels cold because the propellant gas is expanding. An air conditioner works by alternately compressing a gas to make it hot, letting that gas cool, then allowing the gas to expand and get really cold.

Over and over again on our journey, our understanding of gases will be our guide, equally as valid when applied to the hot cores of stars as to a gentle breeze on a summer day. Every time that we talk about an object made of gas, think back to the picture of molecules bouncing around in a box, and remember the commonsense ideas that explain the behavior of the gas.

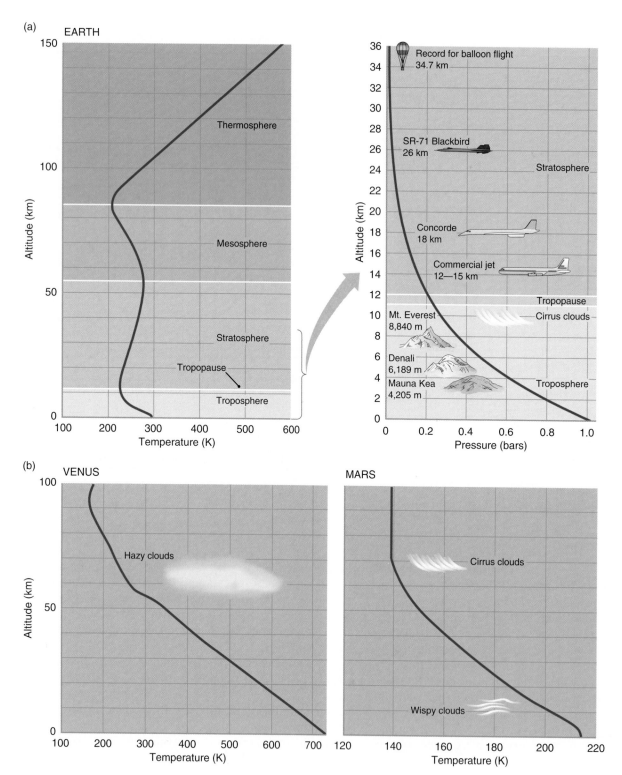

FIGURE 8.6 Earth's atmospheric layers, temperature, and pressure are plotted here as a function of height. Atmospheric temperatures for Venus and Mars are shown for comparison.

CONNECTIONS 8.1

Origins—On Atmospheres and Life

Take a deep breath. The oxygen you inhale sustains you. Breathe out. The carbon dioxide you exhale is food for plant life. If these two gases were to suddenly disappear, most life on Earth would perish.[2] The presence of molecular oxygen and carbon dioxide in our atmosphere is fortunate, but where did these gases come from? As we learned earlier in the chapter, they have notably different origins. Volcanism is a major source of carbon dioxide, and so it occurs quite naturally in our atmosphere. Molecular oxygen is an entirely different matter, though. We would not expect to find it in the atmosphere of a planet like Earth. Oxygen is so chemically reactive it quickly destroys itself by combining with surface minerals to form oxides. As an example, it reacts with iron to produce iron oxide, more commonly known as rust. The reddish surface of Mars is coated with rusted iron-bearing minerals, and this is one reason the Martian atmosphere is almost completely free of oxygen. What then about Earth? As we learned earlier, we owe our terrestrial oxygen to tiny organisms that lived and flourished in ancient seas. Their metabolism produced oxygen as a by-product, much as present-day animal life produces carbon dioxide as a respiratory by-product. Today plants maintain oxygen levels at about 20 percent of our atmospheric total. It is well to remember that without plant life, Earth's oxygen would disappear and all animal life would vanish with it.

[2]Certain microorganisms, called *anaerobes*, live in the total absence of oxygen. In fact most are poisoned by even small amounts of oxygen.

You may have already jumped ahead to an obvious conclusion. One of the surest ways of detecting life on another planet is by finding atmospheric gases that should not otherwise be there. We have already identified molecular oxygen as one of those telltale gases. Ozone is another because it is produced by **photodissociation** and **recombination** of oxygen. Living organisms are the only known source of significant amounts of molecular oxygen or ozone in a planetary atmosphere. Methane is yet another gas that would be unexpected in the atmosphere of an Earth-like planet because it is readily destroyed by photodissociation. The presence of methane means that a source is needed to keep up with photodissociative loss. As with oxygen, the likely source would be biological. Certain microorganisms produce methane as a metabolic by-product. Those bubbles you see rising to the surface of a stagnant pond—sometimes called "swamp gas"—contain biologically produced methane.

So there we have it. Find molecular oxygen, ozone, or methane in the atmosphere of an Earth-like planet and you have a candidate for extraterrestrial life. Is this a realistic expectation? Yes, and it could happen fairly soon. As we learned in Chapter 6, NASA plans to launch the *Terrestrial Planet Finder* (*TPF*) in about a decade. Suppose *TPF* isolates the light from an Earth-like planet and feeds that light into its onboard spectrograph. Suppose further that the spectra reveal substantial amounts of molecular oxygen in the planet's atmosphere. The inescapable conclusion would be that the planet harbors life. Of course the observation would not tell us what kind of life, but it would certainly mark that planet as an object of considerable interest for further study.

high altitudes because there the atmosphere can freely radiate its thermal energy into space. In fact, it would get colder with increasing altitude even faster if it were not for

Convection carries thermal energy upward through the troposphere.

the process of convection. We first encountered convection in our Chapter 7 discussion of the motions of material in Earth's mantle that drive plate tectonics. **Figure 8.7** illustrates how convection carries thermal energy upward

through Earth's atmosphere. At a given pressure, cold air is more dense than warm air. So when cold air encounters warm air, the denser cold air slips under the less dense warm air, pushing the warm air upward. (Rather than "warm air rises" we should perhaps say "cold air sinks.") This convection sets up air circulation between the lower and upper levels of the atmosphere, and this tends to diminish the extremes caused by heating at the bottom and cooling at the top.

Convection also affects the vertical distribution of atmospheric water vapor. The ability of air to hold water in

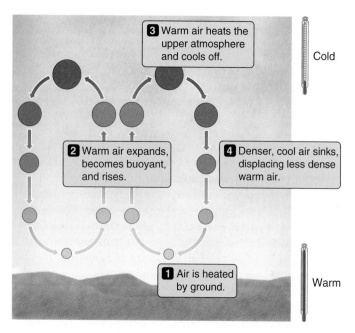

3 Warm air heats the upper atmosphere and cools off.

Cold

2 Warm air expands, becomes buoyant, and rises.

4 Denser, cool air sinks, displacing less dense warm air.

1 Air is heated by ground.

Warm

FIGURE 8.7 Atmospheric convection carries thermal energy from the Sun-heated surface upward through the atmosphere.

the form of vapor depends very strongly on the air temperature. The warmer the air, the more water vapor it can hold. We refer to the amount of water vapor in the air relative to what the air could hold at that temperature as the **relative humidity**. Air that is saturated with water vapor has a rela-

Most atmospheric water stays close to Earth's surface.

tive humidity of 100 percent. As air is convected upward, it cools, limiting its capacity to hold water vapor. If the air temperature decreases to the point where the air can no longer hold all its water vapor, **saturation** occurs. The water begins to condense out in the form of tiny droplets, which in large numbers become visible to us as clouds. When these droplets coalesce to form large drops, convective updrafts can no longer support them, and they fall as rain. For this reason most of the water vapor in Earth's atmosphere stays within 2 km of the surface. At an altitude of 4 km, for example, the Mauna Kea Observatory is higher than approximately one-third of Earth's atmosphere, but it lies above nine-tenths of the atmospheric water vapor. This is important for astronomers who observe the heavens in the infrared region of the spectrum because water vapor strongly absorbs IR light (recall Figure 5.18).

The top of the troposphere, called the **tropopause**, is defined as the height where temperature no longer decreases with increasing altitude (Figure 8.6). This change in atmospheric behavior is caused by heating from absorbed sunlight within the atmospheric layers that lie above the tropopause. And because temperature no longer decreases

with altitude above the tropopause, atmospheric convection also dies out there. The tropopause varies between 10 and 15 km above sea level, depending on latitude, and is highest at the equator.

Above the tropopause and extending upward to an altitude of 50 km above sea level is the **stratosphere**. This is a region in which little convection takes place because the temperature–altitude relationship actually reverses at the

The stratosphere, mesosphere, and thermosphere lie above the troposphere.

tropopause, and the temperature begins to *increase* with altitude. The reason for this is the presence of ozone, which warms the stratosphere by absorbing sunlight.

The region above the stratosphere is called the **mesosphere**. It extends from 50 km to an altitude of about 90 km. In the mesosphere there is no ozone to absorb sunlight, so temperatures once again decrease with altitude. The base of the stratosphere and the upper boundary of the mesosphere are two of the coldest levels in Earth's atmosphere (see Figure 8.6).

At altitudes above 90 km, solar ultraviolet radiation and high-energy particles from the solar wind **ionize** atmospheric atoms and molecules, causing the temperature to once again increase with altitude. We call this region the **thermosphere**, and it is the hottest part of the atmosphere. The temperature can reach 1,000 K near the top of the thermosphere, at an altitude of 600 km. The gases within and beyond the thermosphere are ionized by ultraviolet photons and high-energy particles from the Sun to form a **plasma**. (A plasma is any gas that is made up largely of electrically charged particles rather than only neutral atoms and molecules.) This region of ionized atmosphere is called the **ionosphere**. The ionosphere is important to us in part because it reflects certain frequencies of radio waves back to the ground. Amateur radio operators, for example, are able to communicate with each other around the world by bouncing their signals off the ionosphere.

Earth and its atmosphere are surrounded by a large region filled with electrons, protons, and other charged particles from the Sun that have been captured by Earth's magnetic field. This region, called Earth's **magnetosphere**, has a radius approximately 10 times the radius of Earth, filling a volume over a thousand times the volume of the planet itself. To appreciate Earth's magnetosphere, we need to begin by looking more carefully at magnetic fields and the force they apply to charged particles. Magnetic fields have no effect on charged particles unless the particles are moving. Charged particles are free to move *along* the direction of the magnetic field; but if they try to move *across* the direction of the field, they experience a force that is perpendicular both to their motion and to the magnetic field direction. This force causes them to loop around the direction of the magnetic field, as illustrated in **Figure 8.8(a)**.

It is almost as if charged particles were beads on a string, free to slide along the direction of the magnetic field but unable to cross it.

The picture of how charged particles move in a magnetic field gets even more interesting if the field is pinched together at some point. As particles move into the pinch, they feel a magnetic force that (if conditions are right) pushes them back along the direction from which they came. If charged particles are located in a region in which the field is pinched on both ends, as shown in **Figure 8.8(b)**, then they may bounce back and forth many times. This magnetic

---

Earth's magnetic field traps charged particles from the Sun.

---

field configuration is called a *magnetic bottle.* Earth's magnetic field is pinched together at the two poles and spreads out around the planet. This configuration is like taking many magnetic bottles and bending them over, attaching them to Earth at either end. Earth and its magnetic field are immersed in a constant stream of charged particles from the Sun called the solar wind. When these particles first encounter Earth's magnetic field, the smooth flow is interrupted and they drop suddenly from **supersonic** to **subsonic** speed at a point called the **bow shock.** As they stream by, they are diverted by Earth's magnetic field as a river is diverted around a boulder. As they flow past, some of these charged particles become trapped by Earth's magnetic field,

where they bounce back and forth between Earth's magnetic poles as illustrated in **Figure 8.8(c)**.

An understanding of Earth's magnetosphere is of great practical importance for space travel. Regions in the magnetosphere that contain especially strong concentrations of energetic charged particles, called **radiation belts**, can be very damaging to both electronic equipment and astronauts. Yet we need not leave the surface of the planet to witness beautiful and dramatic effects of the magnetosphere. Disturbances in Earth's magnetosphere can lead to changes in Earth's magnetic field that are large enough to trip power grids, causing blackouts, and to wreak havoc with communications. Earth's magnetic field also funnels energetic charged particles down into the ionosphere in two rings located around the magnetic poles. These charged particles (mostly electrons) collide with atoms such as oxygen, nitrogen, and hydrogen in the upper atmosphere, causing them to glow like the gas in a neon sign. These glowing rings, called **auroras,** can be seen from space (**Figure 8.9(a)**). When viewed from the ground (**Figure 8.9(b)**), they appear as eerie, shifting curtains of multicolored light. People living far from the equator are often treated to spectacular displays of the *aurora borealis* ("northern lights") in the Northern Hemisphere or the *aurora australis* in the Southern Hemisphere. Auroras have also been seen on Venus, Mars, the giant planets, and some moons.

Although our discussion has concentrated on the atmosphere of our own planet, it is important to know that

---

EXCURSIONS 8.2

The Ozone Hole

There is a common misconception confusing the greenhouse effect with the so-called ozone hole. Both are caused by the buildup of certain gases in our atmosphere and both are a cause for concern, but the individual causes and effects are *very* different. Recently the measured amount of ozone in our upper atmosphere has been decreasing, primarily over the polar latitudes in both the Northern and Southern Hemispheres. The decrease appears to be caused by a buildup of atmospheric *halogens*—mostly chlorine, fluorine, and bromine, such as those found in human-made refrigerants. Halogens readily diffuse upward into the stratosphere, where they destroy ozone without themselves being consumed. We call such agents **catalysts**—materials that participate in and accelerate chemical reactions but are not themselves

modified in the process. Because they are not used up, halogens may remain in Earth's upper atmosphere for decades or even centuries.

Why should we be so concerned about the loss of a minor constituent from so high in our atmosphere? Even in tiny amounts, ozone filters out harmful solar ultraviolet radiation, preventing it from reaching Earth's surface. The continuing removal of ozone from the high atmosphere means trouble for terrestrial life as more and more UV radiation reaches the ground. Measured increases in the levels of UV radiation appear to be related to increases in skin cancer in humans, and we do not yet understand the effects it may have on other life forms to which we are inexorably linked. There is good reason to be concerned.

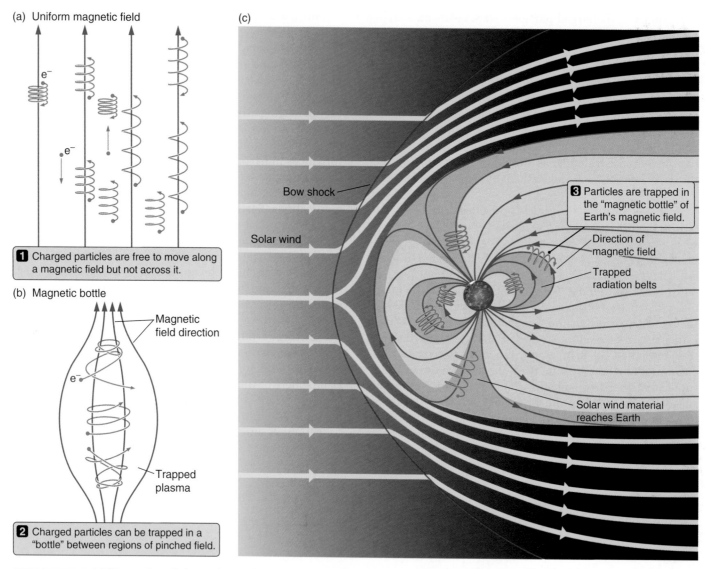

(a) Uniform magnetic field

1 Charged particles are free to move along a magnetic field but not across it.

(b) Magnetic bottle

Magnetic field direction

Trapped plasma

2 Charged particles can be trapped in a "bottle" between regions of pinched field.

(c)

Bow shock

Solar wind

3 Particles are trapped in the "magnetic bottle" of Earth's magnetic field.

Direction of magnetic field

Trapped radiation belts

Solar wind material reaches Earth

FIGURE 8.8 (a) The motion of charged particles, in this case electrons, in a uniform magnetic field. (b) When the field is pinched, charged particles can be trapped in a magnetic bottle. (c) Earth's magnetic field acts like a bundle of magnetic bottles, trapping particles in Earth's magnetosphere.

the structure we have described here is not limited to Earth's atmosphere. The major vertical structural components—troposphere, tropopause, stratosphere, and ionosphere—also exist in the atmospheres of Venus and Mars, as well as in the atmospheres of the giant planets. And as we will see in the following chapter, the magnetospheres of the giant planets are among the largest structures in the Solar System.

Why the Winds Blow

Variation in solar heating is the chief reason for differences in the ground-level temperature of Earth's atmosphere from place to place at similar altitudes and throughout the year.

It is usually warmer in the daytime than at night, warmer in the summer than in winter, and warmer at the equator than in the polar regions. Large bodies of water, such as oceans, also affect atmospheric surface temperatures. As we discovered in Foundations 8.2, heating a gas increases its pressure, which in turn causes it to push into its surroundings. It is because of the horizontal component of these pressure differences that we have winds, and the strength of the winds is closely related to the magnitude of the difference in temperature from place to place. Think about what happens as air in Earth's equatorial regions, heated by the warm surface, begins to rise due to convection. This displaces the air above it, which then has no place to go but toward the poles. Cooling and becoming denser as it moves poleward, the displaced air now descends in the polar re-

Nonuniform solar heating creates our weather.

gions. There it displaces the surface polar air, which is forced back toward the equator, completing the circulation (**Figure 8.10(a)**). This has the effect of keeping the equatorial regions cooler and the polar regions warmer than they otherwise would be. Such planetwide flow of air between equator and poles is called **Hadley circulation**. Global Hadley circulation, it turns out, seldom occurs in planetary atmospheres because other factors break up the planetwide flow into a series of smaller *Hadley cells*. Planet rotation is a major factor here. Most planets and their atmospheres are rotating rapidly, and the Coriolis effects produced by this rotation strongly interfere with Hadley circulation by redirecting the horizontal flow (**Figure 8.10(b)**).

On a rapidly rotating planet, air is not free to flow in just any direction. As we first learned in Chapter 2, when a volume of air starts to move directly toward or away from the poles, the Coriolis effect diverts it into relative motion that is more or less parallel to the planet's equator. This creates winds that blow predominantly in an east–west direction (Figure 8.10(b)). Meteorologists call these **zonal winds**. In general, the more rapid a planet's rotation, the stronger the Coriolis effect and the stronger its zonal winds will be.

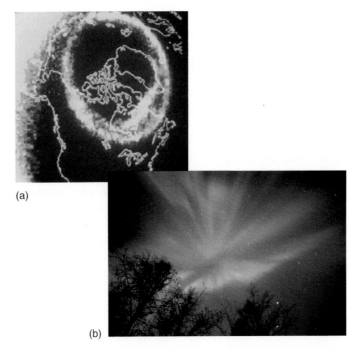

(a)

(b)

FIGURE 8.9 Auroras result when particles trapped in Earth's magnetosphere collide with molecules in the upper atmosphere. (a) An auroral ring around Earth's north magnetic pole, as seen from space. (b) Aurora borealis viewed from the ground—the "northern lights."

FIGURE 8.10 Schematic diagrams of Hadley circulation. (a) The classic Hadley circulation. (b) Hadley flow often breaks up into smaller circulation cells. The poleward–equatorward flow is diverted into zonal flow by the Coriolis effect.

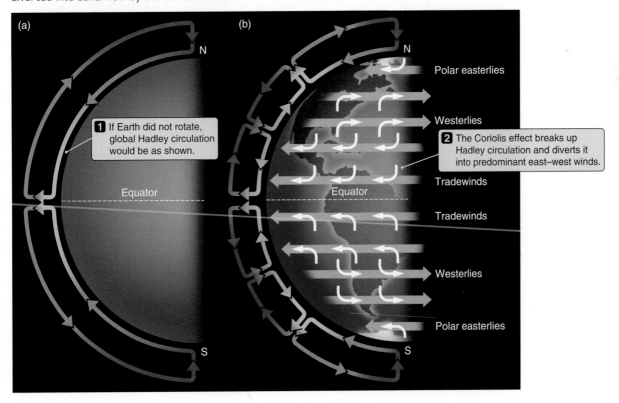

(a)

(b)

1 If Earth did not rotate, global Hadley circulation would be as shown.

2 The Coriolis effect breaks up Hadley circulation and diverts it into predominant east–west winds.

N

Equator

S

N

Polar easterlies

Westerlies

Tradewinds

Equator

Tradewinds

Westerlies

Polar easterlies

S

Zonal winds are often confined to relatively narrow bands of latitude. Between the equator and the poles in most planetary atmospheres, the zonal winds alternate between "easterly," those blowing toward the *west,* and "westerly," those

---

Nonuniform heating together with the Coriolis effect cause east–west zonal winds.

---

blowing toward the *east.* Confusing? Very! This unfortunate terminology is a historical carryover from early terrestrial meteorology, in which winds were labeled not by the direction *toward* which they are blowing but by the direction from which they come.

In Earth's atmosphere, several bands of alternating zonal winds lie between the equator and the poles of both hemispheres. We call this zonal pattern Earth's **global circulation** because its extent is planetwide. The best-known zonal currents are the subtropical trade winds—more or less easterly winds that once carried sailing ships from Europe westward to the New World—and the midlatitude prevailing westerlies that carried them home again.

Embedded within Earth's global circulation pattern are systems of winds associated with large high- and low-pressure regions. We saw in Chapter 2 that a combination of a low-pressure region and the Coriolis effect produces a circulating pattern called *cyclonic motion.* Cyclonic motion is associated with stormy weather, including hurricanes.[3] Similarly, high-pressure systems are localized regions where the air pressure is higher than average. We think of these regions of greater-than-average air concentration as "mountains" of air. Owing to the Coriolis effect, high-pressure regions rotate in a direction opposite to that of low-pressure regions. These high-pressure circulating systems experience *anticyclonic motion* and are generally associated with fair weather.

Thunderstorms Are Integral to the Water Cycle

It takes the absorption of thermal energy to turn liquid water into vapor. Water in Earth's oceans, lakes, and rivers is evaporated by thermal energy acquired from the absorption of sunlight. The water vapor then carries this thermal energy along with it as it circulates throughout the atmosphere. When the water vapor recondenses, it gives up its thermal energy to its surroundings. This is the process that powers hurricanes and thunderstorms (see **Excursions 8.3**). We can

---

[3] Cyclones are regions characterized by rising moist air and are therefore associated with stormy weather. Conversely, anticyclones are vast regions of sinking dry air and thus tend to produce fair weather.

most easily see how it works with a thunderstorm. A *thunderstorm* begins when Earth's surface, heated by the Sun, warms moist air close to the ground (Figure 8.12). The moist air is convected upward, cooling as it gains altitude. Cooling causes the water vapor in the moist air to condense as rain. As it condenses the water vapor gives up its thermal energy to the surrounding air, warming it and thus increasing the strength of the convection. With strong solar heating and an adequate supply of moist air, this self-feeding process can grow within minutes to become a violent thunderstorm. Water, falling back to the surface as rain, eventually returns to the lakes and oceans, wearing down mountains, eroding the soil, and nourishing life as it flows. From the oceans to the air and back again—this is the *water cycle.*

Is Earth Getting Warmer?

Climate is the term we use to define the average state of Earth's atmosphere, including temperature, humidity, winds, and so on. Earth's climate appears to go through lengthy temperature cycles, usually lasting hundreds of thousands of years and occasionally producing shorter cold periods called *ice ages.* These oscillations in the mean global temperature are far smaller than typical geographical or seasonal

---

Climate is the average state of Earth's atmosphere.

---

temperature variations. But Earth's atmosphere is so sensitive to even small temperature changes that it takes a drop of only a few degrees in the mean global temperature to plunge our climate into an ice age. We still do not know the reason for these climate-changing temperature swings. An external influence, such as small changes in the Sun's energy output, may be the cause. Changes in Earth's orbit or the inclination of its rotation axis also have been suggested. Or they may be triggered internally by volcanic eruptions (which can produce global sunlight-blocking clouds or hazes) or long-term interactions between Earth's oceans and its atmosphere.

Many scientists believe that changes in Earth's climate have been accelerating recently and that Earth is becoming warmer. Temperature measurements over the past century show a steady increase in the mean global temperature. But whether this represents the beginning of a long-term change —caused, perhaps, by the buildup of human-made greenhouse gases (**Figure 8.17**)—or merely a temporary short-term cycle is a question still being debated. However, we know that our atmosphere is a delicately balanced mechanism. Tiny changes can produce enormous and often unexpected results. Earth's climate is an example of the sort of complex, chaotic system we will discuss in Chapter 10. To add to the complexity, Earth's climate is intimately tied to ocean tem-

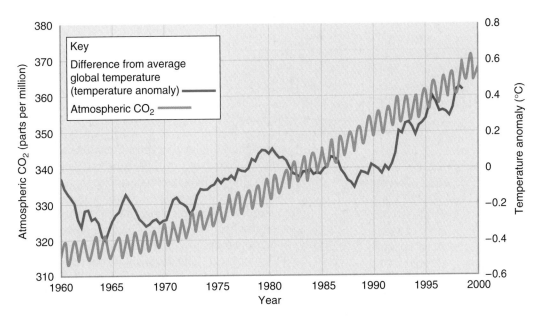

FIGURE 8.17 Annual variations in atmospheric CO_2 are due to seasonal variations in plant life and fossil fuel use, whereas the steady climb is due to human activities. Global temperatures are climbing along with CO_2 concentration. Our climate is delicately balanced, and carbon dioxide plays a vital role in that balance.

peratures and currents. We see examples of this in *El Niño* and *La Niña,* where small shifts in ocean temperature cause much larger global changes in air temperature and rainfall. Recent studies suggest that changes in the flow of the Gulf Stream in the North Atlantic can have very large and unpredictable effects on the climates of North America and Northern Europe—and that these changes may take place

Are we treating our atmosphere irresponsibly?

not over centuries but within a matter of decades! Are we unknowingly jeopardizing our future by meddling with our atmosphere? In a sense, we are much like children playing with matches.

8.4 Venus Has a Hot, Dense Atmosphere

Venus and Earth are similar in many ways—so similar that they might be thought of as sister planets. Indeed, when we used the laws of radiation in Chapter 4 to predict temperatures for the two planets, we predicted that they should be very close. But that was before we considered the greenhouse effect and the role of carbon dioxide in blocking the infrared radiation that a planetary surface radiates. The atmospheric pressure at the surface of Venus is a crushing 92 times greater than at the surface of our own planet, equal to the water pressure at an ocean depth of 900 m. The hull of a World War II era submarine would be crushed on the

Venus has a massive carbon dioxide atmosphere, making it a "poster child" for the greenhouse effect.

surface of Venus. Most (96 percent) of this massive atmosphere is carbon dioxide, with a small amount (3.5 percent) of nitrogen and still lesser amounts of other gases. This thick blanket of carbon dioxide effectively traps the infrared radiation from Venus, driving the temperature at the surface of the planet to a sizzling 737 K, which is hot enough to melt lead. While Earth is a lush paradise, the runaway greenhouse effect has turned Venus into a convincing likeness of hell, an analogy made complete by the presence of choking amounts of sulfurous gases. Venus may be our "sister" planet in many respects, but it will be a *very* long time before humans visit its surface, if ever.

At altitudes between 50 and 80 km (see Figure 8.6), the atmosphere is cool enough for sulfuric acid vapor to react with water vapor to form dense clouds of concentrated sulfuric acid droplets (H_2SO_4). Except for the low-resolution views provided by cloud-penetrating radar, these dense clouds had prevented us from seeing the surface of Venus until the Soviet Union succeeded in landing cameras there in 1975. But it is only with radar that we can globally map the surface of Venus, as so ably done by the spacecraft *Magellan* in the early 1990s.

Sunlight also has difficulty penetrating Venus's dense clouds. Noontime on the surface of Venus is no brighter than a very cloudy day on Earth. High temperatures keep the lower atmosphere of Venus free of clouds and hazes. But though the local horizon could be seen clearly, the distant mountains would not be so clear. Molecules in even a pure gas will scatter light, and the scattering efficiency

Convection Run Amok

THUNDERSTORMS

Those puffy white clouds of a summer day are created when warm moist air is convected upward to cooler atmospheric levels a few kilometers above the surface. Here the moisture condenses out of the air to form the myriad tiny water droplets that we call **cumulus** clouds. For the most part this is a gentle process. But convection can be violent. Summer is also a time for thunderstorms and lightning. The process that creates thunderstorms is the same as that which forms cumulus clouds, but the amount of energy involved is far greater.

Thunderstorms tend to form in the afternoon when solar heating of the ground reaches its maximum. They begin as familiar cumulus clouds; but if the supply of warm moist air at the surface is great enough, the clouds will continue to grow vertically, and we have a case of convection run amok. As more and more moist air rises and the moisture condenses within the cloud, the heat released by condensation continues to warm the surrounding air, forcing convection ever higher. We can easily recognize these **cumulonimbus** clouds, known popularly as *thunderheads*, by their flat, anvil-shaped tops, as seen in **Figure 8.11**. This upper surface visibly marks the tropopause, the level in the atmosphere where convection finally ceases. In the midlatitudes of the continental United States, the tropopause occurs at an altitude of about 12 km above sea level. Convection in the more violent storms can be so strong that the tops of the thunderheads punch right through the tropopause, carrying cloud-forming moisture up into the stratosphere to heights of 20 km or more. The anvil shape occurs when strong stratospheric winds pull ice crystals from the top of the cloud and spread them out horizontally.

For every parcel of air that rises within a thunderstorm another must come back down (see **Figure 8.12**). Downdrafts, including the rain and hail that descend with them, produce ferocious winds that can exceed 100 km/h in the immediate vicinity of the storm. The winds and **turbulence** associated with thunderstorms become so great that even commercial jet aircraft prefer

FIGURE 8.11 Thunderstorms are powered by convection and by thermal energy released as water vapor condenses to form droplets. The "anvil" top is caused by stratospheric winds shearing the top of the convective system.

FIGURE 8.12 Convection within a thunderstorm.

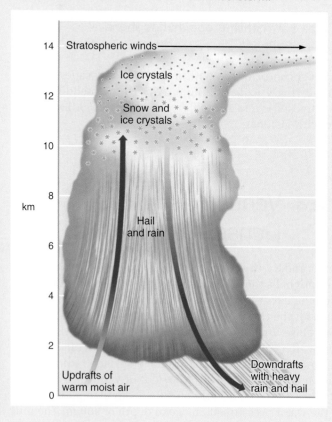

to keep their distance. At its base a thunderstorm may be several kilometers across and travel for tens of kilometers across the landscape. Thunderstorms cause billions of dollars in crop damage every year and are responsible for hundreds of deaths.

LIGHTNING

Lightning is essentially a gigantic electrical spark that results from billions of volts of potential difference—a huge example of the "static electricity" you sometimes generate when walking on a carpet—and is usually associated with thunderstorms and rain. However, lightning can also be created by snow, sand and dust storms, volcanic eruptions, earthquakes, and nuclear explosions. Ice formation in cumulonimbus clouds is a key factor for starting the "electric generator" that produces most lightning. Falling small ice pellets in cumulonimbus clouds become negatively charged by friction as they move through the surrounding air, while small supercooled cloud droplets that strike them bounce off the ice pellets and become positively charged. The supercooled cloud droplets rise on updrafts to the top of the storm cloud while the ice pellets fall and melt in the lower regions of the cloud or, as often is the case, fall all the way to the ground. This creates a difference in electrical potential between the top and bottom of the cloud that can exceed a billion volts. At some point the potential difference between parts of the cloud becomes so great that the electrical resistance of the air breaks down[4] (the molecules become ionized) and a series of lightning bolts flashes between the positively and negatively charged regions of the cloud. We call this cloud-to-cloud lightning. About a third of the time lightning travels between the cloud and the ground. It is here that lightning can become so dangerous. In the United States lightning kills an average of 100 people every year and injures more than 300. Worldwide, lightning strikes occur about 100 times every second.

Lightning bolts are typically 3 to 5 km long and can carry tens of thousands of amperes of current at speeds

[4] Some atmospheric physicists now believe that all lightning is initiated by secondary cosmic ray particles (see Chapter 19), which trigger the electrical breakdown (ionization) of the air along the path of the lightning bolt.

FIGURE 8.13 Lightning bolts carry enormous electric currents, killing or injuring hundreds of people each year in the United States alone.

of 200 km/s (see **Figure 8.13**). On average, the energy of a single lightning bolt could keep a 100-watt lightbulb burning for several months. Although the brilliance of lightning bolts makes them appear huge, they are really no larger in diameter than a quarter or half dollar. As it travels through the air, lightning heats the air to temperatures as high as the surface of the sun, causing the volume of air along the path of the bolt to expand rapidly. The thermal energy causing this rapid expansion of the air is converted into sonic energy, resulting in the familiar boom of thunder.

Lightning also creates powerful electromagnetic waves, some of which you can hear on your AM radio as "lightning static." When the *Pioneer Venus* spacecraft was orbiting our sister planet during the 1980s, its radio receiver picked up many bursts of lightning static—so many that Venus appears to have more lightning activity than our own planet. On Venus, as on Earth, lightning is created in the clouds; but Venus's clouds are so high —typically 55 km—that the lightning bolts never hit the ground. They are all of the cloud-to-cloud type. The *Pioneer Venus* results supported observations made by the Soviet lander *Venera 9*, whose optical spectrometer

(continued on next page)

EXCURSIONS 8.3

had detected flashes of lightning on Venus's dark side five years earlier.

HURRICANES, TORNADOES, AND DUST DEVILS

We first discussed **hurricanes** in Chapter 2 as an example of Coriolis forces acting on air rushing into regions of low atmospheric pressure. But hurricanes, the most powerful of storms, are much more complicated than the simple systems we described earlier. Hurricanes are huge heat engines. They derive their enormous energy from a very common physical phenomenon: the heat of vaporization of water. The conditions for formation must be just right—warm tropical seawater, light winds, and a region of low pressure in which air spirals inward. As warm seawater evaporates, moisture-laden air rises and releases its heat of vaporization as it condenses at cooler levels. (Remember, this is the same process that leads to cumulonimbus thunderstorms.) When the supply of warm seawater is sufficient, a complex of thunderstorms develops. Then, if the winds aloft are light, the complex remains intact and grows in size and strength. Convection ceases at the tropopause, located about 15 km above sea level at tropical latitudes; but the number of individual storm cells in the complex continues to increase. The stage is now set for the birth of a hurricane (see **Figure 8.14**).

FIGURE 8.14 A satellite view of a large hurricane. Hurricane Katrina approaching the Gulf Coast in 2005.

As surface winds driven by the Coriolis effect rush inward to replace the air rising upward in the cumulonimbus complex, the hurricane grows in size and strength. Sustained winds near the center of the storm can reach speeds greater than 300 km/h, causing widespread damage and fatalities if the hurricane moves ashore. In 1900 a hurricane in Galveston, Texas, took 8,000 lives—more than any other natural disaster in U.S. history. Hurricane Andrew caused $12 billion in damage when it hit Florida in 1992. In 2005 Hurricane Katrina devastated much of the Gulf Coast and the city of New Orleans, taking more than a thousand lives and costing more than $200 billion in damage. The eye of a hurricane, on the other hand, is relatively calm and free of clouds. The eye is typically 40 to 50 km wide, whereas the hurricane itself may extend outward for more than 600 km. A hurricane may last for weeks and travel for thousands of kilometers so long as it remains over open ocean. But what happens if the hurricane moves ashore? Over land it loses its principal supply of warm moist air and therefore its source of energy. Without the heat of vaporization of water to feed it, the hurricane eventually fades away.

Tornadoes generally last only a few dozen minutes; but because their energy is so concentrated, they are extremely violent and among the most dangerous and destructive of storms (see **Figure 8.15**). Like their larger hurricane cousins, they are also vortices. They are usually too small to be governed by the Coriolis effect; but the general atmospheric circulation in their vicinity causes many tornadoes, like hurricanes, to rotate counterclockwise in the Northern Hemisphere and clockwise in the Southern Hemisphere.

Tornadoes tend to form in the vicinity of hurricanes and violent thunderstorms where strong, thermally generated updrafts are present. When surrounding air rushes in to replace the rising air in the updraft, it may start spinning, creating a vortex. As the column of rising air extends upward, its diameter shrinks. And because the circulating air must obey the law of conservation of angular momentum, the vortex spins ever faster as it continues to shrink. Wind speeds in the most severe tornadoes have been estimated to reach 800 km/h! The base of an average tornado is about 400 m in diameter. Atmospheric pressure at the base can be extremely low—more than 200 mb lower than the surrounding air. This causes the tornado to act like a gigantic vacuum cleaner, picking up dust, cars, and even buildings. Over the past 50 years tornado-related deaths in the United States have averaged about 90 per year. The debris swept

FIGURE 8.15 A large tornado made visible by the debris it sweeps up. Note the power line structures at the lower right.

up by tornadoes makes them highly visible and reveals their characteristic funnel shape. When a tornado passes over a lake or the ocean, it picks up water and is called a *waterspout*.

Dust devils are similar in structure to tornadoes; but they are generally smaller and less intense, and they usually occur in fair weather. Diameters range from a few meters to a few dozen meters, with average heights of several hundred meters. Like tornadoes, their lifetimes are limited to a few dozen minutes. They form in areas of strong surface heating, usually at the interface between different surface types, such as asphalt and dirt or even irrigated fields and dirt roads. Typically they occur under clear skies and light winds, when the ground can warm the air to temperatures well above the temperatures just above the ground—a very unstable condition. Though their wind speeds seldom exceed 25 m/s, dust devils can still be destructive as they lift dust and other debris into the air. Small structures can be damaged and even destroyed if they have the misfortune of being located in

the path of a strong dust devil. Dust devils are a common sight in the dry deserts of the American Southwest.

We also see dust devils on the arid surface of Mars. The *Viking* orbiters were the first to see them in 1976. In 1997 *Mars Pathfinder* sensed one passing right over it. More recently *Mars Global Surveyor* has spotted a large number of dust devils, made easily visible by the shadows they cast on the Martian surface. Most show dark meandering trails behind them where they have vacuumed up bright surface dust, revealing the dark surface rock that lies beneath the dust (see **Figure 8.16**). Dust devils on Mars, typically higher and wider than their Earth counterparts, reach heights of up to 8 kilometers and have diameters ranging from a few dozen meters to a few hundred meters.

FIGURE 8.16 Meandering dark trails cross a Martian crater where numerous dust devils have removed bright surface dust.

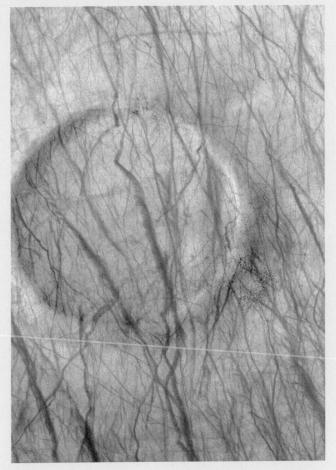

increases sharply with decreasing wavelength. Strong scattering by molecules in the dense atmosphere of Venus would greatly soften any view we might have of distant scenes. We see this same effect, but to a lesser extent, in our own atmosphere. Molecular scattering, always stronger at the shorter wavelengths as discussed in **Excursions 8.4**, causes a loss of contrast and adds a bluish cast to distant terrain. The high atmospheric temperatures also mean that neither liquid water nor liquid sulfurous compounds can exist on the surface of Venus, leaving an extremely dry lower atmosphere with only 0.01 percent of water and sulfur dioxide vapor.

Unlike most planets, Venus rotates on its axis in the opposite sense of its motion around the Sun. Relative to the stars, Venus spins once every 243 Earth days, but a solar day on Venus—the time that it takes for the Sun to return to the same place in the sky—is only 117 Earth days. Regardless, this slow rotation means that Coriolis effects on the atmosphere are small, resulting in a global circulation that is quite close to a classical Hadley pattern (Figure 8.10(a)). Venus is the only planet known to behave in this

---

Surface temperatures on Venus vary little from pole to equator or from day to night.

---

way. Its massive atmosphere is highly efficient in transporting thermal energy around the planet, so the polar regions are only a few degrees cooler than equatorial regions, and there is almost no temperature difference between day and night. Because the Venus equator is nearly in the plane of

---

EXCURSIONS 8.4

Blue Skies, White Clouds, and Red Sunsets

Have you ever noticed the beam from a flashlight or the headlights of a car on a foggy night? As the beam of light shines through the cloud bank, some of the light bounces off tiny water droplets, and you see the light that happens to bounce in your direction. In this way, you "see the beam of light."

When light bounces off small particles in its path, it is called **scattering**. If the particles that are scattering light are much larger than the wavelength of that light, as is the case for water droplets in a fog bank, then photons of all colors are equally likely to be scattered. As a result, the light scattered by a cloud is the same color as the light shining on the cloud. Sunlight is white, so clouds that are illuminated by direct sunlight are also white.

Things get more interesting if the scattering particles are smaller than the wavelength of the light they are interacting with. In such instances, shorter-wavelength photons are more likely to be scattered than are longer-wavelength photons. Molecules in Earth's atmosphere, for example, scatter blue light ($\lambda = 400$ nm) about seven times more effectively than they scatter red light ($\lambda = 650$ nm). When the Sun is high overhead, sunlight follows a short path through the atmosphere and so is relatively unaffected by scattering. Even so, a small fraction of the blue photons in the sunlight are scattered, and when you

look at the sky, your eyes detect the blue photons that were scattered in your direction. As illustrated in **Figure 8.18**, that is why the sky is blue.

This situation changes dramatically as evening approaches. As the Sun drops lower in the sky, the light from the Sun must pass through more air before it reaches you. As the Sun nears the horizon, sunlight passes through hundreds of times more air than it did when the Sun was high overhead. So much of the blue light is scattered away that by the time the light reaches you, the Sun looks orange. Tiny particles of dust and other materials in the atmosphere (which are similar in size to the wavelength of light) scatter away additional blue light, leaving only a glorious red.

The next time you are captivated by the beauty of a sunset, remember also to look up at the deepening blue of the sky overhead. The red sunset is white sunlight minus the blue light that was scattered away. The blue sky is scattered blue light only. The red of sunset, the blue of the sky, and the white of a billowing cloud are three facets of the same gem—a gem called scattering.

FIGURE 8.18 The blue of the sky and the red of a sunset are both due to differences in the way Earth's atmosphere scatters light of different wavelengths.

its orbit, seasonal effects are quite small, producing only negligible changes in surface temperature. (Recall the discussion in Chapter 2 about how the seasons change due to the tilt of Earth's equator relative to the plane of its orbit around the Sun.) Such small temperature variations also mean that wind speeds near the surface of Venus are quite low, typically about a meter per second. High in the rarefied atmosphere, though, where temperature differences can be larger, winds reach speeds of 110 m/s, circling the planet in only four days. The variation of this high-altitude wind speed with latitude can be seen in the chevron shape of the cloud patterns in **Figure 8.19**.

Large variations in the observed amounts of sulfurous compounds in the high atmosphere of Venus suggest to us that the source of sulfur may be sporadic episodes of vol-

canic activity. This strengthens the possibility that Venus remains a volcanically active planet.

8.5 Mars Has a Cold, Thin Atmosphere

Compared to Venus, the surface of Mars is almost hospitable. Because of this, we can confidently expect that humans will eventually set foot on the red planet, quite likely before the end of the 21st century and possibly much sooner. What they will see is a stark, waterless landscape, colored

1 The blue of the sky is short wavelength light scattered from sunlight by Earth's atmosphere, but at midday most sunlight reaches the ground.

2 The red of sunset is sunlight after short-wavelength blue light has been scattered away.

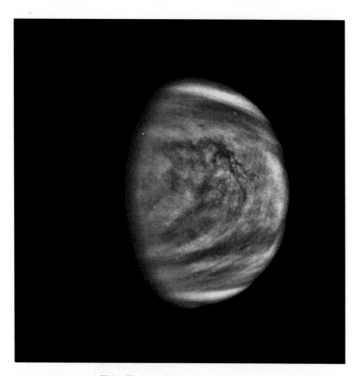

FIGURE 8.19 This *Pioneer Venus* image, taken in ultraviolet light, shows clouds of sulfur compounds in the atmosphere of Venus. Variations in wind speed with latitude cause the clouds to streak out into a "chevron" pattern.

FIGURE 8.20 A view of the dust-filled pink sky of Mars as seen from the *Viking* lander. In the absence of dust, the sky's thin atmosphere would appear deep blue.

reddish by the oxidation of iron-bearing surface minerals. The sky will sometimes be a dark blue but more often a pinkish color caused by windblown dust (**Figure 8.20**). The lower density of the Mars atmosphere makes it more responsive than Earth's to heating and cooling, so temperature extremes are greater. Near the equator at noontime, future astronauts may experience a comfortable 20°C—

The surface of Mars is cold and the air is thin.

---

about the same as a cool room temperature. Nighttime temperatures typically drop to a frigid –100°C, and during the polar night, the air temperature can reach –150°C: cold enough to freeze carbon dioxide out of the air in the form of a dry-ice frost. For human visitors, the low surface pressure will certainly be uncomfortable. With less than 1 percent of our own sea level pressure, the surface of Mars is equivalent to an altitude of 35 km on Earth, far higher than our highest mountain. Like Earth, Mars does have some water vapor in its atmosphere, but the low temperatures condense much of it out as clouds of ice crystals (see Figure 8.6). Early morning ice fog in the lowlands (**Figure 8.21(a)**) and clouds hanging over the mountains will give terrestrial visitors some reminders of their home planet. Although the cold, thin air might be endured, what is seri-

ously lacking and what future astronauts must carry with them is oxygen. Without plants, Mars has only a tiny trace of this life-sustaining gas. Like Venus, the atmosphere of Mars is composed almost entirely of carbon dioxide (95 percent), with a lesser amount of nitrogen (2.7 percent). The near absence of oxygen also means that Mars can have very little ozone, and this lack of ozone allows solar ultraviolet radiation to reach all the way to the surface. To survive on Mars, any surface life forms would have to develop protective layers that could shield against the lethal ultraviolet rays.

The inclination of the Mars equator to its orbital plane is similar to Earth's, so it has similar seasons (see Chapter 2). But the effects are larger for two reasons: Mars varies more in its annual orbital distance from the Sun than does Earth, and the low density of the Mars atmosphere makes it more responsive to seasonal change. The large diurnal, seasonal, and latitudinal temperature differences on Mars

---

Seasonal effects on Mars's climate are larger than Earth's.

---

often create locally strong winds, some estimated to be as high as 100 m/s. For more than a century astronomers have watched the seasonal development of springtime dust storms on Mars. The stronger ones spread quickly and

(a) (b)

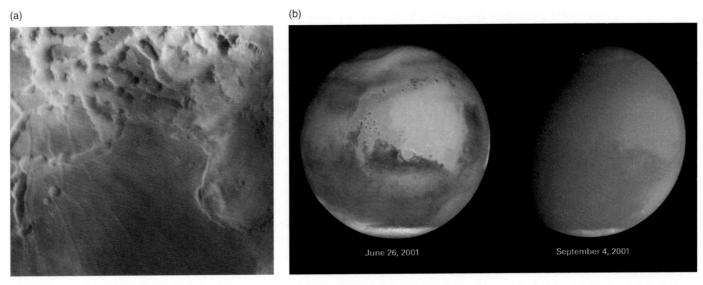

FIGURE 8.21 (a) Patches of early morning water vapor fog forming in canyons on Mars. (b) Hubble Space Telescope images of the development of a global dust storm that enshrouded Mars in September 2001.

June 26, 2001   September 4, 2001

within a few weeks can envelope the entire planet in a shroud of dust (**Figure 8.21(b)**). Such large amounts of windblown dust can take many months to settle out. Seasonal movement of dust from one area to another alternately exposes and covers large areas of dark, rocky surface. This phenomenon led some astronomers of the late 19th and early 20th centuries to believe that they were witnessing the seasonal growth and decay of vegetation. Public imagination carried the astronomers' interpretations a step further—to stories about advanced civilizations on Mars and invasions of Earth by warlike Martians (a theme found in some movies).

Mars likely had a more massive atmosphere in the distant past, and there is geological evidence that liquid water once flowed across its surface, as we saw in Chapter 7; but its low gravity was responsible for the loss of much of this earlier atmosphere. This is a good example of a runaway atmosphere, as discussed earlier in Section 8.2.

8.6 Mercury and the Moon Have Hardly Any Atmosphere

Our story would not be complete if we did not include the kinds of "atmosphere" that exist around the remaining two terrestrial objects, Mercury and the Moon. There is actually little to say: The atmospheres of Mercury and the Moon would hardly be noticed by a visitor from Earth. They are less than a million-billionth (10^{-15}) as dense as our own, and they probably vary somewhat with the strength of the solar wind. Such ultrathin atmospheres can have no effect whatsoever on local surface temperatures. Unless you are one of the few astronomers interested in the interaction between solar wind and solid bodies, you may comfortably ignore the atmospheres of Mercury and the Moon.

Summary

- Primary atmospheres are mostly hydrogen and helium captured from the protoplanetary disk.

- The terrestrial planets lost their primary atmospheres to space.

- Secondary atmospheres are created by volcanic gases and volatiles brought in by impacting comets.

- Plant life is responsible for the oxygen in Earth's atmosphere.

- Nonuniform absorption of solar energy is the cause of all our weather.

- Temperature and pressure decrease with altitude in the tropospheres of Venus, Earth, and Mars.

- The greenhouse effect keeps Earth from freezing but turns the Venus atmosphere into an inferno.

- Earth's magnetosphere shields us from the solar wind.

- Venus has a massive, hot carbon dioxide atmosphere.

- Mars has a thin, cold carbon dioxide atmosphere.

- The Moon and Mercury are almost totally airless.

Seeing the Forest through the Trees

As these words are being written, one of the authors is sitting in his backyard on a pleasant summer morning. As birds fly overhead, he is warmed by the rays of the Sun and cooled by a gentle morning breeze. Such an idyllic setting might seem far from the path of our journey through astronomy, but nothing could be further from the truth. Some of the atoms drifting past arrived at Earth as it formed, later to be expelled into Earth's forming secondary atmosphere by volcanism. The rest, especially volatile materials, arrived in the rain of ice-laden comets that fell on the young Earth. Over the ensuing billions of years, a delicate chemical and physical dance has taken place. The temperature of the young Earth was such that water could condense and fall on the surface as rain. The presence of liquid water served to scrub our atmosphere, absorbing carbon dioxide and locking it up in carbon-bearing rock such as limestone.

Liquid water also provided the bath in which organic chemistry could take place, leading first to molecules with the useful property of being able to reproduce themselves at the expense of other molecules, and later to life itself. Early life on Earth consumed molecules such as carbon dioxide. The waste products formed by this life included what, at the time, was a deadly poison—oxygen. Over the eons the amount of oxygen in our planet's atmosphere increased, and new forms of life evolved to take advantage of its presence. We are among these later life forms. To us the reactive, corrosive gas called oxygen is the breath of life.

In contrast to Earth, Mercury and the Moon were simply too small to hold onto either their primary or their secondary atmospheres, leaving them as airless rocks. Looking inward from Earth toward the Sun, we find a planet that would seem more like our own home. Venus has much the same size and mass as Earth and is a similar distance from the Sun. We might have imagined Venus to be the ideal world for future human colonization. Yet its history has been different from that of Earth. Slightly warmer than the young Earth, the young Venus was too hot for liquid water to pool on its surface, allowing the same greenhouse effect responsible for maintaining Earth's balmy climate to run away with itself. The result is an environment that is hellish beyond even the fevered imagination of Dante Alighieri.

If we step outward from the Sun, we find Mars. Like Venus, Mars is a planet that is not so different from Earth. Although smaller and less massive, it still has strong enough gravity to hold onto its atmosphere. But whereas Venus was too hot, Mars when it was young was too cold. It had a thick atmosphere, and rain fell in torrents. Images of the surface of the planet show flood basins into which the Amazon would have been but a minor tributary. Liquid water on the surface of Mars was too effective at scrubbing the planet's atmosphere of carbon dioxide. The process that prevented Earth from becoming a Venus-like hothouse got out of hand on Mars, and the temperature fell until the planet's atmosphere nearly froze out.

We owe our lives to the thin blanket of atmosphere that covers our planet. At one time we may have felt justified in taking this atmosphere for granted, but after looking around us in the Solar System we have come to appreciate that our world is maintained by the most delicate of balances. Over billions of years, life has shaped the atmosphere of our planet, and today through the activities of humans, life is reshaping our planet's atmosphere once again.

Great political debate surrounds issues such as the release of ozone-destroying chlorofluorocarbons in Earth's atmosphere, as well as the destruction of oxygen-producing plant life and the release of huge amounts of carbon dioxide through the burning of fossil fuels. In the rush for profit and convenience, there are those who say that it is too early to worry about the effects that humans are having on the planet. "Earth is too complex," they say, "and our models too primitive to accurately predict the effect that human activities will have on our atmosphere and climate."

But this burden of proof is incorrectly placed. We know that factors such as the greenhouse effect made our planet what it is today. In the absence of the greenhouse effect, Earth would have an average temperature below the freezing point of water. We also know that human activities are measurably changing the chemical balance of our atmosphere. We are undeniably playing with the knobs that regulate our climate. The only intellectually honest way to look at the situation is to start with the hypothesis that unchecked human activity such as the destruction of rainforests and the burning of fossil fuels will have a significant, adverse impact on the future of our planet. This hypothesis has not been disproved.

Key Terms

atmospheric greenhouse effect, p. 225
greenhouse molecules, p. 226
bar, p. 228
troposphere, p. 229
pressure, p. 230
ideal gas law, p. 230
recombination, p. 233
photodissociation, p. 233
relative humidity, p. 234
saturation, p. 234
tropopause, p. 234
stratosphere, p. 234
mesosphere, p. 234
ionize, p. 234
thermosphere, p. 234
plasma, p. 234
ionosphere, p. 234
magnetosphere, p. 234
bow shock, p. 235
radiation belt, p. 235
aurora, p. 235
catalyst, p. 235
Hadley circulation, p. 237
zonal winds, p. 237
global circulation, p. 238
cumulus, p. 240
cumulonimbus, p. 240
turbulence, p. 240
hurricane, p. 242
tornado, p. 242
dust devil, p. 243
scattering, p. 244

Student Questions

THINKING ABOUT THE CONCEPTS

1. Describe the origins of primary and secondary atmospheres.

2. Primary atmospheres of the terrestrial planets were composed almost entirely of hydrogen and helium. Explain why they contained only these gases and not others.

3. Nitrogen, the principal gas in Earth's atmosphere, was not a significant component of the protostellar disk from which the Sun and planets formed. Where did Earth's nitrogen come from?

4. Contrast the atmospheres and the histories of the atmospheres of Venus, Earth, and Mars.

5. The force of gravity holds objects tightly to the surfaces of the terrestrial planets. Yet atmospheric molecules are constantly escaping into space. Explain how these molecules are able to overcome gravity's grip. How does the mass of a molecule affect its ability to break free?

6. Why is Venus very hot and Mars very cold if both of their atmospheres are dominated by carbon dioxide, an effective "greenhouse" molecule?

7. In what ways does plant life affect the composition of Earth's atmosphere?

8. You check the barometric pressure and find that it is reading only 920 millibars. Two possible effects could be responsible for this low reading. What are they?

9. What is the principal cause of winds in the atmospheres of the terrestrial planets?

10. Describe the solar wind, its origin, and its effects on Earth and society.

11. Global warming appears to be responsible for increased melting of ice in Earth's polar regions.
    a. Why does the melting of Arctic ice, which floats on the surface of the Arctic Ocean, not affect the level of the oceans?
    b. What effect is the melting of glaciers in Greenland and Antarctica having on the level of the oceans?

12. Describe how the atmospheric greenhouse effect works and how it can both help and harm terrestrial life.

13. In 1975 the Soviet Union landed two camera-equipped spacecraft on Venus, giving planetary scientists their first (and only) close-up views of the planet's surface. Both cameras ceased to function after only an hour. What environmental condition most likely led to their demise?

14. Eventually astronomers will have the technical capability to detect Earth-sized planets around other stars. When we find such objects, what observations would indicate with near certainty that they harbor some form of life as we know it?

APPLYING THE CONCEPTS

15. The total mass of Earth's atmosphere is 5×10^{18} kg. Carbon dioxide (CO_2) makes up about 0.06 percent of Earth's atmospheric mass.
    a. What is the mass (kg) of CO_2 in Earth's atmosphere?

b. The annual global production of CO_2 is estimated to be 1.5×10^{13} kg. What annual fractional increase does this represent?

16. The ability of wind to erode the surface of a planet is related in part to the wind's kinetic energy.
    a. Compare the kinetic energy of a cubic meter of air at sea level on Earth (mass = 1.23 kg) moving at a speed of 10 m/s with a cubic meter of air at the surface of Venus (mass = 64.8 kg) moving at 1 m/s.
    b. Compare the terrestrial case with a cubic meter of air at the surface of Mars (mass = 0.015 kg) moving at a speed of 50 m/s.
    c. What would you guess is the reason we do not see more evidence of wind erosion on Earth?

17. Say you seal a rigid container that has been open to air at sea level when the temperature is 0°C (273 K). The pressure inside the sealed container is now exactly equal to the outside air pressure, 10^5 N/m².
    a. What would be the pressure inside the container if it were left sitting in the desert shade where the surrounding air temperature is 50°C (323 K)?
    b. What would be the pressure inside the container if it were left sitting in an Antarctic night where the surrounding air temperature is –70°C (203 K)?
    c. What would you observe in each case if the walls of the container were not rigid?

18. Oxygen molecules (O_2) are 16 times more massive than hydrogen molecules (H_2). Carbon dioxide molecules (CO_2) are 22 times more massive than H_2.
    a. Compare the average speed of O_2 and CO_2 molecules in a volume of air.
    b. Does this ratio depend on air temperature?

19. Using the average density of air at sea level (1.225 kg/m³) and the average mass of Earth's atmosphere above sea level per square meter (1.033×10^4 kg/m²), what would be the total depth of Earth's atmosphere (in km) if its density were the same at all altitudes? (This is called a scale height (H), a useful quantity for comparing Earth's atmosphere with the atmospheres of other planets.)

20. The average surface pressure on Mars is 6.4 millibars. Using Figure 8.6, estimate how high you would have to go in Earth's atmosphere to experience the same atmospheric pressure as you would if you were standing on Mars.

21. Water pressure in Earth's oceans increases by 1 bar for every 10 m of depth. How deep would you have to go to experience pressure equal to the atmospheric surface pressure on Venus?

StudySpace
wwnorton.com/astro21
provides a Study Plan for each chapter that includes a reading outline, animations, keyword flash cards, and gradebook-enabled multiple-choice quizzes. From StudySpace you can also access premium content in the ebook and SmartWork.

Before the starry threshold of Jove's Court
My mansion is.
Above the smoke and stir of this dim spot
Which men call earth.

JOHN MILTON (1608–1674)

Magnificent Saturn looms before the cameras of the *Cassini* spacecraft.

CHAPTER 9

Worlds of Gas and Liquid— The Giant Planets

9.1 The Giant Planets— Distant Worlds, Different Worlds

The four largest planets in our Solar System are Jupiter, Saturn, Uranus, and Neptune, and all have characteristics that clearly distinguish them from the terrestrial, or Earth-like, planets. The most obvious difference is that they are all enormous compared with their terrestrial cousins. Even the smaller of them, Uranus and Neptune, are nearly four times the size of Earth. Another distinguishing feature is their very low density. All are composed almost entirely of light materials such as hydrogen, helium, and water, rather than the rock and metal that make up the terrestrial planets. Collectively we call Jupiter, Saturn, Uranus, and Neptune the **giant planets**, although you may sometimes hear them referred to as the "Jovian planets" after Jupiter, the largest: Jove is another name for Jupiter, the highest-ranking Roman deity.

Even though the giant planets share many characteristics, there are significant differences. Jupiter and Saturn are similar to one another in size and are composed primarily of hydrogen and helium. Uranus and Neptune are also simi-

Jupiter and Saturn are gas giants. Uranus and Neptune are ice giants.

lar in size to one another, and both contain much larger amounts of water and other ices than Jupiter and Saturn. These differences are sufficiently large that some astronomers feel it makes sense to divide the giant planets into two

KEY CONCEPTS

Unlike the solid planets of the inner Solar System, four worlds in the outer Solar System were able to capture and retain gases and volatile materials from the Sun's protoplanetary disk and swell to enormous size and mass. Examining these giant planets, we will discover

- Atmospheres and oceans, but no solid surfaces.

- Two "gas giants," primarily composed of hydrogen and helium, and two "ice giants," primarily composed of water and other volatile materials.

- How changes in temperature and pressure with increasing depth lead to changes in the chemical composition of clouds in giant planet atmospheres.

- Gravitational energy being converted into thermal energy in the interiors of three giant planets, driving strong convection in their atmospheres.

- How the Coriolis effect on these rapidly rotating worlds turns convective motions into powerful winds, huge storms, and planet-spanning bands of multihued clouds.

- Extreme conditions deep within the interiors of the giant planets.

- Brilliant auroras and glowing doughnuts of gas associated with the huge magnetospheres and strong magnetic fields of the giant planets.

(a)

(b)

FIGURE 9.1 The gas giants. (a) Jupiter imaged by *Cassini*. (b) Saturn seen by HST.

(a)

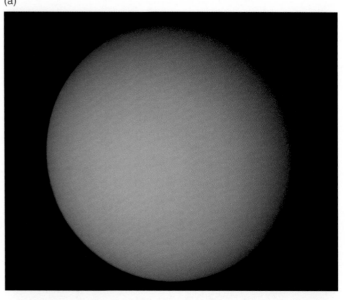

(b)

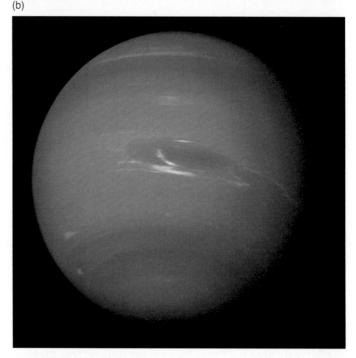

FIGURE 9.2 The ice giants. Uranus (a) and Neptune (b) imaged in visible light by *Voyager*.

classes: Jupiter and Saturn as **gas giants (Figure 9.1)** and Uranus and Neptune as **ice giants (Figure 9.2)**. Other differences between the gas giants and the ice giants[1] will become evident as we more closely examine their physical and chemical characteristics.

The giant planets orbit the Sun far beyond the orbits of Earth and Mars (see **Table 9.1**). The closest to the Sun, and

---

Giant planets orbit much farther from the Sun than do the terrestrial planets.

---

to us as well, is Jupiter, and even it is more than five times as far from the Sun as Earth is. Neptune, the most distant,

---

[1] Remember from Chapter 6 that planetary scientists often refer to these volatile substances as *ices* even when they are in their liquid or gaseous form. Uranus and Neptune may be called *ice giants,* but that does not mean that they are solid, frozen worlds.

is 4½ billion kilometers away, or some 30 times farther from the Sun than we are. To put this distance into perspective: If you were traveling at the speed of a commercial jetliner, it would take you more than 500 years to reach Neptune.

The Sun shines only dimly and provides very little warmth in these remote parts of the Solar System. If we were to journey to Jupiter, we would see the Sun as but a tiny disk, ¹/₂₇ as bright as it appears from Earth. At Neptune the Sun would no longer show a disk but would appear as a brilliant starlike point of light about 500 times brighter than the full Moon in our own sky. How can a point of light be so much brighter than a large disk? It is because the **surface brightness** of the solar disk is 400,000 times greater than the surface brightness of the Moon. Daytime on Neptune is equivalent to a perpetual twilight here on our own planet. With so little sunlight available for warmth, daytime temperatures hover around 123 K at the cloud tops on Jupiter, and they can dip to just 37 K on Neptune's moon Triton.

Two Were Known to Antiquity and Two Were Discovered

Although very far away, the giant planets are so large that all but Neptune can be seen with the unaided eye. Jupiter and Saturn are a familiar sight in the nighttime sky, comparable to the brightest stars. Along with Mercury, Venus, and Mars, they were among the five planets (a Greek word meaning "wandering stars") known to ancient cultures. In contrast, Uranus appears only slightly brighter than the faintest stars visible on a dark night, and Neptune cannot be seen without the aid of binoculars.

Uranus was the first planet to be "discovered," but it was not found until late in the 18th century, more than 170 years after Galileo made the first astronomical observations with a telescope. In 1781 William Herschel, a German–English professional musician and amateur astronomer, came upon it quite by accident. Herschel was producing a catalog of the sky at his home in Bath, England, when he noticed a tiny disk in the eyepiece of his six-inch telescope. At first he thought he had found a comet, but the object's slow nightly motion soon convinced him that it was a new planet beyond the orbit of Saturn. Politically astute, Herschel proposed calling his new planet Georgium Sidus ("George's Star") after King George III of England. Obviously pleased, the monarch rewarded Herschel with a handsome lifetime pension. The astronomical community, however, later rejected Herschel's suggestion, preferring the name Uranus instead. For 65 years Uranus would remain the most distant known planet.

Over the decades that followed Herschel's discovery, astronomers found to their dismay that Uranus was straying from its predicted path in the sky. This aberrant behavior was viewed by mathematicians with grave concern. By then Newton's laws of motion had been firmly established and were the basis for predicting the motion of the newly discovered planet. Could something be wrong with the theory? As a reasonable explanation, a few astronomers suggested that the gravitational pull of some unknown planet might be responsible for this "unacceptable" behavior of Uranus. Using measured positions of Uranus provided by astronomers,

TABLE 9.1

Physical Properties of the Giant Planets

	Jupiter	Saturn	Uranus	Neptune
Orbital radius (AU)	5.2	9.6	19.2	30.0
Orbital period (years)	11.9	29.5	84.0	164.8
Orbital velocity (km/s)	13.1	9.7	6.8	5.4
Mass ($M_\oplus = 1$)	318	95	14.5	17.1
Equatorial diameter (1,000 km)	143	120.5	51.1	49.5
Equatorial diameter ($D_\oplus = 1$)	11.2	9.45	4.0	3.9
Density (water = 1)	1.3	0.7	1.3	1.6
Rotation period (hours)	9.9	10.7	17.2	16.1
Obliquity (degrees)	3.1	26.7	97.8	29.3
Surface gravity (m/s²)	25.1	10.5	9.0	11.2
Escape speed (km/s)	60.2	35.5	21.3	23.5

two young mathematicians, Urbain-Jean-Joseph Le Verrier in France and John Couch Adams in England, independently predicted where the hypothetical planet should be. Although Adams was first to compute his predictions, he was unable to gain the attention of England's Astronomer Royal, so the opportunity for England to gain credit for a second planetary discovery was lost. Meanwhile, Le Verrier was having similar problems convincing French astronomers. It was a German who would finally triumph. Armed with Le Verrier's predictions, Johann Galle began a search at the Berlin Observatory. He found the planet on his first observing night, just where Le Verrier and Adams had predicted it would be. The discovery of Neptune in 1846 became a triumph for mathematical prediction based on physical law—and for the subsequent confirmation of theory by observation.

Neptune became the eighth known planet, and it remains the outermost classical planet in our Solar System. Pluto was discovered in 1930 and was immediately declared to be the ninth planet. However, in 2006, the International Astronomical Union officially reclassified Pluto as a *dwarf planet*. Indeed, Pluto's characteristics are much closer to those of an icy moon or a comet nucleus than to either a giant or a terrestrial planet. For this reason we will discuss Pluto in Chapter 11 when we talk about the moons of the Solar System.

9.2 How Giant Planets Differ from Terrestrial Planets

Historically the vast distances to the giant planets have made them difficult objects for scientific study. All of this changed with the arrival of the space age and with the simultaneous development of more powerful optical and electronic groundbased instruments during the latter decades of the 20th century. Modern instruments on both groundbased telescopes and the Hubble Space Telescope (HST) have provided new insight into the composition and physical structure of the giant planets; but our greatest leaps of knowledge have come from close-up observations made possible by the planetary probes: *Pioneer, Voyager, Galileo,* and *Cassini.* Over the past several decades one or another of these spacecraft has visited each of the four giant planets, sending back a wealth of scientific data with a level of detail that cannot be obtained from the vicinity of Earth. **Figure 9.3** compares the true relative sizes of the giant planets with their relative sizes as seen from Earth. Although groundbased telescopes and the HST have made significant contributions to our knowledge of the giant planets, much of what follows is based on what we have learned from the probes.

The Giant Planets Are Large and Massive

The giant planets represent an overwhelming 99.5 percent of all the nonsolar mass in our Solar System. The vast multitude of other Solar System objects—terrestrial planets, moons, asteroids, and comets—are all included in the remaining 0.5 percent. Jupiter alone is 318 times as massive as Earth and some 3½ times as massive as Saturn, its closest rival. Jupiter, in fact, weighs in at more than twice the mass of all the other planets combined. Uranus and Neptune, the ice giants, are the lightweights among the giant planets, but each still has the equivalent of approximately 15 Earth masses.

How do we measure the mass of the planets? In Chapter 3 we learned how a planet's gravitational attraction can affect the motion of a nearby small body, say one of its moons or a passing spacecraft. We found that the motion of the small body can be accurately predicted if we know the

---

The motions of a planet's moons yield its mass.

---

planet's mass and apply Newton's law of gravitation and Kepler's third law. Seeing how this works, we should now be able to invert the procedure and calculate the planet's mass by observing the motion of a small body. Prior to the space age, we measured a planet's mass by observing the motions of its moons. This, of course, worked only with planets that have moons. (We had to estimate the masses of Mercury and Venus from their size.) The accuracy of those early calculations was limited by how precisely we could measure the positions of the moons with our groundbased telescopes. Planetary spacecraft have now made it possible to measure the masses of planets with unparalleled accuracy. As a spacecraft flies by, the planet's gravity tugs on it and deflects its trajectory. By tracking and comparing the spacecraft's radio signals using several antennae here on Earth, we can detect minute changes in the spacecraft's trajectory, thereby providing a highly accurate measure of the planet's mass.

Jupiter is not only the largest of the eight planets, it is more than a tenth the size of the Sun itself. Saturn is only slightly smaller than Jupiter, with a diameter of 9.5 Earths. Uranus and Neptune are each about 4 Earth diameters across, with Neptune being slightly the smaller of the two.

Stellar occultations let us determine a planet's diameter.

The most accurate measurements of planet sizes have come from observing the length of time it takes a planet to eclipse, or *occult*, a star (**Figure 9.4**). We call these events **stellar occultations**. For example, Newton's laws may tell us that

FIGURE 9.3 *(Top row)* Images of the giant planets, shown to the same physical scale. *(Bottom row)* The same images, scaled according to how the planets would appear as seen from Earth.

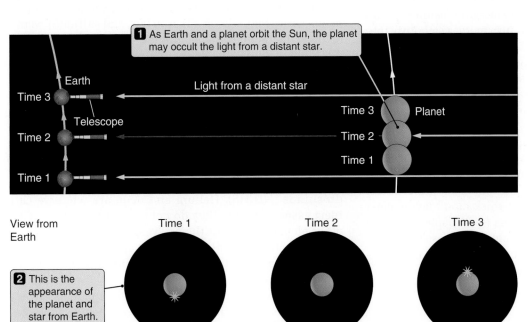

1 As Earth and a planet orbit the Sun, the planet may occult the light from a distant star.

Earth

Light from a distant star

Time 3

Telescope

Time 2

Time 1

Time 3   Planet

Time 2

Time 1

FIGURE 9.4 Occultations occur when a planet, moon, or ring passes in front of a star. Careful measurements of changes in the starlight's brightness and the duration of these changes give information about the size and properties of the occulting object.

View from Earth

Time 1   Time 2   Time 3

2 This is the appearance of the planet and star from Earth.

3 The length of the occultation depends on the known orbital motions and the size of the planet.

Brightness of star

T_1 T_2 T_3

Time ⟶

a particular planet is moving along in its orbit at precisely 25 km/s relative to Earth's own motion, and we observe that this planet takes exactly 2,000 seconds to pass directly in front of a star. The planet then must have a diameter equal to the distance it traveled during the 2,000 seconds, or 50,000 km. The center of the planet rarely passes directly in front of a star, of course; but observations of occultations from several widely separated observatories provide the geometry necessary to calculate both the planet's size and its shape. Occultations of the radio signals transmitted from orbiting spacecraft and images taken by the spacecraft cameras have also provided accurate measures of the sizes and shapes of planets and their moons. You may be surprised to learn that the giant planets are not perfectly round. We will see the reason for this later in the chapter.

We See Only Atmospheres, Not Surfaces

The giant planets are made up mostly or perhaps entirely of gases and liquids. Their characteristic structure is that of a relatively shallow atmosphere merging seamlessly into a deep liquid "ocean," which in turn merges smoothly into a denser liquid or solid core. Although shallow compared with the depth of the liquid layers below, the atmospheres of the giant planets are still much thicker than those of the terrestrial planets—thousands of kilometers as compared to hundreds. As with Venus, only the very highest levels of their atmospheres are visible to us. In the case of Jupiter or Saturn, we see the tops of a layer of thick clouds, the highest of many others that lie below. Although a few thin clouds are visible on Uranus, we find ourselves looking mostly into a clear, seemingly bottomless atmosphere. Atmospheric models tell us that thick cloud layers must lie below, but molecular scattering (see Chapter 8) in the clear part of the atmosphere prevents us from seeing them. Neptune displays a few high clouds with a deep, clear atmosphere showing between them.

Jupiter's Chemistry Is More Like the Sun's Than Earth's

In Chapter 7 we learned that the terrestrial planets are composed mostly of rocky minerals, such as silicates, along with various amounts of iron and other metals. It is true that the atmospheres of the terrestrial planets contain lighter materials, but the masses of these atmospheres—and even of Earth's oceans—are insignificant compared with the total planetary mass. The terrestrial planets are thus very com-

pact with densities ranging from 3.9 (Mars) to 5.5 (Earth) times that of water—they are the densest objects in the Solar System.

In contrast, the giant planets are composed almost entirely of lighter materials such as hydrogen, helium, and water. This makes their densities much lower than the rock-and-metal terrestrial planets. Neptune is the most compact among the giant planets, having a density about 1.6 times that of water. Saturn is the least compact—only 0.7 times the density of water. This means that, placed in a large enough (and deep enough) body of water, Saturn would actually float with 70 percent of its volume submerged. Jupiter and Uranus have densities intermediate between those of Neptune and Saturn.

Jupiter's chemical composition is quite similar to that of the Sun and the rest of the cosmos. Astronomers take the relative abundance of the elements in the Sun as a standard reference, termed **solar abundance**. As illustrated in **Figure 9.5**, hydrogen (H) is the most abundant element, followed by helium (He). Jupiter has about a dozen hydrogen atoms for every atom of helium, which is typical of the universe as a whole. Only 2 percent of its mass is made up of **massive elements** (atoms more massive than helium)—mostly oxygen (O), carbon (C), neon (Ne), nitrogen (N), magnesium (Mg), silicon (Si), iron (Fe), and sulfur (S). Many of these elements have combined chemically with one

---

Jupiter's chemistry is similar to the Sun's.

---

another, and because of its great abundance, most are linked up with hydrogen. Thus atoms of oxygen, carbon, nitrogen, and sulfur have combined with hydrogen to form molecules of water (H_2O), methane (CH_4), ammonia (NH_3), and hydrogen sulfide (H_2S), respectively. More complex combinations produce materials such as ammonium hydrosulfide (NH_4HS). Helium and neon are what we call **inert gases**, meaning they do not normally combine with other elements or with themselves. Most of the iron, the remains of the original rocky planet around which the gas giant grew, and even much of the water has ended up in Jupiter's liquid core.

The chemical compositions of the giant planets are not all the same. Proportionally, Saturn has a somewhat larger inventory of massive elements than Jupiter. In Uranus and Neptune, massive elements are so abundant that they are major compositional components of these two planets. What is especially interesting is that the total mass of massive elements in each of the four giant planets is nearly the same —approximately 10 to 15 $M_\oplus$ (Earth masses). The principal compositional differences among the four of them lie in the amounts of hydrogen and helium they contain. This turns out to be an important clue to the process by which the giant planets formed, a subject we will return to later in the chapter.

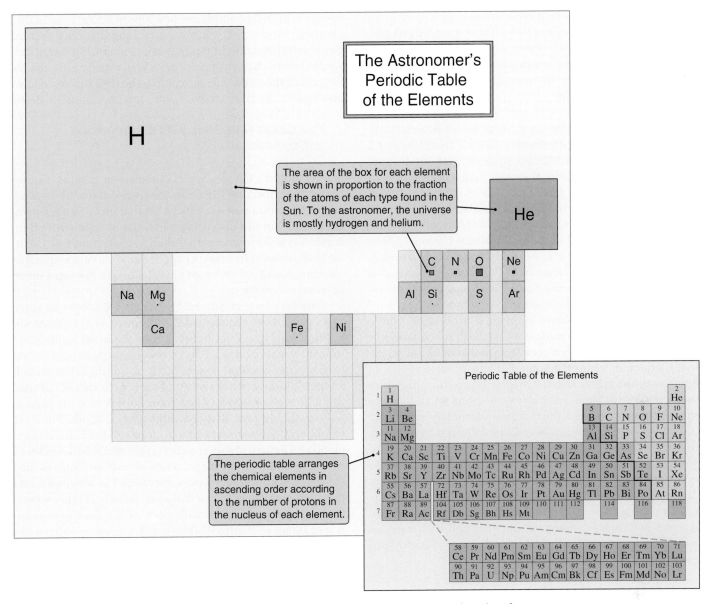

FIGURE 9.5 The periodic table of the elements shows the chemical elements laid out in order of the number of protons in the nucleus of each. But in this "astronomer's periodic table," the universe is made up of mostly hydrogen and helium.

Short Days and Nights on the Giant Planets

Still another distinguishing characteristic of the giant planets is their rapid rotation, meaning that the lengths of their days are short. A day on Jupiter is just under 10 hours long, and Saturn's is only a little longer. Neptune and Uranus have rotation periods of 16 and 17 hours, respectively, giving them days that are intermediate in length between those of Jupiter and Earth.

The rapid rotation of the giant planets distorts their shapes. If they did not rotate at all, these fluid bodies would be perfectly spherical. In Chapter 6 we used the analogy of

Rapid rotation flattens the shapes of the giant planets.

a spinning ball of pizza dough to see why a collapsing, rotating cloud must settle into a disk. The same principle is at work in the rapidly rotating giant planets as well, causing the planets to bulge at their equators and giving them an overall flattened appearance. We call this flattening **oblateness**. Saturn's oblateness is especially noticeable (see Figure 9.3), with an equatorial diameter that is 10 percent greater than its polar diameter. In comparison, the oblateness of Earth is only 0.3 percent.

A planet's obliquity—the inclination of its equatorial plane to its orbital plane—is a major factor in determining the prominence of its seasons. (Recall from Chapter 2 that Earth's obliquity of 23.5° is responsible for our distinct seasons.) The obliquities of the giant planets are shown in Table 9.1. With an obliquity of only 3°, Jupiter has almost no seasons at all. The obliquities of Saturn and Neptune are slightly larger than those of Earth or Mars, giving these planets moderate but well-defined seasons. Curiously, Uranus spins on an axis that is nearly in the plane of its orbit. This creates the appearance from Earth of a planet that is either spinning face on to us or rolling along on its side (or something in between), depending on where Uranus happens to be in its orbit. The obliquity of Uranus is 98°, where a value greater than 90° indicates that the planet rotates in a clockwise direction when seen from above its orbital plane (see Chapter 2). Venus (Chapter 8), Pluto, its moon Charon, and Neptune's moon Triton (Chapter 11) are the only other major bodies that behave this way. Seasons on Uranus are thus extreme, with each polar region alternately experiencing 42 years of continuous sunshine followed by 42 years of total darkness. If we calculate the average amount of sunlight absorbed over an entire orbit, it turns out that the poles are warmed more than the equator, a situation quite different from that of any other planet. Why does Uranus have an obliquity so different from the other planets? As we learned in Chapter 6, many astronomers believe the planet was "knocked over" by the impact of a huge planetesimal near the end of its accretion phase.

9.3 A View of the Cloud Tops

Even when viewed through small telescopes, Jupiter is perhaps the most striking and colorful of all the planets (see Figure 9.1(a)). A dozen or more parallel bands, ranging in hue from bluish gray to various shades of orange, reddish brown, and pink, stretch out across its large, pale yellow disk. Traditionally astronomers call the darker bands *belts* and the lighter ones *zones*. Many small clouds—some dark and some white, some circular and others more oval in shape—appear along the edges of, or within, the belts. The most prominent of these is a large, often brick-red feature in Jupiter's southern hemisphere known as the **Great Red Spot (GRS)**, seen at the lower right in Figure 9.1(a). Oval in shape, with a length of 25,000 km and a width of 12,000 km, the Great Red Spot could comfortably hold two Earths side

by side within its boundaries (see Figure 9.10(a) later in this chapter). No one really knows how long this huge feature has been circulating in Jupiter's atmosphere, but it was first spotted more than three centuries ago, shortly after the invention of the telescope. Since then the GRS has varied unpredictably in size, shape, color, and motion as it drifts

---

The Great Red Spot (GRS) is a giant anticyclone.

---

among Jupiter's clouds. Small clouds seen moving around the periphery of the GRS show that it circulates like a giant **vortex**. Its cloud pattern looks a lot like that of a terrestrial hurricane, but because it rotates in the opposite direction, it exhibits *anticyclonic* rather than *cyclonic* flow (see Chapter 2). Because of its colorful appearance and unpredictable changes, the Great Red Spot has long been a favorite among amateur astronomers.

Our first look at Saturn through a telescope, in some ways even more impressive than Jupiter, is a moment not quickly forgotten (see Figure 9.1(b)). Adorned by its magnificent system of rings, Saturn provides a sight unmatched by any other celestial object. But it is the rings that make it so spectacular; farther away and somewhat smaller in size, the planet itself appears less than half as large as Jupiter. Saturn, like Jupiter, displays atmospheric bands, but their colors and contrasts are much more subdued. Individual clouds on Saturn are seen only rarely from Earth. On these infrequent occasions, large, white cloudlike features may suddenly erupt in the tropics, spread out in longitude, and then fade away over a period of a few months (**Figure 9.6(a)**). Closer views from spacecraft, such as the *Cassini* near-infrared image in **Figure 9.6(b)**, show a wealth of features not visible from Earth.

In even the largest telescopes, Uranus and Neptune appear visually as tiny, featureless disks with a pale bluish green color (similar to the *Voyager* view in Figure 9.2(a)). However, as we will see, infrared imaging reveals a number of individual clouds and belts, giving these distant planets considerably more character.

Observed from close up, the giant planets suddenly appear as real and tangible worlds. The clouds of Jupiter are a landscape of variegated color and intricate formations (Figure 9.1(a)). Time-lapse imaging shows a roiling, swirling giant with atmospheric currents and vortices so complex in nature that we still do not understand the details of how they interact with one another, even after decades of analysis. The GRS alone displays more structure than was visible over all of Jupiter prior to the space age. Dynamically, it also reveals some rather bizarre behavior, such as cloud cannibalism. In a series of time-lapse images, *Voyager* observed a number of Alaska-sized clouds being swept into the GRS vortex. Some of these clouds were carried around the vortex a few times and then ejected, while others were swallowed up and never seen again (**Figure 9.7**). Other smaller clouds

(a)

(b)

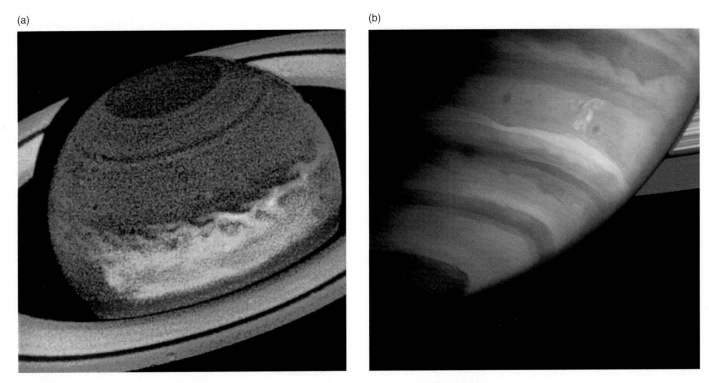

FIGURE 9.6 Tremendous storms are known to occasionally erupt on Saturn. (a) This Hubble Space Telescope image shows a storm that began shortly after the launch of the telescope in 1990. Smaller clouds are seen from closer up. (b) This near-infrared *Cassini* image shows a storm nicknamed "the dragon."

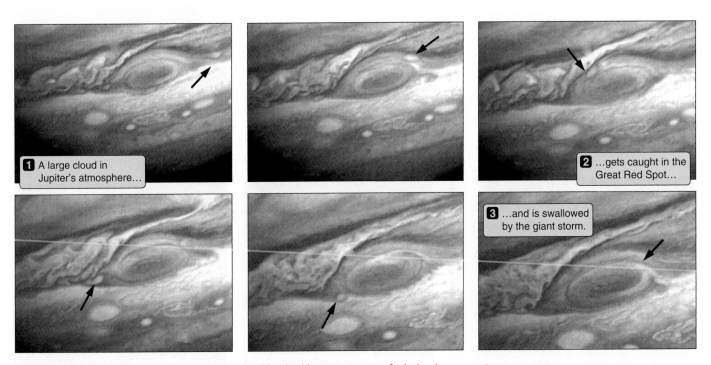

1 A large cloud in Jupiter's atmosphere...

2 ...gets caught in the Great Red Spot...

3 ...and is swallowed by the giant storm.

FIGURE 9.7 This sequence of images, obtained by the *Voyager* spacecraft during its encounter with Jupiter, shows the swirling, anticyclonic motion of Jupiter's Great Red Spot.

(a)

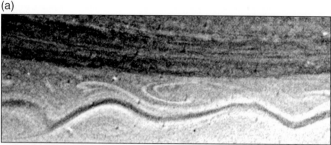

(b)

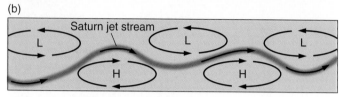

FIGURE 9.8 (a) *Voyager* image of a jet stream in Saturn's northern hemisphere, similar to jet streams in the terrestrial atmosphere. (b) The dynamic relationship between the jet stream and vortices around regions of low and high pressure nested within its peaks and troughs.

with structure and behavior similar to that of the GRS are seen in Jupiter's middle latitudes.

Broad bands and individual clouds show prominently in *Voyager* and *Cassini* images of Saturn's atmosphere. A relatively narrow, meandering band in the midnorthern latitudes encircles the planet in a manner similar to our own terrestrial jet stream (**Figure 9.8**). The largest features are about the size of the continental United States, but many that we see are smaller than terrestrial hurricanes. A small red oval feature, resembling a miniature version of Jupiter's

Great Red Spot, appeared in images taken by *Voyager* in 1980–1981. But 25 years later *Cassini* could find no evidence of Saturn's "mini-Red Spot." It was too small to be seen from

Saturn has jet streams similar to Earth's.

Earth, and we have no idea when or how it vanished. The inability of occasional planetary probes to provide continuous monitoring of the giant planets is one of their more serious shortcomings.

Uranus and Neptune present a rather bland appearance at visible wavelengths. Yet in the near infrared, beyond the spectral range of our eyes, these planets take on a very different appearance, showing limb hazes, bands, and small clouds (**Figure 9.9**). This illustrates how observations made in different spectral regions can add significantly to our understanding of astronomical objects. A few clouds and a weak orange-colored band surrounding the south pole of Uranus are recognizable in HST images (Figure 9.9(a)). Several muted bands appear at lower latitudes. Images from groundbased telescopes using adaptive optics (see Chapter 5) can sometimes show even more detail (Figure 9.9(b)).

A number of small bright clouds appear in *Voyager 2* images of Neptune's atmosphere (see Figure 9.2(b)). In some instances we can see their shadows cast down through the clear upper atmosphere onto a dense cloud layer 75 km below. Some of Neptune's clouds are dark. A large, dark, oval feature in the southern hemisphere (**Figure 9.10(b)**) is reminiscent of Jupiter's Great Red Spot (**Figure 9.10(a)**).

FIGURE 9.9 Hubble Space Telescope images of Uranus (a) and Neptune (b), taken at a wavelength of light that is strongly absorbed by methane. The visible clouds are high in the atmosphere. Uranus's rings show prominently in this image that subdues the brightness of the planet. (c) A groundbased image of Uranus taken four years after (a) in 2004 shows less ring tilt and more of the planet's northern hemisphere. Uranus reached southern autumnal equinox in 2007.

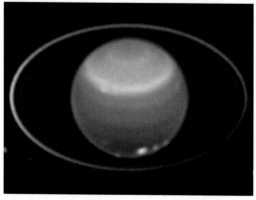

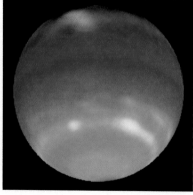

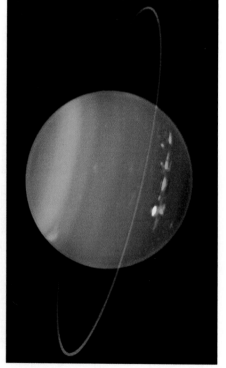

(a)   (b)   (c)

(a) Jupiter's Great Red Spot

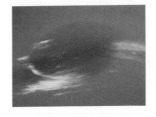

(b) Neptune's Great Dark Spot

(c) Earth

FIGURE 9.10 (a) The Great Red Spot (GRS) on Jupiter and (b) Great Dark Spot (GDS) on Neptune reproduced approximately to scale. Earth is shown for comparison. The Great Red Spot has persisted for centuries, but the Great Dark Spot disappeared between the time *Voyager* flew by Neptune and HST images were obtained.

Predictably, astronomers called it the Great Dark Spot (GDS). However, the Neptune feature was gray rather than red, changed its length and shape more rapidly than the GRS does, and lacked the permanence of the GRS. When the HST trained its corrected optics on Neptune in 1994, the Great Dark Spot had disappeared. It has not been seen since.

9.4 A Journey into the Clouds

Our visual impression of the giant planets is based on our "two-dimensional" view of their cloud tops. Atmospheres, though, have depth. They are three-dimensional structures whose temperature, density, pressure, and even chemical composition vary with height and over horizontal distances. As a rule, atmospheric temperature, density, and pressure all decrease with increasing altitude, although temperature will sometimes reverse itself at very high altitudes (see Chapter 8). The *stratospheres* above the cloud tops appear relatively clear, but closer inspection shows that they contain layers of thin haze that show up best when seen in profile above the limbs of the planets. The composition of the haze particles remains unknown, but we believe that they are **photochemical** —smoglike—products created when ultraviolet sunlight acts on hydrocarbon gases such as methane.

What lies beneath the cloud tops of the giant planets? Imagine riding along in the *Galileo* probe as it descends through Jupiter's atmosphere. At first the only change we note is the expected increase in the outside pressure and temperature as our altitude rapidly decreases. Suddenly we find ourselves passing through dense layers of cloud, sepa-

rated by regions of relatively clear atmosphere. Each of these cloud layers is composed of a different chemical substance. In Earth's atmosphere water is the only substance that can condense into clouds, but the atmospheres of Jupiter and the other giant planets contain a variety of volatiles that can condense at different temperatures and atmospheric pressures. A descent through Jupiter's cloud layers thus becomes a journey that explores the many minor constituents of its atmosphere. Each kind of volatile, such as water, condenses at a particular temperature and pressure, and each therefore forms clouds at a different altitude, as illustrated in **Figure 9.11**. Below the condensation cloud layer, each volatile is freely mixed as a gaseous atmospheric constituent. Above, it is highly depleted. We can see the reason for this: As the

Different volatiles produce different clouds at different heights.

planet's atmosphere convects (a process explained in Connections 7.2), volatile materials are carried upward along with all other atmospheric gases. When a particular volatile reaches an altitude where the temperature is low enough, the condensation process removes most of it from the other gases, leaving it depleted in the air above.

During our descent we find that ammonia has condensed near the top of Jupiter's *troposphere* (see Chapter 8) at a temperature around –140°C (133 K). Next we pass through a layer of ammonium hydrosulfide clouds at a temperature of about –80°C (193 K). Not long after this, information from the probe ends. (While descending slowly via parachute near an atmospheric pressure of 22 bars and a temperature of about 100°C (373 K), the *Galileo* probe failed, presumably because its transmitter got too hot.) What lies below this level in Jupiter's atmosphere must for now be left to our theories and atmospheric models.

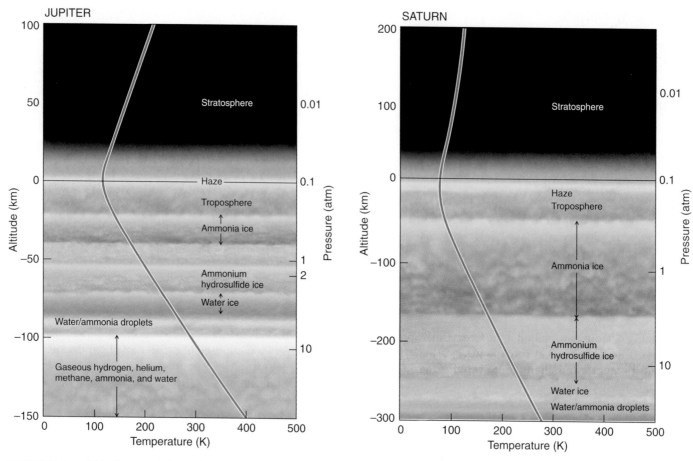

FIGURE 9.11 Volatile materials condense out at different levels in the atmospheres of the giant planets, leading to chemically different types of clouds at different depths in the atmosphere. The 1.0-atm level corresponds to atmospheric pressure at sea level on Earth and is taken as the zero point for altitude.

The distance of each planet from the Sun partly determines its tropospheric temperatures. The farther the planet is from the Sun, the colder its troposphere will be. This now determines the altitude at which a particular volatile, such as ammonia or water, will condense to form a cloud layer on each of the planets (see Figure 9.11). If temperatures are too high, some volatiles may not condense at all. The highest clouds in the frigid atmospheres of Uranus and Neptune are crystals of methane ice. The highest clouds on Jupiter and Saturn are made up of ammonia ice. Methane exists only in gaseous form throughout the warmer atmospheres of Jupiter and Saturn.

In their purest form, the ices that make up the clouds of the giant planets are all white, similar to snow on our own planet. Why, then, are some clouds so colorful, especially Jupiter's? These tints and hues must come from impurities in the ice crystals, similar to the way syrups color "snow cones." Although the identities of these impurities remain unknown, our prime suspects include elemental sulfur and phosphorus and various organic materials produced by the photochemical action of sunlight on atmospheric hydrocarbons. Ultraviolet photons from the Sun, absorbed by molecules of hydrocarbons such as methane, acetylene, and ethane, among others, have enough energy to break these molecules apart. The molecular fragments can then recombine to form complex organic compounds that condense into

Clouds on Jupiter and Saturn are colored by impurities.

solid particles, many of which are quite colorful. Photochemical reactions also occur in our own terrestrial atmosphere. Some of the photochemical products produced close to the ground are rather obvious: We call them "smog."

The upper tropospheres of Uranus and Neptune, unlike those of Jupiter and Saturn, are relatively clear with only a few white clouds—probably methane ice—appearing here and there. Uranus and Neptune are not bluish green because of clouds. Instead they are blue for much the same reason Earth's oceans are blue. Methane gas is much more abundant

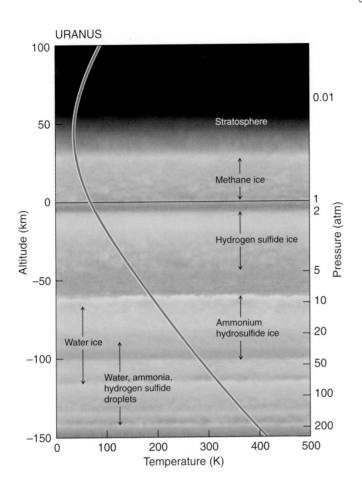

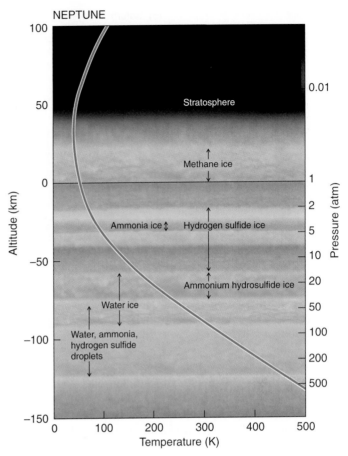

in the atmospheres of Uranus and Neptune than in Jupiter and Saturn. Like water, methane gas tends to selectively absorb the longer wavelengths of light—yellow, orange, and red. This leaves only the shorter wavelengths—green and blue—to be scattered from the relatively cloud-free atmospheres of Uranus and Neptune, giving them a characteristic greenish blue color. We earlier described the atmosphere of Uranus as appearing nearly "bottomless." Molecular scattering also contributes to the bluish color and is so strong in the clear Uranus atmosphere that it completely hides the thick ammonia and water cloud layers that lie far below.

9.5 Winds and Storms—Violent Weather on the Giant Planets

The rapid rotation and resulting strong Coriolis effects (see Chapter 2) in the atmospheres of the giant planets create much stronger zonal winds (see Chapter 8) than we see in the atmospheres of the terrestrial planets, even though less thermal energy is available. **Figure 9.12(a)** shows the zonal wind pattern on Jupiter. On Jupiter the strongest winds are equatorial westerlies, which have been clocked at speeds of up to 550 km/h (150 m/s). (Remember from Chapter 8 that westerly winds are those that blow *from,* not *toward,* the west.) At higher latitudes the winds alternate between easterly and westerly in a pattern that seems to be related to Jupiter's banded structure, but scientists are not sure. Near 20° south latitude, the Great Red Spot appears to be caught between a pair of easterly and westerly currents with opposing speeds of more than 200 km/h. If you think this might imply something about the relationship between zonal flow and vortices, you are right.

The equatorial winds on Saturn are also westerly, but here the wind speed can rise to an impressive 1,700 km/h (470 m/s) (**Figure 9.12(b)**) or drop to as low as 990 km/h (275 m/s). Alternating easterly and westerly winds also occur at higher latitudes; but unlike Jupiter's case, this alternation seems to bear no clear association with Saturn's atmospheric bands. This is but one example of unexplained differences between the giant planets. As mentioned earlier, there is a narrow meandering river of air with alternating crests and troughs (see Figure 9.8(a)). This feature, located

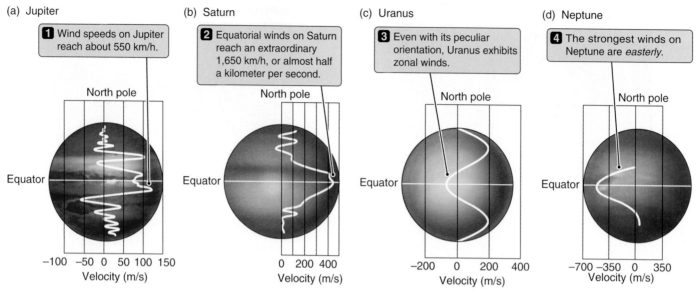

FIGURE 9.12 Strong winds blow the atmospheres of the outer planets, driven by powerful convection and the Coriolis effect on these rapidly rotating worlds. Note that these are typical speeds and not the maximum observed.

near 45° north latitude, is similar to our own terrestrial jet streams, where high-speed winds blow generally from west to east but with alternate wanderings toward and away from the pole. Nested within the crests and troughs of Saturn's jet stream are anticyclonic and cyclonic vortices. They appear remarkably similar in both form and size to terrestrial

Equatorial winds are fast on Jupiter and even faster on Saturn and Neptune.

high- and low-pressure systems, which bring us alternating periods of fair and stormy weather as they are carried along by our terrestrial jet streams. This similarity in jet streams on Saturn and Earth is a good illustration of the many reasons we study other planets—for the ability to compare them to each other and to similar phenomena on Earth. From observing and analyzing similar atmospheric systems on other planets, we often learn more about the way our own weather works.

Our knowledge of global winds on Uranus (**Figure 9.12(c)**) is much poorer than that of the other giant planets because of the relatively few clouds we have been able to see and track. In **Excursions 9.1** we see that winds are measured by observing the motion of clouds. When *Voyager 2* flew by Uranus in 1986, the few clouds we saw were all in the southern hemisphere. The northern hemisphere was in complete darkness at the time. The strongest winds observed were 650 km/h westerlies in the middle to high southern latitudes. No easterly winds were detected on the part of the planet that could be seen. The most important finding was that the winds are zonal. In 1986 the Sun was

shining almost straight down on the south pole of Uranus. Some astronomers had thought earlier that Uranus might have a global wind system very different from that of the other giant planets: Because of its peculiar orientation, Uranus has a "reversed" temperature pattern wherein the poles are warmer than the equator. That the dominant winds on Uranus also turned out to be zonal, as they are on the other giant planets, indicates how dominant the Coriolis effect is and how completely it determines the fundamental structure of the global winds on all the giant planets.

As Uranus moves along in its orbit, spring has come to its northern hemisphere in 2007 and regions previously unseen by modern telescopes are becoming visible. Recent observations by HST and groundbased telescopes show bright cloud bands in the far north extending over 18,000 km in length and revealing wind speeds of up to 825 km/h (230 m/s). As we approach northern summer solstice in the year 2027, we can expect to learn much more about the "unseen northern hemisphere" of Uranus.

As expected, the strongest winds on Neptune occur in the tropics (**Figure 9.12(d)**). The surprise was that they are easterly rather than westerly, with speeds of up to 1,600 km/h. Westerly winds with speeds of 900 km/h and higher were seen in the south polar regions. With wind speeds five times greater than those of the fiercest hurricanes on Earth, Neptune and Saturn are the windiest planets we know of. Summer solstice arrived in Neptune's southern hemisphere in 2005, so much of the north is still in darkness. We'll have to wait a while before we can get a good look at its northern hemisphere. Each season on Neptune lasts for 40 Earth years!

As we learned in Chapter 8, vertical temperature differences can create the localized type of atmospheric motion called *convection*. On the giant planets the thermal energy that drives convection comes from the Sun and also from the hot interiors of the planets themselves. As heating drives air up and down, the Coriolis effect shapes that convection into atmospheric vortices, examples of which are familiar to us on Earth as high-pressure systems, hurricanes, and thunderstorms. On the giant planets convective vortices are

---

Thermal energy drives powerful convection on the giant planets.

---

visible as isolated circular or oval cloud structures. The Great Red Spot (GRS) on Jupiter and the Great Dark Spot (GDS) on Neptune are classic examples. Observations of small clouds distributed within the GRS show that it is an enormous atmospheric whirlpool or eddy, swirling around in a counterclockwise direction with a period of about one week. On a rapidly rotating planet, winds generated by Coriolis effects can be very strong. Clouds circulating around the circumference of the GRS have been clocked at speeds of up to 1,000 km/h. Such weather systems dwarf our terrestrial storms in both size and intensity. Comparable whirlpool-like behavior is observed in many of the smaller oval-shaped clouds found elsewhere in Jupiter's atmosphere, as well as in similar clouds observed in the atmospheres of Saturn and Neptune.

As the atmosphere ascends near the centers of the vortices, it expands and cools. Cooling condenses certain volatile materials into liquid droplets, which then fall as rain (see Chapter 8). As they fall, the raindrops collide with surrounding air molecules, stripping electrons from the molecules and thereby developing tiny electric charges in the air. The cumulative effect of countless falling raindrops is often to generate an electric charge and resulting electric field so great that they break down a conductive path through the atmosphere, creating a surge of current and a flash of lightning (see Excursions 8.3). A single observation of Jupiter's night side by *Voyager 1* revealed several dozen lightning bolts within an interval of three minutes. We estimate the strength of these bolts to be equal to or greater than the "superbolts" that occur in the tops of high convective clouds in the terrestrial tropics. Although various constraints prevented an imaging search for lightning on the other giant planets, radio receivers on *Voyager* picked up lightning static on all of them.

Are atmospheric vortices somehow connected to the global circulation—that is, to zonal flow? We have noted that the narrow zonal jet on Saturn moves with wavelike motion (Figure 9.8(a)). Nestled in each of its crests and troughs are clockwise (anticyclonic) and counterclockwise

---

Vortices created by convection drive the strong zonal winds of the giant planets.

(cyclonic) features, strongly suggesting a dynamic relationship between these systems and the zonal jet. Jupiter's Great Red Spot is situated between pairs of strong zonal winds flowing in opposite directions, as was Neptune's Great Dark Spot. Is this by chance? Similar relationships observed between other isolated vortical clouds and the zonal wind

---

EXCURSIONS 9.1

How Wind Speeds on Distant Planets Are Measured

How do we measure wind speeds on planets that are so far away? It turns out to be surprisingly easy. If we can see individual clouds in their atmospheres, we can measure their winds. As on Earth, clouds are typically carried right along with the local winds. By measuring the positions of individual clouds and noting how much they move during an interval of a day or so, we are able to calculate the local wind speed. We need to know one important additional piece of information though: how fast the planet itself is rotating. This is because we want to measure the speed of the winds with respect to the planet's rotating surface.

In the case of the giant planets, of course, there is no solid surface against which to measure the winds. We must instead assume a hypothetical surface, one that rotates as though it were somehow "connected" to the planet's deep interior. How can we know how rapidly the invisible interior of a planet is rotating? Periodic bursts of radio energy caused by the rotation of a planet's magnetic field tell us how fast the interior of the planet is rotating.

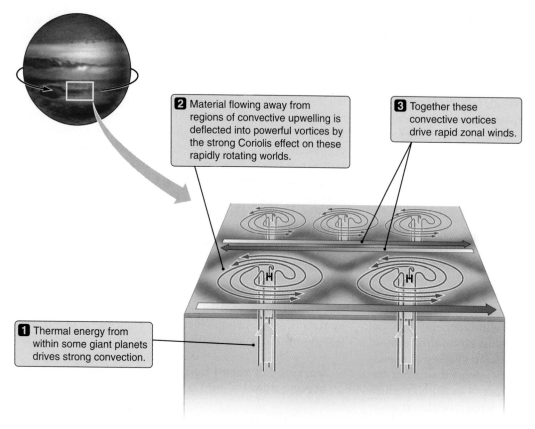

2 Material flowing away from regions of convective upwelling is deflected into powerful vortices by the strong Coriolis effect on these rapidly rotating worlds.

3 Together these convective vortices drive rapid zonal winds.

1 Thermal energy from within some giant planets drives strong convection.

FIGURE 9.13 Material rises within a giant planet due to convection. Anticyclonic flow around the resulting region of high pressure is caused by the Coriolis effect. Such convective vortices drive zonal winds on the giant planets.

flow suggest that it is not. In looking at Figure 9.8(b), we might ask, then, whether the zonal winds are dragging the edges of vortices around or, conversely, the vortices are driving the zonal wind currents. From a careful study of the interaction between vortices and zonal winds, we believe the latter to be true. As shown in **Figure 9.13**, the enormous wind energy developed within numerous convective vortices seems to drive the alternating easterly and westerly zonal winds that characterize the global circulation of the giant planets.

9.6 Some Thermal Energy Comes from Within

We have seen how the uneven heating of a planet, along with the planet's rotation, drives global atmospheric circulation, and how temperature-related differences in pressure from place to place drive winds. In fact, virtually all weather on every planet is driven by the interplays of thermal energy within the planet's atmosphere. On Earth and the other terrestrial planets, the source of this thermal energy is as clear as the Sun shining in a summer sky. Sunlight powers our climate.

This is not a new insight. In Chapter 4 we learned about the equilibrium between the absorption of sunlight and the radiation of infrared light into space, and in Chapter 8 we saw how the resulting equilibrium temperature is modified by the greenhouse effect on Earth and Venus. Yet when we calculate this equilibrium for the giant planets, as we did in

Jupiter has a large internal heat source, as do Saturn and Neptune.

Chapter 4, we find that something seems amiss. According to these calculations, the equilibrium temperature for Jupiter, for example, should be 109 K, but when it is measured we find instead an average temperature of about 124 K. A difference of 15 K might not seem like much, but remember that according to Stefan's Law (see Chapter 4), the energy radiated by an object depends on its temperature raised to the fourth power. Applying this to Jupiter, we get $(T_{\text{true}}/T_{\text{expected}})^4 = (124 \text{ K}/109 \text{ K})^4 = 1.67$. The implications of this are somewhat startling: *Jupiter is radiating roughly two-thirds more energy into space than it absorbs in the form of sunlight.* Almost half of the thermal energy powering Jupiter's weather comes from somewhere other than the Sun. Where else could this extra energy arise but from within the planet itself?

Similarly, the internal energy escaping from both Saturn and Neptune is observed to be almost twice as great as the sunlight that each of them absorbs. Strangely, whatever internal energy may be escaping from Uranus is small compared with the absorbed solar energy.

We have already learned that winds on the giant planets are considerably stronger than on Earth. It is interesting to note that this is true even though less thermal energy is available to drive these winds. Available solar energy per unit area on Jupiter is less than 4 percent of that received by Earth, and on Neptune only 0.1 percent of Earth's. Even with the additional internal energy, the total energy per unit area falls far short of that available to the terrestrial planets. We do not fully understand why these winds should be so strong.

Thermal energy from the hot interiors of the giant planets diffuses slowly outward to warm their upper atmospheres. The rate at which this energy is delivered to the atmosphere depends on both the temperature and the thermal properties of the interior. The most direct path for thermal energy to escape is by convection outward through the liquid and gaseous layers. The final step in this process is seen in the convective vortices that we discussed in the previous section.

With energy continually escaping from the interiors of the giant planets, it is easy to wonder how they have maintained their high internal temperatures over the past 4½ billion years. In **Excursions 9.2** we find that the giant planets are still converting the gravitational potential energy of their creation into thermal energy. This continuous production of thermal energy is sufficient to replace the energy that is escaping from their interiors. In contracting ever so slightly each year, they are continuing the process of their creation.

EXCURSIONS 9.2

Primordial Energy

In Chapter 6 we learned about the collapse of a protostar. A mass of gas collapses under the force of its own self-gravity, and as it does so, it converts its gravitational potential energy to thermal energy. If the mass of the protostar is great enough to form a star, the core warms up to a temperature so high that self-sustaining thermonuclear reactions take place. Once thermonuclear energy becomes available, internal thermal energy can be generated and maintain enough outward pressure to prevent further collapse, thereby stabilizing the newly formed star.

A giant planet, with its smaller mass, does not generate central temperatures that are high enough to result in thermonuclear reactions but is otherwise much like a collapsing protostar. The gaseous planet continues to contract indefinitely, releasing its gravitational potential energy as it shrinks. This is the primary energy source for replacing the internal energy that leaks out of the interior of Jupiter and is probably an important source for the other giant planets as well.

In Saturn's case, and perhaps Jupiter's as well, there is an additional source of internal energy. In Excursions 9.3 we will see that under the right conditions, liquid helium separates from a hydrogen–helium mixture and "rains" downward toward the core. As the droplets of liquid helium sink, they release their gravitational potential energy as thermal energy. Planetary physicists believe that most of Saturn's internal energy and perhaps some of Jupiter's come from this separation of liquid helium.

We can think of this internal energy that lies deep within the giant planets as being primordial. In other words, it is left over from their creation. The giant planets are still contracting and converting their gravitational potential energy into thermal energy today as they did when they first formed, but they are doing it more slowly (see **Connections 9.1**). The annual amount of contraction necessary to sustain their internal temperature is only a tiny fraction of their radius. For Jupiter this is only 1 mm or so—a hundred-billionth of its radius—per year. If this same rate were to continue for the next billion years, Jupiter would shrink by only a thousand kilometers, a little more than 1 percent of its radius.

In the popular literature Jupiter is frequently and erroneously labeled as "a star that failed"—probably because it is gaseous, has a high internal temperature, and releases a large amount of primordial energy. While these statements are true, to imply that it is "almost a star" is most certainly an exaggeration. Jupiter's central temperature is probably not much greater than 20,000 K, whereas the temperatures necessary to initiate the self-sustaining thermonuclear reactions that occur in stars are in the tens of millions of degrees. To achieve such temperatures Jupiter would require a total mass 80 times greater than it actually has. In other words, the least massive stars must still have masses that are approximately 80 times greater than Jupiter's. Jupiter is merely a large planet and never came close to being a star.

CONNECTIONS 9.1

Differentiation in the Gas Giants

In Chapter 7 we learned that denser materials, such as iron and other metals, sank to the centers of the terrestrial planets when they were in an earlier, more molten state. This process, which we call *differentiation*, deposited most of the metals in what became the cores of the terrestrial planets, leaving their mantles and crusts relatively depleted of metals.

The cores of the giant planets did not form in the same manner, however. Most of the material now in their cores was already there in the original bodies that captured hydrogen and helium from the Sun's protoplanetary disk

to form what ultimately became giant planets. Differentiation, however, has occurred and is still occurring in Saturn, and perhaps Jupiter too. In Excursions 9.3 we find that helium tends to condense out of the hydrogen–helium oceans on Saturn. Because these droplets of helium are denser than the hydrogen–helium liquid in which they condense, they sink toward the center of the planet. This tends to enrich helium concentration in the core while depleting it in the upper layers. At the same time the process heats the planet by converting gravitational potential energy to thermal energy.

9.7 The Interiors of the Giant Planets Are Hot and Dense

At depths of a few thousand kilometers, the atmospheric gases of Jupiter and Saturn are so compressed by the weight of the overlying atmosphere that they liquefy. This transition from a gas to a liquid is so subtle as to be hardly noticeable. To put it another way, the *physical* difference between a liquid and a highly compressed, very dense gas is something that could be appreciated only by—well—a physicist. Thus, unlike the well-defined surface between Earth's atmosphere and its oceans, on Jupiter and Saturn there is no clear boundary between the atmosphere and the "ocean" of

> Hydrogen–helium oceans lie at depths of a few thousand kilometers on Jupiter and Saturn.

liquid hydrogen and helium that lies below. The depths of these hydrogen–helium oceans are measured in tens of thousands of kilometers, making them the largest structures within the interiors of any of the giant planets. Uranus and Neptune, as we will see, do not have these oceans of liquid hydrogen and helium.

Figure 9.14 shows the interior structure of the giant planets. At the center of each of the giant planets is a dense, liquid core consisting of a very hot mixture of heavier materials such as water, rock, and metals. Here temperatures

may be in the tens of thousands of degrees, with pressures of tens of **megabars**. As we learned in Chapter 8, a pressure of 1 bar corresponds closely to Earth's atmospheric pressure at sea level. A megabar is a million times as great as 1 bar. For comparison, when the submersible research vessel *Alvin* cruises Earth's ocean bottoms 10,000 meters below the surface, it experiences a surrounding pressure of about 1,000 bars, or 1/1,000 of a megabar. It may seem strange to be talking about water that is still liquid at temperatures of tens of

> Jupiter's core is liquid water and rock at a temperature of about 20,000 K.

thousands of degrees, but there is really nothing peculiar about this. Like a super pressure cooker, the extremely high pressures at the centers of the giant planets prevent the water from turning to steam. The temperature at Jupiter's center is thought to be as high as 20,000 K, and the pressure may reach 70 megabars. Central temperatures and pressures of the other, less massive giant planets are correspondingly lower than those of Jupiter.

Uranus and Neptune Are Different from Jupiter and Saturn

You might suppose that the average densities of the giant planets would tell us how much heavy material they contain. In practice, it is not so simple. In each planet the core mass is roughly the same, perhaps 10 $M_{\oplus}$ for Jupiter and Sat-

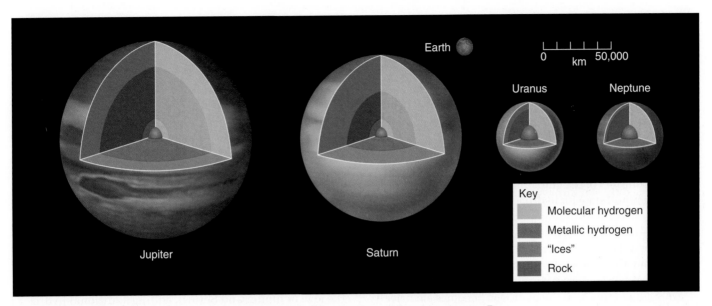

FIGURE 9.14 The central cores and outer liquid shells of the interiors of the giant planets. Even though each core has roughly the same mass (10 Earth masses), only Jupiter and Saturn have significant amounts of the molecular and metallic forms of liquid hydrogen surrounding their cores.

urn and less for Uranus and Neptune. As Table 9.1 shows, Jupiter and Saturn have total masses of 318 and 95 $M_\oplus$, respectively. The heavy materials in their 10 $M_\oplus$ cores, then, contribute very little to their average chemical composition. This means we can think of both Jupiter and Saturn as having approximately the same composition as the Sun and the rest of the universe: about 98 percent hydrogen and helium and only 2 percent of everything else. Why then, with nearly identical compositions, should Jupiter's density be nearly twice as great as Saturn's? In **Excursions 9.3** we find that the internal pressure created by Jupiter's much greater mass compresses its hydrogen and helium and its core to a higher average density than in the core of Saturn. What does this imply about Uranus and Neptune?

If Uranus and Neptune were also of solar composition—that is, if they were made primarily of hydrogen and helium—their average densities would be less than half the density of water, even less than that of Saturn. This is because their lower mass would not be as effective as Saturn's in compressing their hydrogen and helium. But such low average densities are not what we observe. Instead we find Uranus and Neptune to be about twice as dense as Saturn (see Table 9.1). We now have a clear indication that, unlike Jupiter and Saturn, denser materials must dominate the chemical composition of Uranus and Neptune. Is this denser material water or rock? Our observations should be able to tell us. Neptune, the densest of the giant planets, is about 1½ times as dense as uncompressed water and only about half as dense as uncompressed rock. This alone tells us there must be more water than rock. But we must also keep in mind that the high pressures within the interiors of Uranus and Neptune

cause both water and rock to be more dense than in their uncompressed states. Thus water and other low-density ices, such as ammonia and methane, must be the major compositional components of Uranus and Neptune, with lesser amounts of silicates and metals. The total amount of hydrogen and helium in these planets is probably limited to no

Uranus and Neptune are "ice giants" with deep, salty oceans.

more than 1 or 2 $M_\oplus$, with most of it residing in their relatively shallow atmospheres. Based on density alone, as we suggested at the beginning of this chapter, neither Uranus nor Neptune very well fit the description of a "gas giant." It is more appropriate to refer to them as "ice giants."

The water that makes up so much of Uranus and Neptune is probably in the form of deep oceans. Dissolved gases and salts would make these oceans electrically conducting. All of the giant planets have deep layers of a conducting fluid —metallic hydrogen in the case of Jupiter and Saturn, and a saltwater brine in the case of Uranus and Neptune. Currents in these conducting layers are likely the source of the giant planets' intense magnetic fields.

Differences Are Clues to Origins

That each of the giant planets formed around cores of similar mass but turned out so differently is an important clue to their origins. Why do Jupiter and Saturn have so much

EXCURSIONS 9.3

Strange Behavior in the Realm of Ultrahigh Pressure

In the interiors of the giant planets, the transition from a gas to a liquid state for hydrogen and helium is so gradual that there is no well-defined boundary between the two states—as there is, say, between Earth's atmosphere and oceans. When the interior pressure climbs to 4 megabars and the temperature reaches 10,000 K, hydrogen molecules are battered so violently that their electrons are stripped free, and the hydrogen becomes electrically conducting like a liquid metal. This happens at a depth of about 20,000 km in Jupiter's atmosphere and 30,000 km in Saturn's. Uranus and Neptune are less massive, have lower interior pressures, and contain a smaller fraction of hydrogen—conditions that do not favor the formation of metallic hydrogen. Thus their interiors probably contain only a small amount of liquid hydrogen, with little or none of it in a metallic state.

Helium can also be compressed to a liquid, but it does not reach a metallic state under the physical conditions existing in the interiors of the giant planets. At the temperatures found in Jupiter's interior, the liquid helium is mostly dissolved together with the liquid hydrogen. Within Saturn's interior, temperatures are lower, making the helium less soluble. Those who cook know that you can easily dissolve large amounts of sugar in hot water but relatively little when the water is cold. So it is with helium and hydrogen. Physicists believe that some of the helium in Saturn's interior is separating out into tiny droplets, like water from oil, and sinking slowly toward the planet's center. This could explain an observed depletion of helium relative to hydrogen in Saturn's upper atmosphere. And through the release of gravitational potential energy as the helium sinks, the separation process could provide an additional source of internal energy that would not be available in the interiors of other giant planets.

Jupiter's average density is nearly twice as great as Saturn's, even though both have approximately the same composition and are nearly the same size. This is because Jupiter's self-gravity due to its greater mass compresses its material more than Saturn's can. For planets the size of Jupiter and Saturn, adding additional hydrogen does not make them much bigger; it makes them denser instead. We now find ourselves with a curious situation that goes against our intuition: If we were to add much more hydrogen to Jupiter, its diameter would actually get *smaller* rather than larger. A planet with a mass, say, 10 times greater than Jupiter would be about 10 percent smaller. The additional overlying mass creates higher internal pressures, which in turn compress the interior further. The decreased volume caused by increased pressure more than makes up for the increased volume of the additional hydrogen. It turns out that, by chance, Jupiter is almost the largest that any planet can be in this or any other solar system. Planets that are either less massive or more massive will be smaller! Does this mean that stars should also be smaller than Jupiter? They would be except for one important difference. Stars have nuclear reactions going on within their cores and thus have *very* much higher internal temperatures than any planet. These ultrahigh temperatures create enormous internal pressures that better resist gravitational compression. This keeps even the smallest "normal" stars much larger than the largest planets. We say "normal" stars because later, in Chapters 16 and 17, we will introduce you to dying stars that are even smaller than Earth.

---

hydrogen and helium compared with Uranus and Neptune? Why is hydrogen-rich Jupiter so much more massive than hydrogen-rich Saturn? The answers may lie both in the time that it took for these planets to form and in the distribution of material from which they formed. We think that all of the hydrogen and helium in the giant planets was captured from the protoplanetary disk by the strong gravitational at- traction of their massive cores.[2] The much lower hydrogen–helium content of Uranus and Neptune suggests that these cores formed much later than those of Jupiter and Saturn,

---

[2] Although many planetary scientists believe the giant planets formed by core accretion, others have proposed a disk instability process, as we learned in Chapter 6.

at a time when most of the gas in the protoplanetary disk had been blown away by the emerging Sun. Why should the cores of Uranus and Neptune have formed so late? Probably because the icy planetesimals from which they formed were more widely dispersed at their greater distances from the Sun. With more space between planetesimals, it would take longer to build up their cores. Saturn may have captured less gas than Jupiter both because its core formed somewhat later and because less gas was available at its greater distance from the Sun.

9.8 The Giant Planets Are Magnetic Powerhouses

All of the giant planets have magnetic fields that are much stronger than Earth's. Planetary magnetic fields are produced by the motions of electrically conducting liquids deep within planetary interiors. In Jupiter and Saturn magnetic fields are generated in deeply buried layers of metallic hydrogen. In Uranus and Neptune magnetic fields arise in salt brine oceans. Although their origins are complex, we can schematically illustrate the geometry of the magnetic fields of the giant planets as if they came from bar magnets in the interiors of the planets, as shown in **Figure 9.15**.

Jupiter's magnetic field is inclined 10° to its rotation axis, an orientation similar to Earth's, but its axis is displaced about a tenth of a radius from the planet's center. Note that the direction of Jupiter's magnetic field is opposite to that of Earth as defined by where the north end of a compass would point. The total strength of Jupiter's magnetic field is nearly 20,000 times that of Earth's. On the other hand, Jupiter is huge compared to Earth. By the time Jupiter's magnetic field emerges from the cloud tops, the field has dropped

> Jupiter's magnetic field is 20,000 times as strong as Earth's.

to about 4.3 gauss—only 15 times Earth's surface field.

The bar magnet used to simulate Saturn's magnetic field is located almost precisely at the center of Saturn and is almost perfectly aligned with the planet's rotation axis. Saturn's magnetic field is much weaker than Jupiter's, but overall it is still more than 500 times stronger than Earth's. Because Saturn's diameter is much greater than Earth's, the magnetic field strength at its cloud tops is about 0.5 gauss —similar to the strength at Earth's surface. As on Jupiter, a compass would point south on Saturn.

Voyager 2 found that the magnetic field of Uranus is inclined nearly 60° to its rotation axis, and its center is displaced by a third of a radius from its center (see Figure 9.15). Considering the strange spin orientation of Uranus, this did

not come as any great surprise to the *Voyager* scientists. The total strength of the field averages 50 times Earth's field, but the large displacement of Uranus's field from the planet's center causes the field at the cloud tops to vary between 0.1 and 1.1 gauss.

The really big surprise came when *Voyager 2* reached Neptune. The orientation of Neptune's rotation axis is similar to that of Earth, Mars, and Saturn. But Neptune's magnetic field is inclined 47° to its rotation axis, and the center of *this* magnetic field is displaced from the planet's center by more than half the radius—an offset even greater than that of Uranus (see Figure 9.15). The overall strength of Neptune's magnetic field is only half as great as that of Uranus.

The reason for the unusual geometry of the magnetic fields of Uranus and Neptune remains unknown, but it is clearly not related to the orientations of their rotation axes. The displacement of the field is primarily toward Neptune's southern hemisphere, thereby creating a large asymmetry in the field strength at the cloud tops, with 1.2 gauss in the southern hemisphere and only 0.06 gauss in the north.

Giant Planets Have Giant Magnetospheres

Just as Earth's magnetic field traps energetic charged particles to form Earth's magnetosphere, the magnetic fields of the giant planets also trap energetic particles to form mag-

> The magnetospheres of the giant planets are enormous.

netospheres of their own. Although Earth's magnetosphere may have a radius over 10 times that of our planet itself, our magnetosphere is tiny in comparison with the vast clouds of plasma held together by the much more powerful magnetic fields of the giant planets. By far the most colossal of these is Jupiter's magnetosphere. Its radius is as much as 100 times that of the planet itself, or around 7 million kilometers. That is roughly 10 times the radius of the Sun! Although the magnetospheres of the other giant planets are much smaller, even the relatively weak magnetic fields of Uranus and Neptune form magnetospheres that are comparable in size to the Sun.

The solar wind does more than supply some of the particles for a magnetosphere. The pressure of the solar wind also pushes on and compresses a magnetosphere. The size and shape of a planet's magnetosphere can change a great deal depending on how the solar wind is blowing at any particular time. Planetary magnetic fields also divert the solar wind, which flows around magnetospheres the way a stream flows around boulders. Just as a rock in a river creates a wake that extends downstream, the magnetosphere

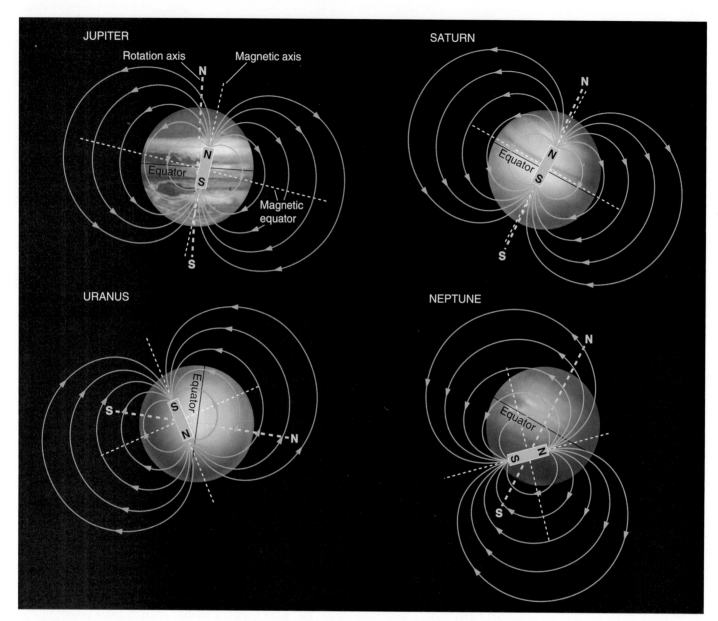

FIGURE 9.15 The magnetic fields of the giant planets can be approximated by the fields from bar magnets offset and tilted with respect to the planets' axes. Compare these with Earth's magnetic field, shown in Figure 7.13(a).

of a planet produces a wake that can extend downstream for great distances. The wake of Jupiter's magnetosphere (**Figure 9.16**) extends over 6 AU outward from the planet, well past the orbit of Saturn. Jupiter's magnetosphere is the largest permanent "object" in the Solar System, surpassed in size only by the tail of an occasional comet. The magnetic wakes of Uranus and Neptune have a curious structure. Because of the tilt and the large displacements of their magnetic fields from the centers of these planets, their magnetospheres wobble as the planets rotate. This causes the wakes of their magnetospheres to twist like corkscrews as they stretch away from the planets.

Magnetospheres Produce Synchrotron Radiation

We need not send spacecraft to the outer Solar System, or even call on telescopes orbiting Earth, to see evidence of the giant planets' magnetospheres. Rapidly moving electrons in planetary magnetospheres spiral around the direction of the magnetic field, and as they do so they emit synchrotron radiation, as discussed in **Foundations 9.1**. If our eyes were sensitive to radio waves, then the second brightest object in the sky would be Jupiter's magnetosphere. The Sun would

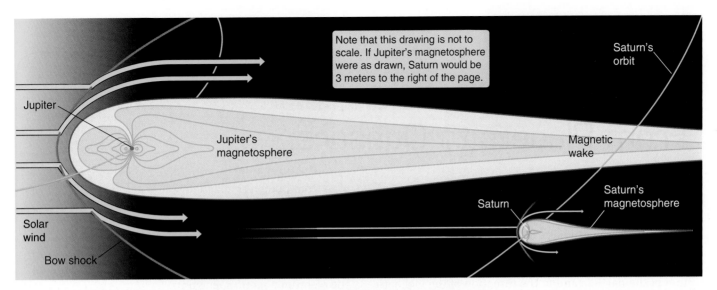

FIGURE 9.16 The solar wind compresses Jupiter's (or any) magnetosphere in the sunward direction and draws it out into a magnetic wake in the direction away from the Sun. Jupiter's wake stretches beyond the orbit of Saturn.

still be brighter, but it would not appear larger: Even at a distance from Earth of 4.2 to 6.2 AU, Jupiter's magnetosphere would still appear roughly twice the size of the Sun in the sky. Saturn's magnetosphere would also be large enough to see, but it would be much fainter than Jupiter's. Even though Saturn has a strong magnetic field, pieces of rock, ice, and dust in Saturn's spectacular rings act like sponges, soaking up magnetospheric particles. Magnetospheric particles typically collide with ring material soon after those particles enter the magnetosphere. With far fewer magnetospheric electrons, there is much less radio emission from Saturn.

We can learn a great deal about planets by studying the synchrotron emission from their magnetospheres. For example, precise measurement of periodic variations in the radio signals "broadcast" by the giant planets tells us the

Radio signals broadcast a planet's true rotation period.

planets' true rotation periods. This is because the magnetic field of each planet is locked to the conducting liquid layers deep within the planet's interior, so the magnetic field rotates with exactly the same period as the deep interior of the planet. Given the fast and highly variable winds that push around clouds in the atmospheres of the giant planets, measurement of radio emission is the only way we have to determine the true rotation periods of the giant planets.

Radiation Belts and Auroras

As a planet rotates on its axis, it drags its magnetosphere around with it. In the enormous magnetosphere of a rap-idly rotating planet like Jupiter, charged particles are swept around at high speeds. These fast-moving charged particles slam into neutral atoms (which do not share the motion of material in the magnetosphere), and the energy released in the resulting high-speed collisions heats the plasma to extreme temperatures. In 1979, while passing through Jupiter's magnetosphere, *Voyager 1* encountered a region of tenuous plasma with a temperature of over 300 million kelvins: 20 times the temperature at the center of the Sun! The density of the plasma (around 10,000 atoms/m^3) was much lower than that of the best vacuum we can produce on Earth, however, so the spacecraft was in no danger. The high temperatures are impressive nonetheless.

Charged particles trapped in planetary magnetospheres are concentrated in certain regions called **radiation belts**. Although Earth's radiation belts are enough for astronauts to worry about, the radiation belts that surround Jupiter are searing in comparison. In 1973 the *Pioneer 11* spacecraft passed through the radiation belts of Jupiter. During its brief encounter, *Pioneer 11* picked up a radiation dose of 400,000

Jupiter has intense radiation belts.

rads, or around 1,000 times the lethal dose for humans. Several of the instruments on board were permanently damaged as a result, and the spacecraft itself barely survived to continue on its journey to Saturn.

In addition to protons and electrons from the solar wind, the magnetospheres of the giant planets also contain large amounts of other elements including sodium, sulfur, oxygen, nitrogen, and carbon. These elements come from several sources, including the planets' extended atmospheres and the moons that orbit within them. The most intense radiation

FOUNDATIONS 9.1

Synchrotron Emission— From Planets to Quasars

In Chapter 4 we found that any time a charged particle experiences an acceleration, the particle emits electromagnetic radiation. The accelerations resulting from forces exerted by magnetic fields are no exception. As we have seen, charged particles moving in a magnetic field experience a force that is perpendicular both to the motion of the particle and to the direction of the magnetic field. This force produces an acceleration, causing the particles to follow helical paths around the direction of the magnetic field. The accelerations also cause the particles to radiate.

If the particles are traveling at a significant fraction of the speed of light, then relativistic effects cause the radiation they emit to be beamed in the direction in which they are traveling. The situation is illustrated in **Figure 9.17**. The resulting radiation is called **synchrotron radiation**, so named because it was first discovered in a type of particle accelerator called a *synchrotron*.

The amount of radiation from a particle depends on the amount of acceleration the particle experiences. For a given amount of force, a less massive particle experiences more acceleration. Because electrons are much less massive than protons or any **ion**, it is the electrons in a magnetized plasma that experience the greatest acceleration. Combining these ideas, we see that it must be the electrons in a magnetized plasma that produce the overwhelming majority of its synchrotron radiation.

The magnetospheres of the giant planets contain energetic electrons moving in strong magnetic fields, so they satisfy the requirements for synchrotron radiation. Synchrotron radiation is unlike thermal (that is, Planck) radiation because instead of being strongly peaked in one part of the electromagnetic spectrum, synchrotron radiation from a single source can range from radio waves to X-rays. The spectrum of synchrotron radiation is determined by the strength of the magnetic field and how energetic the radiating particles are. Synchrotron emis-

sion from planetary magnetospheres is concentrated in the low-energy radio part of the spectrum. This is our first encounter with synchrotron radiation, but it will not be our last. As we move outward into the universe we will find many objects, from quasars to the remnants of supernovae, that emit synchrotron radiation throughout the electromagnetic spectrum.

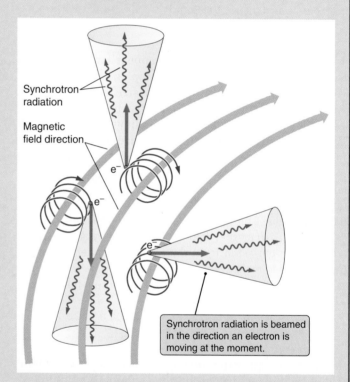

Synchrotron radiation

Magnetic field direction

Synchrotron radiation is beamed in the direction an electron is moving at the moment.

FIGURE 9.17 Rapidly moving charged particles loop around the direction of a magnetic field, giving off electromagnetic radiation as a result of the acceleration they experience. The radiation, called synchrotron radiation, is beamed by relativistic effects in the direction of particle motion.

belt in the Solar System is a toroidal (that is, torus- or doughnut-shaped) ring of plasma associated with Io, the innermost of Jupiter's four Galilean moons. Io is the most volcanically active body in the Solar System. Because of its low surface

gravity and the violence of its volcanism, some of the gases erupting from its interior can escape the moon and become part of Jupiter's radiation belt. As charged particles are slammed into the moon by the rotation of Jupiter's magneto-

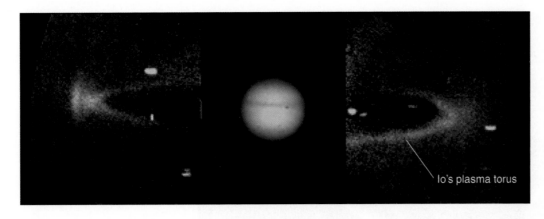

FIGURE 9.18 Atoms from Io, the innermost of Jupiter's Galilean moons, fill a faintly glowing torus of plasma that surrounds Jupiter (center).

sphere, even more material is knocked free of the surface and ejected into space. If the dislodged atoms are electrically neutral, they will continue to orbit the planet in nearly the

Io is a source of magnetospheric particles.

same orbit as the moon from which they escaped. Images of the region around Jupiter, taken in the light of emission lines from atoms of sulfur or sodium, show a faintly glowing torus of plasma supplied by the moon (**Figure 9.18**).

Other moons also influence the magnetospheres of the planets they orbit. The atmosphere of Saturn's largest moon,

Titan, is rich in nitrogen. Leakage of this gas into space is the major source of a plasma torus that forms in Titan's wake. The density of this rather remote radiation belt is highly variable because the orbit of Titan is sometimes within and sometimes outside Saturn's magnetosphere, depending on the strength of the solar wind. When Titan is outside Saturn's magnetosphere, any nitrogen molecules lost from the moon's atmosphere are carried away by the solar wind.

Charged particles spiral along the magnetic field lines of the giant planets, bouncing back and forth between the two magnetic poles, just as they do around Earth. As with Earth, these energetic particles collide with atoms and molecules

FIGURE 9.19 Hubble Space Telescope images of auroral rings around the poles of (a) Jupiter and (b) Saturn. The auroral images were taken in ultraviolet light and superposed on visible light images. (High-level haze obscures the ultraviolet views of the underlying cloud layers.) The bright spot and trail outside the main ring of Jupiter's auroras are the footprint and wake of Io's flux tube.

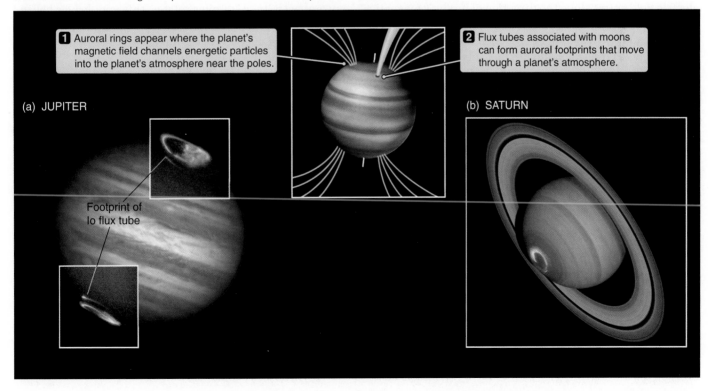

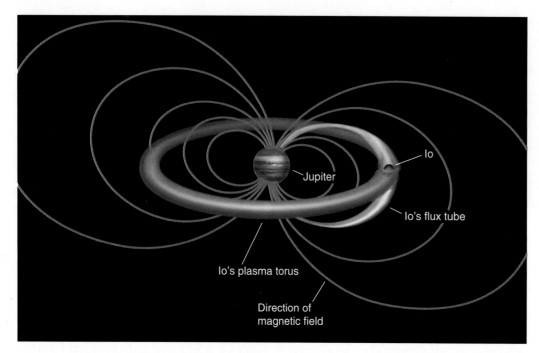

FIGURE 9.20 A diagram showing the geometry of Io's plasma torus and flux tube.

in a planet's atmosphere, knocking them into excited energy states that decay and emit radiation. The results are bright auroral rings (**Figure 9.19**) that surround the poles of the giant planets—just as the *aurora borealis* and *aurora australis* ring the north and south magnetic poles of Earth.

Jupiter's auroras have an added twist that we do not see on Earth, however. As Jupiter's magnetic field sweeps past Io, it behaves like a dynamo, generating an electric potential of 400,000 volts. Electrons accelerated by this enormous electric field spiral along the direction of Jupiter's magnetic field. The result is a magnetic channel, called a **flux tube**, that connects Io with Jupiter's atmosphere near the planet's magnetic poles (**Figure 9.20**). Io's flux tube carries an elec-

tric current of 5 million amperes, amounting to 2 trillion watts of power or about ⅙ of the total power produced by generating stations on Earth. Much of the power generated within the flux tube is radiated away as radio energy. These radio signals are received at Earth as intense bursts of synchrotron radiation. However, a substantial fraction of the energy of particles in the flux tube is deposited into Jupiter's atmosphere. At the very location where Io's flux tube intercepts Jupiter's atmosphere, there is a spot of intense auroral activity. As Jupiter rotates, this spot leaves behind an auroral trail in Jupiter's atmosphere. The footprint of Io's plasma torus, along with its wake, can be seen outside the main auroral ring in Figure 9.19(a).

Summary

- The giant planets are large and massive and less dense than the terrestrial planets.

- Jupiter and Saturn are made up mostly of hydrogen and helium, similar to the Sun.

- Uranus and Neptune contain much larger amounts of "ices" such as water, ammonia, and methane.

- We see only atmospheres, not solid surfaces, on the giant planets.

- Clouds on Jupiter and Saturn are composed of various kinds of ice crystals colored by impurities.

- Powerful convection and the Coriolis effect drive high-speed winds in the upper atmospheres of the giant planets.

- Jupiter, Saturn, and Neptune have large internal heat sources. Uranus seems to have little or none.

- The interiors of the giant planets are very hot and very dense.

- The giant planets have enormous magnetospheres that emit synchrotron radiation.

Seeing the Forest through the Trees

In our discussion of the terrestrial planets a major theme was diversity. We used Earth as the basis for understanding our planetary neighbors, and at the same time we learned from those neighbors what it is about Earth that sets it apart. On this leg of our journey we have come to understand that even the range of conditions from the frozen poles of Mars to the hellish surface of Venus is modest in comparison with what is possible. As the Solar System formed, planets in the cold outer reaches grew more massive than the planets closer to the Sun. They became massive enough to capture and hold onto hydrogen and helium gas that was so abundant in the disk surrounding the young Sun.

As we explored the giant planets, we discovered worlds with scales beyond human experience. In place of Earth's gentle trade winds, we found vast atmospheric bands streaming around the giant planets at up to 1,600 km/h. The most powerful terrestrial hurricanes would be but inconsequential eddies within Jupiter's Great Red Spot, a system so enormous that it could swallow two Earths whole. On Earth we marvel at the sight of a terrestrial thunderstorm towering overhead, lit up by powerful flashes of lightening. Can we begin to imagine the appearance of multihued clouds of water, ammonia, methane, and other compounds, tinted by sulfur, phosphorus, and photochemical smog, as they roil up from within the depths of a giant planet? Can we envision the bolts of lightning that sear through those clouds?

Of course, these analogs of our terrestrial experience only scratch the surface of the differences between the giant planets and Earth. In the case of the terrestrial planets the distinction between planet and atmosphere is clear-cut. On our Earth, clouds billow in a blue sky, adrift over the palpable surface of land and ocean. As we descend into the giant planets, there is no sudden transition to solid or liquid. Instead there is only the steady and inexorable increase in pressure, a smooth transition from the tops of the clouds to a core that in the case of Jupiter may have an absolute temperature almost three times that at the surface of the Sun, and a crushing pressure 70 million times that at the surface of Earth. The higher-than-expected temperatures of the giant planets that we found in Chapter 4 are signposts of these internal differences. Whereas Earth's climate is defined by the interplay between energy from the Sun and the physics and chemistry of our atmosphere and oceans, the fierce weather systems on the giant planets are powered largely by conversion of gravitational energy to thermal energy within their interiors.

Although the bulk of the giant planets dwarfs that of their terrestrial cousins, their influence does not end here. Powerful magnetic fields trap energetic charged particles streaming outward from the Sun, leading to the formation of giant magnetospheres—the largest permanent structures in the Solar System.

The time has come for a brief digression. The systems of moons and rings that surround the giant planets are gravitational playgrounds, rich with phenomena that go far beyond Kepler's laws. Before we take the next step outward, we will pause and return our attention to our old friend, gravity.

Key Terms

giant planet, p. 253
gas giant, p. 254
ice giant, p. 254
surface brightness, p. 255
stellar occultation, p. 256
solar abundance, p. 258
massive element, p. 258
inert gas, p. 258
oblateness, p. 260
Great Red Spot (GRS), p. 260
vortex, p. 260
photochemical, p. 263
megabar, p. 270
radiation belt, p. 275
synchrotron radiation, p. 276
ion, p. 276
flux tube, p. 278

Student Questions

THINKING ABOUT THE CONCEPTS

1. In what manner was Uranus observed to "stray from its path," and how was this used as a clue by Adams and Le Verrier to predict the location of Neptune?

2. Describe the ways in which the giant planets differ from the terrestrial planets.

3. Identify and describe the two subclasses of giant planets, and indicate which planets fall into each subclass.

4. Jupiter's chemical composition is more like the Sun's than Earth's. Yet both planets formed from the same protoplanetary disk. Explain the difference.

5. Astronomers take the unusual position of lumping together all atomic elements other than hydrogen and helium into a single category, which they call "massive elements" or "heavy elements." Why is this a reasonable thing for astronomers to do?

6. Why do the individual cloud layers have different chemical compositions in the atmospheres of the giant planets?

7. Uranus and Neptune, when viewed through a telescope, are distinctly bluish green in color. What are the two reasons for their striking appearance?

8. Jupiter, Saturn, and Neptune all radiate more energy into space than they receive from the Sun. Does this violate the law of conservation of energy? What is the source of the additional energy?

9. The Great Red Spot (GRS) is a long-lasting atmospheric vortex in Jupiter's southern hemisphere. Winds rotate counterclockwise around its center. Is the GRS cyclonic or anticyclonic? Is it a region of high or low pressure? Explain.

10. Jupiter's core is thought to consist of rocky material and ices, all in a liquid state at a temperature of 20,000 K. How can materials such as water and methane be liquid at such high temperatures?

11. When viewed by radio telescopes, Jupiter is the second brightest object in the sky. What is the source of this radiation?

APPLYING THE CONCEPTS

12. The Sun appears 400,000 times brighter than the full Moon in our sky. How far from the Sun (in AU) would you have to go for the Sun to appear only as bright as the full Moon appears in our nighttime sky? Compare your answer with the radius of Neptune's orbit.

13. Uranus occults a star at a time when the relative motion between Uranus and Earth is 23.0 km/s. An observer on Earth sees the star disappear for 37 minutes and 2 seconds and notes that the center of Uranus passed directly in front of the star.
    a. Based on these observations, what value would the observer calculate for the diameter of Uranus?
    b. What could you conclude about the planet's diameter if its center did not pass directly in front of the star?

14. Jupiter is an oblate planet with an average radius of 69,900 km, compared to Earth's average radius of 6,370 km.
    a. Remembering that volume is proportional to the cube of the radius, how many Earths could fit inside Jupiter?
    b. Jupiter is 318 times as massive as Earth. Show that Jupiter's average density is about $\frac{1}{4}$ that of Earth's.

15. Jupiter's equatorial radius is 71,500 km, and its oblateness is 0.065. What is Jupiter's polar radius?

16. The equilibrium temperature for Saturn should be 82 K, but instead we find an average temperature of 95 K. How much more energy is Saturn radiating into space than it absorbs from the Sun?

17. A small cloud in Jupiter's equatorial region is observed to be at 122.0° west longitude in a coordinate system that rotates at the same rate as the deep interior of the planet. (West longitude is measured along a planet's equator toward the west.) Another observation made exactly 10 Earth hours later finds the cloud at 118.0° west longitude. Jupiter's equatorial radius is 71,500 km. What is the observed equatorial wind speed in km/h? Is this an easterly or westerly wind?

18. Ammonium hydrosulfide (NH_4HS) is a molecule in Jupiter's atmosphere responsible for many of its clouds. Using Figure 9.5, calculate the molecular weight of an ammonium hydrosulfide molecule, where the atomic weight of a hydrogen atom is 1.

StudySpace
wwnorton.com/astro21
provides a Study Plan for each chapter that includes a reading outline, animations, keyword flash cards, and gradebook-enabled multiple-choice quizzes. From StudySpace you can also access premium content in the ebook and SmartWork.

Then there are the Tides, so useful to man...
We must be grateful for the Moon's existence
on that account alone.

JAMES NASMYTH (1808–1890)

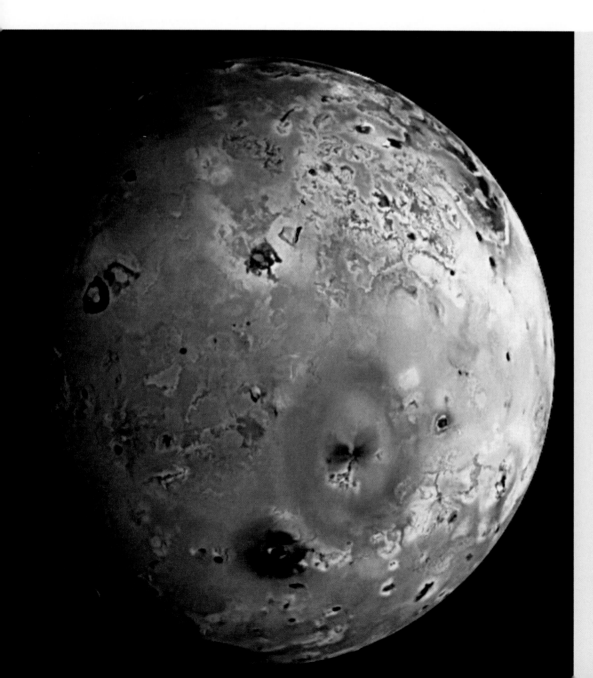

Jupiter's moon, Io, showing
reddish fallout from an
enormous erupting volcano.

CHAPTER 10

Gravity Is More Than Kepler's Laws

10.1 Gravity Once Again

Long before Kepler wrote down the laws describing the motions of the planets about the Sun, or Newton explained and interpreted these motions in terms of the effects of gravity, our ancestors knew that the Moon and the Sun have a direct influence on Earth. Those living near the ocean were especially attuned to these effects. Twice each day—in harmony with the changing position of the Moon in the sky —they saw the seawater rise and then recede once again. They would have noted that this effect, Earth's **tides**, is greatest when the Sun and the Moon are either together in the sky or at opposite extremes (that is, during a full Moon or a new Moon) and is more subdued when the Sun and Moon are separated in the sky by 90° (first quarter Moon or third quarter Moon). Doubtless the unarguable association between the Moon and tides had a great deal to do with the development of superstitions about the power of the Moon over our lives.

Today we understand that tides are the result of the gravitational pull of the Moon and the Sun on Earth. More to the point, tides are the result of *differences* between how hard the Moon and Sun pull on one part of our planet in comparison with their pull on other parts of the planet. So far in our discussion of gravity we have focused on the gravitational interaction between two whole objects. Such interactions explain the motion of the planets about the Sun, the Moon about Earth, and a space shuttle about our globe. Yet gravity is far more than Newton's derivation of Kepler's laws. While concentrating on the elliptical, parabolic, and hyperbolic orbits of one mass about another, we have overlooked many other fascinating and important

KEY CONCEPTS

Newton's law of universal gravitation is simplicity itself. Yet when applied to extended rather than pointlike masses or when acting among three or more objects, this simple rule gives rise to a surprising diversity of phenomena. In this chapter we look beyond Kepler's elegant laws of planetary motion and discover

- Symmetries that allow us to say a great deal about the gravity of an object without actually calculating anything.
- That the gravity within a spherical object is determined only by the mass within a given radius.
- Tides on Earth resulting from the fact that gravity from the Moon and Sun pulls harder on one side of Earth than the other.
- Tidal interactions between planets and moons (including Earth's Moon) that lock a moon's rotation to its orbit.
- Comets that have been shattered by tides, and tortured moons alive with tide-powered volcanism.
- Orbital resonances that nudge asteroids from their orbits and sweep out gaps in planetary rings.
- Chaotic, unpredictable orbits in which the tiniest difference at one moment in time leads to huge differences later on.

manifestations of gravity. We now turn our attention to some of these effects.

10.2 Gravity Differs from Place to Place within an Object

A good place to begin a discussion of the additional effects of gravity is with the way an object interacts gravitationally with *itself.* You can think of Earth, for example, as a collection of small masses, each of which feels a gravitational

---

Self-gravity holds Earth together.

---

attraction toward every other small part of Earth. This gravitational attraction between different parts of Earth is what holds our planet together.

In a certain sense you are one of these small pieces of Earth. You are perhaps slightly more mobile than the average lump of clay; but as far as the gravitational makeup of our planet goes, you serve more or less the same purpose. As you sit reading this book, you are exerting a gravitational attraction on every other fragment of Earth, and every other fragment of Earth is exerting a gravitational attraction on you (see **Figure 10.1(a)**). Your gravitational interaction is strongest with the parts of Earth closest to you. The parts of Earth that are on the other side of our planet are much farther from you, so their pull on you is correspondingly less.

As you know from experience, the net effect of all of these forces is to pull you toward the center of Earth. If you drop a hammer, it falls directly down toward the ground. We can understand why this is so just by thinking about the shape of Earth. Because Earth is nearly spherical, for every piece of Earth pulling you toward your right, a corresponding piece of Earth is pulling you toward your left with just as much force. For every piece of Earth pulling you forward, a corresponding piece of Earth is pulling you backward. Because Earth is nearly **spherically symmetric**, all of these "sideways" forces cancel out, leaving behind an overall force that points toward Earth's center.

Understanding the size of the force is a bit trickier. Some parts of Earth are closer to you while others are farther away, but there must be some "average" or "characteristic" distance between you and each of the small fragments of Earth that is pulling on you. Not surprisingly, this average distance turns out to be the distance between you and the center of Earth. So as illustrated in **Figure 10.1(b)**, *the overall pull that you feel is the same as if all the mass of Earth were concentrated at a single point located at the very center of the planet.* This is true for any spherically symmetric

---

Gravity from a sphere acts as though all of its mass is concentrated at the sphere's center.

---

object. As far as the rest of the universe is concerned, the gravity from such an object behaves as if all the mass of that object were concentrated at a point at its center. We have already made extensive use of this result. It was an implicit assumption in our discussion in Chapter 3 of orbits and Kepler's laws, for example, where we said that the force act-

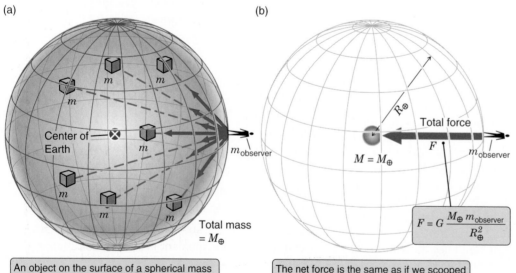

(a)

Center of Earth

m

m

m

m

m

m

m

$m_{observer}$

Total mass $= M_\oplus$

An object on the surface of a spherical mass (such as Earth) feels a gravitational attraction toward each small part of the sphere.

(b)

$R_\oplus$

Total force

F

$M = M_\oplus$

$m_{observer}$

$$F = G \frac{M_\oplus \, m_{observer}}{R_\oplus^2}$$

The net force is the same as if we scooped up the mass of the entire sphere and concentrated it at a point at the center.

FIGURE 10.1 The net gravitational force due to a spherical mass is the same as the gravitational force from the same mass concentrated at a point at the center of the sphere.

(a)

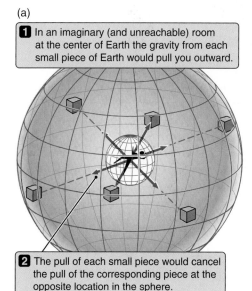

1 In an imaginary (and unreachable) room at the center of Earth the gravity from each small piece of Earth would pull you outward.

2 The pull of each small piece would cancel the pull of the corresponding piece at the opposite location in the sphere.

(b)

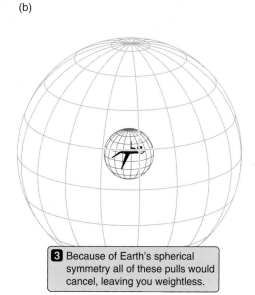

3 Because of Earth's spherical symmetry all of these pulls would cancel, leaving you weightless.

FIGURE 10.2 In an imaginary room at the center of Earth, the gravitational pull from each small piece of Earth would cancel, leaving you weightless.

ing on a space shuttle in orbit about Earth is equal to $GM_\oplus m_{\text{shuttle}}/r^2$, where $M_\oplus$ is the mass of Earth and r is the distance between the center of Earth and the space shuttle.

Gravity Increases Outward from an Object's Center

We now have a way of accounting for the net gravitational attraction of Earth on an object on or above the surface of Earth, but what about the effect of gravity *within* Earth? What is the overall gravitational attraction felt by an object buried deep inside our planet? A good place to begin is by thinking about the gravitational attraction felt by an object at the very center of Earth. Imagine that you were in a room at the center of Earth, as shown in **Figure 10.2**. (Ignore for a moment the crushing pressure or immense temperatures that would surround you!) Every small part of Earth would be exerting a slight gravitational force on you, each of which would be *outward*, away from the center of Earth. Again, Earth's spherical symmetry tells us that overall these gravitational forces must sum to zero. For each part of Earth pulling you outward in one direction, a corresponding piece of Earth on the other side would pull you outward in exactly the opposite direction, as shown in Figure 10.2(a).

Net gravity is zero at the center of Earth.

These two forces would cancel each other so that the net force you would feel from these two small parts of the planet would be zero (Figure 10.2(b)). The same holds true for *each and every* small part of Earth. The net effect of the sum of

all of the gravitational forces acting on an object at the center of Earth is zero! If you were in a room at the center of Earth, you would hang there, surrounded by the enormous mass of a planet but truly weightless. (See **Foundations 10.1** for more discussion of symmetry.)

So much for a room at the center of Earth. Now imagine that your room is located partway out from the center of the planet. To tackle this question we need to think of Earth as being made up of two different pieces, as shown in **Figure 10.3**. The first piece is just a spherical ball containing all the parts of Earth that are closer to the center of Earth than your imaginary room is. You could think of this inner ball as a "planet within a planet." Your room is sitting on the surface of this imaginary sphere. From the discussion in the preceding paragraphs, we know the net effect of the gravity of this inner sphere. It is equivalent to taking all the mass contained within that imaginary ball and placing it at the center of Earth. The force that you would feel from this inner planet is

$$F = \frac{G \times M_{\text{inner}} \times m}{r^2},$$

where M_{inner} is the mass of this inner ball (which depends on where within the planet your room is located), m is your mass, and r is the distance from the center of Earth to your room.

So far so good; but what about the gravitational attraction that you would feel from the *rest* of Earth—the shell of material that surrounds this "inner planet"? The parts of that shell that are closest to you, and so individually pull on you most strongly, are above you. The force from this part of the planet would pull you away from the center of Earth. However, *most* of the mass in the shell is on the side where its gravitational attraction pulls you *toward* the center of Earth. Although this material is farther away from you—which means that each small part of it pulls on you less strongly—there is a good bit more of it than the material immediately over your head. If we were to work it out in detail, we would find that the gravitational attraction from the part of the shell pulling you away from the center of Earth is exactly canceled by the gravitational attraction of the part of the shell that is pulling you toward the center of Earth. The net effect—the overall force—is zero. If you were inside a huge spherical shell, the overall gravitational attraction from the material in that shell would be zero. (This is just a more general case of the "room at the center of Earth" problem.)

Let's put this all together. Regardless of where we are within Earth, we feel a net gravitational force only from the part of the planet that lies deeper within the planet than we are, and that mass acts as if it were all concentrated at the center of Earth. If we want to know what the net gravitational attraction is at any point within a spherical object, all we need to do is determine how much mass is closer to

> At any point, the net gravitational force comes only from the mass closer to a sphere's center.

the center of the sphere than our point of interest. This mass then acts as if it were concentrated at the center of the sphere. This is a general and extremely important result. We have already seen something of this effect in our discussion of the interiors of planets. This result will become even more important as we consider the forces responsible for the structure of stars.

Tides Are Due to Differences in Gravity from External Objects

We have seen that each small part of an object feels a gravitational attraction toward every other small part of the ob-

---

FOUNDATIONS 10.1

The Power of Symmetry

In our discussion of gravity we have managed to arrive at some very interesting results based on nothing but the way one part of an object matches up with another—a property called **symmetry**. We were able to say that the overall gravitational attraction of all parts of Earth must point along a line connecting us with the center of Earth, simply by noting that each sideways pull from a different part of Earth is balanced by a corresponding pull in the other direction. We could make this claim because the distribution of mass to our right as we stand on Earth is just the same as the distribution of mass to our left, which is just the same as the distribution of mass to our front or back. That is to say, Earth is *symmetric.* Similarly, we were able to argue that the net effect of gravity on an object at the center of Earth must be zero because every small force from one part of Earth is balanced by a corresponding force from the corresponding part of Earth on the opposite side of us. Again, no calculation was needed—only an appeal to the symmetry of the distribution of mass within Earth.

If you look around, you will find symmetry everywhere; and when you find it, you will often find a clue to why something works the way it does. The tire on a bicycle rolls because it has *circular symmetry,* meaning that its shape is the same regardless of how it rotates about the axis running through its center. Your image in a mirror is flipped right for left and left for right from your actual appearance—an example of *reflection symmetry.* You hardly notice the difference, however, because your body itself is nearly symmetric between its right and left sides. A child's cubical blocks can be stacked any of six ways because when you rotate them by 90° they still look the same—a special type of *rotational symmetry.* A soccer ball can be kicked in any direction at any time, and will roll any which way—both results of its *spherical symmetry.* An American football makes for a lousy game of soccer but a fantastic spiraling forward pass because of its rotational symmetry about a single axis.

Science progresses by finding and making use of patterns that exist in the universe. Symmetry turns out to be an especially elegant and powerful type of pattern. Physicists and astronomers can often learn a great deal about the properties and behavior of an object or system without making a single calculation, simply by understanding the symmetry of the object or system that they are studying.

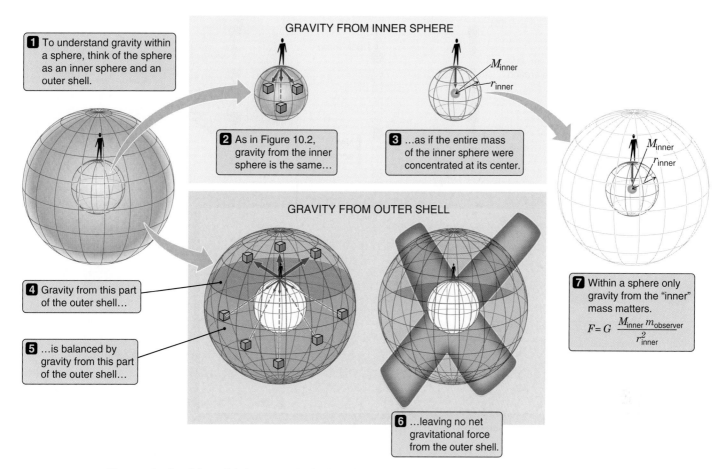

FIGURE 10.3 The gravitational force felt in a room in the interior of a spherical body such as Earth is equal to the force due to the mass closer to the center of the body than the room is.

ject, and that this **self-gravity** differs from place to place. It is also the case that each small part of an object feels a gravitational attraction toward every other mass in the universe, and that these *external* forces differ from place to place within the object as well.

The most notable local example of this involves the effect of the Moon's gravity on Earth. Overall, the Moon's gravity pulls on Earth as if the mass of Earth were concentrated at the planet's center, as we just saw. When astronomers calculate the orbits of Earth and the Moon assuming that the mass of each is concentrated at the center point of each body, they get the right answer. Yet there is more going on than this. The side of Earth that faces the Moon is

The Moon's gravity pulls harder on the side of Earth facing the Moon.

closer to the Moon than the rest of Earth, so it feels a stronger-than-average gravitational attraction toward the Moon. In contrast, the side of Earth facing away from the Moon is *farther* than average from the Moon, so it feels a weaker-than-average attraction toward the Moon. Putting in num-

bers, we find that the pull on the near side of Earth is about 7 percent larger than the pull on the far side of Earth.

Imagine holding three rocks at different altitudes high above the surface of the Moon, as shown in **Figure 10.4(a)**. If you let them go at the same time, they will all fall toward the Moon. However, rock 1, located closest to the Moon and subject to greater acceleration, will fall faster than rocks 2 and 3. Rock 3, located farthest from the Moon, will fall more slowly than either of the other two rocks. As the rocks fall toward the Moon, the separation between them grows. A person falling along with rock 2 would see both rocks 1 and 3 moving away from her. Now suppose the three rocks are connected by springs, as in **Figure 10.4(b)**. As the rocks fall toward the Moon, the differences in the gravitational forces they feel stretch the springs. To someone falling along with rock 2 it would seem as if there were forces pulling rocks 1 and 3 in opposite directions. Exactly the same thing happens with Earth. If we replace the three rocks with different parts of Earth, as shown in **Figure 10.4(c)**, we see that differences in the Moon's gravitational attraction on different parts of Earth try to stretch Earth out along a line pointing toward the Moon.

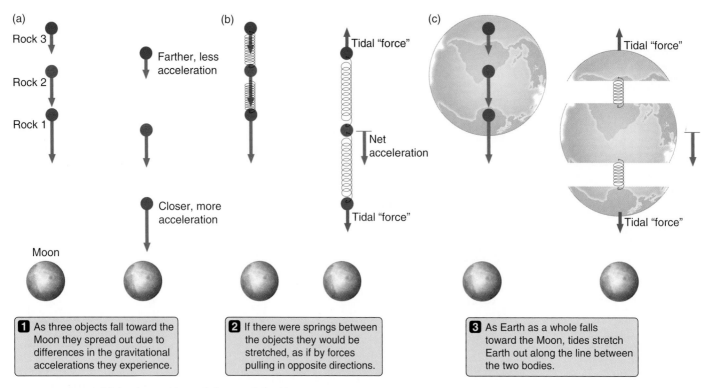

(a)

Rock 3

Rock 2

Rock 1

Moon

Farther, less
acceleration

Closer, more
acceleration

1 As three objects fall toward the
Moon they spread out due to
differences in the gravitational
accelerations they experience.

(b)

Tidal "force"

Net
acceleration

Tidal "force"

2 If there were springs between
the objects they would be
stretched, as if by forces
pulling in opposite directions.

(c)

Tidal "force"

Tidal "force"

Tidal "force"

Tidal "force"

3 As Earth as a whole falls
toward the Moon, tides stretch
Earth out along the line between
the two bodies.

FIGURE 10.4 (a) As three objects fall toward the Moon, they experience different gravitational accelerations. (b) If the three objects were connected by springs, the springs would be stretched as if there were forces pulling outward on each end of the chain. (c) This is the cause of Earth's tides.

We can also think about this situation by applying the idea of relative motion from our discussions of Newton's laws (Chapter 3) and special relativity (Chapter 4). Earth as a whole is constantly falling toward the Moon. The acceleration that Earth experiences is not very great—only about 3.3×10^{-5} m/s². The gravitational acceleration that we feel toward the center of Earth is 300,000 times greater. Yet the gravitational pull of the Moon on Earth is a substantial influence on our planet, causing it to shift more than 9,300 km back and forth over the course of a month. However, we do not personally perceive any sensation from the gravitational attraction of the Moon on Earth. Just as everything in a traveling car shares the motion of the car, everything on Earth falls *together* toward the Moon.

What is *not* exactly the same everywhere on Earth are the "leftover" accelerations—the differences between the gravitational acceleration due to the Moon at any particular location and the average gravitational acceleration acting on Earth as a whole. **Figure 10.5** shows the effect of these leftover accelerations. On the side of Earth closer to the Moon, the actual acceleration is greater than the average acceleration. The result is that 1 kilogram of material on the side of Earth closer to the Moon is pulled *toward* the Moon with a force of 1.1×11^{-6} newtons *relative to the Moon's attractive force on Earth as a whole*. But on the side of Earth

away from the Moon, the actual force is *less than* the average force: The difference between the actual force and the average force points *away from the Moon!* Figure 10.5 shows that there is also a net force squeezing inward on Earth in

Tides stretch out Earth in one direction
and squeeze it in the other.

the direction perpendicular to the line between Earth and the Moon. Together the stretching of tides along the line between Earth and the Moon and the squeezing of the tidal forces perpendicular to this line distort the shape of Earth like a rubber ball caught in the middle of a tug-of-war.

Out of convenience we speak of **tidal stresses** as if there is actually a force pulling the far side of Earth away from Earth's center. However, it is important to remember that the Moon is not pushing the far side of Earth away. Rather, it simply is not pulling on the far side of Earth as hard as it is on the planet as a whole. The far side of Earth is "left behind" as the rest of the planet falls more rapidly toward the Moon.

In Chapter 3 we learned that the strength of the gravitational force between two bodies is proportional to their masses and inversely proportional to the square of the distance between. But the strength of tides caused by one body

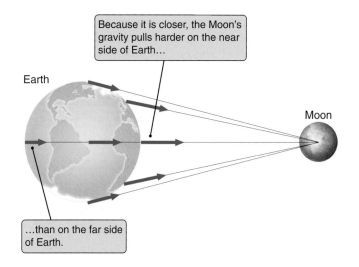

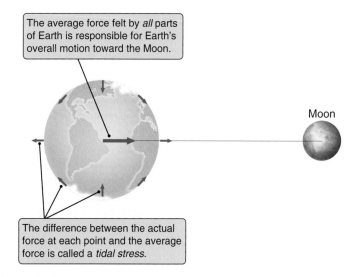

FIGURE 10.5 Tides stretch Earth along the line between the Earth and the Moon but compress the Earth perpendicular to this line.

acting on another is far more complicated. Tidal influence is also proportional to the mass of the tide-raising body, but it is inversely proportional to the *cube* of its distance. (The proof of this is well beyond the level of this text.) So what are the consequences of this inverse cube relationship? Billions of years ago, the Moon was much closer to Earth than it is today. Had the Moon's distance been half of what it is today, oceanic tides would have been *eight* times as large.

Earth's Oceans Flow in Response to Tidal Forces

The tidal acceleration of a mass on either the near side or the far side of Earth is equal to about a ten-millionth of the acceleration due to Earth's gravity. This may not seem like much until we consider the enormous masses involved. Because $F = ma$, a really huge m times a small a results in a large force. In addition, the liquid water covering the majority of Earth's surface is free to move in response to tidal forces. In the idealized case—where the surface of Earth is perfectly smooth and covered with a uniform ocean, and where Earth does not rotate—the **lunar tides** (tides on Earth

---

Earth's oceans have a tidal bulge.

---

due to the gravitational pull of the Moon) would pull our oceans into an elongated **tidal bulge** like that in **Figure 10.6(a)**. The water would be at its deepest on the side toward the Moon and on the side away from the Moon, and at its shallowest at points midway between. Of course, our Earth is *not* a perfectly smooth, nonrotating body covered with

perfectly uniform oceans, and there are many effects that complicate this simple picture. One complicating effect is Earth's rotation. As a point on Earth rotates through the ocean's tidal bulges, that point experiences the ebb and flow of the tides. In addition, friction between the spinning Earth and its tidal bulge drags the oceanic tidal bulge around in the direction of Earth's rotation, as shown in **Figure 10.6(b)**.

Follow along in **Figure 10.6(c)** as we ride through the course of a day. We begin as the rotating Earth carries us through the tidal bulge on the Moonward side of the planet. Because the tidal bulge is not exactly aligned with the Moon, the Moon is not exactly overhead but is instead high in the western sky. When we are at the high point in the tidal bulge, the ocean around us is deeper than average—a condition referred to as *high tide*. About six and a quarter hours later, probably somewhat after the Moon has settled beneath

---

Tides rise and fall twice each day.

---

the western horizon, the rotation of Earth carries us through a point where the ocean is shallowest. It is *low tide*. If we wait another six and a quarter hours, it is again high tide. We are now passing through the region where ocean water is "pulled" (relative to Earth as a whole) into the tidal bulge on the side of Earth that is away from the Moon. The Moon, which is responsible for the tides we see, is itself hidden from view on the far side of Earth. Six and a quarter hours later, probably sometime after the Moon has risen above the eastern horizon, it is low tide. About 25 hours after we started this journey—the amount of time the Moon takes to return to the same point in the sky from which it started —we again pass through the tidal bulge on the near side of the planet. This is the age-old pattern by which mariners have lived their lives for millennia: the twice-daily coming and going of high tide, shifting through the day in lockstep with the passing of the Moon.[1]

The shape of Earth's landmasses and ocean basins is another complicating factor in the tides seen at any particular point. In the open ocean, at a point passing directly "beneath" the Moon as Earth rotates, the difference between the depth of the ocean at low tide and the depth of the ocean at high tide is about a meter. However, the actual tides that are seen along coastlines can be less than this or in some cases much greater than this.

In order to respond to the tidal stresses from the Moon, Earth's oceans must flow around the various landmasses that break up the water covering our planet. Some places, like the Mediterranean Sea and the Baltic Sea, are protected from tides by their relatively small sizes and the narrow passages connecting these bodies of water with the larger ocean. In other places the shape of the land funnels the tidal surge from a large region of ocean into a relatively small area, concentrating its effect. The Bay of Fundy lies between the Canadian provinces of Nova Scotia and New Brunswick. This bay, along with the Gulf of Maine, forms a great basin in which water naturally rocks back and forth with a period of about 13 hours. This is close to the 12½ hour period of the rising and falling of the tides. The characteristics of the basin amplify the tides, sending the water sloshing back and forth like the water in a huge bathtub. **Figure 10.7** shows a picture of one location on the Bay of Fundy at both high and low tide. The average difference between low and high tides on the bay is around 14.5 meters, and under the right conditions this difference can exceed 16.6 meters!

Among the joys we experience as we stroll along the seashores are the tide pools, the natural aquariums teeming with marine life. As the tides roll in and out, they recharge the tide pools with fresh seawater while sometimes exchanging the pool's inhabitants. As you gaze into these microenvironments, think back a billion years or so: Tide pools hold an ancient connection to all plants and creatures now living on solid ground, as you will see in **Connections 10.1**.

The Sun is another complicating factor in Earth's tides. As in the case of the Moon, the side of Earth closer to the Sun is pulled toward the Sun more strongly than the side of Earth away from the Sun. The absolute strength of the Sun's pull on Earth is nearly 200 times greater than the strength of the Moon's pull on Earth. Even so, the Sun is much farther

---

[1] Those who live near the ocean may have noticed that our description of daily tides is often simpler than what you actually observe. For example, high tide does not necessarily occur when the Moon is near its highest point above the horizon. This is because in addition to local shoreline effects (see the following paragraphs), there are oceanwide oscillations, like water sloshing around in a basin. This complicates the simple picture we describe, even in midocean.

away than the Moon, so the Sun's gravitational attraction does not change by much from one side of Earth to the other. As a result, **solar tides** (tides on Earth due to differences in the gravitational pull of the Sun) are only about half as strong as lunar tides. **Figure 10.8** illustrates the interaction between solar and lunar tides. If the Moon and the Sun are lined up with Earth, as occurs at either new Moon or full Moon, then

FIGURE 10.6 Tidal stresses pull Earth and its oceans into a tidal bulge. Earth's rotation pulls its tidal bulge slightly out of alignment with the Moon. As the Earth's rotation carries us through these bulges, we experience the well-known ocean tides.

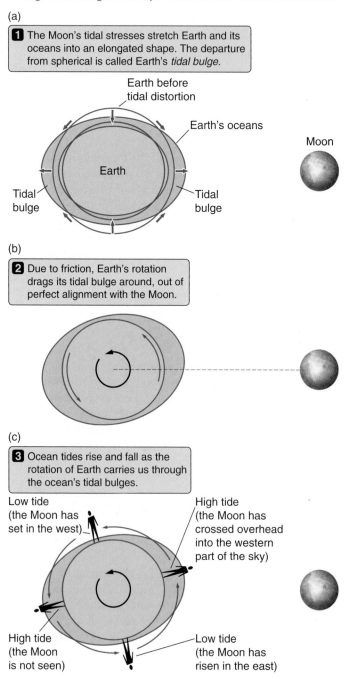

(a)

1 The Moon's tidal stresses stretch Earth and its oceans into an elongated shape. The departure from spherical is called Earth's *tidal bulge*.

Earth before tidal distortion

Earth's oceans

Moon

Earth

Tidal bulge

Tidal bulge

(b)

2 Due to friction, Earth's rotation drags its tidal bulge around, out of perfect alignment with the Moon.

(c)

3 Ocean tides rise and fall as the rotation of Earth carries us through the ocean's tidal bulges.

Low tide (the Moon has set in the west)

High tide (the Moon has crossed overhead into the western part of the sky)

High tide (the Moon is not seen)

Low tide (the Moon has risen in the east)

FIGURE 10.7 The world's most extreme tides are found in the Bay of Fundy, located between Nova Scotia and New Brunswick, Canada. The difference in water depth between low tide (left) and high tide (right) is typically about 14.5 meters and may reach as much as 16.6 meters.

CONNECTIONS 10.1

Origins—Lunar Tides and Life

To ancient Romans and Greeks she was known as Diana or Artemis, goddess of the Moon. From the earliest times to the present, the Moon has occupied a special place in the minds of romantics and scientists alike. Whether we look upon the Moon as an object of nighttime beauty or as our nearest planetary neighbor, few of us are aware that lunar tides played a crucial early role in the drama of life. Directly and indirectly, the Moon imposed its influence on the emergence and evolution of advanced life forms. Life's first home was in the oceans, and instant death was in store for any sea creature having the misfortune to be tossed onto barren land by an errant wave. But there were places of relative safety where life could be exposed briefly to air and land before retreating to the refuge of its natural aquatic environment. These were the tide pools. The twice-daily rise and fall of sea level swept traces of sea life into tide pools, where some spent a few adventurous hours on solid ground. Many probably perished, but a few organisms were successful in making the difficult transition to dry land. The first to take this enormous step were the plants, establishing their dominance on land about 500 million years ago. Animal life followed soon after. Today all creatures of land and air, including ourselves, likely owe their existence to the tides created by the Moon's gravity. Does this mean, as some have claimed, that advanced life can arise only on planets that have a large moon? Perhaps.

The Moon's gravitational grip on Earth has had another probable effect on terrestrial life, albeit one that is a bit more difficult to quantify. The Moon's gravity has held Earth's axial tilt, or *obliquity*, relatively steady between 21.5° and 24.5° over at least the past half billion years. This stability has maintained a smooth and steady climatological transition between our frigid polar regions and warmth of the tropics, allowing for a wide diversity of life. Not so for Mars, though, which has no stabilizing companion. Gravitational tugs from the Sun and planets, among other forces, cause its axial tilt to change chaotically from as little as 0° to as much as 60° over a few tens of millions of years. Without restraint from our Moon, Earth might also have gyrated widely, resulting in an unstable environment for life.

So the next time you gaze at the Moon, look upon it with respect. Quite likely it is one of the reasons you are here!

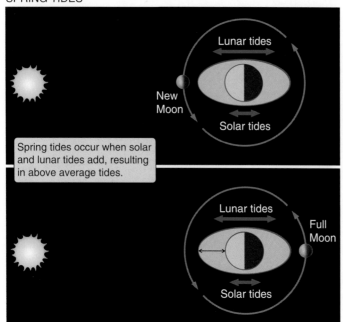

SPRING TIDES

Lunar tides

New Moon

Solar tides

Spring tides occur when solar and lunar tides add, resulting in above average tides.

Lunar tides

Full Moon

Solar tides

NEAP TIDES

Solar tides

Lunar tides

1st quarter

Neap tides occur when solar and lunar tides partially cancel each other.

3rd quarter

Lunar tides

Solar tides

FIGURE 10.8 Earth experiences solar tides that are about half as strong as tides due to the Moon.

the tides on Earth due to the Sun are in the same direction as the tides on Earth due to the Moon. At these times the solar tides reinforce the lunar tides, so the tides are about 50 percent stronger than average. The especially strong tides near the time of new or full Moon are referred to as **spring**

> Solar tides may reinforce or diminish lunar tides.

tides. Conversely, around the first and third quarter Moon, the stretching due to the solar tide is at right angles to the stretching due to the lunar tide. The solar tides pull water into the dip in the lunar tides and pull water away from the tidal bulge due to the Moon. This has the effect of making low tides higher and high tides lower. Such tides, when lunar tides are diminished by solar tides, are called **neap tides**. Neap tides are only about half as strong as average tides and only a third as strong as spring tides.

10.3 Tides Tie an Object's Rotation to Its Orbit

So far in our discussion we have implicitly assumed that the solid body of Earth itself is rigid and that the liquid of Earth's oceans is the only thing that actually moves in response to the tidal stresses from the Moon and Sun. In

reality this is not the case. Earth is somewhat elastic, and like a rubber ball, the solid body of Earth itself is deformed by tidal stresses. The tidal deformation of the solid Earth amounts to a vertical displacement of about 30 cm between high tide and low tide, or roughly a third of the displacement of the oceans.

As Earth rotates through its tidal bulge, the solid body of our planet is constantly being deformed. It takes energy to deform the shape of a solid object. (If you want a practical demonstration of this fact, hold a rubber ball in your hand, and squeeze and release it a few dozen times. While you are shaking your sore hand, imagine the energy that it must take to compress Earth by a third of a meter twice a day!) This energy is converted into thermal energy by *friction* in Earth's interior. This friction opposes and takes energy from the rotation of Earth, causing it to gradually slow. This internal friction within Earth adds to the even greater slowing caused by friction between Earth and its oceans as the planet rotates through the oceans' tidal bulge. Right now Earth's days are getting longer by about 0.0015 second every century.

The Moon Is Tidally Locked to Earth

The Moon has no bodies of liquid to make tides obvious, but it is distorted in the same manner as Earth. In fact, because of Earth's much greater mass and the Moon's smaller radius, the tidal effects of Earth on the Moon are about 20 times as great as the tidal effects of the Moon on Earth. Whereas the average tidal deformation of Earth is about 30 cm, the average tidal deformation of the Moon should be closer to 6 meters. However, what we actually observe on the Moon is a tidal bulge of about 20 meters! This is because the Moon's tidal bulge was "frozen" into its relatively rigid crust at an earlier time when the Moon was closer to Earth and tides were much stronger than they are today. This extreme tidal deformation is the reason why the Moon always keeps the same face toward Earth.

Early in its history the period of the Moon's rotation was almost certainly different from its orbital period. However, as the Moon rotated through its extreme tidal bulge, friction was tremendous, rapidly slowing the Moon's rotation.

Tides lock the Moon's rotation to its orbit around Earth.

After a fairly short time, the period of the Moon's rotation equaled the period of its orbit. When this happened, the Moon no longer rotated *with respect to its tidal bulge*. Instead the Moon and its tidal bulge rotated *together* in lockstep with the Moon's orbit about Earth. With the frictional forces within the Moon gone, the Moon's rotation no longer slowed but instead remained equal to the period of the Moon's orbit about Earth. This is how things remain today as the tidally distorted Moon orbits Earth, always keeping the same face and the long axis of its tidal bulge toward Earth (**Figure 10.9**). The synchronous rotation of the Moon discussed in Chapter 2 is a result of the **tidal locking** of the Moon to Earth.

In addition to their effects on the rotations of the Moon and Earth, tides also influence the orbits of these two objects. Because of its tidal bulge, Earth is not really a spherical object. The material in Earth's tidal bulge on the side nearer the Moon pulls on the Moon more strongly than ma-

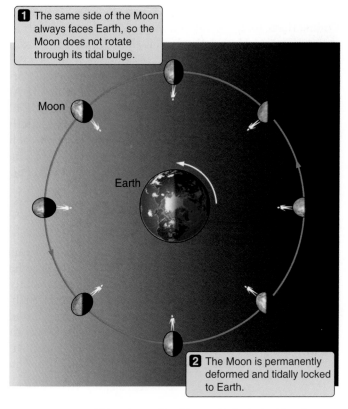

1 The same side of the Moon always faces Earth, so the Moon does not rotate through its tidal bulge.

Moon

Earth

2 The Moon is permanently deformed and tidally locked to Earth.

FIGURE 10.9 Tides due to Earth's gravity lock the Moon's rotation to its orbital period.

terial in the tidal bulge on the back side of Earth. Because the tidal bulge on the Moonward side of Earth "leads" the Moon somewhat, as shown in **Figure 10.10**, the gravitational attraction of the bulge causes the Moon to accelerate slightly along the direction of its orbit about Earth. It is as if the rotation of Earth were dragging the Moon along with it, and in

FIGURE 10.10 Interaction between Earth's tidal bulge and the Moon causes the Moon to accelerate in its orbit and the Moon's orbit to grow.

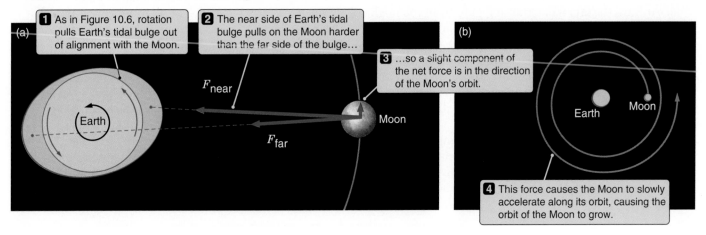

(a)

1 As in Figure 10.6, rotation pulls Earth's tidal bulge out of alignment with the Moon.

2 The near side of Earth's tidal bulge pulls on the Moon harder than the far side of the bulge...

3 ...so a slight component of the net force is in the direction of the Moon's orbit.

F_{near}

Earth

F_{far}

Moon

(b)

Earth   Moon

4 This force causes the Moon to slowly accelerate along its orbit, causing the orbit of the Moon to grow.

a sense this is exactly what is happening. As discussed in **Connections 10.2**, the angular momentum lost by Earth as its rotation slows is exactly equal to the angular momentum gained by the Moon as it accelerates along in its orbit.

The acceleration of the Moon in the direction of its orbital motion causes the orbit of the Moon to grow larger. At present the Moon is drifting away from Earth at a rate of 3.83 cm/year; as the Moon grows more distant, the length of the lunar month increases by about 0.014 second each century. At this rate, within slightly over a billion years the Moon will be far enough away from Earth that a total solar

> The Moon's orbit is growing larger and Earth's rotation is slowing.

eclipse will no longer be possible. If this were to continue long enough (about 50 billion years), Earth would become tidally locked to the Moon, just as the Moon is now tidally locked to Earth. At that point the period of rotation of Earth, the period of rotation of the Moon, and the orbital period of the Moon would all be exactly the same—about 47 of our present days—and the Moon would be about 43 percent farther from Earth than it is today. However, this situation

will never actually occur, or at least not before the Sun itself has burned out.

The effects of tides can be seen throughout the Solar System. Most of the moons in the Solar System are tidally locked to their parent planets, and in the case of Pluto and its largest moon, Charon, each is tidally locked to the other.

Tides and Noncircular Orbits: An Imperfect Match

The tidal locking between a moon and a planet (or between any two objects) can be perfect only if both objects are in circular orbits. Even though *on average* an object in an elliptical orbit may rotate at exactly the same rate at which it moves in its orbit, as the object follows its orbit, it is constantly speeding up and slowing down. Like a spinning top, the Moon rotates on its axis at a steady rate that equals the *average* rate of the Moon's progress around Earth. However, when the Moon is closest to Earth and moving fastest in its orbit, the Moon's orbital motion "gains" a little on its rotation. Conversely, when the Moon is farthest from Earth and

---

CONNECTIONS 10.2

Earth, the Moon, and Conservation of Angular Momentum

In this chapter we learned that the tidal interaction between Earth and the Moon is causing Earth's rotation to slow while it causes the Moon's orbit to grow larger. Given enough time, Earth and the Moon will become tidally locked to each other, with each keeping the same face toward the other. Looking at Figure 10.10 you might imagine that it would be rather difficult to calculate the exact size of the net force causing the Moon's orbit to grow, or the net frictional forces within Earth that slow its rotation. You would be correct. However, we can understand the relationship between these two effects, as well as their end result, without having to worry about any of those difficult details.

As Earth's rotation slows, Earth loses angular momentum. How can this be? When we discussed the formation of stars and planetary systems in Chapter 6, we found that angular momentum is conserved. Where does the angular

momentum that Earth loses as tides slow it down go? The answer is that angular momentum is transferred to the Moon. The angular momentum that the Moon picks up as its orbit grows balances exactly the angular momentum that Earth loses as its rotation about its axis slows down. This may seem like some grand conspiracy, but it is not. The frictional forces in Earth slow its rotation, but they also cause its tidal bulge to be pulled around out of alignment with the Moon. This misalignment in turn causes the Moon's orbit to grow. In the end it *must be* the case that these effects work together to balance Earth's angular momentum loss and the Moon's angular momentum gain. Conservation of angular momentum is a grand pattern that *always* holds true, whether we are talking about the formation of stars, rotations and orbits of planets, a spinning top, or the pinwheel motion of the galaxy itself.

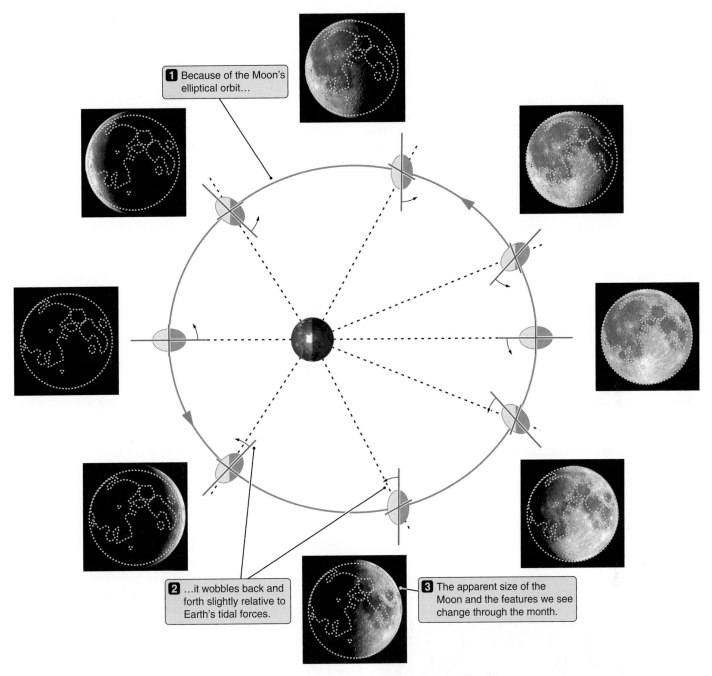

1 Because of the Moon's elliptical orbit…

2 …it wobbles back and forth slightly relative to Earth's tidal forces.

3 The apparent size of the Moon and the features we see change through the month.

FIGURE 10.11 Because of the Moon's elliptical orbit, an observer on Earth sees lunar libration, while an observer on the Moon sees Earth moving slightly back and forth in the sky.

moving most slowly on its orbit, the Moon's orbital motion lags behind its steady rotation. From our perspective on Earth, the face of the Moon appears to be rocking back and forth. This effect is referred to as lunar **libration**. The motions responsible for lunar libration are illustrated in **Figure 10.11**. One consequence of these motions is that the Moon's frozen gravitational bulge is constantly swinging back and forth with respect to the direction of Earth's gravitational pull. Friction within the Moon, working together with the

Tides make orbits more circular in shape.

gravitational interaction between the Moon's tidal bulge and Earth's gravity, is slowly causing the Moon's orbit to become more circular.

The masses of the giant planets are far greater than the mass of Earth, so the tidal stresses they exert on their moons are correspondingly stronger. The tidal effect of Jupiter on its innermost moon, Io, is 250 times greater than the effect of Earth's tides on the Moon. As with most of the moons of the giant planets, Io is tidally locked to Jupiter, rotating once on its axis in exactly the same amount of time that it takes Io to complete one orbit about the planet. Normally we would have expected Io long ago to have settled into a perfectly circular orbit. However, its orbit is constantly being perturbed by the gravity of Jupiter's other moons, which has prevented it from achieving a steady circular orbit. As a result, Io experiences libration, just as the Moon does, but with much more dramatic consequences. As Io's libration forces it to rock back and forth through the moon's tremendous tidal bulge, there is enough frictional heating to keep the interior of Io molten, powering the perpetually active volcanism (**Figure 10.12**) seen on this small world. In Chapter 11 we will learn that Io is the most volcanically active object in the Solar System. The powerhouse behind that activity is tides.

Tidal locking, in which an object's rotation period exactly equals its orbital period, is only one example of **spin-orbit resonance**. Other types of spin-orbit resonance are also possible. For example, Mercury is in a very elliptical orbit about the Sun. As with the Moon, tidal stresses have coupled Mercury's rotation to its orbit. Yet unlike the Moon's synchronous rotation, Mercury is in a 3:2 spin-orbit resonance, spinning on its axis three times for every two trips around the Sun. The period of Mercury's orbit—87.97 Earth days—is exactly one and a half times the 58.64 days that it takes Mercury to spin once on its axis. The forces due to solar tides on Mercury flip Mercury about like a juggler throwing a stick into the air and catching it first by one end, then by the other.

FIGURE 10.12 Volcanoes are constantly erupting on Jupiter's moon Io. The energy to power these volcanoes comes from tides due to Jupiter's gravity.

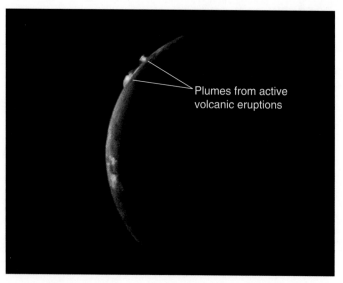

Plumes from active
volcanic eruptions

Tides Can Be Destructive

We normally think of the effects of tides as small compared with the force of gravity holding an object together, yet this is not always the case. Tidal effects act to pull an object apart, which means that when tides become large, they can be extremely destructive. Consider for a moment the fate of a small moon, asteroid, or comet that wanders too close to a massive planet such as Jupiter or Saturn. All objects in the Solar System larger than about a kilometer are held together by their self-gravity. However, the self-gravity of a small object such as an asteroid, a comet, or a small moon is feeble. In contrast, the tidal stresses close to a massive object such as Jupiter can be fierce. If the tidal stresses trying to tear an object apart become greater than the self-gravity trying to hold the object together, the object will begin to break into pieces.

Astronomers have a special name for the distance from the center of a planet at which tidal forces on a second, smaller body just equal that body's self-gravity. It is called the planet's **Roche limit**. For a body having the same density as the planet and having no internal strength, the Roche limit is about 2.45 times the planet's radius. Such an object

---

Tidal forces exceed self-gravity inside the Roche limit.

---

bound together solely by its own gravity can remain intact when it is outside a planet's Roche limit, but not when it is inside the limit.[2] Tidal disruption of small bodies is thought to be the source of the particles that make up the rings of the giant planets. Comparison of the major ring systems around the giant planets shows that most rings lie within their respective planets' Roche limits. In the mid-1990s astronomers were treated to a rare look at tidal disruption in action. Comet Shoemaker-Levy 9 was discovered in March 1993. As astronomers studied the comet, they realized that it had been captured by Jupiter and that in July 1992 the comet had passed within 1.4 Jupiter radii of the planet's center, well within the planet's Roche limit. The comet was broken into pieces by this encounter, so pictures taken by

---

Comet Shoemaker-Levy 9 was shattered by Jupiter's tides.

---

the Hubble Space Telescope (**Figure 10.13(a)**) showed not a single comet, but instead a chain of orbiting fragments. In the summer of 1994 Comet Shoemaker-Levy 9 treated the world to an even more unusual spectacle: One after another the pieces of the comet slammed into Jupiter's atmosphere,

---

[2] Small objects generally have mechanical strength that is much stronger than their own self-gravity. This is why the International Space Station and other Earth satellites are not torn apart even though they orbit well within Earth's Roche limit.

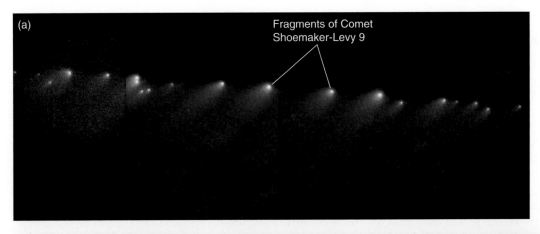

FIGURE 10.13 (a) Comet Shoemaker-Levy 9 was shattered into pieces as it passed within Jupiter's Roche limit in July 1992. (b) Comet West was shattered by solar tides during its 1976 pass through the inner Solar System.

creating enormous fireballs and scars larger than Earth. **Figure 10.13(b)** shows the shattered remains of another comet, Comet West, which was broken apart by the Sun's tides as it passed close to the Sun in 1976.

We have seen that tides are at work throughout the Solar System. In **Connections 10.3** we find they are at work throughout the universe, as well.

10.4 More Than Two Objects Can Join the Dance

We have seen that tidal interactions can lead to simple relationships or resonances between the orbital period of an object and the period of the object's rotation. Resonances can also exist between the orbital periods of two or more bodies orbiting about the Sun or a planet. In Chapter 12 we will study the smaller objects that orbit about the Sun, including the small rocky or metallic worlds called *asteroids*. Most of the asteroids in the Solar System are located in a "belt" between the orbits of Mars and Jupiter. You might imagine that asteroids would be distributed randomly throughout this region. However, if you look at the distribution of the sizes of the orbits of the asteroids in the main belt, shown in **Figure 10.15**, you will see an amazing thing. Rather than asteroids with orbits of all sizes in this region, there are certain-sized orbits that are seldom occupied.

Orbital resonances with Jupiter cause the Kirkwood gaps.

These gaps in the asteroid belt are referred to as the **Kirkwood gaps** after **Daniel Kirkwood** (1814–1895), the American astronomer who first recognized them. Astronomers studying the Kirkwood gaps found a simple relationship between the "missing" orbits and the orbit of Jupiter. For example, none of the asteroids orbiting the Sun have an orbital period equal to half the orbital period of that planet. In fact, all of the Kirkwood gaps in the asteroid belt correspond to orbits that are related to the orbital period of Jupiter by the ratio of two small integers. Such a simple relationship between the periods of the orbits of two or more objects is referred to as an **orbital resonance**. (Note that this is different from spin-orbit resonance, which we discussed in Section 10.3.)

The easiest way to understand the origin of the Kirkwood gaps is to imagine what would happen to an asteroid that started out with an orbital period exactly half that of Jupiter. Follow along in **Figure 10.16** as we watch the orbit of such an asteroid, beginning when the asteroid and Jupiter are at their closest to one another. Because the asteroid is

CONNECTIONS 10.3

Tides on Many Scales

This section of the book concerns itself with the objects that make up our Solar System and the processes that influence them. As such, it is fitting that we have concentrated on the role tides play in the Solar System. On the other hand, we could have discussed tides and their effects at almost any point in the book. Tidal stresses result from differences in gravitational forces from place to place, and such differences are a general consequence of the inverse square law of gravity. *Anytime* two objects of significant size or two collections of objects interact gravitationally, the gravitational forces will differ from one place to another within the objects, giving rise to tidal effects.

We might easily have included a discussion of tides in Chapter 13, where we will find that stars are often members of binary pairs in which two stars orbit each other. Strong tidal interactions between such stars can tidally lock the rotation of each star to its orbital period about its companion, and can even pull material from one star onto the other. When we study this process in Chapter 16, we will learn of the *Roche lobes* of a binary star. These are named after the same **Edouard A. Roche** (1820–1883) who did the pioneering work on the disruption of satellites, and for whom the Roche limit is named.

Under some circumstances in the universe, tides become unimaginably powerful. In the vicinity of very massive, very dense objects the pull of gravity can change by huge amounts over minute distances. In Chapter 17 we will find that the tidal effects near the surface of a black hole can be billions of times stronger than the force of gravity holding us to the surface of Earth. Such tides are enough to rip normal matter apart atom by atom.

Tides continue to play a role at even larger scales. Tidal effects can strip stars from clusters consisting of thousands of stars. Beginning in Chapter 18 we will also learn about *galaxies*—vast collections of hundreds of billions of stars that all orbit each other under the influence of gravity. It often happens that two galaxies pass close enough together to strongly interact gravitationally. When this happens, as in **Figure 10.14**, both galaxies taking part in the interaction can be grossly distorted by tidal effects. Tides even play a role in shaping huge collections of galaxies—the largest known structures in the universe.

The next time you are at the beach "watching the tide roll in," you might take a moment to think about tides and the role they play throughout the universe. Vast collections of billions of stars strewn about millions of light-years of

closer to the Sun than Jupiter is, its orbital velocity is greater than that of Jupiter, so it leaves Jupiter behind as they continue on their respective orbits. By the time the asteroid has completed half of its orbit, Jupiter has completed a fourth of its orbit. By the time the asteroid has gone around its orbit once, Jupiter is halfway. When the asteroid has completed one and one-half orbits, Jupiter has been through three-quarters of its orbit. Only after the asteroid has made two complete orbits and Jupiter has been around the Sun once do the two objects again line up. Courtesy of the relationship between the periods of their two orbits, when Jupiter and the asteroid line up, they are in the same place where they started. As Jupiter and the asteroid continue along in their orbits, they line up again and again at this same location every 11.86 years (the orbital period of Jupiter).

The gravitational force that Jupiter exerts on an asteroid at its closest approach is tiny compared with the gravita-

tional force of the Sun on the asteroid, which is over 360 times stronger. A single close pass between Jupiter and the asteroid does very little to the asteroid's orbit. For an asteroid that is *not* in orbital resonance with Jupiter, the tiny nudges from Jupiter come at a different place in its orbit each time. The effects of these "here and there" nudges average out, and as a result even multiple passes close to Jupiter have little overall effect on the asteroid's orbit. However, as we just saw, for an asteroid with a 2:1 orbital resonance with Jupiter (in other words, with an orbital period equal to half that of Jupiter), the nudge from Jupiter comes at the *same place* in its orbit *every time*. Although no *single* tug has much influence on the asteroid, the *repeated* tugs from Jupiter at the same place add up, changing the asteroid's orbit. This is why there are no asteroids with orbital periods equal to half the orbital period of Jupiter. If we were to put an asteroid in such an orbit, it would not stay there

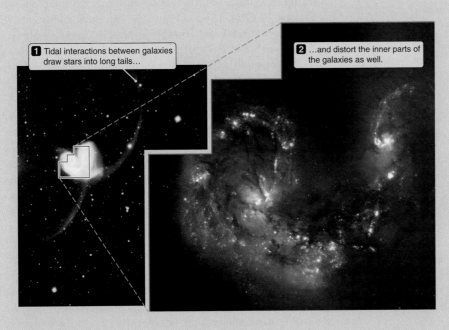

❶ Tidal interactions between galaxies draw stars into long tails…

❷ …and distort the inner parts of the galaxies as well.

FIGURE 10.14 Tidal interactions distort the appearance of two galaxies. The tidal "tails" seen in this image are characteristic of tidal interactions between galaxies.

space, matter shredded on its way into a black hole, gas stripped from a burgeoning star, comets and moons pulled to pieces, planets "flipped" like pancakes—all are the result of the same gravitational effect that is responsible for the gentle twice-daily rise and fall of the waterline on an oceanside dock.

for long. The same phenomenon works for other combinations of orbital periods as well. With a piece of paper and a pencil, you ought to be able to convince yourself that other orbital resonances, such as a 3:1 resonance (in which an asteroid completes three orbits about the Sun in the time that it takes Jupiter to complete one orbit), will also have a similar effect. Asteroids are not found in the Kirkwood gaps because their gravitational interaction with Jupiter prevents them from staying there.

Orbital resonances shape Saturn's rings.

Asteroids are not the only objects in the Solar System that are affected by orbital resonances. Exactly the same phenomenon occurs in the ring systems around the giant planets. In Chapter 11 we will learn of a famous gap in the rings around Saturn called the **Cassini Division**. This gap became famous because of its visibility—it is so large that it could be seen from Earth even with crude, small telescopes. The Cassini Division corresponds to a 2:1 orbital resonance with Saturn's moon Mimas. That is to say, a ring particle located in the Cassini Division would have an orbital period about the planet equal to half the orbital period of Mimas. Just as Jupiter's gravity pulls asteroids out of resonant orbits, leading to the Kirkwood gaps, Mimas pulls ring particles out of the resonant orbits found within the Cassini Division. Orbital resonances and similar effects are responsible for a great deal of the structure in ring systems.

Lagrangian points are orbital resonances formed by the gravity of two objects.

One of the most interesting examples of orbital resonance occurs when two massive bodies (such as a planet and the

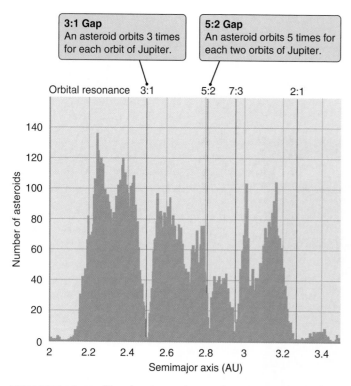

> **3:1 Gap**
> An asteroid orbits 3 times for each orbit of Jupiter.

> **5:2 Gap**
> An asteroid orbits 5 times for each two orbits of Jupiter.

FIGURE 10.15 The plot shows the number of asteroids in the main belt with a given orbital period. The gaps in the distribution of asteroids, called Kirkwood gaps, are caused by orbital resonances with Jupiter.

Sun, or a moon and a planet) move about their common center of mass in circular or nearly circular orbits. In this situation, there are five locations where the combined gravity of the two bodies adds up in such a way that a third, lower-mass object will orbit in lockstep with the other two.

These five locations, called **Lagrangian equilibrium points** or simply Lagrangian points, are named for the French mathematician **Joseph Lagrange** (1736–1813), who first called attention to them.

The exact locations of the Lagrangian points depend on the ratio of the masses of the two primary bodies. The Lagrangian points for the Earth–Moon system are shown in **Figure 10.17(a)**. (Remember that the pattern shown in the figure rotates like a turntable as the two massive objects orbit each other.) Three of the Lagrangian points lie along the line between the two principal objects. These points are designated L_1, L_2, and L_3. These three points are equilibrium points, but they are unstable in the same way that the top of a hill is unstable. A ball placed on the top of a hill will sit there if it is perfectly perched, but give it the slightest bump and it goes rolling down one side. Similarly, an object displaced slightly from L_1, L_2, or L_3 will move away from that point. Thus, while L_1, L_2, and L_3 are equilibrium points, they do not capture and hold objects in their vicinity. Even so, they can be useful. For example, as of this writing, the *SOHO* spacecraft sits at the L_2 point of the Sun–Earth system, where tiny nudges from its onboard jets are enough to keep it orbiting directly between Earth and the Sun, giving a constant, unobscured view of the side of the Sun facing Earth.

The situation is different for the other two Lagrangian points, L_4 and L_5. These two points are located 60° in front and 60° behind the less massive of the two main bodies in its orbit about the center of mass of the system. These are *stable* equilibrium points, like the stable equilibrium that exists at the bottom of a bowl. If you bump a marble sitting at the bottom of a bowl, it will roll around a bit, but it will stay in the bowl. In like fashion, objects near L_4 or L_5 follow elongated, tadpole-shaped paths relative to these gravitational "bowls."

FIGURE 10.16 An asteroid with an orbital period exactly half that of Jupiter would line up with Jupiter at exactly the same place in every other orbit. The repeated gravitational pull from Jupiter at the same point in the asteroid's orbit would nudge the asteroid enough to change its orbital period. There are no asteroids with a 2:1 orbital resonance with Jupiter.

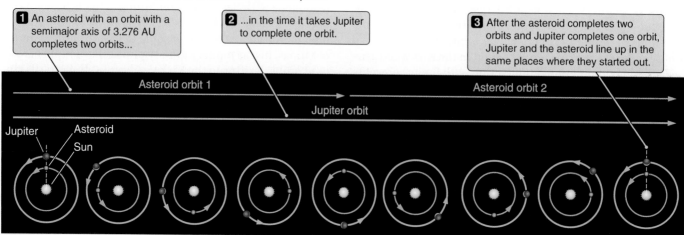

1 An asteroid with an orbit with a semimajor axis of 3.276 AU completes two orbits...

2 ...in the time it takes Jupiter to complete one orbit.

3 After the asteroid completes two orbits and Jupiter completes one orbit, Jupiter and the asteroid line up in the same places where they started out.

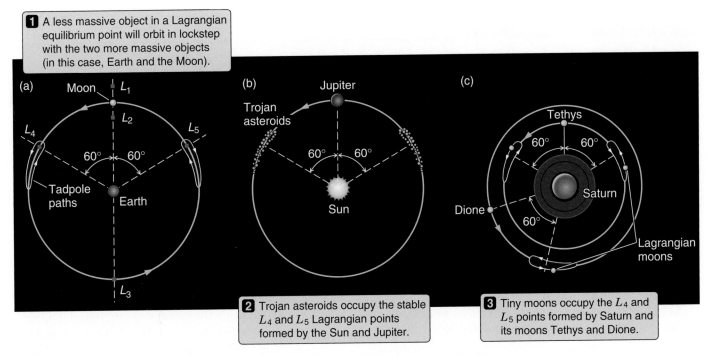

1 A less massive object in a Lagrangian equilibrium point will orbit in lockstep with the two more massive objects (in this case, Earth and the Moon).

(a) Moon L_1 L_2 L_4 L_5 $60°$ $60°$ Tadpole paths Earth L_3

(b) Jupiter Trojan asteroids $60°$ $60°$ Sun

2 Trojan asteroids occupy the stable L_4 and L_5 Lagrangian points formed by the Sun and Jupiter.

(c) Tethys $60°$ $60°$ Saturn Dione $60°$ Lagrangian moons

3 Tiny moons occupy the L_4 and L_5 points formed by Saturn and its moons Tethys and Dione.

FIGURE 10.17 (a) The pattern of Lagrangian equilibrium points of a two-body system, in this case comprised of Earth and the Moon. This entire pattern rotates like a turntable. (b) Trojan asteroids. (c) Small moons occupy Lagrangian points formed by Saturn and Tethys and by Saturn and Dione.

Under the correct circumstances, L_4 and L_5 are able to capture and hold on to passing objects. Objects are often found orbiting about the stable L_4 and L_5 Lagrangian points of two-body systems. For example, the stable Lagrangian points of the Sun–Jupiter system are home to a collection of planetesimals called the **Trojan asteroids (Figure 10.17(b))**. There are also small moons in the L_4 and L_5 Lagrangian points of Tethys and Dione, two moons of Saturn (**Figure 10.17(c)**). A well-known group that advocates human colonization of space calls itself the L_5 *Society* in recognition of the fact that the L_4 and L_5 Lagrangian points of the Earth–Moon system would make excellent locations to park a number of large orbital colonies constructed of material mined on the Moon.

10.5 Such Wondrous Complexity Comes from Such a Simple Force

The tides and resonances that we have discussed in this chapter are but a small sample of the tremendous wealth and variety of phenomena that result from the gravitational interactions between extended objects or collections of more than two objects. We might have mentioned many other such phenomena. The 26,000-year precession of Earth's axis, for example, is a result of the Sun's gravity causing Earth's

Gravity causes the precession of Earth's axis.

axis to wobble like a top's. How a gravitational tug in one direction can cause the axis of a planet (or a top) to move in another direction is a truly fascinating story.

It is with regret that we leave behind the gravitational ballet, but move on we must. Before leaving, however, it is worth pointing out just how broad and deep the general problem of gravitational interactions is. Although the motion of *two* masses orbiting about each other was solved centuries ago by Newton, the general problem of the motion of *three* bodies remains unsolved today! More than a few Ph.D. theses, scholarly journal articles, and monographs in mathematics, astronomy, and physics have been filled with years of work on various aspects of this seemingly simple problem. Even in this day of powerful computers, which can crank out brute force solutions to problems that seemed unapproachable a few decades ago, we remain humble in the face of gravity. As discussed in **Excursions 10.1**, even the gravitational interactions between a few objects such as the planets in our Solar System are *chaotic*. Unbelievably tiny differences in the initial positions and velocities of the planets—differences that are far too tiny

A World of Chaos

Imagine that you head for class but are one second later leaving than you might have been. That single second causes you to get caught at a traffic light, which delays you enough to wind up first in line waiting for a passing train. While you await the train, there is an accident up ahead, and as you sit in the resulting traffic jam, your car's engine overheats. Eventually, while your classmates are listening to an inspiring lecture about orbital motion, you find yourself sitting in a mechanic's waiting room learning the hard way about how much expensive damage you can do to an engine by letting it get too hot. You do not realize it, but the only thing that separated you from a normal day in class was the extra sip of coffee that put you out the door a second later than you might have been. You could never have predicted such a thing would happen. Welcome to the wonderful world of **chaos**.

Chaos is the technical term to describe interrelated systems in which tiny differences at one point in time lead to large differences at later times. Any system capable of exhibiting chaotic behavior is referred to as a **complex system**. Complex systems need not be extremely complicated, like the example in the previous paragraph, but can be seemingly very simple. Newton's law of gravity is simplicity itself, and as long as only two objects are involved, the resulting motions are simple as well. When we think of a planet orbiting the Sun, we think of the regular, repeating elliptical orbits described by Kepler's laws. However, when more than two objects are involved, the resulting motions can be anything but simple and regular.

Figure 10.18(a) shows a calculation of the orbit of Jupiter's moon Pasiphae over an interval of approximately 38 years. Pasiphae orbits far enough from Jupiter that its orbit is strongly perturbed by the Sun's gravity. Rather than a simple ellipse, Pasiphae's orbit looks more like what a 5-year-old might produce if handed a crayon and asked to quickly trace out a circle again and again. Pasiphae's

orbit is an example of chaos. Chaotic orbits are complex, irregular motions in which extremely small differences in an object's position or speed grow into large differences in the object's subsequent motion.

Capture of moons by planets is another example of chaos at work. **Figure 10.18(b)** shows the path of a small planetesimal, such as a comet, as it approaches a planet such as Jupiter. Very tiny differences in the position of the comet lead to very different outcomes. In the illustration there is a gnat's eyelash of difference between the starting point of a path in which the comet impacts the planet, a path in which the comet swings around Jupiter and continues to orbit the Sun, and a path that winds up with the planetesimal orbiting the planet. Each of the giant planets has a handful of moons that revolve about the planet in the wrong direction, indicating they must have been captured in such chaotic events.

Chaos also affects the orbits of planets themselves. Each planet in our Solar System moves under the combined gravitational influence not only of the Sun, but of all the other planets as well. Although these extra influences are small, they are not negligible. Over millions of years they lead to significant differences in the locations of planets in their orbits. While analysis of our Solar System indicates that the orbits of at least eight planets are fairly stable, at least over times of a few billion years, that is not always the case. In many possible planetary systems, chaotic interactions among planets might cause planets to dramatically change their orbits or even be ejected from the system entirely.

In Chapter 6 we discussed the fact that planetary systems have been found around many other stars. Some of these systems contain giant planets that seem much too close to their parent stars to have formed there. Models suggest that these planets might have formed farther out in these systems and then moved inward toward the stars as a result of chaotic interactions. Indeed, events such as the collision between Earth and a Mars-sized body that formed the Moon tell of a more chaotic period early in the history of our own Solar System.

A key characteristic of chaotic systems is that while they are *deterministic*, they are not *predictable*. The law of gravity is well known, and there is in principle no problem with calculating the path that Earth will follow as a result of its ongoing interactions with other planets. It is a perfectly well-determined situation. However, even Earth's orbit is subject to chaos. An uncertainty of only 1 *centimeter* in our knowledge of the position of Earth along its orbit today is enough to make it completely impossible to predict where Earth will be in its orbit 200 million years from now.

Scientists studying the orbits of celestial bodies were among the first to recognize the existence of chaotic be-havior, and in the last decades of the 20th century they began to develop the mathematical tools needed to describe and even manipulate chaos. These tools are finding more and more common applications in our lives. The beating human heart, weather patterns, and traffic control are all examples of complex systems to which these tools have been applied. Pacemakers use these techniques to recognize when the regular beating of a heart is about to give way to the deadly, uncoordinated fluttering called *fibrillation*. A tiny nudge on the part of the pacemaker, slightly changing the timing of a handful of heartbeats before fibrillation begins, is enough to prevent disaster. Chaos—the physics of complexity—is but one of many examples in which studies of the universe have spilled over into better understanding of the immediate world in which we live.

FIGURE 10.18 (a) The chaotic orbit of Jupiter's moon Pasiphae. (b) Tiny differences in the motion of a comet lead to dramatic differences in the comet's fate.

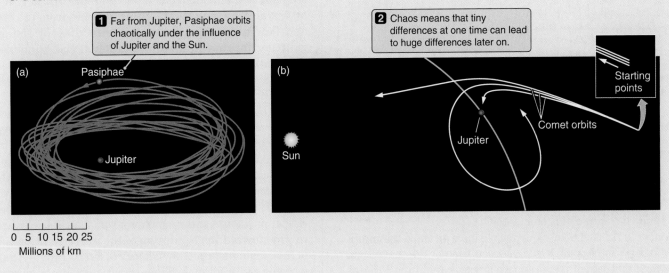

1 Far from Jupiter, Pasiphae orbits chaotically under the influence of Jupiter and the Sun.

(a) Pasiphae

Jupiter

0 5 10 15 20 25
Millions of km

2 Chaos means that tiny differences at one time can lead to huge differences later on.

(b)

Starting points

Comet orbits

Jupiter

Sun

to keep track of with the best computers—can, over time, lead to dramatically different end results.

As we leave this interlude about gravity behind, we can only marvel at the extraordinary complexity and diversity that arise in nature from this deceptively simple force.

What an amazing array of implications, insights, puzzles, and surprises there are within Newton's elegant statement that the force of gravity is proportional to the product of the masses of two objects and inversely proportional to the square of the distance between them!

Summary

- Earth is held together by self-gravity.
- Both the Moon and Sun create tides on Earth.
- As Earth rotates, tides rise and fall twice each day.
- Tides lock the Moon's rotation to its orbit around Earth.
- Tides cause the Moon's orbit to grow and Earth's rotation rate to slow.
- Tides circularize orbits.
- Tidal forces are stronger than self-gravity inside the Roche limit.
- Gravity causes Earth's axis to precess.

Seeing the Forest through the Trees

In Chapter 3 we learned about the birth of modern science. We saw how a few straightforward ideas about forces and the inverse square law of gravity opened a new window on the universe. When Newton used his laws of motion and gravitation to explain the motions of the planets about the Sun, he changed the face of science and the course of history.

On this leg of our journey we have pursued the implications of Newton's insight even further. We have seen how that same inverse square law of gravitation explains not only the simple elliptical orbits of planets about the Sun, but also an amazing wealth of different phenomena. The coming and going of the tides, the synchronous rotation of the Moon, the delicate structure of Saturn's rings—all of these and more are the logical consequences of the fact that gravity is one-fourth as strong when two objects are twice as far apart. The insights that we have arrived at in this chapter will have profound implica-

tions again and again as we continue our journey. For example, when we follow the birth, evolution, and death of stars like our Sun, much of what we will see will be determined by the sizes and masses of the cores of those stars. Because of what we have learned in this chapter, we will be able to say that the force of gravity at the surface of a stellar core is determined by the core alone, regardless of what the rest of the star is doing.

Even more important, this leg of our journey has shown us again the aesthetics of science. If you ask a physicist today about the ultimate goal of physics, you will hear about the desire to explain the way the universe behaves in terms of the fewest and simplest possible physical laws. In Chapter 3 we discovered Newton's laws of motion. In a sense, once we know those laws, everything—from the flight of a bumblebee, to the swirl of a hurricane, to the motions of an Olympic gymnast—is just arithmetic. We began this chapter with the single statement about the inverse square law of gravitation, and went on to tell stories ranging from the ebb and flow of Earth's tides to the shredding of distant galaxies. In Chapter 4 we encountered the principles of quantum mechanics. Earth, wind, fire, water—even thought itself—are but practical applications of these principles.

At the turn of the 21st century, physicists have come to believe that even the basic physical laws that we are using on our journey—Newton's laws, relativity, gravitation, and quantum mechanics—are themselves embodiments of patterns in nature that are more fundamental still. Toward the end of our journey we will learn that today's frontiers of science are located at the two extremes. Theoretical physicists and cosmologists push ever backward in time toward the beginning of the universe in search of the single theory—the "Theory of Everything"—from which all else follows. Some scientists believe we may be just a few decades (or perhaps only a few years) from knowing the shape of such an all-encompassing theory. We may even come to know whether it was possible for the universe to become other than it is. At the same time physicists, astronomers, planetary scientists, biologists, and others are pushing forward with their understanding of the complex world around us. They are developing new tools and approaches, such as the tools describing chaos, to allow them to find the hand of simple physical law in all that we see. So the next time you find yourself looking up at the Moon, gazing on the same face of our sister world that our ancestors saw from time immemorial, think about gravity and tides and the interplay between sister worlds. But at the same time think about what they represent—the glorious complexity of this fascinating universe in which we live, and the beautiful simplicity of physical law from which it all derives.

Key Terms

tide, p. 283
spherically symmetric, p. 284
symmetry, p. 286
self-gravity, p. 287
tidal stress, p. 288
lunar tides, p. 289
tidal bulge, p. 289
solar tides, p. 290
spring tides, p. 292
neap tides, p. 292
tidal locking, p. 293
libration, p. 295
spin-orbit resonance, p. 296
Roche limit, p. 296
Kirkwood gaps, p. 297
orbital resonance, p. 297
Lagrangian equilibrium points, p. 300
Trojan asteroid, p. 301
chaos, p. 302
complex system, p. 302

Student Questions

THINKING ABOUT THE CONCEPTS

1. In our daily lives we encounter many types of symmetry, including rotational, reflection, and bilateral (which is when left and right sides are identical). List some ordinary examples of each of these types of symmetry.

2. The acceleration due to gravity (*g*) at Earth's surface measures 9.832 m/s² at the poles and 9.781 m/s² at the equator (note that both values are close to the value of 9.8 m/s² we use in this text). Give two reasons why the value of *g* is less at Earth's equator than at the poles.

3. The best time to dig for clams along the seashore is when the ocean tide is at its lowest. What phases of the Moon and times of day would be best for clam digging?

4. A pendulum clock, adjusted to run accurately at one geographical location, may not maintain that same accuracy at some other location. Why not?

5. We may have an intuitive feeling for why lunar tides raise sea level on the side of Earth facing the Moon, but why is sea level also raised on the side opposite to the Moon?

6. If lunar tides cannot raise the ocean surface more than 1 m above mean sea level, how can tides as large as 5 to 10 meters occur?

7. If the Moon spins on its axis once every 29½ days relative to Earth, why are there not Earth tides on the Moon rising and falling over this same interval?

8. Tidal friction slows Earth's rotation and causes the Moon to orbit ever farther from Earth. Is this an example of conservation of angular momentum or conservation of energy? Explain your answer.

9. In 1959 a spacecraft launched by the Soviet Union photographed the far side of the Moon, showing us a large part of the Moon's surface that had never before been seen. Yet more than half (59 percent) of the Moon's surface had already been mapped prior to this Soviet achievement. If the Moon always keeps the same face toward Earth, how was this possible?

10. Most commercial satellites orbit Earth well inside the Roche limit. Why are they not torn apart?

11. There is a region in the main asteroid belt (about 3.28 AU from the Sun) where the orbital period is exactly half that of Jupiter's. Yet no asteroids are found there. Explain why.

12. Even our most powerful computers have limited ability to predict the orbits and locations of the various bodies in our Solar System in the far distant future, and no matter how powerful we make our computers, there will always be uncertainties. Why is this so?

APPLYING THE CONCEPTS

13. Assume for the moment that Earth is homogeneous—that is, its density is constant throughout, and that it is spherical in shape. (We know, of course, from Chapter 7 that neither of these assumptions is true.) If you were in a deep well halfway to Earth's center, how would your weight there compare with your weight at Earth's surface?

14. Saturn's small moon Hyperion is in a 4:3 orbital resonance with its largest moon, Titan, which orbits closer to Saturn than does Hyperion. The orbital period of Titan is 15.945 days. What is the orbital period of Hyperion?

15. Tidal influence is proportional to the mass of a disturbing body and is inversely proportional to the *cube* of its distance. Some astrologers claim that your destiny is determined by the "influence" of the planets that are rising above the horizon at the moment of your birth. Compare the tidal influence of Jupiter (mass $= 1.9 \times 10^{27}$ kg; distance $= 7.8 \times 10^{11}$ m) with that of the doctor in attendance (mass $= 80$ kg, distance $= 1$ m).

16. What is Earth's Roche limit for a body having the same density as Earth and no internal strength?

17. The mass of the Sun is 27 million times greater than the mass of the Moon, but the Sun is about 390 times farther away. Show why solar tides on Earth are only about half as strong as lunar tides.

18. Calculate the orbital radius of the Kirkwood gap that is in a 3:1 orbital resonance with Jupiter. Hint: Refer to

Kepler's Third Law in Chapter 3. If you cannot do the calculation, show how you would set it up.

19. The L_4 and L_5 Lagrangian points form equilateral triangles with the Sun and Earth (see Figure 10.17). Should members of the L_5 *Society* ever colonize L_5, how far will they live from their previous homes on Earth?

StudySpace
wwnorton.com/astro21
provides a Study Plan for each chapter that includes a reading outline, animations, keyword flash cards, and gradebook-enabled multiple-choice quizzes. From StudySpace you can also access premium content in the ebook and SmartWork.

Although we are mere sojourners on the surface of the planet, chained to a mere point in space, enduring but for a moment in time, the human mind is not only enabled to number worlds beyond the unassisted ken of mortal eye, but to trace the events of indefinite ages before the creation of our race. . . .

SIR CHARLES LYELL (1797–1875)

Complex structure in Saturn's three bright rings as seen by the *Cassini* spacecraft.

CHAPTER 11

Planetary Moons and Rings, and Dwarf Planets

11.1 Moons and Rings— Galileo's Legacy

In 1610 the Italian astronomer Galileo Galilei observed that Jupiter was accompanied by four "stars" that changed their positions nightly. He quickly realized that, like Earth, Jupiter has moons of its own. We honor his discovery by calling them the *Galilean* moons of Jupiter. Galileo showed that Jupiter and its moons resemble a miniature Copernican planetary system: Just as planets revolve around the Sun, so too do the many moons of Jupiter revolve around it. By the end of the 17th century astronomers had found five moons orbiting around Saturn as well, and this further strengthened their belief in the Copernican system (see Chapter 3). Today we realize that the Solar System abounds with moons. As of June 2006 we count more than 160 observed moons, and there are probably many others—especially in the outer Solar System—that we have not yet found.

If Galileo's discovery of Jupiter's moons was personally satisfying, his other important discovery was decidedly less so. When he first pointed his telescope at Saturn in 1610, the tiny disk seemed to be accompanied by smaller companions on both sides. Unlike the moons of Jupiter that he had found a few months earlier, these features did not move. Galileo was troubled by this because Saturn was like nothing else he had observed. Two years later the "companions" had vanished. Their unexpected disappearance upset the Italian astronomer greatly: He feared his earlier observations had been in error. A few years later the mysterious features reappeared. For more than four decades astronomers puzzled over Galileo's discovery.

KEY CONCEPTS

For centuries, celestial wonders such as Saturn's rings and the Galilean moons of Jupiter delighted those who looked through telescopes. But with the dawn of the space age, robotic explorers traveling through the Solar System have shown us much more than points of light, revealing wondrous, diverse families of worlds orbiting other planets. Among them we will find

- Scores of worlds composed of rock and solid ice, some of which formed with their planets and others that were captured later on.
- Geologically active moons freckled with volcanoes and geysers, and geologically dead moons covered with impact craters.
- Shattered remains of moons and comets forming the rings that surround each giant planet.
- Exquisite, delicate structure in ring systems resulting from subtle gravitational interactions among planets, moons, and ring particles.
- Moons that may harbor deep liquid oceans beneath their ice-covered surfaces and may conceivably provide a home for extraterrestrial life.
- Two frozen, unexplored worlds, Pluto and Charon, orbiting at the boundary between the inner and outer Solar System.

309

In 1655 a 26-year-old Dutch instrument maker, **Christiaan Huygens** (1629–1695), pointed a superior telescope of his own design at Saturn and saw what the astronomers of his day had failed to see. Saturn is surrounded by an apparently continuous, flat **ring**, and as Huygens correctly deduced, the variations in its visibility are caused by changes in the apparent tilt of the ring as Saturn orbits the Sun. What was the nature of this strange ring encircling Saturn? Astronomers assumed it was a solid disk spinning around the planet, an interpretation that lasted for more than a century after Huygens's discovery. That notion began to weaken in 1675 when the great Italian–French astronomer **Jean-Dominique Cassini** (1625–1712) found a gap in the planet's seemingly solid ring. Saturn now appeared to have two rings rather than one, and the gap that separated them became known as the *Cassini Division* that we learned of in the previous chapter.

In 1850 a fainter ring, located just inside the two bright rings, was found independently by English and American observers, giving Saturn a total of three known rings. For illustrators and cartoonists, Saturn had become an icon for depicting all the planets. Yet for more than three and a half centuries, it was the only planet known to have rings.

Most Planets Are Adorned with Moons

The moons of our Solar System are not distributed equally among the planets. Mercury and Venus have none, Earth has one, and Mars has two; thus there are only three moons in the inner part of the Solar System. Among the dwarf planets, Pluto possesses three known moons and Eris has one. All of the remaining moons belong to the giant planets.

Figure 11.1 shows images of many of the major moons in the Solar System, shown to scale. In many ways moons resemble smaller versions of the terrestrial planets. Some, such as our own, are made of rock. Others, especially in the outer Solar System, are mixtures of rock and water ice, with densities intermediate between the two. A few seem to be made almost entirely of ice. Several moons are comparable in size to or even larger than Mercury, while the smallest

Moons are made of rock, ice, or mixtures of both.

known moons would fit within the expanse of a large metropolitan airport. Although most moons are airless, one has an atmosphere denser than Earth's, and several have very low-density atmospheres. Some of the larger moons appear to have differentiated chemically, and three are known to have active volcanoes or geysers.

Our own Moon is not covered in this chapter; we discussed it in Chapter 7 because of its close similarity to the four terrestrial planets of the inner Solar System. On the other hand, Pluto and Eris—nominally dwarf planets—*are* included in this chapter because of their similarity to the icy moons of the outer Solar System.

As discussed in Chapter 6, we believe that many moons were formed at the same time as the planets they orbit, and that they were created in much the same way that the planets themselves grew—from the accumulation of planetesimals orbiting the Sun. In **Excursions 11.1** we follow the formation and evolution of one such moon as it coalesced

Moons that formed together with their planets are called regular moons.

from grains in the disk that orbited the young Jupiter. We call those moons that formed together with their parent planets **regular moons**. Regular moons revolve around their planets in the same direction as the planets rotate and in orbits that lie nearly in the planets' equatorial planes. This is because the debris from which the regular moons formed was orbiting in the planets' equatorial planes and in same direction as the evolving planets were rotating. With few exceptions, regular moons are *tidally locked* to their parent planets. Recall from Chapter 10 that tidal locking causes a body to rotate synchronously with respect to its orbit, as does Earth's Moon. A moon in synchronous rotation around its planet has fixed leading and trailing hemispheres, whose surfaces can appear very different from one another. An especially strange moon is Saturn's Hyperion, seen in **Figure 11.3**. Not only does Hyperion have a peculiar appearance, but its rotation is *chaotic* (see Chapter 10); it tumbles in its orbit with a rotation period and a spin axis orientation that are constantly changing.

Some moons revolve in a direction that is opposite to the rotation of their planets, and some are situated in distant, unstable orbits. These are almost certainly bodies that

Some captured or "irregular" moons have retrograde orbits.

formed elsewhere and were later captured by the planets. We call them **irregular moons**. The largest irregular moon is Neptune's Triton. It orbits Neptune in a **retrograde**, or "backward," direction. Other moons, such as Saturn's Phoebe and Pluto's Charon, also have retrograde orbits. Most of the recently discovered moons of the outer planets are irregular, and many are only a few kilometers across.

We might ask why Mercury and Venus failed to form or capture any moons of their own. But is this really surprising? As we noted earlier, the smaller terrestrial planets tend to have far fewer moons than the giant planets. Mercury and Venus, after all, have only one fewer moon than Earth. Mars apparently captured a couple of asteroids, but Mars is situated adjacent to the main asteroid belt (see Chapter 12). Finally, Earth would be without a moon if not for a cataclysmic collision when the planet was young. What seems

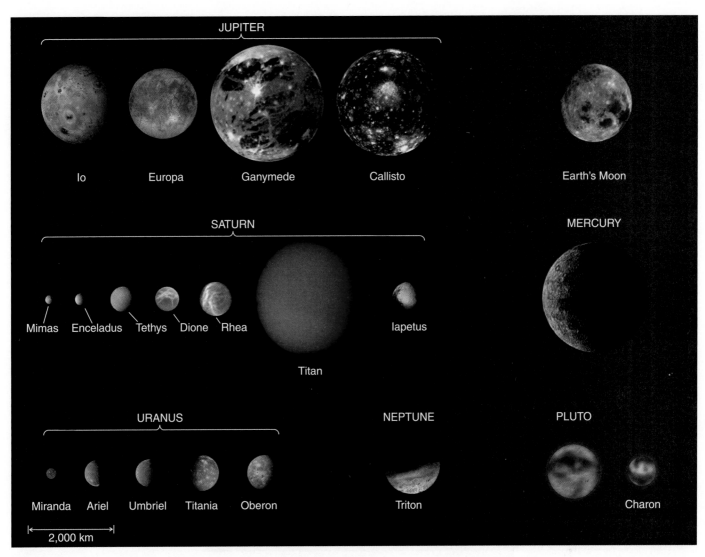

FIGURE 11.1 Images of the major moons in the Solar System obtained by various spacecraft. The images are shown to scale. The planet Mercury and dwarf planet Pluto are shown for comparison. Mars's moons, Phobos and Deimos, are too small to be shown. (Pluto and Charon are artists' conceptions based on ground- and space-based images.)

important is that the larger planets had greater attracting mass and greater amounts of debris around them while they were forming, which is what gave rise to their greater number of moons.

11.2 Rings Surround the Giant Planets

Over the centuries following the discovery of Saturn's rings, exhaustive searches failed to detect rings around any other planet. Papers were published by more than one distin-

guished astronomer explaining why, theoretically, only Saturn could have rings. Then, in the latter part of the 20th century, a new search technique became available: observation of stellar occultations (see Chapter 9). In 1977 a team of American astronomers using the occultation technique to study the atmosphere of Uranus saw brief, minute changes in the brightness of a star as it first approached and then receded from the planet. The interpretation was immediately obvious: Uranus has rings! Over the next several years stellar occultations revealed a total of nine rings surrounding the planet. In 1986 *Voyager 2* imaged two additional Uranus rings, and in 2005 the Hubble Space Telescope recorded two more, bringing the total to 13.

Stellar occultations not only show the existence of rings; they also tell us something about the rings themselves. The

EXCURSIONS 11.1

Formation of a Large Moon

Ganymede, Jupiter's largest moon, is a good example of what we mean by a *regular moon*. Scientists think that a moon like Ganymede formed around young Jupiter in much the same way that the planets formed around the young Sun. Whereas the planets formed from a proto-planetary accretion disk surrounding the Sun, Ganymede grew from the accretion of ice and dust grains that were orbiting in a similar disk surrounding the young, hot proto-Jupiter.

At the distance of Ganymede's orbit from the glowing proto-Jupiter core, temperatures were low enough for grains of water ice to survive, and along with silicate materials, they coalesced to form planetesimals. It took less than a half million years for these planetesimals to accrete and create the moon Ganymede. Heating generated by accretion melted parts of the moon to form an outer water layer, an inner silicate zone, and an ice–silicate core. These layers, however, were not stable. As cooling took place, much of the outer water layer froze to form a dirty ice crust. Most of the denser silicate materials sank to the center to form a core, leaving an intermediate ice–silicate zone (**Figure 11.2**).

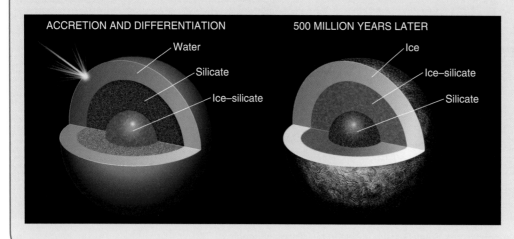

ACCRETION AND DIFFERENTIATION — Water, Silicate, Ice–silicate

500 MILLION YEARS LATER — Ice, Ice–silicate, Silicate

FIGURE 11.2 Diagram showing the evolution of Ganymede's interior. In the final stages of accretion, Ganymede was covered with a global ocean of liquid water, underlain by a zone of silicates and a mixed core of ice and silicates. Five hundred million years later the dense silicates gravitationally had sunk to form a core, and much of the ocean had frozen.

duration of an occultation event is a measure of the width of the ring. The observed decrease in the brightness of a star is an indication of the ring's transparency and therefore of the amount of material it contains. The very brief interruption of starlight as the Uranus rings passed in front of the star showed that they are far too narrow to have been seen in the earlier, unsuccessful searches by more conventional methods.

More ring discoveries were to follow from still another technology: close-up studies by planetary probes. In 1979 cameras on *Voyager 1* recorded a faint ring around Jupiter, and *Pioneer 11* found a narrow ring just outside the bright rings of Saturn.

For a while Neptune seemed to be the only giant planet devoid of rings. Then, in the early to middle 1980s, occultation searches by teams of American and French astronomers began yielding positive but confusing results. Several oc-

cultation events that appeared to be due to rings were seen on only one side of the planet. The astronomers concluded

One of Neptune's rings contains higher-density segments called ring arcs.

that Neptune was surrounded not by complete rings but rather by several arclike ring segments. Only when *Voyager 2* reached Neptune in 1989 did we learn that its rings are indeed complete, and that the **ring arcs** are merely high-density segments within one of its narrow rings. All of Neptune's rings are faint, and with the exception of the ring arcs, they contain too little material to be detected by the stellar occultation technique.

All of the giant planets are now known to have ring systems, although each is unique. The most complex system belongs to what was once thought to be the only ringed planet—spectacular Saturn. Moreover, it turns out that the giant planets are the only planets in our Solar System that have rings; none of the terrestrial ones do.

Now that we have considered this brief history of ring discoveries, what do we know about the rings themselves? We start our discussion of ring structure using Saturn's densely packed, bright rings as an example. Huygens, with his mid-17th-century understanding of physics, believed Saturn's ring to be a solid disk surrounding the planet (as we saw in Section 11.1). This view was challenged in later years, but it was not until the middle of the 19th century

---

Rings are swarms of tiny moons orbiting according to Kepler's laws.

---

that the brilliant Scottish mathematician, James Clerk Maxwell, showed that solid rings would be unstable and would quickly break apart. They must instead consist of countless numbers of small particles, like so many tiny moons in individual orbits around Saturn. What keeps all these small ring particles together? We saw in Chapter 10 how gaps in rings are caused by orbital resonances with satellites. Later in this chapter we will see that there is much more to the gravitational dance performed by the complex system involving the planet, its moons, and the countless individual ring particles.

Kepler's laws dictate that the speed and orbital periods of all ring particles must vary with their distance from the planet, with the closest moving the fastest and having the shortest orbital periods. The orbital periods of Saturn's bright rings range from 5^h45^m at their inner edge to 14^h20^m at the outer one. Ring particles can vary in size from tiny grains to house-sized boulders, and in these densely packed rings of Saturn, all their orbits must be perfectly circular and in precisely the same plane.

To understand why the orbits of particles in dense rings must be so orderly, imagine yourself riding along on a large particle in the middle of Saturn's bright rings. Around you is a swarm of particles of all shapes and sizes, many of them close enough for you to reach out and touch. They are all moving along in almost perfect step with your own, like members of a well-disciplined marching band. We say "almost" because the particles slightly closer to the planet are moving just a bit faster than you are, and those a little farther out are moving just a bit slower, as Kepler's laws demand. The differences in speed are very small. The particle orbiting 1 meter inward from you is moving only a tenth of a millimeter per second faster than you are—a relative speed difference slower than a snail's pace.

Now consider what would happen if that particle were moving in a slightly noncircular orbit, perhaps because it was bumped by one of its other neighbors. Its eccentricity means that it would drift alternately a bit outward and then inward as it completed each orbit. But as it moved outward,

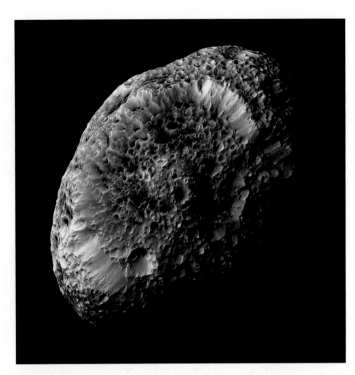

FIGURE 11.3 Saturn's Hyperion, certainly among the strangest-looking moons, rotates chaotically with its rotation period and spin axis constantly changing. The 250-km moon's low density and spongelike texture, seen in this *Cassini* image, suggest that its interior houses a vast system of caverns.

the particle would eventually nudge up against you and be unable to go any farther. With no room for maneuvering, the orbits of errant particles would (and do) quickly become

---

Particles in dense rings must move in circular orbits.

---

circular. The same would be true for any particle in an orbit with even the smallest inclination: During each orbit, the deviant ring fragment would be carried alternately above and below the ring plane. As it approached your particle from, say, above, the mild encounter would quickly remove its out-of-plane motion and bring it into a coplanar (in the same plane) orbit. Saturn's densely packed rings have no place for nonconformists. Like the well-disciplined band, all particles must march along together in unison.

Saturn's Magnificent Rings—A Closer Look

We have chosen the rings of Saturn as our introduction to planetary rings because they alone exhibit such phenomenal complexity. **Figure 11.4** shows the ring systems of the giant planets, including the individual components of Saturn's

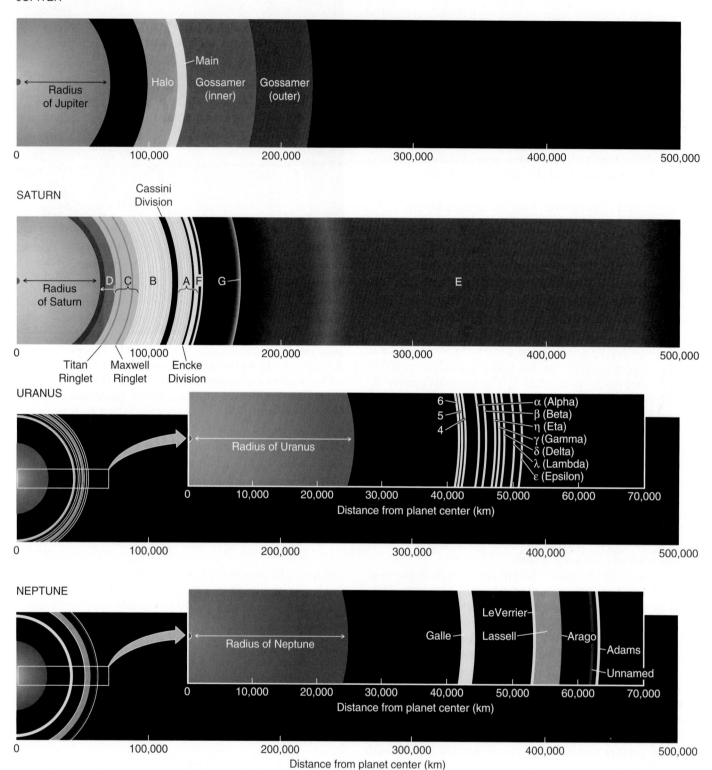

FIGURE 11.4 A comparison of the ring systems of the four giant planets.

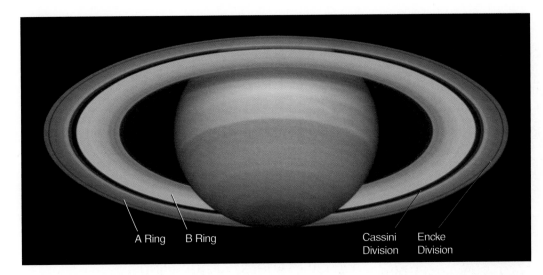

FIGURE 11.5 A Hubble Space Telescope image showing Saturn, the A Ring (the outermost bright ring), the B Ring (the middle bright ring), the Cassini Division (the wide, relatively dark division between the A and B Rings), and the Encke Division (the narrow division near the outer edge of the A Ring).

A Ring   B Ring   Cassini Division   Encke Division

bright ring system and its major divisions and gaps. The most conspicuous are the expansive bright rings, which dominate all photographs of Saturn (**Figure 11.5**). Among the four giant planets, only Saturn has rings so wide and so bright.

Photographs usually show only the two outer and brighter rings, separated by the Cassini Division. The outermost or A Ring is the narrowest of the three bright rings. It has a sharp outer edge and contains several narrow gaps. The conspicuous Cassini Division is so wide (4,700 km) that the

---

"Divisions" and "gaps" in Saturn's rings are not truly empty.

---

planet Mercury would almost fit within it. Astronomers once thought that it was completely empty. In fact, space scientists planning the 1979 encounter of *Pioneer 11* with Saturn gave serious consideration to flying the spacecraft through the Cassini Division to get as close to the planet as possible. Had they carried out this plan, the *Pioneer 11* mission would certainly have ended right there. Images taken the following year by *Voyager 1* show the Cassini Division to be filled with material. Why, then, did astronomers think it was empty? The Cassini Division and many other smaller "gaps" in the bright rings are simply regions with less ring material. In contrast with their denser surroundings, they appear darker and thus empty.

The B Ring is the brightest of Saturn's rings. With a width of 25,500 km, two Earths could fit side by side between its inner and outer edges. Strangely, the B Ring seems to have no gaps at all, at least on the scale of those seen in the other bright rings. The C Ring often fails to show up in normally exposed photographs because of the limited ability of film to record a wide range of brightnesses. At the eyepiece of the telescope, though, this beautifully translucent ring ap-

pears like delicate gossamer and hence is often called the *Crepe Ring*. There is no known gap between the C Ring and either of the adjacent rings. Only an abrupt change in brightness marks the boundary between them. What could cause such a sharp change in the amount of ring material remains an unanswered question. Too dim to be seen next to Saturn's bright disk, the D Ring is a fourth wide ring that was unknown until imaged by *Voyager 1*. It shows more coarse structure than any of the bright rings, and it appears to have no inner edge. The D Ring may extend all the way down to the top of Saturn's atmosphere, where its ring particles would enter and burn up as meteors.

Saturn's bright rings are far from homogeneous. The A and C Rings contain hundreds, and the B Ring thousands, of individual **ringlets**, some only a few kilometers wide (**Figure 11.6**). Each of these ringlets is a narrowly confined

---

Saturn's rings contain thousands of individual narrow rings called ringlets.

---

concentration of ring particles bounded on both sides by regions of relatively little material. Spacecraft images of Saturn's rings turned up many other surprises, as we will see later.

Each time the plane of Saturn's rings lines up with Earth, as it does about every 15 years, the rings all but vanish for a day or so in even the largest telescopes. With the glare of the rings temporarily gone, we can search for undiscovered moons or other faint objects close to Saturn. In 1966 an astronomer looking for moons found weak but compelling evidence for a faint ring near the orbit of Saturn's moon Enceladus. In 1980 *Voyager 1* confirmed the existence of this faint ring, now called the E Ring, and found another closer one known as the G Ring. Both the E and G Rings are diffuse rings with no distinct boundaries. Unlike

FIGURE 11.6 This *Voyager 2* image of the outer B Ring shows so many ringlets and minigaps that it looks like a close-up of an old-fashioned phonograph record. The narrowest features are only 10 km across, at the limit of resolution. Even finer structure was noted during stellar occultations by the rings, as observed by the *Voyager* photometer.

the bright and the narrow rings, diffuse rings do not have well-defined inner and outer edges, nor are they confined to a thin plane.

Although Saturn's bright rings are very wide—more than 62,000 km from the inner edge of the C Ring to the outer edge of the A Ring—they are extremely thin. From our previous discussion, you might guess that they could be no thicker than the diameter of the larger ring particles. But in these densely packed rings, there is simply not enough

---

Saturn's bright rings are exquisitely thin.

---

room to jam all of the particles into the same plane, so they settle down as close as they can get to the ring plane. Saturn's bright rings are thus no more than a hundred meters and probably only a few tens of meters from their lower to upper surfaces. The extremes between their width and their thickness can be difficult to picture, but let's try.

Say you'd like to make a scale model of Saturn and its rings using a basketball to represent Saturn. The basketball is about 20 cm in diameter. You could make the three bright rings out of paper by cutting a circle 45 cm in diameter, with a 25-cm hole cut from the center. To represent the Cassini Division, you could paint a dark stripe 1.5 cm wide and about 12 cm from the outer edge. After mounting the paper ring around the basketball, you have a splendid model of Saturn and its rings. Unfortunately, your model is not completely to scale: The paper rings are more than a thousand times too thick! If you wanted to make your rings from paper similar to that used in this book, the planet would have to be a ball

250 meters in diameter, and the paper rings would have to extend over the length of six football fields! The diameter of Saturn's bright ring system is 10 million times the thickness of the rings.

In diffuse rings, where the separation between particles may be very large, an occasional collision between two particles can cause their orbits to become eccentric, inclined, or both. Because it is unlikely that these disturbed orbits will experience a restoring collision, the particles are likely to remain in their noncircular, noncoplanar orbits. For this reason diffuse rings become spread out and thick, sometimes without any bounds.

Other Planets, Other Rings

Ring structure among the other giant planets is not as diverse as Saturn's. Most rings other than Saturn's are quite narrow, although a few are diffuse. What we see when looking at a ring system depends dramatically on the lighting conditions under which we view the rings. Pebbles and boulders are easiest to see if the light is coming from behind us and reflecting off these relatively large pieces of material. On the other hand, if you have ever tried to drive into sunlight on a dusty day, you may have noticed that particles of dust stand out most strongly when we look *into* the light. As discussed in **Excursions 11.2**, photographers call this effect **backlighting**. When *Voyager* scientists looked at Jupiter's ring with the Sun behind the camera, all they saw was a narrow, faint strand. However, when they looked back toward the Sun while in the shadow of the planet, Jupiter's rings suddenly blazed into prominence. Jupiter's rings are mostly made up not of rocks but of fine dust.

Eleven of the 13 rings of Uranus are quite narrow and widely spaced relative to their widths. Most have widths of only a few kilometers but are many hundreds of kilometers apart (see Figures 11.4 and 11.7(a)). The outermost, called the Epsilon Ring, is the widest of the narrow rings. Its width varies between 20 and 100 km. The 11th ring of Uranus is wide and diffuse, with an undefined inner edge. As with Saturn's D Ring, material in the 11th ring may be spiraling into the top of the Uranus atmosphere. When viewed under backlit conditions by *Voyager 2* (Figure 11.7(b)), the space between Uranus's rings turned out to be filled with dust, much like Jupiter's rings.

Three of Neptune's six rings are narrow, with widths of a few tens of kilometers (see Figure 11.4). They are named after 19th century astronomers who made major contributions to Neptune's discovery. Much of the material in the Adams Ring is clumped together into several arclike segments, with lengths of 4,000 to 10,000 km and a width of about 15 km (**Figure 11.8**). Another ring, as yet unnamed, is 5,800 km wide and lies between the Adams and Le Verrier Rings.

The Backlighting Phenomenon

Photographers often place their models in front of a bright light to highlight their hair, a technique called *backlighting*. Under backlit conditions individual strands of hair shine brightly, creating a halo effect around a model's face. This happens when light falls on very small objects —those with dimensions a few times to several dozen times the wavelength of light. Human hair is near the upper end of this range. Light falling on the strands of hair is not scattered uniformly in all directions; rather, it tends to continue in the direction away from the source of illumination. Very little of the light is scattered off to the side, and almost none back toward the source.

Some of the dustier planetary rings are filled with particles whose sizes are just a few times the wavelength of visible light. To a spacecraft approaching from the direction of the Sun, such rings may be difficult or even impossible to see as in **Figure 11.7(a)**. This is because the tiny ring particles scatter very little sunlight back toward the Sun and the approaching spacecraft. When the spacecraft passes by the planet and looks backward in the general direction of the Sun (**Figure 11.7(b)**), these dusty rings suddenly appear as a circular blaze of light, much like a halo surrounding the nighttime hemisphere of the planet. As illustrated by the images of the rings of Uranus in Figure 11.7, many planetary rings are best seen with backlighting, and some have been observed only under these conditions.

Backlighting has also been remarkably successful in revealing tiny ice crystals erupting from cryovolcanoes on Enceladus, as shown dramatically later in this chapter in Figure 11.17(b).

(a)

(b)

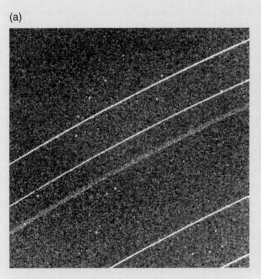
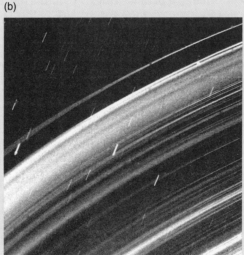

FIGURE 11.7 The appearance of rings depends dramatically on lighting conditions. Here the rings of Uranus (a) appear as narrow, faint bands when viewed with the Sun at our back but (b) burst into dazzling brilliance when illuminated from behind. (Bright lines in the images are stars "streaked out" by the spacecraft motion during exposures.)

Moons Are the Origin of Ring Material

Planetary rings are a wonderland of the sorts of gravitational interactions that we discussed in Chapter 10. Tides, for example, are thought to be largely responsible for much of the material found in planetary rings. If a moon or other planetesimal held together by gravity alone happens to be within the Roche limit of a giant planet, it will be pulled apart by tides. The location of Saturn's bright rings inside the planet's Roche limit, for example, suggests that the rings formed

> Saturn's rings are the remains of icy moons disrupted by tides.

from the breakup of one or more moons or captured planetesimals that were perturbed by larger moons into orbits that took them too close to the planet.

FIGURE 11.8 A *Voyager 2* image of the three brightest arcs in Neptune's Adams Ring. Neptune itself is very overexposed in this image.

If rings are made of material that originated on moons, we might expect to find the composition of a planet's rings mimicking the makeup of the moons that surround the planet. Such is indeed the case. Saturn's bright rings appear bright because they reflect about 60 percent of the sunlight falling on them. We might suspect from their brightness alone that they are made of water ice, and spectral observations confirm this suspicion, clearly showing the distinct signature of water. A slight reddish tint to the rings tells us that they are not made of pure ice but must be contaminated with other materials such as silicates. The icy moons around Saturn or the frozen comets that prowl the outer Solar System could easily provide this material.

Saturn's bright rings are the brightest in the Solar System and are the only ones that we know are composed of water ice. In stark contrast, the rings of Uranus and Neptune are among the darkest objects known in our Solar System. Only 3 percent of the sunlight falling on them is reflected

The rings of Uranus and Neptune are darker than coal.

back into space, making the ring particles blacker than coal or soot. No silicates or similar rocky materials are this dark. However, a number of carbon-rich meteorites are this dark, as is the nucleus of Comet Halley. This suggests that the rings of Uranus and Neptune may be formed largely of similar material, rich in organic compounds. Jupiter's rings are neither as bright as Saturn's nor as dark as those of Uranus and Neptune, suggesting they may be rich in dark silicate materials, like the innermost of Jupiter's small moons.

The jumble of fragments that make up Saturn's rings is easily understood as a product of tidal disruption of a moon or planetesimal, but moons can contribute material to rings in other ways as well. The last looks that *Voyager* had at Jupiter's rings carried hints of things unseen, and so matters were to remain for 20 years. It is ironic that until recently we knew less about Jupiter's rings than any other ring system in the Solar System; yet as a result of *Galileo*'s arrival at the planet, Jupiter's ring system is now among the best observed. **Figure 11.9** shows a *Galileo* image of Jupiter's rings. Jupiter's ring system turns out to be largely the product of the strong gravity of the planet itself, combined with the presence of a handful of small, rocky moons close to the planet. As interplanetary meteoroids are pulled toward Jupiter by its strong gravity, a few of them strike the surface of one of Jupiter's four innermost moons. These moons are so tiny that some of the dust from these impacts is kicked off with speeds in excess of the escape velocities of the moons. This dust provides a steady supply of material for the rings.

The brightest of Jupiter's rings is a narrow strand only 6,500 km across, consisting of material from Metis and Adrastea (Figure 11.9). These two moons orbit in Jupiter's equatorial plane, and the ring they form is thin. Beyond the main ring, however, are the very different Gossamer Rings, so-called because they are extremely tenuous. The Gossamer Rings are supplied by dust from the moons Amalthea and Thebe. Unlike the main ring, the Gossamer Rings are rather thick; the inner Gossamer Ring, associated with Amalthea, is actually located within the outer Gossamer Ring formed of material from Thebe. These rings are so thick because the orbits of the moons that supply the ring material are slightly tilted with respect to Jupiter's equatorial plane. The orbital planes of these moons wobble, as does the orbital plane of Earth's Moon; but instead of taking almost 19 years to complete one wobble (as our Moon does), these moons complete a wobble in only a few months. The Gossamer Rings, made up of material from these wandering moons, are spread as far below and above Jupiter's equatorial plane as the orbits of the satellites that form them.

The innermost ring in Jupiter's system, called the Halo Ring, consists mostly of material from the main ring. As the dust particles in the main ring drift slowly inward toward the planet, they pick up electric charges and are pulled into this rather thick torus, or doughnut-shaped ring, by **electromagnetic forces** associated with Jupiter's powerful magnetic field.

Finally, moons may contribute ring material through volcanism. Volcanoes on Jupiter's moon Io continuously eject sulfur particles into space, many of which drift inward under the influence of pressure from sunlight and find their way into the Jupiter ring. The particles in Saturn's E Ring appear to be ice crystals ejected from the moon Enceladus, which is located in the very densest part of the E Ring.

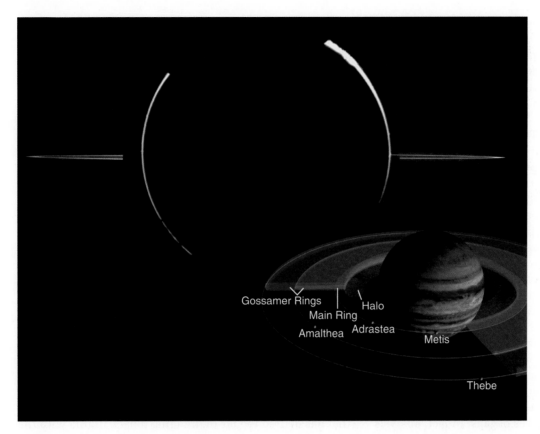

FIGURE 11.9 A backlit *Galileo* view of Jupiter's system of rings and a diagram of the small moons that form them.

Moons Maintain Order and Create Gaps

Planetary rings are ephemeral; they do not have the long-term stability of most Solar System objects. Ring particles are constantly colliding with one another in their tightly packed environment, either gaining or losing orbital energy as they do so. This redistribution of energy can cause particles at the ring edges to leave the rings and drift away,

Rings don't last forever.

aided by nongravitational influences such as the pressure of sunlight. Under these conditions all rings should quickly dissipate. Yet we see that many rings have sharp edges with no visible material appearing in the space beyond. For these rings at least, something is holding most of the particles in place, preventing the rings from quickly dissipating.

The key to the sharp edges and stability of the rings lies with the same sort of orbital resonances discussed in Chap-

Moons maintain sharp ring edges and create gaps in the rings.

ter 10. As moons orbit a planet, their gravity can nudge ring particles that are in resonant orbits, keeping them in line.

Consider, for example, the abrupt outer edge of Saturn's A Ring. There is no visible material beyond this boundary. The ring particles at the edge of the A Ring are in a 7:6 orbital resonance with the co-orbital moons Janus and Epimetheus, meaning that the ring particles make *precisely* seven orbits for every six orbits of Janus and Epimetheus.

Now picture a particle suffering a collision and being forced a little outside the edge of the ring. Each time Janus passes the particle, once out of every six Janus orbits and every seven particle orbits, it does so at the same position in the particle's orbit. And each time the particle receives a small gravitational nudge from the moon in the same direction. The effects build up, and eventually the particle loses enough energy to fall back into the ring. In short, the sharp edge of the ring itself is just like the sharp edge of a resonant gap such as the Cassini Division.

Each time a moon nudges a ring particle inward, the minute amount of energy lost by the particle is absorbed by the more massive moon, causing it to move imperceptibly farther from the ring and the planet. Over time a moon may move so far from the ring edge that it can no longer provide stability to the ring. The ring is then free to dissipate. In the case of Saturn's bright rings, the situation is even more complicated because Janus itself is also in resonance with other moons of Saturn, meaning that all these moons must be pushed away before the rings can dissipate.

Thus the bright rings of Saturn may be much more stable than most rings.

Moons can also influence particles deep within a ring. All that is necessary is for the moon to be massive enough to create a significant gravitational tug on the particles and be in orbital resonance with them. In Chapter 10 we discussed the 2:1 orbital resonance with Saturn's moon, Mimas, that creates the Cassini Division. This is the same mechanism by which Jupiter creates the Kirkwood gaps in the asteroid belt. Such resonances are known to produce some of the gaps that appear in Saturn's bright rings. For example, we know that one of the gaps in the C Ring is caused by a 4:1 resonance between the ring particles and Mimas. Unfortunately, the cause of many—perhaps we should say most —other gaps in Saturn's rings remains unexplained. If they also are produced by resonances, we have yet to identify the source. One possibility is that these gaps are the result of collisions between ring particles. Any collision between two ring particles will cause one of the particles to move to an orbit farther out, and the other to an orbit farther in. Over time this process could sweep some areas clean of ring particles, forming gaps while piling those ring particles up in narrow ringlets between the gaps.

There are also important gravitational interactions between ring particles themselves. These interactions determine the ring shapes at the edges of gaps. In fact, analysis

> The mass of all Saturn's rings combined is about the same as a small icy moon.

of the shapes of ring edges allows us to estimate the masses of the rings. Even though planetary rings can be large and prominent, they account for only the tiniest fraction of the mass of the material around a giant planet. Saturn's bright rings are by far the most massive rings in the Solar System. In fact, they contain more material than all other planetary rings combined. Even so, their total mass is estimated to be less than that of Mimas, a small icy moon of Saturn about 390 km in diameter. The amount of material in the narrow rings is, of course, much less. All of the particles in the largest ring of Uranus, the Epsilon Ring, could be compressed into a single body no more than 20 km across. All of the material in both the Neptune rings and the Jupiter ring could fit into single objects only a few kilometers in diameter.

Other kinds of orbital resonances are possible in ring systems. For example, most narrow rings are caught up in a periodic gravitational tug-of-war with nearby moons.

> Moons that keep narrow rings from spreading are called shepherd moons.

These are called **shepherd moons** in recognition of the way they shepherd a flock of ring particles. Shepherd moons are usually small, are located close to a narrow ring, and often come in pairs, one orbiting just inside and the other just outside the narrow ring. The shepherding mechanism is much like the resonances discussed earlier. A shepherd moon just outside a ring will rob orbital energy from any particles that drift outward beyond the edge of the ring, causing them to move back inward. A shepherd moon just inside a ring will give up orbital energy to a ring particle that has drifted too far in, nudging it back in line with the rest of the ring. In some cases narrow rings are trapped between two shepherd moons in slightly different orbits.

Strange Things among the Rings

Among the many strange rings imaged by *Voyager*, the archetype is clearly Saturn's F Ring, shown in **Figure 11.10**. Images of the ring taken by *Pioneer 11* a year earlier had shown nothing out of the ordinary, but the first high-resolution im-

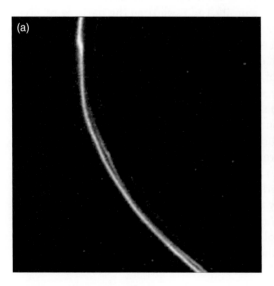

FIGURE 11.10 (a) A *Voyager 1* image of Saturn's "twisted" F Ring. (b) A *Cassini* view of the F Ring and the 85-km moon Pandora.

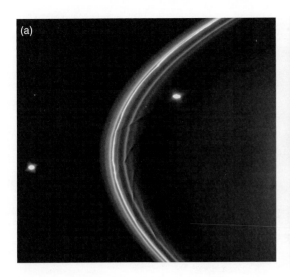

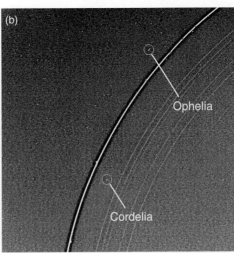

FIGURE 11.11 (a) *Cassini* image of Saturn's F Ring and its shepherd moons, 85-km Pandora (left) and 100-km Prometheus (right). (b) *Voyager 2* image of Uranus's Epsilon Ring with its 40-km shepherds, Cordelia and Ophelia.

ages of the F Ring made by *Voyager* had startled spacecraft scientists staring in disbelief. The ring was separated into several strands that appeared to be intertwined. Some media reporters quickly claimed that the F Ring was "disobeying the laws of physics." This, of course, was not the case; but ready explanations for this seemingly non-Keplerian behavior of the ring particles were not immediately forthcoming. And as if the multiple strands were not perplexing enough, the ring also displayed what appeared to be a number of knots and kinks.

Saturn's F Ring is now understood to be a dramatic example of the action of a pair of shepherd moons. The F Ring is flanked by Prometheus, a moon that orbits 860 km inside the ring, and Pandora, the second moon, which orbits 1,490 km on the outside (**Figure 11.11(a)**). Both moons are irregular in shape, with average diameters of 110 and 90 km

Rings can be distorted by the gravitational influence of nearby moons.

respectively. Because of their relatively large size and proximity, they exert significant gravitational forces on nearby ring particles. The resulting tug-of-war between Prometheus pulling ring particles in its vicinity to larger orbits and Pandora drawing its neighbors into smaller orbits is the cause of the bizarre structure that originally baffled scientists and reporters alike.

The F Ring is not an isolated case. The 360-km-wide Encke Division in the outer part of Saturn's A Ring contains two narrow rings that show knots and gaps (see **Figure 11.12(a)**), structure that must be related to a 20-km diameter moon, Pan, which orbits within the Encke Division. The rings in the Encke Division are unusual but not unique, for they bear some resemblance to the arcs in Neptune's Adams Ring. Arclike segments appear in Uranus's Lambda Ring as well. As yet there is no generally satisfactory explanation for the origin of ring arcs. Recent infrared imaging by the

Hubble Space Telescope shows that the arcs in Neptune's Adams Ring have changed little over the past two decades since they were first seen by *Voyager 2,* but we know nothing about their long-term stability.

If shepherd moons are in eccentric or inclined orbits, they cause the confined ring to also be eccentric or inclined, as is the case for Uranus's Epsilon Ring (**Figure 11.11(b)**). Because ring shepherds can be so small, they often escape detection. According to current theories of ring dynamics, there must be a number of still unknown shepherd moons interspersed among the ring systems of the outer Solar System.

Small moons orbiting within ring gaps can also disturb ring particles along the edges of the gaps. Pan is the cause of a scalloped pattern (**Figure 11.12(b)**) along the Encke Division's inner edge, and **Figure 11.12(c)** catches the 7-km diameter moon Daphnis in the act of disrupting the inner and outer edges of Saturn's 35-km wide Keeler Gap, located near the outer edge of the A Ring.

Various orbital resonances with moons may help guide the orbits of ring particles and delay the dissipation of the rings themselves, but at best this can be only a temporary holding action. All rings eventually face their inevitable fate—total dissipation. It is quite unlikely that the planetary rings we see today have existed in their current form since

Earth doesn't have a ring because it lacks shepherding moons.

the Solar System's beginning. Indeed, it is far more likely that many ring systems may have come and gone over the history of the Solar System. Even our own planet has probably had several short-lived rings at various times during its long history. Any number of comets or asteroids must have passed within Earth's Roche limit[1] and disintegrated

---

[1] Earth's Roche limit is about 25,000 km for rocky bodies and over twice that for icy bodies.

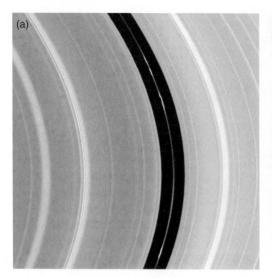

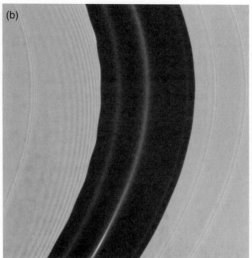

FIGURE 11.12 (a) One of two discontinuous and knotted rings within Saturn's Encke Division. (b) A higher-resolution view of the Encke Division reveals a scalloped pattern along the division's inner edge caused by the small moon Pan. (c) The tiny moon Daphnis disrupts both edges of Saturn's Keeler Gap.

catastrophically into a swarm of small fragments, thereby creating a temporary ring. Yet Earth lacks shepherding moons to provide orbital stability to rings. Interactions between ring particles would have caused such a ring to spread out and dissipate, while the inner parts of a ring around Earth would feel the drag of Earth's extended atmosphere and spiral inward, creating spectacular meteor displays as they fell. A similar absence of small inner moons also prevents Venus and Mercury from keeping rings over geological time scales. We leave Mars off the list for now. Although we know of no ring around Mars, its two tiny moons, Phobos and Deimos, might easily serve to shepherd a collection of orbiting debris. As we continue to explore our Solar System, Mars may surprise us yet.

It is a wild understatement to say that many of the physical properties of planetary rings were unexpected by planetary scientists before spacecraft began surveying the giant planets at close range. We now have come to appreciate that much of this structure—from the narrowest strands in the F Ring to the countless ringlets of the A, B, and C Rings—is the result of the same gravitational dance between moons and rings that we encountered in Chapter 10. Here and in that earlier discussion, we have only scratched the surface of what is possible. Indeed, some of the most extraordinary structure seen in planetary rings is still not well understood, decades after its discovery. In the end we must tip our hats to the seemingly boundless diversity of unpredictable, chaotic behavior that is possible in such complex systems.

Saturn's B Ring has transient radial features called spokes.

Gravitational interactions do not hold a monopoly on strange behavior among ring systems. One of the more puzzling discoveries made by *Voyager 1* at Saturn was the appearance of dozens of dark spokelike features in the outer part of the B Ring (**Figure 11.13**). The **spokes** rotated at the same Keplerian rate as the ring particles, and grew in a radial direction as they circled Saturn. Yet no individual spoke was seen to last for more than half an orbit. This half-orbit survival tells us that the particles in the spokes must be suspended above the ring plane. Why is this obvious?

Any particle that is not in the ring plane must be in an inclined orbit, and it thus has to pass through the plane twice during each orbit of the planet. As the spoke particles try to pass through the densely packed B Ring, they run into the B Ring particles and are absorbed. Such a model can nicely explain what causes the spokes to disappear, but it does not tell us why they appear.

One plausible suggestion links the origin of the spokes with meteoroid impacts on large ring particles. Meteoroids, as they strike these particles, can collide with so much energy that they create an ionized cloud of tiny charged particles—a plasma—that becomes briefly suspended above the ring plane. Before the cloud can return to the ring plane, Saturn's magnetic field causes the charged particles to drift outward, creating the radial spokelike features. Still, questions remain. For example, our model has no explanation for the fact that *Voyager* and *Cassini* have imaged spokes only in the outer part of the B Ring and not in the inner part of the ring or in either of the other two wide rings.

11.3 Moons as Small Worlds

There are several ways we could group the moons of our Solar System. Some schemes are based on the sequence of the moons in their orbits around the parent planets, and others on the sizes or compositions of the moons. For example, earlier in the chapter we saw that a few moons are predominantly rocky objects, some are mostly ices, and most appear to be mixtures of ice and rock.

The scheme we will use here is based on the amount and timing of the moons' geological activity as expressed by the features we see on their surfaces. As we learned when looking at the terrestrial planets and Earth's Moon, surface features observed on planets and moons provide critical clues to their geological history. For example, water ice is a

---

Brightness, structure, and crater density of moon terrains give clues to geological activity.

---

common surface material among the moons of the outer Solar System, and the freshness of that ice tells us something about the ages of those surfaces. Meteorite dust darkens the icy surfaces of moons just as dirt darkens snow late in the season in our own urban areas. In other words, a bright surface often (but not always) means a fresh surface. But as we will see later in this section, the *oldest* surfaces on Jupiter's Europa are bright, whereas the *youngest* are dark.

In addition, as we discussed in Chapter 7, the size and number of impact craters gives us the relative timing of events such as volcanism, and this timing allows us to gauge if and when a moon may have been active in the past. Terrains having a large number of craters are older than those having only a few. Observations of erupting volcanoes, as on Io and Enceladus, are direct evidence that some moons are geologically active today.

FIGURE 11.13 Spokes in Saturn's B Ring appear (a) dark in normal viewing but (b) bright with backlighting, indicating that the spoke particles are very small.

What happens when we apply these age-dating techniques to the moons in the Solar System? We find an immense diversity—some moons have been frozen in time since their formation during the early development of the Solar System, whereas others are even more geologically active than Earth. In our classification scheme of moons, we include four categories of geological activity: (1) definitely active today, (2) possibly active today, (3) active in the past but not today, and (4) apparently not active at any time since their formation.

Geologically Active Moons: Io, Enceladus, and Triton

One of the more spectacular surprises in Solar System exploration was the discovery of active volcanoes on Io, the innermost of the four large Galilean moons of Jupiter. Yet in one of those rare events that happen in science, Io's volcanism was predicted by planetologists just two weeks before its discovery, based on the very same tidal stresses discussed in Chapter 10. Did you ever take a piece of metal and bend it back and forth, eventually breaking it in half?

Io is the Solar System's most volcanically active body.

Touch the crease line and you can burn your fingers! Just like the metal in your hands, the continual flexing of Io's crust caused by the changing strength and direction of the tides generates enough energy to melt parts of the crust. In this way Jupiter's gravitational energy is converted into thermal energy powering the most active volcanism in the Solar System.

As *Voyager* approached Jupiter, images showed that Io's surface is literally covered by volcanic features, including vast lava flows, volcanoes, and volcanic craters. Amazingly, however, pictures from *Voyager* and *Galileo* failed to show a single impact crater, making Io unique among all the solid planetary bodies seen so far in planetary exploration. With a surface so young, Io must be volcanically active indeed, with lava flows and volcanic ash burying impact craters as quickly as they form. Scientists working on the *Voyager* mission discovered just how volcanically active Io is in spectacular and undeniable fashion when postencounter images looking back toward this moon showed explosive volcanic eruptions sending debris hundreds of kilometers above Io's surface! As of this writing, Io has more than 300 known volcanic vents, and over 60 active volcanoes were observed during the *Galileo* mission between 1996 and 2000. The most vigorous eruptions, with vent velocities of up to 1 km/s, spray sulfurous gases and solids as high as 300 km above the surface. Ash and other particles rain onto the surface as far as 600 km from the vents, as can be seen in the opening photograph in Chapter 10. The moon is so active that frequently several huge eruptions are occurring at the same time. One look at an image such as **Figure 11.14** leaves little doubt about the source of the material supplying the plasma torus and Io flux tube we discussed in Chapter 9.

Io's surface, shown in **Figure 11.15**, displays a wide variety of colors—pale shades of red, yellow, orange, and brown. Mixtures of sulfur, sulfur dioxide frost, and sulfurous salts of sodium and potassium on the moon's surface are the likely cause of the colors. Bright patches may be fields of sulfur dioxide snow. You may well wonder how snow can fall from Io's nearly nonexistent atmosphere. According to current understanding, liquid sulfur dioxide must flow beneath Io's surface, held at high pressure by the weight of overlying material. Like water from an artesian spring, this pressurized sulfur dioxide can be pushed out though fractures in the crust, producing sprays of sulfur dioxide snow crystals that travel for up to hundreds of kilometers before settling back to the moon's surface. (To see a similar process in action, just operate a carbon dioxide fire extinguisher. These fire extinguishers contain liquid carbon dioxide at high pressure that immediately turns to "dry ice" snow as it leaves the nozzle.)

Voyager and *Galileo* images show parts of the surface of Io at high resolution. They reveal a variety of plains, irregular craters, and flows, all related to eruption of mostly silicate magmas onto the surface of the moon. They also show high-standing mountains, some nearly twice the height of Mt. Everest, Earth's tallest mountain. Huge structures, some

Silicate magmas dominate Io's volcanism.

65 km across, show multiple calderas and other complex structures telling of a long history of repeated eruptions followed by collapse of the partially emptied magma chambers. Many of the floors are very hot (**Figure 11.16**) and might still contain molten material similar to magnesium-rich lavas that erupted on Earth more than 1.5 billion years ago. It is important to note that volcanoes on Io are spread around the moon in a much more random fashion than on Earth, where tectonic patterns influence the location of volcanism.

Io probably formed at about the same time as the other giant planets' moons. Based on its current volcanic activity, Io's entire mass may have recycled, or turned inside out,

Volcanism may have turned Io inside out several times.

more than once over the past, leading to chemical differentiation. Volatiles such as water and carbon dioxide probably escaped into space long ago, with most heavier materials sinking to the interior to form a core. Sulfur and various sulfur compounds, aided by silicate magmas, are constantly

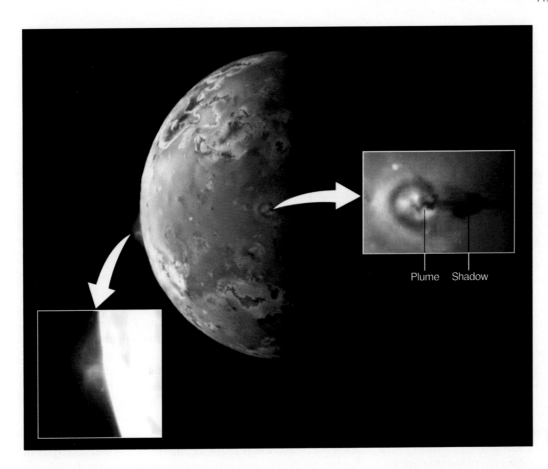

Plume   Shadow

FIGURE 11.14 When this image of Io was obtained by *Galileo*, two volcanic eruptions could be seen at once. The plume of Pillan Patera rises 140 km above the limb of the moon on the left, while the shadow of a 75-km high plume can be seen to the right of the vent of Prometheus, near the moon's terminator.

being recycled, forming the complex surface we see today.

One of Saturn's moons, Enceladus, is only 500 km in diameter. Even so, it shows a wide variety of ridges, faults, and smooth plains—evidence of tectonic processes unexpected for such a small object. Some impact craters, rather than showing crisp features, appear "softened" as though they had "relaxed" by the viscous flow of warm ice. Parts of the moon show no craters at all, indicating recent resurfacing. The orbit of Enceladus is located in the densest part of Saturn's E Ring, and astronomers had long suspected that the moon was providing the ring with a continuous supply of tiny particles, most probably ice crystals.

Confirming evidence of Enceladus activity was provided by *Cassini*. Close-up images of terrain near the moon's south

Water volcanoes on Enceladus are an example of low-temperature cryovolcanism.

pole reveal a cracked and twisted "taffy-like" appearance, as seen in **Figure 11.17(a)**. Temperature measurements of these geologically young "cracks" show them to be warmer than their surroundings. A combination of tidal heating and the decay of radioactive elements within the moon's

FIGURE 11.15 An image of Jupiter's volcanically active moon Io, constructed of images obtained by *Galileo*.

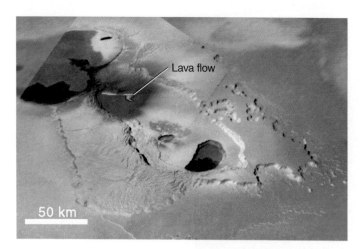

FIGURE 11.16 A *Galileo* image of Io showing regions where lava has erupted. A molten lava flow is shown in false color to make it more visible.

rocky component could generate enough thermal energy to warm the ice and drive it to the surface. This process, called **cryovolcanism**, is similar to normal volcanism but driven by subsurface low-temperature liquids rather than molten rock. *Cassini* showed that active cryovolcanic plumes, like those seen in **Figure 11.17(b)**, are energetic enough to overcome the moon's low gravity, sending tiny ice crystals into space to repopulate particles that are be-

ing continuously lost from the E Ring. We might logically ask why Enceladus is so active while Mimas, a neighboring moon of about the same size and also subject to tidal heating, appears to be quite dead. The answer, unfortunately, remains a mystery.

Triton, the largest moon in the Neptune system, is an irregular moon that must have been captured by its planet. To achieve its present circular, synchronous orbit, it must have experienced extreme tidal stresses from Neptune following its capture. Such stresses would be similar to those on Io and would have generated large amounts of thermal energy. The interior might even have melted, allowing Triton to become chemically differentiated.

Like all moons in the outer reaches of the Solar System, Triton is a cold place. Its surface temperature is only about 38 K. Triton has a thin atmosphere, with a surface composed mostly of ices and frosts of methane and nitrogen (**Figure 11.18**). From the relative lack of craters, we know that the surface is geologically young. Part of Triton is covered with *cantaloupe terrain,* so named because it looks like the skin of a cantaloupe. Irregular pits and hills may represent surface deformation and extrusion of slushy ice onto the surface from the interior. Veinlike features include grooves and ridges that could result from the extrusion of ice along fractures. The rest of Triton is covered with smooth plains of volcanic origin. Irregularly shaped depressions, as wide as 200 km (**Figure 11.19**), formed when mixtures of water, methane, and nitrogen ice melted in the interior of

FIGURE 11.17 *Cassini* images of Enceladus. (a) A twisted and folded surface of deformed ice cracks near the moon's south pole (shown blue in false color) that are warmer than surrounding terrain and appear to be the sources of cryovolcanism. (b) Cryovolcanic plumes in the south polar region are seen spewing ice particles into space.

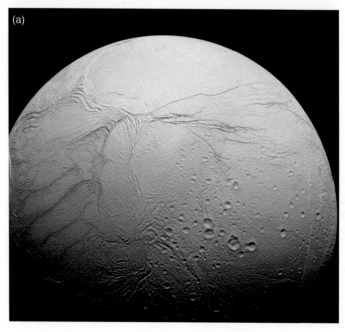

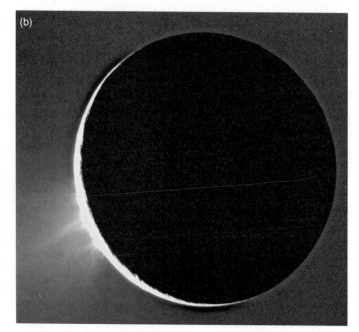

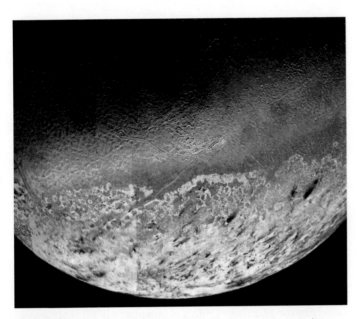

FIGURE 11.18 This *Voyager* mosaic shows various terrains on the Neptune-facing hemisphere of Triton. "Cantaloupe" terrain is visible at the top; its lack of impact craters indicates a geologically younger age than the bright, cratered terrain at the bottom.

Triton and erupted onto the surface, much as rocky magmas erupted onto the lunar surface and filled impact basins on the Moon.

Voyager 2 found four active geyserlike cryovolcanoes on Triton. Each consisted of a plume of gas and dust as much as a kilometer wide rising 8 km above the surface, where the plume was caught by upper atmospheric winds and carried for hundreds of kilometers downwind. How do the eruptions on Triton work? Their association with Triton's southern hemispheric ice cap may provide some clues to

Nitrogen propels Triton's geyserlike volcanism.

the process. We know that nitrogen ice in its pure form is transparent and allows easy passage of sunlight. Thus clear nitrogen ice could create a localized greenhouse effect (see Chapter 8), in which solar energy trapped beneath the ice raises the temperature at the base of the ice layer. A temperature rise of only 4°C would be enough to vaporize the nitrogen ice. As gas is formed, the expanding vapor would exert very high pressures beneath the ice cap. Eventually the ice would rupture and vent the gas explosively into the low-density atmosphere. Dark material, perhaps silicate dust or radiation-darkened methane ice grains, is carried along with the expanding vapor into the atmosphere, from which it subsequently settles to the surface forming dark patches streaked out by local winds, as seen near the bottom of Figure 11.18.

Possibly Active Moons: Europa and Titan

One of the more fascinating objects in the outer Solar System is Jupiter's Europa, a rock world slightly smaller than our Moon but with an outer shell of water. We know that the surface of the water is frozen, but we do not know what lies beneath the ice. Like Io, Europa experiences a continuously changing tidal stress from Jupiter that generates internal energy and possibly volcanism. Some calculations show that the tidal heating may be too small to produce volcanism. On the other hand, sufficient thermal energy might be present to melt the ice (or to keep the water in a liquid state), and there is the possibility of a global ocean below the ice.

Europa's surface is young, with few impact craters, but is deformed by tectonic activity driven by heating from its interior. Regions of chaotic terrain, as shown in **Figure 11.20**, are places where the icy crust has been broken into slabs that have been rafted into new positions. These slabs could be fitted back together like pieces of a jigsaw puzzle. In other

Europa is covered with broken slabs of ice.

areas the crust has split apart and been filled in with new dark material forced to the surface from the interior, in some ways similar to seafloor spreading on Earth (see Chapter 7). With time, frosts cover the dark material, causing a general brightening of the surface. Of the handful of large impact structures preserved on Europa, all are shallow features

FIGURE 11.19 This irregular basin on Triton has been partly filled with frozen water, forming a relatively smooth ice surface. The state of New Jersey could just about fit within the basin's boundary.

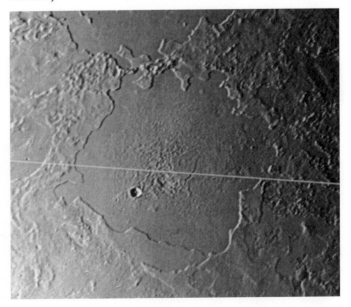

FIGURE 11.20 A high-resolution *Galileo* image of Jupiter's moon Europa, showing where the icy crust has been broken into slabs that, in turn, have been rafted into new positions. These areas of chaotic terrain are characteristic of a thin, brittle crust of ice floating atop a liquid or slushy ocean.

resembling the patterns formed when a rock is dropped into stiff mud. The chaotic terrain, the spreading-like features, the impact structures, and other surface features all suggest that the icy crust of Europa consisted of a thin brittle shell overlying either liquid water or warm, slushy ice at the time when the features formed. Although we do not know whether these are the conditions on Europa today, the geologically young surface holds open that possibility.

Like Mars, Europa is a high-priority target in the search for extraterrestrial life (see **Connections 11.1**). The necessary ingredients for life include liquid water, an energy source, and the presence of organic compounds. A search of the Solar System shows that there are relatively few places where these conditions might be present; Europa is one such place. We know that water is present (at least as ice) and that tidal energy generates thermal energy. Yet the existence of organic materials, the remaining ingredient for life, remains uncertain. However, comets are known to contain organic compounds, and as we will see in Chapter 12, comet nuclei have likely crashed into all of the planets and moons, implanting organic materials throughout the Solar System.

Saturn's moon Titan is larger than Mercury and has a density that suggests a composition of about 45 percent water and 55 percent rocky material. What makes Titan especially remarkable is a thick atmosphere (30 percent denser

A dense, hazy atmosphere obscures Titan's surface.

than Earth's), which generally obscures our view of its surface (**Figure 11.21(a)**). Titan's atmosphere is reminiscent of Los Angeles on a bad day. Like Earth, its atmosphere is mostly nitrogen, but views of Titan's limb show layers in the atmosphere that are probably photochemical hazes, much like smog.

As Titan heated up and differentiated chemically, various ices, including methane, emerged from the interior to form an early atmosphere. Ultraviolet photons from the Sun have enough energy to break the methane molecules apart, a process called *photodissociation*. Methane fragments then recombine to form complex hydrocarbons such as ethane and other organic compounds. Planetary scientists believe that atmospheric ethane behaves much like water in Earth's atmosphere, condensing into a liquid and raining onto the surface. Organic compounds would form tiny particles, creating an organic *smog* and giving Titan's atmosphere its

FIGURE 11.21 *Cassini* images of Saturn's largest moon, Titan. (a) Visible-light imaging shows its orange atmosphere, which is caused by the presence of organic smoglike particles. The purplish haze surrounding the moon is due to short-wavelength scattering of sunlight by small particles. (b) and (c) Infrared light imaging penetrates Titan's smoggy atmosphere and reveals surface features.

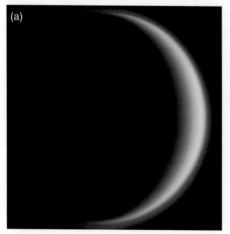
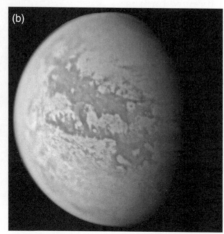
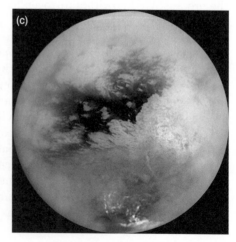

CONNECTIONS 11.1

Origins—Extreme Environments and an Organic Deep Freeze

During a visit in 1835 to the Galápagos Islands in the eastern Pacific Ocean, British naturalist **Charles Darwin** (1809–1882) observed variations in animal and plant life that eventually led to his now legendary **theory of evolution**. Off the coast of the Galápagos, 2,500 meters beneath the ocean's surface, is a form of life that Darwin could never have imagined. Here the Nazca and Pacific plates (see Figure 7.17) grind furiously against one another, creating friction, high temperatures, and seafloor volcanism. Mineral-rich superheated water pours out of *hydrothermal vents*. The surrounding water contains very little dissolved oxygen. No sunlight reaches these depths. Yet in the total darkness of the ocean bottom, life abounds. From tiny bacteria to shrimp to giant clams and tubeworms, sea life not only survives but thrives in this severe environment. In the complete absence of sunlight, the small single-celled organisms at the bottom of the local food chain get their energy from *chemosynthesis*, a process by which inorganic materials are converted into food through the use of chemical energy. Biologists call such life forms *extremophiles*. Robust types of bacteria are found flourishing in the scalding waters of Yellowstone's hot springs, in the bone-dry oxidizing environment of Chile's Atacama Desert, and in the Dead Sea where salt concentrations run as high as 33 percent. Bacteria have even been found in core samples of ancient ice 3,600 m below the surface of the east Antarctic ice sheet. When it comes to harsh habitats, life is amazingly adaptable.

As we have learned in this chapter, Jupiter's moon Europa may contain a shell of liquid water beneath a surface layer of water ice. Some planetologists believe this subsurface sea could be 100 to 150 km deep and contain more water than all of Earth's oceans! How can water exist as a liquid in the frigid realm of the giant planets? Tidal heating, of the kind we encountered in the previous chapter, could pump enough thermal energy into the bottom of Europa's ocean to keep it from freezing. Such an ocean would have some interesting characteristics. We might expect it to be salty due to dissolved minerals and contain an abundance of organic material brought in by impacting comet nuclei. In fact, what we may have on Europa is an environment that is not so different from some of our terrestrial ecological niches that support extremophiles. The universal conditions necessary to create and support life—liquid water, heat, and organic material—could all be present in Europa's oceans.

Europa is not the only moon suspected of having a subsurface ocean. Supporting evidence for Europa's salty sea came when physicists took a close look at *Galileo*'s magnetometer data and found that Europa's magnetic field was variable, indicating an internal electrically conducting fluid. This prompted the scientists to go back and look more closely at the magnetic fields of the other large Jovian moons. To their surprise, Callisto also showed magnetic variability, the signature of a salty ocean. In our search for life elsewhere in the Solar System, Jupiter's large moons should be among our prime targets.

The presence of comet-borne organic material in the oceans of Europa and Callisto cannot yet be confirmed, but there is one place where we *can* find such materials —in Titan's massive atmosphere. Here organic gases reveal themselves through the analytical eyes of the spectrograph. We know that Titan's nitrogen atmosphere contains methane and traces of ethane, propane, ethylene, hydrogen cyanide, carbon monoxide, acetylene, and other compounds of biological interest. For example, five molecules of hydrogen cyanide (HCN) will spontaneously combine to form adenine ($C_5H_3N_4NH_2$), one of the four primary components of DNA and RNA. HCN is also a building block of amino acids, which in turn combine to form proteins. Photodissociation and recombination of these various gases produce complex organic molecules that then rain out onto Titan's surface as a frozen tarry sludge. Biochemists believe that many of these substances are biological precursors, similar to the organic chemistry that preceded the development of life on Earth. For now, these possible clues to the origins of terrestrial life remain locked up in Titan's deep freeze, quietly awaiting analysis by future space missions.

orange hue (see Figure 11.21(a)). But we have a problem here. The photodissociative process would likely remove all atmospheric methane in only 10 million years. So there must be some process for renewing the methane. One likely candidate is cryovolcanism. As on Earth, radioactive decay is an important source of internal heating, and it could cause the release of "new" methane from underground in much the same way that terrestrial volcanism releases water vapor and carbon dioxide into Earth's atmosphere (see Chapter 8). But what about Titan's abundant supply of atmospheric nitrogen? Where did it come from? A likely source is photodissociated atmospheric ammonia, long since depleted by solar ultraviolet photons.

In 2004 the joint NASA–European Space Agency spacecraft *Cassini* began orbiting Saturn with the goal of learning more about this ringed planet and its collection of moons, including Titan (**Figure 11.21(b)**). The spacecraft later released a 320-kg probe, *Huygens*, which plunged through Titan's atmosphere measuring composition, temperature, pressure, and winds and taking pictures as it descended. The atmosphere is mainly nitrogen and methane. Cloud particles contain nitrogen-bearing organic compounds—key components in the production of terrestrial proteins. During its decent *Huygens* encountered 120 m/s winds and temperatures as low as 88 K. As it reached the surface, though, winds had died down to less than 1 m/s, and the temperature had warmed to 112 K. Once on the surface, *Huygens* continued to take pictures and make physical and compositional measurements. *Huygens* scientists described the surface as having characteristics similar to wet or dry

FIGURE 11.22 View of the surface of Titan obtained from the *Huygens* probe, showing a relatively flat surface littered with icy, rounded rocks.

Titan's atmosphere and surface are rich in organic compunds.

---

sand or lightly packed snow. The surface was wet with liquid methane, which evaporated as the probe, heated to 2,000 K during its passage through the atmosphere, landed in the frigid soil. The surface was also rich with other organic compounds, such as cyanogen $(CN)_2$ and ethane (C_2H_6). As shown in **Figure 11.22**, the surface around the landing site is relatively flat and littered with rounded "rocks" of water ice.

Images taken during the descent (see **Figure 11.23(a)**) show terrains reminiscent of Earth, with networks of channels, ridges, hills, and areas that seem to be lake basins. These terrains suggest a sort of *hydrologic cycle* in which methane rain falls to the surface, washes the ridges free of the dark hydrocarbons, then collects into drainage systems that empty into low-lying, liquid methane pools.[2] Stubby dark channels appear to be springs where liquid methane

emerges from the subsurface, while bright curving streaks could be water ice that has oozed to the surface to feed glaciers. Although no liquid methane rain or surface pools were seen in the *Huygens* images, the near absence of impact craters on the surface would point to recent—if not current—hydrologic activity.

As with cloud-covered Venus, an orbiting radar imaging system can pierce Titan's thick cloud cover and expose its hidden surface. *Cassini* has returned intriguing radar views of features resembling terrestrial sand dunes, channels, and impact craters (**Figure 11.23(b)**). Given the dense nitrogen–methane atmosphere and granular material seen on the surface in the *Huygens* images, wind-driven dunes could be composed of snow or ice grains.

In many respects, Titan resembles a primordial Earth, and the presence of organic compounds that could be biological precursors in the right environment makes Titan another high-priority target for continuing exploration (see Connections 11.1).

Formerly Active Moons: Ganymede and Some Moons of Saturn and Uranus

Some moons show clear evidence of past ice volcanism and tectonic deformation, but no current geological activity. Ganymede is the largest moon in the Solar System, even larger than the planet Mercury. The surface is composed of two prominent terrains: a dark, heavily cratered (and there-

---

[2] We are a bit loose with our terminology here. Technically *hydrologic* refers to *water*, as in the terrestrial hydrologic (water) cycle discussed in Chapter 8. On Titan methane assumes the role that water plays on Earth.

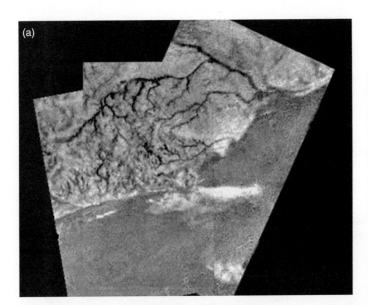

FIGURE 11.23 (a) The surface of Titan viewed from the *Huygens* probe during its descent to the surface. The dark drainage patterns resemble river systems on Earth. (b) A radar image taken from the *Cassini* orbiter, showing an impact crater and its ejecta blanket.

fore ancient) terrain, and a bright terrain characterized by ridges and grooves. Ganymede's low density (1.9 times that of water) indicates that its bulk composition is about half water and half rocky materials. Overall the moon's surface is relatively bright. Even the so-called dark terrains are brighter than the bright areas on Earth's Moon. The high

Ganymede is the Solar System's largest moon.

number of impact craters superposed on the dark terrain reflects the period of intense bombardment during the early history of the Solar System. The most extensive region of ancient dark terrain includes a semicircular area more than 3,200 km across (about the size of Europe) on the leading hemisphere. Furrowlike depressions occurring in many dark areas are among Ganymede's oldest surface features. They may represent surface deformation from internal processes, or they may be relics of impact cratering processes.

Impact craters range up to hundreds of kilometers in diameter, with the larger craters being proportionately more shallow. With time, the icy crater rims deform by viscous (very slow) flow, as might a lump of soft clay, and can ultimately lose nearly all of their varying topography. Such features are seen as flat circular patches of bright terrain, called **palimpsests**, which are characteristic of Ganymede's icy lithosphere (**Figure 11.24(a)**). Palimpsests are found principally in the dark terrain of Ganymede and are believed to be scars left by early impacts onto a thin icy crust overlying water or slush, as illustrated in **Figure 11.24(b)**.

When astronomers first viewed the bright terrain in *Voyager* images, it was thought to represent regions that

had been flooded by water or slush erupted from the interior of Ganymede. *Galileo* images, however, failed to show any indications of such flooding, other than in a few local places. Then how did the bright terrain form? The answer seems to be related to a style of surface formation by tectonic processes not previously considered. In Chapter 7 we discussed how planetary surfaces can be fractured by faults or folded by compression resulting from movements such as those initiated in the mantle. On Ganymede the tectonic processes have been so intense that the fracturing and faulting have completely deformed the icy crust, destroying all signs of older features such as impact craters.

Many other moons show evidence of an early period of geological activity that has resulted in a dazzling array of terrains. A 400-km impact crater scars Saturn's moon Tethys, covering 40 percent of the diameter of the moon itself, while an enormous canyonland wraps at least three-quarters of the way around the moon's equator. Dione's surface shows bright rays that could be frost deposited from the explosive release of interior gases through fractures in the crust. Some craters on Rhea show bright patches on the walls, which may be ice deposits exposed relatively recently by landslides. The trailing hemisphere of Iapetus is bright, reflecting half the light that falls on it, while much of the leading hemisphere is as black as tar. Some planetary scientists believe that the dark material erupts from the interior and then rains onto the surface, creating the wispy, featherlike patterns seen in *Cassini* images (**Figure 11.25**). On the other hand, the fact that these dark deposits appear *only* on the leading hemisphere suggests debris swept up as Iapetus moves along in orbit around Saturn.

(a)

(b)

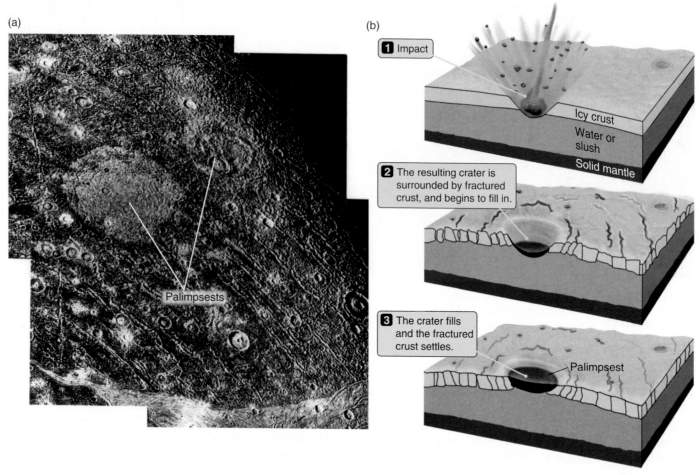

FIGURE 11.24 (a) *Voyager* image of impact scars called palimpsests on Jupiter's moon Ganymede. (b) Palimpsests form as viscous flow smooths out structure left by impacts on icy surfaces.

FIGURE 11.25 *Cassini* images of Iapetus showing the boundary between its bright and dark hemisphere.

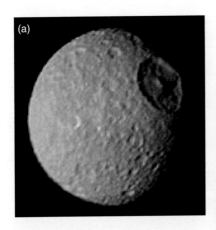

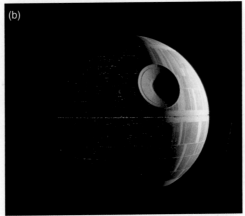

FIGURE 11.26 (a) *Voyager* image of Saturn's moon Mimas and the crater Herschel. (b) The Death Star from *Star Wars.* (The movie was released three years before the image of Mimas was taken.)

Saturn's Mimas, no larger than Ohio, is heavily cratered with deep, bowl-shaped depressions. The most striking feature on Mimas is a huge impact crater on the leading hemisphere. Named Herschel after the German–English astronomer Sir William Herschel, who discovered many of Saturn's moons, the crater is 130 km across, or a third the size of Mimas itself. (**Figure 11.26** illustrates the striking similarity between Mimas and the "Death Star" from the original *Star Wars,* which was released in 1977, three years before

Mimas may have been broken apart and reassembled many times.

the image of Mimas was taken.) It is doubtful that Mimas could have survived the impact of a body much larger than the one that created Herschel. Some astronomers believe that Mimas (and perhaps other small, icy moons as well) was hit many times in the past by objects so large as to fragment the moon into many small pieces. Each time this happened, those individual pieces still in Mimas's orbit would coalesce to reform the moon.

Areas on Uranus's Miranda have been resurfaced by eruptions of icy slush or glacierlike flows (**Figure 11.27**). Other Uranus moons—Oberon, Titania, and Ariel—are covered with faults and other signs of early tectonism. On Ariel, in particular, very old, large craters appear to be missing, perhaps obliterated by earlier volcanism.

Geologically Dead Moons

Geologically dead moons, including Jupiter's Callisto, Uranus's Umbriel, and a large assortment of irregular moons, are those for which there is little or no evidence of internal activity having occurred at any time since their formation. Their surfaces are heavily cratered and show no modification other than the cumulative degradation caused by a long history of impacts.

Callisto is about the size of Mercury. It is also the darkest of the Galilean moons, yet it is still twice as reflective as Earth's Moon. This indicates that it is rich in water ice, but with a mixture of dark, rocky materials. Except for terrains related to large impact events, the surface is essentially uniform, consisting of relatively dark, heavily cratered terrain. High-resolution images reveal that Callisto's surface has been modified by local landslides and places where the small craters have been erased by some unknown process. Its most prominent feature is a 2,000-km, multiringed structure of impact origin named Valhalla (the largest bright feature visible on Callisto's face in Figure 11.1). The impact may have occurred in a relatively thin, rigid crust overlying a fluidlike interior. The fluid mass then rapidly filled the initial crater bowl, leaving only a trace of the impact scar. Geophysical measurements obtained from the *Galileo* spacecraft suggest that Callisto is not differentiated, implying that it never went through a molten phase. On the other hand, the magnetometer aboard *Galileo* returned results suggesting that a liquid ocean could exist beneath the heavily cratered surface, implying that some sort of differentiation has occurred. These observations simply point out that many times in science we are faced with conflict-

FIGURE 11.27 *Voyager* image of fault zones and cratered terrain on Miranda, a 472-km diameter moon of Uranus.

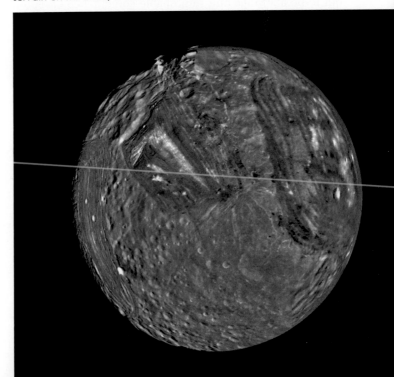

ing ideas that usually can be resolved only with additional measurements or observations.

Umbriel, the darkest of the large Uranus moons, appears uniform in color, reflectivity, and general surface features, indicative of an ancient surface. The real puzzle posed by Umbriel is why it has been dead for so long while the surrounding large moons of Uranus have been so active.

11.4 Pluto and Eris: Dwarf Planets or Gigantic Comets?

Pluto has been an enigma since its discovery. The story begins with what appeared to be unacceptable differences between the observed and predicted orbital positions of Uranus and Neptune throughout the 19th century. Inspired by these apparent discrepancies, astronomers early in the 20th century began a search for the unseen body they believed was perturbing the orbits of Uranus and Neptune. They called it Planet X and estimated that it had six times Earth's mass and was located somewhere beyond Neptune's orbit. Planet X was finally found by the American astronomer **Clyde Tombaugh** (1906–1997) in 1930, not far from its predicted position. It was named Pluto for the Roman god of the underworld.

Over the years that followed, observational evidence began to indicate that the mass of Pluto was far too small to have produced the presumed perturbations in the orbits of Uranus and Neptune. When astronomers reanalyzed the 19th century observations, they found that the orbital discrepancies were erroneous. Pluto's discovery thus turned out to be a strange and improbable coincidence based on faulty data.

Pluto's largest moon, Charon, was found in 1978. By applying Kepler's and Newton's laws to observations of the moon's motion around Pluto, astronomers were finally able to accurately "weigh" the Pluto–Charon system. Its total mass is only about 1/400 that of Earth (**Figure 11.28**). In 2005 two smaller moons were found orbiting this distant body. All three moons appear to be in orbital resonance with one another.

In recent years, Pluto's status as a planet was called into question by many astronomers. The discovery of a distant body, temporarily designated as 2003 UB_{313} and later named Eris, added fuel to the already fiery argument. Eris (see **Figure 11.29**) is even larger than Pluto, so shouldn't it be considered the tenth planet—or should neither be called planets? The question was finally put to rest by the International Astronomical Union (IAU) in August 2006. *Pluto was no longer a planet.* Pluto, along with Eris and the larg-

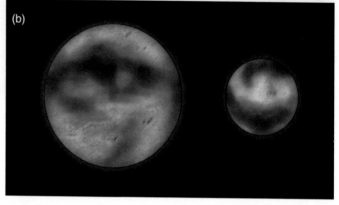

FIGURE 11.28 (a) An HST image and (b) an artist's rendition of Pluto and Charon based on HST images.

est asteroid, Ceres, were now designated as "dwarf planets" (see **Table 11.1**). Not all astronomers were happy with the IAU decision, however, and the question of Pluto's identity remains controversial.

Pluto's eccentric orbit periodically brings it inside Neptune's orbit, which is nearly circular, and from 1989 to 1999 Pluto was closer to the Sun than Neptune. More than 248 Earth years are required for Pluto to complete one orbit. Since its discovery, Pluto has traveled less than a third of its way around the Sun.

Pluto is only two-thirds as large as the Moon and only twice as big as its largest moon, Charon. Because of the similarity in size between Pluto and Charon, we might think of the Pluto–Charon system as a "double planet." Moreover, Pluto and Charon are a dynamically locked pair—the only known example in the Solar System. Both are in synchronous rotation with one another, so that each has one hemi-

---

Pluto has a moon half as big as it is.

---

sphere that always faces the other body and another that never sees the other body (**Figure 11.30**). If immobile inhabitants were living on opposite sides of Pluto and Charon, they would never know of each other's existence. Like Uranus, the plane of Pluto's equator is tipped at almost right angles to its orbital plane.

From observations made with groundbased telescopes and the HST, we know that Pluto's surface contains an icy mixture of frozen water, carbon dioxide, nitrogen, methane, and carbon monoxide. Unlike Pluto, Charon's surface seems to be made up primarily of dirty water ice. Pluto also has a low-density atmosphere similar to that of Neptune's Triton. It is composed primarily of nitrogen and methane and may contain argon or other heavy gases as well. With densities nearly twice that of water, both Pluto and Charon resemble Triton more than most icy moons. They probably consist of a rocky core that makes up about 70 percent of their mass, surrounded by a water–ice mantle. Because of their great distance from Earth, Pluto and Charon were not included in the great program of spacecraft exploration that took

Pluto is a mixture of rock and ice and has a thin methane atmosphere.

place late in the 20th century. We thus know little about the properties of their surfaces and nothing at all about their geological history. However, a planetary spacecraft called *New Horizons* was launched in 2006 and should reach the "double planet" in 2015.

It was the 2005 announcement that Eris is even larger than Pluto that intensified the argument over Pluto's status as a planet and made Eris's then temporary name, 2003 UB$_{313}$, a household word. The highly eccentric orbit of Eris carries

FIGURE 11.29 This distant dwarf planet Eris (shown within the white circle) has physical characteristics very similar to those of Pluto, but is slightly larger. Following its discovery, Eris was temporarily known as 2003 UB$_{313}$.

it from 37.8 AU at perihelion to 97.6 AU at aphelion, with an orbital period of 557 years. By chance it was found near its farthest point from the Sun, making it the most distant known object in the Solar System. Because of its great distance at this time, it now appears about 100 times fainter

TABLE 11.1

Physical Properties of the Dwarf Planets

	Ceres	Pluto	Eris
Orbital radius (AU)	2.7	39.5	67.7
Orbital period (years)	4.6	248.1	557
Orbital velocity (km/s)	17.9	4.7	3.4
Mass ($M_\oplus = 1$)	0.00016	0.0021	0.0025?
Equatorial diameter (1,000 km)	950	2306	2400?
Equatorial diameter ($D_\oplus = 1$)	0.07	0.18	0.19
Density (water $= 1$)	2.1	2.0	?
Rotation period (hours)	9.1	10.7	>8?
Obliquity (degrees)	4.0	119.6	?
Surface gravity (m/s^2)	0.27	0.58	?
Escape speed (km/s)	0.51	1.2	?

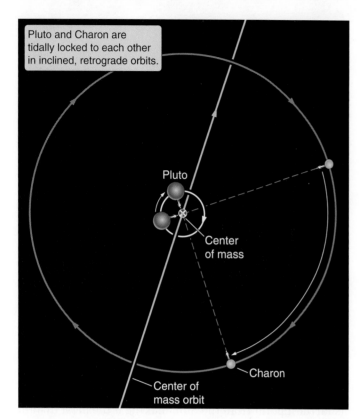

FIGURE 11.30 A diagram showing the doubly synchronous rotation and revolution in the Pluto–Charon system.

than Eris and Pluto. And like Pluto–Charon, many KBOs are known to form binary pairs. Some planetologists would classify Pluto not as a dwarf planet but as a large member among the family of Kuiper Belt objects. Yet one characteristic sets Pluto aside from all the known KBOs and makes it seem more planetlike. *Pluto has an atmosphere.* For this and historical reasons we have included Pluto and Charon in our discussion of planets and moons.

11.5 Ceres: Dwarf Planet or Large Asteroid?

On New Year's Day, 1801, a Sicilian astronomer named Giuseppe Piazzi found a bright object between the orbits of Mars and Jupiter. Piazzi named the new object Ceres. When Piazzi discovered Ceres, he thought he might have found a hypothetical "missing planet." But as more objects were discovered orbiting between Mars and Jupiter, astronomers realized that Ceres was a new kind of Solar System object, which they called *asteroids*. With a diameter of about 950 km, Ceres (see **Figure 11.31**) is larger than most moons and smaller than any planet. It contains about a third of the total mass in the asteroid belt. For more than two centuries, Ceres would be known as the largest asteroid.

FIGURE 11.31 Ceres, once known as the largest asteroid, is now called a dwarf planet. Note its spherical shape, one of the criteria that gives Ceres its dwarf planet status. The nature of the bright spot is unknown, but it reveals the rotation of Ceres as seen in this set of four images taken by HST over an interval of 2 hours and 20 minutes.

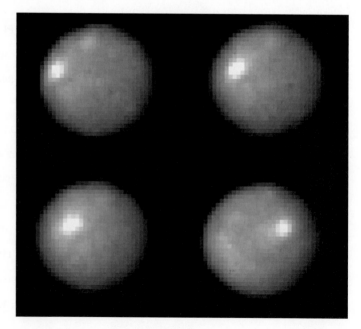

than Pluto. Although somewhat larger, Eris resembles Pluto in many respects. Both are attended by a relatively large moon, although Pluto has two other smaller moons. Both have frozen methane on their surface. And while Eris does not show evidence of a Pluto-like atmosphere today, the icy volatiles on its surface suggest that it may develop one when it comes closest to the Sun in the year 2257. There are also differences. Pluto's orbit has a higher eccentricity and a higher inclination than any of the classical planets. But the orbit of Eris is even more eccentric and even more highly inclined to the ecliptic plane than Pluto's.

Pluto's inclusion in the list of objects we call planets was little more than an accident of history. In the chapter that follows we will learn that in the outermost parts of the disk of the Solar System, beyond the realm of the planets, countless frozen icy planetesimals orbit. These objects, called *Kuiper Belt objects*, or KBOs,[3] are comet nuclei. We now know of several KBOs that are only slightly smaller[4]

---

[3] Many astronomers refer to KBOs as TNOs, trans-Neptunian objects.
[4] With a few exceptions, the sizes of KBOs cannot be measured directly and must be estimated. Although brightness and approximate distance are known, their albedos are uncertain, making estimates of their size correspondingly uncertain. Nevertheless, reasonable limits for the albedos of KBOs can set maximum and minimum values for their size.

According to the 2006 IAU decision, to be given planetary status, a body must meet three criteria (see Appendix 8). It was the third criterion, namely that a planet must be able to clear smaller neighboring bodies from its orbital surroundings, that disqualified Pluto as a planet, and put it into the dwarf planet category. Curiously, this same definition raised the status of Ceres from asteroid to a dwarf planet. NASA's *Dawn* mission, scheduled to be launched in 2007, will explore Ceres and Vesta, once the second largest asteroid and now the largest.

Summary

- All four giant planets are surrounded by rings.

- Saturn's ring system is the most complex.

- Rings are composed of tiny particles moving around their parent planets in Keplerian orbits.

- Rings may be transient features held in place by moons.

- The moons of the outer Solar System are composed of rock and ice.

- Some moons were formed along with their parent planets; others were captured later.

- A few moons are geologically active; most are dead.

- Jupiter's moon Io is the most volcanically active body in the Solar System.

- Jupiter's moon Europa may contain an enormous subsurface ocean.

- Dwarf planets Pluto and Eris more closely resemble giant comet nuclei.

Seeing the Forest through the Trees

We normally think of the Solar System as consisting of the planets we learned about in grade school. Yet there are far more moons in our Solar System than there are planets, and each moon is a unique world in its own right. There was no way in 1610 for Galileo to know that the four points of light he discovered circling giant Jupiter were the flagships of a vast armada of strange and amazing worlds. The diversity of the solid worlds of the inner Solar System, while remarkable, is nothing compared with the diversity of the worlds that surround Jupiter, Saturn, Uranus, and Neptune. Some of these worlds are frozen remnants of the time long ago when the giant planets formed at the centers of their own swirling disks. Others are testament to chaotic events in which passing objects were captured in gravitational interactions between planets and other moons. At the other extreme are moons such as Io, a world so geologically active that its surface is remaking itself through volcanism as we watch.

These moons are not the uniform worlds, with iron-and-nickel cores and silicate mantles, found close to the Sun. Instead they, along with Pluto, are assembled from the diverse mix of metals, rocks, and ices that existed in solid form in the outer reaches of the disk that surrounded the newly formed Sun nearly 4.6 billion years ago. As such, these bodies offer important clues about the history of our Solar System, allowing us to test our ideas of what that young Solar System must have been like. These worlds may even give us clues about the history of life in the universe. The fractured ice flows that make up the surface of Europa may well cover a vast, deep ocean, warmed and enriched by geothermal vents not unlike those that dot the floors of Earth's oceans. When scientists finally realized that Mars and Venus were not as Earth-like as once imagined, prospects for finding life elsewhere in the Solar System seemed dim. Now, at the opening of the 21st century, we are turning our attention with fresh excitement and hope to these small worlds that circle far from the Sun.

Systems of moons may offer a fascinating future for geologists; but they also represent gravitational playgrounds in which the simple two-body interactions described by Kepler's laws give way to complex, chaotic interactions among many different objects. The tidal stresses we learned about in Chapter 10 also rip and pull at these worlds, locking them into synchronous orbits and in some cases churning their interiors into seething cauldrons of geological activity. Sometimes these tidal stresses are so great that a moon or passing comet pays the ultimate price and is ripped into tiny fragments. Born from these violent events are the majestic rings that circle Saturn and the less spectacular but equally fascinating systems of rings surrounding the other giant planets. The extraordinary complexity of these rings is also a testament to Newton's remarkable inverse square law of gravitation as planets, moons, and rings perform their gravitational dance. These interactions can even distort rings into structures so remarkable that they seemed all but impossible until revealed by the eyes of the *Voyager* and *Cassini* spacecraft.

It is with sadness that we leave this realm of moons and rings so quickly; these worlds deserve much more attention than we can give them on our journey. As we might do on a package tour promising to show us Europe in three days, we look at the marvels through the window, glance at our guidebooks, and promise to return someday when we have more time. Planetary scientists themselves share our wistful regret. Our glances from passing spacecraft at the moons and rings surrounding the giant planets are really only enough to whet our appetites. No doubt lifetimes of wonder and surprise remain as we continue to explore the smaller worlds of our Solar System.

There may be more moons than there are planets in the Solar System, but planets, moons, and rings do not complete the accounting of objects adrift within the gravitational realm of the Sun. Still remaining are the uncounted swarms of asteroids and comets that orbit within the domain of the planets and also stretch halfway to the nearest star. In the next stage of our journey we will discover that far from being insignificant, this flotsam and jetsam carries with it the most direct record of the history of our Solar System, providing essential pieces to the puzzle of the origin and history of life on Earth.

Key Terms

ring, p. 310
regular moon, p. 310
irregular moon, p. 310
retrograde, p. 310
ring arc, p. 312
ringlet, p. 315
backlighting, p. 316
electromagnetic force, p. 318
shepherd moon, p. 320
spoke, p. 322
cryovolcanism, p. 326
theory of evolution, p. 329
palimpsest, p. 331

Student Questions

THINKING ABOUT THE CONCEPTS

1. Explain the difference between regular and irregular moons.

2. Identify the three moons known to be geologically active. Name three geologically dead moons.

3. Identify and explain two possible mechanisms that can produce planetary ring material.

4. Will the particles in Saturn's bright rings eventually stick together to form one solid moon orbiting at the mean distance of all the ring particles? Explain your answer.

5. Describe and explain a mechanism that keeps planetary rings from dissipating.

6. Why does Earth not have a ring?

7. Explain the process that drives volcanism on Jupiter's moon Io.

8. Could there be tidal heating effects on a moon located in a perfectly circular orbit around a planet? Think carefully about this and explain your answer.

9. Discuss evidence that supports the idea that Europa might have a subsurface ocean of liquid water.

10. What is cryovolcanism? Name two moons that have cryovolcanoes.

11. In certain ways Titan resembles a frigid version of Earth or Venus. Explain the similarities.

12. What evidence do we have that Ganymede and several other geologically dead moons were once active?

13. Describe ways in which Pluto differs significantly from the classical Solar System planets.

14. Pluto, now designated as a dwarf planet, continues to be a source of controversy among astronomers. Discuss arguments for and against its being listed as a planet. What is your opinion and why?

APPLYING THE CONCEPTS

15. The inner and outer diameters of Saturn's B Ring are 184,000 km and 235,000 km, respectively. If the average thickness of the ring is 10 m and the average density is 150 kg/m^3, what is the mass of Saturn's B ring?

16. The mass of Saturn's small icy moon Mimas is 3.8×10^{19} kg. How does this mass compare with the mass of Saturn's B Ring? Why is this comparison meaningful?

17. Particles at the very outer edge of Saturn's A Ring are in a 7:6 orbital resonance with the moon Janus. If the orbital period of Janus is 16^{h}41^m, what is the orbital period of the outer edge of Ring A?

18. Planetary scientists have estimated that Io's extensive volcanism could be covering the moon's surface with lava and ash to an average depth of up to 3 mm per year.
    a. Io's radius is 1,815 km. If we model Io as a sphere, what are its surface area and volume?
    b. What is the volume of volcanic material deposited on Io's surface each year?
    c. How many years would it take for volcanism to perform the equivalent of depositing Io's entire volume on its surface?
    d. How many times might Io have "turned inside out" over the age of the Solar System?

19. Enceladus has mass of 8.4×10^{19} kg and a radius of 250 km. Using the formula for escape velocity (see Chapter 3), calculate the minimum speed at which ice crystals from the moon's cryovolcanoes must be traveling in order to escape to Saturn's E Ring.

20. Our Moon has a diameter of 3,474 km and orbits at an average distance of 384,400 km. At this distance it subtends an angle just slightly larger than ½ degree in our sky. Pluto's moon Charon has a diameter of 1,186 km and orbits at a distance of 19,600 km from the dwarf planet.
    a. Compare the appearance of Charon in Pluto's skies with the Moon in our own skies.
    b. Describe where in the sky Charon would appear as seen from various locations on Pluto.

StudySpace
wwnorton.com/astro21
provides a Study Plan for each chapter that includes a reading outline, animations, keyword flash cards, and gradebook-enabled multiple-choice quizzes. From StudySpace you can also access premium content in the ebook and SmartWork.

Almost in the center of it, above the Prechistenka Boulevard, surrounded and sprinkled on all sides by stars . . . shone the enormous and brilliant comet of 1812—the comet which was said to portend all kinds of woes and the end of the world.

LEO TOLSTOY (1828–1910), *WAR AND PEACE*

Comet Hyakutake approached close to Earth in 1996.

CHAPTER 12

Asteroids, Meteorites, Comets, and Other Debris

12.1 Ghostly Apparitions and Rocks Falling from the Sky

The Sun, Moon, and planets are usually the brightest, most obvious members of our Solar System, but they are not all there is. Like an ocean afloat with flotsam and jetsam, our Solar System swarms with smaller members that might totally escape our attention until one of them washes ashore. It would be a mistake to assume this planetary debris is unimportant, however. Comets, asteroids, and meteorites provide astronomers with some of their most important clues about how our Solar System formed and evolved.

Throughout human history, the sudden and unexpected appearance of a bright comet has elicited fear, wonder, and superstition. Early cultures viewed these spectacular visitors as omens. Comets were often seen as dire warnings of disease, destruction, and death, but sometimes as portents of victory in battle or as heavenly messengers announcing the impending birth of a great leader. The earliest records of comets date from as long ago as the 23rd century B.C. Mediterranean and Far Eastern literatures are especially full of references and popular superstitions about comets.

Meteorites have also engendered fascination and awe for as long as humankind has watched the heavens. Nearly all ancient cultures venerated these rocks from the sky. Early Egyptians preserved meteorites along with the remains of their pharaohs, Japanese placed them in Shinto shrines, and ancient Greeks worshipped them. In 1492—the same year Columbus voyaged to the New World—the townsfolk of Ensisheim (now in France) watched as a 172-kg stone

KEY CONCEPTS

Asteroids and comets are tiny in comparison to planets; yet these objects, and their fragments that fall to Earth as meteorites, have told us much of what we know about the early history of the Solar System. As we explore these bits of interplanetary flotsam and jetsam, we will discover

- Small, irregular worlds called asteroids that are made of rock and metal.
- That some asteroids are primitive while others are differentiated or are pieces of larger, differentiated bodies.
- Different types of meteorites that are fragments of these varieties of asteroids.
- That most asteroids orbit between Mars and Jupiter, but some have orbits that cross Earth's.
- Comet nuclei—pristine, icy planetesimals—adrift in the frozen outer reaches of the Solar System.
- Spectacular active comets that are warmed by the Sun as they dive through the inner Solar System.
- Meteor showers that occur when Earth passes through a comet's trail of debris.
- World-jarring meteorite impacts that still occur today and have played a vital role in shaping the history of life on Earth.

FIGURE 12.1 A woodcut of the meteorite fall at Ensisheim in 1492.

fell to Earth. The event was recorded in a famous woodcut (**Figure 12.1**), and for a while the meteorite was enshrined in a local church. Despite numerous eyewitness accounts of meteorite falls, however, many were slow to accept that these peculiar rocks actually came from far beyond Earth. In 1807 two Yale professors investigating the meteorite that landed in Weston, Connecticut, concluded that the object truly did drop from the sky. Thomas Jefferson, third president of the United States and an enlightened scientist, is rumored to have responded, "I would rather believe that two Yankee professors would lie, than that stones fall from heaven." But as the evidence continued to mount, it became impossible to ignore. By the early 1800s scientists had documented so many meteorite falls that their true origin was indisputable. Today hardly a year passes without a recorded meteorite fall, including some that have smashed into cars, houses, pets, and (occasionally) people.

A third type of interplanetary debris was not discovered until the age of telescopes. As we learned in Chapter 11, in 1801, Giuseppe Piazzi found a bright object between the orbits of Mars and Jupiter and named the new object Ceres. But more objects were then discovered between the orbits of Mars and Jupiter. Because these new objects appeared in astronomers' eyepieces as nothing more than faint points of light, William and Carolyn Herschel (the brother–sister pair of English amateur astronomers who discovered Uranus) named them *asteroids*, a Greek word meaning "star-like." As we begin the 21st century we have found more than 330,000 asteroids, of which about 130,000 have well-determined orbits. Discoveries continue at the rate of about 5,000 per month.

Apart from their role as fascinating curiosities and reminders that there is more to heaven and Earth than meets the eye, this interplanetary debris plays a very significant role in our quest to discover humankind's cosmic origins. We now know that meteorites, asteroids, and comet nuclei are fragments of our Solar System's ancient past. Like ar-

cheologists who assemble the history of civilizations from fossil bones and shards of pottery, planetary scientists work

Comets and asteroids provide scientific clues to our past.

to piece together the story of our Solar System from these relics of a time when the Sun, Earth, and planets were young. Fortunately, although scientists have had to reconstruct the puzzle of the early Solar System from the pieces they find, on our journey in this book we have the luxury of starting with the results of their work. We begin by describing what the completed puzzle looks like and then talk about the pieces and how they fit together.

12.2 Asteroids and Comets: Pieces of the Past

We began our discussion of the Solar System in Chapter 6 with the story of its history. Our Solar System was born 4.6 billion years ago inside an enormous interstellar cloud of gas and dust. At the same time that our Sun was becoming a star, tiny dust grains of primitive material were sticking together to form swarms of planetesimals. Some kinds of materials, such as iron–nickel alloys and rock, were able to stick together even in the hot inner parts of the protoplanetary disk that surrounded the Sun. Those planetesimals that formed near the young Sun were composed mostly of these heat-resistant materials. Other materials—such as ices of water, methane, ammonia, and various organic (carbon-containing) compounds—could coalesce only in the outer parts of the disk where temperatures were much lower. In time, many of these planetesimals stuck together to build still larger bodies, the moons and planets. When the planet-building process finally ended, some of the original planetesimals remained in orbit around the Sun.

Most asteroids are relics of rocky or metallic planetesimals that originated in the region between the orbits of Mars and Jupiter. Although early collisions between these planetesimals created several bodies large enough to differentiate, Jupiter's tidal disruption prevented them from forming

Asteroids are relics of rocky planetesimals that formed between the orbits of Mars and Jupiter.

a single moon-sized planet. As they orbit the Sun, asteroids continue to collide with one another, producing small fragments of rock and metal. Most meteorites are pieces of these asteroidal fragments that have found their way to Earth and crashed to its surface.

Icy planetesimals surviving today form the solid cores, or *nuclei,* of comets. Although asteroids can survive in the inner Solar System, comet nuclei cannot. Today most comet nuclei are preserved in the frigid regions of space beyond the planets. Sometimes, however, something hap-

Comet nuclei are remnants of icy planetesimals that formed far from the Sun.

pens to cause one of these icy bodies to change direction and dive inward toward the domain of the inner planets. When a comet nucleus approaches too close to the Sun, solar heating unglues its surface structure, gradually breaking it up into tiny clumps of dusty ice. When Earth passes through swarms of dusty ice from a disintegrating comet nucleus or pebbles from the breakup of an asteroid, the particles can burn up in our atmosphere, creating an atmospheric phenomenon we refer to as a **meteor** (see **Excursions 12.1**).

Planetesimals that became part of the planets or their moons were so severely modified by planetary processes that

EXCURSIONS 12.1

Meteors, Meteoroids, Meteorites, and Comets

Meteors are familiar to all of us. Just stand outside for a few minutes on a starry night, away from bright city lights, and you will almost certainly be rewarded by a glimpse of a "falling star," a meteor. Most of us realize we are not witnessing the demise of a real star but are instead seeing a piece of Solar System debris entering our atmosphere. Few are aware that it is seldom the actual debris we see, but rather a trail of heated atmospheric gas glowing from the friction caused by a small unseen object entering at speeds of up to about 70 km/s. A meteor, then, is an *atmospheric phenomenon.*

The small[1] solid body that creates the meteor is called a **meteoroid**. These small objects are of either cometary or asteroidal origin. Cometary fragments are typically less than a centimeter across and have about the same density as cigarette ash. What they lack in size and mass, they make up for in speed. A 1 g meteoroid (about half the mass

[1] By definition, meteoroids are those objects whose size lies within a range of 100 μm to 100 m. Particles smaller than this are considered *interplanetary dust,* whereas larger objects are called *planetesimals.*

of a dime) entering Earth's atmosphere at 50 km/s has a kinetic energy comparable to that of an automobile cruising along at the fastest highway speeds. Before plunging into our atmosphere, the meteoroid may have been orbiting the Sun for millions of years after being chipped from an asteroid or left behind by a disintegrating comet nucleus. Most meteoroids are so small and fragile that they burn up completely before reaching Earth's surface.

A meteoroid large enough to survive all the way to the ground is called a **meteorite**. All meteoroids are pieces of Solar System debris. The larger pieces that survive the plunge through Earth's atmosphere are usually fragments of asteroids. Most of the smaller pieces that burn up before reaching the ground are cometary fragments.

Comets appear as fuzzy patches of light in the nighttime sky. They are bodies of ice, dust, and gas that shine by reflected sunlight. Comets frequently remain visible in our skies for weeks at a time, with tails that can be 100 million km long and stretch halfway across our sky. In contrast, a meteor may streak across 100 km of our atmosphere and last at most a few seconds.

nearly all information about their original physical condition and chemical composition has been hopelessly lost. On the other hand, asteroids and comet nuclei constitute an ancient and far more pristine record of what the early Solar System was like. If you visit your local planetarium or science museum, you may find a meteorite on display that you can walk right up to and touch. Do so with respect. The meteorite under your hand may be older than Earth itself. Some meteorites contain tiny grains of material that predate the formation of our Solar System. These grains include diamonds and carbon compounds that originated as material ejected from dying stars.

12.3 Meteorites: A Chip Off the Old Asteroid Block

Most asteroids are so small compared with planets that they appear as nothing but unresolved points of light through telescopes on Earth. Historically, this has made it difficult to learn much about them directly. Even determining gross properties such as size, shape, and rate of spin has required sophisticated analysis of telescopic, spacecraft, and radar observations. Despite this challenge, we

> Meteorites are pieces of asteroids. They tell us about the bodies they came from.

probably know more about the structure and composition of asteroids than any Solar System object other than Earth! As asteroids orbit the Sun, they occasionally collide with each other, chipping off smaller rocks and bits of dust. Occasionally one of these fragments is captured by Earth's gravity and survives its fiery descent through Earth's atmosphere as a meteor, to be picked up and added to someone's collection of meteorites.

Thousands of meteorites reach the surface of Earth every day, but only a tiny fraction of these are ever found and identified as such. The best meteorite hunting in the world is in Antarctica. Meteorites are no more likely to fall in Antarctica than anywhere else, but in Antarctica they are far easier to distinguish from their surroundings. Antarctic snowfields serve as a huge net for collecting meteorites. Glaciers then carry the meteorites along, concentrating them in regions where the glacial ice is eroded by wind. The meteorites are left lying on the surface for collectors to pick up. The big advantage of hunting on Antarctic ice is that in many places the *only* stones to be found on the ice are meteorites. Antarctic meteorites have also spent their time on Earth literally in the deep freeze; many are thus relatively well preserved, showing little weathering or contamination from terrestrial dust or organic compounds.

Meteorites are extremely valuable because they are samples of the same relatively pristine material from which asteroids are made. Also, unlike planets or asteroids themselves, we can take meteorites into the laboratory and study them using sophisticated equipment and techniques. Scientists compare them to rocks found on Earth and the Moon and contrast their structure and chemical makeup with those of rocks studied by spacecraft that have landed on Mars and Venus. Meteorites are also compared with asteroids and other objects on the basis of what colors of sunlight they reflect and absorb.

Meteorites are normally grouped into three categories based on the kinds of materials they are made of and the

> There are three classes of meteorites: stony (including chondrites and achondrites), iron, and stony-iron.

degree of differentiation they experienced within their parent bodies. **Figure 12.2** shows cross sections of these several meteorite types.

Over 90 percent of meteorites are **stony meteorites**, which are similar to terrestrial silicate rocks. A stony meteorite can be recognized by the thin coating of melted rock that forms as it passes through the atmosphere. Many stony meteorites contain small round spherules called **chondrules**, which range in size from that of sand grains up to marble-sized objects. Chondrules are all held together by a matrix of finer-grained material, much as the gravel in concrete is held together by a matrix of cement. Stony meteorites containing chondrules are called **chondrites**, whereas those that do not contain chondrules are called **achondrites**. Unlike chondrites, achondrites have crystals that appear to have formed in the same place and at the same time.

The second major category of meteorites, **iron meteorites**, is the easiest to recognize. Their surfaces have a melted and pitted appearance generated by frictional heating as they streaked through the atmosphere. (The meteorite in your local museum is probably an iron meteorite.) Even so, many are never found, either because they land in water or simply because no one happens to notice and recognize them for what they are. As an interesting side note, one of the more pleasant surprises from the Mars rover *Opportunity* was the discovery of an iron meteorite lying right there in plain sight on the Martian surface (**Figure 12.3**). Not only was its appearance typical of iron meteorites found on Earth, but its placement on the smooth, featureless plains made it instantly recognizable. Chemical analysis from the rover instruments showed it to be composed mostly of iron and nickel.

The final category is the **stony-iron meteorites**, which consist of a mixture of rocky material and iron–nickel alloys. Stony-iron meteorites are relatively rare.

One of the most interesting stories that meteorites have to tell about the Solar System is its age. In Chapter 7 we

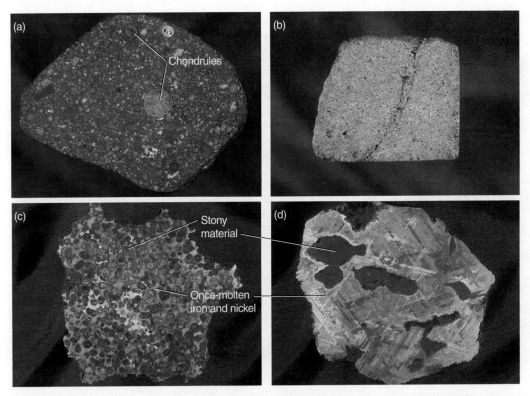

FIGURE 12.2 Cross sections of several kinds of meteorites: (a) a chondrite, (b) an achondrite, (c) a stony-iron meteorite, and (d) an iron meteorite. Chondrites and achondrites are stony meteorites.

learned that radioactive materials locked up in solids provide a kind of clock that begins ticking the moment the material solidifies. By comparing how much of a radioactive substance remains in a rock with how much of the radioactive material's decay products have accumulated, we can tell how long this clock has been ticking and therefore how long the rock has been solid. When we apply this technique to meteorites, nearly all turn out to be about 4.54 billion years old. One type of meteorite stands out with particular interest. **Carbonaceous chondrites**, chondrites that are rich in carbon, are thought to be the very building blocks of the Solar System. Although these meteorites cannot be dated directly because they are

---

Meteorites tell us the age of the Solar System.

---

composed of aggregates of many individual grains, indirect measurements suggest they are about 4.56 billion years old. This is our best measurement of the time that has passed since the formation of the Solar System.

The diversity among meteorites and asteroids reflects the range of physical conditions under which they were formed. For example, laboratory experiments simulating formation show that chondrules formed at higher temperatures than the finer-grained minerals that surround them. Chondrules must once have been molten droplets that rapidly cooled to form the crystallized spheres we see today. Meteorites thus hold clues to the physical conditions existing in the Solar System at the time they were formed.

Not surprisingly, the different classes of meteorites are related to different classes of asteroids. As we learned in Chapter 6, the objects that we find in the Solar System today are the result of small objects sticking together to form

FIGURE 12.3 A basketball-size iron meteorite lying on the surface of Mars, imaged by the Mars exploration rover *Opportunity*.

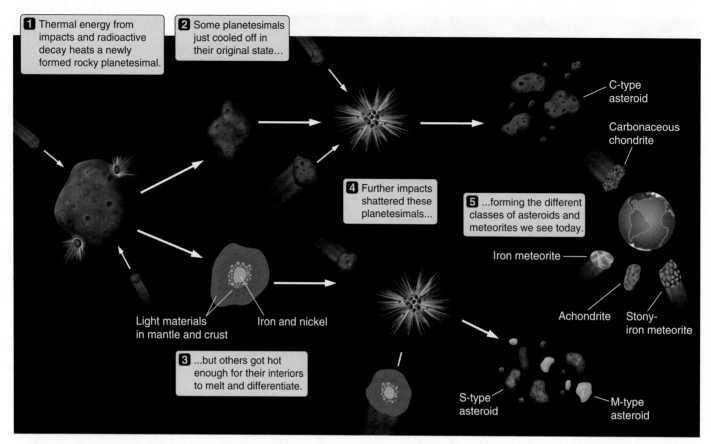

FIGURE 12.4 The fate of a rocky planetesimal in the young Solar System depends on whether it gets large and hot enough to melt and differentiate, as well as the impacts it experiences. Different histories lead to the different types of asteroids and meteorites found today.

larger objects. This process is called *accretion*. Refer to **Figure 12.4** as we follow the possible fates of a planetesimal. As was the case in the interior of the terrestrial planets, large amounts of energy were released in collisions as planetesimals accreted smaller objects. Additional thermal

C-type asteroids never differentiated.

energy was released as radioactive elements inside the planetesimals decayed. Some planetesimals never reached the temperatures needed to melt their interiors. These planetesimals—aggregations of the primordial material from which they formed—simply cooled off, and they have since remained pretty much as they were when they formed. These are the **C-type asteroids**, which are composed of primitive material that we believe has largely been unmodified since the origin of the Solar System, almost 4.6 billion years ago.

In contrast to the C-type asteroids, some planetesimals were heated enough by impacts and radioactive decay to cause them to melt. Once the interior of a planetesimal melted, it differentiated, just as all the planets and large

moons did. Denser matter such as iron sank to the center of the planetesimal while lower-density material such as compounds of calcium, silicon, and oxygen floated toward the surface and combined to form a mantle and crust of silicate rocks. **S-type asteroids** (and the stony achondritic meteorites that come from them) may be pieces of the mantles and crusts of such differentiated planetesimals. They are chemically more similar to igneous rocks found on Earth than to primitive chondrites and C-type asteroids. This means that they were hot enough at some point to lose their carbon compounds and other volatile materials to space. The large interlocking crystals that make up achondrites are also like

S- and M-type asteroids may be pieces of larger differentiated bodies.

those seen in rocks that slowly cooled from molten material. Similarly, **M-type asteroids** (from which iron meteorites come) are easy to understand as fragments of the iron- and nickel-rich cores of one or more differentiated planetesimals that shattered into small pieces during collisions with other planetesimals. Indeed, slabs cut from iron meteorites show large interlocking crystals characteristic of iron that cooled very slowly from molten metal. The rare stony-iron meteorites may come from the transition zone between the stony mantle and the metallic core of such a planetesimal.

A few planetesimals in the region between the orbits of Mars and Jupiter seem to have evolved toward becoming tiny planets before being shattered by collisions. They even became volcanically active, complete with eruption of lava flows onto their surfaces. Vesta (now the largest known asteroid, with a diameter of 525 km) seems to be a piece of such a highly evolved planetesimal. Vesta does not fit well into any of the main asteroid groups, but its spectrum is strikingly similar to the reflection spectrum of a peculiar group of meteorites. These meteorites, which are probably pieces of Vesta, look like rocks taken from iron-rich lava flows on Earth and the Moon. The main difference is that they are *only* 4.5 billion years old, a bit younger than most meteorites. It is hard to escape the conclusion that early in the history of the Solar System, some planetesimals were large enough to become differentiated and to develop volcanism. But rather than going on to form planets, these planetesimals broke into pieces in collisions with other planetesimals.

The story of asteroids and meteorites is a great success of planetary science. A vast wealth of information about this diverse collection of objects has come together into a single consistent story of planetesimals growing and possibly becoming differentiated, then being shattered by subsequent collisions. The story is even more satisfying because it, in turn, fits so well with the even grander story of how most planetesimals were accreted into the planets and their moons.

But what of the "missing planet" that Piazzi thought he found in 1801? Was there once a single planet between orbits of Mars and Jupiter that shattered in a tremendous collision? Although this may be a wonderfully romantic notion, the evidence says that events did not happen that way. A more detailed look at the wide variety of chemical properties and orbits of today's asteroids seems to indicate that they are fragments of at least a dozen or more parent bodies, rather than of a single planet. As discussed in Chapter 10, we now understand that it was the "stirring" of this region by Jupiter's gravity that prevented the asteroids from coalescing into a planet.

As we stress throughout this book, patterns are extremely important in science—not only when they are followed, but also when they are broken. Some types of meteorites fail to follow the patterns just discussed. Whereas most achondrites have ages of 4.5 billion to 4.6 billion years, some members of one group are less than 1.3 billion years old. Some are chemically and physically similar to the soil and the atmospheric gases that were measured on Mars by the *Viking* landers. The similarities are so strong that most planetary

> **To hold certain meteorites is to hold a piece of Mars.**

scientists believe these meteorites are pieces of Mars that were knocked into space by asteroidal impacts.[2] This means we have pieces of another planet we can study in laboratories here on Earth. The Martian meteorites support the general belief that much of the surface of Mars is covered with iron-rich volcanic materials.

If pieces of Mars have reached Earth from its orbit almost 80 million kilometers beyond our own, we might expect that pieces of our companion Moon would find their way to Earth as well. Indeed, another group of meteorites bears striking similarities to samples returned from the Moon. Like the meteorites from Mars, these meteorites are chunks of the Moon that were blasted into space by impacts and later fell to Earth.

12.4 Asteroids Are Fractured Rock

With a few possible exceptions, asteroids are too small for their self-gravity to have pulled them into a spherical shape. Some asteroids imaged by spacecraft or by Earth-based radar have highly elongated shapes, suggesting objects that either are fragments of larger bodies or were created haphazardly from collisions between smaller bodies. The masses of a number of asteroids have been measured by noting the effect of their gravity on spacecraft passing nearby. Knowing the mass and the size of an asteroid allows us to determine its density. The densities of these asteroids range between 1.3 and 3.5 times the density of water. Those at the lower end of this range are considerably less dense than the meteorite fragments they create. How can asteroids be less dense than the rock that likely came from them? Planetologists think that some of them are shattered heaps of rubble, with large voids between the fragments. Once again, this is what we would expect of objects that were assembled from smaller objects and then suffered a history of violent collisions.

Asteroids rotate just as planets and moons do, although the rotation of irregularly shaped asteroids is more of a tumble than a spin. A day on a typical asteroid is about 9 hours long, although the rotation periods of some asteroids

---

[2] In 1996 a NASA research team announced that a particular Martian meteorite (ALH84001) showed physical and chemical evidence of early life on Mars. These conclusions have since been challenged by many scientists and are now generally discredited.

are as short as 2 hours, and others are longer than 40 Earth days. How can we measure the rotation period of an object that appears starlike in a telescope? Unless the asteroid is

---

A day on a typical asteroid is only 9 hours long.

---

perfectly round, and we know that very few are, we measure rotation periods by watching changes in their brightness as they alternately present their broad and narrow faces to Earth.

Asteroids are found throughout the Solar System. Most orbit the Sun in several distinct zones, with the majority residing between the orbits of Mars and Jupiter in the **main asteroid belt**. The main belt asteroids are not distributed uniformly within the belt. As we discussed in Chapter 10, several empty regions, the Kirkwood gaps, have been depleted by perturbations from Jupiter.

The main belt contains at least 1,000 objects larger than 30 km (about the size of Washington, D.C.), of which about 200 are larger than 100 km. Small asteroids are more difficult to find than large ones, but there are many more of them. Taking this into account, we estimate that there are about 30 times as many asteroids larger than 30 km as there are asteroids larger than 100 km. Similarly, there seem to be about 30 times as many asteroids over 10 km as there are asteroids larger than 30 km. By assuming that this pattern continues for smaller and smaller bodies, we estimate that there may be as many as 10 million asteroids larger than 1 km in the main asteroid belt. However, while there are a great number of asteroids, they account for only a tiny fraction of the matter in the Solar System. If all of the asteroids were combined into a single body, it would be about a third the size of Earth's Moon.

Figure 12.5 shows orbits of several asteroids that are representative of their classes. In Chapter 10 we discussed the Trojan asteroids that occupy the L_4 and L_5 Lagrangian points of the Sun–Jupiter system. Asteroids whose orbits bring them close to Earth's orbit are called **near-Earth asteroids**. Members of one group, called the **Amors**, cross the orbit of Mars but do not cross Earth's orbit. Members of

---

Apollo and Aten asteroids cross Earth's orbit and could collide with us one day.

---

two other groups, the **Atens** and the **Apollos**, have orbits that cross Earth's orbit (see Figure 12.5). The difference between these two populations is that Apollos also cross the orbit of Mars, while the Atens remain inside Mars's orbit. Atens and Apollos occasionally collide with Earth or the Moon. Astronomers estimate that there are more than 3,500 Atens and Apollos with diameters larger than a kilometer. As we discovered in Connections 7.1, such collisions are geologically important and have dramatically altered the history of life on Earth.

12.5 Asteroids Viewed Up Close

Until the space age, scientists had no good idea of what asteroids looked like. As the 1980s drew to a close planetary scientists were about to have their theories put to the test. Would the first close-up views of asteroids pull the rug out from under our carefully constructed story of asteroids, meteorites, and the early Solar System? (If so, it would not be the first time a closer examination of some phenomenon revealed that people's ideas about it were wrong!) Fortunately, as spacecraft began to return detailed information about the asteroids they visited, those data placed our ideas about the history of the Solar System on even more solid footing.

In 1991 the *Galileo* spacecraft was following a circuitous path through the inner Solar System, gaining momentum from encounters with Earth and Venus to help send it on its journey to Jupiter. While flying through the main asteroid belt, it got close enough to the asteroid Gaspra to get dozens of good images. Gaspra is an S-type asteroid, about 12 by 20 by 11 km. The *Galileo* images show it to be cratered and irregular in shape. Faint, groovelike patterns on its surface may be fractures resulting from the same impact that chipped Gaspra from a larger planetesimal. Images taken in several different wavelengths of light show distinctive colors on Gaspra's surface, implying Gaspra is covered with a variety of different types of rock.

Later in its mission, *Galileo* made another pass through the main belt and this time sent back images of the asteroid Ida (**Figure 12.6**). Ida orbits in the outer part of the main asteroid belt, and it too is an S-type asteroid. *Galileo* flew so close to Ida that it could see details as small as 10 m —about the size of a small house or cottage. Ida is shaped like a *croissant,* 56 km long and 24 by 21 km across. Like Gaspra, Ida shows the scars of a long history of impacts with smaller bodies. Just as planetary scientists use the number and sizes of craters to estimate the ages of the surfaces of planets and moons (as we saw in Chapter 7), the cratering on Ida indicates that its surface is a billion years old, twice the age estimated for Gaspra. Also like Gaspra, the *Galileo* images of Ida revealed the presence of fractures in the asteroid. The fractures seen in the two asteroids indicate that they must be made of relatively solid rock. (You can't "crack" a loose pile of rubble.) This is an important confirmation of the idea—crucial to our story of asteroids and meteorites —that some asteroids must be pieces chipped from larger, solid objects.

Several features on Ida appear to be landslides, indicating that the surface of Ida must be covered with a "soil" of sorts. Other features are clearly rocks, some larger than football fields, exposed on the surface of the asteroid. Together these data show that Ida's gravity was able to hold onto some of the material thrown about when smaller objects hit its surface. This hardly seems worth mentioning

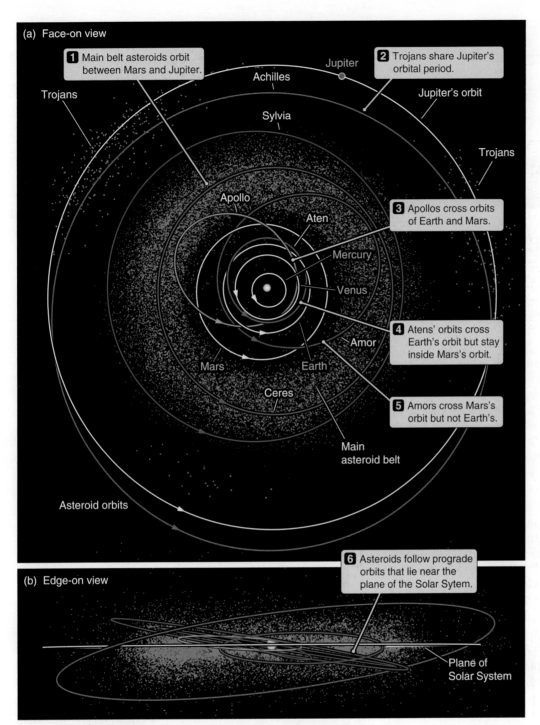

(a) Face-on view

1 Main belt asteroids orbit between Mars and Jupiter.

2 Trojans share Jupiter's orbital period.

Trojans

Achilles

Jupiter

Sylvia

Jupiter's orbit

Trojans

Apollo

3 Apollos cross orbits of Earth and Mars.

Aten

Mercury

Venus

4 Atens' orbits cross Earth's orbit but stay inside Mars's orbit.

Amor

Mars

Earth

Ceres

5 Amors cross Mars's orbit but not Earth's.

Main asteroid belt

Asteroid orbits

6 Asteroids follow prograde orbits that lie near the plane of the Solar Sytem.

(b) Edge-on view

Plane of Solar System

FIGURE 12.5 Blue dots show the locations of known asteroids at a single point in time. Most families of asteroids take their names from prototype members of the family. For example, Apollo family asteroids cross the orbits of Earth and Mars, like the asteroid Apollo. Most asteroids, such as Sylvia, are main belt asteroids.

until you consider that Ida's surface gravity is feeble—only a thousandth as strong as that of Earth. The escape velocity from Ida is only about 20 meters per second—about half the speed of a good fastball pitch.

Ida's gravity has managed to hold onto more than a bit of soil—*Galileo* images show a tiny moon orbiting about the asteroid. Ida's moon, called Dactyl, is only 1.4 km across, and like nearly all Solar System objects with solid surfaces,

its surface is cratered from impacts. Although planetary scientists had long speculated that asteroids might themselves have moons, Dactyl was the first such object to be discovered. But is Dactyl really a moon? To planetary scientists, any significant Solar System body orbiting a larger body other than the Sun is called a *moon.* Dactyl may not be the stuff poetry is written about, but by the planetary scientist's definition of moon, it counts. As it turns out,

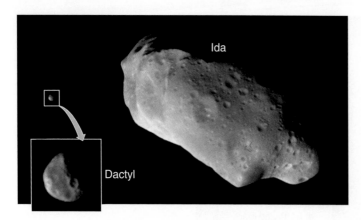

FIGURE 12.6 A *Galileo* spacecraft image of the asteroid Ida with its tiny moon, Dactyl (also shown enlarged in inset).

Ida's feeble gravity holds a tiny moon called Dactyl.

Dactyl is hardly unique. Moons have now been found around more than four dozen asteroids.

Since *Galileo*'s visits to Gaspra and Ida, spacecraft have visited four other asteroids: Mathilde, Eros, Braille, and Itokawa. This last object was seen only briefly in 1999 during the flyby of *Deep Space 1*. It was found to have a highly reflective surface, similar to Vesta, leading to speculation that Braille could be a chunk knocked from Vesta by a collision sometime in the past.

The *NEAR Shoemaker*[3] spacecraft was designed to do more than make a brief encounter with an asteroid—in early 2000 it was placed into orbit around the asteroid Eros to begin long-term observations of this S-type object. **Figure 12.7** shows a number of images of Eros obtained by *NEAR Shoemaker*. As a near-Earth asteroid, Eros is one of the 800 known objects whose orbits bring them within 1.3 AU of the Sun. Eros measures about 33 by 13 by 13 km and, like its counterparts Gaspra and Ida, shows a surface with grooves, rubble, and impact craters, including a crater 8.5 km across. However, the scarcity of smaller craters suggests that its surface could be younger than Ida's. One of the most important discoveries made by *NEAR Shoemaker* was analysis of X-rays emitted by the asteroid as it was being irradiated by X-rays from an eruption on the Sun. The X-ray/gamma-ray spectrometer aboard *NEAR Shoemaker* detected emission lines from a number of elements, allowing us to measure the chemical composition of the asteroid. These observations confirmed the similarity in composition between Eros and primitive meteorites. At the end of

[3]*NEAR* stands for *Near Earth Asteroid Rendezvous*. The spacecraft was renamed *NEAR Shoemaker* in March 2000 in honor of Gene Shoemaker, a pioneer of the study of comets and asteroids who died in an automobile accident in 1997 while studying impact craters in the Australian outback.

its mission, *NEAR Shoemaker* became the first spacecraft to land on an asteroid, when on February 12, 2001, the orbiter was gently crashed into the surface of Eros. On its way in, *NEAR Shoemaker* obtained images showing features as small as a few centimeters.

On its way to Eros, *NEAR Shoemaker* flew past the asteroid Mathilde and provided our first information about a C-type body. Mathilde was found to measure 66 by 48 by 44 km and to have a surface only half as reflective as charcoal. Its color properties suggest a composition of materials such as carbonaceous chondrules. However, estimates of the overall density of Mathilde are about 1,300 kg/m^3 (1.3 times the density of water), only half that of carbonaceous chondrite meteorites measured in the laboratory. This implies that Mathilde is "porous" inside, as we would expect if it is composed of chunks of rocky material stuck together by the accretion process with open spaces between the chunks. The surface of Mathilde is dominated by craters, the largest of which is more than 33 km across. The great number of craters means that Mathilde probably dates back to the very early history of the Solar System.

The spectrum of sunlight reflected by Mars's tiny moons, Phobos and Deimos (**Figure 12.8**), is similar to that of C-type asteroids, and many scientists believe these moons must be asteroidlike objects that were somehow captured by Mars. Images of Phobos and Deimos returned by *Mariner 9* in 1971 showed them to be irregular shaped objects with cratered surfaces, but planetary scientists got an even better look at

FIGURE 12.7 Images of the asteroid Eros obtained by the *NEAR Shoemaker* spacecraft.

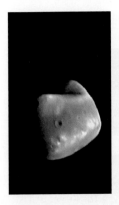

FIGURE 12.8 *Viking* orbiter and *Mars Express* images of Mars's two tiny moons: Deimos *(left)* and Phobos *(right)*.

the two moons a few years later. NASA controllers used the *Viking* orbiters to take close-up views of both moons, commanding *Viking 2*'s orbiter to whiz by within 28 km of Deimos and raising a few concerns about a possible unplanned "impact experiment" in Mars orbit! Fortunately the delicate maneuver came off without a hitch, and the spacecraft sent back

Phobos and Deimos may be asteroids captured by Mars.

images showing features as small as a meter. While the *Viking* images of Phobos and Deimos were spectacular, it is still not clear that they answered the question at hand: Are Phobos and Deimos really asteroids? Controversy about this question has never been put to rest. Some scientists argue that it is unlikely that Mars could have "captured" two asteroids and that Phobos and Deimos must somehow have evolved together with Mars. A third possibility is that the bodies must have been parts of a much larger body that was fragmented by a collision early in the history of Mars.

Once called the "vermin of the skies," asteroids have now achieved considerable respectability in scientific circles. Not only do they hold fascinating and unique clues about how the Solar System formed, but they also play a part in the continuing (albeit slow) evolution of the Solar System. And if an asteroid were to get *too* close to Earth, as happens from time to time, it would quickly become a *very* respectable topic for conversation!

12.6 The Comets: Clumps of Ice

As with many astronomical objects, our concept of comets changed markedly during the 16th and 17th centuries, when we looked upon nature with an increasingly refined and rational eye. In much earlier times comets were regarded simply as ghostly apparitions, pale luminous patches or streaks of light in the nighttime sky that would mysteriously appear, remain for a few days or weeks, and then vanish. Until the end of the Middle Ages, the Aristotelian view of comets prevailed, regarding them as atmospheric phenomena rather than astronomical objects. The popular beliefs that comets were very nearby and that their tails contained poisonous vapors contributed to the fear and panic that comets frequently generated.

A major change in our view came in the 16th century thanks to Tycho Brahe. Tycho reasoned that if comets were atmospheric phenomena like clouds, as Aristotle had sup-

Comets are astronomical objects, *not* atmospheric phenomena.

posed, then their appearance and location in the sky should be very different to observers located many miles away from each other. However, when Tycho compared sightings of comets made by observers at several different sites, he found no evidence of such differences, and he concluded that comets must be at least as far away as the Moon. Lest we become arrogant in our faith in humankind's rationality, it is worth noting that the proof that comets are far outside Earth's atmosphere did not completely dispel the fear they evoke. As recently as 1910, the news that Earth would pass through the tail of Halley's Comet was accompanied by widespread apprehension throughout Europe and the United States. And during the approach of Comets Kohoutek in 1973 and Hale-Bopp[4] in 1997, many otherwise rational people experienced similar anxieties, regrettably fueled by certain opportunistic writers who exploit human fear and superstition for power or profit.

The Abode of the Comets

When we think of a comet, we think of a spectacular display of the sort produced by Comet Hyakutake in 1996, as seen in the opening photograph of this chapter. But for most

Swarms of comets spend most of their lives far beyond the planets.

of their lives comets are merely icy planetesimals—*comet nuclei*—adrift in the frigid outer reaches of the Solar System. Comet nuclei put on a show only when they come deep enough into the inner Solar System to suffer destructive heating from the Sun. When they are close enough to show the effects of solar heating, we call them **active comets**. Comet nuclei spend most of their lives as nothing more than

---

[4]Believing that Comet Hale-Bopp was accompanied by a UFO that would carry their souls away to a "better place," 39 members of the cult Heaven's Gate committed mass suicide in March 1997.

icy planetesimals—small clumps of primordial material that managed to escape the planet-building process—that follow orbits that keep them far beyond the planets in the outermost reaches of the Solar System. Drifting in these distant regions, most are much too small and far away to be seen and counted by telescopes on Earth, so no one really knows their total number. Estimates range as high as a trillion (10^{12}) comet nuclei—more than all the stars in the Milky Way Galaxy. Despite this, we have seen hardly more than a thousand, perhaps only a billionth of the presumed total.

We know where comets reside by observing their orbits as they pass through the inner Solar System. Based on these studies, they seem to fall into two distinct groups: the *Kuiper*

The Kuiper Belt and the Oort Cloud are reservoirs of comets.

Belt comets and the *Oort Cloud* comets. These two populations of comet nuclei are named for scientists **Gerard Kuiper** (1905–1973) and **Jan Oort** (1900–1992), who first proposed their existence in the mid–20th century.[5]

The extent of the realm of comets is illustrated in **Figure 12.9**. Comet nuclei from the **Kuiper Belt** orbit the Sun within a disk-shaped region that begins just beyond the orbit of Neptune and extends outward to perhaps a thousand astronomical units from the Sun. The innermost part of the Kuiper Belt, which seems to end somewhat abruptly at about 50 AU, appears to contain tens of thousands of icy planetesimals, which we call **Kuiper Belt objects** (**KBOs**). As we learned in the previous chapter, the larger KBOs are well over a thousand kilometers across. This makes them larger than Ceres and similar in size to Pluto. Like the asteroids, some KBOs are attended by moons, and at least one has as many as three. Astronomers have already found more than a thousand KBOs, but many smaller ones must also be there, just beyond the reach of our largest telescopes. Although they are too far from the Sun to be "active," it seems likely that these planetesimals are comet nuclei.

One of the larger known KBOs, Quaoar (pronounced kwa-whar), is also one of the few whose size we have been able to measure. Hubble Space Telescope observations show a tiny disk with a diameter of about 1,250 km, slightly larger than Pluto's Charon. By knowing its apparent brightness, distance, and size, we can easily calculate Quaoar's albedo. It turns out to be 0.12, which is more reflective than the nuclei of those comets that have entered the inner Solar System but far less than that of Pluto. Quaoar's remote location and pristine condition have apparently allowed some bright ices

[5] Some planetary scientists refer to the Kuiper Belt as the Edgeworth-Kuiper Belt in honor of Kenneth Edgeworth (1880–1972), an Irish engineer who predicted a reservoir of comets beyond the planets in a paper published in 1943, several years before the predictions of Kuiper and Oort.

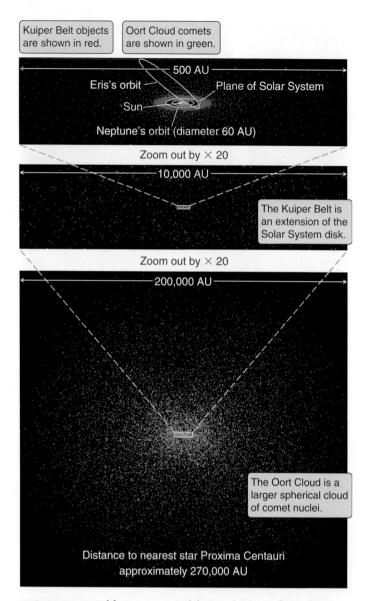

FIGURE 12.9 Most comet nuclei near the inner Solar System form an extension to the disk of the Solar System called the Kuiper Belt. The spherical Oort Cloud is far larger and contains many more comets.

to survive on its surface. Quaoar orbits between 42 AU and 45 AU from the Sun, and in this frigid region of the Solar System, its surface temperature is only 50 K.

Sedna is an interesting object that remains outside the inner Kuiper Belt. Its highly elliptical orbit takes it from 75 AU to more than 800 AU from the Sun. In such an extended orbit, Sedna requires *10,500 years* to make a single trip around the Sun. When discovered in 2004, Sedna was about 90 AU from the Sun and getting closer. It will reach its closest point in 2076. Because of its great distance, Sedna's size cannot be measured directly, even by the HST. If its albedo were the same as Quaoar's, Sedna would be the largest KBO.

Right now our knowledge of the chemical and physical properties of KBOs remains meager at best. Because of their great distance, we know very little about their composition and surface properties and absolutely nothing about their geology. This may change in the not too distant future, however. Following its encounter with Pluto in 2015, the *New Horizons* spacecraft is scheduled to continue outward into the Kuiper Belt, where it will be maneuvered to fly close to one or more KBOs.

Unlike the flat disk of the Kuiper Belt, the **Oort Cloud** is a spherical distribution of comet nuclei that are much too remote to be seen by even the most powerful telescopes. We know the size and shape of the Oort Cloud because comet nuclei from the Oort Cloud approach the inner Solar System from seemingly random directions in the sky, following orbits that bring them in from as far as 50,000 AU from the Sun, or about a fifth of the way to the nearest stars!

We can think of the Kuiper Belt and Oort Cloud as enormous reservoirs of icy planetesimals that now and then fall into the inner Solar System. But why do they sometimes leave their frigid haven and take what will probably be a fatal plunge inward toward the Sun? One idea is that comet nuclei run into each other from time to time, knocking each other from their orbits. But even with the huge number of nuclei that the Oort Cloud must contain, the immense volume of this region means that the distances between comet nuclei are great. The average separation between an Oort Cloud object and its nearest neighbor must be at least 10 AU—as great as the distance between the Sun and Saturn—making collisions between comet nuclei *extremely* uncommon events. It seems more likely that gravitational perturbations by objects beyond the Solar System—such as stars and interstellar clouds—disturb the orbits of comet nuclei, kicking the nuclei in toward the Sun.

By looking at the distribution of stars around us, astronomers estimate that every 5 million to 10 million years, one star or another, traveling in nearly the same orbit around the Milky Way as does the Sun, passes within about 100,000 AU of the Sun, perturbing the orbits of comet nuclei. The gravitational attraction from huge clouds of dense interstellar gas concentrated in the plane of the Milky Way Galaxy[6] might also stir up the Oort Cloud as the Sun passes back and forth through the galactic plane during its 220-million-year orbital journey around the center of our galaxy. These disturbances from beyond the Solar System have little effect on the orbits of the planets and other bodies in the inner Solar System. Inner Solar System objects are close enough to the Sun that external disturbances never exert more than a tiny fraction of the gravitational force of the Sun. In the distant Oort Cloud, however, things are quite different. Oort Cloud comet nuclei are so far from the Sun, and the Sun's gravitational force on them is so feeble, that they are barely

bound to the Sun at all. The tug of a slowly passing star or interstellar cloud can compete with the Sun's gravity, significantly changing the orbits of Oort Cloud objects. If the interaction adds to the orbital energy of a comet nucleus, it may move outward to an even more distant orbit, or perhaps escape from the Sun completely to begin an eons-long odyssey through interstellar space. On the other hand, a comet nucleus that loses orbital energy falls inward. Some of these come all the way into the inner Solar System, where they may appear briefly in our skies as active comets before returning once again to the Oort Cloud.

Unlike comet nuclei in the Oort Cloud, those in the Kuiper Belt are packed close enough together to interact gravitationally from time to time. In such events, one nucleus gains energy while the other loses it. The "winner" may gain enough energy to be sent into an orbit that reaches far beyond the boundary of the Kuiper Belt. It seems likely that the Oort Cloud was populated in just this way (by comet nuclei ejected from the inner edge of the Kuiper Belt in what is now the Uranus–Neptune region). The experience could prove fatal for the "loser" if enough orbital energy is lost for it to fall inward toward the Sun.

Anatomy of an Active Comet

Figure 12.10 shows a diagram of a fully active comet. The small object at the center of the comet—the icy planetesimal itself—is the comet nucleus. The nucleus is by far the smallest component of a comet, but it is the source of all the mass that we see stretched across the skies as the comet nears the Sun. Comet nuclei are anywhere from a few dozen meters to several hundred kilometers across. We might

---

An active comet has a tiny nucleus at its heart.

---

formally describe comet nuclei as "planetesimals of modest size composed of a mixture of various ices of volatile compounds, organics, and dust grains loosely packed together to form a porous conglomerate"; but mercifully a shorter description is possible. In the middle of the 20th century astronomer **Fred Whipple** (1906–2004) offered an elegant way to sum all of this up in just two words: comets are "dirty snowballs."

Unlike asteroids, which have been through a host of chemical and physical changes as a result of collisions, heating, and differentiation, most comet nuclei have been preserved over the past 4.6 billion years by the "deep freeze" of the outer Solar System. Comet nuclei are made of the most nearly pristine material remaining from the formation of the Solar System.

As a comet nucleus nears the Sun, sunlight heats its surface, turning volatile ices into gases, which then stream

---

[6]Interstellar clouds and the structure of the Milky Way Galaxy will be discussed in Chapters 15 and 19, respectively.

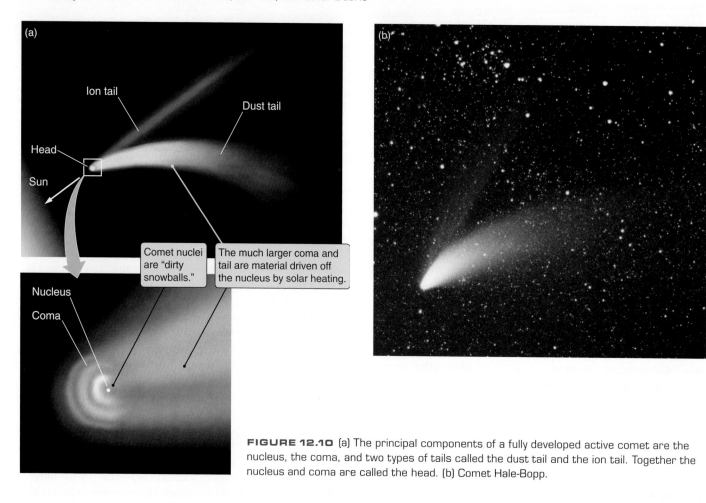

FIGURE 12.10 (a) The principal components of a fully developed active comet are the nucleus, the coma, and two types of tails called the dust tail and the ion tail. Together the nucleus and coma are called the head. (b) Comet Hale-Bopp.

away from the nucleus, carrying embedded dust particles along with them. This process of conversion from solid to gas is called **sublimation**. An example much closer at hand illustrates this process. Dry ice (frozen carbon dioxide) does not melt like water ice but instead turns directly into carbon dioxide gas. Water ice melts.[7] Dry ice sublimates—that is why we call it "dry." Set a piece of dry ice out in the sun on a summer day, and you will get a pretty good idea of what happens to a comet.

The gases and dust driven from the nucleus of an active comet form a nearly spherical atmospheric cloud around

> **The coma is an extended cometary atmosphere surrounding the nucleus.**

the nucleus called the **coma** (plural *comae*). The nucleus and the inner part of the coma are sometimes referred to collectively as the comet's **head**. Pointing from the head of

[7] Water ice also sublimates in certain environments. For example, an ice cube left in a frost-free freezer will eventually disappear as water molecules slowly break free from its surface.

the comet in a direction more or less away from the Sun are long streamers of dust, gas, and ions called the **tail**.

A Visit to Comet Halley

Typical comet nuclei at typical distances appear so small that even the most powerful telescopes see them only as points of light. Even when a rare comet happens to approach close to Earth, we are still out of luck because the nucleus remains hidden within the dusty coma that surrounds it. Although astronomers have studied the gases that make up the comae and tails of comets with great interest for centuries, direct observations of comet nuclei have remained an elusive prize.

This all changed in 1986 when Halley's Comet made its "once in a lifetime" trip into the inner Solar System, providing a rare opportunity to send space probes to observe a nucleus at close range. This opportunity was so rare because, despite the fact that even relatively pristine comet nuclei are frequent visitors to the inner Solar System, we seldom have enough advance warning of a comet's visit or

sufficiently detailed knowledge of its orbit to mount a successful mission to intercept it. But although planetary scientists knew the orbit of Halley extremely well, sending
probes to the comet still turned out to be a difficult task as

Halley's Comet comes around once in a lifetime.

space missions go. The problem was that the orbit of Comet
Halley is *retrograde*—it goes around the Sun in a direction
opposite to that of Earth and the other planets. As a result,
the closing speed between an Earth-launched spacecraft
and the comet would be 68 km/s, or about 75 times the speed
of a bullet from a high-powered rifle. Observations close to
the nucleus would have to be made very quickly, and there
would be the always-present danger of high-speed collisions
with debris from the nucleus.

Despite the difficulties, working in cooperation with
each other, three different space agencies rose to the challenge of sending a five-spacecraft armada to meet Comet
Halley as it passed through the inner Solar System. The
whole operation was rather like driving down the highway
at 70 miles per hour and trying to thread a needle held out
of the window by someone in the oncoming lane; but in the
end the mission was a glorious success. Much of what we
know about comet nuclei and the innermost parts of the
coma was learned from data sent back by this armada of
Soviet, European, and Japanese spacecraft.

Images taken by these spacecraft showed the nucleus of
Halley to be an irregular peanut shell–shaped object measuring 8 by 8 by 14 km (**Figure 12.11(a)**). As the Soviet *VEGA*
and the European *Giotto* spacecraft passed close by the nucleus, they observed **jets** of gas and dust moving at speeds
of up to 1,000 m/s from its surface, far exceeding the ability
of its feeble gravity to contain the material. Here was dramatic evidence that the gas and dust that make up a comet's
coma and tail come directly from the nucleus. But instead
of streaming out from the whole surface of the nucleus, the
spacecraft images showed that material was coming from
several hot spots, or jets, that covered only about a tenth of
the surface. Ninety percent of the surface was "quiet" at the
time of the observations. This is most easily understood if
the surface of the nucleus is covered by a thick crust of rubble, with the escaping gas coming from beneath this crust
through a number of small cracks or fissures.

By observing the jets of material streaming away from
the nucleus of Halley, planetary scientists estimated that
20,000 kg of gas and 10,000 kg of dust were being lost by the
nucleus each second. During its 1986 passage around the
Sun, Comet Halley must have lost at least 100 billion (10^{11})
kilograms of material. But these jets also indirectly allowed
planetary scientists to measure another fundamental property of this comet. Just like the engines on a jet airplane,
the jets of material streaming away from the nucleus push

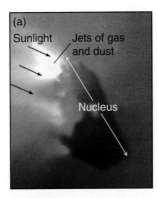
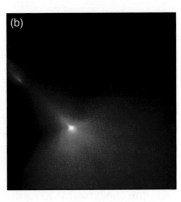

FIGURE 12.11 (a) The nucleus of Comet Halley as imaged by
the *Giotto* spacecraft in 1986. (b) A Hubble Space Telescope
image of the coma of Comet Hyakutake. (Images are false color.)

on it, subtly altering its orbit. By observing the jets flowing
away from the nucleus, scientists could tell how hard the
jets were pushing on it. And by observing the comet's orbit,
they learned how the nucleus was responding to these
shoves. Using these observations, scientists were then able

Comet Halley loses 100 billion kg each time it comes around.

to apply Newton's laws of motion, which require the change
in the momentum of the nucleus to balance the change in
momentum of the material streaming away. They thus estimated the mass of the nucleus to be about 3×10^{14} kg. This
mass, divided by the volume as measured from the images
of the nucleus, tells us that the nucleus is only about a fourth
as dense as water. This is more like a loosely packed pow-
dery snow than normal ice. As suggested by Whipple years
earlier, the nucleus of Comet Halley must be a very fragile
object that is porous throughout. Even with a mass loss of
100 billion kg of material during each encounter, this is less
than 0.1 percent of its total mass, so Halley's Comet should
still be entertaining terrestrial onlookers hundreds of generations from now.

The surface of the nucleus is also much darker than expected. It reflects only 3 percent of the sunlight falling on
it, making it even blacker than coal or black velvet. The
"dirty snowball" at the heart of Comet Halley is not only
dirty, it is among the *darkest known objects* in the Solar
System. This low reflectivity means that comet nuclei, or at

The nucleus of Comet Halley is blacker than coal.

least the nucleus of Halley, must be rich in organic matter
that is far more complex than simple hydrocarbons such as
methane, propane, butane, or octane. Simple organic compounds such as these form clear liquids and ices at low

temperatures. In contrast, more complex hydrocarbons such as tar are usually very dark. The implication is that very complex organic matter was present as dust in the disk around the young Sun, perhaps even in the interstellar cloud from which the Solar System formed.

The coma of a comet is a tenuous and temporary atmosphere surrounding the nucleus, consisting of gas and dust that are driven from the nucleus by solar heating. **Figure 12.11(b)** shows an image of the coma of Comet Hyakutake obtained with the Hubble Space Telescope. As a comet nucleus falls from the outer Solar System inward toward the Sun, the coma begins to form at about the time the nucleus passes within the orbit of Jupiter. (Although from our vantage point so close to the heart of the Solar System Jupiter may seem far removed from the Sun's warmth, from the comet's perspective the solar warming at 5 AU is hundreds to many millions of times greater than it is in the Kuiper Belt or the Oort Cloud.) As the nucleus approaches still closer to the Sun, rapid sublimation of ices near its surface can cause the coma to swell to a diameter of a million kilometers, three times the distance between Earth and the Moon. But while the coma is huge compared with the nucleus, it is still more tenuous than the very best vacuums we can produce on Earth.

VEGA and *Giotto* entered the coma of Comet Halley when they were still nearly 300,000 km from the nucleus. In addition to cameras, these spacecraft also carried instruments capable of measuring the size and chemical composition of the dust particles in Halley's coma. These experiments found that the dust grains escaping from the nucleus were generally between 0.01 and 10 mm in size. This dust is very fine—finer even than the dust responsible for that annoying thin film that collects on your car and ruins its shine. Chemically the dust from Comet Halley was found to consist of a mixture of light organic substances (combinations of hydrogen, carbon, nitrogen, and oxygen) and heavier rocky material (combinations of magnesium, silicon, iron, and oxygen). Instruments on the spacecraft designed to study the gases surrounding the nucleus found it consisted of about 80 percent water (H_2O) and 10 percent carbon monoxide (CO), with smaller amounts of carbon dioxide (CO_2), methane (CH_4), ammonia (NH_3), and hydrogen cyanide (HCN).

Visits to Comets Borrelly, Wild 2, and Tempel 1

The early years of the 21st century have witnessed dramatic progress in our understanding of comet nuclei. Three spacecraft were sent by NASA to visit three Jupiter family comets. Such comets, as we will learn later in the chapter, are escapees from the Kuiper Belt that have been captured by Jupiter into short-period orbits. It may come as no surprise to learn that all three comet nuclei appear vastly different when studied up close.

In 2001 NASA's *Deep Space 1* spacecraft flew within 2,200 km of Comet Borrelly's nucleus, penetrating a surrounding coma of dust and gas. The comet's 8-km, potato-shaped nucleus was seen spewing jets of gas from fissures concentrated on just one side, similar to the asymmetry observed on Comet Halley's nucleus. Comet Borrelly's tar-black surface is among the darkest seen on any Solar System object and, much to the surprise of scientists, showed no evidence of water ice. Was Comet Borrelly's crust too old and depleted of its volatiles? Astronomers wanted a look at a more pristine comet, and that opportunity was soon to come.

In 2004 NASA's *Stardust* spacecraft flew within 235 km of the nucleus of Comet Wild 2 (pronounced "Vilt Two"), snapping dozens of images, analyzing gas and dust streaming from its surface, and collecting dust samples that were returned to Earth in January 2006. Astronomers were especially interested in this comet because it is a newcomer to the inner Solar System. Comet Wild 2 had previously resided in the frigid region between the orbits of Jupiter and Uranus, but a close encounter with Jupiter in 1974 perturbed its orbit, bringing this relatively pristine body closer to the Sun as it travels back and forth in its new orbit between Jupiter and Earth. At the time of *Stardust*'s encounter with Wild 2, the comet had made only five trips around the Sun in its new orbit, compared with more than 100 visits by Comet Halley.

Wild 2's nucleus is about 5 km across, and unlike other comets visited earlier by spacecraft (Comets Halley and Borrelly), its shape is more spherical. Images taken by *Stardust* show more than two dozen active gas jets. Some were as large as 100 m in diameter and carried within them surprisingly large chunks of surface material. (A few particles, as large as a bullet, penetrated the outer layer of the spacecraft's protective shield.) The surface of Wild 2 appears covered with craterlike features (**Figure 12.12**). Although these depressions do not have the normal impact characteristics, they may in fact be impact craters that have been modified by ice sublimation, small landslides, and erosion by jetting gas. The larger depressions are more than 1.5 km across. Some show flat floors, suggesting a relatively solid interior beneath a porous surface layer.

How do we know what really lies beneath a comet's surface? To find the answer, astronomers came up with a novel approach. They proposed to fire a massive bullet at very high speed into the nucleus of a comet and observe what happens. In 2005 a 370-kg impacting projectile, launched by NASA's *Deep Impact* spacecraft, hit the nucleus of Comet Tempel 1 at a speed of more than 10 km/s. The kinetic energy of the impact was equivalent to 5 tons of TNT, sending 10 thousand tons of subsurface debris flying off into space at speeds of 50 m/s (**Figure 12.13**). A camera, mounted on the projectile, continued to snap photos of its

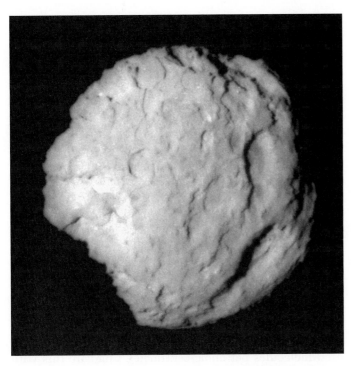

FIGURE 12.12 The nucleus of Comet Wild 2 imaged by the *Stardust* spacecraft.

FIGURE 12.13 A 370-kg projectile impacts Comet Tempel 1 at a speed of 10.2 km/s. This image, taken 67 s after impact, shows subsurface cometary material being ejected into space by the force of the impact.

target (**Figure 12.14**) until both were vaporized by the impact. Observations of the event were made locally by the *Deep Impact* spacecraft and back at Earth by a multitude of orbiting and groundbased telescopes. Spectra indicated the presence of water, carbon dioxide, hydrogen cyanide, iron-bearing minerals, and a host of complex organic molecules. Analysis of the debris showed that the comet's outer layer is composed of fine dust with a consistency of talcum powder. No large chunks were seen flying from the impact site. Beneath the dust are layers made up of water ice and organic materials. The nucleus has a density only 0.6 times that of water and is perhaps best described as a porous rubble pile held weakly together by self-gravity rather than the mechanical strength of its constituent materials. One surprise for scientists was the presence of well-formed impact craters (Figure 12.14), which had been absent on close-up images of Comets Borrelly and Wild 2. Why should some comet nuclei have fresh impact craters and others none? This remains a question that planetary scientists have yet to answer.

Comets Have Two Types of Tails

The tail, which is the largest and most spectacular part of a comet, is also the "hair" for which comets are named. (*Comet* comes from the Greek word *kometes,* which means "hairy

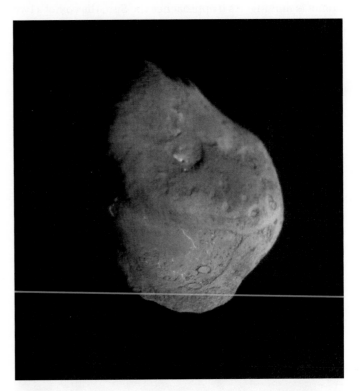

FIGURE 12.14 The surface of Comet Tempel 1 seen just before impact by the *Deep Impact* projectile. The impact occurred between the two 370-m diameter craters located near the bottom of the image. The smallest features appearing in this image are about 5 m across.

one.") Active comets have two different types of tails, as shown in Figure 12.10. One type of tail is called the **ion tail**. Many of the atoms and molecules that make up a comet's coma are ions. Because they are electrically charged, ions in the coma feel the effect of the solar wind, the stream of charged particles that blows continuously away from the Sun. The solar wind pushes on these ions, rapidly accelerating them to speeds of more than 100 km/s—far greater than the orbital velocity of the comet itself—and sweeps them out into a long wispy structure. Because the particles that make up the ion tail are so quickly picked up by the solar wind, ion tails are usually very straight and point from the head of the comet directly away from the Sun.

Dust particles in the coma can also have a net electric charge and feel the force of the solar wind. In addition, sunlight itself exerts a force on cometary dust. But dust particles are much more massive than individual ions, so they are accelerated more gently and do not reach such high relative speeds as the ions. As a result, the dust particles are unable to keep up with the comet, and the **dust tail** often appears to gently curve away from the head of the comet as the dust particles are gradually pushed from the comet's orbit in the direction away from the Sun.

Figure 12.15 shows the tails of a comet at various points in its orbit. Remember that both types of tails always point *away from the Sun,* regardless of the direction in which the comet is moving. As it approaches the Sun, the comet's two tails trail behind its nucleus. But like the flag on a sailboat sailing with the wind, the comet's tails extend *ahead* of the nucleus as it moves outward from the Sun.

> Comet tails *always* point away from the Sun.

Comet tails are fascinating structures that vary greatly from one comet to another. Some comets, such as Comets West and Halley, display both types of tails simultaneously, while for reasons we do not understand some comets produce no tails at all! A tail often forms as the comet crosses the orbit of Mars, where the increase in solar heating drives gas and dust away from the nucleus at a prodigious rate. In October 1965 early morning risers were held spellbound by the sight of Comet Ikeya-Seki's glowing tail (**Figure 12.16**) extending 150 million kilometers across the sky—a length as great as the distance between Earth and the Sun. For a brief few weeks Comet Ikeya-Seki achieved temporary splendor as the longest object in the entire Solar System.

The gas in a comet's tail is even more tenuous than the gas in its coma, with densities of no more than a few hundred molecules per cubic centimeter. Compare this with Earth's atmosphere, which at sea level contains more than 10^{19} molecules per cubic centimeter. Dust particles in the tail are typically about a micrometer in diameter, about the size of smoke particles. One astronomer's often-repeated quip that "all of the material in a long comet tail could be packed into a suitcase" is less an exaggeration than you might imagine. When Earth passed through the tail of Comet Halley in 1910, there were no observable effects. In contrast to the widespread predictions of disaster, it was all "much ado about nothing."

FIGURE 12.15 The orientation of the dust and ion tails at several points in a comet's orbit. The ion tail points directly away from the Sun while the dust tail curves along the comet's orbit.

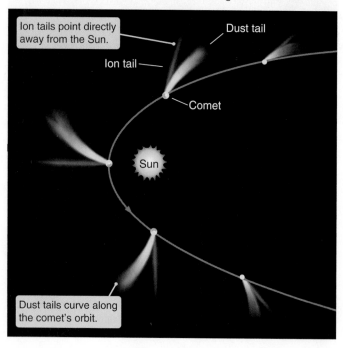

Ion tails point directly away from the Sun.

Dust tail

Ion tail

Comet

Sun

Dust tails curve along the comet's orbit.

The Orbits of Comets

How long a dying nucleus will survive depends on its orbital period and **perihelion** (closest point to the Sun) distance—in other words, how frequently it passes by the Sun and how close it comes. Each passage takes its toll of ice and dust. By convention, comet orbits are generally referred to as *short-period* or *long-period orbits*. The division between the two is somewhat arbitrarily set at a period of 200 Earth years. Comets with periods of less than 200 years are called **short-period comets**. The total number of short-period comets known today is about 330. Comets with orbital periods longer than 200 years are termed **long-period comets**. The total number of long-period comets observed to date is about 2,000, with an average of six new ones being discovered each year.

The orbits of nearly all comets are highly elliptical, in sharp contrast to the relatively circular orbits of the planets. Only a few comets have orbital eccentricities that are less than Pluto's, whose orbit is less circular than any of the clas-

FIGURE 12.16 Comet Ikeya-Seki as it appeared in the predawn hours in October 1965. Soon afterward, just before it passed behind the Sun, the head of the comet became so bright that it was visible in broad daylight. When it reappeared a few days later it had split into two separate components.

sical planets. With a little thought, we might have anticipated this result. When a comet nucleus first enters the inner Solar System, it *must* be on a very elongated orbit because one end

> The orbits of most comets are very elliptical.

of the orbit is close to the Sun while the other end is in the distant parts of the Solar System. Because of this, we might expect all comets seen in the inner Solar System to have *extremely* elliptical orbits and extremely long orbital periods

that carry them again and again back to the Oort Cloud or the Kuiper Belt. The surprise is that more than a hundred short-period comets remain relatively close to the Sun. How do these denizens of the frigid hinterlands become resident aliens in the realm of the planets?

The key to solving this mystery lies in an interesting difference between the orbits of long- and short-period comets. **Figure 12.17** shows the orbits of a number of comets. Whereas long-period comets come diving into the inner Solar System from all directions, short-period comets have orbits that are strongly concentrated in the ecliptic plane. Most have orbital inclinations less than Pluto's, which has a higher inclination than any of the classical planets. Because their orbits are nearly in the plane of the ecliptic and cross the orbits of many of the planets, short-period comets frequently get close enough to a planet for the planet's gravity to change the comet's orbit about the Sun in the sorts of chaotic encounters discussed in Chapter 10. Presumably short-period comets originated in the Kuiper Belt; but as they fell in toward the Sun, they were forced into their present short-period orbits by gravitational encounters with planets.

Short-Period Comets Figure 12.17(a) shows the orbits of a number of short-period comets in the inner Solar System. Nearly all short-period comets have **prograde** orbits. That is, they orbit the Sun in the same direction as the planets. Comet Halley, with its retrograde orbit, is an exception. This, along with the concentration of short-period comets in the plane of the Solar System, is important evidence supporting our theory of the origin of short-period comets and how they were captured. The disklike Kuiper Belt appears to share the prograde rotation of the rest of the Solar System (except the Oort Cloud). Because Kuiper Belt comet nuclei start out on prograde orbits, they are more likely to be captured into short-period prograde orbits as well. Also, the encounter between a planet and a comet nucleus on a prograde orbit will last longer than the encounter between a planet and a comet nucleus on a retrograde orbit. This means that a prograde comet's orbit will be more strongly influenced by such an encounter than the orbit of a retrograde comet will be. Jupiter, the most massive of the planets, is the one most likely to affect the orbit of a Kuiper Belt comet. There are about 270 known short-period comets, referred to as the *Jupiter family,* that all have their **aphelion** (farthest point from the Sun) near Jupiter's orbit and travel in prograde orbits around the Sun.

Halley's Comet is the brightest and most famous of the short-period comets. It also has a very special place in the history of science because it was the first comet whose return was predicted. In 1705 Edmund Halley was studying the orbits of comets employing the gravitational laws that had recently been discovered by his friend and colleague, Isaac Newton. (Halley was instrumental in encouraging Newton to publish his work on orbits and gravitation—work

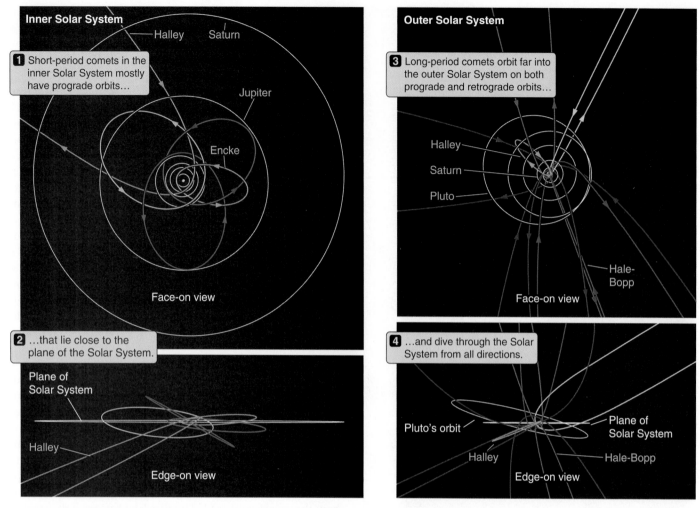

FIGURE 12.17 Orbits of a number of comets in face-on and edge-on views of the Solar System. Populations of short- and long-period comets have very different orbital properties.

that became the foundation for the science of physics.) Halley noted that a bright comet that had appeared in 1682 had an orbit remarkably similar to those of comets seen in 1531 and 1607. He concluded that all three were actually one and the same, and he made the daring prediction that this comet would return in 1758. Unfortunately Halley did not live to see it. His comet reappeared on Christmas Eve in the very year he had predicted. Astronomers quickly named it Halley's Comet and heralded it as a triumph for the genius of both Newton and Halley.

Comet Halley's highly elongated orbit takes it from perihelion, about halfway between the orbits of Mercury and Venus, out to aphelion beyond the orbit of Neptune. Astronomers and historians have now identified sightings of the comet that go back to 467 B.C. Halley's Comet has an average period of 76 years, and many of us mark the "once in our lifetime" when we are fortunate enough to see it. Mark Twain is famous for saying that he came in with the comet in 1834 and he would go out with the comet in 1910

—a promise that he kept. For the authors of this book, our opportunity to see Halley's Comet came in 1986. Although that appearance was not especially spectacular compared to the one in 1910—in 1986 Halley's Comet and Earth were on opposite sides of the Sun when the comet put on its display—seeing the ghostly sight of Halley is an experience we will all remember. Halley will reach aphelion in 2024 and then begin its long journey back to the inner Solar System, becoming visible to the naked eye once again in 2061.

Long-Period Comets Of the known long-period comets, more than 600 have well-determined orbits. Some have orbital periods of hundreds of thousands or even millions of years, and their nuclei spend almost all their time in the frigid, outermost regions of the Solar System. Orbits of a few long-period comets are shown in Figure 12.17(b). Unlike the mostly prograde, ecliptic plane orbits of the short-period comets, long-period comets split about evenly into prograde and retrograde and fall into the inner Solar System from all

Some long-period comets come by only once in a million years.

directions. These are the comets that tell us of the existence of the Oort Cloud. Because of their very long orbital periods, these comets cannot make more than a single appearance throughout the entire course of recorded history and, in most cases, in all of human history.

Long-period comets differ from short-period comets in another way as well. With a few exceptions, Halley's Comet among them, the nuclei of short-period comets have been badly "worn out" by their repeated exposure to heating by the Sun. As the volatile ices are driven from a nucleus, some of the dust and organics are left behind on the surface. As this covering builds up, it slows down cometary activity. (Envision how, as a pile of urban—and therefore dirty—snow melts, the dirt left behind is concentrated on the surface of the snow.) That is why most short-period comets create little excitement. In contrast, long-period comets are usually relatively pristine. More of their supply of volatile ices still remains close to the surface of the nucleus, and they can produce a truly magnificent show. A half dozen or so long-period comets arrive each year. Most pass through the inner Solar System at relatively large distances from Earth or Sun and never become bright enough to attract much public attention.

Comet Ikeya-Seki (Figure 12.16) is a member of a family of comets called **sungrazers**, comets whose perihelia are located very close to the surface of the Sun. Many sungrazers fail to survive even a single pass by our local star. Ikeya-Seki became so bright as it neared perihelion in 1965 that it was visible in broad daylight, only two solar diameters away from the noontime Sun, and when it reappeared from behind the Sun, it had been split into two pieces. Sungrazers sometimes come in groups, with successive comets following in nearly identical orbits. It seems likely that each member of such a group started as part of a single larger nucleus that broke into pieces during an earlier perihelion passage.

The closing years of the 20th century witnessed two spectacular long-period comets sporting long, beautiful tails: Hyakutake in 1996 (chapter opening photograph) and Hale-Bopp in 1997 (Figure 12.10(b)). Both were widely seen by the viewing public, Comet Hale-Bopp being perhaps the most observed comet ever. It was bright enough to spot even in urban nighttime skies and dazzled those who were fortunate enough to see it far from city lights.

Comet Hyakutake is not very large as comets go, with a nucleus only a few kilometers across. The comet's awesome appearance was primarily a consequence of its unusually close trajectory past Earth. It approached within a mere 15 million kilometers less than two months after its discovery by Japanese amateur astronomer Yuji Hyakutake.

Comet Hale-Bopp, on the other hand, is a huge comet, with a nucleus perhaps as large as 40 km in diameter. It was an especially important scientific object for professional astronomers because it was discovered far from the Sun near Jupiter's orbit two years before its perihelion passage. This extended the total time available to study its development as it approached the Sun and provided ample opportunity to plan the important observations as it neared perihelion. Warmed by the Sun, the nucleus produced large quantities of gas and dust, as much as 300 tons of water per second, with lesser amounts of carbon monoxide, sulfur dioxide, cyanogen, and other gases. As this book goes to press Comet Hale-Bopp is still being observed with large telescopes. It remains active and continues to show a tail even though it is now beyond the orbit of Uranus. Comet Hale-Bopp will continue its outward journey for well over a thousand years. It will not return to the inner Solar System until sometime around the year 4377.

When will the next bright comet like Hyakutake or Hale-Bopp come along? On average a spectacular comet appears about once per decade, but it is all a matter of chance. It might be many years from now—or it could happen tomorrow.

12.7 Collisions Still Happen Today

As we studied the inner planets and the moons of the outer planets, we learned that almost all hard-surfaced objects in the Solar System still bear the scars of a time when the Sun and planets were young and tremendous impact

In 1994 much of the world watched as a comet crashed into Jupiter.

events were common occurrences. The collision of Comet Shoemaker-Levy 9 with Jupiter in 1994 (see **Excursions 12.2**) focused attention on the fact that although such impacts are far less frequent today than they once were, they still happen. Nearly every major observatory in the world observed this spectacular and dramatic event, as did the Hubble Space Telescope and *Galileo* spacecraft. The "Jupiter comet crash" was also a landmark event in the history of the public's access to fast-breaking scientific events. Occurring when the World Wide Web was growing in popularity, the impact's latest images were downloaded daily by millions of people around the world, scientists and laypeople alike. (Although such access may be commonplace to you now, it was heady and exciting stuff to the authors back in 1994.)

In addition to providing a convincing demonstration that major collisions still occur in the Solar System today, the collision of Shoemaker-Levy 9 with Jupiter also let planetary

Comet Shoemaker-Levy 9

From a wealth of ground- and space-based observations, planetary scientists have pieced together the events leading to the collision of Comet Shoemaker-Levy 9 with Jupiter. We have already encountered Comet Shoemaker-Levy 9 twice on our journey: first in Chapter 7 in our discussion of the devastation caused by impacts, and later in Chapter 10 when we learned of the destructive nature of tides. Early in the 20th century a comet nucleus from the Kuiper Belt was perturbed from its path and sent on

an orbital journey that carried it close to Jupiter. In 1992 it passed so close to the planet that tidal stresses broke it into two dozen major fragments, which subsequently spread out along more than 7 million km of its orbit. The trajectory carried the fragments around for one more two-year orbit about the planet.

Then in July 1994 the entire string of fragments crashed into Jupiter. Over a week's time one after another of the fragments, each traveling at 60 km/s, plunged through

FIGURE 12.18 Illustration of events following the impact of a fragment of Comet Shoemaker-Levy 9 on Jupiter.

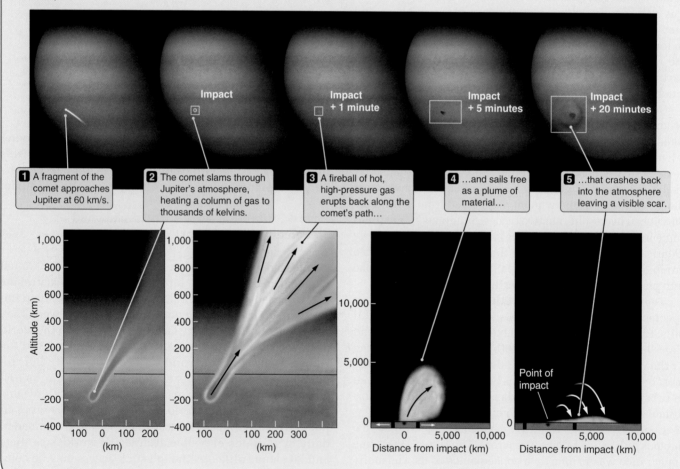

Jupiter's stratosphere. Even though the impacts occurred just behind the limb of the planet, where they could not be observed from Earth, the *Galileo* spacecraft was on its way to Jupiter and was able to image some of the impacts. In addition, astronomers using groundbased telescopes and the HST could see immense plumes rising from the impacts to heights of more than 3,000 km above the cloud tops at the limb. The debris in these plumes then rained back onto Jupiter's stratosphere, causing ripple effects like pebbles thrown into a pond. **Figure 12.18** shows the sequence of events that took place as each fragment of the comet slammed into the giant planet. Sulfur and carbon compounds released by the impacts formed giant, Earth-sized scars in the atmosphere that persisted for months (**Figure 12.19**) and were visible even through small amateur telescopes.

FIGURE 12.19 (a) Ground based infrared images of Jupiter showing the hot glowing scars left by fragments of Comet Shoemaker-Levy 9. (b) Although the fragments struck Jupiter on the back side, these HST images show the fireball from one fragment rising above the limb of the planet. (c) HST images of the evolution of the scar left by one fragment of the comet.

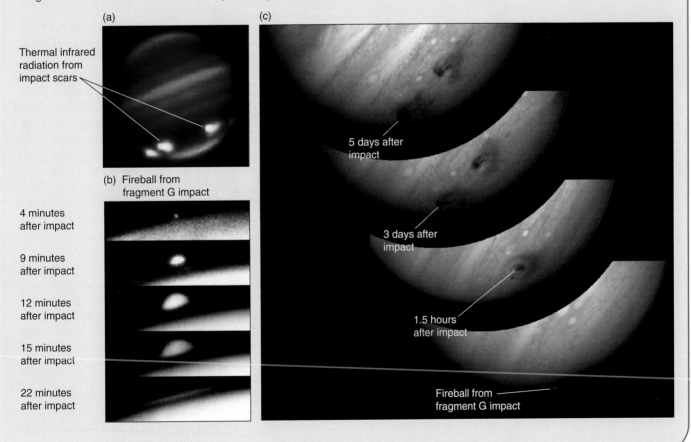

scientists study the role that planetary atmospheres play in such events. In Chapter 7 we found that the dense atmosphere of Venus caused some objects that collided with the planet to break up before reaching the surface, creating clusters of craters in a pattern similar to a shotgun blast or sometimes leaving only a dark "splotch" on the surface with no crater at all. The Shoemaker-Levy 9 collision also gave direct evidence that the tremendous amounts of energy released in such impacts can create huge atmospheric fireballs. Having seen the effects of Shoemaker-Levy 9's collision with Jupiter, it is easy to imagine the global firestorms on Earth that accompanied the impact that occurred at the **Cretaceous–Tertiary boundary** 65 million years ago (see Excursions 7.1). **Connections 12.1** discusses several ways that such impacts affected the course of life on Earth.

What we learned from the Jupiter comet crash also helps explain an event that has puzzled scientists for the past century. The Tunguska River flows through a remote region of western Siberia, inhabited mostly by reindeer and a few reindeer herders. In the summer of 1908 the region was

CONNECTIONS 12.1

Origins—Comets, Asteroids, and Life

One of the great questions of modern astronomy involves understanding how life on Earth is connected with processes at work in the greater cosmos. In a certain sense, this is not so different a quest from that pursued by many of the world's great religions and philosophies. At most of the stops along our journey through the universe, we find threads of the cosmic tapestry of which terrestrial life is a part, and this stop is no exception.

For example, it is possible that without the existence of comet nuclei, water on Earth might have been less plentiful than it is today. We think that some of Earth's water was contributed during impacts of icy planetesimals (in other words, comet nuclei) early in the history of the Solar System. How could this have come about? The icy planetesimals likely condensed from the protoplanetary disk surrounding the young Sun and grew to their present size near the orbits of the giant planets. These planetesimals subsequently suffered strong orbital disturbances from those same planets. In such interactions about half of the planetesimals would be flung outward to form the Kuiper Belt and Oort Cloud and half inward toward the Sun. Some of the objects flung toward the Sun would likely have hit Earth, the largest planet in the inner Solar System. Because most of the mass in comet nuclei appears to be in the form of water ice, it is likely that some of Earth's current water supply came from this early bombardment. Water, as we know, is essential to life.

Yet the existence of comets can threaten life on Earth as well. It seems certain that occasional collisions of comet nuclei and asteroids with Earth have resulted in widespread devastation of Earth's ecosystem and the extinction of many species. Passing stars or galactic plane crossings may send showers of comet nuclei into the inner Solar System every 10 million to 100 million years —intervals similar to those between mass extinctions found in fossil records of life on our planet. Although such events certainly qualify as global disasters for the plants and animals alive at the time, they also represent global opportunities for new life forms to evolve and fill the niches left by species that did not survive. Such a collision probably played a central role in ending the 180-million-year reign of dinosaurs and provided our ancient mammalian ancestors an opportunity for world dominance.

In our study of comets we may also have found a key to the chemical origins of life on Earth. Comets turn out to be rich in complex organic material—the chemical basis for terrestrial life—and cometary impacts on the young Earth may have played a role in chemically seeding the planet. Also, because comets are believed to be pristine samples of the material from which the Sun and planets formed, it means that organic material must be widely distributed throughout interstellar space. This conclusion, which is supported by radio telescope observations of vast interstellar clouds throughout our galaxy, might have significant implications as we consider the possibility of life elsewhere in the universe.

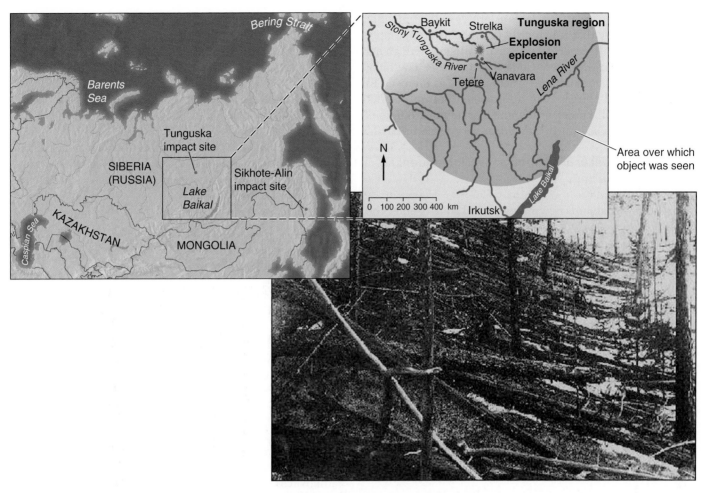

FIGURE 12.20 A large region of forest near the Tunguska River in Siberia was flattened by an atmospheric explosion in 1908 caused by the impact of a small asteroid or comet.

blasted with the energy equivalent of 2,000 of the atomic bombs dropped over Hiroshima. **Figure 12.20** shows a map of the region, along with a photo of the destruction caused by the blast. Eyewitness accounts included the destruction of dwellings, the incineration of reindeer (including one herd

In 1908 a small asteroid or comet smashed into Siberia.

of 700), and the deaths of at least five people. Although trees were burned or flattened for over 2,150 square kilometers—an area greater than metropolitan New York City—no crater was left behind!

For many years scientists and nonscientists alike speculated about the cause of the Tunguska event. Ideas included such fanciful notions as a collision with a mini–black hole that passed through Earth, to a collision with an object made of *antimatter,* to an act of aggression by aliens (a favorite of the supermarket tabloids). It now seems fairly clear that the Tunguska event was the result of a tremendous high-altitude explosion that occurred when a small asteroid or comet nucleus hit Earth's atmosphere, ripped apart, and formed a fireball before reaching Earth's surface. This is just the sort of event seen on Jupiter and hypothesized to account for the "splotches" seen on Venus. Recent expeditions to the Tunguska area have recovered resin from the trees blasted by the event. Chemical traces in the resin suggest that the impacting object may have been a stony asteroid.

The Tunguska impact and the collision of Shoemaker-Levy 9 with Jupiter are truly sobering events when you stop to think about them. We know of several impacts occurring in the Solar System within a human lifetime, and more than one of them involved our planet![8] We know enough about

[8] At 10:38 A.M. on February 12, 1947, yet another planetesimal struck Earth, this time in the Sikhote-Alin region of eastern Siberia. Composed mostly of iron, it had an estimated diameter of about 100 m and broke into a number of fragments before hitting the ground, leaving a cluster of craters and widespread devastation. Witnesses reported a fireball brighter than the Sun and sound that was heard from 300 km away.

the distribution of relatively large asteroids in the inner Solar System to say that it is highly improbable a populated area on Earth will experience a major collision with an asteroid within our lifetimes.[9] Comets and smaller asteroids, however, are less predictable. There may be as many as 10 million asteroids larger than a kilometer, but only about 130,000 have well-known orbits, and most of the unknowns are too small to see until they come very close to Earth.

Comet or asteroid impacts are infrequent but devastating.

Several previously unknown long-period comets enter the inner Solar System each year. If one happens to be on a collision course with Earth, we might not notice it until just a few weeks or months before impact. For example, Comet Hyakutake was discovered only two months before passing near Earth, and a potentially destructive asteroid that just missed Earth in 2002 was not discovered until three days *after* its closest approach! Although this has become a favorite theme of science fiction disaster stories (*Lucifer's Hammer* by Larry Niven and Jerry Pournelle being a favorite of several of the authors of this text), Earth's geological and historical record suggests that actual impacts by large bodies are infrequent events. Does this mean we need not lose any sleep over a possible collision with a large comet or asteroid? Probably, but remember that even though the probability may be small, the consequences of such an event are enormous. Just ask the dinosaurs.

12.8 Solar System Debris

Although some comets and asteroids meet the spectacular fates of Comet Shoemaker-Levy 9 or the Tunguska planetesimal, most go out with more of a fizzle than a bang. Comet nuclei that enter the inner Solar System generally disintegrate within a few hundred thousand years as a result of their repeated passages close to the Sun. Asteroids have much longer lives but still are slowly broken into pieces from occasional collisions with each other. Disintegration of comet nuclei and asteroid collisions are the source of most of the debris that fills the inner part of the Solar System.

---

[9]The asteroid Apophis will pass about 31,000 km above Earth's surface on April 13, 2029 (Friday the 13th), and will be bright enough to be seen with the naked eye. Depending on its exact miss distance in 2029, it could be perturbed into an orbit that would bring it into a collision course with Earth on April 13, 2036. Although the probability of a collision remains small at this time, the blast from such an impact would release energy equivalent to 60,000 of the atomic bombs dropped over Hiroshima.

tem. As Earth and other planets move along in their orbits, they continuously sweep up this fine debris. This is the source of most of the meteoroids that Earth encounters. When they burn up in our atmosphere, these meteoroids

---

Comet nuclei and asteroids are the source of Solar System debris.

---

become the meteors that you can see streaking across the sky on any clear dark night. As discussed in Excursions 12.1, only when they are large enough to reach Earth's surface do we call them meteorites.

Meteoroids and Meteors

Some 100,000 kg of meteoritic debris is swept up by Earth every day, and what does not burn up (mostly particles smaller than 100 μm) eventually settles to the ground as fine dust. We can demonstrate that meteors are atmospheric phenomena using the same method Tycho Brahe did to figure

---

Meteors are atmospheric phenomena.

---

out that comets are not. If the same meteor is captured on photographs taken from two different locations, it appears to be in different places as seen against the background of stars. We can also measure meteor heights by radar because radio waves bounce off the ionized gas in meteor trails just as they bounce off the metal in an airplane or an automobile. Using these techniques, we find that the altitudes of meteors are between 50 and 150 km. For comparison, commercial jet aircraft typically fly at heights of about 10 km.

Fragments of asteroids are much denser than cometary meteoroids. If an asteroid fragment is large enough—about the size of your fist—it can survive all the way to the ground to become one of the meteorites discussed in Section 12.3. The fall of a 10-kg meteoroid can produce a fireball so bright that it lights up the night sky more brilliantly than the full Moon. Such a large meteoroid, traveling many times faster than the speed of sound, may create a sonic boom heard hundreds of kilometers away. It may even explode into multiple fragments as it nears the end of its flight, becoming a **bolide**. Some fireballs glow with a brilliant green color, caused by metals in the meteoroid that created them.

Meteor Showers and Comets

Standing under a dark nighttime sky with a clear horizon, you can expect to see about a dozen meteors per hour on any night of the year. These are called **sporadic meteors**, and they occur as Earth sweeps up random bits of cometary and asteroidal debris in its annual path around the Sun. However, if you happen to be meteor watching on the night of August 11, for instance, you may see four to five times this number. If you pay close attention, you might also notice that nearly all of the meteors seem to be coming from the same region of the sky. This phenomenon is called a **meteor shower**. We call the particular meteor shower that peaks in mid-August the **Perseids** because the trails they leave all point back to the constellation Perseus.

Meteor showers happen when Earth's orbit crosses the orbit of a comet. Bits of dust and other debris released by a comet nucleus as it rounds the Sun remain in orbits of their own that are similar to the orbit of the nucleus itself. When Earth crosses through this concentration of cometary debris, the result is a meteor shower. Because the meteoroids

---

Meteor showers occur when Earth passes through cometary debris.

---

that are being swept up are all in similar orbits, they all enter our atmosphere moving in the same direction. As a result, the paths of all shower meteors are parallel to one another. But just as the parallel rails of a railroad track appear to vanish to a single point in the distance, as in **Figure 12.21**, our perspective makes all the meteors appear to originate from the same point in the sky. This point is called the shower's **radiant**.

The Perseids are the result of Earth crossing the orbit of Comet Swift-Tuttle. Although spread out along the comet's orbit, the debris is more concentrated in the vicinity of the comet itself. In 1992 Comet Swift-Tuttle returned to the inner Solar System for the first time since its discovery in 1862. An exceptional Perseid meteor shower resulted, with counts of up to 500 meteors per hour.

There are more than a dozen comets whose orbits come close enough to Earth's to produce annual meteor showers. Around November 16 each year, Earth passes almost directly through the orbit of Comet Tempel-Tuttle, a short-period comet with an orbital period of 33.2 years. We call the meteor shower that Tempel-Tuttle produces the **Leonids**. In most years the Leonids fail to produce much of a show because this comet distributes little of its debris around its orbit. However, in 1833 and 1866 Comet Tempel-Tuttle was not far away when Earth passed through its orbit. The Leonid showers in those two years were so intense that meteors filled the sky with as many as 100,000 per hour (**Figure 12.22**). In 1900, one comet orbit later, nothing out of the ordinary happened. Again in 1933 the Leonids were disappointing. Perturbations of Comet Tempel-Tuttle by Jupiter had moved the orbit of the comet's nucleus slightly away from Earth's, causing a sharp decrease in shower strength. What Jupiter took away, though, it gave back. Further perturbations of the comet's orbit caused a spectacular Leonid shower in 1966 that may have produced as many as a half million meteors per hour! The Leonid shower put on less spectacular but

FIGURE 12.21 (a) Meteors appear to stream away from the radiant of the Perseid meteor shower. These streaks are actually parallel paths that appear to emerge from a vanishing point, like the railroad tracks in (b).

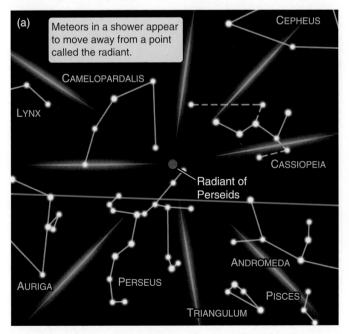

FIGURE 12.22 An engraving of the 1833 Leonid shower seen in France. One of the authors of this text recalls witnessing the 1966 shower, in which the sky was covered with meteor trails —perhaps as many as a half million per hour!

still impressive shows between 1999 and 2003, when several thousand meteors per hour were seen (**Figure 12.23**).

Zodiacal Dust

In the same way that we can "see" sunlight streaming through an open window by observing its reflection from motes of dust drifting in the air, we can see the sunlight reflected off tiny **zodiacal dust** particles that fill the inner parts of the Solar System close to the plane of the ecliptic. On a clear, moonless night, not long after the western sky has gotten dark, this dust is visible as a faint column of light slanting upward from the western horizon along the path of the ecliptic, as seen in **Figure 12.24**. This band, called the

zodiacal light, can also be seen in the eastern sky before dawn. With good eyes and an especially dark night, you may be able to follow the zodiacal dust band all the way across the sky. In its brightest parts, the zodiacal light can be several times brighter than the Milky Way, for which it is sometimes mistaken.

Like meteoroids, zodiacal dust is a mixture of cometary debris and ground-up asteroidal material. The dust grains are roughly a millionth of a meter in diameter, the size of smoke particles. In the vicinity of Earth there are only a few particles of zodiacal dust in each cubic kilometer of space.

> Zodiacal dust is cometary debris and ground-up asteroidal material.

The total amount of zodiacal dust in the entire Solar System is estimated to be 10^{16} kg, equivalent to a solid body about 25 km across, or roughly the size of a large comet nucleus. Grains of zodiacal dust are constantly being lost as they are swept up by planets or pushed out of the Solar System by

FIGURE 12.23 Leonid meteors seen in 2001.

FIGURE 12.24 Zodiacal light shines in the eastern sky before dawn .

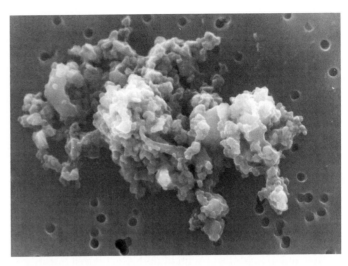

FIGURE 12.25 A 10-μm diameter cluster of interplanetary dust grains collected in Earth's stratosphere by a NASA U-2 airplane.

the pressure of sunlight. Such interplanetary dust grains have been recovered from Earth's upper atmosphere by very high-flying aircraft (**Figure 12.25**). If not replaced by new dust from comets, all zodiacal dust would be gone within a brief span of 50,000 years.

In the infrared region of the spectrum, thermal emission from the band of warm zodiacal dust makes it one of the brightest features in the sky. It is so bright that astronomers wanting to observe faint infrared sources are frequently hindered by its foreground glow.

Summary

- Asteroids are small Solar System bodies made of rock and metal.

- The orbits of most asteroids are located between the orbits of Mars and Jupiter.

- Some asteroids cross Earth's orbit and are potentially dangerous.

- Comets are small icy planetesimals that reside in the frigid regions beyond the planets.

- Comets that venture into the inner Solar System are warmed by the Sun, producing an atmospheric coma and a long tail.

- Very large asteroids or comets striking Earth create enormous explosions that can wipe out most of terrestrial life.

- Meteoroids are small fragments of asteroids and comets.

- Meteor showers occur when Earth passes through a trail of cometary debris.

- When a meteoroid enters Earth's atmosphere, frictional heat causes the air to glow, producing a phenomenon called a meteor.

- A meteoroid that survives to a planet's surface is called a meteorite.

Seeing the Forest through the Trees

Long before their true nature was understood, comets were granted great significance by humankind. These spectacular celestial displays were taken as omens of great events, and more often than not, as harbingers of the end of the world. How ironic that they have turned out instead to be messengers from the time when the world was born.

Anyone who has read a mystery novel knows that sometimes things that seem least significant at first glance turn out to hold the crucial clues to the biggest questions. In our study of the Solar System, two questions rise above all others: How did the Solar System form, and what is its history? Planets may dominate the environment around the Sun, but planets keep imperfect records of the earliest days of the Solar System. The violence of the planet formation process and eons of geological activity together effectively erase most clues to their origin. The best samples of the early Solar System that have survived to the present day are instead the smallest bodies —asteroids and especially comets, frozen denizens of the outer reaches of the Sun's influence. Buried within these dirty snowballs are grains that predate even the Solar System itself. These tiny bits of solid material formed in the atmospheres of stars and in material blasted into space in tremendous stellar explosions, survived for a time in the vast reaches between the stars, participated in the collapse of the interstellar cloud that would become the Solar System, and wound up embedded within the small bodies that are the most numerous citizens of our planetary system. The discovery of such grains provides a direct link between our existence and our origins in the stars. How remarkable and fortuitous it is that these very pieces of our Solar System's past are delivered to our doorstep, falling to Earth as meteorites to be picked up from a cornfield in the Midwest or a glacier in the Antarctic, and then deciphered using the most advanced tools modern science has to offer.

Grains that predate the Sun make up only the tiniest fraction of the material found in meteorites. Most of this material was instead formed along with the Solar System itself. This material was vaporized in the violence that accompanied the formation of the disk around the young Sun and then condensed again into solid form. Some asteroids and all comets are formed directly from this pristine material. Meteorites broken off from such bodies provide a window into the conditions that existed at that time. Other asteroids instead carry clues about how planets formed. Iron and stony-iron meteorites, for example, are pieces broken off from what were once larger, differentiated bodies. Pick up a stony-iron meteorite, and you hold in your hand a snapshot, frozen in time, of the processes that shaped the world.

But comets and asteroids are far more than scientific curiosities; they have played a major role in shaping the history of life on our planet. Your body is largely water, and it is quite likely that much of that water (along with the water that covers the surface of our world) arrived on Earth billions of years ago as volatile-rich comets slammed into the surface of our young world. Across the ages, occasional impacts of comets and asteroids on Earth have dramatically altered the planet's climate for a time and redirected the course of life's flow. Intelligent life on Earth descended from mammals rather than dinosaurs only because of a cosmic fluke—the impact of a comet, 65 million years ago, in the region that is now the Yucatán Peninsula. The awe-inspiring fireballs that accompanied the impact of Comet Shoemaker-Levy 9 on Jupiter and the devastation of a remote corner of Siberia by a small piece of a comet or asteroid that hit Earth's atmosphere in 1908 are reminders that there is a grain of truth in humankind's deep-seated superstitions about these objects. Sometimes the appearance of a comet *does* mean the end of the world as we know it. It has happened in the past, and unless we develop technology to prevent such events (not an easy task), it *will* happen again in the future.

However, as we look to the future of human exploration and utilization of the Solar System, asteroids and comets may play a vital and positive role as ready-made way stations and caches of raw materials that we would need if we were to expand beyond our home planet. The history and destiny of our kind, as well as the course of our intellectual journey through the universe, are inextricably tied to this flotsam and jetsam adrift in interplanetary space.

So far on our journey we have spent our time digging in our own backyard. The Solar System may dwarf the scales of our everyday lives, but it is vanishingly small compared with the universe. Just as we found that we could understand our own planet and its history only by putting it within the context of the Solar System, we are unable to understand the Sun and the worlds of our Solar System without placing them within the context of the broader universe. The starting point on the next leg of our journey begins as we gaze at the myriad stars that fill the night sky and wonder about what we see there.

Key Terms

meteor, p. 343
meteoroid, p. 343

Student Questions

THINKING ABOUT THE CONCEPTS

1. How does the composition of an asteroid differ from that of a comet nucleus?

2. Most meteorites (pieces of S-type and M-type asteroids) are 4.54 billion years old. Carbonaceous chondrites (pieces of C-type asteroids), however, are 20 million years older. What determines the time of "birth" of these pieces of rock? What does this tell you about the history of their parent bodies?

3. Why are most asteroids very irregular in shape? What does this tell you about their history?

4. Most asteroids are found between the orbits of Mars and Jupiter, but astronomers are especially interested in the relative few whose orbits cross that of Earth. Why?

5. Describe the differences among meteoroids, meteors, and meteorites.

6. How could you and a friend, armed only with your cell phones and a knowledge of the night sky, prove conclusively that meteors are an atmospheric phenomenon?

7. Suppose you found a rock that has all of the characteristics of a meteorite. You take it to a physicist friend who confirms that it is a meteorite but says that radioisotope dating indicates an age of only a billion years. What might be the origin of this meteorite?

8. What are the differences between a comet and a meteor in terms of their size, distance, and how long they remain visible?

9. Name the three parts of a comet. Which part is the smallest? Which is the most massive?

10. Kuiper Belt objects (KBOs) are actually comet nuclei. Why do they not display comae and tails?

11. Comets contain substances closely associated with the development of life, such as water (H_2O), ammonia (NH_3), methane (CH_4), carbon monoxide (CO), and hydrogen cyanide (HCN). Which of these are organic compounds?

12. In 1910 Earth passed through the tail of Halley's Comet. Among the various gases in the tail was hydrogen cyanide, deadly to humans. Yet nobody became ill from this event. Why not?

13. Picture a comet leaving the vicinity of the Sun and racing toward the outer Solar System. Does its tail point backward toward the Sun or forward in the direction the comet is moving? Explain your answer.

APPLYING THE CONCEPTS

14. One recent estimate concludes that nearly 800 meteorites with mass greater than 100 g (massive enough to cause personal injury) strike the surface of Earth each day. Assuming that you present a target of 0.25 m² to a falling meteorite, what is the probability that you will be struck by a meteorite during your 100-year lifetime? Note that the surface area of Earth is approximately 5×10^{14} m².

15. Electra is 182-km diameter asteroid accompanied by a small moon orbiting at a distance of 1,350 km in a

circular orbit with a period of 3.92 days. Refer back to Chapter 3:

a. What is the mass of Electra?

b. What is Electra's density?

16. The orbital periods of Comets Encke, Halley, and Hale-Bopp are 3.3 years, 76 years, and 2,380 years, respectively.

a. What are the semimajor axes (in astronomical units, or AU) of the orbits of these comets?

b. Assuming negligible perihelion distances, what are the maximum distances from the Sun (in AU) reached by Comets Halley and Hale-Bopp in their respective orbits?

c. Which would you guess is the most pristine comet among the three? Which is the least? Explain your reasoning.

17. A cubic centimeter of the air you breath contains about 10^{19} molecules. A cubic centimeter of a comet's tail may typically contain 10 molecules per cubic centimeter. Calculate how large a cube of comet tail material it would take to hold 10^{19} molecules.

18. The total number of asteroids larger than 1,000 m that cross Earth's orbit (Atens and Apollos asteroids) is currently estimated to be about 3,600, with five times as many having diameters larger than 500 m. Assuming this progression remains constant for still smaller asteroids, how many would there be with diameters larger than 125 m? (Note that the impact on Earth of any asteroid larger than 100 m in diameter could cause major, widespread damage.)

19. A one-megaton hydrogen bomb releases 4.2×10^{15} joules (J) of energy. Compare this with a 10-km diameter comet nucleus ($m = 5 \times 10^{14}$ kg) hitting Earth at a speed, v, of 20 km/s. Recall from Chapter 4 that $E_k = \frac{1}{2}mv^2$ (where E_k is the kinetic energy in J, m is in kg, and v is in m/s).

20. The estimated amount of zodiacal dust in the Solar System remains constant at approximately 10^{16} kg. Yet zodiacal dust is constantly being swept up by planets or removed by the pressure of sunlight.

a. If all the dust would disappear (at a constant rate) over a span of 30,000 years, what must be the average production rate in kg/s?

b. Is this an example of static or dynamic equilibrium? Explain your answer.

StudySpace
wwnorton.com/astro21

provides a Study Plan for each chapter that includes a reading outline, animations, keyword flash cards, and gradebook-enabled multiple-choice quizzes. From StudySpace you can also access premium content in the ebook and SmartWork.

PART III

Stars and Stellar Evolution

It is stern work, it is perilous work to
    thrust your hand in the sun
And pull out a spark of immortal flame
    to warm the hearts of men.

JOYCE KILMER (1886–1918)

A sunset in New York City.

CHAPTER 14

A Run-of-the-Mill G Dwarf: Our Sun

14.1 The Sun Is More Than Just a Light in the Sky

How wonderful, after a long cold night, to see the light and feel the warmth of the rays of the Sun. Energy from the Sun is responsible for daylight, for our weather and seasons, and for terrestrial life itself. No object in nature has been more revered or more worshipped than the Sun. In fact, the modern symbol for the Sun, ⊙, is nothing other than the ancient Egyptian hieroglyph for the Sun god Ra. Yet although the Sun may have dominated human consciousness since the dawn of our species, the discussion in Chapter 13 offers a very different perspective. To an astronomer at the opening of the 21st century, the Sun is the prototype for main sequence stars. With a middle-of-the-road spectral type of G2, the Sun is all but indistinguishable from billions of other stars in our galaxy. It is also the star against which all other stars are measured. The mass of the Sun, the size of the Sun, the luminosity of the Sun—these basic properties of the Sun provide the yardsticks of modern astronomy.

The Sun may be run-of-the-mill as far as stars go, but that makes it no less awesome an object on a human scale. The mass of the Sun, 1.99×10^{30} kilograms, is over 300,000 times that of Earth. The Sun's radius of 696,000 km is over 100 times that of Earth. At a luminosity of 3.85×10^{36} watts, the Sun produces more energy in a second than all of the electrical power plants on Earth could generate in 10 million years. The Sun is also the only star we can study at close range. Much of the detailed information that we know about stars has come only by studying our local star. Even though the Sun has been intensively studied for quite some time,

KEY CONCEPTS

To most humans the Sun is the most important object in the heavens. It lights our days, warms our planet, and provides the energy for life. But to astronomers the Sun is a typical main sequence star, located conveniently nearby for detailed study. In this chapter, as we take a closer look at our local star, we will learn about

- The balances between pressure and gravity and between energy generation and loss that determine the structure of the Sun.

- Fusion of hydrogen to helium, and how mass is efficiently converted into energy in the Sun's core.

- The different ways that energy moves outward from the Sun's core toward its surface.

- Physical models of the Sun's interior and how they are tested using observations of solar neutrinos and seismic vibrations on the surface of the Sun.

- The structure of the Sun's atmosphere, from its 5,770 K photosphere to its 1 million K corona.

- Sunspots, flares, coronal mass ejections, and other consequences of magnetic activity on the Sun.

- Eleven- and 22-year cycles in solar activity.

- The solar wind streaming away from the Sun.

- How solar activity affects Earth.

the wealth of phenomena displayed by the Sun will keep solar astronomers busy for many years to come.

In the previous leg of our journey, we looked at the gross physical properties of distant stars, including their mass, luminosity, size, temperature, and chemical composition. Now, as we turn our attention toward our own local star, we face more fundamental questions. How does the Sun work? Where does it get its energy? Why does it have the size, temperature, and luminosity that it does? How has it been able to remain so constant over the billions of years since the Solar System formed? In short, we now confront the question "What is a star?"

14.2 The Structure of the Sun Is a Matter of Balance

In Chapter 7 we tackled the question of how we know about the interior of Earth, despite the fact that no machine has ever done more than scratch the surface of our planet. The answer was a combination of physical understanding, detailed computer models, and clever experiments that test the predictions of those models. The task of exploring the interior of the Sun is much the same. As with Earth, the structure of the Sun is governed by a number of physical processes and relationships. Using our understanding of physics, chemistry, and the properties of matter and radiation, we express these processes and relationships as mathematical equations. High-speed computers are then used to simultaneously solve these equations and arrive at a model of the Sun. One of the great successes of 20th century astronomy was the successful construction of a physical model of the Sun that agrees with our observations of the mass, composition, size, temperature, and luminosity of the real thing.

Our current model of the interior of the Sun is the culmination of decades of work by thousands of physicists and astronomers. Understanding the details that lie within this model is the work of lifetimes. Even so, the essential ideas underlying our understanding of the structure of the Sun are found in a few key insights. In turn, these insights can be summed up in a single statement: *The structure of the Sun is a matter of balance.*

The first key balance within the Sun is the balance between pressure and gravity illustrated in **Figure 14.1**. The Sun is a huge ball of hot gas. If gravity were stronger than pressure within the Sun, the Sun would collapse. Likewise, if pressure were stronger than gravity, the Sun would blow itself apart. We have seen this balance before and have given it a name—*hydrostatic equilibrium* (see Foundations 7.2). Hydrostatic equilibrium sets the pressure at each point within a planet and determines the atmospheric pressure

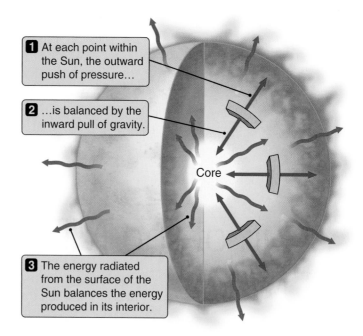

1 At each point within the Sun, the outward push of pressure…

2 …is balanced by the inward pull of gravity.

Core

3 The energy radiated from the surface of the Sun balances the energy produced in its interior.

FIGURE 14.1 The structure of the Sun is determined by balances between forces and in the outward flow of energy.

at Earth's surface. Hydrostatic equilibrium says that the pressure at any point within the Sun's interior must be just enough to hold up the weight of all the layers above that point. If the Sun were not in hydrostatic equilibrium, then

The Sun is in hydrostatic equilibrium.

forces within the Sun would not be in balance, so the surface of the Sun would *move*. The Sun today is the same as it was yesterday and the day before. That is all the observation we need to infer that the interior of the Sun is in hydrostatic equilibrium.

Hydrostatic equilibrium becomes an even more powerful concept when combined with what we know about the way gases behave. As we move deeper into the interior of the Sun, the weight of the material above us becomes greater, and hence the pressure must increase. As we have learned before (see Foundations 8.2), in a gas, higher pressure means higher density and/or higher temperature. **Figure 14.2(a)** shows how conditions vary as distance from the center of the Sun changes. As we go deeper into the Sun, the pressure climbs; as it does, the density and temperature of the gas climb as well.

A second fundamental balance within the Sun is a balance of energy. Stars like the Sun are remarkably stable objects. Geological records show that the luminosity of the Sun has remained nearly constant for billions of years. In fact, the very existence of the main sequence says that stars do not change much over the main part of their lives. To remain in balance, the Sun must produce just enough en-

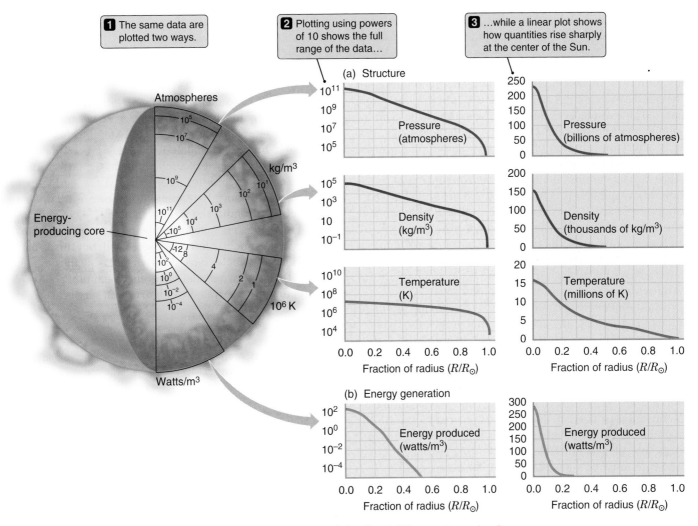

1 The same data are plotted two ways.

2 Plotting using powers of 10 shows the full range of the data...

3 ...while a linear plot shows how quantities rise sharply at the center of the Sun.

(a) Structure

(b) Energy generation

FIGURE 14.2 A cutaway figure showing the interior structure of the Sun. (a) Temperature, density, and pressure increase toward the center of the Sun. (b) Energy is generated in the Sun's core.

Solar energy production must balance what is radiated away.

ergy in its interior each second to replace the energy that is radiated away by its surface each second. This is a new type of balance, one we have not dealt with before. Understanding the balance of energy within the Sun requires thinking about how energy is generated in the interior of the Sun, and how that energy finds its way from the interior to the Sun's surface, where it is radiated away.

The Sun Is Powered by Nuclear Fusion

One of the most basic questions facing the pioneers of stellar astrophysics was where the Sun and stars get their energy. The answer to this question came not from astronomers' telescopes, but from theoretical work and the laboratories

of nuclear physicists. At the heart of the Sun lies a nuclear furnace capable of powering the star for billions of years.

The nucleus of most hydrogen atoms consists of a single proton. Nuclei of all other atoms are built from a mixture of protons and neutrons. Most helium nuclei, for example, consist of two protons and two neutrons. Most carbon nuclei consist of six protons and six neutrons. Protons have a

Atomic nuclei are held together by the strong nuclear force.

positive electric charge, and neutrons have no net electric charge. Like charges repel, so all of the protons in an atomic nucleus must be pushing away from each other with a tremendous force because they are so close to each other. If electric forces were all there was to it, the nuclei of atoms would rapidly fly apart—yet atomic nuclei *do* hold together. We conclude that there must be some other force in nature, even stronger than the electric force that "glues" the protons and neutrons in a nucleus together. That force, which acts only over short distances, is called the **strong nuclear force**.

The strong nuclear force is indeed a very powerful force. It would take an enormous amount of energy to pull apart the nucleus of an atom such as helium into its constituent parts. The reverse of this process says that when you assemble an atomic nucleus from its component parts, this same enormous amount of energy is released. The process of combining two less massive atomic nuclei into a single more massive atomic nucleus is referred to as **nuclear fusion**. Many nuclear processes are possible, and as we continue our study of stars, we will find that a wide range of nuclear reactions can occur in stars. In the Sun, as in all main sequence stars, the only significant process going on is the fusion of hydrogen to form helium—a process often referred to as **hydrogen burning** (even though it has nothing to do with fire in the usual sense of the word).

Main sequence stars get their energy by fusing hydrogen atoms together to make helium. To judge the effectiveness of this reaction, we can make use of one of the key results of special relativity (Chapter 4): the equivalence between mass and energy. Mass can be converted to energy and energy can be converted to mass, with Einstein's famous equation $E = mc^2$ providing the exchange rate between the two. Comparing the mass of the *products* of a reaction with the mass of the *reactants* tells us what fraction of the original mass was turned into energy in the process. The mass of four separate hydrogen atoms is 1.007 times greater than the mass of a single helium atom; so when hydrogen fuses to make helium, 0.7 percent of the mass of the hydrogen is converted to energy.

Conversion of 0.7 percent of the mass of the hydrogen into energy might not seem very efficient—until we compare it with other sources of energy and discover that it is millions of times more efficient than even the most efficient chemical reactions. Fusing a *single gram* of hydrogen into helium releases about 6×10^{11} joules of energy, which is enough to boil all of the water in about 10 average backyard swimming pools. Scale this up to converting roughly 600

Nuclear fusion is a very efficient source of energy.

---

million metric tons of hydrogen into helium every second (with 4 million metric *tons* of matter converted to energy in the process), and you have our Sun. The sunlight falling on Earth may be responsible for powering almost everything that happens on our planet, but it amounts to only about a hundred-billionth of the energy radiated by our local star. The Sun has been burning hydrogen at this prodigious rate for 4.6 billion years, during which time it has converted about half of the hydrogen at its center into helium. A favorite theme of science fiction is the fate that awaits Earth when the Sun dies, but we need not worry anytime soon.

The Sun is only about halfway through its 10-billion-year lifetime as a main sequence star.

Whether a ball rolling down hill, a battery discharging itself through a lightbulb, or an atom falling to a lower state by emitting a photon, most systems in nature tend to seek the lowest-energy state available to them. Going from hydrogen to helium is a big ride downhill in energy, so we might imagine that hydrogen nuclei would naturally tend to fuse together to make helium. Fortunately for us, however, a major roadblock stands in the way of nuclear fusion. The strong nuclear force responsible for binding atomic nuclei together can act only over very short distances—10^{-15} meter or so, or about a hundred-thousandth the size of an atom. To get atomic nuclei to fuse, they must be brought close enough to each other for the strong nuclear force to assert itself, but this is hard to do. All atomic nuclei have positive electric charges, which means that any two nuclei will repel each other. This electric repulsion, illustrated in **Figure 14.3**, serves as a barrier

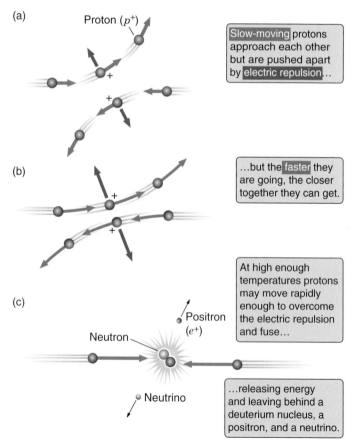

FIGURE 14.3 Atomic nuclei are positively charged and so repel each other. If two nuclei are moving toward each other, the faster they are going, the closer they will get before veering away, as shown in (a) and (b). (c) At the temperatures and densities found in the centers of stars, thermal motions of nuclei are so energetic that nuclei can overcome this electric repulsion, so fusion takes place.

(a) Proton (p^+)

Slow-moving protons approach each other but are pushed apart by electric repulsion...

(b) ...but the faster they are going, the closer together they can get.

At high enough temperatures protons may move rapidly enough to overcome the electric repulsion and fuse...

(c) Positron (e^+)

Neutron

Neutrino

...releasing energy and leaving behind a deuterium nucleus, a positron, and a neutrino.

against nuclear fusion. Fusion cannot take place unless this electric barrier is somehow overcome.

As shown in Figure 14.2(b), energy in the Sun is produced in its innermost region, called the Sun's *core*. Conditions in the Sun's core are extreme. Matter at the center of the Sun has a density about 150 times the density of water (which is 1,000 kg/m³), and the temperature at the center of the Sun is about 15 million K. The thermal motions of atomic nuclei in the Sun's core are tens of thousands times more energetic than the thermal motions of atoms at room temperature. As illustrated in Figure 14.3(c), under these conditions atomic nuclei slam into each other hard enough to overcome the electric repulsion between them and allow short-range nuclear forces to act. The hotter and denser a

Hydrogen fuses to helium in the core of the Sun.

gas, the more of these energetic collisions will take place each second. For this reason, the rate at which nuclear fusion reactions occur is extremely sensitive to the temperature and the density of the gas. Half of the energy produced by the Sun is generated within the inner 9 percent of the Sun's radius, or less than 0.1 percent of the volume of the Sun (Figure 14.2(b)).

There are several reasons that hydrogen burning is the most important source of energy in main sequence stars. Hydrogen is the most abundant element in the universe, so it offers the most abundant source of nuclear fuel at the beginning of a star's lifetime. Hydrogen burning is also the most efficient form of nuclear fusion, converting a larger fraction of mass into energy than any other type of reaction. But the most important reason why hydrogen burning is the dominant process in main sequence stars

is that hydrogen is also the easiest type of atom to fuse. Hydrogen nuclei—protons—have an electric charge of +1. The electric barrier that must be overcome to fuse protons is the repulsion of a single proton against another. Compare this to the force required, for example, to get two

Hydrogen burns mostly via the proton–proton chain.

carbon nuclei close enough to fuse. To fuse carbon we must overcome the repulsion of six protons in one carbon nucleus pushing against the six protons in another carbon nucleus. The resulting force is proportional to the product of the charges of the two atomic nuclei, making the repulsion between two carbon nuclei 36 times stronger than that between two protons. For this reason, hydrogen fusion occurs at a much lower temperature than any other type of nuclear fusion. In the cores of low-mass stars such as the Sun, hydrogen burns primarily through a process called the **proton–proton chain**. The dominant branch of the proton–proton chain is illustrated in **Figure 14.4** and discussed in **Foundations 14.1**.

Energy Produced in the Sun's Core Must Find Its Way to the Surface

Some of the energy released by hydrogen burning in the core of the Sun escapes directly into space in the form of neutrinos (see Foundations 14.1), but most of the energy goes instead into heating the solar interior. To understand the structure of the Sun we must understand how thermal energy is able to move outward through the star. The nature

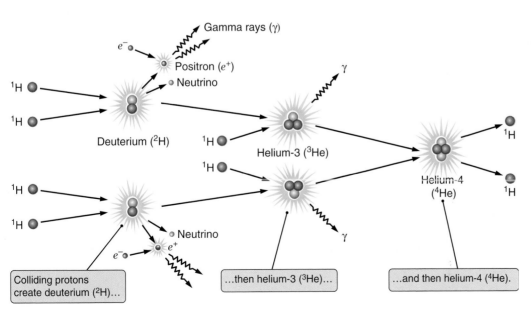

FIGURE 14.4 The Sun and all main sequence stars get their energy by fusing the nuclei of four hydrogen atoms together to make a single helium atom. In the Sun, about 85 percent of the energy produced comes from the branch of the proton–proton chain shown here.

of **energy transport** within a star is one of the key factors determining the star's structure.

Thermal energy can be transported from one place to another by a number of methods. Pick up a bucket of hot coals and carry it from one side of the room to the other, and you have transported thermal energy. A common way in which energy is transported in our everyday lives is by **thermal conduction**. For example, hold one end of a metal rod while you put the other end into a fire. Soon the end of the rod that is in your hand becomes too hot to hold. Thermal conduction occurs as the energetic thermal vibrations of atoms and molecules in the hot end of the rod cause their

cooler neighbors to vibrate more rapidly as well. However, while thermal conduction is the most important way energy is transported in *solid* matter, it is typically ineffective in a *gas*. Thermal conduction is unimportant in the transport of energy from the core of the Sun to its surface. Thermal energy is instead carried outward from the center of the Sun by two other mechanisms: radiation and convection.

The transport of energy from one place to another by **radiative transfer** involves photons moving from hotter regions to cooler regions, carrying energy with them. Imagine a hotter region of a star sitting next to a cooler region, as shown in **Figure 14.5**. Recall from our study of radiation in

FOUNDATIONS 14.1

The Proton–Proton Chain

Hydrogen burning in the Sun and other low-mass stars takes place through a series of nuclear reactions called the proton–proton chain. There are three different "branches" to the proton–proton chain. The most important of these (Figure 14.4), responsible for about 85 percent of the energy generated in the Sun, consists of three steps. In the first step, two hydrogen nuclei fuse. In the process, one of the protons is transformed into a neutron by emitting a positively charged particle called a **positron** and another type of elementary particle called a **neutrino**. The conversion of a proton into a neutron by the emission of a positron and a neutrino is one variety of a process referred to as **beta decay**.

The positron is expelled at a great velocity, carrying away some of the energy released in the reaction. Electrons and positrons have opposite electric charges, so they attract each other. As a result, our expelled positron will soon collide with one of the many electrons moving freely about in the center of the Sun. But the positron is the **antiparticle** of the electron, and as we'll learn later in Chapter 21, when particle and antiparticle collide they annihilate each other, with their total mass being converted into energy. Thus the annihilation of electrons and positrons in the Sun's core produces energy in the form of gamma ray photons. These photons carry part of the energy released when the two protons fused, thereby helping to heat the surrounding gas. The neutrino, on the other hand, is a very elusive particle. Its interactions with matter are so feeble that its most likely fate is to escape from the Sun without further interactions with matter.

The new atomic nucleus formed by the first step in the proton–proton chain consists of a proton and a neutron. This is the nucleus of a heavy isotope of hydrogen called *deuterium,* or ^{2}H. The second step of the proton–proton chain occurs when another proton slams into the deuterium nucleus, fusing with it to form the nucleus of a light isotope of helium, ^{3}He, consisting of two protons and a neutron. The energy released in this step is carried away as a gamma ray photon. The third and final step in the proton–proton chain occurs when two ^{3}He nuclei collide and fuse together, producing an ordinary ^{4}He nucleus and ejecting two protons in the process. The energy released in this step shows up as kinetic energy of the helium nucleus and two ejected protons.

This dominant branch of the proton–proton chain can be written symbolically as

$$^1\text{H} + {}^1\text{H} \rightarrow {}^2\text{H} + e^+ + \nu,$$

followed by

$$e^+ + e^- \rightarrow \gamma + \gamma$$

$$^2\text{H} + {}^1\text{H} \rightarrow {}^3\text{He} + \gamma$$

$$^3\text{He} + {}^3\text{He} \rightarrow {}^4\text{He} + {}^1\text{H} + {}^1\text{H}$$

Here the symbols are e^- for an electron, e^+ for a positron, ν (the Greek letter "nu") for a neutrino, and γ ("gamma") for a gamma ray photon.

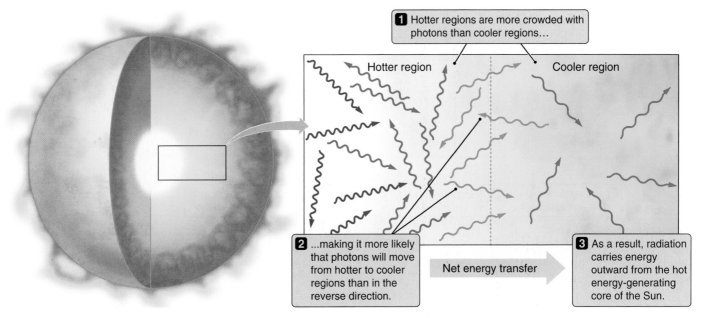

1 Hotter regions are more crowded with photons than cooler regions...

Hotter region   Cooler region

2 ...making it more likely that photons will move from hotter to cooler regions than in the reverse direction.

Net energy transfer

3 As a result, radiation carries energy outward from the hot energy-generating core of the Sun.

FIGURE 14.5 Higher-temperature regions deep within the Sun produce more radiation than lower-temperature regions farther out. Although radiation flows in both directions, more radiation flows from the hot regions to the cooler regions than from the cooler regions to the hot regions. In this way radiation carries energy outward from the inner parts of the Sun.

Chapter 4 that the hotter region will contain more (and more energetic) photons than the cooler region. (There will be a Planck spectrum of photons in both regions, so the total energy carried by photons will be proportional to the fourth power of the temperature in each region.) More photons will

Radiation carries energy from hotter regions to cooler regions.

move by chance from the hotter ("more crowded") region to the cooler ("less crowded") region than in the reverse direction. Thus there is a net transfer of photons and photon energy from the hotter region to the cooler region; and in this way radiative transfer carries energy from hotter regions to cooler regions.

If the temperature differs by a large amount over a short distance, then the concentration of photons will differ sharply as well, which favors rapid radiative energy transfer. The transfer of energy from one point to another by radiation also depends on how freely radiation can move from one point to

Opacity impedes the outward flow of radiation.

another within a star. The degree to which matter impedes the flow of photons through it is referred to as **opacity**. The opacity of material depends on many things including the density of material, its composition, its temperature, and the wavelength of the photons moving through it.

Radiative transfer is most efficient in regions where opacity is low. In the inner part of the Sun, where temperatures are high and atoms are ionized, opacity comes mostly from the interaction between photons and free electrons (electrons not attached to any atom). Here opacity is relatively low, and radiative transfer is capable of carrying the energy produced in the core outward through the star. The region in which radiative transfer is responsible for energy transport extends 71 percent of the way out toward the surface of the Sun. This region (see **Figure 14.6**) is referred to as the Sun's **radiative zone**. Even though the opacity of the radiative zone is relatively low, photons are still able to travel only a short distance before interacting with matter. The path that a photon follows is so convoluted that on average it takes the energy of a gamma ray photon produced in the interior of the Sun about 100,000 years to find its way to the outer layers of the Sun. Opacity serves as a blanket, holding energy in the interior of the Sun and letting it seep away only slowly.

From a peak of 15 million K at the center of the Sun, the temperature falls to about 100,000 K at the outer margin of the radiative zone. At this temperature atoms are no longer completely ionized, which increases the opacity. As the opacity becomes greater, radiation becomes less efficient in carrying energy from one place to another. The energy that is flowing outward through the Sun "piles up." The physical sign that energy is piling up is that the *temperature gradient*—that is, how rapidly temperature drops with increasing distance from the center of the Sun—becomes

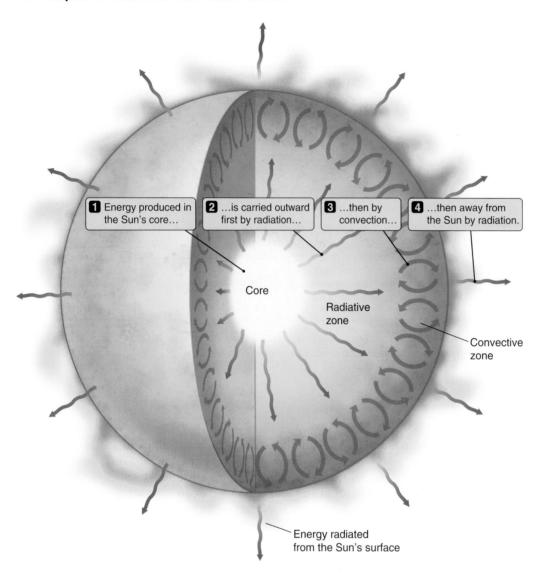

1 Energy produced in the Sun's core…

2 …is carried outward first by radiation…

3 …then by convection…

4 …then away from the Sun by radiation.

Core

Radiative zone

Convective zone

Energy radiated from the Sun's surface

FIGURE 14.6 The interior structure of the Sun is divided into zones on the basis of where energy is produced and how it is transported outward.

very steep. (Radiative transfer carries energy from hotter regions to cooler regions, smoothing out temperature differences between them. As the opacity increases, radiation becomes less effective in smoothing out temperature differences, so temperature differences between one region and another become greater.)

As we move farther toward the surface of the Sun, radiative transfer becomes so inefficient (and the temperature

In the outer part of the Sun, energy is carried by convection.

gradient so steep) that a different way of transporting energy takes over. Like a hot-air balloon, cells (or packets) of hot gas become buoyant and rise up through the lower-temperature gas above them, carrying energy with them. Thus convection begins. Just as convection carries energy from the interior of planets to their surfaces, or from the Sun-heated surface of Earth upward through Earth's atmosphere, convection also plays an important role in the trans-

port of energy outward from the interiors of many stars, including the Sun. The solar **convective zone** extends from the outer boundary of the radiative zone out to just below the visible surface of the Sun.

In the outermost layers of stars, radiation again takes over as the primary way that energy is transported. (This must be the case. After all, it is radiation that transports energy from the outermost layers of a star off into space.) Even so, the effects of convection can be seen as a perpetual roiling of the visible surface of the Sun.

What If the Sun Were Different?

At this point we have seen the pieces that go into calculating a model of the interior of the Sun. To put these pieces together, we again stress that the key is *balance*. The temperature and density in the core of the model Sun must be just right to produce the same amount of energy as would be radiated away by a star of the same size and surface temperature as the Sun. The temperature and density at each point within the model Sun must be just right so that transport of energy away from the core by radiation and convection just balances the amount of energy produced by fusion in the core. The density, temperature, and pressure of the model Sun must vary from point to point in such a way that the outward push of pressure is everywhere balanced by the inward pull of gravity. Finally, the whole model must depend on only two things: the total mass of gas from which the star is made, and the chemical composition of that gas.

Let us restate that last idea to drive the point home. *In order to be successful, our model of the Sun must start with no more information than the known mass and chemical composition of the real Sun, and from these it must correctly predict all the other observed properties of the real Sun.* The amazing thing is that our model does exactly that. Computer modeling shows that a ball of gas containing one solar mass with the same composition as our Sun can have only *one* structure and still satisfy all the different balances just listed *at the same time.* Our model predicts what the size, temperature, and luminosity of the resulting star should be—predictions that agree remarkably well with the observed properties of the real Sun.

To better understand *why* there is only one possible structure that a 1 $M_{\odot}$ star can have, we can "what-if" the Sun. For example, we might ask, "What if a star had the same mass, surface temperature, and composition as the Sun, but was somehow larger than the Sun? What properties would such a hypothetical star have? Specifically, what would happen to the balance between the amount of energy generated within such a hypothetical star and the amount of energy that it radiates away into space?" Follow along in **Figure 14.7** as we consider what would happen if the Sun were "too large."

We begin with the second part of this balance. Because our hypothetical star would have more surface area than the Sun, it would be able to more effectively radiate its energy into space. To keep a one-solar-mass star inflated to a size larger than the Sun would require that the star be more luminous than the Sun.

But now let us consider what is going on in the interior of our hypothetical star. Because the star is larger than the Sun but contains the same amount of mass as the Sun, the force of gravity at any point within our hypothetical star would be less than the force of gravity at the corresponding location within the Sun. (This is a result of the inverse square law of gravitation. If the radius R is larger in our hypothetical star, then $1/R^2$ must be smaller.) Weaker gravity means that the weight of matter pushing down on the interior of our hypothetical star would be less than in the Sun. Because according to hydrostatic equilibrium the pressure at any point within a star is equal to the weight of overlying matter, the

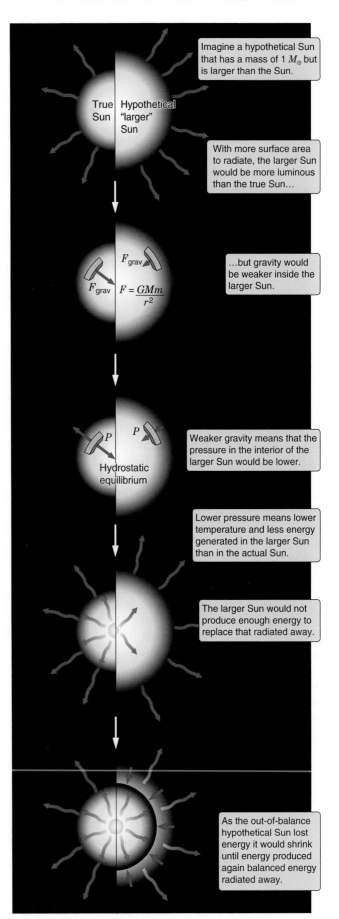

FIGURE 14.7 A star like the Sun can have only the structure that it has. Here we imagine the fate of a Sun with too large a radius.

pressure at any point in the interior of our hypothetical star would be less than the pressure at the corresponding point in the Sun. This would affect the amount of energy that our star produces. The proton–proton chain runs faster at higher temperature and density, so the lower pressure in the interior of hypothetical star means that less energy would be generated there than in the core of the Sun.

But wait a minute—there is a contradiction here. We found that our hypothetical star would have to be more luminous than the Sun, but at the same time it would be producing less energy in its interior than the Sun does. This violates the balance that must exist in any stable star between the amount of energy generated within the star and

---

If the Sun were any different, it would not be in balance.

---

the amount of energy radiated into space. Our hypothetical star cannot exist! Stated another way, even if we could somehow magically pump up the Sun to a size larger than it actually is, it would not remain that way. Less energy would be generated in its core, while more energy would be radiated away at its surface. The Sun would be out of balance. As a result, the Sun would lose energy, the pressure in the interior of the Sun would decline, and the Sun would shrink back toward its original (true) size.

We could do the same thought experiment the other way around, asking what would happen if the Sun were smaller than it actually is. With less surface area it would radiate less energy. At the same time the Sun's mass would be compacted into a smaller volume, driving up the strength of gravity and therefore the pressure in its interior. Higher pressure implies higher density and temperature, which in turn would cause the proton–proton chain to run faster, increasing the rate of energy generation. Again there is a contradiction—an imbalance. This time, with more energy being generated in the interior than is radiated away from the surface, pressure in the Sun would build up, causing it to expand toward its original (true) size.

In the end, there is only one possible Sun. If a star contains one solar mass of material of solar composition, there is only one structure of that star that maintains all of the balances that must be maintained. Our stable, reliable Sun is the result.

14.3 The Standard Model of the Sun Is Well Tested

The standard model of the Sun correctly predicts such global properties of the Sun as its size, temperature, and luminosity. This is a remarkable feat, but the model predicts much more than these properties. In particular, the standard model of the Sun predicts exactly what nuclear reactions should be occurring in the core of the Sun, and at what

---

Neutrinos escape freely from the core of the Sun.

---

rate. The nuclear reactions that make up the proton–proton chain produce copious quantities of neutrinos. As we have noted, neutrinos are extremely elusive beasts—so elusive that almost all of the neutrinos produced in the heart of the Sun travel freely through the outer parts of the Sun and on into space as if the Sun were not there. The core of the Sun lies buried beneath 700,000 km of dense, hot matter, seemingly buried forever away from our view. Yet the Sun is *transparent* to neutrinos.

It may take thermal energy produced in the heart of the Sun 100,000 years to find its way to the Sun's surface, but the solar neutrinos streaming through you as you read these words were produced by nuclear reactions in the very heart of the Sun only $8\frac{1}{3}$ minutes ago. If we could find a way to capture and analyze these neutrinos, think of what we might learn! In principle, neutrinos offer a direct window into the very heart of the Sun's nuclear furnace.

Using Neutrinos to Observe the Heart of the Sun

Turning the promise of neutrino astronomy into reality turns out to be a formidable technical challenge. The same property of neutrinos that makes them so exciting to astronomers—the fact that their interaction with matter is so feeble that they can escape unscathed from the interior of the Sun—also makes them notoriously difficult to observe. Suppose we wanted to build a neutrino detector capable of stopping half of the neutrinos falling on it. Our hypothetical detector would need the stopping power of a piece of lead a light-year thick! Yet despite the difficulties, neutrinos offer such a unique and powerful window into the Sun that they are worth going to great lengths to try to detect.

Fortunately, the Sun produces a truly enormous number of neutrinos. As you lie in bed at night, about 400 trillion solar neutrinos pass through your body each second, having already passed through Earth. With this many neutrinos about, a neutrino detector does not have to be very efficient to be useful. Several methods have been devised to measure neutrinos from the Sun and from other astronomical sources such as supernova explosions, and a number of such experiments are under way. These experiments have successfully detected neutrinos from the Sun, and in so doing they have provided crucial confirmation that nuclear fusion reactions indeed are responsible for powering the Sun.

However, as with many good experiments, measurements of solar neutrinos raised new questions while answering others. After their initial joy at confirming that the Sun really is a nuclear furnace, astronomers became troubled that there seemed to be only about a third to a half as many solar neutrinos as predicted by solar models. The difference between the predicted and measured flux of solar neutrinos was referred to as the **solar neutrino problem**.

Understanding the solar neutrino problem was an area of very active research. One possible explanation was that our understanding of the structure of the Sun was somehow wrong. This seemed unlikely, however, because of the many other successes of our solar model. A second possibility was that our understanding of the neutrino itself was incomplete. The neutrino was long thought to have zero mass (like photons) and to travel at the speed of light. However, if neutrinos actually do have a tiny amount of mass, then theories from particle physics predict that solar neutrinos should *oscillate* (alternate back and forth) among three different kinds or *flavors*—the *electron, muon,* and *tau neutrinos,* as shown in **Figure 14.8**. Because only one of these, the electron neutrino, could interact with the atoms in the earlier neutrino detectors (described in **Tools 14.1**), neutrino oscillations provided a convenient explanation for why we saw only about a third of the expected number of neutrinos. And, as seen in Figure 14.8, electron neutrinos should also change flavor as they interact with solar material during their escape from the Sun.

After several decades of work on the solar neutrino problem, this last idea has won out. Work currently under way at high-energy physics labs, nuclear reactors, and neutrino telescopes around the world is showing that neutrinos *do* have a nonzero mass, and this work has uncovered evidence of neutrino oscillations.

FIGURE 14.8 If neutrinos have mass, they should oscillate among the three types of neutrinos (electron, muon, and tau). In sequences (1) through (5) a neutrino oscillates between electron and muon types. Changes from one type to another can also take place in the presence of matter at a certain density found within the Sun. In (6) an electron neutrino created in the Sun's core converts (7) to a muon neutrino, which then arrives at Earth. Early neutrino detectors would not have recognized this muon neutrino.

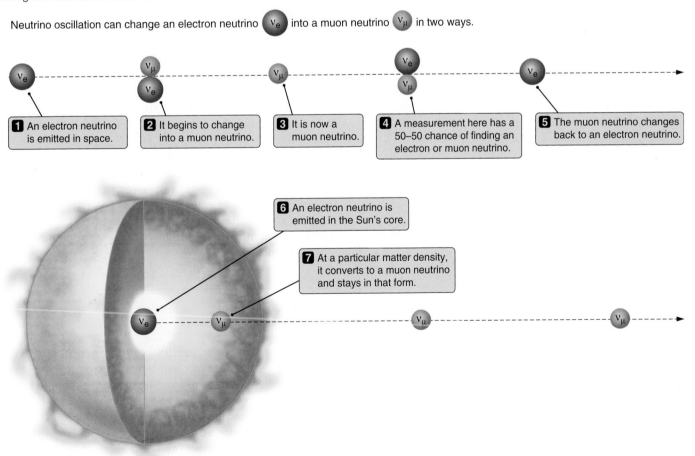

Neutrino Astronomy

A neutrino telescope hardly fits anyone's expectation of what a telescope should look like. The first experiment designed to detect solar neutrinos consisted of a cylindrical tank filled with 100,000 gallons of dry cleaning fluid—C_2Cl_4, or perchloroethylene—buried 1,500 meters deep within the Homestake gold mine in Lead, South Dakota. A tiny fraction of neutrinos passing through this fluid interact with chlorine atoms, causing the reaction

$$^{37}Cl + \nu \rightarrow {}^{37}Ar + e^-$$

to take place. The ^{37}Ar formed in the reaction is a radioactive isotope of argon. The tank must be buried deep within Earth to shield the detector from the many other types of radiation capable of producing argon atoms. The argon is flushed out of the tank every few weeks and measured.

The Homestake detector (**Figure 14.9(a)**) began operations in 1965. Over the course of two days, roughly 10^{22} (10 billion trillion) solar neutrinos pass through the Homestake detector. Of these, on average only *one* neutrino interacts with a chlorine atom to form an atom of argon. Even so, this is enough of a signal to measure, and the effects of solar neutrinos are seen. Since then over a dozen and a half neutrino detectors have been built, each using different reactions to detect neutrinos of different energies. For example, the Soviet–American Gallium Experiment (SAGE) and the European GALLEX experiment use reactions involving conversion of gallium atoms into germanium atoms ($^{71}Ga + \nu \rightarrow {}^{71}Ge + e^-$) to detect solar neutrinos. (SAGE uses 60 tons of metallic gallium. GALLEX uses 30 tons of gallium in the form of gallium chloride. These two experiments account for most of the world's current supply of gallium.)

Among the more ambitious detectors capable of detecting all three flavors of neutrinos are the Super Kamiokande detector and the Sudbury Neutrino Observatory (SNO). Super Kamiokande is located in an active zinc mine 2,700 m under Mount Ikena, near Kamioka, Japan. It is a 50,000-ton tank of ultrapure water, surrounded by 13,000 photomultiplier tubes capable of registering extremely faint flashes of light. When a neutrino interacts with an atom in the tank, a faint conical flash of blue light is produced. This flash is seen by some of the photomultipliers. **Figure 14.9(b)** shows a picture of Super Kamiokande, while **Figure 14.9(c)** shows a map of the flash of light from a single neutrino. SNO uses 1,000 tons of heavy water (D_2O) contained in a 12-m sphere surrounded by light detectors (**Figure 14.9(d)**) and buried 2,000 m deep in an active nickel mine near Sudbury, Ontario.

Neutrino telescopes observe neutrinos produced in the heart of the Sun, allowing us to directly observe the results of the nuclear reactions going on there. While these observations have provided crucial confirmation that stars are powered by nuclear reactions, they have also challenged our models of the solar interior and led us to change our ideas about the nature of the neutrino itself (as is evident from the discussion of the solar neutrino problem in the text). In addition to solar neutrinos, a number of experiments detected neutrinos from supernova 1987A. As we will see in Chapter 17, this was an explosion marking the end of the life of a massive star located 160,000 light-years away in a small galaxy called the Large Magellanic Cloud. Neutrino astronomy was one of the great innovations of 20th century astronomy and is certain to have many applications in the 21st century.

Helioseismology Observes Echoes from the Sun's Interior

In Chapter 7 we found that models of Earth's interior predict how density and temperature change from place to place within our planet. These differences affect the way pressure waves travel through Earth, bending the paths of these waves. Models of Earth's interior are tested by comparing measurements of seismic waves from earthquakes with model predictions of how seismic waves should travel through the planet.

The same basic idea has now been applied to the Sun. Detailed observations of motions of material from place to place across the surface of the Sun show that the Sun vibrates or "rings," something like a struck bell. Compared

FIGURE 14.9 Neutrino "telescopes" do not look much like visible-light telescopes. (a) The Homestake neutrino detector is a 100,000-gallon tank of dry cleaning fluid located deep in a mine in South Dakota. (b) The Super Kamiokande detector (shown while being filled) is a tank containing 50,000 tons of pure water surrounded by about 13,000 photomultiplier tubes that record flashes of light from reactions within the tank. (c) A map of the flash of light from a single neutrino detected by the Super Kamiokande detector. (d) The Sudbury neutrino detector, buried 2 km deep in a Canadian nickel mine.

to a well-tuned bell, however, the vibrations of the Sun are very complex, with many different frequencies of vibrations occurring simultaneously. These motions are echoes of what lies below. Just as geologists use seismic waves from earthquakes to probe the interior of Earth, solar physicists use the surface oscillations of the Sun to test our understanding of the solar interior. This science, new to the later years of the 20th century, is called **helioseismology**. Like neutrino astronomy, helioseismology has created quite a stir among astronomers by letting us "see" into the invisible heart of the Sun (**Figure 14.10**).

Observing the "ringing" motion of the surface of the Sun is a difficult business. To detect the disturbances of helioseismic waves on the surface of the Sun, astronomers must use instruments capable of measuring Doppler shifts of less than 0.1 m/s while detecting changes in brightness

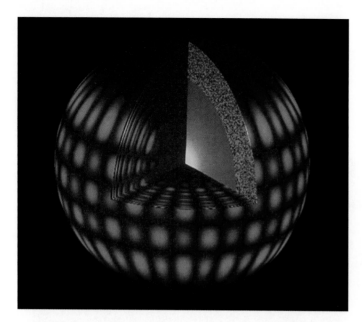

FIGURE 14.10 The interior of the Sun rings like a bell as helioseismic waves move through it. This figure shows one particular "mode" of the Sun's vibration. Red shows regions where gas is traveling inward, blue where gas is traveling outward. We can observe these motions by using Doppler shifts.

To interpret helioseismology data, scientists compare the strength, frequency, and wavelengths of observed vibrations with predicted vibrations calculated from models of the solar interior. This technique provides a powerful test of our understanding of the solar interior, and it has led both to some surprises and to improvements in our models.

Helioseismology confirms the predictions of solar models.

For example, some scientists had proposed that the solar neutrino problem might be solved if there were less helium in the Sun than generally imagined—an explanation that was ruled out by analyzing the waves that penetrate to the core of the Sun. Helioseismology also showed that the value for opacity used in early solar models was too low. This led astronomers to recalculate the location of the bottom of the convective zone. Both theory and observation now put the base of the convective zone at 71.3 percent of the way out from the center of the Sun, with an uncertainty in this number of less than half a percent! The amazing agreement between the predictions of computer models of the Sun and the temperature and density structure within the Sun measured by helioseismology is truly remarkable.

of only a few parts per million at any given location on the Sun. In addition, there are tens of millions of different wave motions possible within the Sun. Some waves travel around the circumference of the Sun, providing information about the density of the upper convection zone. Other waves travel through the interior of the Sun, revealing the density structure of the Sun close to its core. Still others travel inward toward the center of the Sun until they are bent by the changing solar density and return to the surface. All of these wave motions are going on at the same time. Sorting out this jumble requires computer analysis of long, unbroken strings of solar observations.

Helioseismology studies of the Sun got a huge boost in the closing years of the 20th century with two very successful projects. The Global Oscillation Network Group, or *GONG*, is a network of six solar observation stations spread around the world. With this network, solar astronomers are able to observe the surface of the Sun approximately 90 percent of the time. The other project is the Solar and Heliospheric Observatory, or *SOHO* spacecraft, which is a joint mission between NASA and the European Space Agency. By orbiting at the L_1 Lagrangian point of the Sun–Earth system (see Chapter 10), *SOHO* orbits in lockstep with Earth at a location approximately 1,500,000 km (0.01 AU) from Earth on a line directly between Earth and the Sun. *SOHO* carries a complement of 12 scientific instruments designed to monitor the Sun and measure the solar wind upstream of Earth. *SOHO* observations have dramatically improved our detailed knowledge of the Sun.

14.4 The Sun Can Be Studied Up Close and Personal

The Sun is a large ball of gas, and so it has no solid surface in the sense that Earth does. Instead it has the kind of surface that a cloud on Earth does, or the "surface" of one of the Jovian planets in our Solar System. To help you understand such a surface, imagine a fog bank. The surface of the fog bank is a gradual thing—an illusion really. Imagine watching some people walking into a fog bank. When they disappear from view, you would say that they were definitely inside the fog bank, even though they never passed through a noticeable boundary. The apparent surface of the Sun is defined by the same effect. Light from the surface of

The apparent surface of the Sun is called the photosphere.

the Sun can escape into space, and so we can see it. Light from below the surface of the Sun cannot escape directly into space, and so we cannot see it. The Sun's surface is referred to as the solar **photosphere**. (*Photo* means light; the photosphere is the place light comes from.) There is no instant when you can say that you have suddenly crossed the surface of a fog bank, and by the same token there is no instant when we suddenly cross the photosphere of the Sun.

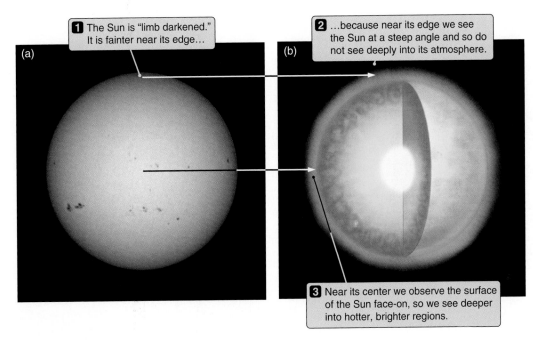

1 The Sun is "limb darkened." It is fainter near its edge…

2 …because near its edge we see the Sun at a steep angle and so do not see deeply into its atmosphere.

3 Near its center we observe the surface of the Sun face-on, so we see deeper into hotter, brighter regions.

FIGURE 14.11 (a) When viewed in visible light, the Sun appears to have a sharp outline, even though it has no true surface. The center of the Sun appears brighter while the limb of the Sun is darker, an effect known as limb darkening. (b) Looking at the middle of the Sun allows us to see deeper into the Sun's interior than looking at the edge of the Sun does. Because higher temperature means more luminous radiation, the middle of the Sun appears brighter than its limb.

The surface of the Sun—the photosphere—has an **effective temperature**[1] of 5,770 K and ranges from 6,600 K at the bottom to 4,400 K at the top. It is a zone about 500 km thick, across which the density and opacity of the Sun increase sharply. The reason the Sun appears to have a well-defined surface and a sharp outline when viewed from Earth (*never* look at the Sun directly!) is because this zone is relatively shallow; 500 km does not look very thick when viewed from a distance of 150 million km.

Look at a photograph of the Sun such as **Figure 14.11(a)**, and notice that the Sun appears to be fainter near its edges than near its center. This effect, called **limb darkening**, is an artifact of the structure of the Sun's photosphere. (The *limb* of a celestial body is the outer border of its visible disk.) The cause of limb darkening is illustrated in **Figure 14.11(b)**. Near the edge of the Sun you are looking through the photosphere at a steep angle. As a result, you do not see as deeply into the interior of the Sun as when looking directly down through the photosphere near the center of the Sun's disk. The light you see coming from near the limb of the Sun is from a layer in the Sun that is shallower and hence cooler and fainter.

The Solar Spectrum Is Complex

In one sense the surface of the Sun may be an illusion, but in another sense it is not. The transition between "inside" the Sun and "outside" the Sun is quite abrupt. In the outer-most part of the Sun, the density of the gas drops very rapidly with increasing altitude. This is the region we refer to as the Sun's **atmosphere**. **Figure 14.12** shows how the pressure and temperature change across the atmosphere of the Sun. The Sun's atmosphere is where all visible solar phenomena take place.

Very nearly all the radiation from below the Sun's photosphere is absorbed by matter and cannot escape—it is trapped. This is exactly our definition from Chapter 4 for the conditions under which blackbody radiation is formed. We should not then be surprised that the radiation able to leak out of the Sun's interior has a spectrum very close to being a Planck (blackbody) spectrum. This is why in Chapter 13 we were able to understand much about the physical properties of stars by applying our understanding of blackbody radiation.

As we look in more detail at the structure of the Sun, however, this simple description of the spectra of stars begins to go wrong. The fact that light from the solar photosphere must escape through the upper layers of the Sun's atmosphere affects the spectrum that we see. In Chapter 13 we discussed the presence of absorption lines in the spectra of stars. Now we can take a closer look at how these absorption lines form. As photospheric light travels upward through the solar atmosphere, atoms in the solar atmosphere absorb light at discrete wavelengths. Because from our perspective the Sun is so much brighter than any other star, its spectrum can be studied in far more detail. Today specially designed telescopes and high-resolution spectrographs have been built specifically to study the light from the Sun. The amazing structure present in the solar spectrum can be seen in **Figure 14.13**. Absorption lines from over 70 elements have been identified. Analysis

---

[1] The effective temperature of the photosphere is the temperature at which the Sun appears to radiate.

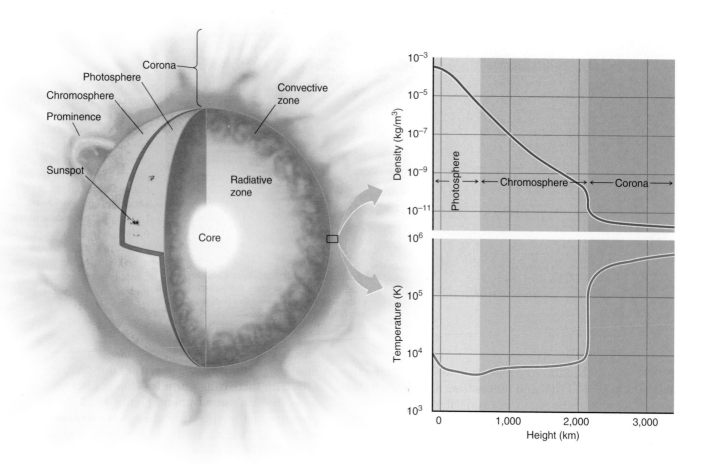

FIGURE 14.12 The components of the Sun's atmosphere, along with plots of how the temperature and density change with height at the base of the Sun's atmosphere.

of these lines forms the basis for much of our knowledge of the solar atmosphere, including the composition of the Sun, and is the starting point for our understanding of the atmospheres and spectra of other stars.

The Sun's Outer Atmosphere: Chromosphere, Corona, and Solar Wind

As we move upward through the Sun's photosphere, the temperature continues to fall, reaching a minimum of about 4,400 K at the top of the photosphere. At this point the temperature trend reverses and slowly begins to climb, rising to about 6,000 K at a height of 1,500 km above the top of the photosphere. This region above the photosphere is called the **chromosphere** (see **Figure 14.14(a)**). The chromosphere was discovered in the 19th century during observations of total solar eclipses (**Figure 14.14(b)**). The chromosphere is seen most strongly as a source of emission lines, especially

The Sun's chromosphere lies above the photosphere.

the Hα line from hydrogen. In fact, it is from the deep red color of the Hα line that the *chromosphere* (the place where color comes from) gets its name. It was also from a spectrum of the chromosphere that the element helium was discovered in 1868. Helium is named after *helios,* the Greek word for "Sun."[2]

At the top of the chromosphere, across a transition region that is only about 100 km thick, the temperature suddenly soars (Figure 14.12). Above this transition lies the outermost region of the Sun's atmosphere, called the **corona**, in which temperatures reach 1 million to 2 million K. The corona is probably heated by magnetic waves and magnetic fields in much the same way the chromosphere is, but why the temperature changes so abruptly at the transition be-

[2] It is rather remarkable that helium, the second most common element in the universe, was discovered in the Sun before it was identified on Earth!

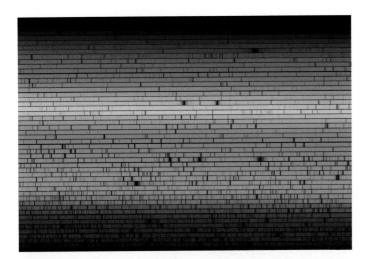

FIGURE 14.13 A high-resolution spectrum of the Sun, stretching from 400 nm (lower left corner) to 700 nm (upper right corner), showing a wealth of absorption lines.

> **The corona has a temperature of millions of kelvins.**

tween the chromosphere and the corona is not at all clear. Since ancient times the Sun's corona has been known—visible during total solar eclipses as an eerie outer glow stretching a distance of several solar radii beyond the Sun's surface (see **Figure 14.14(c)**). Because it is so hot, the solar corona is a strong source of X-rays. Atoms in the corona are also highly ionized. The spectrum of the corona shows

emission lines from ionic species such as Fe^{13+} (iron atoms from which 13 electrons have been stripped) and Ca^{14+} (calcium atoms from which 14 electrons have been stripped).

Solar Activity Is Caused by Magnetic Effects

Virtually all of the structure seen in the atmosphere of the Sun is imposed on the gas by the Sun's magnetic field. High-resolution images of the Sun, such as **Figure 14.15**, show *coronal loops* that make the Sun look as though it were covered with matted, tangled hair. This fibrous or rope-like texture in the chromosphere is the result of magnetic structures called *flux tubes,* much like the flux tube that connects Io with Jupiter (see Chapter 9). Magnetic fields are responsible for much of the structure of the corona as well. Recall from Chapter 8 that "hot" atoms are able to escape from the tops of planetary atmospheres. Shouldn't this happen on the Sun as well? Well, in this case, the answer is yes and no. The corona is far too hot to be held in by the Sun's gravity, but over most of the surface of the Sun coronal gas is confined instead by magnetic loops with both ends firmly anchored deep within the Sun. The magnetic field in the corona acts almost like a network of rubber bands that coronal gas is free to slide along but cannot cross. In contrast, about 20 percent of the surface of the Sun is covered by an ever-shifting pattern of **coronal holes**. These are apparent in X-ray images of the Sun (**Figure 14.16**) as dark regions, indicating that they are cooler and lower in density than

FIGURE 14.14 (a) This image of the Sun, taken in Hα light during a transit of Venus, shows structure in the Sun's chromosphere. The planet Venus is seen in silhouette against the disk of the Sun. (b) The chromosphere seen during a total eclipse. (c) This eclipse image shows the Sun's corona, consisting of million-kelvin gas that extends for millions of kilometers beyond the surface of the Sun.

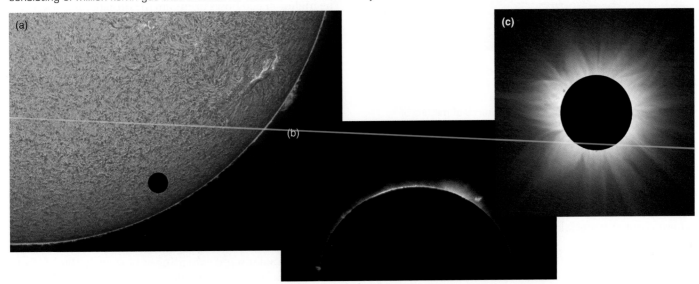

FIGURE 14.15 A close-up image of the Sun showing the tangled structure of coronal loops.

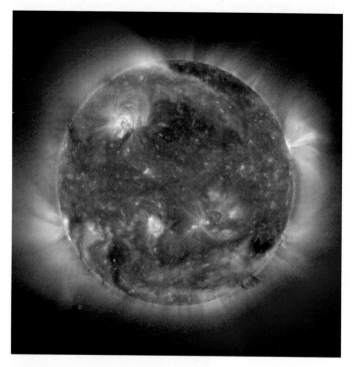

FIGURE 14.16 X-ray images of the Sun show a very different picture of our star than images taken in visible light. The brightest X-ray emission comes from the base of the Sun's corona, where gas is heated to temperatures of millions of kelvins. This heating is most powerful above magnetically active regions of the Sun. Also visible are dark coronal holes.

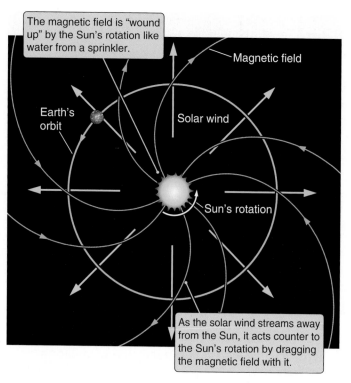

FIGURE 14.17 The solar wind streams away from active areas and corona holes on the Sun. As the Sun rotates, the solar wind picks up a spiral structure, much like the spiral of water that streams away from a rotating lawn sprinkler.

their surroundings. Coronal holes are large regions where the magnetic field points outward, away from the Sun, and where coronal material is free to stream away into interplanetary space.

We have encountered this flow of coronal material away from the Sun before. This is the same *solar wind* responsible for shaping the magnetospheres of planets (Chapters 8 and 9) and for blowing the tails of comets away from the Sun (Chapter 12). The relatively steady part of solar wind consists of lower-speed flows, with velocities of around 350 km/s, and higher-speed flows, with velocities up to about 700 km/s.

A "wind" blows away from the Sun.

The higher-speed flows originate in coronal holes. Depending on their speed, particles in the solar wind take about two to five days to reach Earth. Frequently, two to five days after a coronal hole passes across the center of the face of the Sun, there is an increase in the speed and density of the solar wind reaching Earth. The solar wind drags the Sun's magnetic field along with it. The magnetic field in the solar wind gets "wound up" by the Sun's rotation, as shown in **Figure 14.17**. This gives the solar wind a spiral structure, something like a stream of water from a rotating lawn sprinkler.

The effects of the solar wind are felt throughout the Solar System. As we have seen, the solar wind causes the tails of comets, shapes the magnetospheres of the planets, and provides the energetic particles that power Earth's spectacular auroral displays. Using space probes, we have been able to observe the solar wind extending out to nearly 100 AU from the Sun. But the solar wind does not go on forever. The far-

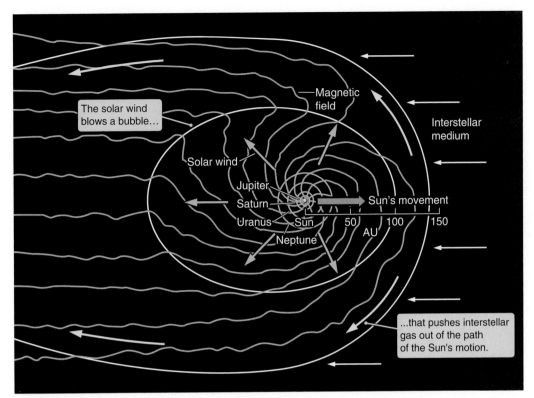

The solar wind blows a bubble…

…that pushes interstellar gas out of the path of the Sun's motion.

Magnetic field

Interstellar medium

Solar wind

Jupiter
Saturn
Uranus — Sun
Neptune

Sun's movement

50   100   150
AU

FIGURE 14.18 The solar wind streams away from the Sun for about 100 AU, until it finally piles up against the pressure of the interstellar medium through which the Sun is traveling. At the beginning of the 21st century, the *Voyager 1* spacecraft is approaching the expected location of this boundary.

ther it gets from the Sun, the more it has to spread out. Just like radiation, the density of the solar wind follows an inverse square law. At a distance of around 100 AU from the Sun, the solar wind is assumed no longer to be powerful enough to push the **interstellar medium** (the gas and dust that lie between stars in a galaxy and that surround the Sun) out of the way. There the solar wind will stop abruptly, "piling up" against the pressure of the interstellar medium. **Figure 14.18** shows this region of space over which the wind from the Sun holds sway. Sometime within the next decade the *Voyager 1* spacecraft is expected to cross the outer edge of this boundary and begin sending back our first "on the scene" measurements of true interstellar space.[3]

The best-known features on the surface of the Sun are relatively dark blemishes in the solar photosphere, called **sunspots**, that come and go over time. Early telescopic observations of sunspots made during the 17th century led to the discovery of the Sun's rotation, which has an average period of about 27 days as seen from Earth and 25 days relative to the stars. Observations of sunspots also show that the Sun, like Saturn, rotates more rapidly at its equator than it does at higher latitudes. This effect, referred to as **differential rotation**, is possible only because the Sun is a large ball of gas rather than a solid object. Sunspots appear dark, but only in contrast to the brighter surface of the Sun. Sun-

[3] The truth is, we really don't know just how far from the Sun this boundary lies, but we *will* know it when we see it.

Sunspots are cooler than their surroundings.

spots are about 1,500 K cooler than their surroundings. What does this tell us? Think back to Stefan's Law in Chapter 4. The flux from a blackbody is proportional to the fourth power of the temperature. Let's take round numbers for the temperature of a typical sunspot and the surrounding photosphere as 4,500 K and 6,000 K, respectively. Stefan's Law informs us that the surface brightness of a sunspot is about

$$\left(\frac{4,500}{6,000}\right)^4 = 0.32$$

or about ⅓ the brightness of the surrounding photosphere.

Figure 14.19 is a photograph of a sunspot group, and **Figure 14.20** shows the remarkable structure of one of these blemishes on the surface of the Sun. Each sunspot consists of an inner dark core called the **umbra**, which is surrounded by a less dark region called the **penumbra**. The penumbra

Magnetic fields cause sunspots.

shows an intricate radial pattern, reminiscent of the petals of a flower. Observations of sunspots show that they are magnetic in origin, with magnetic fields thousands of times greater than the magnetic field at Earth's surface. Sunspots occur in pairs that are connected by loops in the magnetic

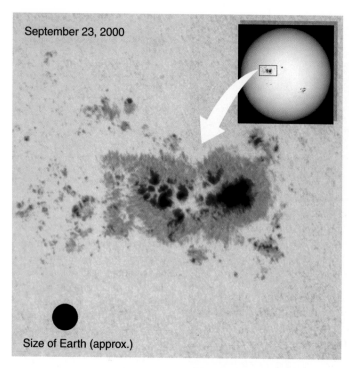

September 23, 2000

Size of Earth (approx.)

FIGURE 14.19 This false-color *SOHO* image shows a group of sunspots. The darkest part of sunspots are called umbrae. The surrounding areas, which look like petals on a flower, are the penumbrae. Sunspots are magnetically active regions that are cooler than the surrounding surface of the Sun.

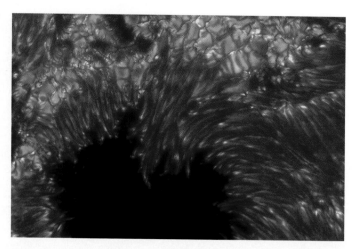

FIGURE 14.20 A very high-resolution view near the central dark region of a large sunspot. Nearly two Earths could fit side by side in this image.

field, much like the shape of a horseshoe. Sunspots range in size from about 1,500 km across up to complex groups that may contain several dozen individual spots and span 50,000 km or more. The largest sunspot groups are so large that they can be seen with the naked eye and have been noted since antiquity. (*But do not look directly at the Sun! More than one solar observer has paid the price of blindness for a glimpse of a naked-eye sunspot.* **Figure 14.21** shows a number of ways to look at the Sun in safety. Direct viewing through welder's glass is also safe.)

Individual sunspots are ephemeral. Although a few sunspots last for 100 days or longer, half of all sunspots come and go in about 2 days, and 90 percent are gone within 11 days. The amount of sunspot activity and the locations on the Sun where sunspots are seen change over time in a pronounced 11-year pattern called the **sunspot cycle (Figure 14.22(a))**. At the beginning of a cycle, sunspots begin to appear at solar latitudes of around 30° to the north and south

Sunspots come and go in an 11-year cycle.

of the solar equator. Over the following years the regions where most sunspots are seen move toward the equator as the number of sunspots increases and then declines. As the last few sunspots near the equator are seen, sunspots again

begin appearing at middle latitudes, and the next cycle begins. A diagram (**Figure 14.22(b)**) showing the number of sunspots at a given latitude plotted against time has the appearance of a series of opposing diagonal bands and is often referred to as the sunspot "butterfly diagram."

Telescopic observations of sunspots date back for almost 400 years, and there were naked-eye reports of sunspots even before that. **Figure 14.23** shows the historical record of sunspot activity. Although astronomers often speak of the 11-year sunspot cycle, the cycle is neither perfectly periodic nor especially reliable. The time between peaks in the number of sunspots actually varies between about 9.7 and 11.8 years. The number of spots seen during a given cycle fluctuates as well. Some cycles are real monsters. There have also been times when sunspot activity has disappeared almost entirely for extended periods. The most recent extended lull in solar activity, called the **Maunder minimum**, lasted from 1645 to 1715. Normally there would be six peaks in solar activity in 70 years, but virtually no sunspots were seen during the Maunder minimum.

In the early 20th century George Ellery Hale was the first to show that the 11-year sunspot cycle was actually half of a 22-year magnetic cycle during which the direction of the Sun's magnetic field reverses from one 11-year sunspot cycle to the next. In one sunspot cycle the leading sunspot in each pair tends to be a north magnetic pole, whereas the trailing sunspot tends to be a south magnetic pole. In the next sunspot cycle this polarity is reversed: The leading spot in each pair is a south magnetic pole. The transition between these two magnetic polarities occurs near the peak of each sunspot cycle (Figure 14.22(c)). The predominant theory for what causes this magnetic cycle involves a *dynamo* in the interior of the Sun, much like the dynamos that generate the magnetic fields of the planets.

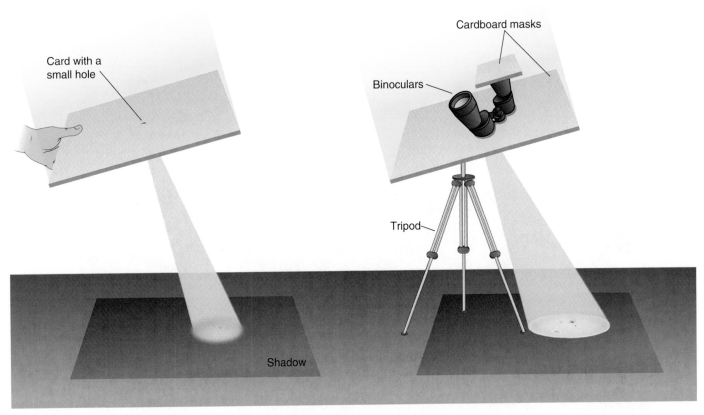

FIGURE 14.21 There are a number of ways of safely viewing sunspots on the Sun or observing a partial solar eclipse. Never look at the Sun directly (except during totality of a total eclipse)!

The effects of magnetic activity on the Sun are felt throughout the Sun's photosphere, chromosphere, and corona. Sunspots are only one of a host of phenomena that follow the Sun's 22-year cycle of magnetic activity. The peaks of the cycle, referred to as **solar maxima**, are times of intense activity, as can be seen in the ultraviolet images of the Sun in Figure 14.22(d). Although sunspots are darker-than-average features, they are often accompanied by a brightening of the solar chromosphere that is seen most clearly in the light of emission lines such as Hα. The magnificent loops seen arching through the solar corona are also anchored in solar active regions. These include solar **prominences** such as those shown in **Figure 14.24.** Prominences are magnetic flux tubes of relatively cool (5,000 to 10,000 K) gas extending through the million-kelvin gas of the corona. Although most prominences are relatively quiescent, others can erupt out through the corona, towering a million kilometers or more over the surface of the Sun and ejecting material into the corona at velocities of 1,000 km/s.

Figure 14.25(a and b) shows a picture of a **solar flare** erupting from a sunspot group. Solar flares are the most energetic form of solar activity. These are violent eruptions in which tremendous amounts of magnetic energy are released over the course of a few minutes to a few hours. Flares can heat gas to temperatures of 20 million K, and they are the source of intense X-ray and gamma ray radiation. Hot plasma is seen to move outward from flares at speeds that can reach 1,500 km/s. These events, called **coronal mass ejections (Figure 14.25(c and d))**, send powerful bursts of energetic particles outward through the Solar System. Magnetic effects in flares can accelerate subatomic particles to almost the speed of light.

Solar Activity Affects Earth

The amount of solar radiation at Earth's distance from the Sun is, on average, 1.35 kilowatts per square meter.[4] Satellite measurements of the amount of radiation coming from the Sun (**Figure 14.26**) show that this value can vary by as much as 0.2 percent over periods of a few weeks as

---

[4] Photons from the Sun provide an important alternative source of energy.

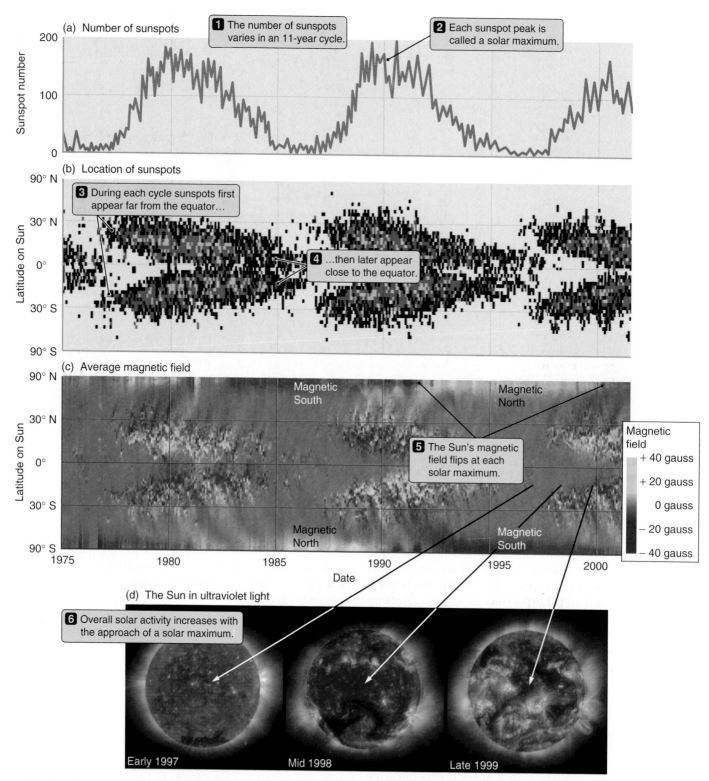

FIGURE 14.22 (a) The number of sunspots versus time for the last few solar cycles. (b) The solar butterfly diagram, showing the fraction of the Sun covered at each latitude. (c) The Sun's magnetic field flips every 11 years. (d) The approach of a solar maximum is apparent in the *SOHO* images taken in ultraviolet light.

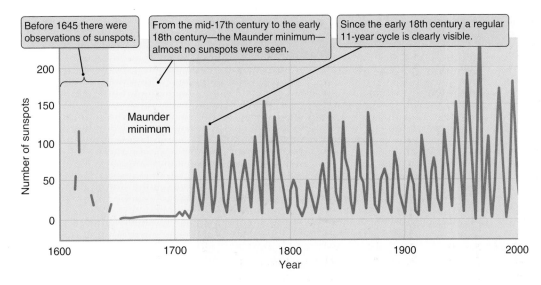

Before 1645 there were observations of sunspots.

From the mid-17th century to the early 18th century—the Maunder minimum—almost no sunspots were seen.

Since the early 18th century a regular 11-year cycle is clearly visible.

Maunder minimum

FIGURE 14.23 Sunspots have been observed for hundreds of years. In this figure the 11-year cycle in the number of sunspots (half of the 22-year solar magnetic cycle) is clearly visible. Sunspot activity varies greatly over time. The period from the middle of the 17th century to the early 18th century, when almost no sunspots were seen, is called the Maunder minimum.

dark sunspots and bright spots in the chromosphere move across the disk. Overall, however, the increased radiation from active regions on the Sun more than makes up for the reduction in radiation from sunspots. On average the Sun seems to be about 0.1 percent brighter during the peak of a solar cycle than it is at its minimum.

Solar activity has many effects on Earth. Solar active regions are the source of most of the Sun's extreme ultraviolet and X-ray emission. This energetic radiation heats Earth's upper atmosphere and during periods of solar ac-

tivity causes Earth's upper atmosphere to swell. When this happens, it can significantly increase the atmospheric drag on spacecraft orbiting near Earth, causing their orbits to decay. Prior to the launch of the Hubble Space Telescope, many people working on the project were concerned that because it was to be launched into the teeth of a solar maximum, the telescope might not survive for long. One reason for repeated shuttle visits to the HST was to "reboost" it up to its original orbit to make up for the slow decay in its orbit caused by drag in the rarefied outer parts of Earth's

FIGURE 14.24 Solar prominences are magnetically supported arches of hot gas that rise high above active regions on the Sun. The inset shows detail at the base of a large prominence.

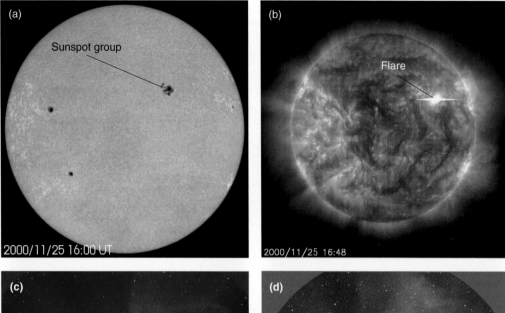

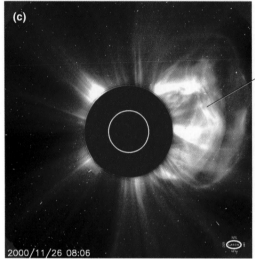

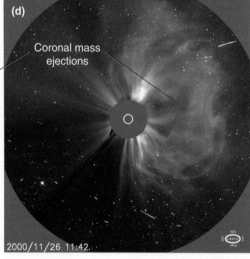

FIGURE 14.25 In this sequence of false-color images taken by the *SOHO* spacecraft a large sunspot group (a) emits a powerful flare, shown (b) in ultraviolet light. The resulting coronal mass ejection is seen streaming away from the Sun in (c) and (d). (The dark circular masks in the last two frames obscure the Sun's disk, allowing *SOHO* to observe the corona. The white circles indicate the size of the Sun.)

atmosphere. As this book goes to press, NASA's plans for a shuttle reboost visit to HST remain in doubt. Left unattended, Earth's atmospheric drag will eventually bring HST back to Earth, although not in one piece.

Earth's magnetosphere is the result of the interaction between Earth's magnetic field and the solar wind. Increases in the solar wind accompanying solar activity, especially coronal mass ejections directed at Earth, can disrupt Earth's

> Solar storms cause auroras and disrupt electric power grids on Earth.

magnetosphere in ways that are obvious even to nonscientists. Spectacular auroras can accompany such events, as can magnetic storms that have been known to disrupt electrical power grids, causing blackouts across large regions. In fact, the first clear evidence that Earth's environment was affected by solar magnetic activity was the observed rela-

tionship between sunspot activity and terrestrial events such as auroras and perturbations in Earth's magnetic field. Solar flares can have an especially dramatic effect on the environment within the Solar System. Energetic particles accelerated in solar flares pose one of the greatest dangers to human exploration of space.

Over the years people have attempted to relate sunspot activity to phenomena ranging from the performance of the stock market to the mating habits of Indian elephants. Few of these supposed relationships have withstood serious scrutiny, however. An especially interesting idea is that variations in Earth's climate might be related to solar activity. We have certainly seen that solar activity affects Earth's upper atmosphere, and it might not be too much of a reach to imagine that it could affect weather patterns as well. It has also been suggested that variations in the amount of radiation from the Sun might be responsible for variations in Earth's climate in the past. Although the causes of global

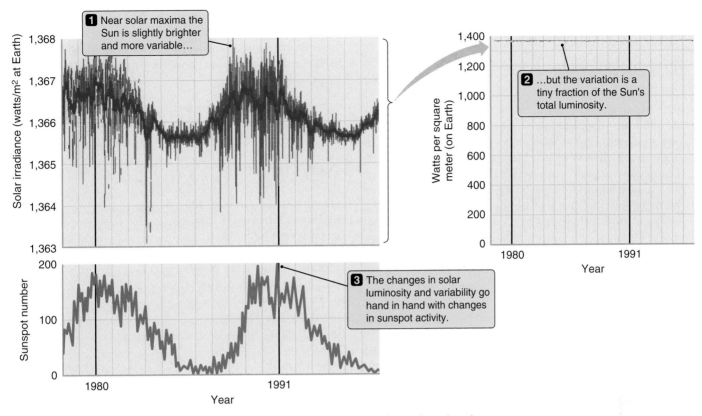

FIGURE 14.26 Measurements taken by satellites above Earth's atmosphere show that the amount of light from the Sun changes slightly over time.

climatic change remain controversial, models suggest that observed variations in the Sun could account for only about 0.1 K differences in Earth's average temperature—less than the effects due to the ongoing buildup of carbon dioxide in Earth's atmosphere. On the other hand, triggering the onset of an ice age may require a drop in global temperatures of only around 0.2 to 0.5 K, so the quest for a definite link between solar variability and changes in Earth's climate persists.

Summary

- Nuclear reactions, converting hydrogen to helium, are the source of the Sun's energy.

- Energy created in the Sun's core moves outward to the surface, first by radiation and then by convection.

- Neutrinos are elusive, almost massless particles that interact only very weakly with other matter.

- The apparent surface of the Sun is called the photosphere.

- Sunspots are photospheric regions that are cooler than their surroundings.

- The Sun's outermost region is called the corona.

- The temperature of the Sun's atmosphere ranges from about 6,000 K near the bottom to a million degrees at the top.

- Sunspots reveal the 11- and 22-year cycles in solar activity.

- Material streaming away from the Sun's corona creates the solar wind.

- Solar storms can produce terrestrial auroras and disrupt power grids.

Seeing the Forest through the Trees

The universe contains more stars than there are grains of sand on all of Earth's beaches. Even "nearby" stars are so distant that at best we see them as faint points of light in the night sky. Only one star is close enough for its complex structure to be studied in great detail. That star is a maelstrom of fierce activity, fueled by nuclear fires burning deep within its interior. Storms rage across its surface. Giant flares burst forth, arching millions of kilometers into space before collapsing back again. That star heats a huge and rarefied corona to millions of kelvins and blows a bubble in interstellar space large enough to swallow our entire Solar System. That star—a main sequence G2 star, different in no way from main sequence G2 stars throughout our galaxy and the universe—is the star we call our Sun.

No machine of human manufacture has ever visited the interior of the Sun, just as no machine has more than scratched the surface of Earth. Even so, our understanding of the interior of the Sun is remarkably complete. Like everything else in the universe, the Sun is governed by the laws of physics. Computer models based on those physical laws show that there is only one possible structure that a 1 $M_\odot$ ball of gas with the chemical composition of the Sun can have while remaining in balance. According to those models, the Sun draws its energy from the fusion of hydrogen to make helium in its core. That energy, carried from the Sun's core to its surface by radiation and convection, arrives at just the rate needed to replace the energy radiated away into space. All of this takes place while maintaining a delicate standoff in the battle between the inward pull of gravity and the outward push of pressure fueled by the nuclear fires in the Sun's interior. Models of the Sun make a host of detailed predictions, ranging from the density and temperature at every point within the Sun's interior to the nuclear reactions that take place in its core. Over the past decades the predictions of models of the Sun have been tested in ever more sophisticated and detailed ways, and those predictions have been confirmed. At the opening of the 21st century, it is fair to say that we *know* what the interior of the Sun is like.

Our knowledge of the Sun also forms the cornerstone of our knowledge of all stars everywhere. Starting with a successful model of the Sun, we can now go on to ask a host of other questions. What happens if we increase the mass of the model star? What happens if we change its chemical abundances? What happens if we run the model forward in time until all of the hydrogen at the star's center is gone? How do we assemble something that looks like the Sun in the first place? Our study of the Sun has set the stage for the next leg of our journey as we turn our attention to how stars form, how they differ from one another, how they change with time, and how they die.

Key Terms

strong nuclear force, p. 406
hydrogen burning, p. 406
proton–proton chain, p. 407
energy transport, p. 408
thermal conduction, p. 408
radiative transfer, p. 408
positron, p. 408
neutrino, p. 408
beta decay, p. 408
antiparticle, p. 408
opacity, p. 409
radiative zone, p. 409
convective zone, p. 410
helioseismology, p. 415
photosphere, p. 416
effective temperature, p. 417
limb darkening, p. 417
atmosphere, p. 417
chromosphere, p. 418
coronal hole, p. 419
interstellar medium, p. 421
sunspot, p. 421
differential rotation, p. 421
umbra, p. 421
penumbra, p. 421
sunspot cycle, p. 422
Maunder minimum, p. 422
solar maximum, p. 423
prominence, p. 423
solar flare, p. 423
coronal mass ejection, p. 423

Student Questions

THINKING ABOUT THE CONCEPTS

1. Engineers and physicists dream of solving the world's energy supply problem by constructing power plants that would convert globally plentiful hydrogen to helium. Our Sun seems to have solved this problem. On

Earth, what is a major obstacle to this environmentally clean solution?

2. Explain the proton–proton chain process in which the Sun generates energy by converting hydrogen to helium.

3. In the proton–proton chain process, the mass of four protons is slightly greater than the mass of a helium nucleus. Explain what this difference in mass means for a star like the Sun.

4. What experiences have you had with energy transmitted to you by radiation alone? By convection alone? By conduction alone? (You are exposed to all three every day.)

5. If neutrinos have mass, as now seems to be the case, they cannot travel at the speed of light. Explain why not.

6. Why are neutrinos so difficult to detect?

7. Discuss the "solar neutrino problem" and how this problem was solved.

8. What techniques do you find in common between how we probe the internal structure of the Sun and the internal structure of Earth?

9. Explain the relationship between the 11-year sunspot cycle and the 22-year magnetic cycle.

10. Solar flares and magnetic storms can adversely affect Earth's power grids, radio communication, and the health of humans in orbit. How might observations of the Sun help us forecast such events?

APPLYING THE CONCEPTS

11. The Sun shines by converting mass into energy according to Einstein's well-known relationship, $E = mc^2$. Show that if the Sun produces 3.78×10^{26} J of energy per second, it must convert 4.2 million metric tons (4.2×10^9 kg) of mass per second into energy.

12. Assume the Sun has been producing energy at a constant rate over its lifetime of 4.5 billion years (1.4×10^{17} s).
    a. How much mass has it lost creating energy over its lifetime?

b. The present mass of the Sun is 2×10^{30} kg. What fraction of its present mass has been converted into energy over the lifetime of the Sun?

13. Imagine the source of energy in the interior of the Sun were to abruptly change.
    a. How long would it take before a neutrino telescope could detect the event?
    b. When would a visible-light telescope see evidence of the change?

14. On average, how long does it take particles in the solar wind to reach Earth from the Sun if they are traveling at 300 km/s?

15. Sunspots appear relatively dark because their surface temperature typically is 1,500 K cooler than the surrounding photosphere (5,780 K). Compare the surface brightness (power radiated per square meter) of a dark sunspot to that of the surrounding photosphere.

16. The hydrogen bomb represents humankind's efforts to duplicate processes going on in the core of the Sun. The energy released by a 5-megaton hydrogen bomb is 3×10^{14} J.
    a. How much mass did Earth lose each time a 5-megaton hydrogen bomb was exploded?
    b. This textbook, *21st Century Astronomy*, has a mass of about 1 kg. If we converted all of its mass into energy, how many 5-megaton bombs would it take to equal that energy?

StudySpace
wwnorton.com/astro21
provides a Study Plan for each chapter that includes a reading outline, animations, keyword flash cards, and gradebook-enabled multiple-choice quizzes. From StudySpace you can also access premium content in the ebook and SmartWork.

There is a theory which states that if ever anyone discovers exactly what the universe is for and why it is here, it will instantly disappear and be replaced by something even more bizarre and inexplicable.

There is another which states that this has already happened.

DOUGLAS ADAMS (1952–2001)

The search for our origins is one of the grand themes of modern science.

CHAPTER 21

The Origin of Structure

21.1 Whence Structure?

Throughout the ages there have been as many different ideas about the origin of the universe as there have been cultural traditions. Thoughts about what the universe was like once upon a time have been part of the mythologies and traditions of all great civilizations. We live at a remarkable moment in history, when the nature of the early universe has moved from philosophical speculation to the realm of scientific fact. All that we have to do to see the early universe is point our microwave telescopes at the sky. The glow of the cosmic background radiation is there for anyone with the appropriate technology to see.

The early universe was an extraordinary place—an expanding fireball that was far more uniform than the blue of the bluest sky on the clearest day. How different that universe was from the universe we see about us today! Today's universe is one of stars and galaxies, viewed from the surface of a planet with oceans and mountains and uncountable species of living things. The contrast between these two realities cries out for explanation. How did we get from there to here? We come to one of the most philosophically intriguing questions in the history of human thought: What is the origin of structure?

KEY CONCEPTS

The universe that emerged from the Big Bang was incredibly uniform, wholly unlike today's universe of galaxies, stars, and planets. As our journey comes to its end, we tackle face-on the question that has been with us all along: Where does structure in the universe come from? Here we find that complex structure is a natural, unavoidable consequence of the action of physical law in an evolving universe:

- Matter and the fundamental forces of nature "froze out" of the uniformity of the expanding and cooling universe moments after the Big Bang.

- Galaxies formed as slight ripples in the dark matter emerging from the Big Bang, which then collapsed under the force of gravity, pulling in normal matter as well.

- Galaxies were drawn together by gravity to form the large groupings of galaxies we see today.

- Much lies ahead for Earth and for the universe.

- Like planets, stars, and galaxies, life is another form of structure that evolved through the action of the physical processes that shape the universe—the universe may teem with life.

- Evolution of structure through the action of physical law is the lynchpin that ties all of modern science together and brings sense to what we see.

21.2 Galaxies Form Groups, Clusters, and Larger Structures

Just as stars and clouds of glowing gas show us the structure of our galaxy, it is the distribution of galaxies themselves that shows us the structure of our universe. And just as gravity holds galaxies together, giving them their shape, it is gravity that shapes the universe itself. No galaxy exists in utter isolation. The vast majority of galaxies are parts of gravitationally bound collections of galaxies. The smallest and most common of these are called **galaxy groups**. A galaxy group is an irregular collection of a few dozen galaxies, most of them dwarf galaxies. Our Milky Way is a member of the **Local Group**, which consists of two giant spirals (the Milky Way and the Andromeda Galaxy), along with more than 30 smaller dwarf galaxies in a volume of space about 4 Mly in diameter (see **Figure 21.1**). Close to 98 percent of the galaxy mass in the Local Group resides in just its two giant galaxies.

Larger systems of galaxies, which can consist of hundreds of galaxies and often have a more regular structure than groups, are called **galaxy clusters**. Galaxy clusters are larger than groups, typically occupying a volume of space 10 to 15 Mly across; but clusters and groups are similar in composition. Giant galaxies dominate the mass of the far more numerous dwarf galaxies. One interesting difference among galaxy groups and clusters is that while spiral galaxies are common in most systems, elliptical galaxies are common in only about a quarter of groups and clusters. The Virgo cluster (**Figure 21.2(a)** and Figure 21.1), located 53 Mly from Earth, is an example of the first type of cluster, whereas the Coma cluster (**Figure 21.2(b)** and Figure 21.1) is an example of a cluster that is dominated by giant elliptical and S0 galaxies.

Clusters and groups of galaxies themselves bunch together to form enormous **superclusters**, which contain tens of thousands or even hundreds of thousands of galaxies and span regions of space typically more than 100 Mly in size. Our Local Group is part of the Virgo supercluster, which also includes the Virgo cluster.

Hubble's Law is a powerful tool for mapping the distribution of galaxies in space. Using Hubble's Law, all that we need to determine the distance to a galaxy is a single spec-

---

Galaxy redshift surveys measure distances to large numbers of galaxies.

---

trum from which we can measure the galaxy's redshift. Although it is easy in principle to measure the redshift of a galaxy, in practice this can be a time-consuming business. The first redshifts were measured from spectra recorded on photographic plates. Exposures of several hours were required to capture the feeble signal, and results rolled in at the breakneck pace of one or two redshifts per night of painstaking observation. As of 1975 redshifts had been measured for only about 1,000 of the hundred billion or so galaxies we can see. Since that time there has been a happy marriage of large telescopes, new instruments (such as CCD detectors and spectrographs capable of observing many galaxies at once), and powerful computers that can process large amounts of data in rapid, automated fashion. By 1990 the redshifts of over 10,000 galaxies had been measured; and by the time you read this book, that number will be well over 1 million galaxy redshifts.

The first large redshift survey was conducted by the Harvard Center for Astrophysics, which in 1986 presented

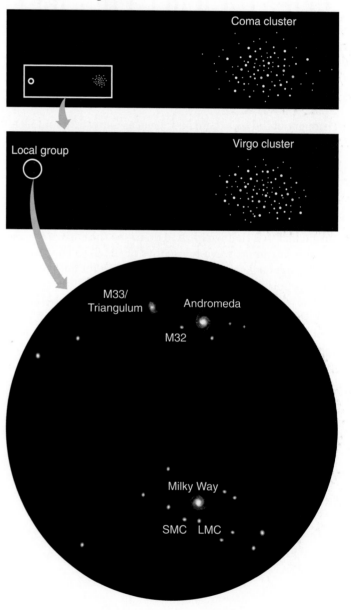

FIGURE 21.1 The Local Group of galaxies and the Virgo and Coma clusters of galaxies.

FIGURE 21.2 (a) The Virgo cluster contains a large fraction of spiral galaxies. (b) The larger, richer Coma cluster is dominated by elliptical galaxies.

the astronomical community with a "slice of the universe." **Figure 21.3(a)** shows what this slice looked like. The observations show that the clusters and superclusters, rather than being scattered randomly through space, are linked together in an intricate network of "filaments" and "walls." The concentrations of galaxies in turn surround large voids, regions of space that are largely devoid of galaxies. These voids represent some of the largest "structures" seen in the universe. Though the voids may seem like "a whole lot of nothing," we do not know that they are empty of *matter*—only

that they are largely empty of galaxies. Clusters and superclusters are located within the walls and filaments. This structure is not peculiar to the "nearby" universe. Subsequent surveys have looked at much larger volumes of space.

Large-scale structure fills the universe.

Figure 21.3(b) shows the results of one such survey conducted using the Anglo-Australian Telescope in Siding Spring, Australia. A more recent survey, the Sloan Digital Sky Survey (SDSS), is shown in **Figure 21.3(c)**. Out as far as our observatories can currently measure, the universe has a porous structure reminiscent of a sponge. Together, galaxies and the larger groupings in which they are found are referred to as **large-scale structure**.

21.3 Gravity Forms Large-Scale Structure

Large-scale structure cries out for explanation. Clearly it is the consequence of well-defined physical processes, but what processes? A number of ideas have been proposed. For example, early on it was suggested that voids were the result of huge expanding blast waves from tremendous explosions that might have occurred in the early universe. The correct answer has turned out to be less fanciful, but far more satisfying. Large-scale structure is the fingerprint of our old friend gravity.

At the close of the previous chapter we learned that the slight ripples in the glow of the cosmic background radiation probably result from quantum mechanical variations

Ripples in the universe were the seeds of structure.

that imprinted structure on the early universe at the time of inflation. These variations provided the seeds from which galaxies and collections of galaxies grew. In our discussion of star formation in Chapter 15 we learned about gravitational instabilities. As illustrated in Figure 15.14, we begin with a molecular cloud with clumps inside it, and gravity then causes those clumps to collapse faster than their surroundings. If conditions are right, then what began as relatively minor variations in the density of a cloud are turned by gravity into stars. If you replace "molecular cloud" with "universe" and "star" with "galaxy," you will have a good starting point for understanding the way galaxies formed from slight inhomogeneities in the distribution of matter following recombination.

It is one thing to say that galaxies and larger structures formed from gravitational instabilities that began with slight

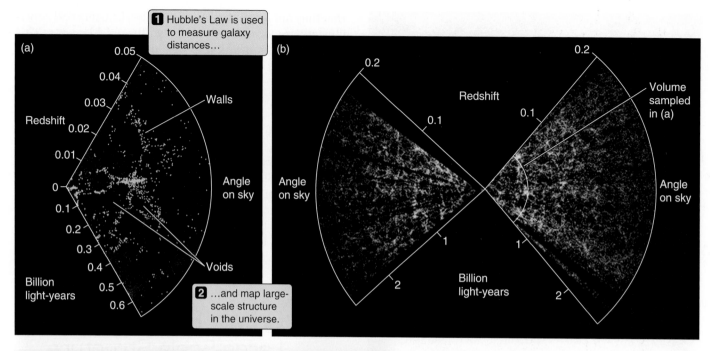

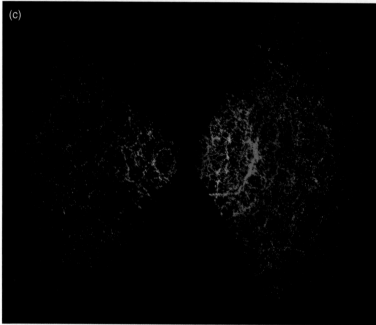

FIGURE 21.3 Redshift surveys use Hubble's Law to map the universe. (a) The Harvard-Smithsonian Center for Astrophysics redshift survey, called "A Slice of the Universe," was the first to show that clusters and superclusters of galaxies are part of even larger-scale structures. (b) The 2dF Galaxy Redshift Survey shows similar structures at even larger distances. (c) The Sloan Digital Sky Survey map of the universe extends outward to a distance of about 2 billion light-years. Shown here is a sample of 67,000 galaxies that lie near the plane of the Earth's equator.

irregularities in the early universe. It is quite another to turn this statement into a real scientific theory with testable predictions. To make that step, we return to the same basic

Models show how gravity turns early seeds into large-scale structure.

technique we have used to look into the centers of planets and stars and to answer many other questions about things we cannot observe directly. We combine the ideas we want to test with the laws of physics, construct a model, and then

compare the predictions of that model with observations of the universe.

To build a model of the formation of large-scale structure, we have to begin by making three key choices. First we have to decide what universe we are going to model. That is, what values of Ω_{mass} and Ω_Λ (see Chapter 20) are we going to assume? These are important in part because they determine how rapidly the universe expands and therefore how difficult it is for gravity to overcome this expansion in some region. The more rapidly a universe is expanding, or the less mass it contains, the more difficult it will be for gravity to pull material together into galaxies and larger-scale structures.

The second thing we need to know to construct our model is what the early bumps in the density of the universe looked like. How large and how concentrated were these early bumps? There are several ways to approach this question. One such method uses observations of variations in the CBR. Both COBE and WMAP (see Figure 20.17 and the opening photograph for Chapter 20) provide a good picture of structure in the CBR, from which we can infer what the early inhomogeneities in the universe must have looked like. Future observations such as those from the European Space Agency's proposed Planck mission will do even better.

A different way to approach this question is to look at models of inflation, which make predictions about the structure that will emerge following this episode of rapid expansion. These are among the predictions of the inflation model that will be tested in the years to come. These are especially important predictions to test because they tie together the large-scale structure of today's universe with our most basic

---

Smaller structures form first. Larger structures form later.

---

ideas about what the universe was like in the briefest instant after the Big Bang. While we await final answers about the details of structure, we know enough to say that the early universe was more irregular on smaller (galaxy-sized) spatial scales than it was on the scales of the clusters, superclusters, filaments, and voids seen in Figure 21.3. An immediate consequence of this fact is that smaller structures (such as galaxies and even subgalactic clumps) formed first, whereas larger structures took more time to form. The idea of small structure forming first and larger structure forming later is referred to as **hierarchical clustering**. Hierarchical clustering has become one of the most important themes in our growing understanding of how structure in the universe formed.

The third thing we need if we are to model the formation of large-scale structure is a complete list of the types and amounts of ingredients the early universe was made of. Specifically, we need to know the balance between radiation, normal matter, and dark matter. We also need to make some choices about the nature of the dark matter that we use in our model. We'll discuss this further in a bit.

Once we have made these three choices—the shape of the universe we are modeling, the way matter inhomogeneities developed, and the nature and mixture of different forms of matter we are using—the rest is physics and calculations: We simply need to "turn the crank," as the expression goes. As we discuss later, there are some real difficulties in carrying out this strategy. For example, our uncertainties about how gas turns into stars make it difficult to compare our models with the real universe. Despite technical dif-

ficulties, we are aiming to answer this key question: What choices lead to a model universe that most resembles the real universe in which we live?

Galaxies Formed Because of Dark Matter

Observations of nonuniformity in the CBR made with COBE and WMAP show variations of about 1 part in 100,000. Models clearly show that such tiny variations at the time of recombination (when the universe was about half a million years old) are far too slight to explain the structure we see in today's universe. Gravity is simply not strong enough to grow galaxies and clusters of galaxies from such poor "seeds." These models indicate that for ripples in the density of the universe to collapse to form today's galaxies, the density of those ripples must have been at least 0.2 percent greater than the average density of the universe at the time of recombination.

If normal luminous matter in the early universe had been clumped at this level, then the variations in the CBR today would be at least 30 times larger than what is observed. At first glance this discrepancy might seem to be a horrible problem with our understanding of the origin of structure in the universe, but it has turned out instead to be a crucial result that ties together a number of pieces of the puzzle. The theme of this story is "dark matter."

Dark matter first appeared on our journey back in Chapter 18 as an ad hoc construction—an annoyance, really—used by astronomers to explain the oddly flat rotation curves of spiral galaxies. When we turn our attention to clusters of galaxies, we find that dark matter dominates normal matter on those scales as well. Now we find that irregularities in normal matter in the early universe were too slight to form galaxies. So where are we going to turn for an answer to the question of how structure in the universe *did* form? You guessed it: dark matter.

In our discussion of primordial nucleosynthesis we learned that dark matter cannot be made of normal matter consisting of neutrons, protons, and electrons. If it were, then it would have affected the formation of chemical elements in the early universe. The abundances of several isotopes of the least massive elements would be quite different

---

Dark matter provides the seeds for observed structure.

---

from what we find in nature. Dark matter must be something else—something that has no electric charge (so it does not interact with electromagnetic radiation) and that interacts only feebly with normal matter. Clumps of such dark matter in the early universe would not interact with radiation or normal matter, so they would not be seen directly when looking at the CBR. This unseen dark matter saves the day for modeling the formation of galaxies and clusters of galaxies.

What is so different about the behavior of dark matter and normal matter in the early universe? First, pressure waves and radiation did a very good job of smoothing out ripples in the distribution of normal matter in the early universe. Feebly interacting dark matter would be immune to these processes, so clumps of dark matter survived long after bumps in the normal matter had been smoothed out, as illustrated in **Figure 21.4**. Second, the dark matter in these clumps does not glow, so we do not see it directly when we look at the cosmic background radiation. Although these clumps of dark matter do cause slight gravitational redshifts in the light coming from normal matter, the resulting variations fit well with current observations of the CBR. Thus clumps of dark matter could have (and, according to our theories of the very early universe, *should* have) existed in the early universe, even though we are unable to see evidence of them directly.

Just because we cannot see clumps of dark matter in the early universe does not mean they had no effect. In the same way that dark matter dominates the gravitational fields of today's galaxies and clusters, so too did the mass of dark matter control the growth of gravitational instabilities in the early universe.

There Are Two Classes of Dark Matter

Here is the story of galaxy formation in a nutshell. Dark matter in the early universe was much more strongly clumped than normal matter. Within a few million years after recombination, these dark matter clumps pulled in the surrounding normal matter. Later, gravitational instabilities caused these clumps to collapse. The normal matter in the clumps went on to form visible galaxies. This story seems plausible enough, but the details of how this happened depend greatly on the properties of dark matter itself. Even though we do not yet know exactly what dark matter in the universe is made of, we can talk about two broad classes of dark matter based on how it behaves.

One possibility is that dark matter consists of feebly interacting particles that are moving about relatively slowly, like the slowly moving atoms and molecules in a cold gas.

Cold dark matter consists of relatively massive, slowly moving particles.

For obvious reasons this type of dark matter is called **cold dark matter**. There are several candidates for cold dark matter. It is possible that cold dark matter consists of tiny black holes that might have been produced in the early universe. Few physicists and cosmologists favor this idea. Most think

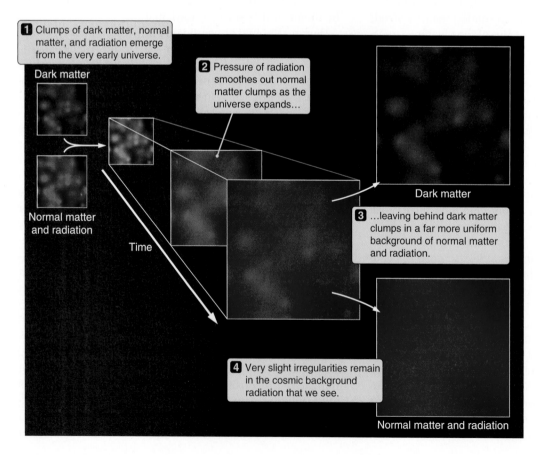

1 Clumps of dark matter, normal matter, and radiation emerge from the very early universe.

Dark matter

2 Pressure of radiation smoothes out normal matter clumps as the universe expands…

Normal matter and radiation

Time

Dark matter

3 …leaving behind dark matter clumps in a far more uniform background of normal matter and radiation.

4 Very slight irregularities remain in the cosmic background radiation that we see.

Normal matter and radiation

FIGURE 21.4 Radiation pressure and other processes in the early universe smoothed out variations in normal matter, but irregularities in the dark matter survived to become the seeds of galaxy formation.

instead that cold dark matter consists of some unknown elementary particle. One candidate is the **axion**, a hypothetical particle first proposed to explain some observed properties of neutrons. Even though axions would have very low mass, they would have been produced in great abundance in the Big Bang. Another candidate is the **photino**, an elementary particle related to the photon. Some theories of particle physics predict that these particles exist and have a mass of around 10,000 times that of the proton. Our state of knowledge might soon change quickly: Photinos might be detected in current particle accelerators, and experiments are under way to search for axions and photinos that are trapped in the dark matter halo of our galaxy.

If the first class of dark matter is called cold dark matter, then the other class of dark matter is called—are you ready?—**hot dark matter**. Hot dark matter consists of particles that are moving very rapidly. Neutrinos are one example of hot dark matter that we know exists. We have seen that

Hot dark matter consists of less massive, rapidly moving particles.

neutrinos interact with matter so feebly that they are able to flow freely outward from the center of the Sun, passing through the dense overlying layers of matter as if they were not there. There is no question that the universe is filled with neutrinos. Calculations indicate that around 300 million of these cosmic relics of the Big Bang fill each cubic meter of space. Preliminary measurements of the neutrino mass obtained at the Super Kamiokande detector in Japan and the Sudbury Neutrino Observatory in Canada suggest that neutrinos might account for as much as 5 percent of the mass of the universe. Although this percentage is not enough to account for all of the dark matter in the universe, it may be enough to have had a noticeable effect on the formation of structure.

The different effects of cold and hot dark matter on structure formation have to do with differences in the way the two types of dark matter cluster in a gravitational field. The faster an object is moving, the harder it is for gravity to hold onto it. This is the same basic idea that we used to explain why hydrogen and helium are able to escape Earth's atmo-

Cold dark matter forms galaxies.

sphere while molecules such as oxygen and nitrogen are not. It is also the same idea we used to explain why the hot gas in elliptical galaxies extends far beyond their visible boundaries. Slow-moving particles are more easily corralled by gravity than fast-moving particles, so particles of cold dark matter clump together more easily into galaxy-sized structures than do particles of hot dark matter. The result of the calculations of galaxy formation is clear: To account for the formation of today's galaxies, we need cold dark matter.

A Galaxy Forms within a Collapsing Clump of Dark Matter

We can best see how models of galaxy formation work by following the events predicted by the models step by step. Let us consider the case of a universe made up of 90 percent cold dark matter and 10 percent ordinary matter, clumped together in a manner consistent with observations of the cosmic background radiation. On the scale of an individual galaxy, the effect of the cosmological constant is so small it can be ignored.

Figure 21.5(a) shows the state of affairs of one clump of dark matter at the time of recombination. The dark matter is less uniformly distributed than normal matter, but overall the distribution of matter is still remarkably uniform. By a few million years after recombination (**Figure 21.5(b)**), the universe of our model calculation has expanded severalfold. Spacetime is expanding, so the clump of dark matter is also expanding. However, the clump of dark matter is not expanding as rapidly as its surroundings because its self-gravity has slowed down its expansion. The clump now stands out more with respect to its surroundings. Another change that has taken place is that the gravity of the dark matter clump has begun to pull in normal matter as well. By the stage shown in **Figure 21.5(c)**, normal matter is clumped in much the same way as dark matter.

A ball thrown in the air will slow, stop, and then fall back to Earth. In like fashion, the clump of dark matter will stop expanding when its own self-gravity slows and then stops its initial expansion. By the time the universe is around a billion years old, the clump of dark matter has

Clumps of dark matter can collapse only so far.

reached its maximum size and is beginning to collapse (Figure 21.5(c)). The collapse of the dark matter clump stops when the clump is about half its maximum size, however, because the particles making up the cold dark matter are moving too rapidly to be pulled in any closer (**Figure 21.5(d)**). The clump of cold dark matter is now given its shape by the orbits of its particles, in the same way that an elliptical galaxy is given its shape by the orbits of the stars it contains.

Unlike dark matter (which cannot emit radiation), the normal matter in the clump is able to radiate away energy and cool. As the gas cools, it loses pressure; and as it loses pressure, it collapses. Models show that small concentrations of normal matter within the dark matter clump collapse under their own gravity to form clumps of normal matter that range from the sizes of globular clusters to the sizes of dwarf galaxies. These clumps of normal matter then fall inward toward the center of the dark matter clump, as shown in **Figure 21.5(e)**. We discussed a similar chain of events in Chapter 15: the collapse of a molecular cloud core on its way to becoming a star. It is important to note that,

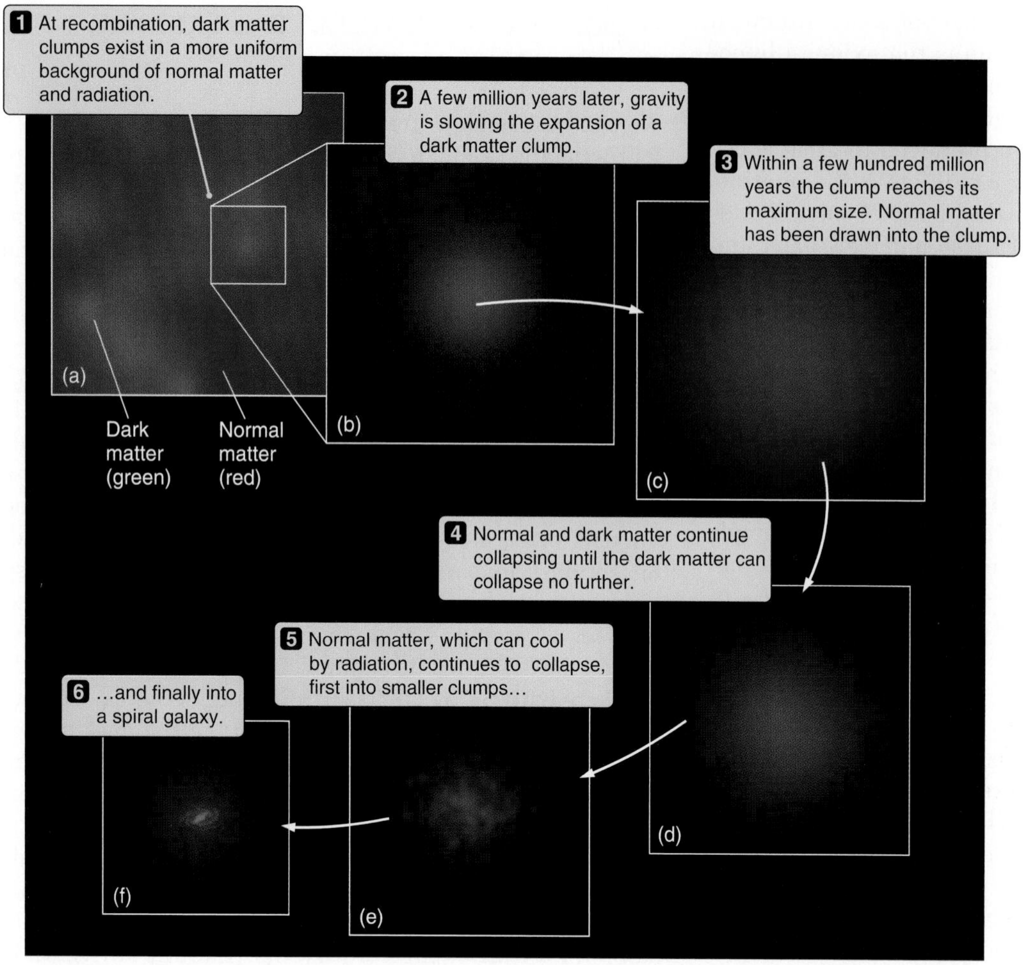

1 At recombination, dark matter clumps exist in a more uniform background of normal matter and radiation.

2 A few million years later, gravity is slowing the expansion of a dark matter clump.

3 Within a few hundred million years the clump reaches its maximum size. Normal matter has been drawn into the clump.

(a)

Dark matter (green)   Normal matter (red)

(b)

(c)

4 Normal and dark matter continue collapsing until the dark matter can collapse no further.

5 Normal matter, which can cool by radiation, continues to collapse, first into smaller clumps…

6 …and finally into a spiral galaxy.

(f)

(e)

(d)

FIGURE 21.5 Stages in the formation of a spiral galaxy from the collapse of a clump of cold dark matter.

according to models, gas in our universe can cool quickly enough to fall in toward the center of the dark matter clump only if the clump has a mass of 10^8 to 10^{12} $M_\odot$. This is just

---

Normal matter cools and falls toward the center of the dark matter halo.

---

the range of masses of observed galaxies! This agreement between theory and observations is an important success of the cold dark matter theory of galaxy formation.

As discussed in **Connections 21.1**, there are many similarities between the collapse of a molecular cloud to form stars and the collapse of a clump in the early universe to form galaxies. Another process that we saw at work in star formation also comes into play at this point. The clumps of matter collapsing to form galaxies do not exist in isolation. They have been tugged on by the gravity of neighboring clumps and have been pushed around by the pressure waves that ran through the young universe, smoothing out its structure. As a result, each protogalactic clump has a little bit of rotation when it begins its collapse. As normal

matter falls inward toward the center of the dark matter clump, this rotation forces much of the gas to settle into a rotating disk (**Figure 21.5(f)**), just as the collapsing cloud that was to become our Sun settled first into an accretion disk. And just as the accretion disk around the Sun provided the raw materials for the planets of our Solar System, the disk formed by the collapse of each protogalactic clump becomes the disk of a spiral galaxy.

Searching for Signs of Dark Matter

Dark matter has come to play a very important role in our understanding of the universe. We might reasonably be uncomfortable having our models of galaxy formation rely so heavily on something we cannot see directly. Fortunately, dark matter shows itself in a variety of ways. The flat rotation curves of spiral galaxies, for example, are compelling evidence for the existence of extended halos comprised of dark matter. Similarly, the ability of elliptical galaxies to hold

Parallels between Galaxy and Star Formation

As you read about galaxy formation, you might find it enlightening to think back to our discussion of star formation in Chapter 15. Both processes involve the gravitational collapse of vast clouds to form denser, more concentrated structures. To help you with the comparison, we list here a few of the similarities and differences between the two:

Gravitational instability: In both star and galaxy formation, the collapse begins with a gravitational instability. Regions only slightly denser than their surroundings are pulled together by their own self-gravity. As the matter in these regions becomes more compact, gravity becomes stronger, and the collapse process snowballs. One key difference is that for a galaxy to form, the dark matter clump must collapse rapidly enough to counteract the overall expansion of the universe itself.

Fragmentation: In both cases the original cloud separates into smaller pieces as a result of the gravitational instability. However, the order of fragmentation differs between star and galaxy formation. In molecular clouds large regions begin to collapse first, then fragment further to form individual stars. In contrast to this "top-down" process, galaxy formation is "bottom-up": Smaller structures collapse first and then merge together to form galaxies and eventually assemblages of galaxies.

Compression, heating, and thermal support: As an interstellar molecular cloud collapses, its temperature climbs and the pressure in the cloud increases. The higher pressure would eventually be enough to prevent further collapse except that the cloud core is able to radiate away thermal energy. That is the bright infrared radiation that allows us to see star-forming cores. Compare this with galaxy formation: As a dark matter clump collapses, its temperature

also climbs, and it too reaches a point where there is a balance between gravity and the thermal motions of the dark matter. However, dark matter is *not* able to radiate away energy, so once this balance is reached, the collapse of the dark matter is over. Only the normal matter within the cloud of dark matter is able to radiate away thermal energy and continue collapsing. That is why normal matter collapses to form galaxies, while the dark matter remains in much larger dark matter halos.

As galaxies form, dark matter remains in extended halos. Dark matter is too hot to settle into galaxy disks. It is far too hot to become concentrated into even smaller structures such as molecular clouds or to take part in the formation of stars. Dark matter may be the dominant form of matter in the universe, and it may determine the structure of galaxies; but dark matter can never collapse enough to play a role in the processes that shape stars, planets, or the interstellar medium.

Angular momentum and the formation of disks: Conservation of angular momentum is responsible for the formation of disk galaxies, just as conservation of angular momentum is responsible for the formation of the accretion disks around young stars. The Milky Way and the Solar System are both flat for the same reasons. The origin of the angular momentum is different, though. Turbulent motions within the molecular cloud precursors of stars produce the net angular momentum for a stellar disk, whereas gravitational interactions with nearby clumps lead to the angular momentum of the Milky Way.

The end product: Once a stellar accretion disk forms, most of the matter moves inward and is collected into a star. In contrast, much of the matter in a spiral galaxy remains in the disk.

onto hot gas convincingly demonstrates that they contain far more mass than can be accounted for by their stars alone.

We can use a similar approach to look for evidence of dark matter on larger scales. Galaxy clusters are filled with extremely hot (10 million to 100 million K) gas, which occupies the space between galaxies (**Figure 21.6**). Even though this gas is *extremely* tenuous, the volume of space that it occupies is enormous: The mass of this hot gas can be up to five times the mass of all the stars in that cluster. X-ray spectra show that this gas contains significant amounts of massive elements that must have formed in stars. This chemically enriched gas has been either blown out of galaxies in winds driven by the energy of massive stars, or stripped from galaxies during encounters with neighboring galaxies. Were it not for the gravity of the dark matter filling the volume of the cluster, this hot gas would have dispersed long ago.

We can also look at the motions of the cluster's galaxies (which behave like very large "atoms" in this context) and again ask, How strong must gravity be to hold this cluster together? Again, the answer is that the total mass of the clusters, including dark matter, must be about 10 times greater than the normal matter they contain.

A third way to look for dark matter relies on the predictions of Einstein's general theory of relativity, which states that mass distorts the geometry of spacetime, causing even light to bend near a massive object. In particular, light from a distant object is bent by the gravity of a galaxy or cluster of galaxies, so that images of the distant object can be seen magnified on either side of the intervening galaxy or cluster. The result is a **gravitational lens (Figure 21.7(a))**.

We have already encountered gravitational lensing in our discussion of MACHOs in Excursions 19.2, where we found that lenses can make background objects appear brighter. When considering the more complex three-dimensional geometry of gravitational lenses, we find that

Gravitational lenses are one way to measure the masses of galaxy clusters.

lenses can show multiple images of background objects, and that these magnified images are often drawn out into an arclike appearance. The greater the gravitational lensing, the greater the mass that must be in the cluster. **Figure 21.7(b)** shows an image of a galaxy cluster that is acting as a gravitational lens for a number of background galaxies. Analysis of such images allows us to determine the mass of the lensing cluster.

Regardless of how we measure the masses of galaxy clusters—by looking at the motions of their galaxies, by

Dark matter dominates the mass of galaxy groups and clusters.

measuring their hot gas, or by using them as gravitational lenses—the results are the same. By mass, galaxy clusters, like the galaxies they are made of, are dominated by the dark matter they contain.

(a) X-rays

1 For gravity to hold onto this hot X-ray emitting gas…

1 Mly

(b) Visible light

2 …this galaxy cluster must be more massive than the galaxies we see.

3 Galaxy clusters are mostly dark matter.

FIGURE 21.6 (a) An X-ray image of a cluster of galaxies shows that the cluster is full of hot tenuous gas. (b) An image of the same field of view taken in visible light. (The two brightest objects are foreground stars.)

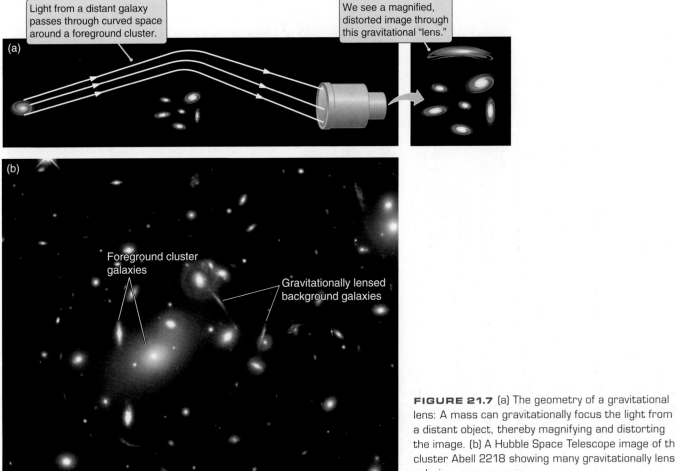

Light from a distant galaxy passes through curved space around a foreground cluster.

We see a magnified, distorted image through this gravitational "lens."

(a)

(b)

Foreground cluster galaxies

Gravitationally lensed background galaxies

FIGURE 21.7 (a) The geometry of a gravitational lens: A mass can gravitationally focus the light from a distant object, thereby magnifying and distorting the image. (b) A Hubble Space Telescope image of the cluster Abell 2218 showing many gravitationally lensed galaxies, seen as arcs.

Mergers Play a Large Role in Galaxy Formation

The picture of galaxy formation given in the previous sections is much cleaner and more idealized than what happened in reality. In our hierarchical clustering picture, smaller structures collapsed first. If there was substructure within the original galaxy-sized dark matter clump, that substructure would itself have collapsed into subgalaxy-sized objects before the galaxy as a whole finished forming. This gives us a way to understand the existence of such objects as the globular star clusters and dwarf companion galaxies near the Milky Way and other galaxies. These objects formed inside the larger, cold dark matter clump when the luminous matter in the clump was still settling toward what would become the disk of the galaxy.

It is also likely that the same larger clump often produced more than one galaxy, and that these galaxies later interacted or even merged. When two spiral (or protospiral) galaxies merged, all sorts of commotion were likely. The tidal interactions between the galaxies and the collisions be-

tween gas clouds in the galaxies probably triggered many regions of star formation throughout the combined system.

The merging of galaxies also answers another puzzle. We have seen how spiral galaxies could form in the early universe, but what about the formation of elliptical galax-

Ellipticals form from the mergers of spirals.

ies? Ellipticals are now thought to result from the merger of two or more spiral galaxies. **Figure 21.8** shows a computer simulation of such a merger. If the merging galaxies were not originally spinning in the same direction, then the resulting merged galaxy loses its disklike character. The dark matter halos of the galaxies merge, and the stars eventually settle down into the bloblike shape of an elliptical galaxy. As we might expect on the basis of this picture, elliptical galaxies are known to be more common in dense clusters where mergers are likely to have been more frequent.

Such events also played a major role in feeding supermassive black holes, which themselves must have formed

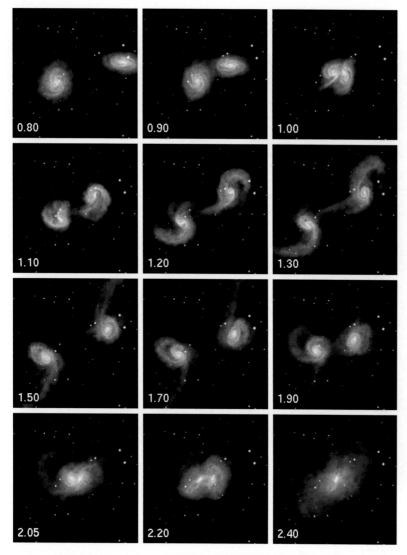

FIGURE 21.8 In this computer simulation, two spiral galaxies merge to form an elliptical galaxy. Numbers indicate time since the interaction began.

When we look at the bright, active galaxies that populate the early universe, it is important to remember that what we actually see is only the tip of the iceberg. The luminous galaxies are small in comparison with the dark matter clumps that surround them. These clumps are still found in today's universe as the invisible dark matter halos responsible for the flat rotation curves of spiral galaxies.

Observations Help Fill Gaps in the Models

We would like to be able to turn our models of collapsing cold dark matter clumps into detailed models of what galaxies in the young universe should look like, but we have not yet reached such a level of sophistication. One major problem is that we lack a good understanding of how stars form in young galaxies. We know that most of the normal matter in a galaxy winds up as stars; but when and where do those stars form, and how long does it take? Although we can say a great deal about how individual stars form in our galaxy, as we did in Chapter 15, we do not yet have theories or models that predict such important pieces of the puzzle as what fractions of stars should have what masses. Nor do we clearly understand the differences between star formation in the early universe and the star formation going on around us today. In a universe devoid of massive elements there were no dense, dusty molecular clouds from which stars might form, and no spiral disks in which such clouds might congregate. Instead the first stars must have formed from the collapse of clouds of hydrogen and helium gas within the overall clump of matter of which they were a part.

Although a detailed understanding of star formation in a forming galaxy is in the future, we can still make a number of factual statements about when and where stars form in a collapsing galaxy. These clues come from the study of our

---

Star formation began very early on in collapsing galaxies.

---

very early on in the collapse of galaxies and protogalaxies. The same mergers and interactions that formed the giant galaxies that we see today also provided the fuel to power the quasars and other AGNs that were common in the early universe.

With galaxies crashing together, supermassive black holes forming, quasars flooding the young universe with intense radiation and powerful jets, and star formation running amok, galaxy formation in the young universe must have been a violent, messy business. This conclusion has clear consequences for what we should expect to see when looking back to a time in the history of the universe when galaxies were actively forming. Rather than the well-formed spirals and ellipticals that dominate today's universe, the early universe should have contained many clumpy, irregular objects. As expected, images of the young universe such as the Hubble Space Telescope image shown in Figure 18.1 are filled with lumpy, distorted objects that are still in the process of settling in.

own galaxy. The fact that the atmospheres of even the oldest halo stars in the Milky Way contain some amount of massive elements tells us that a significant amount of star formation must have taken place *very* early during the collapse of the Milky Way. If we could closely inspect a collapsing protogalactic clump in the stages depicted in Figure 21.5(d) and (e), we would expect to see stars already forming and supernovae exploding from place to place in the collapsing halo. The halo stars that we see today must have formed while the Milky Way was still collapsing, before the gas they formed from coalesced into the disk. In contrast, disk stars (which all have relatively high abundances of massive elements) formed from gas that was enriched by early generations of halo stars before it settled into the galaxy's disk.

Thus while we cannot say exactly how stars formed in the early universe, we can say that star formation must have been going on throughout the process of galaxy formation. The pattern of stellar ages and massive element abundances evident in our galaxy today fits naturally with our current models in which galaxies formed from collapsing clumps of cold dark matter.

Uncertainties in our understanding of star formation complicate quantitative comparisons between models of galaxy formation and observations of the early universe. Fortunately, though, when we talk about structure on scales much larger than galaxies, star formation becomes less of an issue. To understand the formation of clusters and larger structures, we only need worry about the way that galaxy-sized clumps of matter themselves fall together under the force of gravity.

Figure 21.9 shows the results of one computer simulation that follows the motions of millions of clumps of dark matter as they fall through space under their mutual gravitational attraction. The scale of this simulation is so large that we cannot follow the outcome of individual galaxies,

Clusters, filaments, and voids form after galaxies form.

but instead are looking at the overall distribution of dark matter. The simulation shows that the smaller-scale structures develop first. Over the first few billion years, dark matter falls together into structures comparable in size to today's clusters of galaxies. Only later do the spongelike filaments, walls, and voids become well defined.

The similarities between the results of these models and observations of large-scale structure, such as those in Figure 21.3, are quite remarkable. Not any model will do, however. Only model universes with certain combinations of shape, mass, nature of ripples, type of dark matter, and values for the cosmological constant will produce structure similar to what we actually see. It is a very important result that models beginning with assumptions consistent with our knowledge of the early universe wind up predicting the formation of large-scale structure similar to what we see in today's universe.

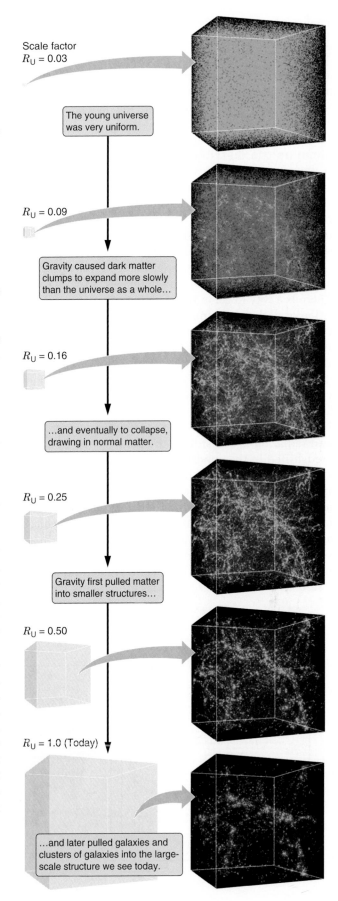

FIGURE 21.9 A computer simulation of the formation of very large-scale structure in a universe filled with cold dark matter.

Peculiar Velocities Trace the Distribution of Mass

In our comparison between models and observations, there is an important caveat. The models show where the *mass* is, whereas our observations show where the *light* is. We have

---

The distribution of light may not be the same as the distribution of mass.

---

already noted that our understanding of star formation in the early universe is incomplete at best. What if a dark matter clump formed and collapsed, but for some reason the normal matter associated with that clump did not produce stars? Such clumps could still contain a significant amount of mass and could even affect the geometry of the universe, but would themselves not be seen in images of the sky.

There is a way to get at the question of the overall distribution of dark matter more directly. If the clumping of galaxies is a result of the action of gravity, then we should be able to look at how galaxies are moving and see them falling together. In fact, measurements of the way galaxies fall together should allow us to infer the gravitational field they are experiencing, and hence the overall distribution of dark matter.

In the previous chapter we learned that the cosmic background radiation is blueshifted in one direction in space and redshifted in the opposite direction, telling us of our peculiar velocity relative to the CBR. Peculiar velocities of galaxies other than our own are difficult to measure. They require us to accurately determine the distances to galaxies using standard candles, and then to use those distances along with Hubble's Law to predict what the redshifts of those galaxies should be. Comparison of observed redshifts with redshifts predicted on the basis of Hubble expansion then tells us how fast these galaxies are moving with re-

spect to the cosmic background radiation (at least along our line of sight).

Figure 21.10(a) shows one such map of the peculiar velocities inferred for galaxies in our part of space. If we use observations of peculiar velocities to map concentrations of mass in this part of the universe, we get a picture that looks like two mountains surrounded by foothills and valleys, as shown in **Figure 21.10(b)**. Unfortunately, these two "mountains" of mass, which exert the greatest gravitational pulls on our own galaxy, are located in regions of the sky that are heavily obscured by the Milky Way's dusty disk.

Peculiar velocities tell us how dark matter is distributed.

The largest pull that we are experiencing comes from a region dubbed the Great Attractor, so-called because of its gravitational tug.

The picture painted in this chapter of the formation of large-scale structure is almost certainly correct in its broad outlines. Even so, uncertainties in star formation, our understanding of the exact nature of dark matter, and our knowledge of the shape of the universe limit the amount of detail in any comparisons we can make between models and observations. As the 21st century opens, this is a field of very active research. Several projects such as the Sloan Digital Sky Survey are currently collecting large amounts of data on the distribution and redshifts of galaxies, and satellites such as WMAP and the ESA's proposed Planck mission can dramatically improve our knowledge of the structure in the CBR. Larger and more sophisticated models are being built all the time, while studies of star-forming regions in our galaxy and in galaxies with lower massive element abundances are continuing to improve our understanding of this important process. There is also hope that new theories and laboratory experiments might identify

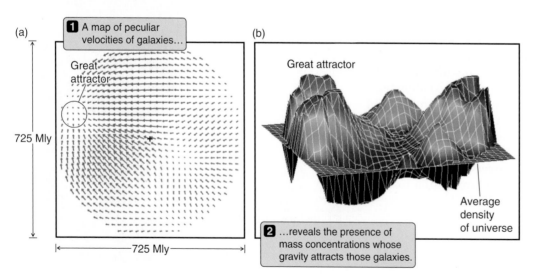

(a)   **1** A map of peculiar velocities of galaxies...

Great attractor

725 Mly

|← 725 Mly →|

(b)

Great attractor

Average density of universe

2 ...reveals the presence of mass concentrations whose gravity attracts those galaxies.

FIGURE 21.10 (a) A map of peculiar velocities of galaxies in our neighborhood. Arrows show the velocities of material lying in the plane of our own local supercluster. Motions seem to be converging onto the position of the Great Attractor. The black cross marks the location of the Milky Way Galaxy. (b) A map of the mass distribution inferred from observations of peculiar velocities.

the particles that make up dark matter. A more complete understanding of the formation of galaxies and large-scale structure in the universe will not likely come in a single "eureka" moment, but will instead arrive in bits and pieces throughout the first part of the 21st century as a result of advances in all of these different areas.

21.4 The Earliest Moments

It is easy to look at galaxies and stars and say, "Aha, structure!" But there is more to the origin of structure than simply the clumping of matter. The forces that govern the behavior of matter and energy in the universe are themselves a kind of structure. There are four fundamental forces in nature, and everything in the universe is a result of their action. Chemistry and light are products of the *electromagnetic force* acting between protons and electrons in atoms and molecules. The energy produced in fusion

---

There are four fundamental forces in nature.

---

reactions in the heart of the Sun comes from the *strong nuclear force* that binds together the protons and neutrons in the nuclei of atoms. Beta decay of nuclei, in which a neutron decays into a proton, an electron, and an antineutrino, is governed by the **weak nuclear force**. Finally there is our old friend *gravity,* which has played such a major role at every point along our journey. How these forces—these physical laws—came into being is part of the history of the universe as well.

May the Forces Be with You

In Chapter 4 we spoke of electromagnetism using the electric and magnetic fields, but we also spoke of the quantum mechanical description of light as a stream of particles called photons. Because there is only one reality, both of these descriptions of electromagnetism have to coexist. The branch of physics that deals with this reconciliation is called **quantum electrodynamics** or **QED**.

QED treats charged particles almost as if they were baseball players engaged in an endless game of catch. As the baseball players throw and catch baseballs, they experience forces. Similarly, in QED, charged particles "throw" and "catch" an endless stream of virtual photons, as illustrated in **Figure 21.11**. Earlier we grappled with the idea that quantum mechanics is a science of possibilities rather than certainties. QED describes the electromagnetic interaction between two charged particles by averaging over all of the

possible ways that the particles could throw photons back and forth. The result is a force that, over large scales, acts like the classical electric and magnetic fields described by Maxwell's equations. Physicists speak of the electromag-

---

In QED, photons carry the electromagnetic force.

---

netic force being "mediated by the exchange of photons." As is always the case with quantum mechanics, the world described by QED is hard to picture. Even so, QED is one of the most accurate, well-tested, and precise branches of physics. As of this writing, not even the tiniest measurable difference between the predictions of the theory and the outcome of an actual experiment has been found.

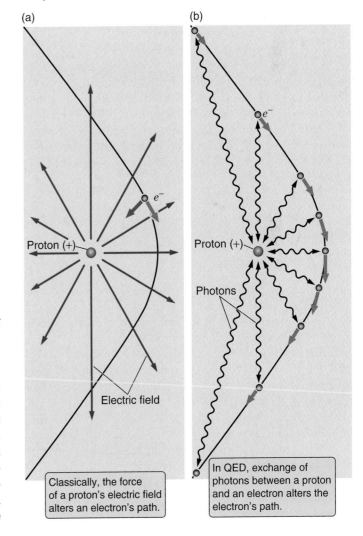

FIGURE 21.11 (a) The classical view of an electron being deflected from its course by the electric field from a proton. (b) According to quantum electrodynamics, the interaction is properly viewed as an ongoing exchange of photons between the two particles.

(a)

Proton (+)

e^-

Electric field

Classically, the force of a proton's electric field alters an electron's path.

(b)

e^-

Proton (+)

Photons

In QED, exchange of photons between a proton and an electron alters the electron's path.

The central idea of QED—forces mediated by the exchange of carrier particles—provides a template for understanding two of the other three fundamental forces in

The weak nuclear and electromagnetic forces combine in electroweak theory.

nature. The electromagnetic and weak nuclear forces have been combined into a single theory called **electroweak theory**. This theory predicts the existence of three particles—labeled the W^+, W^-, and Z^0—that mediate the weak nuclear force. The essential predictions of electroweak theory were confirmed in the 1980s when these particles were identified in laboratory experiments.

The strong nuclear force is described by a third theory, called **quantum chromodynamics** or **QCD**. In this theory particles such as protons and neutrons are composed of more fundamental building blocks, called **quarks**, that are bound together by the exchange of another type of carrier particle dubbed **gluons**. Together electroweak theory and QCD are referred to as the **standard model** of particle physics. A deeper investigation of the standard model must

Electroweak theory + QCD = The standard model of particle physics.

await another journey. Here we leave the discussion by pointing out that the standard model is able to explain all the currently observed properties of matter and has made many predictions that were subsequently confirmed by laboratory experiments. The standard model does not exclude the neutrino from having mass, but the existence of neutrino mass would have an explanation lying outside the standard model.

A Universe of Particles and Antiparticles

For modern theories of particle physics to make any sense, every type of particle in nature must also have an alter ego—an *antiparticle*—that is the opposite of that particle in every way described by quantum mechanics. For the electron there is the antielectron, otherwise known as the positron. For the proton there is the antiproton, for the neutron the antineutron, and so on down the list. One fascinating property of these particle–antiparticle pairs is that if you bring such a pair together, the two particles will annihilate each other.

When a particle–antiparticle pair annihilates, the mass of the two particles is converted into energy in accord with Einstein's special theory of relativity ($E = mc^2$). For example, in **Figure 21.12(a)** an electron and a positron annihilate each other, and the energy is carried away by a pair of photons.

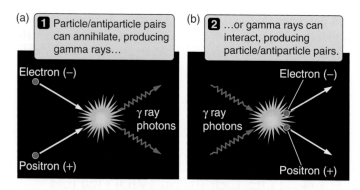

FIGURE 21.12 (a) An electron and a positron annihilate, creating two gamma ray photons that carry away the energy of the particles. (b) In the reverse process, pair creation, two gamma ray photons collide to create an electron–positron pair.

(This is the idea behind *Star Trek*'s "antimatter" engines.) When we run time backward (as we are allowed to do in particle physics), we see two high-energy photons colliding with each other, as in **Figure 21.12(b)**, creating in their place an electron–positron pair. This is an example of how an energetic event can create a particle and its corresponding antiparticle—a process called **pair creation**.

In principle, *any* type of particle and its antiparticle can be created in this way. The only limitation comes when there is not enough energy available to supply the rest mass of the particles being created. A specific example helps to show how this works. An electron (or positron) has a mass of $m_e = 9.11 \times 10^{-31}$ kg, which corresponds to an energy ($E = m_e c^2$) of 8.20×10^{-14} joules. If two gamma ray photons with a combined energy greater than 16.40×10^{-14} J ($= 2 \times m_e c^2$) collide, then the two photons may disappear and leave an electron–positron pair behind in their place. If the photons have more than the necessary energy, then the extra energy goes into the kinetic energy of the two newly formed particles.

Now we apply this idea to a hot universe awash in a bath of Planck radiation. Using Wien's Law from Chapter 4 [$\lambda_{peak} = (2{,}900\ \mu m\ K)/T$] and the expression for photon energy ($E = hc/\lambda$), we can show that when the universe was less than about 100 seconds old and had a temperature greater than a billion kelvins, it was filled with photons that had enough energy to create electron–positron pairs. Under

The early universe was filled with photons, electrons, and positrons.

these conditions, photons were constantly colliding, creating electron–positron pairs; and electron–positron pairs were constantly annihilating each other, creating pairs of gamma ray photons. The whole process reached an equilibrium, determined strictly by temperature, in which pair creation and pair annihilation exactly balanced each other. Rather than being filled only with a swarm of photons, at this time the universe was filled with a swarm of photons, electrons, and positrons, as illustrated in **Figure 21.13(a)**.

The Frontiers of Physics

Physics has given us the tools to understand all the structures we have seen so far on our journey. There are still many gaps in what we know, but we are confident that everything from planets to stars to galaxies can be understood by applying our current knowledge of the four fundamental forces. At the turn of the 21st century, most of astronomy is a struggle with the complexity of the universe, rather than a shortcoming of our understanding of the fundamental laws governing matter, energy, and spacetime. However, as we push further back toward the Big Bang itself, the nature of the game changes. We need new physical theories.

In Foundations 10.1 we explored the power of symmetry—the idea that we can learn a lot about nature just by thinking about the way one part of something matches up with another. There we were talking about the gravitational forces that hold planets together, but other kinds of symmetry exist as well. In the process of pair creation there is a symmetry between matter and antimatter. For every particle created, its antiparticle is created as well.

As the universe cooled, there was no longer enough energy to support the creation of particle pairs, so the swarm of particles and antiparticles that filled the early universe annihilated each other and were not replaced. When this happened, every electron should have been annihilated by a positron. Every proton should have been annihilated by an antiproton. This was almost the case, but not quite. For every electron in the universe today, there were 10 billion

> For every 10 billion positrons there were 10 billion and one electrons.

and one electrons in the early universe, but only 10 billion positrons. This one part in 10 billion excess of electrons over positrons meant that when electron–positron pairs finished annihilating each other, some electrons were left over—enough to account for all the electrons in all the atoms in the universe today (**Figure 21.13(b)**).

If the standard model of particle physics were a complete description of nature, then the one-part-in-10-billion imbalance between matter and antimatter would not have been there in the early universe. The symmetry between matter and antimatter would have been complete. No matter at all would have survived into today's universe, and we would not exist. The fact that you are reading this page demonstrates that something more needs to be added to the model.

According to current ideas, the symmetry between matter and antimatter may be broken in a theory that joins the

> GUTs unify the strong and electroweak interactions.

electroweak and strong nuclear forces together in much the same way that electroweak theory unified our understanding

FIGURE 21.13 (a) A swarm of electrons, positrons, and photons in the very early universe. For every 10 billion positrons, there were really 10 billion and one electrons. (b) After these particles annihilated, only the one electron was left.

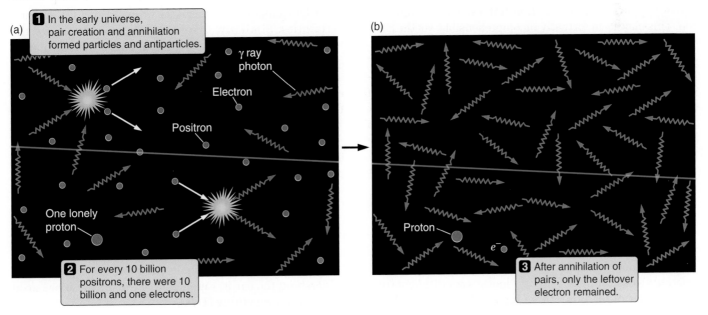

of electromagnetism and the weak nuclear force. Such a theory, which combines three of the four fundamental forces into a single grand, unified force, is referred to as a **grand unified theory** or **GUT**.

Many possible grand unified theories exist, and they make many predictions about the world. The problem is that the particles that mediate GUTs are so massive that it takes enormous amounts of energy to bring them into existence—roughly a trillion times as much energy as can be achieved in today's particle accelerators! Even so, some predictions of GUTs are testable with current technology. For

---

GUTs predict that even the proton will decay.

---

example, GUTs predict that protons should be unstable particles that, given enough time, will decay into other types of elementary particles. This is a *very* slow process. Over the course of your life, GUTs predict that there is about a 1 percent chance that *one* of the 10^{28} or so protons in your body will decay. As of this writing, proton decay has yet to be observed. As we speak, however, large arrays of detectors are peering into huge tanks of water, waiting to see the signature of such an event. Perhaps your local news stand holds a story heralding the confirmation of this central prediction of GUTs.

The particles that mediate GUTs may be beyond the reach of today's high-energy physics labs; but when the universe was *very* young (younger than about 10^{-35} second) and *very* hot (hotter than about 10^{27} K), there was enough energy available for these particles to be freely created. During this time, the distinction between the electromagnetic, weak, and strong nuclear forces had not yet come into being. There was only the one grand, unified force. Welcome to the era of GUTs.

During this era of GUTs, the apparent size of the entire universe was less than a trillionth the size of a single proton. This may seem virtually incomprehensible, yet the basic ideas needed to understand this universe are in place. However, as we move backward in time, there is one threshold we have yet to cross. The story of advances in our understanding of physical law has been a story of unification—of the electromagnetic and weak forces into the electroweak theory, and then of these and the strong nuclear force into a GUT. But this program is incomplete. How does gravity fit into this scheme?

General relativity provides a beautifully successful description of gravity that correctly predicts the orbits of planets, describes the ultimate collapse of stars, and even allows

---

Gravity does not fit into the GUT picture.

---

us to calculate the structure of the universe. Yet general relativity's description of gravity "looks" very different from our theories of the other three forces. Rather than talking

about the exchange of photons or gluons or other carrier particles, general relativity talks instead about the smooth, continuous canvas of spacetime upon which events are painted.

We might be tempted to say, "Oh well. Gravity works one way and the other forces work another way, and that is how the universe happens to be." In practice that is exactly what we do when we call on quantum mechanics to tell us about the properties of atoms, then use relativity to describe the passage of time or the expansion of the universe. Even the era of GUTs is described perfectly well by treating gravity as a separate force. However, as we push back even closer to the moment of the Big Bang, this happy coexistence between relativity and quantum mechanics turns instead into a brutal confrontation.

Toward a Theory of Everything

When the universe was younger than about 10^{-42} second old, the density of the universe was incomprehensibly high. The universe was so small that 10^{60} universes would have fit into the volume of a single proton! Under these extreme conditions, quantum mechanical fluctuations in the matter and radiation making up the universe involved immense amounts of mass—so much mass that quantum fluctuations made mincemeat out of spacetime. Rather than a smooth sheet, spacetime was a quantum mechanical froth. General relativity fails to describe this early universe in ways that are reminiscent of the failure of Newtonian mechanics to describe the structure of atoms. An electron in an atom

---

In the Planck era the whole universe was a quantum mechanical froth.

---

must be thought of in terms of probabilities rather than certainties. Similarly, there is no unique history for the earliest moments after the Big Bang. This era in the history of the universe is referred to as the **Planck era**, signifying that we can understand the structure of the universe itself during this period only by using the ideas of quantum mechanics.

The conflict between the continuous and the discrete —between general relativity and quantum mechanics— brings us to the current limits of human knowledge. The physics that we know can take us back to a time when the universe was a millionth of a trillionth of a trillionth of a trillionth of a second old, but to push back any further we need something new. We need a theory that combines general relativity and quantum mechanics into a single theoretical framework unifying all four of the fundamental forces. Here we have reached the holy grail of modern physics. To understand the earliest moments of the universe, we need a **theory of everything (TOE)**.

A successful theory of everything would do more than unify general relativity with quantum mechanics. It would tell us which of the possible GUTs is correct and would provide an answer for the nature of dark matter. A successful theory of everything would also necessarily answer several outstanding issues in cosmology, including the how,

Superstring theory is a possible theory of everything.

when, and why of inflation, and the physics underlying the possible cosmological constant adding to the expansion of the universe. Physicists are currently grappling with what a TOE might look like. The current contender for the title is **superstring theory**. Here elementary particles are viewed not as points but as tiny loops called *strings*. A guitar string vibrates in one way to play an F, another way to play a G, and yet another way to play an A. According to superstring theory, different types of elementary particles are like different "notes" played by vibrating loops of string.

Superstring theory in principle provides a way to reconcile general relativity and quantum mechanics, but there is a price to pay for this success. To make superstring theory work, we have to imagine that these tiny loops of string are vibrating in a universe with 10 spatial dimensions! (Adding time to the list would make our universe 11-dimensional.)

Superstring theory predicts 11 dimensions, but we experience only 4 of them.

How can that be, when we clearly experience only three spatial dimensions? Whereas the three spatial dimensions that we know spread out across the vastness of our universe, the other seven dimensions predicted by string theory wrap tightly around themselves (**Figure 21.14**), extending no further today than they did a brief instant after the Big Bang.

To better appreciate how this bizarre notion works, imagine what it would be like to live in a three-dimensional universe (like the one we experience) in which one of those dimensions extended for only a tiny distance. Living in such a universe would be like living within a thin sheet of paper that extended on for billions of light-years in two directions but was far smaller than an atom in the third. In such a universe we would easily be aware of length and width—we could move in those directions at will. In contrast, we would have no freedom to move in the third dimension at all, and might not even recognize that the third dimension existed. Perhaps our only inkling of the true nature of space would come from the fact that in order to explain the results of particle physics experiments, we would have to assume that particles extended into a third, unseen dimension. In like fashion, if superstring theory is correct, we see three spatial dimensions extending on possibly forever, but we are unaware of the fact that each point in our three-dimensional space actu-

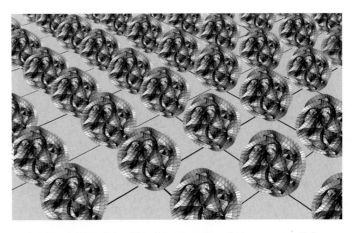

FIGURE 21.14 It is difficult to visualize what seven spatial dimensions wrapped up into structures far smaller than the nucleus of an atom would be like. Here such geometries are projected onto the two-dimensional plane of the paper.

ally has a tiny but finite extent in seven other dimensions at the same time!

Superstring theory is only a pale shadow of the sort of well-tested theories that we have made use of throughout this book. In some respects superstring theory is no more than a promising idea providing direction to theorists searching for a TOE. It is worth noting that we will probably never be able to build particle accelerators that allow us to directly search for the most fundamental particles predicted by a TOE. The energies required are simply too high. Fortunately, however, nature has provided us with the ultimate particle accelerator—the Big Bang itself! The structure of the universe that we see around us today is the observable result of that grand experiment.

At first glance particle physics and cosmology might seem to have almost nothing in common. Particle physics is the study of the quantum mechanical world that exists on the tiniest scales imaginable, whereas cosmology is the study of the changing structure of a universe that extends for billions of light-years and probably much farther. Yet the last quarter of the 20th century saw the boundary between these two fields fade and eventually disappear as cosmologists and particle physicists came to realize that the structure of the universe and the fundamental nature of matter are two sides of the same scientific coin.

One of the most intriguing questions that a successful TOE may answer is whether the universe could have been different. Do we live in only one of many possible universes,

Is ours the only possible universe?

each with different physical laws? Or are the physical laws that govern our universe the *only* consistent set of physical laws that could exist? We do not know the answer yet, but possibly within your lifetime we will.

Order "Froze Out" of the Cooling Universe

We started our discussion of the fundamental forces of nature by pointing out that these forces themselves represent a type of structure in the universe. Just as galaxies and stars are structure that condensed out of the uniformity of the Big Bang, so too are the four fundamental forces structure that condensed out of the uniformity of the theory of everything.

Figure 21.15 illustrates the origin of structure in the evolving universe. The first 10^{-43} second after the Big Bang is described by the TOE, when the physics of elementary particles and the physics of spacetime were one and the same. As the universe expanded and cooled, gravity parted ways with the forces described by the GUT. Spacetime took on the properties described by general relativity. Inflation may also have been taking place at this time.

As the universe continued to expand and its temperature fell further, less and less energy was available for the creation of particle–antiparticle pairs. When the particles responsible for GUT interactions could no longer form, the

---

First, atomic nuclei and then atoms formed as the universe cooled.

---

strong force split off from the electroweak force. One might speak of this transition as the strong force "freezing out" of the GUT because this and other similar transitions are reminiscent of the phase change that occurs as water changes to ice, and molecules become more constrained in their motions. Somewhere along the line, as the unity of the original TOE was lost, the symmetry between matter and antimatter was broken. As a result, the universe ended up with more matter than antimatter.

The next big change took place when the particles responsible for unifying the electromagnetic and weak nuclear forces froze out, leaving these two forces independent of each other. The four fundamental forces of nature that govern today's universe had now come into their own. The temperature of the universe had fallen to a chilly 10^{16} K, and a 10-trillionth of a second had ticked off the cosmic clock. It was a full minute or two before the universe cooled to the billion-kelvin mark, below which not even pairs of electrons and positrons could form.

Although the universe was now too cool to form additional particles and their antiparticles, it was still hot enough for the thermal motions of protons to overcome the electric barriers between them, allowing nuclear reactions to take place. These reactions formed the least massive elements, including helium, lithium, beryllium, and boron, but could not create more massive elements.

Big Bang nucleosynthesis came to an end by the time the universe was 5 minutes old, and the temperature of the universe had dropped below about 800 million K. The density of the universe at this point had fallen to only about a 10th that of water. Normal matter in the universe now consisted of atomic nuclei and electrons, awash in a bath of radiation. So the universe remained for the next several hundred thousand years, until finally the temperature dropped so far that electrons were able to combine with atomic nuclei to form neutral atoms. We have encountered this event before. This was the era of recombination, which we see directly when we look at the cosmic background radiation.

21.5 Life Is Another Form of Structure

As we near the end of our journey, let's briefly appreciate the distance we have covered. We have followed the origin of structure in the universe from the earliest instants after the Big Bang, when the fundamental forces of nature came to be, through to formation of the galaxies and other large-scale structure that is visible today. Earlier on our journey, we watched as stars (including our Sun) formed from clouds of gas and dust within these galaxies, and planets (including our Earth) formed around those stars. We learned of the geological processes that shaped early Earth into the planet we see around us today. In short, we have traced the origin of structure from the instant the universe came into existence up until the modern day. To help put this into perspective, **Excursions 21.1** shows how the major events in the history of the universe would fit into a single 24-hour day that began with the Big Bang and ended today. Yet there is one piece that we have left out. Although the focus of our journey is 21st century *astronomy,* no discussion of how structure evolved in the universe would be complete without some consideration of the origin of the particular type of structure we refer to as *life* (**Figure 21.16**).

Evolution Is Unavoidable

Imagine that just once during the first few hundred million years after the formation of Earth, a single molecule formed by chance somewhere in Earth's oceans. That molecule had a very special property: Chemical reactions between that

---

To create life, only one self-replicating molecule needed to form by chance.

---

molecule and other molecules in the surrounding water resulted in that molecule making a copy of itself. Now there were two such molecules. Being duplicates of the original, chemical reactions would produce copies of each of these molecules as well, making four. Four became 8, 8 became 16,

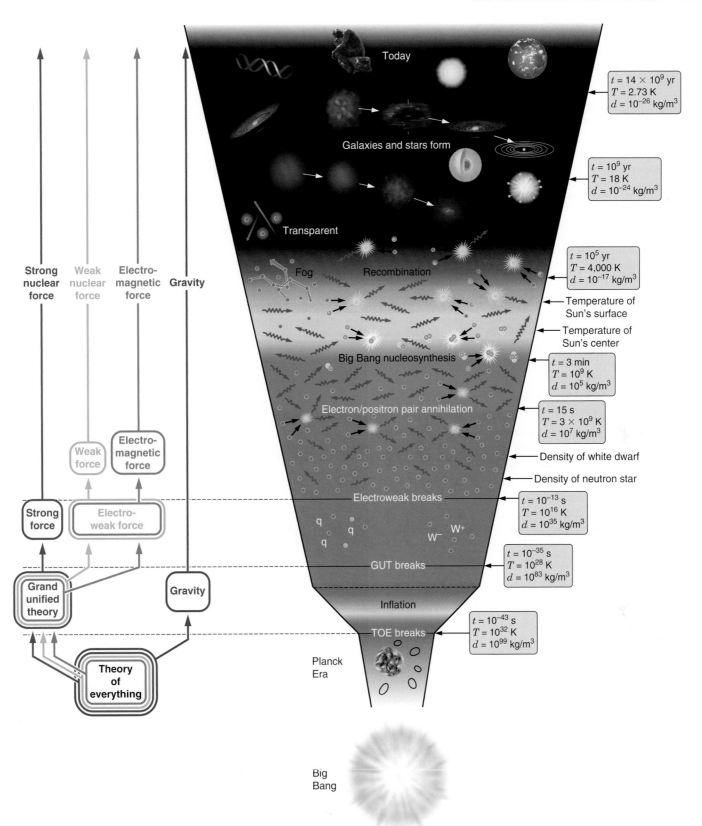

FIGURE 21.15 Eras in the evolution of the universe. As the universe expanded and cooled following the Big Bang, it went through a series of phases determined by what types of particles could be created freely at that temperature. Later the structure of the universe was set by the gravitational collapse of material to form galaxies and stars, and by the chemistry made possible by elements formed in stars.

Forever in a Day

We have seen that the events that ultimately led to our existence here as inhabitants of planet Earth stretched out over many billions of years, intervals that even astronomers have difficulty visualizing. We can sometimes get a better grasp of such enormous spans of time by compressing them into much shorter intervals with which we have day-to-day experience. Try then to imagine the age of the universe and those important events associated with our origins as if they were all taking place within a single day. Our cosmic day begins at the stroke of midnight:

12:00:00 A.M. The embryonic universe is a mixed broth, minute specks of matter suspended in a vast soup of radiation. The entire universe exists only as an extraordinarily hot bath of photons and a zoo of elementary particles.

12:00:02 A.M. It is now just 2 seconds after midnight. All of the early eras in the history of the universe have passed. The fundamental forces have frozen out, Big Bang nucleosynthesis has formed the universe's original complement of atomic nuclei, and things have cooled down enough for atomic nuclei to combine with electrons to produce neutral atoms. Both normal and dark matter are now available to create galaxies and stars, but that process will take some time.

12:40 A.M. The first stars, quasars and galaxies appear. At some point—we are not sure exactly when—our own galaxy becomes visible as star formation begins. Throughout the cosmic day, stars will continue to form. The more massive stars each go through their brief life cycles in only 5 to 10 seconds of our imaginary 24-hour clock. Each massive star shines briefly, creates its heavy elements, and then disperses this material throughout interstellar space as it dies in a violent supernova explosion. Stars similar to our Sun go through less dramatic life cycles, each lasting about 16 cosmic hours. Stars with mass less than 0.8 $M_\odot$ will last for several dozen cosmic hours and will survive beyond the end of our cosmic day.

4:00 P.M. Our Solar System forms out of a giant cloud of gas and dust. Collapse of the cloud's protostellar core, followed by the appearance of the Sun and the planets—including Earth—all take place within the span of a single cosmic minute.

4:05 P.M. A Mars-size planetesimal crashes into Earth, forming the Moon.

---

16 became 32,... and so on. By the time the original molecule had copied itself just 100 times, over a *million trillion trillion* (10^{30}) of these molecules existed. That is 100 million times more of these molecules than there are stars in the observable universe!

Chemical reactions are never perfect. Sometimes when a copy is attempted, it is not an exact duplicate of its predecessor. The imperfection in the attempted copy is called a **mutation**. Most of the time such an error would be devastating, leading to a molecule that could no longer reproduce at all; but occasionally a mutation would actually help. It would lead to a molecule that was *more* successful in duplicating itself than the original. Even if imperfections in the copying process cropped up only once every 100,000 times a molecule reproduced itself, and even if only 1 out of 100,000 of these errors turned out to be beneficial, that still meant that after only 100 generations there were a hundred million trillion (10^{20}) errors that, by blind luck, improved on the original design. Copies of each of these improved molecules would inherit the change. These molecules would have happened upon a form of **heredity**—the ability of one generation of structure to pass on its characteristics to future generations.

In this way, as these molecules continued to interact with their surroundings and make copies of themselves, they split into many different varieties. Eventually these descendants of our original molecule became so numerous that the building blocks they needed in order to reproduce became scarce. Faced with this scarcity of resources, varieties of molecules that were more successful than others in reproducing themselves became more numerous. Varieties that could break down other varieties of self-replicating molecules and use them as raw material were especially

Success breeds success.

5:20 P.M. The first primitive life appears on Earth. It evolves into the simplest life forms: unicellular organisms such as bacteria, cyanobacteria, and archaebacteria.

8:40 P.M. More complex single-cell organisms appear, making it possible for multicellular life to develop.

9:30 P.M. The first multicellular organisms (fungi) appear on dry land.

11:00 P.M. Multicellular organisms become abundant. This paves the way for still larger and more complex life forms.

11:20 P.M. The first animals (fish) make the transition from ocean to dry land, a major step toward our existence.

11:35 P.M. The first dinosaurs make their appearance. Various small animals appear as well, but they remain dominated by larger life forms.

11:53:10 P.M. A large comet or asteroid crashes into Mexico's Yucatán Peninsula. Seventy percent of all species on Earth (including the dinosaurs) suddenly vanish. In the minutes that follow, the mammals, being more adaptable to the changed environment, gain prominence.

11:59:25 P.M. Our earliest human ancestors finally appear on the plains of Africa just 35 seconds before the end of our cosmic day.

11:59:59.8 P.M. Modern humans arrive with only a fifth of a second to spare—a fraction of a heartbeat before the day's end.

12:00:00 A.M. Just as our cosmic day draws to a close, will a worldwide catastrophe occur? Will 21st century humans permanently scar the face of the planet, driving many of its life forms into oblivion by polluting the atmosphere and poisoning the land, the rivers, and the oceans? Or will 21st century humans finally break free of their gravitational bondage to their planetary home and begin to claim the rest of the Solar System as their own? In comparison to all that came before, what will happen in the next century will occur in a blur—less than a blink of the eye.

successful in this world of limited resources. Competition and predation had entered the picture. After a few generations, certain varieties of molecules came to dominate the mix, while less successful varieties became less and less common. This competition, in which better-adapted molecules thrive and less well-adapted molecules die out, is referred to as **natural selection**.

Four billion years is a long time—enough time for the combined effects of heredity and natural selection to shape the descendants of that early self-copying molecule into a huge variety of complex, competitive, successful structures. Geological processes on Earth, such as sedimentation, have preserved a fossil record of the history of these structures, as illustrated in **Figure 21.17**. Among these descendants are "structures" capable of thinking about their own existence and unraveling the mysteries of the stars.

The molecules of DNA (deoxyribonucleic acid) that make up the chromosomes in the nuclei of the cells throughout your body are direct descendants of those early self-duplicating molecules that flourished in the oceans of a young Earth. Although the game played by the molecules

Evolution is inevitable.

of DNA in your body is far more elaborate than the game played by those early molecules in Earth's oceans, the fundamental rules remain the same. We now realize that this process is inevitable: Any system that combines the elements of heredity, mutation, and natural selection *must* and *will* evolve.

In *The Selfish Gene* (1976), Richard Dawkins points out that, when viewed from a purely utilitarian perspective, a human being is a machine whose "purpose" is to produce copies of its genetic material. The remarkable thing about humanity is that our intelligence (itself a very powerful tool for survival in a competitive world) has also allowed us to

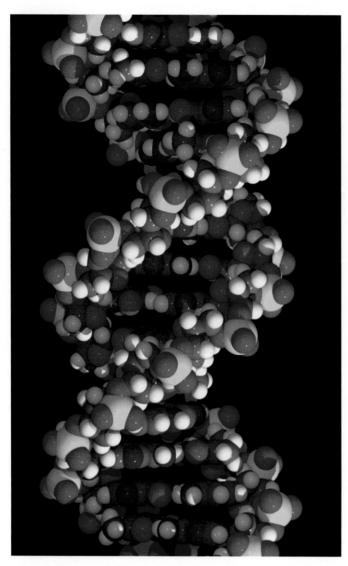

FIGURE 21.16 DNA is the blueprint for life.

develop more noble pursuits (for example, astronomy) than those dictated by our basic biochemical imperative.

Are We Alone?

The story of the evolution of life cannot be separated from the story of 21st century astronomy. We know that we live in a universe full of stars and that systems of planets orbit many and probably most of those stars. We also know that there is nothing mysterious about the origin of life or the processes that cause it to evolve. In fact, as discussed in **Connections 21.2**, the evolution of life on Earth is but one of the many examples that we have encountered of the emergence of structure in an evolving universe. This leads naturally to one of the more profound questions that we ask about the universe. Has life arisen elsewhere, and is it intelligent? Are we alone?

Humans have already made preliminary efforts to "reach out and touch someone." The *Pioneer 11* spacecraft, which will probably spend eternity drifting through interstellar space, carries the plaque shown in **Figure 21.19** describing ourselves to any future interstellar traveler who might happen to find it. This may not be the most efficient way to make contact with the universe, but it is a significant gesture nonetheless. A somewhat more practical effort was made in 1974 when the kilometer-wide dish of the Arecibo radio telescope was used to beam the message shown in **Figure 21.20** toward the star cluster M13. (If someone from M13 answers, we will know in 48,000 years!)

The first serious effort to search for intelligent extraterrestrial life was made by astronomer **Frank Drake** (1930–) in 1960. Drake used what was then astronomy's most powerful radio telescope to listen for signals from two nearby stars. Although his search revealed nothing unusual, it prompted him to develop an equation that still bears his name. The **Drake equation** is a prescription for calculating the likelihood that intelligent civilizations exist beyond our own Solar System.

The Drake equation estimates the number of advanced civilizations in the Milky Way.

In its basic form, the Drake equation is a prescription for calculating N, the number of technologically advanced civilizations in our galaxy today:

$$N = F f_p n_{pm} f_l f_c L$$

The six factors on the right side of the equation relate to the conditions that must be met for a civilization to exist:

- F is the total number of stars in our galaxy. We know this to be several hundred billion stars.

- f_p is the fraction of stars that form planetary systems. Questions remain, but we do know that planets form as a natural by-product of star formation and that many, perhaps most, other stars have planets. For this calculation we will assume that f_p is between 0.5 and 1.

- n_{pm} is the average number of planets and moons in each planetary system capable of supporting life. If such planets are like Earth, they will have to be of the terrestrial type and orbit within a "life zone," where temperatures are neither too hot nor too cold. If our own Solar System is a good example, there would be one or two in each planetary system. (In addition to Earth, Mars appears capable of supporting life, whether or not it presently exists there.) Although we now know that many planetary systems do not resemble our own, it is possible that moons orbiting giant planets near stars could also harbor life. For purposes of calculation, we put the value of n_{pm} between 0.1 and 2.

FIGURE 21.17 Fossils record the history of the evolution of life on Earth.

- f_l is the fraction of planets and moons capable of supporting life on which life actually arises. Remember that just a single self-replicating molecule may be enough to get the ball rolling. Many biochemists now believe that if the right chemical and environmental conditions are present, then life *will* develop. If they are correct, then f_l is close to 1. We will use a range for f_l of 0.01 to 1.

- f_c is the fraction of those planets harboring life that eventually develop technologically advanced civilizations. With only one example of a technological civilization to work with, f_c is hard to estimate. Intelligence is certainly the kind of survival trait that might often be strongly favored by natural selection. On the other hand, on Earth it took 4 billion years—half the expected lifetime of our star—to evolve tool-building intelligence. The correct value for f_c might be close to 1, or it might be closer to 1 in 1,000.

- L is the likelihood that any such civilization exists *today*. This is certainly the most difficult factor of all to estimate because it depends on the long-term stability of advanced civilizations. We have had a technological civilization on Earth for around 100 years, and during that time have developed, deployed, and used weapons with the potential to eradicate our civilization and render Earth hostile to life for many years to come. We have also so degraded our planet's ecosystem that many respectable biologists and climatologists wonder if we are nearing the brink. Do all technological civilizations destroy themselves within a thousand years? A thousand years is only one 10-millionth of the lifetime of our star.

On the other hand, if most technological civilizations learn to use their technology for survival rather than self-destruction, perhaps they instead will survive for hundreds of millions of years. For our calculation we will use a range for L of between 10^{-7} (for civilizations that live 1,000 years) and 10^{-2} (for civilizations that live 100 million years).

As illustrated in **Figure 21.21**, the conclusions we draw based on the Drake equation depend a great deal on the assumptions we make. Using the most pessimistic of our estimates, the Drake equation sets the likelihood of finding a technological civilization in our galaxy at 1 percent, or 1 chance in 100. If this is correct, then we are in all likelihood the *only* technological civilization in the Milky Way.

The nearest advanced civilization may be as far away as 30 Mly...

In fact, this would suggest that only 1 in 100 galaxies would contain a technological civilization *at all*. Such a universe would still be full of intelligent life. With a hundred billion galaxies in the observable universe, even these pessimistic assumptions would mean that there are a billion technological civilizations out there somewhere. On the other hand, we would have to go a *very* long way (30 Mly or so) to find our nearest neighbors.

At the other extreme, what if we take the most optimistic numbers, assuming that intelligent life arises and survives everywhere it gets the chance? The Drake equation then says that there should be 40 million technological

Origins—Life, the Universe, and Everything

Popular discussions of the origin of life almost always wind up struggling with the question of how it is that complex, highly ordered structures such as living things could have emerged from a simpler, more disordered past. Place a drop of ink in a glass of water and watch what happens. The ink spreads out, diffusing through the water, until the only sign that the ink is there is the fact that the water is a different color. The order represented by the discrete drop of ink naturally fades away as the ink spreads out through the water. Yet no matter how long we watch, we will never see that drop of ink spontaneously reassemble itself.

Physicists discuss the degree of order of a system using the concept of **entropy**, which is a measure of the number of different ways a system could be rearranged and still appear the same. A neatly ordered system (such as the drop of ink and the glass of clear water) has low entropy, whereas a disordered system (the glass of inky water, which can be stirred or turned and still look the same) has higher entropy. The **second law of thermodynamics** says that, left on its own, an isolated system will always move toward higher entropy—that is, from order toward disorder. This is common sense. As time goes by we are more likely to find a system in a state that is, well, more likely.

In light of this inescapable march toward disorder dictated by the second law, how can ordered structure emerge spontaneously? Creationist claims that the origin of life flies in the face of the second law of thermodynamics are about as common as reports of Elvis sightings in supermarket tabloids. Yet such claims make a crucial

mistake: They focus attention on one player while ignoring the rest of the game.

We have all seen what happens when we set a glass of ice water out on a hot, humid day (**Figure 21.18**). Water vapor from the surrounding air condenses into drops of liquid water on the surface of the cold glass. This is amazing: We have just watched as ordered structure (drops of water) spontaneously emerged from disorder (water vapor in the air). It is almost as if we saw the drop of ink reassemble itself. This simple, everyday event appears to violate the second law of thermodynamics… but it does not.

To understand why, we need to step back and look at more than just the drops of water on the glass. When the drops condensed, they released a small amount of thermal energy that slightly heated both the glass and the surrounding air. Heating something increases its entropy. The decrease in entropy due to the formation of the drops is more than made up for by the increase in the entropy of their surroundings. Ordered structure spontaneously emerged, but *overall*, disorder increased.

It is important to note that energy can be used to reduce entropy at one location while increasing entropy somewhere else. An air conditioner is a good example. Electric energy produced at a power plant is used to pump thermal energy from inside a home and dump that thermal energy outside. When an air conditioner is run in reverse to provide heating, it is referred to as a "heat pump," but "entropy pump" might be a more accurate description. As you sit in your armchair on a summer day, all you immediately notice is that when the air

civilizations in our galaxy alone! In this case, our nearest neighbors may be "only" 40 or 50 light-years away. If scientists in that civilization are listening to the universe with

…or as near as 40 light-years.

their own radio telescopes, hoping to answer the question of life in the universe for themselves, then as you read this page they may be puzzling over an episode of *I Love Lucy*.

It is fascinating to speculate about the implications of the Drake equation. If there are civilizations around every corner, cosmically speaking, then why have we not heard from them? Perhaps they are not interested in talking to the new kid on the block, or perhaps the fact that we know of no other civilizations simply means that there are none nearby.

If we did run across another technologically advanced civilization, what would it be like? Looking back at the conditioner comes on, the temperature drops. Entropy decreases inside your house. If you look at the system as a whole, on the other hand, including the heating of air by the outside coils and the entropy produced by the burning of coal or natural gas during the production of electricity, then you see that turning on the air conditioning causes an overall *increase* in entropy.

This idea is especially important when we consider the origin of life. A living thing, whether an amoeba or a human being, represents a huge local increase in order. However, no violation of the second law of thermodynamics is involved. In our everyday lives, the food we eat gives us the energy we need to stave off the relentless advance of entropy. Viewed even more broadly, the evolution of life on Earth was powered primarily by energy striking Earth in the form of sunlight. A local increase in order on Earth (such as you) is "paid for" in the end by the much greater decrease in order that accompanies thermonuclear fusion in the heart of the Sun. Order emerges in localized regions within a system, but the second law of thermodynamics is obeyed overall.

The unifying theme of this chapter and, in many ways, this entire book has been understanding the origin of structure. The answer we have found is clear. Whether we are discussing the freezing out of matter and the fundamental forces in the young universe; the gravitational collapse of stars, planets, and galaxies; the evolution of life; or water beading up on the outside of a cold glass—ordered structure *does* emerge spontaneously as an unavoidable consequence of the action of physical law.

FIGURE 21.18 On a humid day, water condenses into droplets on the surface of a cold glass. The second law of thermodynamics does not mean that ordered structure cannot spontaneously emerge.

Drake equation, it is highly unlikely that we have neighbors nearby unless civilizations typically live for many thousands or even millions of years. In this case, any civilization that we encountered would almost certainly have been around for much longer than we have. Having survived that

Any civilization we discover will almost certainly be advanced.

long, would its members have learned the value of peace, or would they have developed strategies for controlling pesky neighbors? Movie theaters and the science fiction shelves of libraries and bookstores are filled with amusing and thoughtful stories that explore what life in the universe might be like (**Figure 21.22**). For the moment let's set aside such speculation to look at the question as scientists. Given this fascinating question, how might we go about finding a real answer?

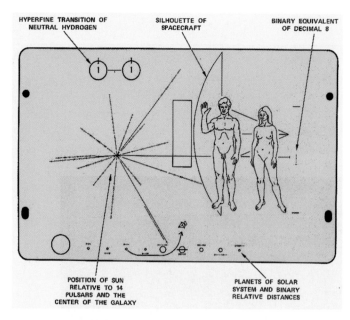

HYPERFINE TRANSITION OF NEUTRAL HYDROGEN

SILHOUETTE OF SPACECRAFT

BINARY EQUIVALENT OF DECIMAL 8

POSITION OF SUN RELATIVE TO 14 PULSARS AND THE CENTER OF THE GALAXY

PLANETS OF SOLAR SYSTEM AND BINARY RELATIVE DISTANCES

FIGURE 21.19 The plaque included with the *Pioneer 11* probe, which left the Solar System to travel through the millennia in interstellar space.

The Search for Signs of Intelligent Life in the Universe

One question we may be able to answer using observations of our own Solar System is, What is the likelihood of life originating at all? When the *Viking* landers failed to discover life on Mars in 1976, hopes faded for the presence of other life on worlds orbiting the Sun. Since that time, however, optimism has been renewed. A better understanding of the history of Mars indicates that at one time the planet was wet and warm, and many scientists believe that fossil life or even living microbes may be buried under the planet's surface.

Even more exciting are discoveries in the outer Solar System. There is strong evidence that life on Earth may have originated deep under the oceans, where geothermal vents (**Figure 21.23**) provided the thermal and chemical energy needed for life to gain a toehold. If we are correct that geological activity churns the floor of a deep ocean on Jupiter's moon Europa, for example, then this could be an excellent place to find life. One other example is all we need. If life arose independently *twice* in the same planetary system, then we would have no choice but to conclude that f_1 in the Drake equation is close to 1 and that life is ubiquitous throughout the universe.

Another way to search for intelligent life is to simply turn our ears to the sky and listen. Drake's original project of listening for radio signals from intelligent life around two nearby stars has grown over the years into a much more elaborate program that is referred to as the Search for Extraterrestrial Intelligence, or **SETI**. Scientists from around the world have thought carefully about what strategies might be useful for finding life in the universe. Most of these have focused on the idea of using radio telescopes to listen for signals from space that bear an unambiguous signature of an intelligent source. Some have listened intently at certain

SETI listens for radio signals from other civilizations.

"magic" frequencies, such as the frequency of the interstellar 21-cm line from hydrogen gas. The assumption behind this technique is that if a civilization wanted to be heard, they would tune their broadcasts to a channel that astronomers throughout the galaxy should be listening to. More recent searches have made use of advances in technology to record as broad a range of radio signals from space as possible, and then use computers to search these databases for types of regularity in the signals that might suggest that they are intelligent in origin.

Unlike much astronomical research, SETI is funded mostly through private rather than governmental means, and SETI researchers have been very ingenious at finding ways to accomplish as much as possible with limited resources. One especially clever idea, coming out of the SETI Institute in Mountain View, California, uses the underutilized computing power of thousands of personal computers around the world to analyze the institute's data. SETI screen saver programs installed on worldwide desktops download radio observations from the SETI Institute over the Internet, analyze these data while the computers' owners are off living their lives, and then report the results of their searches back to the institute. It is fun to think that the first sign of intelligent life in the universe might be found by a computer sitting on a table in the corner of your living room!

A number of SETI projects are on the drawing board. One is called the Allen Telescope Array (ATA), named after the cofounder of Microsoft, Paul Allen, who provided much of the initial financing for the project. The ATA (see **Figure 21.24**) will consist of a "farm" of hundreds of small, inexpensive radio dishes like those used to capture sig-

FIGURE 21.20 The message we beamed toward the star cluster M13. A reply may be forthcoming in 48,000 years.

nals from orbiting communication satellites. The ATA, a joint venture between the SETI Institute and the University of California, will use sensitive modern receiver technology to search the sky 24 hours a day, seven days a week, for signs of intelligent life. Just as your brain can sort out sounds coming from different directions, this array of radio telescopes will be able to determine the direction a signal is coming from, allowing it to listen to many stars at the same time. Over several years' time the ATA is expected to survey as many as a million stars, hoping to find a civilization that has sent a signal in our direction. If reality is anything like the more optimistic of the assumptions we used in evaluating the Drake equation, this project will stand a good chance of success.

If SETI finds even *one* nearby civilization in our galaxy, then the message will be clear. If we find a *second* technological civilization in our own small corner of the universe, then it means that the universe as a whole must literally be teeming with intelligent life. SETI may not be in the mainstream of astronomy, and the likelihood of its success may be difficult to predict, but its potential payoff is enormous. Few discoveries would do more to change our understanding of ourselves than certain knowledge that we are not alone.

21.6 The Future, Near and Far

We have used our understanding of physics and cosmology to look back through time and watch as structure formed throughout the universe. Now, as our journey nears its end, we look toward the future and contemplate the fate that awaits Earth, humanity, and the universe as a whole. Beginning close to home, around 5 billion years from now, the Sun will end its long period of relative stability. Shedding its identity as the passive, benevolent star that has nurtured life on Earth for nearly 4 billion years, the Sun will expand to become a red giant and later an asymptotic giant branch (AGB) star, swelling to hundreds of times its present size. The giant planets, orbiting outside the extended red giant atmosphere, should survive the Sun's cranky old age in some form. Even so, they will suffer the blistering radiation from a Sun grown thousands of times more luminous than it is today.

The terrestrial planets will not be so lucky. Some and perhaps all of the worlds of the inner Solar System will be engulfed by the expanding Sun. Just as an artificial satellite is slowed by drag in Earth's tenuous outer atmosphere and eventually falls to the ground in a dazzling streak of white-hot light, so too will a terrestrial planet caught in the Sun's atmosphere be consumed by the burgeoning star. If this is Earth's fate, our home world will leave no trace other than

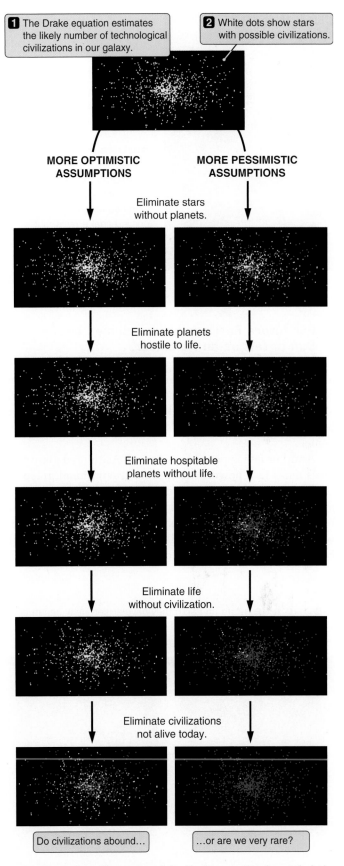

FIGURE 21.21 Simulation of the Drake equation for optimistic and pessimistic assumptions about the factors affecting the prevalence of intelligent life in the universe.

FIGURE 21.22 The classic 1951 film *The Day the Earth Stood Still* portrayed intelligent extraterrestrials who had no interest in our internal affairs, but promised destruction of Earth if we carried our violent ways into space.

Far future Earth will be consumed by the Sun or left as an icy cinder.

---

a slight increase in the amount of massive elements in the Sun's atmosphere. As the Sun loses more and more of its atmosphere in an AGB wind, our atoms may be expelled back into the reaches of interstellar space from which they came.

Another fate is possible, however. As the red giant Sun loses mass in a powerful wind, its gravitational grasp on the planets will weaken, and the orbits of both the inner and outer planets will spiral outward. If Earth moves out far enough, it may survive as a seared cinder, orbiting the white dwarf that the Sun will become. Barely larger than Earth and with its nuclear fuel exhausted, the white dwarf

Sun will slowly cool, eventually becoming a cold, inert sphere of degenerate carbon, orbited by what remains of its retinue of planets. The ultimate outcome for our Earth —consumed in the heart of the Sun or left behind as a frigid burned rock orbiting a long-dead white dwarf—is not yet certain. In either case Earth's status as a garden spot will be at an end.

The Fate of Life on Earth

Life on our planet will not survive long enough to witness the Sun's departure from the main sequence. Well before that cataclysmic event takes place, the luminosity of our heretofore benevolent star will begin to rise. As solar luminosity increases, so will temperatures on all the planets, including our own. Eventually Earth's temperatures will climb so high that all animal and plant life will perish. Models of the Sun's evolution are still too imprecise to predict with certainty when that fatal event will occur, but the end of all terrestrial life may be only 1 or 2 billion years away. That is, of course, a comfortably distant time from now, but it is well short of the Sun's departure from the main sequence.

Yet it is far from certain that the descendants of today's humanity will even be around a billion years from now. Some of the threats that await us come from beyond Earth. For the remainder of the Sun's life, the terrestrial planets, including Earth, will continue to be bombarded by asteroids and comets. Perhaps a hundred or more of these impacts will involve kilometer-sized objects, capable of causing the kind of devastation that eradicated the dinosaurs (and most other species) 65 million years ago. Although these events may create new surface scars, they will have little effect on

FIGURE 21.23 Life on Earth may have arisen near ocean geothermal vents like the one at left. Similar environments might exist elsewhere in the Solar System. Life around vents (right) is powered by geothermal rather than solar energy.

FIGURE 21.24 When complete, the Allen Telescope Array, seen here in an artist's view, will listen for evidence of intelligent life from as many as a million stellar systems.

the integrity of Earth itself. Earth's geological record is filled with such events, and each time they happen, life manages to recover and reorganize.

It seems likely, then, that some form of life will survive to see the Sun begin its march toward becoming a red giant. On the other hand, individual species do not necessarily fare so well when faced with cosmic cataclysm. If the descendants of humankind survive, it will be because we chose to become players in the game by changing the odds of such planetwide biological upheavals. We are rapidly developing technology that could allow us to detect most threatening asteroids and modify their orbits well before they can strike Earth. Comets are more difficult to guard against because long-period comets appear from the outer Solar System with little warning. To offer protection, defense capabilities would have to be in place, ready to be

To survive, humanity must learn to manage the threat of impacts.

used on very short notice. We have been slow to take such threats seriously. Although impacts from kilometer-sized objects are infrequent, objects a few dozen meters in size, carrying the punch of a several-megaton bomb, strike Earth about once every 100 years. Perhaps an explosion like the 1908 Tunguska blast occurring over New York or Paris would be enough to convince us that such precautions are worthwhile.

We might protect ourselves from the fate of the dinosaurs, but in the long run the descendants of humanity will either leave this world or die out. Planetary systems surround other stars, and all that we know tells us that many

other Earth-like planets should exist throughout our galaxy. Colonizing other planets is currently the stuff of science fiction, but if our descendants are ultimately to survive the death of our home planet, off-Earth colonization must become science fact at some point in the future.

But though humankind may soon be capable of protecting Earth from life-threatening comet and asteroid impacts, in other ways we are our own worst enemy. We are poison-

Humanity is the worst threat to its own survival.

ing the atmosphere, the water, and the land that form the habitat for all terrestrial life. As our population grows unchecked, we are occupying more and more of Earth's land and consuming more and more of its resources, while sending thousands of species of plants and animals to their extinction each year. At the same time, human activities are dramatically affecting the balances of atmospheric gases. The climate and ecosystem of Earth constitute a finely balanced, complex system capable of exhibiting chaotic behavior. The fossil record shows that Earth has undergone sudden and dramatic climatic changes in response to even minor perturbations. Such drastic changes in the overall balance of nature would certainly have consequences for our own survival. When politics is added to the mix, even more immediate dangers await. For the first time in human history, we possess the means to unleash nuclear or biological disasters that could threaten the very survival of our species. In the end, the fate of humanity will depend more than anything on whether we accept stewardship of ourselves and of our planet.

The Future of Our Expanding Universe

So much for our planet. What about the universe itself? As we have seen, if the mass of the universe is large enough and the cosmological constant small enough, gravity will win in the end. Hubble expansion will eventually reverse, and the universe will collapse back into a state resembling that of the young, hot universe. Galaxies, stars, planets, molecules, atoms, and subatomic particles—all might cease to exist as matter is replaced by pure energy. Perhaps from such a state, a new universe would emerge. At the dawn of the 21st century, however, few cosmologists see such a "Big Crunch" in the future of the universe. It appears that our universe will expand forever, perhaps even at an ever-accelerating pace. Does this mean the universe will go on without end? Yes, but not in the form that we see today.

In 1997 Fred Adams and Gregory Laughlin of the University of Michigan published their calculations of the great eras, past and future, in the history of the universe. During the first era—the first 500,000 years after the Big Bang and before

---

Other eras await our universe in the far future.

---

recombination—the universe was a swarm of radiation and elementary particles. Today we live during the second era, the Era of Stars, but this too will end. Some 100 trillion (10^{14}) years from now the last molecular cloud will collapse to form stars, and a mere 10 trillion years later the least massive of these stars will evolve to form white dwarfs.

Following the Era of Stars, most of the normal matter in the universe will be locked up in degenerate stellar objects: brown dwarfs, white dwarfs, and neutron stars. During this Era of Degeneracy, the occasional star will still flare up as ancient substellar brown dwarfs collide, merging to form low-mass stars that burn out in a short trillion years or so. However, the main source of energy during this era will come from the decay of protons and neutrons and the annihilation of particles of dark matter. Even these processes will eventually run out of fuel. In 10^{39} years white dwarfs will have been destroyed by proton decay, and neutron stars will have been destroyed by the beta decay of neutrons.

As the Era of Degeneracy comes to an end, the only significant concentrations of mass left will be black holes. These will range from black holes with the masses of single stars to greedy monsters that grew during the Era of Degeneracy to have masses as large as those of galaxy clusters. During the period that follows, the Era of Black Holes, these black holes will slowly evaporate into elementary particles through the emission of Hawking radiation. A black hole

with a mass of a few solar masses will evaporate into elementary particles in 10^{65} years, whereas galaxy-sized black holes will evaporate in around 10^{98} years. By the time the universe reaches an age of 10^{100} years, even the largest of the black holes will be gone. A universe vastly larger than ours will contain little but photons with colossal wavelengths, neutrinos, electrons, positrons, and other waste products of black hole evaporation. The Dark Era will have arrived as the universe continues to expand forever into the long, cold, dark night of eternity. This may be the final victory of **entropy**—the **heat death** of the universe.

From our perspective, the universe of the far distant future sounds like an extremely dull and lifeless place—but will that necessarily be so? Imagine for a moment that in-

---

We cannot predict what structure might arise in the far future.

---

telligent life evolved amid the swarm of free quarks and gluons that filled the universe immediately after inflation. Such organisms would have been far smaller than today's atoms and perhaps would have lived out their lives in 10^{-40} second or so. To such organisms the universe—all three meters of it—would have seemed incomprehensibly vast. These creatures and their entire civilization would have had to evolve, live, and die out in a millionth of a trillionth of a trillionth of the time that it takes for a single synapse in our brains to fire. If such creatures ever calculated the conditions in *today's* universe, they would have recoiled in horror. They would have imagined a frozen time when the temperature of the universe was only a thousandth of a trillionth of a trillionth of what they knew—a time when most matter had ceased to exist altogether, and the tiny fraction that remained was spread out over a volume of space 10^{75} times greater than that of their universe. In short, such creatures would have looked forward and seen *our* universe as the frozen, desolate future. It seems doubtful they could have foreseen the existence of stars, galaxies, planets, and intelligent creatures for whom a single thought took longer than a billion trillion times the entire history of the universe they knew.

Now turn the tables, and think *forward* to a time when the universe is 10^{50} times older than it is today. Who can say that there will not be life then as well? Perhaps they will be organisms of magnetic fields and tenuous electron plasmas, spread across countless trillions of light-years of space, whose lives unfold over untold eons of time. For such organisms, if they ever exist, it will be *we* who are the impossible creatures, alive for the briefest of instants, still immersed in the momentary fireball of the Big Bang.

Summary

- Galaxies reside in groups, clusters, and larger structures, all of which formed after galaxies did.

- Structure formed in the universe through the gravitational collapse of cold dark matter inhomogeneities arising from the early universe.

- Observed galaxies come from complex mergers, with the visible gas in galaxies cooling and falling inward to form the visible stars, which are surrounded by a dark matter halo.

- Understanding the very earliest moments in the universe requires that we also understand how the four fundamental forces of nature were all unified into one basic phenomenon at early times.

- Life may be an inevitable consequence of the formation of a single molecule that evolves through natural selection to more complex structures.

- Other advanced civilizations may be as close as 40 light-years or as far away as 30 million light-years, and we are searching for signals from such civilizations using SETI.

Seeing the Forest through the Trees

As our journey nears its end, we have reached high ground, a vantage point from which we can look back and survey the terrain we have covered. From this perspective it becomes clear that from the beginning, our journey has been guided by a single underlying quest: to understand how we came to be. Whether we are talking about the formation of galaxies, the origin and evolution of stars, the history of our Solar System, the geology of our own world, the evolution of life, the changes that shaped the universe itself during its earliest moments, or the eras of the far future, 21st century astronomy (the science, as well as the book) is organized around the desire to better understand the origin of structure in the universe.

The past century has seen remarkable strides toward this goal, and along the way a pattern has emerged. Look back on the quotation opening Chapter 16. What statement remains true in all circumstances? "And this, too, shall pass away." Structure is ephemeral. The universe is not about destination. The universe is a place of process and change. The things that "matter" are the things that happen along the way!

When you see your breath fog up on a cold winter day, think about elementary particles freezing out of the early universe, giving rise not only to matter but to the fundamental forces of nature. As you watch the clouds build before a thunderstorm, think of galaxies coalescing within halos of dark matter. Glance at a crystalline snowflake as it lands on the sleeve of your coat, and in your mind's eye see a star that condensed out of clouds of interstellar gas and dust. The origin of structure is everywhere around us and is no less remarkable for its familiarity.

As it is for galaxies and stars and planets, so too it is for ourselves. A recurring theme on our journey of discovery has been the systematic dismantling of conceptual walls that in our minds separated us from the larger universe. Humanity does not stand outside the processes that shape the universe. We are instead one more variety of the structure to which the universe has given birth—a way station on a long road of evolving structure stretching back 14 billion years. Few single words are capable of eliciting as much emotional reaction from some people as the word *evolution*. Yet the public controversy surrounding evolution cannot change the scientific standing of this theory as one of the best tested and most successful theories in all of science. As stated by Stephen J. Gould, "The theory of evolution is in about as much trouble as the theory that Earth revolves around the Sun." By any reasonable standards of scientific knowledge, evolution is a fact.

Opponents of evolutionary theory often act as if evolution were a tiny piece of science that can simply be cast aside without doing violence to the rest. Having traveled the journey of *21st Century Astronomy,* you should see clearly how absurd such a claim is. Modern astronomy would simply cease to be were it stripped of our understanding of the origin and evolution of galaxies, stars, planets, and every other component of the universe that we observe. Everything we see says that the universe formed 14 billion years ago, not 6,000. Cosmology is the ultimate evolutionary science—the science of the origin and evolution of the universe itself. During the 20th century geology became the science of the evolution of the surface of Earth, while planetary science applied our understanding of terrestrial geology to understanding the evolution of other planets and their moons. The most fundamental questions in physics concern the origin and evolution of physical laws. So too is the case in modern biology, which simply makes no sense until it is organized around the theme of evolution by natural selection.

Evolution is anything but a scientific appendix that can be harmlessly removed. It is the backbone and central nervous system of modern science. In *Darwin's Dangerous Idea* (1995) Daniel Dennett speaks of the concept of evolution as "universal acid: it eats through just about every traditional concept, and leaves in its wake a revolutionized worldview." As the 21st century gets under way, evolution of structure is *the* unifying theme that ties the breadth of modern science together into a beautiful, powerful, comprehensive whole. If we tried to pull the thread of evolution out of this tapestry, the whole cloth would unravel before our eyes.

Key Terms

galaxy group, p. 608
Local Group, p. 608
galaxy cluster, p. 608
supercluster, p. 608
large-scale structure, p. 609
hierarchical clustering, p. 611
cold dark matter, p. 612
photino, p. 613
hot dark matter, p. 613
weak nuclear force, p. 621
quantum electrodynamics (QED) , p. 621
electroweak theory, p. 622
quantum chromodynamics (QCD) , p. 622
quark, p. 622
gluon, p. 622
standard model, p. 622
pair creation, p. 622
grand unified theory (GUT) , p. 624
Planck era, p. 624
theory of everything (TOE) , p. 624
superstring theory, p. 625
mutation, p. 628
heredity, p. 628
Drake equation, p. 630
entropy, p. 632
second law of thermodynamics, p. 632
heat death, p. 638

Student Questions

THINKING ABOUT THE CONCEPTS

1. Of the four fundamental forces in nature, which does not depend on electric charge?

2. Suppose you could view the early universe at a time when galaxies were first forming. How would it be different from the universe we see today?

3. What are the basic differences between a grand unified theory (GUT) and the theory of everything (TOE)?

4. As clumps containing cold dark matter and normal matter collapse, they heat up. When a clump collapses to about half its maximum size, the increased thermal motion of particles tends to inhibit further collapse. Whereas normal matter can overcome this effect and continue to collapse, dark matter cannot. Explain the reason for this difference.

5. How do astronomers use the following to measure the amount of dark matter contained in a cluster of galaxies?
   a. Motions of individual members of the cluster.
   b. Extremely hot gas that fills the intergalactic space within the cluster.
   c. Gravitational lensing by the cluster.

6. Imagine that there are galaxies in the universe composed mostly of dark matter with relatively few stars or other luminous normal matter. If this were true, how might we learn of the existence of such galaxies?

7. Why is dark matter so essential to the galaxy formation process?

8. If we should eventually find life on Europa, what would this tell you about the probability of finding life on worlds around other stars?

9. A few scientists believe we may be the only advanced life in the galaxy today. If this were indeed the case, which factors in the Drake equation would have to be extremely small?

10. The second law of thermodynamics says that the entropy (a measure of disorder) of the universe is always increasing. Yet living organisms exist by creating order from disorder. Why does this not violate the second law of thermodynamics?

APPLYING THE CONCEPTS

11. The proton and antiproton each have the same mass, $m_p = 1.67 \times 10^{-27}$ kg. What is the energy (in joules) of each of the two gamma rays created in a proton–antiproton annihilation?

12. Suppose you brought together a gram of ordinary matter hydrogen atoms (each composed of a proton and an electron) and a gram of antimatter hydrogen atoms (each composed of an antiproton and a positron). Keeping in mind that two grams is less than the mass of a dime:
   a. Calculate how much energy (J) would be released as the ordinary matter and antimatter hydrogen atoms annihilated one another.

b. Compare this with the energy released by a one-megaton hydrogen bomb (1.6×10^{14} J).

13. Excursions 21.1 ("Forever in a Day") takes events spread out over enormous intervals of time and compresses them into the more comprehensible interval of a single 24-hour day. Make your own "Life in a Day" by compressing all the important events of your lifetime into a single day, starting with your birth at the stroke of midnight and continuing to the present at the end of the day.

14. Consider an organism Beta that, because of a genetic mutation, has a 5 percent greater probability of survival than its nonmutated form, Alpha. Alpha has only a 95 percent probability (p_r) of reproducing itself compared to Beta. After n generations, Alpha's population within the species would be $S_p=(p_r)^n$ compared to Beta's. Calculate Alpha's relative population after 100 generations. (You may need a scientific calculator or help from your instructor to evaluate the quantity 0.95^{100}.)

15. To fully appreciate the power of heredity, mutation, and natural selection, consider Alpha's relative population (from Question 14) after 5,000 generations in a case where Beta has a mere 0.1 percent survivability advantage over Alpha.

16. Make your own best guess at the six factors on the right side of the Drake equation to estimate the number of technologically advanced civilizations in our galaxy.

17. The Drake equation estimates the number of technologically advanced civilizations in our galaxy. Instead estimate the most optimistic and most pessimistic number of systems containing life (whether advanced or not) in our galaxy.

18. The kinetic energy of an object is given by $KE = mv^2/2$. A 100-meter radius spherical rock having a mass of 10^{10} kg strikes the Earth with a speed of 20 km/sec. How much energy does it deposit on Earth compared to a one-megaton hydrogen bomb (1.6×10^{14} J)?

StudySpace
wwnorton.com/astro21
provides a Study Plan for each chapter that includes a reading outline, animations, keyword flash cards, and gradebook-enabled multiple-choice quizzes. From StudySpace you can also access premium content in the ebook and SmartWork.

Sometimes the light's all shining on me
Other times I can barely see
Lately it occurs to me
What a long strange trip it's been.

ROBERT HUNTER (1938–)

Epilogue: We Are Stardust in Human Form

The Long and Winding Road

Go out at night and look at the stars, and feel the same sense of wonder and awe that our kind has always felt at the sight. Take it in, be amazed at the majestic canopy overhead, and let your imagination roam, just as our ancestors have done for thousands of years. As you do, drift back down the road we have followed. Reflect on all we have come to know about the universe, and on how much more magnificent the heavens are than our ancestors ever could have imagined. Our journey has been more than a description of the universe and what it contains. It has been a travelogue of the struggle and triumph of the human mind and spirit. Ever since humans first recognized that the patterns shaping our lives are echoed in the sky, we have searched for the threads connecting us to the cosmos and have sought to understand our place in it. We have the privilege of being among the first generations of humans to find those threads and to learn real answers to those age-old questions.

We have no way to count how many different stories have been told about the heavens, but we do know that for thousands of years most of those stories shared a common foundation. One cornerstone was the belief that Earth occupies a special place in the scheme of things. In our minds we were at the center of Creation, fixed and immovable, and all that we saw was present only to give meaning to our existence. A second cornerstone of this traditional worldview was the belief that the heavens are fundamentally different from Earth. To our ancestors our world was the realm of the ordinary and mundane—a terrestrial existence built from Aristotle's earth, wind, fire, and water. In contrast, the heavens had their own separate reality. There we saw a realm of gods and angels, of mysticism and magic, of the perfect and unchanging fifth element.

So it remained for thousands of years until, at the dawn of the Renaissance, a Polish monk dared to challenge the wisdom of the ages and to think the unthinkable. Reviving a notion that had been discarded long before by the Greeks, Nicolaus Copernicus allowed himself to imagine that perhaps it was the motion of Earth, rather than the motions of the Sun and stars, that shaped the passage of the days and the years. In so doing he not only conceptually dislodged Earth from its moorings, but he also broke the shackles that had for so long constrained the human mind. What began as a crack in the foundation of our preconceptions would in the end turn that old view of the world to rubble. In its place we would construct an edifice of knowledge that has given us dominion over our world and carried our thoughts to the edges of the universe.

Copernicus was one of a succession of people with great minds who could not rest without first picking at the loose threads of the ideas in which they had been raised to believe. Water runs into the cracks in a slab of granite and freezes, expanding and pushing the cracks open, exposing the flaws in the rock. As the seasons come and go, the imposing boulder stands no chance in the face of this persistent onslaught. In like fashion, the persistent questioning and probing of great minds would eventually shatter the reign of enforced ignorance and entrenched authority. We have met a few of these great minds on our journey; there were many others. They were of different nationalities, different upbringings, and different dispositions and beliefs. But their work shared a common theme: The answers to questions about the world come not from the pronouncements of authority or the prejudices taught in childhood, but from observing nature itself and thinking carefully, honestly, and openly about what we see.

In Newton's famous thought experiment, a ball fired from an imaginary cannon moves rapidly enough that it falls around the world in a circle. Such a "cannon" could not be built with the technology available in Newton's day. That would have to await the launch of *Sputnik* hundreds of years later. Yet although Newton could not make his cannon reality, he did not have to. All he had to do was look at the sky and watch as the Moon traced out its monthly path. Newton realized that the force holding the Moon in its orbit about Earth is the same force that gives us weight and guides the path of a ball thrown into the air. In fact, all Newton had to do to see his cannonball was look out across the English countryside—all of us ride Newton's cannonball as the force of the Sun's gravity holds Earth in its yearly orbit.

This thought experiment was an important step in Newton's work. Eventually it led him to invent calculus, which he used to calculate the motions of the planets, making predictions that were confirmed by Kepler's empirical laws. The philosophical and scientific significance of Newton's thought experiment goes far deeper, however: It signifies the final collapse of the barriers that humankind had placed between Earth and the heavens. With Newton's insight we came to see the heavens as part of the world around us—made of the same substance and shaped by the same physical laws. It is ironic that for knowledge to progress, we had to turn the early cornerstones of our thinking upside down. The modern foundation of our understanding of the universe, the cosmological principle, is the literal negation of those early beliefs: Not only is Earth *not* the center of the universe, but Earth occupies no special place in the universe *at all*. The heavens are *not* "the other," but are instead "the same." We can know the heavens by going into terrestrial laboratories and learning about the nature of matter and energy and radiation, then applying this knowledge to careful observations of a universe of stars, planets, and galaxies that is governed by physical law.

In our journey we have followed the trail of discovery that grew from this profound change in our understanding. We have watched as stars and planets formed, as stars lived out their lives and died, and as galaxies coalesced out of the primordial fireball of the Big Bang. We have followed our physics back to the very briefest of instants after the event that brought space and time into existence, and we have seen the hints of theories that may in our lifetimes carry us the final step. There is no doubting the wonder of what we have seen. Yet for you there is another aspect to our journey that, in a practical sense, should be even more significant. While learning *about* the universe, you have also come to better appreciate *how we know* those—things and in so doing have found a powerful, workable definition for what it means "to know."

According to Richard Paul, director of the Foundation for Critical Thinking in Dillon Beach, California, the three most common standards that people apply to knowledge even today are "it is true because I believe it," "it is true because we believe it," *and* "it is true because I want to believe it." Of course none of these has anything to do with what really *is* true. To learn about the universe and our world, we have had to set aside these notions, which blur the line between reality and fantasy, and replace them with a tough, unforgiving, and very different standard: "It is *provisionally* true because we have worked very hard to show that it is false, but so far have failed." This standard alone places reality itself in front of our parochial ideas and beliefs. It puts what *is* true ahead of what we would *like* to be true. Only by testing the falsifiable predictions of our theories about the world have we learned to push aside the comfortable notions that for so long prevented us from truly seeing our world and our universe.

We Are Stardust in Human Form

As witnessed by the obelisks of Stonehenge or the ruins of a Mayan pyramid, humans have always built temples to the stars. We still build temples to the stars today. They are seen as an array of radio telescopes spread across the high desert of New Mexico, or a city of domes atop the summit of a dormant Hawaiian volcano, or a satellite telescope carried into orbit and subsequently repaired by space shuttle astronauts, or a tiny rover crawling across the surface of Mars. These modern temples are the legacy of insights by Copernicus, Kepler, Galileo, Newton, Einstein, Hubble, and countless others.

The discoveries that pour forth from these modern-day temples stretch the mind and stir the imagination. We have walked on the Moon and have come to see the planets not as points of light in the sky, but as worlds as rich and complex as our own. We have looked at the remnants of stars that exploded long ago and have peered into eerie columns of glowing interstellar gas within which new stars are being born. We have gazed back in time at galaxies forming when the universe was young, and we have even learned to recognize the birth of the universe itself in the faint glow of the cosmic background radiation. As we contemplate those wonders, the words from act I, scene V, of Shakespeare's *Hamlet* seem almost frighteningly appropriate. As Hamlet faces the challenges and revelations brought by the ghost of his father, Horatio cries out:

"O day and night, but this is wondrous strange!"

To this comes Hamlet's immortal reply:

And therefore as a stranger give it welcome.
There are more things in heaven and earth, Horatio,
Than are dreamt of in your philosophy.
There are more things in heaven and earth, Horatio, than
are dreamt of in your philosophy

Here is a message to shout back through the ages. At the start of our journey we asked the most basic question about what we see in the sky—"How big is it?"—and the answers were enough to expose the comedy of humanity's ancient conceits. Traveling at the speed of a modern jetliner, it would take us over 5 million years to cross the distance to even the nearest star beyond our Sun. Even so, we live in a galaxy containing hundreds of billions of such stars, which are themselves outnumbered by other galaxies filling a universe that may stretch on forever. Using the speed of light as our yardstick, we have come to realize that Earth—the world of our birth and the stage on which all of human history has been played—is to the expanse of the observable universe as a single snap of our fingers is to the aeons that have transpired since time itself came into existence, roughly 14 billion years ago.

As we stare at the images that have come to symbolize 21st century astronomy and consider what they show, it is easy to understand why some people recoil from these insights. "Wouldn't it be nice," they say, "if we could just go back to imagining that Earth is only 6,000 years old, and that humanity occupies a special place at the pinnacle of Creation?" Indeed, if this were where our story ended—with the fact of our seeming insignificance in the universe—we might *all* long for an excuse to retreat into ignorance. Fortunately, this is not where our story ends. Rather, this is where our true journey of discovery begins. Although modern science may have shattered our egotistical views about our exalted place in the scheme of things, it has also offered us a wonderful new appreciation and understanding of ourselves to fill that void.

When we look at distant galaxies, the light we see is produced by stars like our Sun. Each of those stars formed when a cloud of interstellar gas and dust collapsed under the same force of gravity that guided Newton's cannonball. As each of those clouds collapsed, it spun faster and faster, obeying the same laws of motion that accelerate the spin of an Olympic skater as she pulls her arms and leg ever more tightly to her body. This spin prevented those clouds from collapsing directly into stars, forcing them instead to settle into flat rotating disks. We see such disks today when we look at the newest generation of stars. Inside those disks, grains of dust stick together to make larger grains, which stick together to make still larger grains—the beginning

of a bottom-up process that culminates with planetesimals crashing together to make worlds. Our Earth is one such world. We have come to view our Sun, Earth, and Solar System as products of natural processes still going on around us today. As we watch new generations of stars form and search for the planets that surround them, we are witnessing a replay of the birth of our own world, 4.6 billion years ago. We have found our roots in the stars.

Go out at sunset on an evening when a waxing crescent Moon hangs low above the western horizon, and several planets stretch out across the darkening sky. The plane of the ecliptic is there in front of you, and your mind's eye might even envision the flat, rotating accretion disk from which our Solar System formed. Once you realize what you are looking at, the cradle of our world and ourselves hangs there in the night sky for all to see.

Looked at in this way, the sunset takes on a whole new significance. It will never be the same. Yet even the majesty of the planets spread across the sky fails to capture the intimacy of our connection with the universe. In the most basic sense, the question "What are we?" is easily answered. Humans and all other terrestrial life are an organized assemblage of various organic molecules, most of which are very complex. Counting the numbers of atoms in our bodies, we are approximately 60 percent hydrogen, 26 percent oxygen, 11 percent carbon, and 2 percent nitrogen, with a small fraction of metals and other heavy elements mixed in. Over the course of our journey we have witnessed the history of those atoms. A very long time ago—roughly 14 billion years—something wonderful happened. The universe came into being, and time began. From an infinitesimally small volume of concentrated energy, the universe expanded, growing in size at the speed of light. Within the first few minutes, particles of solid matter, including protons and electrons, condensed out of this dense, primordial ball of energy. Nuclear reactions caused some of the protons to fuse into other light nuclei. Several hundred thousand years later, when the universe had cooled to a temperature of a few thousand kelvins, those nuclei combined with electrons to form atoms. Of those atoms, roughly 90 percent were hydrogen atoms and 10 percent were helium atoms. There were traces of lithium, beryllium, and boron as well, but that was all that existed in the way of normal luminous matter as the universe expanded past the threshold of recombination.

The hydrogen atoms in our bodies date back to this early time in the history of the universe, but what of the rest? Having taken our journey of discovery, you know the answers. As the universe emerged from the Big Bang, clumps of dark matter began to collapse under the force of gravity, pulling normal matter along with it. Within these collapsing protogalaxies the first generations of stars formed—nuclear furnaces powered by the fusion of those original hydrogen atoms into increasingly massive elements. Carbon, oxygen, silicon, sulfur—elements all the way up to iron and nickel were formed in those stellar infernos. As those first generations of stars ended their lives, they blasted this nuclear ash back into the reaches of interstellar space. Nucleosynthesis did not end with fusion, however. In the extreme environments of supernovae, free neutrons were captured by the products of fusion, building more massive elements still. Atoms of copper, zinc, tin, silver, and gold—all the way up to the most massive naturally occurring element, uranium—were formed and expelled into space. Here were the chemical elements to fill the periodic table and to build the compounds of life. As early protogalaxies merged to form early galaxies, more generations of stars continued to enrich the universe with the fruits of their alchemy. As early galaxies settled into the well-ordered ellipticals and spirals of today's universe, still more generations of stars came and went, adding to the chemical richness of the universe. When our Solar System appeared on the scene 4.6 billion years ago, it formed from interstellar material that carried the chemical building blocks of planets and of life—atoms produced both in the Big Bang and in the hearts of generations of stars that had lived and died during the 9 billion years that had transpired since the universe began. "What are we, and how did we get here?" We are stardust in human form.

The Future Arrives Every Day

While traveling the highways and back roads of 21st century astronomy, we have come to see our world and ourselves in a very different light. Even so, nothing we have seen has changed the most basic circumstances of our day-to-day existence. Earth remains our world, our home. The hopes and dreams we humans feel are no less real today than they were a thousand years ago. In 1968 Stanley Kubrick and Arthur C. Clarke collaborated to make the film *2001: A Space Odyssey*. This provocative piece of speculative fiction captured the imagination of a generation, and the year 2001 came to signify the future. That future is now here: yet little of our modern-day life is recognizable in those cinematic prognostications dating from only three decades ago. It is ironic that while we can forecast the future of the Sun with great accuracy and can even calculate the fate of the universe itself, our vision of our own future is so much less certain.

These words were first drafted on the afternoon of December 31, 2000. At that moment in the lives of your authors, roughly half the surface of Earth remained in the 20th century, while the other half of the world had witnessed the beginning not only of a new century but also of a new millennium. As the dividing line between the two swept across Europe and out into the Atlantic Ocean, it was just another moment in the long dance of Earth as it spins on its axis and falls around the Sun. The Sun took no notice of the moment as it continued its orbit around the center of the Milky Way Galaxy, which itself is but a speck in an expanding universe. Yet while that December afternoon had no objective physical significance, it was loaded with symbolic meaning for our species. The year of Kubrick and Clarke's mind-bending tale was at hand. Though the details of their vision have turned out to be incorrect, there is no question that we live at an amazing moment in history—a moment when science has for the first time allowed us to truly see the universe beyond ourselves and offers the promise of showing us the universe within ourselves, as well.

There is no denying the fact that we live in an evolving universe—a place of ongoing and unending change that continues to shape humanity as certainly as it shapes the cosmos itself. The grim prospects mentioned in the closing sections of Chapter 21 are real. Saying that they do not exist or choosing to push them from our minds will not make them go away. Yet at the same time modern medicine, food production, transportation, and a thousand other technologies that push back the ancient scourges of humanity are equally real. Differences among people remain, but modern communication offers the hope of spreading understanding. Meanwhile, a picture of Earthrise above the lunar horizon taken by the *Apollo 8* astronauts (**Figure E.1**) hangs forever as part of the human experience and perspective, showing us as nothing else could that Earth is a tiny fragile island to be cherished. Whether we like it or not, humanity is a single, interrelated, interdependent global village that will face the future together—or not at all.

At the dawn of the 21st century, science has shown us the wonders of the universe and at the same time has given us the knowledge and power to shape our world and choose our future. The future of humanity may depend entirely on how well we treat Earth and ourselves over the few decades and centuries ahead. If 21st century astronomy does nothing else, it forces us to change our perspective. As we study the laws that govern the workings of atoms, we learn something about the conditions of our own lives: The future is not yet written. As we use our telescopes to stare at galaxies 10 billion light-years away, collecting light from stars that died billions of years before our Sun was even born, manifest destiny seems a pretty silly concept. There are no guarantees that things will work out in the end for one tiny world or for the species to which it gave birth. If we choose to destroy our world, either through direct action or simple neglect, so be it. The universe as a whole will carry on in sublime indifference to our fate.

This is not, however, a message of despair. Instead it is a message of hope, responsibility, and maturity. We have the power to make our Earth a paradise or to leave our children's children to cope with a world choking from our shortsighted excess. The choices are ours, whether we want them or not. Borrowing from the book of Genesis, we have truly tasted of the tree of knowledge. As we stand at the beginning of the 21st century, we face a future filled with choices: but one choice we are not allowed is refusing to acknowledge responsibility for our own destiny.

As you read these words your authors' moment of introspection at the threshold between the second and third millennia has long passed. Even so, the moment when you read these words is not so different from that New Year's Eve of the year A.D. 2000. Although the symbolism of your moment may not be as palpable as that of an instant when the world passes a major milestone in humanity's accounting of events, your moment is a milestone nonetheless. It is one of a stream of moments that define your life as the possibilities of the future cross, inexorably, into the unchangeable reality of the past. That moment is there for you to use as you will. The course of the future is yours to shape.

With that thought, we come to the end of our journey. We, the authors, hope this journey has helped to open your eyes to the wonders of the world and the universe around you. Even more, we hope this journey has given you pause to reflect on who we are as humans and on our place in the larger reality in which we find ourselves. Finally, we hope this journey has changed the shape of the way you think—not only about the sights you see in the night sky, but also about the events of your daily life. If any or all of these hopes are fulfilled, then the journey will have been worth taking—worthy of your time and thought and of ours.

FIGURE E.1 This image of Earth taken from lunar orbit by the *Apollo 8* astronauts forever changed our understanding of our Earth and ourselves.

Mathematical Tools

Working with Proportionalities

Most of the mathematics in *21st Century Astronomy* involves proportionalities—statements about the way that one physical quantity changes when another quantity changes. Here we offer a practical guide to working with proportionalities.

To use a statement of proportionality to compare two objects, begin by turning the proportionality into a ratio. For example, the price of a bag of apples is proportional to the weight of the bag:

$$\text{Price} \propto \text{Weight}.$$

Here the symbol "$\propto$" is pronounced "is proportional to." What this means is that the ratio of the prices of two bags of apples is equal to the ratio of the weights of the two bags:

$$\text{Price} \propto \text{Weight} \quad means \quad \frac{\text{Price of A}}{\text{Price of B}} = \frac{\text{Weight of A}}{\text{Weight of B}}.$$

Work a specific example: Suppose bag A weighs two pounds, while bag B weighs one pound. That of course means that bag A will cost twice as much as bag B. We can turn our proportionality into this equation:

$$\frac{\text{Price of A}}{\text{Price of B}} = \frac{\text{Weight of A}}{\text{Weight of B}} = \frac{2\ \text{lb}}{1\ \text{lb}} = 2.$$

So the price of bag A is two times the price of bag B.

Let's work another, more complicated, example. In Chapter 13 we discuss the relationship between the luminosity, brightness, and distance of stars. The luminosity of a star—the total energy the star radiates each second—is proportional to the star's brightness times the square of its distance:

$$\text{Luminosity} \propto \text{Brightness} \times \text{Distance}^2.$$

What this proportionality means is that if we have two stars—call them A and B—then

$$\frac{\text{Luminosity of A}}{\text{Luminosity of B}} = \frac{\text{Brightness of A}}{\text{Brightness of B}} \times \left(\frac{\text{Distance of A}}{\text{Distance of B}}\right)^2.$$

If we use the symbols L, b, and d to represent luminosity, brightness, and distance, respectively, this becomes

$$\frac{L_A}{L_B} = \frac{b_A}{b_B} \times \left(\frac{d_A}{d_B}\right)^2.$$

As an example, suppose Star A appears twice as bright in the sky as Star B, but Star A is located 10 times as far away as Star B. Compare the luminosities of the two stars. Because we know that

$$\text{Luminosity} \propto \text{Brightness} \times \text{Distance}^2$$

we write

$$\frac{\text{Luminosity of A}}{\text{Luminosity of A}} = \frac{\text{Brightness of A}}{\text{Brightness of B}} \times \left(\frac{\text{Distance of A}}{\text{Distance of B}}\right)^2$$

$$= \frac{2}{1} \times \left(\frac{10}{1}\right)^2 = 200.$$

Star A is 200 times as luminous as Star B.

A final note: In our original example we said that the price of a bag of apples is proportional to the weight of the bag, which allowed us to say that a two-pound bag of apples costs twice as much as a one-pound bag of apples. This is a statement about the way the world *works*. A two-pound bag of apples costs *more* than a one-pound bag, not *less* than a one-pound bag. To actually figure out how much a bag of apples will cost, we need another piece of information: the price per pound. The price per pound is an example of a *constant of proportionality*. Wrapped up in the price per pound is all sorts of information about the cost of growing apples, the cost of transporting them, how the market is at the moment, the profit the grocer needs to make, and so forth. In other words, this constant of proportionality is a statement about the way the world *is*.

Proportionalities help us understand how the world works and let us compare one object to another. Constants of proportionality allow us to calculate real values for things. In *21st Century Astronomy* it is usually the "understanding"—the proportionality itself—that we care about.

Scientific Notation

Astronomy is a science both of the very large and the very small. The mass of an electron, for example, is

$$0.00000000000000000000000000000009109 \text{ kg}$$

whereas the distance to a galaxy far, far away might be around

$$100,000,000,000,000,000,000,000,000 \text{ m}.$$

All it takes is a quick glance at these two numbers to see why astronomers, like most scientists, make heavy use of scientific notation to express numbers.

Powers of 10

Our number system is based on powers of 10. Going to the left of the decimal place,

$$10 = 10 \times 1,$$

$$100 = 10 \times 10 \times 1,$$

$$1,000 = 10 \times 10 \times 10 \times 1,$$

and so on. Going to the right of the decimal place.

$$0.1 = \frac{1}{10} \times 1,$$

$$0.01 = \frac{1}{10} \times \frac{1}{10} \times 1,$$

$$0.001 = \frac{1}{10} \times \frac{1}{10} \times \frac{1}{10} \times 1,$$

and so on for as long as we care to continue. In other words, each place to the right or left of the decimal place in a number represents a power of 10. For example, 1 million can be written

$$1 \text{ million} = 1,000,000$$
$$= 1 \times 10 \times 10 \times 10 \times 10 \times 10 \times 10.$$

That is to say, 1 million is 1 times six factors of 10. Scientific notation combines these factors of 10 in convenient short-

hand. Rather than writing out all six factors of 10, instead we combine them in easy shorthand using an exponent:

$$1 \text{ million} = 1 \times 10^6$$

which means 1 times six factors of 10.

Moving to the right of the decimal place, each step we take *removes* a power of 10 from the number. One millionth can be written

$$1 \text{ millionth} = 1 \times \frac{1}{10} \times \frac{1}{10} \times \frac{1}{10} \times \frac{1}{10} \times \frac{1}{10} \times \frac{1}{10}.$$

We are removing powers of 10, so we express this using a negative exponent. We write

$$\frac{1}{10} = 10^{-1}$$

and

$$1 \text{ millionth} = 1 \times 10^{-1} \times 10^{-1} \times 10^{-1} \times 10^{-1} \times 10^{-1} \times 10^{-1}$$

$$= 1 \times 10^{-6}$$

Returning to our earlier examples, the mass of an electron is 9.109×10^{-31} kg, whereas the distant galaxy is located 1×10^{26} away: these are *much* more convenient ways of writing this information. *Notice that the exponent in scientific notation gives you a feeling for the size of a number at a glance.* The exponent of 10 in the electron mass is –31, which tells us instantly that it is a very small number. The exponent of 10 in the distance to a remote galaxy, +26, tells us immediately that it is a very large number. This exponent is often called the *order of magnitude* of a number. When you see a number written in scientific notation while reading *21st Century Astronomy* (or elsewhere), just remember to look at the exponent to better understand what the number is telling you.

Scientific notation is also convenient because it makes multiplying and dividing numbers easier. Again, let's look at an example. Two billion times eight thousandths can be written

$$2,000,000,000 \times 0.008$$

but it is more convenient to write these two numbers using scientific notation as

$$(2 \times 10^9) \times (8 \times 10^{-3}).$$

This can be regrouped as

$$(2 \times 8) \times (10^9 \times 10^{-3})$$

The first part of the problem is just $2 \times 8 = 16$. The more interesting part of the problem is the multiplication in the right parentheses. The first number, 10^9, is just shorthand for $10 \times 10 \times 10 \ldots$ nine times. That is, it represents nine factors of 10. The second number stands for three factors of 1/10—or removing three factors of 10 if you prefer to think of it that way. All together, that makes $9 - 3 = 6$ factors of 10. In other words,

$$10^9 \times 10^{-3} = 10^{9-3} = 10^6.$$

Putting the problem together,

$$(2 \times 10^9) \times (8 \times 10^{-3}) = (2 \times 8) \times (10^9 \times 10^{-3})$$

$$= 16 \times 10^6.$$

By convention, when a number is written in scientific notation, only one digit is placed to the left of the decimal point. In this case, there are two. However, 16 is 1.6×10, so we can add this two factor of 10 to the exponent at right, making the final answer

$$1.6 \times 10^7.$$

Dividing is just the inverse of multiplication. Dividing by 10^3 means removing three factors of 10 from a number. Using the previous number.

$$(1.6 \times 10^7) \div (2 \times 10^3) = (1.6 \div 2) \times (10^7 \div 10^3)$$
$$= 0.8 \times 10^{7-3}$$
$$= 0.8 \times 10^4.$$

This time we have only a zero to the left of the decimal point. To get the number into proper form, say that $0.8 = 8 \times 10^{-1}$, giving

$$0.8 \times 10^4 = (8 \times 10^{-1}) \times 10^4 = 8 \times 10^3.$$

Adding and subtracting numbers in scientific notation is somewhat more difficult because each number has to be written as a value times the *same* power of 10 before they can be added or subtracted. However, almost all modern calculators have scientific notation built in. They keep up with the powers of 10 for you. If you do not have such a calculator you may want to buy one and learn to use it before tackling the mathematical problems in this book. There are more examples on our Web site that you can use to better learn how to work with scientific notation.

Significant Figures

In the previous example we actually broke some rules in the interest of explaining how powers of 10 are treated in scientific notation. The rules that we broke involve the *precision* of the numbers we are expressing. In everyday speech we might say, "The store is a kilometer away," by which we probably mean that the store is *roughly* a kilometer away. If it turned out to be 0.8 km, or 1.2 km, it is unlikely that the recipient of our directions would quibble. But when expressing quantities in science, it is extremely important to know not only of the value of a number, but also how precise that value is.

The most complete way to keep track of the precision of numbers is to actually write down the uncertainty in the number. For example, if we know the distance to the store (call it d) is between 0.8 km and 1.2 km, we can write

$$d = 1.0 \pm 0.2 \text{ km}$$

where the symbol "$\pm$" is pronounced "plus or minus." In this example, d is between $1.0 - 0.2 = 0.8$ km and $1.0 + 0.2 = 1.2$ km. This is an unambiguous statement about the limitations on our knowledge of the value of d: but carrying along the formal errors with every number we write would be cumbersome at best. Instead we keep track of the approximate precision of a number by using *significant figures*.

The convention for significant figures works like this: We assume the number we write has been rounded from a number that had one additional digit to the right of the decimal point. If we say that some quantity d, which might represent the distance to the store, is "1.", what we mean is that d is close to 1. It is likely not as small as 0., and it is likely not as large as 2. If we say instead

$$d = 1.0$$

then we mean that d is likely not 0.9 and is likely not 1.1. It is roughly 1.0 to the nearest tenth. The greater the number of significant figures, the more precisely the number is being specified. For example, 1.00000 is *not* the same number as 1.00. The first number, 1.00000, represents a value that is probably not as small as 0.99999 and is probably not as large as 1.00001. The second number, 1.00, represents a value that is probably not as small as 0.99 nor as large as 1.01.

When we carry out mathematical operations, significant figures are important. For example, $2.0 \times 1.6 = 3.2$. It does *not* equal 3.20000000000. The product of two numbers cannot be known to any greater accuracy than the numbers themselves! As a general rule, when we multiply and divide, the answer should have the same number of significant figures as the less precise of the numbers being multiplied or divided. In other words, $2.0 \times 1.602583475 = 3.2$. Because all we know is that the first factor is probably closer to 2.0 than to 1.9 or 2.1, all we know about the product is that it is between about 3.0 and 3.4. It is 3.2. It is not 3.205166950 (*even if that is the answer your calculator gives!*). The rest of the digits to the right of 3.2 just do not mean anything.

When we add and subtract, the rules are a bit different. If one number has a significant figure with a particular place value but another number does not, their sum or difference cannot have a significant figure in that place value. In other words,

1,045.

+1.34567

1,046.

The answer is 1,046., *not* 1,046.34567. Again, the extra digits to the right of the decimal place have no meaning because 1,045. is not known to that accuracy.

There is always a fly in the ointment, and this is no exception. What is the precision of the number 1,000,000? As it is written, the answer is unclear. Are all those zeros really significant, or are they placeholders? If we write the number in scientific notation, on the other hand, there is never a question. Instead of 1,000,000 we write 1.0×10^6 for a number that is known to the nearest hundred thousand or so: or we write 1.00000×10^6 for a number that is 1 million to the nearest 10.

So our earlier example would have been more correct had we said

$$(2.0 \times 10^9) \times (8.0 \times 10^{-3}) = 1.6 \times 10^7.$$

Algebra

There are many branches of mathematics. The branch that tells us about the relationships between quantities is called *algebra*. If you are reading this book, you have almost certainly taken an algebra class: but you are not alone if you feel a little review is in order. Basically, algebra begins by using symbols to represent quantities. For example, we might write the distance you travel in a day as d. As it stands, d has no value. It might be 10,000 miles. It might be 30 feet. It does, however, have *units*—in this case the units of distance.

The average speed at which you travel is equal to the distance you travel divided by the time you take. If we use the symbol v to represent your average speed and the symbol t to represent the time you take, then instead of writing out, "Your average speed is equal to the distance you travel divided by the time taken," we can write

$$v = \frac{d}{t}.$$

The meaning of this algebraic expression is exactly the same as the sentence quoted before it, but it is much more concise. As it stands, v, d, and t still have no specific values. There are no numbers assigned to them yet. However, this expression tells us what the relationship between those numbers will be when we *do* look at a specific example. For example, if

you go 500 km ($d = 500$ km) in 10 hours ($t = 10$ hours), this expression tells you that your average speed is

$$v = \frac{d}{t} = \frac{500 \text{ km}}{10 \text{ hours}} = 50 \text{ km/hour}.$$

Notice that the units in this expression act exactly like the numerical values. They are just multiplicative factors. When we say "500 km" what we really mean is "500 $\times$ kilometers." Likewise, 10 hours means "10 $\times$ hours." When we divide the two we find that the units of v are kilometers divided by hours, or km/hr (pronounced kilometers per hour).

We introduced algebra as shorthand for expressing relations between quantities, but it is far more powerful than that. Algebra provides rules for manipulating the symbols used to represent quantities. We begin with a bit of notation for *powers* and *roots*. When we talk about raising a quantity to a power, we mean multiplying the quantity by itself some number of times. For example, if S is a symbol for something (anything), then S^2 (pronounced "S squared" or "S to the second power") means $S \times S$, and S^3 (pronounced "S cubed" or "S to the third power") means $S \times S \times S$. Suppose S represents the length of the side of a square. The area of the square is given by

$$\text{Area} = S \times S = S^2.$$

If $S = 3$ m, then the area of the square is

$$S^2 = 3 \text{ m} \times 3 \text{ m} = 9 \text{ m}^2$$

(pronounced 9 square meters). It should be obvious why raising a quantity to the second power is called "squaring" the quantity. We could have done the same thing for the sides of a cube and found that the volume of the cube is

$$\text{Volume} = S \times S \times S = S^3$$

If $S = 3$ m, then the volume of the cube is

$$S^3 = 3 \text{ m} \times 3 \text{ m} \times 3 \text{ m} = 27 \text{ m}^3$$

(pronounced 27 cubic meters). Again, it is clear why raising a quantity to the third power is called "cubing" the quantity.

Roots are the reverse of this process. The square root of a quantity is the value that, when squared, gives the original quantity. The square root of 4 is 2, which means that $2 \times 2 = 4$. The square root of 9 is 3, which means that $3 \times 3 = 9$. Similarly, the cube root of a quantity is the value that, when cubed, gives the original quantity. The cube root of 8 is 2, which means that $2 \times 2 \times 2 = 8$. Roots are written with the symbol $\sqrt{\ }$. For example, we write

$$\sqrt{9} = 3$$

for the square root of 9 and

$$\sqrt[3]{8} = 2$$

for the cube root of 8. If the volume of a cube is $V = S^3$, we can also write

$$S = \sqrt[3]{V} = \sqrt[3]{S^3}.$$

Roots can also be written as powers. Powers and roots behave exactly like the exponents of 10 in our discussion of scientific notation. (They had better: The exponents used in scientific notation are just powers of 10.) For example, if a, n, and m are all algebraic quantities, then

$$a^n \times a^m = a^{n+m}, \quad \text{and} \quad \frac{a^n}{a^m} = a^{n-m}.$$

(To see if you understand all this, explain why the square root of a can also be written $a^{\frac{1}{2}}$ and the cubed root of a can be written $a^{\frac{1}{3}}$.)

Some of the rules of algebra are summarized next. These are really no more than the rules of arithmetic applied to the symbolic quantities of algebra. The important thing is this: So long as we apply the rules of algebra properly, then the relationships among symbols we arrive at through our algebraic manipulations remain true for the physical quantities those symbols represent.

Here we summarize a few algebraic rules and relationships. In this summary, a, b, c, n, m, r, x, and y are all algebraic quantities:

Associative rule:

$$a \times b \times c = (a \times b) \times c = a \times (b \times c)$$

Commutative rule:

$$a \times b = b \times a$$

Distributive rule:

$$a \times (b + c) = (a \times b) + (a \times c)$$

Cross multiplication:

$$\text{If } \frac{a}{b} = \frac{c}{d}, \text{ then } ad = bc.$$

Working with exponents:

$$\frac{1}{a^n} = a^{-n} \qquad a^n a^m = a^{n+m}$$

$$\frac{a^n}{a^m} = a^{n-m} \qquad (a^n)^m = a^{n \times m} \qquad \left(\frac{a}{b}\right)^n = \frac{a^n}{b^n}$$

Equation of a line with slope m and y-intercept b:

$$y = mx + b$$

Equation of a circle with radius r centered at $x = 0$, $y = 0$:

$$x^2 + y^2 = r^2$$

Angles and Distances

The farther away something is, the smaller it appears. This is common sense and everyday experience. In astronomy, where we seldom get to walk up to the object we are studying and measure it with a meterstick, our knowledge about the sizes of things usually depends on knowing the relationship between the size of an object, its distance, and the angle it covers in the sky.

The natural way to measure angles is using a unit called *radians*. As shown in **Figure A1.1(a)**, the size of an angle in radians is just the length of the arc subtending the angle divided by the radius of the circle. In the figure, the angle $x = S/r$ radians.

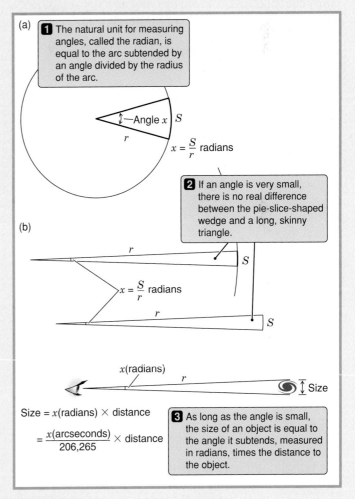

(a)

1 The natural unit for measuring angles, called the radian, is equal to the arc subtended by an angle divided by the radius of the arc.

—Angle x S
r
$x = \dfrac{S}{r}$ radians

2 If an angle is very small, there is no real difference between the pie-slice-shaped wedge and a long, skinny triangle.

(b)

r S
$x = \dfrac{S}{r}$ radians
r S

x(radians)
r Size

$\text{Size} = x(\text{radians}) \times \text{distance}$

$= \dfrac{x(\text{arcseconds})}{206{,}265} \times \text{distance}$

3 As long as the angle is small, the size of an object is equal to the angle it subtends, measured in radians, times the distance to the object.

FIGURE A1.1

Because the circumference of a circle is 2π times the radius, $C = 2\pi r$, a complete circle has an angular measure of $(2\pi r)/r = 2\pi$ radians. In more conventional angular measure, a complete circle is 360°: so we can say that

$$360° = 2\pi \text{ radians}$$

or that

$$1 \text{ radian} = \frac{360°}{2\pi} = 57.2958°$$

When talking about stars and galaxies, we often use seconds of arc to measure angles. A degree is broken into 60 minutes of arc, each of which is broken into 60 seconds of arc—so there are 3,600 seconds of arc in a degree. That means

$$3,600 \, \frac{\text{arcseconds}}{\text{degree}} \times 57.2958 \, \frac{\text{degrees}}{\text{radian}} = 206,625 \, \frac{\text{arcseconds}}{\text{radian}}$$

If the angle is small enough (which it usually is in astronomy), there is precious little difference between the pie slice just described and a long skinny triangle with a short side of length S (see **Figure A1.1(b)**). So, if we know the distance d to an object, and we can measure the angular size x of the object, then the size of the object is just

$$S = x \text{ (in radians)} \times d = \frac{x \text{ (in degrees)}}{57.2958 \text{ degrees/radian}} \times d$$

$$= \frac{x \text{ (in arcseconds)}}{206,265 \text{ arcseconds/radian}} \times d,$$

which is all we need to turn our knowledge of the angular size and the distance to an object into a measurement of the object's physical size. (Our Web site has plenty of examples of such calculations.)

Circles and Spheres

To round out our mathematical tools, here are a few useful formulas for circles and spheres. The circle or sphere in each case has a radius r.

$$\text{Circumference}_{\text{circle}} = 2\pi r$$

$$\text{Area}_{\text{circle}} = \pi r^2$$

$$\text{Area}_{\text{circle}} = 4\pi r^2$$

$$\text{Volume}_{\text{sphere}} = \frac{4}{3}\pi r^3$$

Physical Constants and Units

Fundamental Physical Constants

Constant	Symbol	Value
Speed of light in a vacuum	c	2.99792×10^8 m/s
Universal gravitational constant	G	6.673×10^{-11} N m^2/kg^2
Planck constant	h	6.62607×10^{-34} J s
Electric charge of electron or proton	e	1.60218×10^{-19} C
Boltzmann constant	k	1.38065×10^{-23} J/k
Stefan-Boltzmann constant	σ	5.67040×10^{-8} W/(m^2 k^4)
Mass of electron	m_e	9.10938×10^{-31} kg
Mass of proton	m_p	1.67262×10^{-27} kg

Source: National Institute of Standards and Technology (http://physics.nist.gov).

Unit Prefixes

Prefix[a]	Name	Factor[b]
n	nano-	10^{-9}
μ	micro-	10^{-6}
m	milli-	10^{-3}
k	kilo-	10^3
M	mega-	10^6
G	giga-	10^9
T	tera-	10^{12}

These prefixes ([a]), when appended to a unit, change the size of the unit by the factor ([b]) given. For example, 1 km (kilometer) is 10^3 meters.

Units and Values

Quantity	Fundamental Unit	Values
Length	meter (m)	Radius of Sun ($R_\odot$) = 6.96265×10^8 m astronomical unit (AU) = 1.49598×10^{11} m 1 astronomical = 149,598,000 km Light-year (ly) = 9.4605×10^{15} m 1 light-year = 6.324×10^4 AU 1 parsec (pc) = 3.261 ly = 3.0857×10^{16} m 1 m = 3.281 feet
Volume	meters^3 (m^3)	1 m^3 = 1,000 liters = 264.2 gallons
Mass	kilogram (kg)	1 kg = 1,000 g Mass of Earth ($M_\oplus$) = 5.9736×10^{24} kg Mass of Sun ($M_\odot$) = 1.9891×10^{30} kg
Time	seconds (s)	1 hour (h) = 60 minutes (min) = 3,600 s Solar day (noon to noon) = 86,400 s Sidereal day (Earth rotation period) = 86,164.1 s Tropical year (equinox to equinox) = 365.24219 days = 3.15569×10^7 s Sidereal year (Earth orbital period) = 365.25636 days = 3.15581×10^7 s
Speed	meters/second (m/s)	1 m/s = 2.236 miles/h 1 km/s = 1,000 m/s = 3,600 km/h $c = 3.00 \times 10^8$ m/s = 300,000 km/s
Acceleration	meters/second^2 (m/s^2)	Gravitational acceleration on Earth (g) = 9.78 m/s^2
Energy	joules (J)	1 J = 1 kg m^2/s^2 1 megaton = 4.19×10^{15} J
Power	watt (W)	1 W = 1 J/s Solar luminosity ($L_\odot$) = 3.827×10^{26} W
Force	newton (N)	1 N = 1 kg m/s^2 1 pound (lbs) = 4.448 N 1 N = 0.22481 pounds (lbs)
Pressure	newtons/meter^2 (N/m^2)	Atmospheric pressure at sea level = 1.013×10^5 N/m^2 = 1.013 bar
Temperature	kelvins (K)	Absolute zero = 0 K = $-273.15°$C = $-459.67°$F

Sources: National Space Science Data Center (2002); *Observer's Handbook 2002*, Rajiv Gupta (Royal Astronomical Society of Canada, 2001); National Institute of Standards and Technology (2002).

Periodic Table of the Elements

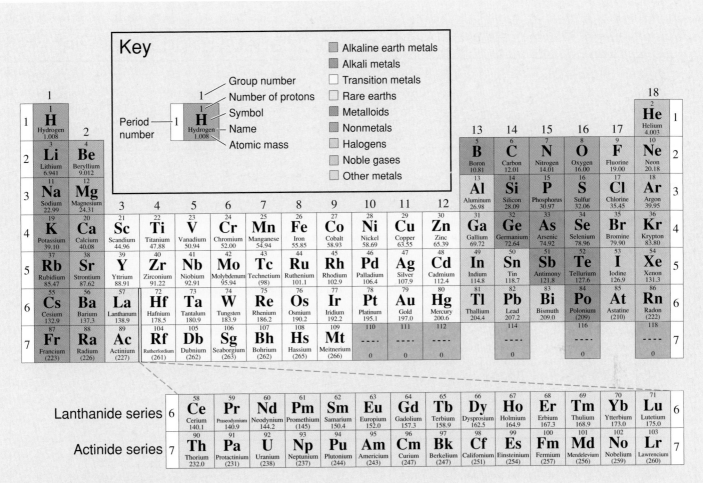

Key

- Group number
- Number of protons
- Period number
- Symbol
- Name
- Atomic mass

- Alkaline earth metals
- Alkali metals
- Transition metals
- Rare earths
- Metalloids
- Nonmetals
- Halogens
- Noble gases
- Other metals

Lanthanide series 6														
58 Ce Cerium 140.1	59 Pr Praseodymium 140.9	60 Nd Neodynium 144.2	61 Pm Promethium (145)	62 Sm Samarium 150.4	63 Eu Europium 152.0	64 Gd Gadolium 157.3	65 Tb Terbium 158.9	66 Dy Dysprosium 162.5	67 Ho Holmium 164.9	68 Er Erbium 167.3	69 Tm Thulium 168.9	70 Yb Ytterbium 173.0	71 Lu Lutetium 175.0	6

Actinide series 7														
90 Th Thorium 232.0	91 Pa Protactinium (231)	92 U Uranium (238)	93 Np Neptunium (237)	94 Pu Plutonium (244)	95 Am Americium (243)	96 Cm Curium (247)	97 Bk Berkelium (247)	98 Cf Californium (251)	99 Es Einsteinium (254)	100 Fm Fermium (257)	101 Md Mendelevium (256)	102 No Nobelium (259)	103 Lr Lawrencium (260)	7

Sources: Los Alamos National Laboratory; National Institute of Standards and Technology.

Properties of Planets, Dwarf Planets, and Moons

Physical Data for Planets and Dwarf Planets

Planet	Equatorial Radius km	R/R$_\oplus$	Mass kg	M/M$_\oplus$	Average Density (relative to water[a])	Rotation Period (days)	Tilt of Rotation Axis (relative to orbit)	Surface Gravity (relative to Earth[b])	Escape Velocity (km/s)	Average Surface Temperature (K)
Mercury	2,440	0.383	3.30×10^{23}	0.055	5.427	58.64	0.01°	0.378	4.3	440 (100 725)[d]
Venus	6,052	0.949	4.87×10^{24}	0.815	5.243	243.02[c]	177.36°	0.907	10.36	737
Earth	6,378	1.000	5.97×10^{24}	1.000	5.515	1.000	23.45°	1.000	11.19	288 (183 331)[d]
Mars	3,397	0.533	6.42×10^{23}	0.107	3.933	1.0260	25.19°	0.377	5.03	210 (133 293)[d]
Ceres	950	0.075	9.60×10^{20}	0.0002	2.100	9.075	3.0°	0.27	0.51	200
Jupiter	71,492	11.209	1.90×10^{27}	317.83	1.326	0.4136	3.13°	2.364	59.5	165
Saturn	60,268	9.449	5.68×10^{26}	95.16	0.687	0.4440	26.73°	0.916	35.5	134
Uranus	25,559	4.007	8.68×10^{25}	14.537	1.270	0.7183[c]	97.77°	0.889	21.3	76
Neptune	24,764	3.883	1.02×10^{26}	17.147	1.638	0.6713	28.32°	1.12	23.5	58
Pluto	1,195	0.187	1.25×10^{22}	0.0021	1.750	6.387[c]	122.53°	0.059	1.1	40
Eris	1,200	0.188	1.5×10^{22} (est.)	0.0025 (est.)	?	8.?	?	?	?	30

[a]The density of water is 1,000 kg/m^3.
[b]The surface gravity of Earth is 9.78 m/s^2.
[c]Venus, Uranus, and Pluto rotate opposite to the directions of their orbits. Their north poles are south of their orbital planes.
[d]Where given, values in parentheses give extremes of recorded temperatures.

Orbital Data for Planets and Dwarf Planets

Planet	Mean Distance from Sun (A[a]) 10⁶ km	AU	Orbital Period (P) (sidereal years)	Eccentricity	Inclination (relative to ecliptic)	Average Speed (km/s)
Mercury	57.91	0.387	0.2408	0.2056	7.005°	47.87
Venus	108.2	0.723	0.6152	0.0067	3.395°	35.02
Earth	149.6	1.000	1.000	0.0167	0.000°	29.78
Mars	227.9	1.524	1.8809	0.0935	1.850°	24.13
Ceres	413.9	2.767	4.6027	0.097	9.73°	17.88
Jupiter	778.6	5.204	11.8618	0.0489	1.304°	13.07
Saturn	1,433.5	9.582	29.4566	0.0565	2.485°	9.69
Uranus	2,872.5	19.201	84.0106	0.0457	0.772°	6.81
Neptune	4,495.1	30.047	164.7856	0.0113	1.769°	5.43
Pluto	5,869.7	39.236	247.6753	0.2488	17.142°	4.72
Eris	10,123	67.668	557.	0.4418	44.187°	3.44

[a]A is the semimajor axis of the planet's elliptical orbit.
Sources: National Space Science Data Center, *Astronomical Almanac.*

Properties of Selected Moons[a]

Planet	Moon	Orbital Properties P (days)	A (10³ km)	Physical Properties R (km)	M (10²⁰ kg)	Density[b] (water = 1)
Earth (1 moon)	Moon	27.32	384.4	1,737.4	735	3.34
Mars (2 moons)	Phobos	0.32	9.38	13.5 × 10.8 × 9.4	0.0001	2.0
	Deimos	1.26	23.46	7.5 × 6.1 × 5.5	0.00002	1.7
Jupiter (63 known moons)	Metis	0.29	127.97	20	0.00096	2.8
	Amalthea	0.50	181.30	131 × 73 × 67	0.0717	1.8
	Io	1.77	421.60	1,815	894	3.55
	Europa	3.55	670.90	1,569	480	3.01
	Ganymede	7.16	1,070	2,631	1,480	1.94
	Callisto	16.69	1,883	2,403	1,080	1.86
	Himalia	250.57	11,480	93	0.0956	2.8
	Pasiphae	735[c]	23,500	25	0.0019	2.9
	Callirrhoe	759[c]	24,100	4.3	0.00001	2.6

(continued)

Properties of Selected Moons[a]

Planet	Moon	Orbital Properties		Physical Properties		
		P (days)	A (10^3 km)	R (km)	M (10^{20} kg)	Density[b] (water = 1)
Saturn (56 known moons)	Pan	0.58	133.58	20	0.00003	—
	Prometheus	0.61	139.35	$72.5 \times 42.5 \times 32.5$	0.0027	0.7
	Pandora	0.63	141.70	$57 \times 42 \times 31$	0.0022	0.7
	Mimas	0.94	185.52	196	0.38	1.17
	Enceladus	1.37	238.02	250	0.84	1.24
	Tethys	1.89	294.66	530	7.55	1.21
	Dione	2.74	377.40	560	10.5	1.43
	Rhea	4.52	527.04	765	24.9	1.33
	Titan	15.95	1,222	2,575	1,350	1.88
	Hyperion	21.28	1,481	$205 \times 130 \times 110$	0.177	1.4
	Iapetus	79.33	3,561	735	18.8	1.21
	Phoebe	550.48[c]	12,952	107	0.04	0.7
	Paaliaq	686.9	15.200	11	0.0001	2.3
Uranus (27 known moons)	Cordelia	0.34	49.75	20	0.0004	1.3
	Miranda	1.41	129.78	236	0.64	1.15
	Ariel	2.52	191.24	579	12.7	1.56
	Umbriel	4.14	265.97	585	12.7	1.52
	Titania	8.71	435.84	789	34.9	1.70
	Oberon	13.46	582.60	761	30.3	1.64
	Setebos	2.225[c]	17,418	15	0.0002	1.5
Neptune (16 known moons)	Naiad	0.29	48.0	$48 \times 30 \times 26$	0.002	1.3
	Larissa	0.55	73.6	$108 \times 102 \times 84$	0.05	1.3
	Proteus	1.12	117.6	$210 \times 208 \times 202$	0.5	1.3
	Triton	5.88[c]	354.8	1,353	214	2.07
	Nereid	360.14	5,513.40	170	0.3	1.5
Pluto (3 moons)	Charon	6.39	19.60	593	16.2	1.85
Eris	Dysnomia	14	36	?	?	?

[a]Innermost, outermost, largest, and/or a few other moons for each planet.
[b]The density of water is 1,000 kg/m^3.
[c]Irregular moon (has retrograde orbit).

Nearest and Brightest Stars

Stars within 15 Light-Years of Earth

Name[a]	Visual[b] Brightness (Sirius = 1,000,000)	Distance (ly)	Spectral Type[c]	Visual[b] Luminosity (Sun = 1.000)
Sun	1.29×10^{16}	1.58×10^{-5}	G2V	1.000
Alpha Centauri C (Proxima Centauri)	10.3	4.23	M5.5V	0.00006
Alpha Centauri B	76,600	4.40	K1V	0.5
Alpha Centauri A	263,000	4.40	G2V	1.6
Barnard's star (dbl)[d]	40	6.0	M4Ve	0.0004
CN Leonis	1.1	7.8	M5.5	0.00002
BD +36-2147	260	8.3	M2	0.0055
Sirius A	1,000,000	8.6	A1V	23
Sirius B	110	8.6	wdDA	0.0025
BL Ceti A	2.6	8.8	M5.5	0.00006
BL Ceti B	1.7	8.8	M5.5	0.00004
V1216 Sagittarii	19	9.7	M3.5V	0.0005
HH Andromedae	3.2	10.3	M5var	0.0001
Epsilon Eridani (dbl)	8,500	10.5	K2V	0.28
Lacaille 9352	300	10.7	M0.5	0.010
FI Virginis	9.3	10.9	M4	0.0003
EZ Aquarii	3.1	11.1	M5.5	0.0001
61 Cygni A	2,200	11.4	K5V	0.084
61 Cygni B	990	11.4	K7V	0.039
Procyon A	184,000	11.4	F5IV-V	7.4
Procyon B	14	11.4	wdDA	0.00054

(continued)

Stars within 15 Light-Years of Earth

(continued) Name[a]	Visual[b] Brightness (Sirius = 1,000,000)	Distance (ly)	Spectral Type[c]	Visual[b] Luminosity (Sun = 1.000)
Gleise 227-046B	34	11.5	M3.5	0.0014
Gleise 227-046A	69	11.7	M3V	0.0028
Groombridge 34 A	150	11.7	M1	0.0062
Groombridge 34 B	9.5	11.7	M3.5	0.00039
DX Cancri	0.31	11.8	M6	0.00001
Epsilon Indi (dbl)	3,500	11.8	K4.5V	0.15
Tau Ceti	10,500	11.9	G8V	0.45
YZ Ceti	3.8	12.1	M4.5V	0.00017
Luyten's star	30	12.4	M3.5	0.0014
Kapteyn's star	75	12.8	M0	0.0037
AX Microscopii	550	12.9	M1/M2V	0.027
Kruger 60A	38	13.1	M2V	0.0020
Kruger 60B	7.9	13.1	M6V	0.0004
V577 Monocerotis A	9.3	13.4	M4.5V	0.0005
V577 Monocerotis B	0.38	13.4	—	0.00002
CD 25 10553A (dbl)	5.4	13.9	M1	0.0003
Gliese 153-058	24	13.9	M3.5	0.0014
FL Virginis A	1.6	14.2	M5	0.0001
FL Virginis B	1.1	14.2	M7	0.00007
Gliese 267-025	98	14.2	M1.5	0.0060
HIP 15689 (dbl)	3.6	14.4	—	0.0002
Van Maanen 2	2.9	14.4	wdDG	0.0002
TZ Arietis	3.3	14.6	M4.5	0.0002
LHS 288	0.74	14.7	M	0.00005
CD 25 10553B	3.9	14.7	M1.5	0.0003
LP 731-58	0.15	14.8	M6.5	0.00001
Gliese 240-063 (dbl)	57	14.8	M3	0.0037
CD 46 11540	46	14.8	M2.5	0.0030

[a]Stars may carry many names, including common names (such as Sirius), names based on their prominence within a constellation (such as Alpha Canis Majoris, another name for Sirius), or names based on their inclusion in a catalog (such as BD + 36 – 2147). Addition of letters A, B, and so on, or superscripts indicates membership in a multiple-star system.

[b]Brightness and luminosity in these tables refer only to radiation in "visual" light.

[c]Spectral types such as M3 are discussed in Chapter 13. Other letters or numbers provide additional information. For example, V after the spectral type indicates a main sequence star, whereas III indicates a giant star.

[d](dbl) means an unresolved double star.

The 52 Brightest Stars in the Sky

Name	Common Name	Visual Brightness (Sirius = 1,000,000)	Distance (ly)	Spectral Type	Visual Luminosity (Sun = 1.000)
Sun	Sun	1.29×10^{16}	1.58×10^{-5}	G2V	1.000
Alpha Canis Majoris	Sirius	1,000,000	8.60	A1V	23
Alpha Carinae	Canopus	506,000	310	F0II	15,000
Alpha Bootis	Arcturus	270,000	36.7	K1.5IIIFe-0.5	110
Alpha^1 Centauri	Rigel Kentaurus	263,000	4.40	G2V	1.6
Alpha Lyrae	Vega	254,000	25.3	A0Va	50
Alpha Aurigae	Capella	242,000	42	G5IIIe+G0III	130
Beta Orionis	Rigel	233,000	770	B8Ia:	43,000
Alpha^1 Canis Minoris	Procyon A	184,000	11.4	F5IV-V	7.4
Alpha Eridani	Achernar	171,000	144	B3Vpe	1,100
Alpha Orionis	Betelgeuse	164,000	430	M1-2Ia-Iab	9,300
Beta Centauri	Hadar	149,000	530	B1III	13,000
Alpha Aquilae	Altair	128,000	16.8	A7V	11
Alpha Tauri	Aldebaran	119,000	65	K5+III	160
Alpha Scorpii	Antares	108,000	600	M1.5Iab-Ib+B4Ve	12,000
Alpha Virginis	Spica	106,000	260	B1III-IV+B2V	2,200
Beta Geminorum	Pollux	91,200	33.7	K0IIIb	32
Alpha Piscis Austrinus	Fomalhaut	89,500	25.1	A3V	17
Beta Crucis	Becrux	82,400	350	B0.5III	3,200
Alpha Cygni	Deneb	82,400	3,000	A2Ia	270,000
Alpha^1 Crucis	Acrux A	76,600	320	B0.5IV	2,400
Alpha^2 Centauri	Alpha Centauri B	76,600	4.40	K1V	0.5
Alpha Leonis	Regulus	75,200	77	B7V	140
Epsilon Canis Majoris	Adhara	65,500	430	B2II	3,800
Gamma Crucis	Gacrux	58,100	88	M3.5III	140
Lambda Scopii	Shaula	58,100	700	B2IV+B	8,900
Gamma Orionis	Bellatrix	57,500	240	B2III	1,100
Beta Tauri	El Nath	57,000	131	B7III	300
Beta Carinae	Miaplacidus	55,500	111	A2IV	210
Epsilon Orionis	Alnilam	54,500	1,300	B0Ia	30,000
Alpha^2 Crucis	Acrux B	53,000	800	B1V	11,000

(continued)

The 52 Brightest Stars in the Sky

(continued)

Name	Common Name	Visual Brightness (Sirius = 1,000,000)	Distance (ly)	Spectral Type	Visual Luminosity (Sun = 1.000)
Alpha Gruis	Al Na'ir	52,500	101	B7IV	170
Epsilon Ursae Majoris	Alioth	51,000	81	A0pCr	100
Gamma2 Velorum	Suhail	50,600	840	WC8+O9I	11,000
Alpha Persei	Mirfak	50,100	590	F5Ib	5,400
Alpha Ursae Majoris	Dubhe	50,100	124	K0IIIa	240
Delta Canis Majoris	Wezen	47,900	1,800	F8Ia	47,500
Epsilon Sagittarii	Kaus Australis	47,400	145	B9.5III	300
Epsilon Carinae	Avior	47,000	630	K3III+B2:V	5,800
Eta Ursae Majoris	Benetnasch	47,000	101	B3V	150
Theta Scorpii	Sargas	46,600	270	F1II	1,100
Beta Aurigae	Menkalinan	45,300	82	A2IV	94
Alpha Trianguli Australis	Atria	44,500	415	K2IIb-IIIa	2,400
Gamma Geminorum	Alhena	44,100	105	A0IV	150
Alpha Pavonis	Peacock	43,700	183	B2IV	450
Delta Velorum	—	42,900	80	A1V	84
Beta Canis Majoris	Murzim	42,100	500	B1II-III	3,200
Alpha Geminorum	Castor	42,100	52	A1V	35
Alpha Hydrae	Alphard	42,100	177	K3II-III	410
Alpha Arietis	Hamal	41,300	66	K2-IIICa-1	55
Alpha Ursae Minoris	Polaris (Pole Star)	40,600	430	F7:Ib-II	2,300
Sigma Sagittarii	Nunki	40,600	224	B2.5V	630

Sources for Appendix 5: *The Hipparcos and Tycho Catalogues*, 1997, European Space Agency SP-1200; *Bright Star Catalogue*, 5th rev. ed. (Hoffleit, 1991), NSSDC Astronomical Data Center; *Burnham's Celestial Handbook* (Dover, 1983).

Observing the Sky

The purpose of this appendix is to provide enough information so you can make sense of a star chart or list of astronomical objects, as well as find a few objects in the sky.

Celestial Coordinates

In Chapter 2 we discuss the celestial sphere—the imaginary sphere with Earth at its center upon which celestial objects appear to lie. A number of different coordinate systems are used to specify the positions of objects on the celestial sphere. The simplest of these is the *altitude–azimuth coordinate system*. The altitude–azimuth coordinate system is based on the "map" direction to an object (the object's azimuth, with north = 0°, east = 90°, south = 180°, and west = 270°) combined with how high the object is above the horizon (the object's altitude, with the horizon at 0° and the zenith at 90°). For example, an object that is 10° above the eastern horizon has an altitude of 10° and an azimuth of 90°. An object that is 45° above the horizon in the southwest is at altitude 45°, azimuth 225°.

The altitude–azimuth coordinate system is the simplest way to tell a friend where in the sky to look at the moment, but it is not a good coordinate system for cataloging the positions of objects. The altitude and azimuth of an object are different for each observer, depending on their position on Earth, and are constantly changing as Earth rotates on its axis. If we need to specify the direction to an object in a way that is the same for everyone, we need a coordinate system that is fixed relative to the celestial sphere. The most common such coordinates are called *celestial coordinates*.

Celestial coordinates are illustrated in **Figure A6.1.** Celestial coordinates are much like the traditional system of latitude and longitude used on the surface of Earth. On Earth, latitude specifies how far you are from Earth's equator, as discussed in Chapter 2. If you are on Earth's equator, your latitude is 0°. If you are at Earth's North Pole, your latitude is 90° north. If you are at Earth's South Pole, your latitude is 90° south.

The latitudelike coordinate on the celestial sphere is called *declination,* often signified with the Greek letter δ (delta). The celestial equator has $\delta = 0°$. The north celestial pole has $\delta = +90°$. The south celestial pole has $\delta = -90°$. (See Chapter 2 if you need to refresh your memory of the celestial equator or celestial poles.) Declination is usually expressed in degrees, minutes of arc, and seconds of arc. For example, Sirius, the brightest star in the sky, has $\delta = -16° \, 42' \, 58''$, meaning that it is located not quite 17° south of the celestial equator.

On Earth, east–west position is specified by longitude. Lines of constant longitude run north–south from one pole to the other. Unlike latitude, for which the equator provides a natural place to call "zero," there is no natural starting point for longitude, so we just have to invent one. By arbitrary convention, the Royal Greenwich Observatory in Greenwich, England, is defined to lie at a longitude of 0°. On the celestial sphere the longitudelike coordinate is called *right ascension,* often signified with the Greek letter α (alpha). Unlike longitude, there *is* a natural point on the celestial sphere to use as the starting point for right ascension—the vernal equinox, or the point at which the ecliptic crosses the celestial equator with the Sun moving from the southern sky into the northern sky. The vernal equinox defines the line of right ascension at which $\alpha = 0°$. The autumnal equinox, located on the opposite side of the sky, is at $\alpha = 180°$.

Normally right ascension is measured in units of time rather than in degrees. It takes Earth 24 hours (of sidereal time) to rotate on its axis, so the celestial sphere is broken into 24 hours of right ascension, with each hour of right ascension corresponding to 15°. Hours of right ascension are then subdivided into minutes and seconds of time. Right ascension increases going to the east. The right ascension of Sirius, for example, is $\alpha = 06^h \, 45^m \, 08.9^s$, meaning that Sirius is about 101° (that is, $06^h \, 45^m$) east of the vernal equinox. Time is a natural unit for measuring right ascension because time naturally tracks the motion of objects due to Earth's rotation on its axis. If stars on the meridian at a certain time have 6^h, then an hour later the stars

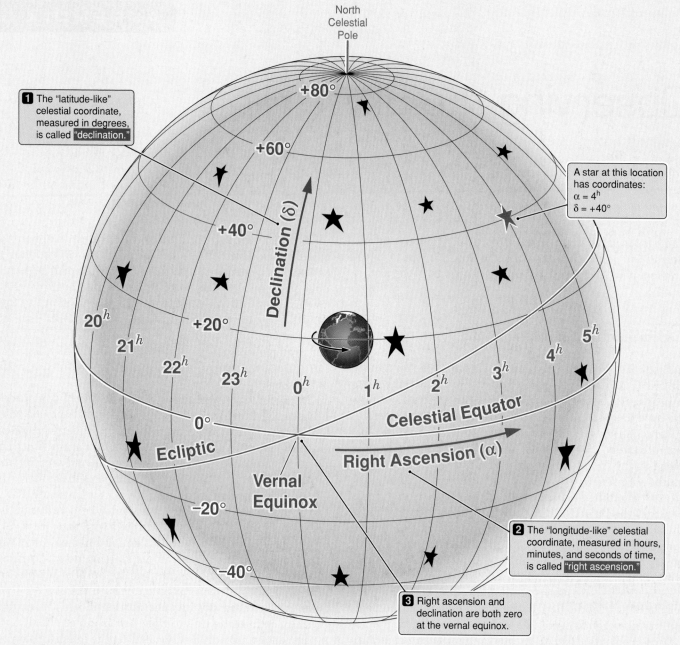

North
Celestial
Pole

1 The "latitude-like" celestial coordinate, measured in degrees, is called "declination."

+80°

+60°

A star at this location has coordinates:
$\alpha = 4^h$
$\delta = +40°$

+40°

Declination (δ)

20^h +20°

21^h

22^h

23^h 0^h 1^h 2^h 3^h 4^h 5^h

0° Celestial Equator

Ecliptic Right Ascension (α)

Vernal Equinox

−20°

2 The "longitude-like" celestial coordinate, measured in hours, minutes, and seconds of time, is called "right ascension."

−40°

3 Right ascension and declination are both zero at the vernal equinox.

FIGURE A6.1

on the meridian will have 7^h, and an hour after that they will have $\alpha = 8^h$. The *local sidereal time*, or "star time," at your location right now is equal to the right ascension of the stars that are on your meridian at the moment. Because of Earth's motion around the Sun, a sidereal day is about 4 minutes shorter than a solar day: so local sidereal time constantly gains on solar time. At midnight on September 23 the local sidereal time is 0^h. By midnight on December 22 local sidereal time has advanced to 6^h. On March 21

local sidereal time at midnight is 12^h. Local sidereal time at midnight on June 21 is 18^h.

Putting this all together, right ascension and declination provide a convenient way to specify the location of any object on the celestial sphere. Sirius is located at $\alpha = 06^h 45^m 08.9^s$, $= -16° 42' 58''$, which means that at midnight on December 22 (local sidereal time = 6^h) you will find Sirius about 45^m east of the meridian, not quite 17° south of the celestial equator.

There is just one final caveat. As we discussed in Chapter 2, the directions of the celestial equator, celestial poles, and vernal equinox are constantly changing as Earth's axis wobbles like the axis of a spinning top. In Chapter 2 we called this 26,000-year wobble the "precession of the equinoxes," meaning that the location of the equinoxes is slowly advancing along the ecliptic. So when we specify the celestial coordinates of an object, we need to specify the date at which the positions of the vernal equinox and celestial poles were measured. By convention, coordinates are usually referred to with the position of the vernal equinox on January 1, 2000. A complete, formal specification of the coordinates of Sirius would then be $\alpha(2000)06^h\ 45^m\ 08.9^s$, $\delta(2000)-16°\ 42'\ 58''$, where the "2000" in parentheses refers to the equinox of the coordinates.

Constellations and Names

Although it is certainly possible to specify exactly any location on the surface of Earth by giving its latitude and longitude, it is usually convenient to use a more descriptive address. We might say, for example, that one of the coauthors of this book works near latitude 37° north, longitude 122° west: but it would probably mean a lot more to you were we to say that George Blumenthal works in Santa Cruz, California.

Just as the surface of Earth is divided into nations and states, the celestial sphere is divided into 88 *constellations,* the names of which are often used to refer to objects within their boundaries (see the star charts in **Figure A6.2**, starting on the next page). The brightest stars within the boundaries of a constellation are referred to using a Greek letter combined with the name of the constellation. For example, the star Sirius is the brightest star in the constellation Canis Major (the Great Dog) and so is referred to as α *Canis Majoris.* The bright red star in the northeastern corner of the constellation of Orion is referred to as α *Orionis,* also known as Betelgeuse. Rigel, the bright blue star in the southwest corner of Orion, is also called β *Orionis.*

Astronomical objects can take on a bewildering range of names. For example, the bright southern star Canopus, also known as α *Carinae* (the brightest star in the constellation of Carina) has no fewer than 34 different names, most of which are about as memorable as "SAO 234480" (number 234,480 in the Smithsonian Astrophysical Observatory catalog of stars).

You may have noticed a slight diffrence in the way a constellation is spelled when it becomes part of a star's name. Thus we see that Sirius is called α *Canis Majoris,* not α *Canis Major;* Rigel is referred to as α *Orionis,* not α *Orion;* and Canopus becomes α *Carinae,* not α *Carina.* This is because

we use the Latin genitive or possessive case with star names where, for example, *Orionis* means "of *Orion.*"

Astronomical Magnitudes

Throughout the text we refer to the brightness of objects: but when discussing the appearance of an object in the sky, astronomers normally speak instead of the object's *magnitude.* The system of astronomical magnitudes dates back to the Greek astronomer Hipparchus, who when classifying stars ordered them according to their "rank." First–rank stars were the brightest stars in the sky, second–rank stars were the next brightest, and so on. The faintest stars that could be seen were called sixth–rank stars. When methods were devised to actually measure the brightness of stars, it was discovered that first–rank stars were typically about 100 times as bright as sixth–rank stars and that the steps from one rank to the next were *logarithmic.* That is a first–rank star was typically about 2.51 *times* as bright as a second–rank star, whereas a second–rank star was typically about 2.51 *times* as bright as a third–rank star.

This rough classification of stellar "rank" was formalized into the modern system of astronomical magnitudes. A difference of five magnitudes between the brightness of two stars (say a star with $m = 6$ and a star with $m = 1$), corresponds to a hundredfold difference in brightness. Notice that the magnitude scale is backward—the *greater the magnitude, the fainter the object.*

If five steps in magnitude correspond to a factor of 100 in brightness, then one step in magnitude must correspond to a factor of $100^{1/5} = 10^{2/5} = 2.512\ldots$ in brightness ($100^{1/5} \times 100^{1/5} \times 100^{1/5} \times 100^{1/5} \times 100^{1/5} = 100$). The relationship between brightness and magnitude is most easily written using common or base-10 logarithms. If star 1 has a brightness of b_1 and star 2 has a brightness of b_2, then the difference in magnitude $m_2 - m_1$ between the two stars is

$$m_2 - m_1 = -2.5\ log_{10}\frac{b_2}{b_1}.$$

To convert from magnitude differences to brightness ratios, divide by −2.5 (that is, multiply by −0.4) and raise 10 to the resulting power:

$$\frac{b_2}{b_1} = 10^{-0.4 \times (m_2 - m_1)}$$

The last thing we need to set the magnitude scale is an object that we define to have a magnitude of zero. Many different magnitude scales exist, but perhaps the most common scale for visual magnitudes defines the star Vega as having a magnitude of 0.

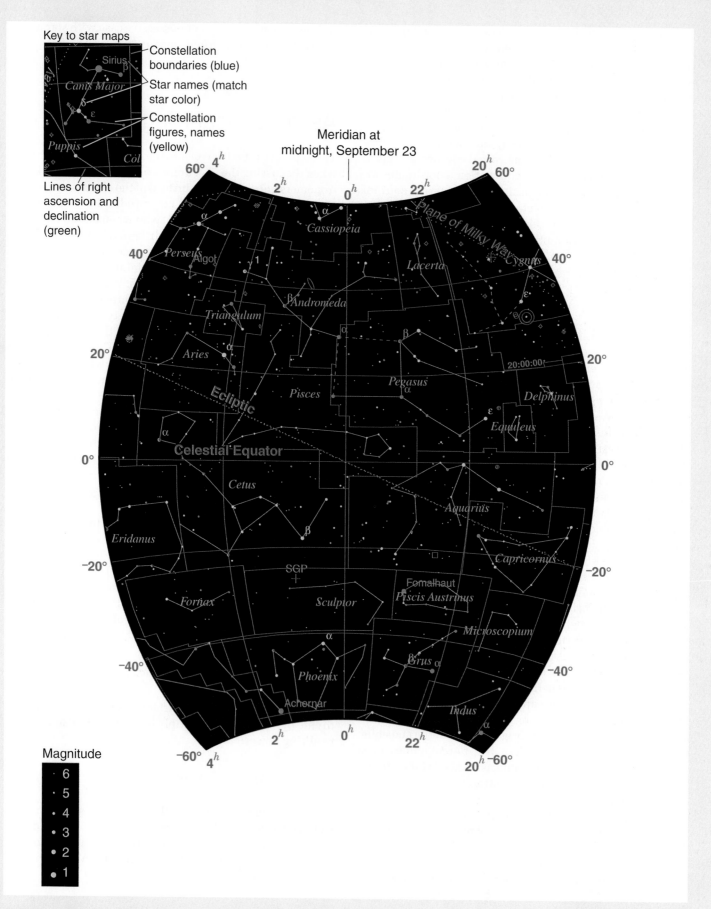

FIGURE A6.2(A) The sky from right ascension 20ʰ to 4ʰ and declination −60° to +60°.

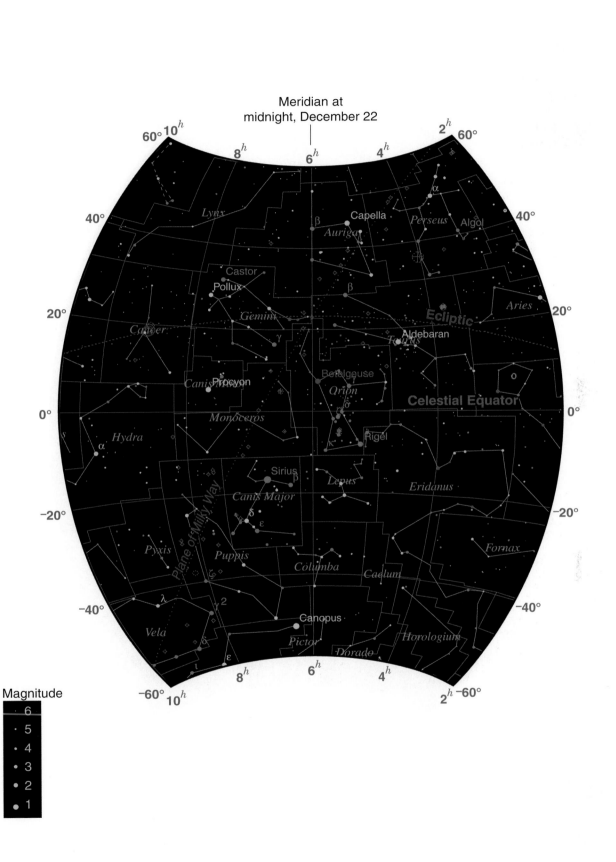

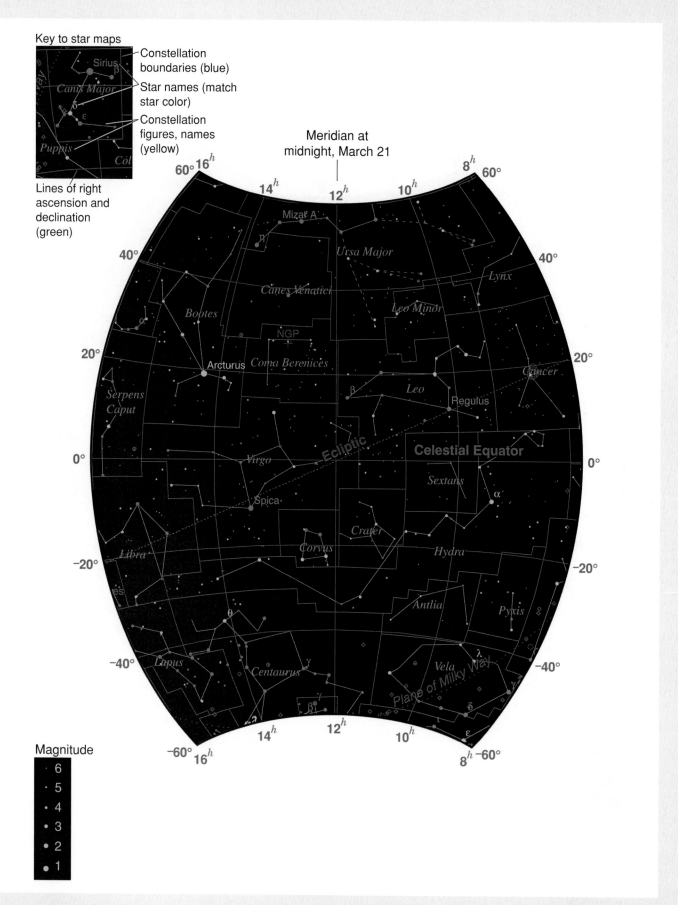

Key to star maps

Constellation boundaries (blue)

Star names (match star color)

Constellation figures, names (yellow)

Lines of right ascension and declination (green)

Meridian at midnight, March 21

Magnitude

FIGURE A6.2(C) The sky from right ascension 8ʰ to 16ʰ and declination −60° to +60°.

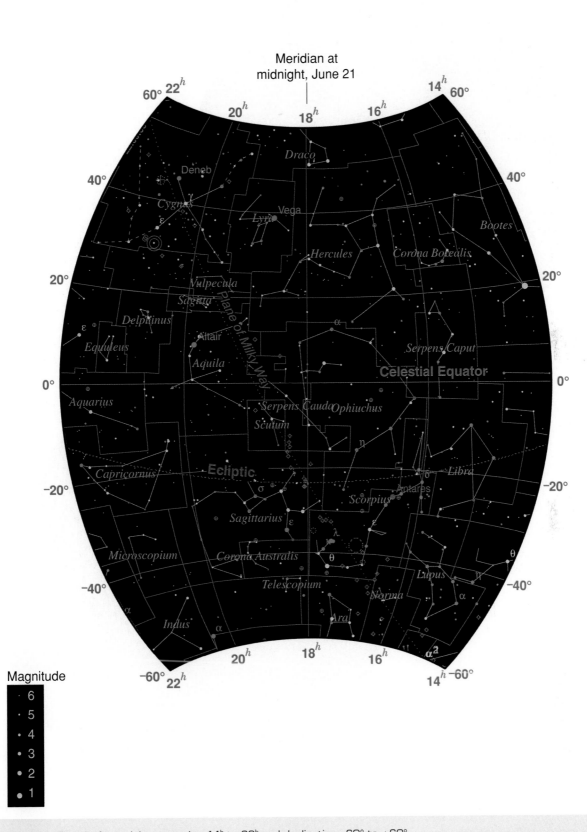

FIGURE A6.2(D) The sky from right ascension 14^h to 22^h and declination −60° to +60°.

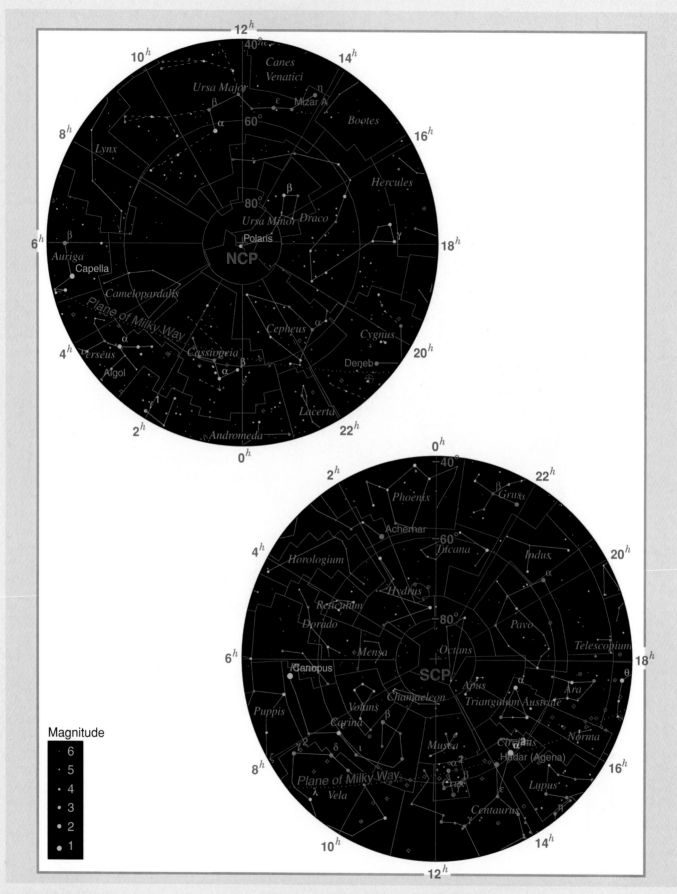

FIGURE A6.2(E) The regions of the sky north of declination +40° and south of declination −40°.

A final note: In Chapters 13 and onward, we used "colors" based on the ratio of the brightness of a star as seen in two different parts of the spectrum. The "b_B/b_V color," for example, was just the ratio of the brightness of a star seen through a blue filter, divided by the brightness of a star seen through a red filter. Normally astronomers instead discuss the "B-V color" of a star, which is equal to the difference between a star's blue magnitude and its visual magnitude. We can use the expression previous for a difference in magnitude to write

$$\text{B-V color} = m_B - m_V = -2.5 \, log_{10} \frac{b_B}{b_V}.$$

So a star with a b_B/b_V color of 1 has a B-V color of 0. A star with a b_B/b_V color of 1.4 has a B-V color of -0.37. Notice that, as with magnitudes, B-V colors are "backward." The bluer a star the greater its b_B/b_V color, but the less its B-V color.

Although the system of astronomical magnitudes and colors is actually convenient in many ways—which is why astronomers continue to use it—it can also be confusing to new students, especially students for whom math is a foreign language. Having seen something of how astronomical magnitudes work, we think you may agree with our choice to discuss the brightness of stars rather than their magnitudes throughout the text! However, if you do need to use the magnitude system (for example, to read star charts, make sense of a popular astronomy article, or complete a lab assignment), just remember three things and you will probably get by:

1. The greater the magnitude, the *fainter* the object.

2. A magnitude less means about two and a half times brighter.

3. The brightest stars in the sky have magnitudes of less than 1, while the faintest stars that can be seen with the naked eye on a dark night have magnitudes of about 6.

Uniform Circular Motion and Circular Orbits

In Chapter 3 (see Section 3.5 and Figure 3.15) we discuss the motion of an object moving in a circle at a constant speed. This motion, called uniform circular motion, is the result of centripetal force always acting toward the center of the circle. The key question when thinking about uniform circular motion is "how hard do I have to pull to keep the object moving in a circle?" Part of the answer to this question is pretty obvious—the more massive an object is, the harder it will be to keep it moving on its circular path. According to Newton's Second Law, $F = ma$, or in this case, the centripetal force equals the mass times the centripetal acceleration. The larger the mass, the greater the force required to keep it moving in its circle.

The centripetal force needed to keep an object moving in constant circular motion also depends on two other quantities: the speed of the object and the size of the circle. The faster an object is moving, the more rapidly it has to change direction to stay on a circle of a given size. The second quantity that influences the needed acceleration is the radius of the circle. The smaller the circle, the greater the pull needed to keep it on track. You can understand this by looking at the motion. A small circle requires a continuous "hard" turn, whereas a larger circle requires a more gentle change in direction. It takes more force to keep an object moving faster in a smaller circle than it does to keep the same object moving more slowly in a larger circle. (To get a better feeling for how this works, think about the difference between riding in a car that is taking a tight curve at high speed and a car that is moving slowly around a gentle curve.)

To arrive at the circular velocity and other results discussed in Chapter 3, we need to turn these intuitive ideas about uniform circular motion into a quantitative expression for exactly how much centripetal acceleration is needed to keep an object moving in a circle with radius r at speed v. **Figure A7.1** shows a ball moving around a circle of radius r at a constant speed v at two different times. The centripetal acceleration that is keeping the ball on the circle is a. Remember that the acceleration is always directed toward the center of the circle, whereas the velocity of the ball is always perpendicular to the acceleration. The ball's velocity and its acceleration are always at right angles to each other. As the object moves around the circle, the direction of motion and the direction of the acceleration change together in lockstep.

In the figure we have drawn two triangles. In triangle 1 we show the velocity (speed and direction) at each of the two times. The arrow labeled Δv connecting the heads of the two velocity arrows shows how much the velocity changed between time 1 and time 2. This change is the effect of the centripetal acceleration. If we imagine that points 1 and 2 are very close together—so close that the direction of the centripetal acceleration does not change by much between the two—then we can say that the centripetal acceleration equals the change in the velocity divided by the time between the two, $\Delta t = t_2 - t_1$. So we have $\Delta v = a\Delta t$.

In triangle 2 we do a similar thing. Here the arrow labeled Δr indicates the change in the position of the ball between time 1 and time 2. Again, if we imagine that the time between the two points is very short, we can say that Δr is equal to the velocity times the time, or $\Delta r = v\Delta tv$.

The line between the center of the circle and the ball is always perpendicular to the velocity of the ball. So if the direction of the ball's velocity changes by an angle α, then the direction of the line between the ball and the center of the circle must also change by the same angle α. In other words, triangles 1 and 2 are *similar triangles*. They have the same *shape*. If the triangles are the same shape, the ratio of two sides of triangle 1 must equal the ratio of the two corresponding sides of triangle 2. Using this fact we can write

$$\frac{a\Delta t}{v} = \frac{v\Delta t}{r}.$$

If we divide out the Δt from both sides of the equation, then cross-multiply, we obtain

$$ar = v^2,$$

which after dividing both sides of the equation by r becomes

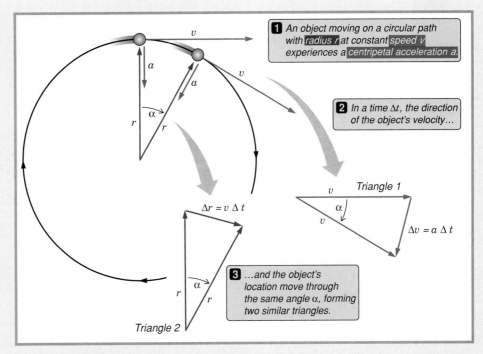

FIGURE A7.1 *Similar triangles used to find the centripetal force needed to keep an object moving at a constant speed on a circular path.*

$$a_{\text{centripetal}} = \frac{v^2}{r}.$$

We have added the subscript "centripetal" to a to signify that this is the centripetal acceleration needed to keep the object moving in a circle of radius r at speed v. The centripetal force required to keep an object of mass m moving on such a circle is then

$$F_{\text{centripetal}} = ma_{\text{centripetal}} = \frac{mv^2}{r}.$$

Circular Orbits

In the case of an object moving in a circular orbit there is no string to hold the ball on its circular path. Instead this force is provided by gravity.

Think about an object with mass m in orbit about a much larger object with mass M. The orbit is circular, and the distance between the two objects is given by r. The force needed to keep the smaller object moving at speed v in a circle with radius r is given by the previous expression for $F_{\text{centripetal}}$. The force actually provided by gravity (see Chapter 3) is

$$F_{\text{gravity}} = G\frac{Mm}{r^2}.$$

If gravity is responsible for holding the mass in its circular motion, then it should be true that $F_{\text{gravity}} = F_{\text{centripetal}}$. That is, if mass m is moving in a circle under the force of gravity, the force provided by gravity *must* equal the centripetal force needed to explain that circular motion. Setting our two expressions for $F_{\text{centripetal}}$ and F_{gravity} equal to each other gives

$$\frac{mv^2}{r} = G\frac{Mm}{r^2}.$$

All that remains is a bit of algebra. Dividing out the m on both sides of the equation and multiplying both sides by r gives

$$v^2 = G\frac{M}{r}.$$

Taking the square root of both sides then brings us to the desired result:

$$v_{\text{circular}} = \sqrt{\frac{GM}{r}}$$

This is the "circular velocity" we presented in Chapter 3. It is the velocity at which an object in a circular orbit *must* be moving. If the object were not moving at this velocity, then gravity would not be providing the needed centripetal force, and the object would not move in a circle.

IAU 2006 Resolutions: Definition of a Planet in the Solar System, and Pluto

24 August 2006, Prague
SOURCE: International Astronomical Union

Resolutions

Resolution 5 is the principal definition for the IAU usage of "planet" and related terms.

Resolution 6 creates for IAU usage a new class of objects, for which Pluto is the prototype. The IAU will set up a process to name these objects.

Resolution 5

Definition of a Planet in the Solar System

Contemporary observations are changing our understanding of planetary systems, and it is important that our nomenclature for objects reflect our current understanding. This applies, in particular, to the designation "planets." The word "planet" originally described "wanderers" that were known only as moving lights in the sky. Recent discoveries lead us to create a new definition, which we can make using currently available scientific information.

The IAU therefore resolves that planets and other bodies, except satellites, in our Solar System be defined into three distinct categories in the following way:

(1) A "planet"[1] is a celestial body that (a) is in orbit around the Sun, (b) has sufficient mass for its self-gravity to overcome rigid body forces so that it assumes a hydrostatic equilibrium (nearly round) shape, and (c) has cleared the neighbourhood around its orbit.

(2) A "dwarf planet" is a celestial body that (a) is in orbit around the Sun, (b) has sufficient mass for its self-gravity to overcome rigid body forces so that it assumes a hydrostatic equilibrium (nearly round) shape[2], (c) has not cleared the neighbourhood around its orbit, and (d) is not a satellite.

(3) All other objects,[3] except satellites, orbiting the Sun shall be referred to collectively as "Small Solar-System Bodies."

Resolution 6

Pluto

The IAU further resolves:

Pluto is a "dwarf planet" by the above definition and is recognized as the prototype of a new category of Trans-Neptunian Objects.[4]

---

[1] The eight planets are: Mercury, Venus, Earth, Mars, Jupiter, Saturn, Uranus, and Neptune.

[2] An IAU process will be established to assign borderline objects into either dwarf planet and other categories.

[3] These currently include most of the Solar System asteroids, most Trans-Neptunian Objects (TNOs), comets, and other small bodies.

[4] An IAU process will be established to select a name for this category.

Glossary

A

aberration of starlight The apparent displacement in the position of a star due to the finite speed of light and Earth's orbital motion around the Sun.

absolute zero The temperature at which thermal motions cease. The lowest possible temperature. Zero on the kelvin temperature scale.

absorption The capture of electromagnetic radiation by matter.

absorption line An intensity minimum in a spectrum due to absorption of electromagnetic radiation at a specific wavelength determined by the energy levels of an atom or molecule.

acceleration The rate at which the speed and/or direction of an object's motion is changing.

accretion The process by which smaller bodies coalesce to form larger ones.

accretion disk A flat, rotating disk of gas and dust surrounding a condensed mass, such as a young stellar object, a forming planet, or a collapsed star in a binary pair.

achondrite A stony meteorite that does not contain chondrules.

active comet A comet nucleus that approaches close enough to the Sun to show signs of activity, such as the production of a coma and tail.

active galactic nucleus (AGN) A highly luminous, compact galactic nucleus whose luminosity may exceed that of the rest of the galaxy.

adaptive optics Electro-optical systems that largely compensate for image distortion caused by Earth's atmosphere.

AGB star A star on the asymptotic giant branch.

AGN See *active galactic nucleus.*

albedo The fraction of electromagnetic radiation incident on a surface that is reflected by the surface.

alpha particle A He⁴ nucleus consisting of two protons and two neutrons. Alpha par-

ticles are given off in the type of radioactive decay referred to as *alpha decay*—hence the name.

Alpher, Ralph (1921–) The physicist who worked with George Gamow to predict the existence of the cosmic background radiation.

Amors A Family of asteroids with orbits that cross the orbit of Mars but not that of Earth.

amplitude In a wave, the maximum excursion from equilibrium. For example, in a water wave the amplitude is the vertical distance from crests to the undisturbed water level.

angular momentum A conserved property of a rotating or revolving system whose value depends on the velocity and distribution of its mass.

annular solar eclipse A solar eclipse that occurs when the apparent diameter of the Moon is less than that of the Sun, leaving a visible ring of light (annulus) surrounding the dark disk of the Moon.

Antarctic Circle The circle on Earth with latitude 66.5° south, marking the limit of the Antarctic region.

antimatter Matter made from antiparticles.

antiparticle An elementary particle of antimatter identical in mass but opposite in charge and all other properties to its corresponding ordinary matter particle.

aperture The clear diameter of a telescope's objective lens or primary mirror.

aphelion (pl. **aphelia**) The point in a solar orbit that is farthest from the Sun.

Apollos A group of asteroids whose orbits cross the orbits of both Earth and Mars.

arcminute See *minute of arc.*

arcsecond See *second of arc.*

Arctic Circle The circle on Earth at latitude 66.5° north, marking the limit of the Arctic region.

Aristotle (384–322 B.C.) The ancient Greek philosopher whose observations of the natural world later became the basis of modern science.

asteroid A primitive rocky or metallic body (planetesimal) that has survived planetary accretion. Asteroids are parent bodies of meteoroids. Asteroids are also known as minor planets.

asteroid belt The region between the orbits of Mars and Jupiter that contains most of the asteroids in our Solar System.

astrobiology An interdisciplinary science combining astronomy, biology, and geology to study life in the cosmos.

astrology The belief that the positions and aspects of stars and planets influence human affairs and characteristics as well as terrestrial events.

astronomical seeing A measurement of the degree to which Earth's atmosphere degrades the resolution of a telescope's view of astronomical objects.

astronomical unit (AU) The average distance from the Sun to Earth: approximately 150 million kilometers.

astronomy The scientific study of planets, stars, galaxies, and the universe as a whole.

astrophysics The application of physical laws to the understanding of planets, stars, galaxies, and the universe as a whole.

asymptotic giant branch (AGB) In the H-R diagram, a separate "branch" that goes from the horizontal branch toward higher luminosities and lower temperatures, asymptotically approaching and then rising above the red giant branch.

Atens A group of asteroids whose orbits cross Earth's orbit, but not the orbit of Mars.

atmosphere The gravitationally bound, outer gaseous envelope surrounding a planet, moon, or star.

atmospheric greenhouse effect A warming of planetary surfaces produced by atmospheric gases that transmit optical solar radiation but partially trap infrared radiation. Compare with *greenhouse effect.*

atmospheric windows   A regions of the electromagnetic spectrum in which radiation is able to penetrate a planet's atmosphere.

atom   The smallest piece of an element that retains the properties of that element. Each atom is composed of a nucleus (neutrons and protons) surrounded by a cloud of electrons.

aurora   Emission in the upper atmosphere of a planet from atoms that have been excited by collisions with energetic particles from the planet's magnetosphere.

autumnal equinox   (a) The point where the ecliptic crosses the celestial equator, with the Sun moving from north to south. (b) The day, around September 23, on which the Sun appears at this location, which is the first day of autumn (fall) in the Northern Hemisphere.

axion   A hypothetical elementary particle first proposed to explain certain properties of the neutron and now considered a candidate for *cold dark matter*.

B

backlighting   Illumination from behind a subject as seen by an observer. Fine material such as human hair and dust in planetary rings stands out best when viewed under backlighting conditions.

bar   A unit of pressure. One bar is equivalent to 10^5 newtons per square meter—approximately equal to Earth's atmospheric pressure at sea level.

barred spiral   A spiral galaxy with a bulge having an elongated, barlike shape.

basalt   Gray to black volcanic rock, rich in iron and magnesium.

b_B/b_V color   Measurement of the color of an object on the basis of the ratio of its brightness in blue light (b_B) to its brightness in "visual" (or yellow-green) light (b_V).

beta decay   The decay of a neutron into a proton by emission of an electron (beta ray) and an antineutrino, or the decay of a proton into a neutron by emission of a positron and a neutrino.

Bethe, Hans (1906–2005)   The physicist who won the Nobel Prize for his work on the source of stellar energy.

Big Bang   The event that occurred about 14 billion years ago that marks the beginning of time and the universe.

Big Bang nucleosynthesis   The formation of low-mass nuclei (H, He, Li, Be, B) during the first few minutes following the Big Bang.

binary star   A system in which two stars are in gravitationally bound orbits about their common center of mass.

binding energy   The minimum energy required to separate an atomic nucleus into its component protons and neutrons.

bipolar outflow   Material streaming away in opposite directions from either side of the accretion disk of a young star.

blackbody   An object that absorbs and can reemit all electromagnetic energy it receives.

blackbody spectrum   See *Planck spectrum*.

black hole   An object so dense its escape velocity exceeds the speed of light: a *singularity* in spacetime.

blueshift   The Doppler shift toward shorter wavelengths of light from an approaching object. Compare with *redshift*.

Bohr model   A model of the atom, proposed by Neils Bohr in 1913, in which a small positively charged nucleus is surrounded by orbiting electrons, similar to a miniature solar system.

Bohr, Neils (1885–1962)   The Danish theoretical physicist best known for his discovery of the quantization of atomic states.

bolide   A very bright, exploding meteor.

Boltzmann's constant (k)   The constant that relates the temperature of a gas to the average kinetic energy of the molecules of the gas.

bound orbit   A closed orbit in which the velocity is less than the escape velocity.

bow shock   (a) The boundary at which the speed of the *solar wind* abruptly drops from supersonic to subsonic in its approach to a planet's *magnetosphere;* the boundary between the region dominated by the solar wind and the region dominated by a planet's magnetosphere. (b) The interface between strong collimated gas and dust outflow from a star and the interstellar medium.

Brahe, Tycho (1546–1601)   The foremost astronomical observer before the invention of the telescope. Tycho provided the data that permitted Kepler to deduce the motions of the planets around the Sun.

brightness   The apparent intensity of light from a luminous object. Brightness depends on both the luminosity of a source and its distance. Units at the detector: watts per square meter (W/m²).

brown dwarf   A "failed" star that is not massive enough to cause hydrogen fusion in its core.

bulge   The central region of a spiral galaxy that is similar in appearance to a small elliptical galaxy.

C

Cannon, Annie Jump (1863–1941)   The American astronomer who systematically classified the spectral types of stars in the Henry Draper Catalog, the first catalog to list stellar spectral types for 250,000 stars.

carbonaceous chondrite   A primitive stony meteorite that contains chondrules and is rich in carbon and volatile materials.

carbon–nitrogen–oxygen cycle   See *CNO cycle*.

carbon star   A cool *red giant* or AGB star that has an excess of carbon in its atmosphere.

Cassini Division   The largest gap in Saturn's rings, discovered by Jean-Dominique Cassini in 1675.

Cassini, Jean-Dominique (1625–1712)   The Italian astronomer (nationalized as French in 1673) who discovered four satellites of Saturn as well as the division within Saturn's ring that bears his name.

catalyst   An atomic and molecular structure that permits or encourages chemical and nuclear reactions but does not change its own chemical or nuclear properties.

CBR   See *cosmic background radiation*.

CCD   See *charge-coupled device*.

celestial equator   The imaginary great circle that is the projection of Earth's equator onto the celestial sphere.

celestial sphere   An imaginary sphere with celestial objects on its inner surface and Earth at its center. The celestial sphere has no physical existence but is a convenient tool for picturing the directions in which celestial objects are seen from the surface of Earth.

Celsius (C)   The arbitrary temperature scale with 0°C at the freezing point of water and 100°C at the boiling point of water at sea level. Defined by Anders Celsius (1701–1744). Also known as the Centigrade scale.

center of mass   The location associated with an object system at which we may regard the entire mass of the system as being concentrated. The point in any isolated system that moves according to Newton's first law of motion.

centripetal acceleration   The acceleration of an object directed toward the center of curvature of its motion.

centripetal force   A force directed toward the center of curvature of an object's curved path.

Cepheid variable   An evolved high-mass star with an atmosphere that is pulsating, leading to variability in the star's luminosity and color.

Chandrasekhar limit The upper limit on the mass of an object supported by electron degeneracy pressure—approximately 1.4 $M_\odot$.

chaos Behavior in complex, interrelated systems in which tiny differences in the initial configuration of a system result in dramatic differences in the system's later evolution.

charge-coupled device (CCD) A common type of solid-state detector of electromagnetic radiation that transforms the intensity of light directly into electric signals.

chemistry The study of the composition, structure, and properties of substances

chondrite A stony meteorite containing chondrules.

chondrule A spherical inclusion of rapidly cooled melt found inside some meteorites.

chromatic aberration A detrimental property of a lens in which rays of different wavelengths are brought to different focal distances from the lens.

chromosphere The region in the Sun's atmosphere located between the *photosphere* and the *corona.*

circular velocity The orbital velocity needed to keep an object moving in a circular orbit.

circumpolar That part of the sky, near either celestial pole, that can always be seen above the horizon from a specific location on Earth.

classical mechanics The science of applying Newton's laws to the motion of objects.

climate The state of an atmosphere averaged over an extended time.

closed universe A finite universe with a curved spatial structure such that the angles of a triangle always exceed 180°.

CNO cycle (carbon–nitrogen–oxygen cycle) One of the ways in which hydrogen burning (the fusion of four hydrogen atoms to form one helium atom) can take place. See also *triple-alpha process.*

cold dark matter Dark matter particles that move slowly enough to be gravitationally bound even in the smallest galaxies.

coma (pl. **comae**) The nearly spherical cloud of gas and dust surrounding the nucleus of an active comet.

comet A complex object consisting of a small solid, icy nucleus, an atmospheric halo, and a tail of gas and dust.

comet nucleus (pl. **nuclei**) A primitive planetesimal, composed of ices and *refractory materials* that has survived planetary accretion. The "heart" of a comet, containing nearly the entire mass of the comet. A "dirty snowball."

comparative planetology The study of planets through comparison of their chemical and physical properties.

complex system An interrelated system capable of exhibiting chaotic behavior. See also *chaos.*

composite volcano A large, cone-shaped volcano.

compound lens A lens made up of two or more elements of differing refractive index, the purpose of which is to minimize *chromatic aberration.*

conservation law A physical law stating that the amount of some physical quantity (such as *energy or angular momentum*) of an isolated system does not change over time.

conservation of angular momentum The physical law stating that the amount of angular momentum of an isolated system does not change over time.

conservation of energy The conservation law stating that in an isolated, closed system, the total energy does not change.

constant of proportionality The multiplicative factor by which one quantity is related to another.

constellation An imaginary image formed by patterns of stars; any of 88 defined areas on the celestial sphere used by astronomers to locate celestial objects.

constructive interference A state in which the amplitudes of two intersecting waves reinforce one another. See also *destructive interference, interference.*

continental drift The slow motion (centimeters per year) of Earth's continents relative to each other and to Earth's mantle.

continuous radiation Electromagnetic radiation with intensity that varies smoothly over some range of wavelengths.

convection The transport of thermal energy from the lower (hotter) to the higher (cooler) layers of a fluid by motions within the fluid driven by variations in buoyancy.

convective zone A region within a star in which energy is transported outward by convection.

Copernicus, Nicolaus (1473–1543) The Polish monk who first proposed the model of a Sun-centered Solar System to the Western world.

core (a) The innermost region of a planetary interior. (b) The innermost part of a star.

core accretion A process for forming giant planets, whereby large quantities of surrounding hydrogen and helium are gravitationally captured onto a massive rocky core.

Coriolis effect The apparent displacement of objects in a direction perpendicular to their true motion as viewed from a rotating frame of reference. On a rotating planet, different latitudes rotating at different speeds cause this effect.

corona The hot, outermost part of the Sun's atmosphere.

coronal hole A lower-density region in the solar corona containing "open" magnetic field lines along which coronal material is free to stream into interplanetary space.

coronal mass ejection An eruption on the Sun that ejects hot gas and energetic particles at much higher speeds than are typical in the solar wind.

cosmic background radiation (CBR) Isotropic microwave radiation from the celestial sphere having a 2.73 K Planck spectrum. The CBR is understood as residual radiation from the Big Bang.

cosmic ray A very fast-moving particle (usually an atomic nucleus) that resides in the disk of our galaxy.

cosmological constant A constant introduced into general relativity by Einstein that characterizes an extra, repulsive force in the universe due to the vacuum of space itself.

cosmological principle The (testable) assumption that the same physical laws that apply here and now also apply everywhere and at all times, and that there are no special locations or directions in the universe.

cosmological redshift (z) The redshift that results from the expansion of the universe rather than from the motions of galaxies or gravity (see *gravitational redshift*).

cosmology The study of the large-scale structure and evolution of the universe as a whole.

Crab Nebula The remnant of the Type II supernova explosion witnessed by Chinese astronomers in A.D. 1054.

crescent Phases of the Moon, Mercury or Venus in which the object appears less than half illuminated by the Sun.

Cretaceous period The interval in Earth's history from 146 million to 65 million years ago.

Cretaceous–Tertiary (K-T) boundary The boundary between the Cretaceous and Tertiary periods in Earth's history;

corresponds to the time of the impact of an asteroid or comet and the extinction of the dinosaurs.

critical mass density   The value of the mass density of the universe that, ignoring any cosmological constant, is just barely capable of halting expansion of the universe.

crust   The relatively thin, outermost, hard layer of a planet, which is chemically distinct from the interior.

cryovolcanism   Low temperature volcanism in which the *magmas* are composed of motter *ices* rather than rocky material. See also *volcanism*.

C-type asteroid   An asteroid made of material that has largely been unmodified since the formation of the Solar System; the most primitive type of asteroid.

cumulonimbus   A towering cumulus-type cloud associated with thunderstorms, also known as a *thunderhead*.

cumulus   A large puffy cloud with a flat base.

Curtis, Heber D. (1872–1942)   The Lick Observatory astronomer who argued, in the Great Debate of 1920, that galaxies are island universes. See also *Shapley, Harlow*.

cyclonic motion   The rotation of a weather system resulting from the Coriolis effect as air moves toward a region of low atmospheric pressure.

Cygnus X-1   A binary X-ray source and probable black hole.

D

dark matter   The matter existing in galaxies and in groups and clusters of galaxies that does not emit or absorb electromagnetic radiation; thought to comprise most of the mass in the universe.

dark matter halo   The centrally condensed, greatly extended dark matter component of a galaxy that contains up to 95 percent of the galaxy's mass.

Darwin, Charles (1809–1882)   The English naturalist whose publication, *The Origin of Species*, opened the way to the modern theory of evolution.

daughter element   An element resulting from radioactive decay of a more massive parent element.

decay   (a) The process of a radioactive nucleus changing into its daughter element. (b) An atom or molecule dropping from a higher-energy state to a lower-energy state.

density   The measure of an object's mass per unit of volume. Units: kg/m^3.

destructive interference   A state in which the amplitudes of two intersecting waves cancel one another. See also *constructive interference, interference*.

differential rotation   Rotation of different parts of a system at different rates.

differentiation   The process by which materials of higher density sink toward the center of a molten or fluid planetary interior.

diffraction   The spreading of a wave after it passes through an opening or past the edge of an object.

diffraction limit   The limit of a telescope's angular resolution caused by diffraction.

dispersion   The separation of rays of light into their component wavelengths.

distance ladder   A sequence of techniques for measuring cosmic distances; each method is calibrated using the results from other methods that have been applied to closer objects.

Doppler effect   The change in wavelength of sound or light due to the relative motion of the source toward or away from the observer.

Doppler redshift   See *redshift*.

Doppler shift   The change in observed wavelength in a wave from a source moving toward or away from an observer.

Drake equation   A prescription for estimating the number of intelligent civilizations existing elsewhere.

Drake, Frank (1930–)   The American astronomer best known for his formulation of the *Drake equation*.

dust devil   A small tornadolike column of air containing dust or sand.

dust tail   A type of comet tail consisting of dust particles that are pushed away from the comet's head by radiation pressure from the Sun.

dwarf galaxy   A small galaxy with a luminosity ranging from 1 million to 1 billion solar luminosities.

dwarf planet   A body with characteristics similar to that of a classical planet except that it has not cleared smaller bodies from the neighboring regions around its orbit. Compare with *planet*.

dynamic equilibrium   A situation in which the configuration of a system does not change even though the components of that system are in motion.

dynamo   A device that converts mechanical energy into electrical energy in the form of electric currents and magnetic fields. The *dynamo effect* is thought to create magnetic fields in planets and stars by electrically charged currents of material flowing within their cores.

E

eccentricity (e)   A measure of the departure of an ellipse from circularity; the ratio of the distance between the two foci of an ellipse to its major axis.

eclipse season   Times during the year when the line of nodes is sufficiently close to the Sun for eclipses to occur.

eclipsing binary   A binary system in which the orbital plane is oriented such that the two stars appear to pass in front of one another as seen from Earth.

ecliptic   (a) The apparent annual path of the Sun against the background of stars. (b) The projection of Earth's orbital plane onto the celestial sphere.

effective temperature   The temperature at which a black body, such as a star, appears to radiate.

Einstein, Albert (1879–1955)   The eminent 20th century German physicist whose theories about relativity and the properties of light changed how we view the physical universe.

ejecta   (a) Material thrown outward by the impact of an asteroid or comet on a planetary surface, leaving a crater behind. (b) Material thrown outward by a stellar explosion.

electric field   A field that is able to exert a force on a charged object, whether at rest or moving.

electric force   The force exerted on a charged particle by an electric field.

electromagnetic force   The force, including both electric and magnetic forces, that acts on electrically charged particles. One of four fundamental forces of nature. The force mediated by photons.

electromagnetic radiation   A traveling disturbance in the electric and magnetic fields caused by accelerating electric charges. In quantum mechanics, a stream of photons. Light.

electromagnetic spectrum   The spectrum made up of all possible frequencies or wavelengths of electromagnetic radiation, ranging from gamma rays through radio waves and including the portion our eyes can use.

electromagnetic wave   A wave consisting of oscillations in the electric field strength and the magnetic field strength.

electron (e⁻)   A fundamental particle having a negative charge of 1.6×10^{-19} coulomb, a rest mass of 9.1×10^{-31} kg, and mass-equivalent energy of 8×10^{-14} joules.

electron degenerate   Describes the state of material compressed to the point where electron density reaches the limit imposed by the rules of quantum mechanics.

electroweak theory The quantum theory that combines descriptions of both the electromagnetic force and the weak nuclear force.

element One of 92 naturally occurring substances (such as hydrogen, oxygen, and uranium) and more than 20 human-made ones (such as plutonium). Each element is chemically defined by the specific number of protons in the nuclei of its atoms.

elementary particle One of the basic building blocks of nature, such as the *electron* and the *quark*.

ellipse A conic section produced by the intersection of a plane with a cone by passing the plane through the cone at some angle to the axis other than 0° or 90°.

elliptical galaxy A galaxy with a circular to elliptical outline on the sky containing almost no disk and a population of old stars (abbreviated E galaxy).

emission The release of electromagnetic energy when an atom, molecule, or particle drops from a higher-energy state to a lower-energy state.

emission line An intensity peak in a spectrum due to sharply defined emission of electromagnetic radiation in a narrow range of wavelengths.

empirical science Based primarily on observations and experimental data. Descriptive rather than based on theoretical inference.

energy The conserved quantity that gives objects and systems the ability to do work. Units: joules (J).

energy transport The transport of energy from one location to another. In stars, energy transport is mostly carried out by radiation or convection.

entropy A measure of the disorder of a system related to the number of ways a system can be rearranged without affecting its appearance.

equator The great circle on the surface of a body midway between its poles. The equatorial plane passes through the center of the body and is perpendicular to its rotation axis.

equilibrium The state of an object in which physical processes balance each other so that its properties or conditions remain constant.

equinox ("equal night") (a) One of two positions on the ecliptic where it intersects the celestial equator. (b) Either of the two times of year at which the Sun is at one of these two positions. At this time, night and day are of the same length everywhere on Earth. See also *autumnal equinox, vernal equinox*.

equivalence principle The principle stating that there is no difference between a reference frame that is freely floating through space and one that is freely falling within a gravitational field.

escape velocity The minimum velocity needed for an object to achieve a parabolic trajectory and thus permanently leave the gravitational grasp of another mass.

Eta Carinae An extremely massive luminous star that underwent an episode of great mass loss during the 1840s.

event A particular location in *spacetime*.

event horizon The effective "surface" of a *black hole*. Nothing inside this surface—not even light—can escape from a black hole.

evolutionary track The path that a star follows across the H-R diagram as it evolves through its lifetime.

excited state An energy level of a particular atom, molecule, or particle, higher than its ground state.

extrasolar planet A planet orbiting a star other than the Sun.

F

Fahrenheit (F) The arbitrary temperature scale with 32°F at the melting point of water and 212°F at the boiling point of water at sea level. Defined by Gabriel Fahrenheit (1686–1736).

fault A fracture in the crust of a planet or moon along which blocks of material can slide.

filter An instrument element that transmits a limited wavelength range of electromagnetic radiation. For the optical range such elements are typically made of different kinds of glass and take on the hue of the light they transmit.

first quarter Moon The phase of the Moon as seen from Earth in which only the western half of the Moon is illuminated by the Sun; occurs about a week after a new Moon.

fissure A fracture in the planetary lithosphere from which magma emerges.

flatness problem The surprising result that the sum of Ω_{mass} plus Ω_{Λ} is extremely close to unity in the present-day universe; equivalent to saying that it is surprising the universe is so flat.

flat rotation curve A rotation curve of a spiral galaxy in which rotation rates do not decline in the outer part of the galaxy, but remain relatively constant to the outermost points.

flat universe An infinite universe whose spatial structure obeys Euclidean geometry, such that the sum of the angles of a triangle always equals 180°.

flux The total amount of energy passing through each square meter of a surface each second. Units: watts per square meter (W/m^2).

flux tube Tubelike structures in a plasma, such as the atmosphere of the Sun, or the interaction of Io and Jupiter, that are "bundled" together by magnetic fields.

flyby A spacecraft that first approaches and then continues on past a planet or moon. Flybys can visit multiple objects, but they remain in the vicinity of their targets only briefly. See also *orbiter*.

focal length The optical distance between a telescope's objective lens or primary mirror and the plane (called the focal plane) on which the light from a distant object is focused.

focal plane The plane, perpendicular to the optical axis of a lens or mirror, in which an image is formed.

focus (pl. **foci**) (a) One of two points that define an ellipse. (b) A point in the *focal plane* of a telescope.

fold A location where rock is bent or warped.

Foucault, Jean-Bernard-Leon (1819–1868) The French physicist who was the first to demonstrate that Earth rotates using the periodic motions of a pendulum. A pendulum that demonstrates Earth's rotation is called a *Foucault pendulum* in his honor.

Foucault pendulum A pendulum used to demonstrate the rotation of Earth.

frame of reference A frame against which an observer measures positions, motions, and the like.

free fall The motion of an object when the only force acting on it is gravity.

frequency The number of times per second that a periodic process occurs. Unit: hertz (Hz), 1/s.

full Moon The phase of the Moon as seen from Earth in which the near side of the Moon is fully illuminated by the Sun; occurs about two weeks after a new Moon.

G

galactic cluster See *open cluster*.

galaxy A gravitationally bound system that consists of stars and star clusters, gas, dust, and dark matter; typically greater than 1,000 light-years across; and recognizable as a discrete, single object.

galaxy cluster   A large, gravitationally bound collection of galaxies containing hundreds to thousands of members; typically 10 to 15 Mly across.

galaxy group   A small, gravitationally bound collection of galaxies containing from several to a hundred members; typically 4 to 6 Mly across.

Galileo Galilei (1564–1642)   The first person to use a telescope to make useful discoveries about the heavens and to develop the concept of inertia.

gamma ray   Electromagnetic radiation with higher frequency, higher photon energy, and shorter wavelength than all other types of electromagnetic radiation.

Gamow, George (1904–1968)   The Ukrainian-American theoretical physicist who, with his student Ralph Alpher, first predicted the existence of the *cosmic background radiation*.

gas giant   A giant planet formed mostly of hydrogen and helium.

general relativistic time dilation   The verified prediction that time passes more slowly in a gravitational field than in the absence of a gravitational field. See *time dilation*.

general relativity   See *general theory of relativity*.

general theory of relativity   Einstein's theory explaining gravity as the distortion of spacetime by massive objects; also known simply as *general relativity*. This theory deals with all types of motion; for contrast, see *special relativity*.

geodesic   The path an object will follow through *spacetime* in the absence of external forces.

geology   The study of the composition, behavior, and history of solid bodies, including Earth, the terrestrial planets, and the moons of the Solar System.

giant galaxy   A galaxy with luminosity greater than about a billion solar luminosities.

giant molecular cloud   An interstellar cloud composed primarily of molecular gas and dust, having hundreds of thousands of solar masses.

giant planet   One of the largest planets in the Solar System, typically 10 times the size and many times the mass of the terrestrial planets and lacking a solid surface.

gibbous   Phases of the Moon, Mercury, or Venus in which the object appears more than half illuminated by the Sun.

global circulation   The overall, planet-wide circulation pattern of a planet's atmosphere.

globular cluster   A spherically symmetric, highly condensed cluster of stars, containing tens of thousands to a million members.

gluon   The particle that carries (or, equivalently, mediates) interactions due to the *strong nuclear force*.

gradation   The leveling of a planet's surface through weathering, erosion, transpiration, and deposition of rock debris by water, wind, and gravity.

grand unified theory (GUT)   A unified quantum theory that combines the *strong nuclear, weak nuclear,* and *electromagnetic forces* but does not include *gravity*.

granite   Rock that is cooled from magma and is relatively rich in silicon and oxygen.

grating   A *diffraction* grating. An optical surface containing many narrow, closely and equally spaced parallel grooves or slits that create *dispersion* of reflected or transmitted light.

gravitational lens   A massive object that gravitationally focuses the light of a more distant object to produce multiple brighter, magnified, possibly distorted images.

gravitational lensing   The bending of light by gravity.

gravitational potential energy   The potential energy that an object has solely due to its position within a gravitational field.

gravitational redshift   The shifting to longer wavelengths of the wavelength of radiation from an object deep within a gravitational well.

gravity   (a) The mutually attractive force between massive objects. (b) An effect arising from the bending of *spacetime* by massive objects. (c) One of four fundamental forces of nature.

gravity wave   Wave in the fabric of *spacetime* emitted by accelerating masses.

great circle   Any circle on a sphere that has as its center the center of the sphere. The celestial equator, the meridian, and the ecliptic are all great circles on the sphere of the sky, as is any circle drawn through the zenith.

Great Red Spot (GRS)   The giant, oval, brick-red anticyclone seen in Jupiter's southern hemisphere.

greenhouse effect   The solar heating of air in an enclosed space, such as a closed building or car, resulting primarily from the inability of the hot air to escape. Compare with *atmospheric greenhouse effect*.

greenhouse molecule   One of a group of atmospheric molecules such as carbon dioxide that are transparent to visible radiation but absorb infrared radiation.

Gregorian calendar   The modern calendar. A modification of the Julian calendar decreed by Pope Gregory XIII in 1582. By this time the less accurate Julian calendar had developed an error of 10 days over the 13 centuries since its inception.

ground state   The lowest possible energy state for a system or part of a system, such as an atom, molecule, or particle.

GUT   See *grand unified theory*.

Guth, Alan (1947–)   The physicist who first proposed that the universe underwent an early, rapid *inflation*, which resolves both the flatness and horizon problems.

H

Hadley circulation   A simplified, and therefore uncommon, atmospheric global circulation that carries thermal energy directly from the equator to the polar regions of a planet.

half-life   The time it takes half a sample of a particular radioactive element to decay to a daughter element.

halo   The spherically symmetric, low-density distribution of stars and dark matter that defines the outermost regions of a galaxy.

harmonic law   Another name for *Kepler's third law* of planetary motion.

Hawking, Stephen (1942–)   The British physicist/astrophysicist who has done much to further our understanding of the universe and the phenomena surrounding black holes.

Hawking radiation   Radiation from a black hole.

Hayashi track   The path that a protostar follows on the H-R diagram as it contracts toward the main sequence.

head   The part of a comet that includes both the nucleus and the inner part of the coma.

heat death   The possible eventual fate of an open universe, in which entropy has triumphed and all energy- and structure-producing processes have come to an end.

heavy element   See *massive element*.

Heisenberg uncertainty principle   The physical limitation that the *product of the position and the momentum* of a particle cannot be smaller than a well-defined value, *Planck's constant* (h).

Heisenberg, Werner (1901-1976) The German theoretical physicist and Nobel laureate who was one of the founders of quantum mechanics. He is perhaps best known for his discovery of the *Heisenberg uncertainty principle*.

helioseismology The use of solar oscillations to study the interior of the Sun.

helium flash The runaway explosive burning of helium in the degenerate helium core of a *red giant* star.

Herbig-Haro (HH) object A glowing, rapidly moving knot of gas and dust that is excited by bipolar outflows in very young stars.

heredity The process by which one generation passes on its characteristics to future generations.

hertz (Hz) A unit of frequency equivalent to cycles per second.

Hertz, Heinrich (1857−1894) The 19th century physicist who supplied the first experimental data confirming Maxwell's predictions about electromagnetic radiation.

Hertzsprung, Einar (1873−1967) The Danish astronomer who, along with Henry Norris Russell, defined the main diagnostic tool for studying the properties of stars, in which we graph stellar luminosity versus stellar spectral type (or temperature, or color). The *Hertzsprung-Russell (H-R) diagram* is named in his honor.

Hertzsprung-Russell diagram See *H-R diagram*.

HH object See *Herbig-Haro object*.

hierarchical clustering The "bottom-up" process of forming large-scale structure. Small-scale structure first produces groups of galaxies, which in turn form clusters, which then form superclusters.

high-velocity star A star belonging to the halo found near the Sun, distinguished from disk stars by moving far faster and often in the direction opposite to the rotation of the disk and its stars.

homogeneous In cosmology, describes a universe in which observers in any location would observe the same properties.

horizon The boundary that separates the sky from the ground.

horizon problem The puzzling observation that the cosmic background radiation is so uniform in all directions, despite the fact that widely separated regions should have been "over the horizon" from each other in the early universe.

horizontal branch A region on the H-R diagram defined by stars burning helium to carbon in a stable core.

hot dark matter Particles of dark matter that move so fast that gravity cannot confine them to the volume occupied by a galaxy's normal luminous matter.

hot Jupiter A large, Jovian-type *extrasolar planet* located very close to its parent star.

hot spot A place where hot plumes of mantle material rise near the surface of a planet.

H-R diagram The Hertzsprung-Russell diagram, which is a plot of the luminosities versus the surface temperatures of stars. The evolving properties of stars are plotted as tracks across the H-R diagram.

H II region A region of interstellar gas that has been ionized by UV radiation from nearby hot massive stars.

Hubble constant (H_0) The constant of proportionality relating the recession velocities of galaxies to their distances. See also *Hubble time*.

Hubble, Edwin P. (1889−1953) The American astronomer who discovered the true nature of galaxies and that the universe is expanding. The *Hubble Space Telescope* is named after him.

Hubble's law The law stating that the speed at which a galaxy is moving away from us is proportional to the distance of that galaxy.

Hubble Space Telescope (HST) A 2.4-m telescope placed in Earth orbit in 1990.

Hubble time An estimate of the age of the universe from the inverse of the Hubble constant, $1/H_0$.

hurricane A large tropical cyclonic system circulating counterclockwise in the Northern Hemisphere and clockwise in the Southern Hemisphere. Hurricanes can extend outward from their center to more than 600 km and generate winds in excess of 300 km/h. Hurricanes are also known as *cyclones* or *typhoons*.

Huygens, Christiaan (1619−1695) Among the founders of mechanics and optics; the Dutch physicist first to recognize that Saturn has rings.

hydrogen burning The release of energy from the nuclear fusion of four hydrogen atoms into a single helium atom.

hydrogen shell burning Fusion of hydrogen in a shell surrounding a stellar core that may be either degenerate or fusing more massive elements.

hydrosphere The portion of Earth that is largely liquid water.

hydrostatic equilibrium The condition in which the weight bearing down at some point within an object is balanced by the pressure within the object.

hypothesis A well-thought-out idea, based on scientific principles and knowledge, that makes testable predictions.

hypothetical Proposed but not yet tested.

I

ice The solid form of a volatile material; sometimes the volatile material itself, regardless of its physical form.

ice giant A giant planet formed mostly of the liquid form of volatile substances.

ideal gas A gas in which all collisions between individual atoms or molecules are like collisions between billiard balls or marbles.

ideal gas law The relationship between pressure (P), number density of particles (n), and temperature (T) expressed as $P = nkT$, where k is Boltzmann's constant.

igneous activity The formation and action of molten rock or magma.

impact crater The scar of the impact left on a solid planetary or moon surface by collision with another object.

index or refraction (n) The ratio of the speed of light in a vacuum, c, to the speed of light in an optical medium.

inert gas A gaseous element that combines with other elements only under conditions of extreme temperature and pressure—such as helium, neon, and argon.

inertia The tendency for objects to retain their state of motion.

inertial frame of reference A reference frame that is not accelerating. In general relativity, a reference frame that is falling freely in a gravitational field.

inflation An extremely brief phase of ultra-rapid expansion of the very early universe. Following inflation, the standard Big Bang models of expansion apply.

infrared (IR) radiation Electromagnetic radiation occurring in the spectral region between those of visible light and microwaves.

instability strip A region of the H-R diagram containing stars that pulsate with a periodic variation in luminosity.

integration time The time interval over which photons are collected and added up in a detecting device.

intensity Amount per second per unit area. In the case of electromagnetic radiation: W/m^2.

intercloud gas   A low-density region of the interstellar medium that fills the space between interstellar clouds.

interference   The interaction of two sets of waves pruducing high and low intensity, depending on whether their amplitudes reinforce or cancel. See also *constructive interference* and *destructive interference*.

interferometer   A group or array of separate but linked optical or radio telescopes whose overall separation determines the angular resolution of the system.

interferometric array   See *interferometer*.

intermediate-mass star   Star with mass between about 3 $M_\odot$ and 8 $M_\odot$.

interstellar cloud   A discrete, high-density region of the interstellar medium made up mostly of atomic or molecular hydrogen and dust.

interstellar dust   Small particles or grains (0.01 to 10 μm) of matter, primarily carbon and silicates, distributed throughout interstellar space.

interstellar extinction   The dimming of visible and ultraviolet light by interstellar dust.

interstellar medium   The gas and dust that fill the space between the stars within a galaxy.

inverse square law   The rule that a quantity or effect diminishes with the square of the distance from the source.

ion   An *atom* or *molecule* that has lost or gained one or more *electrons*.

ionize   The process by which electrons are stripped free from an *atom* or *molecule*, resulting in free *electrons* and a positively charged atom or molecule.

ionosphere   A layer high in Earth's atmosphere in which most of the atoms are ionized by solar radiation.

ion tail   A type of comet tail consisting of ionized gas. Particles in the ion tail are pushed directly away from the comet's head in the antisolar direction at high speeds by the solar wind.

iron meteorite   A metallic meteorite composed mostly of iron–nickel alloys.

irregular galaxy   A galaxy without regular or symmetric appearance.

irregular moon   A moon that has been captured by a planet. Some revolve in the opposite direction from the rotation of the planet, and many are in distant, unstable orbits.

isotopes   Forms of the same element with differing numbers of neutrons.

isotropic   In cosmology, a universe whose properties observers find to be the same in all directions.

J

jansky (*Jy*)   The basic unit of flux density. Units: W/m²/Hz.

Jansky, Karl (1905-1950)   The American physicist and radio engineer whose discovery of radio emissions from the *Milky Way Galaxy* led to the birth of radio astronomy.

jet   (a) A stream of gas and dust ejected from a comet nucleus by solar heating. (b) A collimated linear feature of bright emission extending from a *protostar or active galactic nucleus*.

joule (J)   A unit of energy or work. 1 J = 1 newton meter.

K

Kant, Immanuel (1724–1804)   The 18th century German philosopher who hypothesized that the nebulae seen by Messier and Herschel were "island universes," separate from our own.

kelvin (K)   The temperature scale using Celsius-sized degrees, but with 0 K defined as *absolute zero* instead of the melting point of water. Defined by William Thompson, better known as Lord Kelvin (1824–1907).

Kepler, Johannes (1571–1630)   The discoverer of the elliptical shape of planetary orbits and how orbital period is related to average distance from the Sun.

Kepler's first law   The law stating that planets move in orbits of elliptical shapes with the Sun at one focus.

Kepler's laws   The three rules of planetary motion inferred by Johannes Kepler from the data acquired by Tycho Brahe.

Kepler's second law   The law stating that a line drawn from the Sun to a planet sweeps out equal areas in equal times as the planet orbits the Sun. Also called the *law of equal areas*.

Kepler's third law   The relationship between the period of a planet's orbit and its distance from the Sun. Also called the *harmonic law*.

kinetic energy   The energy of an object due to its motions. KE = 1/2 mv^2. Units: joules (J).

Kirkwood, Daniel (1814-1895)   The American astronomer who discovered resonant gaps in the asteroid belt, now known as *Kirkwood gaps*.

Kirkwood gap   A gap in the main asteroid belt related to orbital resonances with Jupiter.

Kuiper Belt   A disk-shaped population of comet nuclei extending from Neptune's orbit to perhaps several thousand AU from the Sun.

Kuiper Belt object (KBO)   An icy planetesimal (comet nucleus) that orbits within the Kuiper Belt beyond the orbit of Neptune. Also known as a *Trans-Neptunian Object* (TNO).

Kuiper, Gerard Peter (1905-1973)   The Dutch-American planetary astronomer who discovered CO_2 in the atmosphere of Mars and pioneered airborne infrared observing. Kuiper is well known for his suggestion that a belt of trans-Neptunian objects, now known as the *Kuiper Belt*, is the source of *short-period comets*.

L

Lagrange, Joseph (1736–1813)   Italian-French mathematician who contributed to the theory of celestial mechanics.

Lagrangian equilibrium point   One of five points of equilibrium in a system consisting of two massive objects in nearly circular orbit around a common center of mass. Only two (L_4 and L_5) represent stable equilibrium. A third smaller body located at one of the five points will move in lockstep with the center of mass of the larger bodies.

lambda peak (λ_{peak})   The wavelength at which the thermal radiation from an object is most intense. Literally, the peak of the Planck spectrum as specified by Wien's Law.

lander   An instrumented spacecraft designed to land on a planet or moon. See also *rover*.

large-scale structure   Observable aggregates on the largest scales in the universe, including galaxy groups, clusters, and superclusters.

latitude   The angular distance north (+) or south (–) from the equatorial plane of a nearly spherical body.

law of conservation of angular momentum   The physical law stating that a system's total angular momentum remains constant, unless acted on by an external torque.

law of equal areas   See *Kepler's second law*.

law of gravitation   See *universal law of gravitation*.

leap year   A year that contains 366 days. Leap years occur every four years when the year is divisible by four, correcting for the accumulated excess time in a normal year, which is approximately 365¼ days long.

Leavitt, Henrietta (1868–1921) The American astronomer who discovered that periods of variation of Cepheid stars are related to their luminosities by studying the variable stars in the Large and Small Magellanic Clouds.

Leonids A November meteor shower associated with the dust debris left by comet Tempel-Tuttle.

libration The apparent wobble of an orbiting body that is tidally locked to its companion (such as Earth's Moon) resulting from the fact that its orbit is elliptical rather than circular.

light All electromagnetic radiation, which comprises the entire *electromagnetic spectrum*.

light-year (ly) The distance light travels in one year—about 9 trillion kilometers.

limb darkening The darker appearance caused by increased atmospheric absorption near the limb of a planet or star.

line of nodes (a) A line defined by the intersection of two orbital planes. (b) The line defined by the intersection of Earth's equatorial plane and the plane of the ecliptic.

Lippershey, Hans (1570-1619) The German-Flemish spectacle maker who is generally credited with the discovery of the optical telescope.

lithosphere The solid, brittle part of Earth (or any planet or moon), including the crust and the upper part of the mantle.

lithospheric plate A separate piece of Earth's lithosphere capable of moving independently. See *continental drift, plate tectonics*.

Local Group The small group of galaxies of which the Milky Way and the Andromeda Galaxy are members.

longitudinal wave A wave that oscillates parallel to the direction of the wave's propagation.

long-period comet A comet with an orbital period greater than 200 years.

look-back time The time that it has taken the light from an astronomical object to reach Earth.

low-mass star A star with a main sequence mass of less than about 3 $M_\odot$.

luminosity The total flux emitted by an object. Unit: watts (W).

luminosity–temperature–radius relationship A relationship among these three properties of stars indicating that if any two are known, the third can be calculated.

luminous matter Matter in galaxies—including stars, gas, and dust—that emits electromagnetic radiation.

lunar eclipse An eclipse that occurs when the Moon is partially or entirely in Earth's shadow.

lunar tide The component of Earth's tides due to the Moon.

M

MACHO "Massive compact halo object," such as brown dwarfs, white dwarfs, and black holes, which are candidates for being considered dark matter. See also *WIMP*.

magma Molten rock often containing dissolved gases and solid minerals.

magnetic field A field that is able to exert a force on a moving electric charge. See *electromagnetic force*.

magnetic force A force associated with, or caused by, the relative motion of charges.

magnetosphere The region surrounding a planet filled with relatively intense magnetic fields and plasmas.

magnitude A system used by astronomers to describe the brightness or luminosity of stars. The brighter the star, the smaller its magnitude.

main asteroid belt See *asteroid belt*.

main sequence The strip on the H-R diagram where most stars are found. Main sequence stars are fusing hydrogen to helium in their cores.

main sequence lifetime The amount of time a star spends on the main sequence, fusing hydrogen into helium in its core.

main sequence turnoff The location on the H-R diagram of a single-aged stellar population (such as a star cluster) where stars have just evolved off the main sequence. The position of the main sequence turnoff is determined by the age of the stellar population.

mantle The solid portion of a rocky planet that lies between the crust and the core.

mare (pl. **maria**) A dark region on the Moon, composed of basaltic lava flows.

mass (a) Inertial mass is the property of matter that resists changes in motion. (b) Gravitational mass is the property of matter defined by its attractive force on other objects. According to general relativity the two are equivalent.

massive element (a) In astronomy, refers to all elements more massive than helium. (b) In other sciences (and sometimes also

in astronomy), refers to the most massive elements in the periodic table, such as uranium and plutonium. Also called heavy element.

mass transfer The transfer of mass from one member of a binary star system to its companion. Mass transfer occurs when one of the stars evolves to the point that it overfills its *Roche lobe*, so that its outer layers are pulled toward its binary companion.

mathematics The science and language of patterns; the language of science. Mathematics provides the tools used by scientists to understand, describe, and predict the patterns found in nature.

matter Objects made of particles that have mass, such as protons, neutrons, and electrons; anything that occupies space and has mass.

Maunder minimum The period 1645–1715, when there were few sunspots.

Maxwell, James Clerk (1831–1879) The 19th century Scottish physicist who summarized all of classical electromagnetism, including electromagnetic radiation, with a set of four equations.

megabar A unit of pressure equal to 1 million *bars*.

mega-light-year (Mly) A unit of distance equal to 1 million *light-years*.

meridian The imaginary arc in the sky running from the horizon at due north through the zenith to the horizon at due south. The meridian divides the observer's sky into eastern and western halves.

mesosphere A portion of Earth's atmosphere that lies above the stratosphere.

Messier, Charles (1730–1817) An early French comet hunter; published the first catalog of bright, diffuse celestial objects to prevent their being confused with comets.

Messier's Catalog The catalog made by Charles Messier in the late 18th century in which he identifies 103 bright, diffuse-appearing objects visible from France, all of them either star clusters, planetary nebulae, H II regions, or galaxies; the catalog was later expanded to 110 objects.

meteor The incandescent trail produced by a small piece of interplanetary debris as it travels through the atmosphere at very high speeds.

meteorite A meteoroid that survives to reach a planet's surface.

meteoroid A small cometary or asteroidal fragment ranging in size from 100 μm to 100 m. When entering a planetary atmosphere, the meteoroid creates a *meteor,* which is an atmospheric phenomenon.

meteor shower   A larger than normal display of meteors, occurring when Earth passes through the orbit of a disintegrating comet, sweeping up its debris.

micrometer (μm) 10^{-6} meter; a unit of length used for measuring the wavelength of electromagnetic radiation. Also called *micron.*

micron   See *micrometer.*

microwave radiation   Electromagnetic radiation occurring in the spectral region between those of infrared radiation and radio waves.

Milky Way Galaxy   The *galaxy* in which our Sun and Solar System reside.

minute of arc (′)   A unit for measuring angles. A minute of arc is 1/60 of a degree of arc. Also called *arcminute.*

Mly   See *mega-light-year.*

modern physics   A term usually used to refer to those physical principles, including relativity and quantum mechanics, developed since Maxwell's equations were published.

molecular cloud   An interstellar cloud composed primarily of molecular hydrogen.

molecular cloud core   A dense clump within a molecular cloud that forms as the cloud collapses and fragments. Protostars form from molecular cloud cores.

molecule   Generally, the smallest particle of a substance that retains its chemical properties and is composed of two or more atoms. A very few types of molecules, such as helium, are composed of single atoms.

momentum   The product of the *mass* and *velocity* of a particle. Units: kg m/s.

moon   A less massive satellite orbiting a more massive object. Moons are found around planets, dwarf planets, asteroids, and KBOs.

M-type asteroid   An asteroid once part of the metallic core of a larger, differentiated body that has since been broken into pieces; mostly made of iron and nickel.

mutation   In biology, an imperfect reproduction of self-replicating material.

N

natural selection   The process by which forms of structure, ranging from molecules to whole organisms, that are best adapted to their environment become more common than less well-adapted forms. See also *Theory of Evolution.*

nature   The word frequently used by scientists to denote the physical universe. Note the difference from other common uses.

neap tide   An especially weak tide that occurs around the time of the first or third quarter Moon when the gravitational forces of the Moon and the Sun on Earth are at right angles to each other, thus producing the least pronounced tides.

near-Earth asteroid   An asteroid whose orbit brings it close to the orbit of Earth. See also *near-Earth object.*

near-Earth object (NEO)   An asteroid, comet, or large meteoroid whose orbit intersects Earth's orbit.

nebula (pl. nebulae)   A cloud of interstellar gas and dust, either illuminated by stars (bright nebula) or seen in silhouette against a brighter background (dark nebula).

neutrino   A very low-mass, electrically neutral particle emitted during beta decay. (Neutrinos interact with matter only very feebly and so can penetrate through great quantities of matter.)

neutrino cooling   The process in which thermal energy is carried out of the center of a star by neutrinos rather than by electromagnetic radiation or convection.

neutron (n^0)   A subatomic particle having no net electric charge, and a rest mass and rest energy nearly equal to that of the proton.

neutron star   The neutron degenerate remnant left behind by a Type II supernova.

new Moon   The phase of the Moon that occurs when the Moon is between Earth and the Sun, and we see only the side of the Moon not being illuminated by the Sun.

newton (N)   The force required to accelerate a 1-kg mass at a rate of 1 m/s². Units: kg m/s².

Newton's first law of motion   The law stating that an object will remain at rest or will continue moving along a straight line at a constant speed until an unbalanced force acts on it.

Newton, Sir Isaac (1642–1727)   The English discoverer and codifier of the laws describing the motion of objects and of the law of gravity; in many respects, the founding father of modern science.

Newton's laws   See *Newton's first law of motion, Newton's second law of motion,* and *Newton's third law of motion.*

Newton's second law of motion   The law stating that if an unbalanced force acts on a body, the body will have an accelera-

tion proportional to the unbalanced force and inversely proportional to the object's mass. The acceleration will be in the direction of the unbalanced force.

Newton's third law of motion   The law stating that for every force, there is an equal and opposite force.

nonthermal radiation   Any form of electromagnetic radiation, such as *synchrotron radiation,* that does not originate from thermal energy.

normal matter   Matter in galaxies that emits and absorbs electromagnetic radiation. Compare with *dark matter.*

north celestial pole   The northward projection of Earth's rotation axis onto the celestial sphere.

North Pole   The point in the Northern Hemisphere where Earth's rotation axis intersects the surface of Earth.

north star   See *Polaris.*

nova (pl. novae)   A stellar explosion that results from runaway *nuclear fusion* in a layer of material on the surface of a white dwarf in a binary system.

nuclear fusion   The combination of two less massive atomic nuclei into a single more massive atomic nucleus. Release of energy by fusion of low-mass elements is often referred to as *nuclear burning.*

nuclear burning   Release of energy by fusion of low-mass elements.

nucleosynthesis   The formation of more massive atomic nuclei from less massive nuclei, either in the Big Bang (Big Bang nucleosynthesis) or in the interiors of stars (stellar nucleosynthesis).

nucleus (pl. nuclei)   (a) The dense, central part of an atom. (b) The central core of a galaxy, comet, or other diffuse object.

O

oblateness   The flattening of an otherwise spherical planet or star caused by its rapid rotation.

obliquity   The inclination of a celestial body's equator to its orbital plane.

observational uncertainty   The fact that real measurements are never perfect; all observations are uncertain by some amount.

Occam's razor   The principle that the simplest hypothesis is the most likely; named after William of Occam (circa 1285–1349), the medieval English cleric to whom the idea is attributed.

Oort Cloud   A *spherical* distribution of comet nuclei stretching from beyond the

Kuiper Belt to more than 50,000 AU from the Sun.

Oort, Jan (1900–1992) The Dutch radio astronomer who first suggested that a distant, spherical swarm of objects, now known as the *Oort Cloud*, is the source of *long-period comets*.

opacity A measure of how effectively a material blocks the radiation going through it.

open cluster Group of a few dozen to a few thousand stars that formed together in the disk of a spiral galaxy. Also called *galactic clusters*.

open universe An infinite universe with a negatively curved spatial structure (much like the surface of a saddle) such that the sum of the angles of a triangle is always less than 180°.

orbit The path taken by one object moving around another object under the influence of their mutual gravitational or electric attraction.

orbital resonance A situation in which the orbital periods of two objects are related by a ratio of small integers.

orbiter A spacecraft placed in orbit around a planet or moon. See also *flyby*.

organic A substance, not necessarily of biological origin, containing the element carbon.

origins The subject relating to the genesis of the universe and life.

P

pair creation The production of a particle–antiparticle pair from a source of electromagnetic energy.

paleomagnetism The record of Earth's magnetic field as preserved in rocks.

palimpsest A flat, circular feature in icy lithospheres, believed to be a scar of an ancient impact.

parallax The apparent shift in the position of one object relative to another object, caused by the changing perspective of the observer. In astronomy, the displacement in the apparent position of a nearby star caused by the changing location of Earth in its orbit.

parent element A radioactive element. See *daughter element*.

parsec (pc) The distance to a star with a parallax of 1 arcsecond using a base of 1 AU. One parsec is approximately 3.26 light-years.

partial solar eclipse Any solar eclipse in which the observer is outside the path of totality.

peculiar velocity The motion of a galaxy relative to the overall expansion of the universe.

pendulum A mass supported by a string, wire, or rod that is free to swing back and forth in a gravitational field.

penumbra (pl. **penumbrae**) (a) The outer part of a shadow, where the source of light is only partially blocked. (b) The region surrounding the umbra of a sunspot. The penumbra is cooler and darker than the surrounding surface of the Sun but not as cool or dark as the umbra of the sunspot.

penumbral lunar eclipse A lunar eclipse in which the Moon passes through the penumbra of Earth's shadow.

Penzias, Arno (1933–) The Bell Laboratories physicist who, together with Robert Wilson, in the early 1960s, discovered the cosmic background radiation, for which they won the Nobel Prize in physics.

perihelion (pl. **perihelia**) The closest point to the Sun on a solar orbit.

period The time it takes for a regularly repetitive process to complete one cycle.

period–luminosity relationship The relationship between the period of variability of a pulsating variable star, such as a Cepheid or RR Lyrae variable, and the luminosity of the star. Longer-period Cepheid or RR Lyrae variables are more luminous than their shorter-period cousins.

Perseids A prominent August meteor shower associated with the dust debris left by comet Swift-Tuttle.

phase One of the various appearances of the sunlit surface of the Moon or a planet caused by the change in viewing location of Earth relative to both the Sun and the object.

photino An elementary particle related to the *photon*. One of the leading candidates for the cold dark matter.

photochemical A chemical reaction driven by the absorption of electromagnetic radiation.

photodisintegration The disintegration of atomic nuclei by absorption of high energy radiation. For example, the splitting of iron into helium by the absorption of gamma rays.

photodissociation The breaking apart of molecules into smaller fragments or individual atoms by the action of photons.

photoelectric effect An effect whereby electrons are emitted from a substance illuminated by photons above a certain critical frequency.

photon A discrete unit or particle of electromagnetic radiation; a *quantum of light*. The energy of a photon is equal to Planck's constant (h) times the frequency (f) of its electromagnetic radiation: $E_{photon} = h \times f$. The particle that mediates the electromagnetic force.

photosphere The apparent surface of the Sun as seen in visible light.

physical laws Broad statements that predict some aspect of how the physical universe behaves and that are supported by many empirical tests.

physics The scientific study of the fundamental principles that govern the behavior of matter and energy, and the application of those principles to understanding and predicting natural phenomena.

pixel The smallest picture element in a digital image array.

Planck era The early time, just after the Big Bang, when the universe as a whole must be described with quantum mechanics.

Planck, Max (1858–1947) The German physicist who first postulated the quantum nature of electromagnetic radiation.

Planck's constant (h) The constant of proportionality between the energy of a photon and the frequency of the photon. This constant defines how much energy a single photon of a given frequency or wavelength has. Value: $h = 6.63 \times 10^{-34}$ joule-second.

Planck spectrum The spectrum of electromagnetic energy emitted by a blackbody per unit area per second, which is determined only by the temperature of the object. Also called *blackbody spectrum*.

planet (a) A large body that orbits the Sun or other star that shines only by light reflected from the Sun or star. (b) In the Solar System, a body that orbits the Sun, has sufficient mass for self-gravity to overcome rigid body forces so that it assumes a spherical shape, and has cleared smaller bodies from the neighborhood around its orbit. Compare with *dwarf planet*.

planetary nebula The expanding shell of material ejected by a dying AGB star. A planetary nebula glows from flourescence caused by intense ultraviolet light coming from the hot, stellar remnant at its center.

planetary system A system of planets and other smaller objects in orbit around a star.

planetesimal A primitive body of rock and ice, 100 m or more in diameter, that combines with others to form planets.

plasma A gas composed largely of charged particles but that also may include some neutral atoms.

plate tectonics   The geological theory concerning the motions of lithospheric plates, which in turn provides the theoretical basis for continental drift.

Polaris   The star that currently is the closest bright star to the north celestial pole.

positron   A positively charged subatomic particle; the *antiparticle* of the electron.

power   The rate at which work is done or at which energy is delivered. Unit: watts or joules/second (W or J/s).

precession of the equinoxes   The slow change in orientation between the ecliptic plane and the celestial equator caused by the wobbling of Earth's axis.

pressure   Force per unit area. Units: newtons per square meter (N/m²) or bars.

primary atmosphere   An atmosphere, composed mostly hydrogen and helium, that forms at the same time as its host planet.

primary mirror   The principal optical mirror in a reflecting telescope, which determines the telescope's light-gathering power and resolution.

primary wave   A longitudinal seismic wave in which the oscillations involve compression and decompression parallel to the direction of travel (that is, a pressure wave).

principle   A general idea or sense about how the universe is that guides us in constructing new scientific theories. Principles can be testable theories.

prograde   (a) Describes rotational or orbital motion of a moon that is in the same sense as the planet it orbits. (b) Describes the counterclockwise orbital motion of Solar System objects as seen from above Earth's orbital plane. See also *retrograde*.

prominence   An archlike projection above the solar photosphere often associated with a sunspot.

proportional   Two things are proportional if their ratio is a constant.

proton (*p* or *p*⁺)   A fundamental particle having a positive electric charge of 1.6×10^{-19} coulomb, a mass of 1.67×10^{-27} kg, and a rest energy of 1.5×10^{-10} joules.

proton–proton chain   One of the ways in which hydrogen burning—the fusion of four hydrogen atoms to form a helium atom—can take place. This is the most important path for hydrogen burning in low-mass stars such as the Sun.

protoplanetary disk   The remains of the accretion disk around a young star from which a planetary system may form.

protostar   A young stellar object that derives its luminosity from the conversion of gravitational energy to thermal energy, rather than from nuclear reactions in its core.

protostellar disk   The accretion disk that forms as an interstellar cloud collapses on its way to becoming a star.

pulsar   A rapidly rotating neutron star that beams radiation into space in two searchlight-like beams. To a distant observer, the star appears to flash on and off, earning its name.

pulsating variable star   A variable star that undergoes periodic radial pulsations.

Q

QCD   See *quantum chromodynamics*.

QED   See *quantum electrodynamics*.

quantized   Describes a quantity that exists as discrete, irreducible units.

quantum chromodynamics (QCD)   The quantum mechanical theory describing the strong nuclear force and its mediation by gluons.

quantum efficiency   The fraction of photons falling on a detector that actually produces a response in the detector.

quantum electrodynamics (QED)   The quantum theory that describes the electromagnetic force and its mediation by photons.

quantum mechanics   The branch of physics that deals with the quantized and probabilistic behavior of atoms and subatomic particles.

quantum of light   The discrete particle of light we call a *photon*.

quark   The building block of protons and neutrons.

quasar   The most luminous of the active galactic nuclei (AGNs), seen only at great distances from our galaxy; shortened from "quasi-stellar radio" source.

Quaternary Period   The period in Earth's history from 1.8 million years ago through today.

R

radial velocity   The component of velocity that is directed toward or away from the observer.

radian   The angle at the center of a circle subtended by an arc equal to the length of the circle's radius. Therefore, 2π radians equals 360° and 1 radian equals approximately 59°.296.

radiant   The direction in the sky from which the meteors in a meteor shower seem to come.

radiation belt   A toroidal ring of high-energy particles surrounding a planet.

radiative transfer   The transport of energy from one location to another by electromagnetic radiation.

radiative zone   A region in the interior of a star through which energy is transported outward by radiation.

radio galaxy   An elliptical galaxy having very strong emission (10^{35} to 10^{38} watts) in the radio part of the electromagnetic spectrum.

radiometric dating   Use of radioactive decay to measure the ages of materials such as minerals.

radio telescope   An instrument for detecting and measuring radio-frequency emissions from celestial sources.

radio wave   Electromagnetic radiation in the extreme long-wavelength region of spectrum, beyond the region of microwaves.

ray   (a) A beam of electromagnetic radiation. (b) A bright streak emanating from a young impact crater.

recombination   (a) The combining of ions and electrons to form neutral atoms. (b) An event early in the evolution of the universe in which hydrogen and helium nuclei combined with electrons to form neutral atoms. The removal of electrons caused the universe to become transparent to electromagnetic radiation.

reddening   The effect by which stars and other objects, when viewed through interstellar dust, appear redder than they actually are. Reddening is caused by blue light being more strongly absorbed and scattered than red light.

red giant   A low-mass star that has evolved beyond the main sequence and is now fusing hydrogen in a shell surrounding a degenerate helium core.

red giant branch   A region on the H-R diagram defined by low-mass stars evolving from the main sequence toward the *horizontal branch*.

redshift   The shifting of the wavelength of light to longer wavelengths by any of several effects including Doppler shifts, gravitational redshift, or cosmological redshift. Compare with *blueshift*.

reflecting telescope   A telescope that uses mirrors for collecting and focusing incoming electromagnetic radiation to form an image in their focal planes. The size of a reflecting telescope is defined by the diameter of the first mirror from which incoming light is reflected (called the *primary mirror*).

reflection   The redirection of a beam of light incident on the surface between two media having different refractive indices. If the surface is flat and smooth, the angle of incidence equals the angle of reflection.

refracting telescope A telescope that uses objective lenses to collect and focus light.

refraction The bending of a beam of light when it crosses the boundary between two media having different refractive indices.

refractory material Material that remains solid at high temperatures. Compare with *volatile material*.

regular moon A moon that formed together with the planet it orbits.

relative humidity The amount of water vapor held by a volume of air at a given temperature compared (stated as a percentage) to the total amount of water that could be held by the same volume of air at the same temperature.

relative motion The difference in motion between two individual frames of reference.

relativistic Pertaining to physical processes that take place in systems traveling at nearly the speed of light or located in the vicinity of very strong gravitational fields.

relativistic beaming The effect created when material moving at nearly the speed of light beams the radiation it emits in the direction of its motion.

remote sensing The use of images, spectra, radar, or other techniques to measure the properties of an object from a distance.

resolution The ability of a telescope to separate two point sources of light. Resolution is determined by the telescope's *aperture* and the *wavelength* of light it receives.

rest wavelength The wavelength of light we see coming from an object at rest with respect to the observer.

retrograde (a) Describes rotation or orbital motion of a moon that is in the opposite sense to the rotation of the planet it orbits. (b) Describes the clockwise orbital motion of Solar System objects as seen from above Earth's orbital plane. See also *prograde*.

ring An aggregation of small particles orbiting a planet or star. The rings of the four giant planets of the Solar System are composed variously of silicates, organic materials, and ices.

ring arc A discontinuous region of higher density within an otherwise continuous, narrow ring.

ringlet A narrow subdivision within a larger feature called a *ring*.

Roche, Edouard A. (1820–1883) The French astronomer who systematized the effects of tidal stress on the structure of moons and planets.

Roche limit The distance at which a planet's tidal forces exceed the self-gravity of a smaller object, such as a moon, asteroid, or comet, causing the object to break apart.

Roche lobe The hourglass or figure eight–shaped volume of space surrounding two stars, which constrains material that is gravitationally bound by one or the other.

Roentgen, W. C. (1845–1923) The German discoverer of X-rays, for which he won the first Nobel Prize in physics in 1901.

Rømer, Ole (1644–1719) The Danish astronomer who first estimated the speed of light by timing eclipses of Jupiter's moons when Earth was at different distances from Jupiter.

rotation curve A plot showing how the orbital velocity of stars and gas in a galaxy changes with radial distance from the galaxy's center.

rover A remotely controlled instrumented vehicle designed to traverse and explore the surface of a terrestrial planet or moon. See also *lander*.

RR Lyrae variable A variable giant star whose regularly timed pulsations are good predictors of its luminosity. RR Lyrae stars are used for distance measurements to globular clusters.

Russell, Henry Norris (1877–1957) The American astronomer who, together with Einar Hertzsprung, defined the main diagnostic tool for studying the properties of stars. The *Hertzsprung-Russell (H-R) diagram* is named in his honor, as is the *Vogt-Russell theorem*.

Rutherford, Ernest (1871–1937) The New Zealand nuclear physicist, often considered the "father" of nuclear physics, who pioneered the orbital theory of the atom.

S

satellite (a) An object in orbit about a more massive body. (b) A moon.

saturation An atmosphere reaching the point at which it can hold no more water (or other substance) in the gas phase.

scale factor (R_U) A dimensionless number proportional to the distance between two points in space. The scale factor increases as the universe expands.

scattering The random change in the direction of travel of photons, caused by

their interactions with molecules or dust particles.

science The search for the rules that govern the behavior of the universe and the explanation of observed phenomena in terms of those rules.

scientific method The formal procedure—including hypothesis, prediction, and experiment or observation—used to test (attempt to falsify) the validity of scientific hypotheses and theories.

scientific notation The standard expression of numbers with one digit (which can be zero) to the left of the decimal point and multiplied by 10 to the exponent required to give the number its correct value. Example: $2.99 \times 10^8 = 299,000,000$.

secondary atmosphere A planetary atmosphere that formed—as a result of volcanism, comet impacts, or some other process—sometime after the planet formed.

secondary crater A crater formed from ejecta by a primary impact.

secondary mirror A small mirror placed on the optical axis of a reflecting telescope that returns the beam back through a small hole in the primary mirror, thereby shortening the mechanical length of the telescope.

secondary wave A transverse seismic wave.

second law of thermodynamics The law stating that the entropy or disorder of an *isolated* system always increases as the system evolves.

second of arc (") A unit used for measuring very small angles. A second of arc is 1/60 of a minute of arc, or 1/3,600 of a degree. Also called *arcsecond*.

seismic wave A vibration due to earthquakes, large explosions, or impacts on the surface that travels through a planet's interior.

seismometer An instrument that measures the amplitude and frequency of seismic waves.

self-gravity The gravitational attraction between all the parts of the same object.

semimajor axis Half of the longer axis of an ellipse.

SETI The Search for Extraterrestrial Intelligence project, which uses advanced technology combined with radio telescopes to search for evidence of intelligent life elsewhere in the universe.

Seyfert, Carl (1911–1960) The American astronomer who discovered AGNs in spiral galaxies. *Seyfert galaxies* are named in his honor.

Seyfert galaxy   A type of spiral galaxy with an active galactic nucleus (AGN) at its center, first discovered in 1943 by Carl Seyfert.

Shapley, Harlow (1885–1972)   The Harvard College Observatory astronomer who first showed that our galaxy is nearly 300,000 light-years in size. Shapley was the participant in the Great Debate who argued that our galaxy alone encompassed the entire universe. See also *Curtis, Heber*.

shepherd moon   A moon that orbits close to rings and gravitationally confines the orbits of the ring particles.

shield volcano   A volcano formed by very fluid lava flowing from a single source, spreading out from that source.

short-period comet   A comet with an orbital period of less than 200 years.

sidereal   The orbital or rotational period of an orbit measured with respect to the stars.

sidereal time   Time based on the true rotation period of Earth. The 24-hour sidereal day used by astronomers is approximately four minutes shorter than a solar day.

silicate   One of the family of minerals composed of silicon and oxygen in combination with other elements.

singularity   (a) The point where a mathematical expression or equation becomes meaningless, such as the denominator of a fraction approaching zero. (b) A black hole.

Slipher, Vesto (1875–1969)   The Lowell Observatory astronomer who first made measurements of the redshifts of galaxies, supplying most of the data used by Edwin Hubble to discover that our universe is expanding.

small solar system body   A collective term that includes most asteroids, comets, and most objects orbiting beyond Neptune.

solar abundance   The relative amount of an element detected in the atmosphere of the Sun, expressed as the ratio of the number of atoms of that element to the number of hydrogen atoms.

solar day   The 24-hour period of Earth's axial rotation that brings the Sun back to the same local meridian.

solar eclipse   Blocking of all or part of the Sun by the Moon.

solar flare   Explosive events on the Sun's surface associated with complex sunspot groups and strong magnetic fields.

solar maximum (pl. **maxima**)   The time, occurring about every 11 years, when the Sun is at its peak activity, meaning that sunspot activity and related phenomena (such as prominences, flares, and coronal mass ejections) are at their peak.

solar neutrino problem   The historical observation that only about a third as many neutrinos as predicted by theory seemed to be coming from the sun.

Solar System   The gravitationally bound system made up of the Sun, planets, dwarf planets, moons, asteroids, comets, and KBOs, and their associated gas and dust.

solar tide   The component of Earth's tide due to the Sun's gravity. Also see *tide* and *lunar tide*.

solar time   Time based on the apparent rotation period of Earth. A 24-hour solar day is the average interval between two successive passages of the Sun across the local meridian.

solar wind   The stream of charged particles emitted by the Sun that flows at high speeds through interplanetary space.

solstice ("sun standing still")   (a) One of the two most northerly and southerly points on the ecliptic. (b) The time of year when the Sun is at one of these two points.

south celestial pole   The southward projection of Earth's rotation axis onto the celestial sphere.

South Pole   The location in the Southern Hemisphere where Earth's rotation axis intersects the surface of Earth.

spacetime   The four-dimensional continuum in which we live, and which we experience as three spatial dimensions plus time.

special relativity   The consequences of the speed of light being a constant for nonaccelerating frames of reference; discovered by Albert Einstein. Compare with *general theory of relativity*.

spectral type   A classification system for stars based on the presence and relative strength of absorption lines in their spectra. Spectral type is related to the surface temperature of a star.

spectrograph   A device that spreads out the light from an object into its component wavelengths.

spectrometer   A spectrograph in which the spectrum is generally recorded by electronic means. See *spectrograph*.

spectroscopic parallax   Use of the spectroscopically determined luminosity and the observed brightness of a star to determine the star's distance.

spectroscopy   The study of electromagnetic radiation from an object in terms of its component wavelengths.

spectrum   (a) The intensity of electromagnetic radiation as a function of wavelength. (b) Waves sorted by wavelength.

speed   The rate of change of an object's position with time without regard to the *direction* of movement. Units: m/s, km/h. See *velocity*.

spherically symmetric   Describes an object whose properties depend only on distance from the object's center, so that the object has the same form viewed from any direction.

spin-orbit resonance   A relationship between the orbital and rotation periods of an object such that the ratio of their periods can be expressed by simple integers.

spiral density wave   A stable spiral-shaped change in the local gravity of a galactic disk that can be produced by periodic gravitational kicks from neighboring galaxies or from nonspherical bulges and bars in spiral galaxies.

spiral galaxy   A galaxy of Hubble type "S" class, with a discernible disk in which large spiral patterns exist.

spoke   One of several narrow radial features seen occasionally in Saturn's B ring. Spokes appear dark in backscattered light and bright in forward, scattering light, indicating that they are composed of tiny particles. Their origin is not well understood.

sporadic meteor   A meteor not associated with a specific meteor shower.

spreading center   A zone from which two tectonic plates diverge.

spring tide   An especially strong tide that occurs near the time of a new or full Moon, when lunar tides and solar tides reinforce each other.

stable equilibrium   An equilibrium condition in which the system returns to its former condition after a small disturbance. Compare with *unstable equilibrium*.

standard candle   An object whose luminosity either is known or can be predicted in a distance-independent way, so its brightness can be used to determine its distance via the inverse square law of radiation.

standard model   The theory of particle physics that combines electroweak theory with quantum chromodynamics to describe the structure of known forms of matter.

star   A luminous ball of gas that is held together by gravity. A normal star is powered by nuclear reactions in its interior.

star cluster   A group of stars that all formed at the same time and in the same general location.

static equilibrium A state in which the forces within a system are all in balance so that the system does not change. Compare with *dynamic equilibrium.*

Stefan–Boltzmann constant (σ) The proportionality constant that relates the flux emitted by an object to the fourth power of its absolute temperature.

Stefan, Josef (1835–1893) The Austro-Hungarian physicist best known for his discovery of the power law of radiation from a black body.

Stefan's Law Gives the amount of electromagnetic energy emitted from the surface of a body, summed over the energies of all photons of all wavelengths emitted, which is proportional to the fourth power of the temperature of the body.

stellar mass loss The loss of mass from the outermost parts of a star's atmosphere during the course of its evolution.

stellar occultation An event in which a planet or other Solar System body moves between the observer and a star, eclipsing the light emitted by that star.

stellar population A group of stars with similar ages, chemical compositions, and dynamical properties.

stereoscopic vision The way an animal's brain combines the different information from its two eyes to perceive the distances to objects around it.

stony-iron meteorite A meteorite consisting of a mixture of silicate minerals and iron–nickel alloys.

stony meteorite A meteorite composed primarily of silicate minerals, similar to those found on Earth.

stratosphere The atmospheric layer immediately above the troposphere. On Earth it extends upward to an altitude of 50 km.

strong nuclear force The attractive short-range force between protons and neutrons that holds atomic nuclei together; one of the four fundamental forces of nature, mediated by the exchange of *gluons.*

S-type asteroid An asteroid made of material that has been modified from its original state, likely as the outer part of a larger, differentiated body that has since been broken into pieces.

subduction zone A region where two tectonic plates converge, with one plate sliding under the other and being drawn downward into the interior.

subgiant A giant star smaller and lower in luminosity than normal giant stars of the same spectral type. Subgiants evolve to become giants.

subgiant branch A region of the H-R diagram defined by stars that have left the main sequence but have not yet reached the *red giant branch.*

sublimation The process of a solid becoming a gas without first becoming a liquid.

subsonic Moving within a medium at a speed slower than the speed of sound in that medium.

summer solstice The time of year in the Northern Hemisphere (about June 21) when the Sun is at its most northerly distance from the celestial equator.

sungrazer A comet whose perihelion is within a few solar diameters of the surface of the Sun.

sunspot A cooler, transitory region on the solar surface produced when loops of magnetic flux break through the surface of the Sun.

sunspot cycle The approximate 11-year cycle during which sunspot activity increases and then decreases. This is one-half of a full 22-year cycle in which the magnetic polarity of the Sun first reverses, then returns to its original configuration.

supercluster A large conglomeration of galaxy clusters and galaxy groups; typically more than 100 million light-years in size and containing tens of thousands to hundreds of thousands of galaxies.

superluminal motion The appearance (though not the reality) that a jet is moving faster than the speed of light.

supermassive black hole A black hole of 1,000 solar masses or more that resides in the center of a galaxy, and whose gravity powers active galactic nuclei (AGNs).

supernova A stellar explosion resulting in the release of tremendous amounts of energy, including the high-speed ejection of matter into the interstellar medium. See *Type I supernova* and *Type II supernova.*

supersonic Moving within a medium at a speed faster than the speed of sound in that medium.

superstring theory The theory that conceives of particles as strings in 11 dimensions of space and time; the current contender for a *theory of everything (TOE).*

surface brightness The amount of electromagnetic radiation emitted or reflected per unit area.

surface wave Seismic wave that travels on the surface of a planet or moon.

symmetry (a) The property that an object has if the object is unchanged by rotation or reflection about some point, line, or plane. (b) In theoretical physics, the correspondence of different aspects of physical laws or systems, such as the symmetry between matter and antimatter.

synchronous rotation The case in which the period of rotation of a body on its axis equals its period of revolution in its orbit around another body. A special type of spin-orbit resonance.

synchrotron radiation Radiation from electrons moving at close to the speed of light as they spiral in a strong magnetic field; named because this kind of radiation was first identified on Earth in particle accelerators called synchrotrons.

synodic A period or time related to apparent changes in celestial objects with respect to the Sun as seen from a planet. See also *sidereal.*

S0 galaxy A galaxy with a bulge and a disklike spiral, but smooth in appearance like ellipticals.

T

tail A stream of gas and dust swept away from the coma of a comet by the solar wind and by radiation pressure from the Sun.

tectonism Deformation of the lithosphere of a planet.

telescope The basic tool of astronomers. Working over the entire range from gamma rays to radio, astronomical telescopes collect and concentrate electromagnetic radiation from celestial objects.

temperature A measure of the average kinetic energy of the atoms or molecules in a gas, solid, or liquid.

terrestrial planet An Earth-like planet, made of rock and metal and having a solid surface.

Tertiary Period The period in Earth's history from 65 million until 1.8 million years ago.

theoretical (a) Using mathematics to formalize rules or laws from ideas and/or empirical results. (b) Using rules and laws to explain observed phenomena or predict phenomena not yet seen.

theoretical model A detailed description of the properties of some object or system in terms of known physical laws or theories. Often a computer calculation of predicted properties based on such a description.

theory A well-developed idea or group of ideas that are consistent with known physical laws and make testable predictions about the world. A very well-tested theory may be called a physical law, or simply a fact.

Theory of Everything (TOE) A theory that unifies all four fundamental forces of nature: strong nuclear, weak nuclear, electromagnetic, and gravitational forces.

Theory of Evolution First published in 1859 by Charles Darwin, the theory maintains that variation or evolution within species occurs randomly, and that the survival or extinction of an organism is determined by its ability to adapt to its environment. Modern theory involves both evolutionary change and the concept of natural selection as the agent of change. See also *natural selection*.

thermal conduction The transfer of energy in which the thermal energy of particles is transferred to adjacent particles by collisions or other interactions. The transport of energy by thermal conduction is most important in solids.

thermal energy The energy that resides in the random motion of atoms, molecules, and particles, by which we measure their temperature.

thermal equilibrium The state in which the rate at which thermal energy emitted by an object is equal to the rate at which thermal energy is absorbed.

thermal motion The random motion of atoms, molecules, and particles that gives rise to thermal radiation.

thermal radiation Electromagnetic radiation resulting from the random motion of the charged particles in every substance.

thermosphere That portion of Earth's atmosphere between the exosphere and the mesosphere.

third quarter Moon The phase of the Moon in which the eastern half of the Moon as viewed from Earth is illuminated. The third quarter Moon occurs about one week after the full Moon.

tidal bulge A distortion of a body resulting from tidal stresses.

tidal locking Synchronous rotation of an object caused by internal friction as the object rotates through its tidal bulge.

tidal stress Stress due to differences in the gravitational force of one mass on different parts of another mass.

tide (a) The deformation of a mass due to differential gravitational effects of one mass on another because of the extended size of the masses. (b) On Earth, the rise and fall of the oceans as Earth rotates through a tidal bulge caused by the Moon and Sun.

time dilation The relativistic "stretching" of time.

TOE See *Theory of Everything*.

Tombaugh, Clyde William (1906–1997) The American astronomer best known for his discovery of the planet Pluto in 1930.

topographic relief The differences in elevation from point to point on a planetary surface.

tornado A violent rotating column of air, typically 75 m across with 200 km/h winds. Some tornadoes can be more than 3 km across, and winds up to 500 km/h have been observed.

total lunar eclipse A lunar eclipse in which the Moon passes through the umbra of Earth's shadow.

total solar eclipse The type of eclipse that occurs when Earth passes through the umbra of the Moon's shadow so that the disk of the Sun is completely blocked by the Moon.

transform fault The actively slipping segment of a fracture zone.

Trans-Neptunian Object (TNO) Another name for a *Kuiper Belt object*.

transverse wave A wave in which the oscillations are perpendicular to the direction of the wave's propagation.

triple-alpha process The nuclear fusion reaction that combines three helium nuclei, or "alpha particles," together into a single nucleus of carbon. See also *CNO cycle*.

Trojan asteroid One of a group of asteroids orbiting in the L_4 and L_5 Lagrangian points of Jupiter's orbit.

tropical year The time between one crossing of the vernal equinox and the next. Due to precession of the equinoxes, a tropical year is slightly shorter than the time that it takes for Earth to orbit once about the Sun.

tropics The region on Earth between 23.5° south latitude and 23.5° north latitude and in which the Sun appears directly overhead twice during the year.

tropopause The top of a planet's troposphere.

troposphere The convection-dominated layer of a planet's atmosphere. On Earth the atmospheric region closest to the ground within which most weather phenomena take place.

T Tauri star A young stellar object that has dispersed enough of the material surrounding it to be seen in visible light.

tuning fork diagram The two-pronged diagram showing Hubble's classification of galaxies into ellipticals, S0s, spirals, barred S0s and spirals, and irregular galaxies.

turbulence The random motion of blobs of gas within a larger cloud of gas.

Type I supernova A supernova explosion in which no trace of hydrogen is seen in the ejected material. Most supernovae of this type are thought to be the result of runaway carbon burning in a white dwarf star onto which material is being deposited by a binary companion.

Type II supernova A supernova explosion in which the degenerate core of an evolved massive star suddenly collapses and rebounds.

U

ultraviolet (UV) radiation Describes electromagnetic radiation with frequencies and photon energies greater than those of visible light but less than those of X-rays, and wavelengths shorter than those of visible light but longer than those of X-rays.

umbra (a) The darkest part of a shadow, where the source of light is completely blocked. (b) The darkest, innermost part of a sunspot.

unbalanced force The nonzero net force acting on a body.

unbound orbit An orbit in which the velocity is greater than the escape velocity.

unified model of AGN A model in which many different types of activity in the nuclei of galaxies are all explained by accretion of matter around a supermassive black hole.

uniform circular motion Motion in a circular path at a constant speed.

universal gravitational constant (G) The constant of proportionality in the universal law of gravity. Value: $G = 6.673 \times 10^{-11}$ Nm^2/kg^2.

universal law of gravitation The gravitational force between any two objects is proportional to the product of their masses and inversely proportional to the square of the distance between them.

universe All of space and everything contained therein.

unstable equilibrium An equilibrium state in which a small disturbance will cause a system to move away from equilibrium.

V

vacuum A region of space which contains very little matter. However, in quantum mechanics and general relativity, even a perfect vacuum has physical properties.

variable star A star with varying luminosity. Many periodic variables are found within the instability strip on the H-R diagram.

velocity The rate and direction of change of an object's position with time. Units: m/s, km/h. Compare with *speed*.

vernal equinox (a) The point where the ecliptic crosses the celestial equator, with the Sun moving from south to north. (b) The day on which the Sun appears at this location; the first day of spring in the Northern Hemisphere (about March 21).

virtual particle A particle which, according to quantum mechanics, comes into momentary existence. According to theory, fundamental forces are mediated by the exchange of virtual particles.

visual binary A binary star in which the two stars can be seen individually from Earth.

Vogt-Russell theorem A theorem that states that the structure of a star is completely determined by its mass and chemical composition.

volatile material Material that remains gaseous at moderate temperature, sometimes referred to as an "ice." Compare with *refractory material*.

volcanic dome Steep-sided, dome-shaped volcanic feature, produced by viscous lavas.

volcanism The occurrence of volcanic activity on a planet or moon.

vortex (pl. vortices) Any circulating fluid system: (a) An atmospheric anticyclone or cyclone. (b) A whirlpool or eddy.

W

waning The changing phases of the Moon as it becomes less fully illuminated between full Moon and new Moon as seen from Earth.

watt (W) A measure of *power*. Units: J/s.

wave A disturbance moving along a surface or passing through a space or a medium.

wavefront The imaginary surface of an electromagnetic wave, either plane or spherical, oriented perpendicular to the direction of travel.

wavelength The distance on a wave between two adjacent points having identical characteristics. The distance a wave travels in one period. Unit: m.

waxing The changing phases of the Moon as it becomes more fully illuminated between new Moon and full Moon as seen from Earth.

weak nuclear force The force underlying some forms of radioactivity and certain interactions between subatomic particles. It is responsible for radioactive beta decay and for the initial proton–proton interactions that lead to nuclear fusion in the Sun and other stars. One of the four fundamental forces of nature.

Wegener, Alfred (1880–1930) The multidisciplinary German scientist who in 1915 first proposed the theory of continental drift, for which he was generally ridiculed. His theory finally became widely accepted in the 1950s. See also *plate tectonics*.

weight The force equal to the mass of an object times the local acceleration due to gravity; or (in general relativity) the force equal to the mass of an object times the acceleration of the reference frame in which the object is observed.

Whipple, Fred (1906–2004) The American planetary astronomer who proposed the "dirty snowball" model of comet nuclei.

white dwarf The stellar remnant left at the end of the evolution of a low-mass star. A typical white dwarf has a mass of $0.6\ M_\odot$, and a radius about that of Earth: it is made of nonburning, electron degenerate carbon.

Wien's Law A relationship describing how the peak wavelength, and therefore the color, of electromagnetic radiation from a glowing black body changes with temperature. Also known as Wien's Displacement law. See *lambda peak*.

Wilson, Robert (1936–) The Bell Laboratories physicist who, together with Arno Penzias, in the early 1960s discovered the cosmic background radiation, for which they won the Nobel Prize in physics.

WIMP "Weakly interacting massive particle." A hypothetical massive particle that interacts through gravity but not with electromagnetic radiation and is a candidate for dark matter. See also *MACHO*.

winter solstice The time of year (about December 22) when the Sun is at its most southerly distance from the celestial equator. The first day of Northern Hemisphere winter.

X

X-ray Electromagnetic radiation having frequencies and photon energies greater than those of ultraviolet light but less than those of gamma rays, and wavelengths shorter than those of UV light but longer than those of gamma rays.

X-ray binary A binary system in which mass from an evolving star spills over onto a collapsed companion such as a neutron star or black hole. The material falling in is heated to such high temperatures that it glows brightly in X-rays.

Y

year The time it takes Earth to make one revolution around the Sun. A solar year is measured from equinox to equinox. A sidereal year, Earth's true orbital period, is measured relative to the stars.

Z

zenith The point on the celestial sphere located directly overhead from an observer.

zodiac The 12 constellations lying along the plane of the ecliptic.

zodiacal dust Particles of cometary and asteroidal debris less than 100 μm in size that orbit the inner Solar System close to the plane of the ecliptic. Compare with *meteoroid* and *planetesimal*.

zodiacal light A band of light in the night sky caused by sunlight reflected by zodiacal dust.

zonal wind The planetwide circulation of air that moves in directions parallel to the planet's equator.

Credits

These pages contain credits for all 21 chapters in the complete volume of *21st Century Astronomy*, Second Edition. The Solar System Edition does not include Chapters 13 and 15–20. The Stars and Galaxies Edition does not include Chapters 6–12.

Part I Opener STScI. **Part II Opener** NASA/JPL. **Part III Opener** NASA/ESA/STScI/J. Hester and P. Scowen (Arizona State University). **Part IV Opener** Jason T. Ware/Photo Researchers, Inc.

Chapter 1

Opener Courtesy of Anthony Ayiomamitis (http://www.perseus.gr). **1.1** Craig Lovell/Corbis. **1.2** Dr. Bernard G. Lindsay, Physics and Astronomy Department, Rice University. **1.3a & b** Neil Ryder Hoos; **1.3c** Owen Franken/Corbis; **1.3d** George Hall/Corbis; **1.3e** PhotoDisc; **1.3f & g** American Museum of Natural History; **1.3h** NASA/JPL/Caltech. **1.4** Bettmann/Corbis. **1.5 left** NASA/STScI/Arizona State University/Hester/Ressmeyer/Corbis; **1.5 right** Ron Watts/Corbis. **1.6a** Michael J. Tuttle (NASM/NASA); **1.6b** NSSDC. **1.7** NRAO/AUI, James J. Condon, John J. Broderick, and George A. Seielstad. **1.8** ©1969 by The New York Times Co. Reprinted by permission. **1.9 top left** Bettmann/Corbis; **1.9 top right** *Special Relativity: The M.I.T. Introductory Physics Series* by A. P. French. ©1968, 1966 by Massachusetts Institute of Technology. Used by permission of W. W. Norton & Company, Inc; **1.9 center** Seth Joel/Corbis; **1.9 bottom left** Musée d'Orsay, Paris, France. Photo: Réunion des Musées Nationaux/Art Resource, NY. Photograph by Herve Lewandowski; **1.9 bottom right** Robbie Jack/Corbis. **1.10** Bettmann/Corbis. **1.11** Four seasons photographs, Jim Schwabel/Index Stock Imagery/PictureQuest. **1.12** NON SEQUITUR by Wiley Miller. Dist. by Universal Press Syndicate. Reprinted with permission. All rights reserved.

Chapter 2

Opener Jon Arnold Images/Alamy. **2.1** British Library Maps. **2.7 left** Pekka Parviainen/Photo Researchers, Inc.; **2.7 right** David Nunuk/Photo Researchers, Inc. **2.19 top left** AFP/Getty Images; **2.19 top right** AP Images; **2.19 bottom left** Lynn Goldsmith/Corbis; **2.19 bottom right** Lindsay Hebberd/Corbis. **2.24** Roger Ressmeyer/Corbis. **2.26** Courtesy of Nick Quinn. **2.27** Courtesy of Koen van Gorp, www.koenvangorp.be **2.28a** Reuters New Media Inc./Corbis. **2.28b** Photograph ©2001 by Fred Espenak, courtesy of www.MrEclipse.com.

Chapter 3

Opener Richard Cummins/Corbis. **3.11** Courtesy of Alan Bean.

Chapter 4

Opener Courtesy David Malin. **4.8** Eyewire Collection/Getty Images. **4.28** NASA/NSSDC/GSFC.

Chapter 5

Opener NASA/JPL. **5.1** Junenoire Photography. **5.2** Gianni Tortoli/Photo Researchers, Inc. **5.3** Jim Sugar/Corbis. **5.4a** Yerkes Observatory photographs: Richard Dreiser; **5.4b** W. M. Keck Observatory/CARA. **5.5** ESO Telescope Systems Division. **5.8b** Howard Voss. **5.10b** Howard Voss. **5.11** Howard Voss. **5.13c** Steve Percival/Photo Researchers, Inc. **5.17** Keck Observatory. **5.18 left to right** NASA; NASA/CXC/SAO; Graphic courtesy Orbital Sciences Corp; NOAO/AURA/NSF; NASA/JPL/Caltech; Joint Astronomy Center in Hilo, Hawaii; NRAO VLA Image Gallery; The National Radio Astronomy Observatory, Green Bank; Dave Finley, Courtesy National Radio Astronomy Observatory and Associated Universities, Inc. **5.19a** Dr. Jeremy Burgess/Photo Researchers, Inc.; **5.19b** Hulton-Deutsch Collection/Corbis. **5.20** NASA/IPAC/NED/JPL/Caltech/SSDS. **5.21b** Jean-Charles Cuillandre (CFHT). **5.22a** Courtesy of Ed Grafton; **5.22b** Steve Larson/University of Arizona. **5.24a** Roger Ressmeyer/Corbis; **5.24b** David Parker/Photo Researchers, Inc. **5.25** Image courtesy of NRAO/AUI. **5.26** ESO Telescope Systems Division. **5.27** NASA/NSSDC. **5.28** NASA/JPL. **5.29** Fermilab/Photo Researchers, Inc. **5.30** Cern/Photo Researchers, Inc. **5.31** Frank Summers, Space Telescope Science Institute; Chris Mihos, Case Western Reserve University; Lars Hernquist, Harvard University.

Chapter 6

Opener Caltech/NASA/JPL. **6.1a** STScI photo: Karl Stapelfeldt (JPL); **6.1b** D. Padgett (IPAC/Caltech), W. Brandner (IPAC), K. Stapelfeldt (JPL) and NASA; **6.1c** D. Padgett (IPAC/Caltech), W. Brandner (IPAC), K. Stapelfeldt (JPL) and NASA; **6.1d** D. Padgett (IPAC/Caltech), W. Brandner (IPAC), K. Stapelfeldt (JPL) and NASA. **6.2** Photograph by Pelisson, SaharaMet. **6.4** Reuters/Corbis. **6.11** NASA/JPL/Northwestern University.

Chapter 7

Opener NASA/GRIN. **7.1** Stockli, Nelson, Hasler, Goddard Space Flight Center/NASA. **7.2** USGS Hawaiian Volcano Observatory. **7.3b** NASA/JSC. **7.4** Montes De Oca & Associates. **7.5** Photograph by D.J. Roddy and K.A. Zeller, USGS, Flagstaff, AZ. **7.6** Don Davis. **7.7** NASA/JSC. **7.8** NASA/JPL/Caltech. **7.9** Dennis Flaherty/Photo Researchers, Inc. **7.10** NASA/JPL/Caltech. **7.13b** Grant Heilman Photography, Inc. **7.14** Photo courtesy of Ron Greeley. **7.19** NASA/JSC. **7.20** NSSDC/NASA. **7.21** NASA/Magellan Image/JPL. **7.23** NASA/JSC.

Chapter 16

Opener NASA, ESA and H.E. Bond (STScI). **16.11a** Minnesota Astronomical Society; **16.11b** NASA and The Hubble Heritage Team (STScI/AURA). **16.12 top left** Dr. Raghvendra Sahai (JPL) and Dr. Arsen R. Hajian (USNO), NASA and The Hubble Heritage Team (STScI/AURA); **16.12 center left** A. Hajian (USNO) et al., Hubble Heritage Team (STScI/AURA), NASA; **16.12 bottom left** J. P. Harrington and K. J. Borkowski (University of Maryland), HST, NASA; **16.12 top right** NASA, A. Fruchter and the ERO team (STScI); **16.12 center right** Anglo-Australian Telescope, photograph by David Malin; **16.12 bottom right** Bruce Balick (University of Washington), Vincent Icke (Leiden University, The Netherlands), Garrelt Mellema (Stockholm University), and NASA.

Chapter 17

Opener NASA/HST/ASU/J. Hester et al. **17.7** STScI/NASA/Arizona State University/ Hester/Ressmeyer/Corbis. **17.10** Anglo-Australian Observatory, photographs by David Malin. **17.11a** Joachim Trumper, Max-Planck-Institut für extraterrestrische Physik; **17.11b** Courtesy Nancy Levenson and colleagues, American Astronomical Society; **17.11c** Jeff Hester, Arizona State University, and NASA. **17.15a** European Southern Observatory, ESO PR Photo 40f199; **17.15b & c** Jeff Hester, Arizona State University, and NASA.

Chapter 18

Opener NASA/ESA/STScI/Photo Researchers, Inc. **18.1** NASA, ESA, S. Beckwith (STScI) and the HUDF Team. **18.3** Courtesy David Malin; **18.3 top right** Canada-France-Hawaii Telescope/J.-C. Cuillandre/Coelum. **18.4** The Hubble Heritage Team (AURA/STScI/NASA). **18.6** NOAO/AURA/NSF. **18.7** Courtesy David Malin. **18.8** C. Howk (JHU), B. Savage (University of Wisconsin), N. A. Sharp (NOAO)/WIYN/NOAO/NSF. **18.10 top left** George Jacoby, Bruce Bohannan, Mark Hanna/NOAO/AURA/NSF; **18.10**

top right European Southern Observatory (ESO). **18.11 bottom left** Courtesy David Malin; **18.11 bottom right** Courtesy David Malin. **18.12** R. Windhorst and D. Burstein (Arizona State University). **18.13 left** Todd Boroson/NOAO/AURA/NSF; **18.13 right** Richard Rand, University of New Mexico. **18.15** Courtesy David Malin. **18.16b** Z. Frei, Institute of Physics, Eotvos University, Hungary. **18.17** Images courtesy G. Fabbiano, Harvard-Smithsonian Center for Astrophysics. **18.18** J. Bahcall (Institute for Advanced Study), M. Disney (University of Wales) and NASA. **18.19** NRAO/Alan Bridle et al. **18.22b** Courtesy Bill Keel, University of Alabama, and NASA; **18.22c** Holland Ford, STScI/Johns Hopkins University, and NASA. **18.22d** R. P.van der Marel, STScI, F. C. van den Bosch, University of Washington, and NASA. **18.23** F. Owen, NRAO, with J. Biretta, STScI, and J. Eilek, NMIMT. **18.24** Holland Ford, STScI/Johns Hopkins University; Richard Harms, Applied Research Corp.; Zlatan Tsvetanov, Arthur Davidsen, and Gerard Kriss at Johns Hopkins University; Ralph Bohlin and George Hartig at STScI; Linda Dressel and Ajay K. Kochhar at Applied Research Corp. in Landover, MD; and Bruce Margon from the University of Washington, Seattle/ NASA. **18.25** John Biretta, STScI. **18.26** Halton Arp/Caltech.

Chapter 19

Opener Gordon Garradd 1996. **19.1a** Axel Mellinger; **19.1b** C. Howk and B. Savage (University of Wisconsin); N. Sharp (NOAO)/WIYN, Inc. **19.2** U.S. Patent Office Library. **19.3** Hubble Heritage Team (AURA/STScI/NASA). **19.4** NOAO/ AURA/NSF. **19.5** Courtesy Jeff Hester. **19.7** Courtesy J. Binney. **19.9** Anglo-Australian Observatory. **19.12** The Electronic Universe Project. **19.16a** Carl Heiles, UC Berkeley; **19.16b** C. Howk and B. Savage (University of Wisconsin), N. Sharp (NOAO). **19.18** NRAO. **19.19** Anglo-Australian Observatory, photograph by David Malin.

Chapter 20

Opener NASA/WMAP Science Team. **20.2** TIMEPIX/Getty Images. **20.3 top** Courtesy

of Jeff Hester; **20.3 bottom** NASA William C. Keel (University of Alabama, Tuscaloosa). **20.5** WIYN/NOAO/NSF, WIYN Consortium, Inc. **20.14** Lucent Technologies' Bell Labs. **20.15** Ann and Rob Simpson. **20.17a & b** NASA COBE Science Team; **20.17c** NASA/ WMAP Science Team.

Chapter 21

Opener NASA/JPL/Origins. **21.2a** Anglo-Australian Observatory; **21.2b** Omar Lopez-Cruz and Ian Shelton/NOAO/AURA/NSF. **21.3** Max Tegmark/SDSS Collaboration; **21.6** AAO/David Malin. **21.7** NASA, A. Fruchter and the ERO team (STScI, ST-ECF). **21.8** Images produced by Donna Cox Smithsonian Institution and Motorola Corporation. **21.14** From *The Elegant Universe* by Brian Greene. ©1999 by Brian R. Greene. Used by permission of W. W. Norton & Company, Inc. **21.16** Kenneth EWard/Biografx/Photo Researchers, Inc. **21.17 top left** Mark A. Schneider/Visuals Unlimited; **21.17 middle left** K. Sandved/ Visuals Unlimited; **21.17 bottom left** Ken Lucas/Visuals Unlimited; **21.17 top right** Science VU/National Museum of Kenya/ Visuals Unlimited; **21.17 middle right** Ken Lucas/Visuals Unlimited; **21.17 bottom right** Ken Lucas/Visuals Unlimited. **21.18** Jeff J. Daly/Visuals Unlimited. **21.19** NASA. **21.20** NASA. **21.21** F. Drake (UCSC) et al., Arecibo Observatory (Cornell, NAIC). **21.22** FOXCLIPS. **21.23 bottom left** Dr. Michael Perfit, University of Florida, Robert Embley/NOAA; **21.23 bottom right** Woods Hole Oceanographic Institution, Deep Submergence Operations Group, Dan Fornari. **21.24** Isaac Gary.

Epilogue NASA/Manned Spacecraft Center/ Image # 68-HC-870.

Appendix 8 The International Astronomical Union/Martin Kornmesser.

Every attempt has been made to contact the permission holders of each image. If there is a change or correction, please contact the Photo Permissions Department at W. W. Norton & Company, 500 Fifth Avenue, New York, NY 10110.

Index

This Index contains page references for all 21 chapters in the complete volume of *21st Century Astronomy*, Second Edition. The Solar System Edition does not include pages 374–401 and 430–605. The Stars and Galaxies Edition does not include pages 161–372.